T0136030

MINERALOGY OF ARIZONA, FOURTH EDITION

Mineralogy
of Arizona

RAYMOND W. GRANT
RONALD B. GIBBS
HARVEY W. JONG
JAN C. RASMUSSEN
STANLEY B. KEITH

MAPS AND ILLUSTRATIONS
BY JOHN CALLAHAN

THE UNIVERSITY OF
ARIZONA PRESS
TUCSON

The University of Arizona Press
www.uapress.arizona.edu

ISBN-13: 978-0-8165-4358-8 (hardcover)
ISBN-13: 978-0-8165-4357-1 (paperback)

Cover design by Leigh McDonald
Cover photo: Wulfenite, Rowley Mine, Maricopa County, 4 cm high specimen, Evan and
Melissa Jones Collection, photo by Jeff Scovil
Designed and typeset by Sara Thaxton in 10.5/15 Warnock Pro with Interstate Condensed

FM FREEPORT-McMoRan

Publication of this book was made possible in part by financial assistance from
Freeport-McMoRan.

Library of Congress Cataloging-in-Publication Data
Names: Grant, Raymond W., author.
Title: Mineralogy of Arizona / Raymond W. Grant [and 4 others] ; maps and illustrations by
    John Callahan.
Description: Fourth edition. | Tucson : University of Arizona Press, 2022. | Includes
    bibliographical references and index.
Identifiers: LCCN 2021041195 | ISBN 9780816543588 (hardcover) | ISBN 9780816543571
    (paperback)
Subjects: LCSH: Minerals—Arizona.
Classification: LCC QE375.5.A6 G73 2022 | DDC 549.9791—dc23
LC record available at https://lccn.loc.gov/2021041195

Dedicated to the memory of John W. Anthony,
Sidney A. Williams, and Richard A. Bideaux

What is presented here would not be
possible without their previous work and their
pursuit of the study of Arizona minerals.

CONTENTS

ACKNOWLEDGMENTS

This edition has benefited greatly from the labors of many people who have supplied information about mineral occurrences, provided valuable review, and assisted in many other ways:

Richard Graeme for his many contributions and suggestions
Dave Fanger and George V. Polman for helpful review of chapter 2
Ken Rippere for critical reading of the manuscript
Dr. Hexiong Yang and Dr. Robert Downs and the RRUFF project at
    the University of Arizona

We are grateful for the major funds we received to support the inclusion of color photography in this edition from Freeport-McMoRan.
We are thankful to the following individuals and organizations who have also helped support the inclusion of color photography:

Arizona Mineral Minions
John Callahan
Earth Science Museum
John Ebner
Flagg Mineral Foundation
Mark Hay
Robert W. Jones
Mineralogical Society of Arizona
Tony Potucek
Prescott Gem and Mineral Club
Paula and Les Presmyk

Joseph Ryan
SRK Consulting Inc.
Tucson Gem and Mineral Society
Frank Valenzuela
Chris Whitney-Smith
Will and Pam Wilkinson
Dr. Hexiong Yang

MINERALOGY OF ARIZONA, FOURTH EDITION

Introduction

Many new mineral species have been documented in Arizona since the third edition of *Mineralogy of Arizona* was published in 1995. This new edition adds to the extensive work of the previous authors, who built the foundation on which this new effort rests. This edition is dedicated to these men: John Anthony, Sidney Williams, and Richard Bideaux.

The field of mineralogy has undergone many changes since the third edition. Mineralogical nomenclature has changed, and now some species have been relegated to group names (e.g., monazite group); others have had element designations added, as in euxenite-(Y); and still others have been discredited for various reasons. The new analytical equipment available today has made it possible to more easily subdivide previous valid species, often based on the major element present. This is illustrated by the zeolite minerals, where chabazite is now a group name with five species who differ only in the major element present: Ca, K, Mg, Na, or Sr. Older references are unaware of the new nomenclature, and many occurrences are noted where the specific species as now defined is not presently known. This new edition lists a total of 992 valid mineral species in Arizona, which is an increase of 182 species compared with the third edition. Several mineral species found in the third edition have been discredited and are no longer found in the occurrences section. Many are discussed in some detail in appendix E. Fifty-one minerals were removed from the third edition because they were either discredited as a mineral species by the International Mineralogical Association (IMA), changed to a group name, or renamed by changes in nomenclature as mineral groups and relationships were redefined.

A review of the occurrences in Arizona of important gemstone and lapidary materials is found in chapter 1. These are an important part of the state's mineral heritage. Arizona turquoise is world famous, as are many other unique gem materials from Arizona.

Fluorescent minerals are widely distributed in Arizona, and chapter 2 discusses the occurrences and characteristics of the most notable Arizona fluorescent minerals. The reasons for and the types of florescence are discussed in detail, including a list of known Arizona minerals that fluoresce.

Chapter 3 lists all the presently known mineral species found in Arizona with their notable occurrences. Important data from the earlier edition have been retained, and new information about additional species and occurrences has been added. The RRUFF project in the Geosciences Department at the University of Arizona has been a significant source of information in updating this section (http://rruff.info). In addition, several extensive databases, such as mindat.org, are now available on the internet and are a great source of information. The mineral occurrence section has also benefited from the work and observations of the many dedicated mineral collectors in Arizona. Inevitably, not all reported occurrences can be included here because of space limitations, overall significance, and the need for some verification. Mineral occurrence descriptions include references to published works or existing institutional collections so that additional information can be obtained by the reader. Sometimes occurrences are passed on verbally by reputable sources and have been included where the occurrence is beyond doubt. The third edition included a number of occurrences that are unverifiable due to the passage of time; they have been retained and noted as coming from that edition.

In this new edition, chapter 4 contains a discussion of the use of metallic mineral districts in the text and on the maps. The previous edition used mining districts as part of the locality description. These political and geographic district names have been replaced with a more descriptive term, *metallic mineral district*. This term shows the geochemical and mineralogical affinity of the mineral districts as well as the known or estimated age of mineralization. These designations are based on the commodities produced and known geologic mapping, age dating, and mineral occurrences. This information will enable the reader to learn more about the geology of the district and to possibly predict the minerals that could occur in similarly categorized districts. Where possible, the metallic mineral district for each mineral locality is given for the mineral occurrence in chapter 3.

History of Mineralogy in Arizona

As Arizona mineralogy has evolved over the years, the number of species described from the state has risen from 18 in Blake's 1866 report to 992 in this

TABLE I.1 Publications of Arizona mineralogy

Year	Title	Author	Years from previous work
1866	*Annotated Catalogue*	Blake	—
1909	*Minerals of Arizona*	Blake	43
1910	*Mineralogy of Arizona*	Guild	1
1915	*Directory of Arizona Minerals*	Willis	5
1941	*Minerals of Arizona*	Galbraith	26
1947	*Minerals of Arizona* (2nd ed.)	Galbraith	6
1958	*Mineralogical Journeys in Arizona*	Flagg	11
1959	*Minerals of Arizona* (3rd ed.)	Galbraith and Brennan	1
1977	*Mineralogy of Arizona*	Anthony, Williams, and Bideaux	18
1982	*Checklist of Arizona Minerals*	Grant	5
1982	*Mineralogy of Arizona* (2nd ed.)	Anthony, Williams, and Bideaux	—
1995	*Mineralogy of Arizona* (3rd ed.)	J. W. Anthony et al.	13
2007	*Checklist of Arizona Minerals* (2nd ed.)	Grant	12
2022	This volume	Grant et al.	15

new edition. Over the past 156 years, fourteen summaries on the mineralogy of Arizona have been published (table I.1). The first published list of the minerals found in Arizona was made by William P. Blake (1866) and was part of a report for California and Arizona.

William Blake was the state mineralogist for California in 1866. From 1895 to 1910, he was a professor of geology and mining and the director of the School of Mines at the University of Arizona. In 1898 he also became the territorial geologist of Arizona. In 1909, Blake published the first complete work on Arizona minerals, *Minerals of Arizona: Their Occurrence and Association with Notes on Their Composition, a Report to the Hon. J. H. Kibby, Governor of Arizona*. His list had 102 mineral species for Arizona.

A year later, in 1910, Frank N. Guild authored *The Mineralogy of Arizona*, which was published privately by the Chemical Publishing Company of Easton, Pennsylvania. Guild was a professor of chemistry and mineralogy at the University of Arizona. His list had 130 species, and it is interesting that these two works were published a year apart. Guild had Blake's list, as he has two references to Blake (1909). But he did not include several minerals that Blake listed, such as bismuthinite, cuprotungstite, troilite, trona, tungstite, and zinkenite. We can only imagine the relationship between these authors and their competing mineral lists in 1909 and 1910.

In 1915–16 and in 1916–17, Charles Willis, who was director of the Arizona State Bureau of Mines, published *Directory of Arizona Minerals* and

Mineralogy of Useful Minerals in Arizona. He had a short list of seventy-five minerals of economic importance but included eight additional minerals from the earlier lists: arsenopyrite, carnotite, celestite, dufrenosyite, dyscrasite, freieslebenite, jamesonite, and volborthite.

It was twenty-six years until the next state mineralogy was published. In 1941, Frederick W. Galbraith, chairman of the Department of Geology at the University of Arizona and curator of the Mineral Museum, authored *Minerals of Arizona* as Arizona Bureau of Mines Bulletin 149, listing 267 species. He wrote a revised edition six years later, in 1947, as Arizona Bureau of Mines Bulletin 153, adding 17 species to the list. In the introduction, Galbraith writes, "Since the first edition of Bulletin 149 was published in 1941, Volume I of the revised Dana's system has been published. A tremendous amount of new data on the elements, sulfides, sulfosalts, and oxides has been made available, and based on this data the groupings and order of these minerals have been radically changed."

A third edition of *Minerals of Arizona* was published in 1959 by Galbraith and Daniel J. Brennan. Brennan was a graduate student at the University of Arizona and later a geologist for Shell Oil Company at Oklahoma City when the book was published. This time it was released as the University of Arizona Bulletin vol. 30, number 2, by the University of Arizona Press and listed 403 species for Arizona. *Minerals of Arizona* was published again in 1970 as Arizona Bureau of Mines Bulletin 181, but no changes were made from the 1959 edition.

In 1958, Arthur L. Flagg wrote *Mineralogical Journeys in Arizona*, a book he described as "primarily for rockhounds," but he also gave a list of more than 400 mineral species for Arizona. He had six other special lists of minerals, such as pegmatite minerals, contact-metamorphic minerals, and secondary minerals.

The next series of mineralogies for Arizona were the three editions of the *Mineralogy of Arizona* by John Anthony, Sidney Williams, and Richard Bideaux, with Raymond Grant added on the third edition. The first edition was published in 1977, and it added 224 species to the list for a total of 605 species. The second edition in 1981 had 45 additional species in a supplement bound into the end of the book. The main text was not changed. The third edition in 1995 added another 187 species. This brought the total species for Arizona to 810. (This number is less than the total if you add all the additional species to Galbraith's list. This is because some species were discredited and others dropped from the list.) Grant published two editions of the *Checklist of Arizona Minerals.* The first in 1982 had 655 species, and the second in 2007 added 61 additional species to the list from the 1995 *Mineralogy of Arizona* for a total of 870.

Figure I.1 highlights the evolution of known Arizona minerals since lists were first started in 1866. The number of known minerals has steadily increased, and recently, 182 additional species have been added to the list of Arizona minerals since the third edition (see appendix B). Appendix B includes those species that are new occurrences and does not include those whose names have been changed by revised nomenclature. Some species were lost from the third edition as a result of nomenclature changes or being discredited.

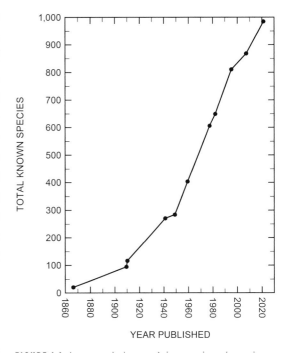

FIGURE I.1 Increase in known Arizona mineral species since 1866.

Arizona State Symbols

Arizona has many state symbols, and those of special interest to mineral collectors and mineralogists are the state gem, fossil, metal, and mineral. Official state symbols are suggested, researched, and drafted into legislation for the ultimate approval of the governor.

The first of these symbols to be created is the official state gem, turquoise. Legislation was signed into law by the governor in 1974. Turquoise is widespread throughout the copper-mining areas in Arizona and has been used in jewelry for centuries. Large-scale open-pit copper mining brought an increase in the amount of available high-quality turquoise for use in jewelry. It has been and continues to be an important and valuable commodity in Arizona.

In 1988, the governor signed into law legislation making the petrified wood of *Araucarioxylon arizonicum*, a prehistoric conifer, the official state fossil. The beautiful colors found in the petrified wood from Petrified Forest National Park are well known throughout the world. A large industry developed mining petrified wood found outside the park and preparing it for decorative use.

In 2015, copper was proclaimed the state metal. Legislation was drafted after a fourth-grade class at Copper Creek Elementary School had the idea and reached out to their representative about it. The governor signed the legislation into law, saying, "A crucial driver of our economy, copper is represented on our state seal and is one of Arizona's '5 C's' along with climate, cattle, citrus and cotton."

Wulfenite became the official state mineral in 2017. Arizona needed a mineral that was known to museums around the world and unrivaled for its beauty. No place on earth produced more spectacular specimens of wulfenite than Arizona, including many mines that for decades produced the finest wulfenite specimens in the world.

Public Mineral Collections in Arizona

There are excellent mineral collections available for public viewing in several museums in Arizona. These museums are a resource for mineral collectors to learn more about the wide variety of minerals that occur in the state. They also offer exhibits and experiences related to Arizona history, culture, and wildlife.

Of the many mineral collections, both small and large, scattered throughout the state, three stand out. These collections are available throughout the year. All three maintain their own specific focus on Arizona mineralogy. The University of Arizona Alfie Norville Gem and Mineral Museum (formerly the University of Arizona Mineral Museum) collection focuses on Arizona and Mexico as well as the world at large. The Arizona-Sonora Desert Museum specializes in minerals of the Sonoran Desert, which includes the southeastern corner of California, the southern portion of Arizona, and Sonora and Baja California, Mexico. The Bisbee Mining and Historical Museum focuses on the history of the mines and town of Bisbee. All three collections are highly recommended when planning a visit to Arizona.

The new **University of Arizona Alfie Norville Gem and Mineral Museum** (gemandmineralmuseum.arizona.edu) is located in the old Pima County Courthouse building at 115 North Church Avenue in downtown Tucson. The original museum was opened in 1892 as the Museum of Geology and Mineralogy on the university campus. The collections have grown dramatically and today comprise more than twenty thousand specimens in the main collections, plus more than seven thousand specimens in the micromount collection. The new museum is much larger and now occupies the basement and half of the first floor of the courthouse. Exhibits take you through Mineral Evolution, where you travel through time to see how earth's minerals formed; followed by the Arizona Gallery, with displays of the best of Arizona's minerals and a re-created azurite stope; then into the Crystal Lab and Gem Galleries to see how crystals form and how to identify minerals; and finally you come to displays of the best of gem minerals, gemstones in jewelry, and gem identification. The basement houses collection storage, specimen preparation facilities, meeting rooms, classrooms, and a research lab. The new museum also includes an auditorium for educational and special programs.

Another fine collection is found at the **Arizona-Sonora Desert Museum** (desertmuseum.org), which is only a thirty-minute drive from downtown Tucson, on the west side of the Tucson Mountains. The address is 2021 N. Kinney Rd., Tucson, Arizona 85843. The focus of the museum is to preserve the heritage of the Sonoran Desert region, including the mineral heritage. The mineral collection was begun in the 1950s and the first curator was hired in 1972. The collection has blossomed from then and contains minerals found in Arizona, as well as from Sonora and Baja California, Mexico. Currently, the collection holds more than sixteen thousand specimens, including several important micromount collections. The mineral hall in the Earth Science Center has the best specimens on display, which includes several reconstructed crystal pockets. The display is accessed through an impressive manmade "natural cave."

Portions of the collection are often displayed at regional gem and mineral shows, especially at the Tucson Gem and Mineral Show each February. The museum engages in numerous educational outreach programs and holds the annual Mineral Madness sale just prior to the Tucson Gem and Mineral Show. A visit to Tucson should include a visit to the Desert Museum. The museum grounds include extensive gardens featuring the native plants and animals found in the Sonoran Desert.

The **Bisbee Mining and Historical Museum** (bisbeemuseum.org) also has a fine mineral collection. The museum is at No. 5 Copper Queen Plaza in Bisbee and is open daily. It has the finest collection of the "classic" Bisbee minerals in Arizona. Three collections form the bulk of the museum's holdings: the M. J. Cunningham collection, the Lem Shattuck collection, and specimens on loan from the National Museum of Natural History (NMNH), including many from the James Douglas collection. The well-arranged displays were designed by a team from the NMNH. A natural cave display is a great educational tool. This collection must be seen if you have an interest in the amazing minerals from the Bisbee mines.

The **Arizona Mining, Mineral and Natural Resources Education Museum** (ammnre.arizona.edu) was recently established by the University of Arizona to promote Arizona's natural resource heritage and future through education, research, and outreach. The museum is currently in the development and renovation stage at the historic Polly Rosenbaum building at 1502 W. Washington Street, Phoenix, home of the former Arizona Mining and Mineral Museum, which closed in 2011. The new museum will house and showcase the former museum's diverse collection, which dates back to the 1884 Arizona Territorial Fair and includes more than twenty thousand specimens from Arizona and around the world. The new museum will also highlight the historical and social contributions of Arizona's natural resource industries: mining, mineral, timber, livestock, and agriculture. The University of Arizona's vision for the museum is to create a destination where people can continue to learn and experience the value of our natural resources through interactive galleries, workshops, and public spaces. An opening date will be set after the completion of major building renovation and the installation of displays. Information about progress and the expected future opening can be found at the website.

1 Gemstones and Lapidary Materials of Arizona

Gemstones and lapidary materials can be defined as those minerals used for jewelry or ornamental purposes. They can be used in their natural state or be worked by the lapidarist through cutting and polishing. Almost any mineral can be used in jewelry, but this chapter concentrates on the occurrence of minerals from Arizona that are commonly used as gemstones or lapidary materials.

Arizona is one of the leading states in the nation in the production of semiprecious gemstones. Millions of dollars' worth of turquoise, petrified wood, and peridot have been mined in Arizona. Smaller amounts of other lapidary material have been collected, and there are literally hundreds of localities around the state where collectors can find good lapidary material.

Unfortunately, the precious gems (diamond, sapphire, ruby, and emerald) are not known to occur as gem minerals in Arizona. There are occurrences of corundum and beryl in the state, but they are not found with the qualities of the precious gems.

More information about each of these gem minerals can be found in chapter 3 along with additional localities and references of interest. Many of these gemstone occurrences have been mined commercially, and many others are well known to the mineral and lapidary clubs. A few of the localities are mysteries because the exact location is unknown. These are from references found in the literature but are not specific or give locations not readily identified. They are given in this chapter in italics and are included here with the idea that additional research and fieldwork by collectors might lead to finding interesting material. Finally, many remote areas in Arizona still have the potential to produce interesting gemstones.

One of the earliest works with information about Arizona gemstones is George Kunz's 1890 book, *Gems and Precious Stones of North America*. He lists garnet, olivine, quartz, turquoise, azurite, malachite, and petrified wood from Arizona. In 1917, Frank Culin authored one of the first publications related to gemstones in Arizona, *Gems and Precious Stones of Arizona*. He lists about a dozen minerals that he considered gems, with turquoise, garnet, peridot, and chrysocolla being the most important.

Azurite

Arizona has hundreds of azurite localities, and azurite from many of Arizona's copper mines and prospects has been used as lapidary material. The large open-pit and underground copper mines in Arizona have contributed the bulk of azurite production and have put large amounts of material on the market. The mines at Bisbee, in Cochise County, have produced large quantities of gem-quality azurite. Masses of azurite up to several tons were found. A polished azurite from the Junction shaft in Bisbee is featured on the U.S. mineral stamp issued in 1992.

At Morenci, in Greenlee County, gem-quality azurite was also commercially recovered. The best material from Morenci included combinations of azurite with malachite eyes when stalactitic formations were cut across the grain and when banded material was cut parallel to the grain.

The Blue Ball Mine, in Gila County, has produced thousands of azurite nodules up to a few inches in diameter, which are often hollow and lined with sparkling crystals. They are split and used in jewelry (Grant 1989).

FIGURE 1.1 Azurite-cementing breccia, San Xavier North Mine, Pima County, polished butterfly, 5.6 cm high.

Beryl

Beryl is found at many places in Arizona, in pegmatites, alpine veins, and tungsten deposits. Most of this beryl is opaque and not gem quality. But a few localities have produced facet-quality aquamarine. The best known is the Bella Donna claim in the Sierrita Mountains in Pima County, where beryl occurs as beautiful blue-green crystals in quartz (Galbraith and Brennan 1970). Stones up to forty carats have been faceted from aquamarine crystals from the Sierrita Mountains (Muntyan 2010). Another location with potential gem material is on Groom Peak in Mohave County (Nyle Niemuth, pers. comm., 2014). Recent activity has produced some aquamarine from the Santa Teresa Mountains. This material is not facet quality but can be used in jewelry.

Calcite

Calcite in the form of travertine, which is sometimes called Mexican onyx or onyx, has been mined and collected at many localities in Arizona. The travertine is usually deposited by springs, and much of it was formed during the Ice Age, when the climate in Arizona was wetter.

The most colorful travertine in Arizona comes from a deposit just north of Mayer, Yavapai County, a popular spot for collectors. It has been mined commercially, and some was shipped to Omaha, Nebraska, to make ornamental items. The travertine has red, yellow, brown, and white bands. Another location is the white travertine deposit about ten miles north of Kingman in the Stockton Hills. This location was claimed in the past by the Kingman lapidary club. Other well-known travertine locations include one about ten miles south of Ash Fork, in Yavapai County, and one three miles north of Seven Springs, near Cave Creek, in Maricopa County.

Marble, the metamorphic rock made up of calcite, is present at multiple localities in southern Arizona, where Paleozoic limestone formations have been metamorphosed. Arizona Bureau of Mines Bulletin 180 (Stanley B. Keith 1969) has the following statement about Arizona marble: "Most of the marble is highly fractured, contains crosscutting igneous dikes, or has color variations which limit the size of blocks or the amounts that have uniform color. In general, the localities are not good sources of large blocks of high-quality, finished construction stone, but are a source of smaller monument and decorative facing stone." This reference lists thirteen potential commercial marble locations in Arizona. Total production is small, amounting to only several thousand tons over the years. One of these locations is in Hewitt Canyon, near Queen Valley, Pinal County. It has pale pink-, green-, and white-striped marble and has been used for lapidary arts.

Chalcopyrite

Chalcopyrite is the most common mineral mined for copper in Arizona. It has been marketed as a cutting material called Apache gold or Geronimo's gold. A large quantity of this material came from the United Verde Mine at Jerome, in Yavapai County, where it occurred as bright-yellow chalcopyrite in a black schist matrix. Almost all the copper mines in Arizona are a potential source for chalcopyrite to be used as cutting material.

Chrysocolla

Chrysocolla is one of the most common secondary copper minerals found in Arizona, and it occurs in every county in the state. Occurring in various shades of blue, most of it is very friable and soft, and therefore it is not good lapidary material. It can be stabilized with epoxy to make a useful gemstone. Also, chrysocolla often acts as a coloring agent for chalcedony, which is referred to as gem chrysocolla, gem silica, chrysocolla chalcedony, or chrysocolla aventurine, and it is a valuable gemstone. Earlier writers also referred to the chrysocolla-stained chalcedony as chrysoprase (Guild 1910), but this name should be used for green chalcedony colored by nickel. When small crystals of clear quartz cover the chrysocolla, it is referred to as *drusy chrysocolla.*

Chrysocolla chalcedony has been found at many copper mines in the state, but the best material is from mines in the Globe-Miami area in Gila County. The Keystone Mine, the Inspiration Mine, and the Old Dominion Mine have all produced good gem-quality material. Blake (1909) writes the following about the material from the Keystone Mine: "It varies in color from bright to pale blue, bluish-green and nearly apple green, resembling some chrysoprase. It bears some resemblance to the greenish colored chalchuite (turquoise), but is more brilliant and translucent. It takes a high polish, is hard and when well cut, en cabochon, and mounted forms a pleasing jewel." The best blue transparent material is rare and commands a high price. The Ray Mine, in Pinal County, has produced good chrysocolla chalcedony and a large quantity of drusy chrysocolla. Literally tons of gem chrysocolla were mined at the Bagdad Mine in Yavapai County in the 1980s and 1990s (Wallace 2012; Muntyan 2015a).

FIGURE 1.2 Gem chrysocolla, Ray Mine, Pinal County, 14.3 ct faceted stone in ring with rough chrysocolla.

Corundum

Corundum is not a common mineral in Arizona, and only one locality has produced lapidary material: the Ruby No. 1 claim southwest of Kingman, in Mohave County, where gray to black single crystals up to three inches long occur in biotite schist. A dome cut on the *c*-axis shows pink/lavender chatoyancy and, rarely, a weak star. These stones are not true gem rubies because of their very poor color.

A corundum locality that needs investigation is the one described by Galbraith and Brennan (1970) of blue, red, and white material in a pegmatite dike with andalusite from Grand Wash Cliffs, Red Lake District, Mohave County. It is not known if this material is of gem quality.

Dumortierite

Dumortierite is a relatively rare mineral found in metamorphic rocks, and the pale-violet to blue massive material, mixed with quartz, is a good lapidary material. It was first found as float material along the Colorado River near Clip, north of Yuma. The main deposit is about three miles southwest of Quartzsite. Here it occurs in schist with several other minerals, including kyanite and andalusite.

Feldspar

There are small pieces of facet-grade anorthite, a variety known as labradorite, at Red Mountain, twenty-five miles northwest of Flagstaff. Red Mountain is an eroded cinder cone in the Coconino National Forest with a well-marked trail. Augite crystals and pieces of labradorite can be picked up from the eroded and weathered volcanic rock. Other basaltic volcanic rocks around the state also contain labradorite.

Two elusive gem feldspar localities in Arizona have been described. Arthur Flagg (1958) has the following entry: "On the slopes of the Trincheras Mountains are cleavage fragments of moonstone, a variety of feldspar. No effort has been made to discover better material in place; it may be there." This locality is unknown as there are no mountains officially named Trincheras in Arizona, unless this was a local name and not an official name. Sinkankas (1959) describes "andesine and andesine sunstone showing bright coppery reflections . . . reported from the Altar Mountains within the Apache Reservation near Globe." No record can be found for any Altar Mountains in Arizona.

Garnet

Several different garnet minerals are found in Arizona, including andradite, grossular, pyrope, and spessartine, but the facet-grade pyrope is the desired gem material. In Arizona it is found in Apache County on the Navajo Reservation in serpentinized ultramafic microbreccias. These rocks are found at Garnet Ridge and Buell Park. Native Americans and early settlers collected the pyrope because it could be faceted into good red stones up to several carats and was often sold as Arizona ruby.

FIGURE 1.3 Pyrope, Navajo Nation, Apache County, 13.9 ct faceted stone with rough pyrope.

George Kunz, in *Gems and Precious Stones of North America*, published in 1890, writes about the pyrope: "They have been found on ant-hills and near the excavations made by scorpions, having been taken there by the busy occupants as obstructions to the erection of their galleries and chambers." No mining has been recorded for the pyrope. Pieces are picked up on the surface, where they have weathered out of the rock.

The andradite garnet from Stanley Butte is well known to mineral collectors. It is a very distinctive green color with a metallic appearance. Cabochons and carvings exhibiting a metallic sheen have been made from this material.

K. A. Phillips and Boyd (1988) report the following from pegmatites in Arizona: "Wine red gem quality spessartites (spessartine) up to an inch in diameter have been found, but reports of such occurrences are very rare." No exact location was given. Reports of uvarovite from Arizona have not been verified, and whether any of the green garnets were gem quality is unknown. Schlegel (1957) reports uvarovite from Graham Mountain (Mt. Graham?) in Graham County. B. E. Thomas (1953) reports it from the Tennessee Mine, Cerbat Mountains, Mohave County.

Malachite

Malachite, like azurite and chrysocolla, is a very common secondary copper mineral found at numerous localities in Arizona. The best malachite has bands of various shades of green with a chatoyant pattern formed by compact acicular crystals when polished. Bisbee and Morenci have both produced quantities of good lapidary material. Some of the material from both localities is the banded mixture of malachite and azurite. The Miami area mines also have good malachite and malachite mixed with chrysocolla.

Olivine

Peridot is the name used for the gemstone variety of the mineral forsterite, also known as olivine, and in many of the older references it is called chrysolite. There are two main occurrences of peridot in Arizona. The first material was collected with pyrope garnet from the serpentinized ultramafic microbreccias at Garnet Ridge and Buell Park, Apache County (Kunz 1890). The grains of peridot were picked up while collecting the pyrope garnet. Blake (1909) reports that olivine "occurs in the country of the Navajo Indians, who collect the stones and sell to local traders. The gem is also known as peridot and locally as Job's tears." Guild (1910) writes: "It is found in beautiful tints near Talkai, and is frequently collected, together with the red garnets with which they are frequently found, by the Indians and prospectors. A crystal from this locality, exhibited at the World's Fair, at Portland, in 1905, after being cut, was a beautiful gem of 25 and ¾ carats."

FIGURE 1.4 Olivine variety peridot, Peridot Mesa, San Carlos Apache Reservation, Gila County, 15 ct faceted stone.

The most abundant Arizona olivine is found as nodules in basalt flows. There are multiple localities around the state, but the one at Peridot Mesa on the San Carlos Apache Reservation, in Gila County, stands out from all the others, with an amazing amount of olivine, which is mined for gem material. The volcanic eruptions of basalt that contain the olivine happened between one and four million years ago, and these nodules came from the mantle at an estimated depth of fifty to one hundred miles beneath the earth's surface. Stones up to thirty-five carats have been faceted from this material. Peridot Mesa supplies the majority of the world's supply of gem peridot. Sinkankas (1997) reports: "Much of Arizona peridot finds its way to the Hawaiian Islands, where it is set in jewelry and sold to tourists, with the implication that it is native to the Islands."

Opal

Opal forms as a late mineral in many geologic environments in Arizona. Gemquality precious opal is restricted to Tertiary silica-rich volcanic rocks. There are many claims for precious opal north of Ruby, in Santa Cruz County. They

FIGURE 1.5 Opal, Ruby, Santa Cruz County; largest cabochon is 27 × 33 mm, with slab of rough opal.

have been referred to as the Jay R claim and the Arizona Blue Fire Opal claim. Located about four miles east-northeast of Ruby, they have been worked sporadically since the early 1970s. They produce a blue opal with play of colors. Currently, the company Southern Skies Opal is marketing precious blue opal from Arizona (Lurie 2012).

A second occurrence of precious opal is in Morgan City Wash, west of Lake Pleasant, in Maricopa County. This locality is not well known and has produced only a few specimens.

Opal as a replacement for asbestos is reported from the Blue Mule Asbestos Mine in Bear Canyon, in Gila County (Sinkankas 1997). Several tons of material were recovered from the mine dump, and Sinkankas (1997) gives this description: "when replacement of the asbestos [by opal] is complete, the chatoyancy of the asbestos is faithfully preserved, and very handsome cabochons can be cut from the material in dark red, brown, pale brown, yellow, white, and dark green hues."

Quartz

Quartz is one of the most common minerals in the earth's crust, and because it has a hardness of seven, it makes a good gemstone. It also takes on a wide variety of forms and colors, many of which were considered distinct minerals in the past. These include agate, chalcedony, jasper, flint, prase, chrysoprase, and chert. There are also many local names for varieties of quartz, such as all the different names given to agates. Some of the more important varieties of Arizona quartz will be described separately below: chalcedony and agate, amethyst, jasper and chert, and petrified wood. These are all quartz, and in which category the material fits is somewhat arbitrary.

Clear facet-grade quartz is found at many localities in Arizona. *Arizona diamonds* resemble Herkimer diamonds, with high clarity, and are found at the Diamond Point area in Gila County.

Clear quartz with inclusions such as rutile, chlorite, and other minerals make the most interesting gemstones. Sinkankas (1959) reports about the Crystal Peak locality south of Quartzsite: "In 1956, collectors at the well-known Crystal Peak locality in Yuma County (now La Paz County), found a pocket of quartz crystals which contain numerous silky inclusions of an unidentified nature and from which exceptionally handsome chatoyant cabochons and cat's-eyes have been cut. The author can recall of no chatoyant quartz from any locality in the world in which the inclusions are so fine and so brilliantly reflective."

The Santa Teresa Mountains, in Graham County, have produced a lot of quartz, including good quality smoky, clear, and pink crystals. Occasional crystals from this area contain needles of rutile. Several areas in the Huachuca Mountains, Cochise County, have produced large crystals with various inclusions, including green chlorite. Rose quartz is not abundant in Arizona, but a deposit on Sycamore Creek in Maricopa County was mined for ornamental stone. It is generally a pale pink and not a great lapidary material.

Quartz, Variety Amethyst

Amethyst is quite common as a gangue mineral in many mineral deposits in Arizona, and many localities have gem material. It is commonly cut into cabochons and slabs, although several localities have produced facet-grade stones.

Facet-grade material occurs in the McConnico District located southwest of Kingman, in Mohave County. Guild (1910) reports about amethyst: "In the McConnico District it is found in Precambrian rocks, sometimes in samples of great beauty. One crystal is reported to have been sold to Tiffany & Co., of New York City, for $59.00." Schrader (1908) reports from the McConnico District: "Amethyst associated with some stibnite and galena, is found about one-fourth of a mile northeast of Boulder Spring, in the foothills, probably as a vein pegmatite dike. This amethyst-bearing property is owned by Kinsman and Solomon. It is developed by a 65-foot shaft and has produced some good stones." There is no recent information about this locality.

The most famous Arizona amethyst locality is the Four Peaks Amethyst Mine or Arizona Amethyst Mine in Maricopa County. The early history of the

FIGURE 1.6 Quartz variety amethyst, Four Peaks, Maricopa County, 80 ct faceted stone with rough amethyst.

location is vague. Jim McDaniels is given credit for finding the deposit in the early 1900s (J. Lowell and Rybicki 1976). There is also a legend that the Spanish found it in the 1700s and that there is amethyst from Four Peaks in the Spanish crown jewels. The property was patented in 1942 and has been worked intermittently until the present day. It has produced large quantities of amethyst, but only a small percentage has the rich red-violet color for which the deposit is famous (Korwin 2013).

Quartz, Varieties Chalcedony and Agate

Chalcedony and agate are the common names used for fibrous fine-grained quartz. Normally people use *agate* for the banded form, although material like fire agate and moss agate are exceptions. Chalcedony and agate localities are scattered over most of Arizona.

FIGURE 1.7 Quartz variety agate, Fourth of July Butte, Maricopa County, cut and polished nodule, 5.5 cm high.

Usually, chalcedony is colorless or white and not a valuable lapidary material. It can be found everywhere in the state where there are Tertiary silica-rich volcanic rocks. When chalcedony forms the shape of roses, they can be used in jewelry, and the chalcedony usually has a bright-green fluorescence under shortwave ultraviolet light, making it of interest to fluorescent mineral collectors. When there is a coloring agent present, chalcedony makes a good gem material. When nickel is the coloring agent, chalcedony is called chrysoprase. In the Plomosa Mountains, north of Interstate 10, Raymond Perry had a chrysoprase mine in the 1960s. The chalcedony was colored green by the nickel mineral willemseite (Melchiorre 2017).

Fire agate is chalcedony with layered inclusions of an iron oxide mineral, which allows light to be diffracted into the colors of the rainbow when the stone is cut correctly. Arizona has several well-known fire agate localities. One is near Ed's Camp east of Oatman, in Mohave County. This is a pay-to-dig location, in the Black Mountains, and it is possible to look at other places along this range for fire agate. There are several collecting areas for fire agate in eastern Arizona near Safford, in Graham County, and near Clifton, in Greenlee County. The Bureau of Land Management has designated two rockhound collecting areas for collecting fire agate. The Black Hills rockhound area is about eighteen miles north of Safford, and the Round Mountain area is southeast of Duncan.

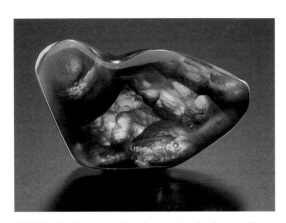

FIGURE 1.8 Fire agate, San Carlos Apache Reservation, Gila County, 17 ct polished stone.

Arizona has produced some extremely fine agates. The deposits are small, and none has produced large quantities of material. Floyd Getsinger did a summary of the agate locations around Arizona in the Lapidary Journal publication *The Agates of North America* (Getsinger 1966). Another good source of information is the book *Agates*, which includes fifty-eight pages of photographs of agates from forty localities in Arizona (McMahan 2016).

One area currently being worked for agate is Mulligan Peak, north of Clifton, in Greenlee County. The agates show fine banding and are found in purple, lilac, and gray colors. The area is remote, and there are some valid mining claims for agate in the area.

Quartz, Varieties Jasper and Chert

The names *jasper* and *chert* are used for opaque fine-grained quartz with a more granular structure. Chert is very common in northern Arizona, as it forms beds and nodules in several of the limestone formations in the region. Both the Kaibab and Redwall Limestones have abundant chert. As the limestone weathers, the chert is accumulated and covers the surface. Unfortunately, most of it is pale gray to white and of little interest, but in some localities, it is red, and near Payson, Gila County, there is an alternating gray- and white-layered chert called zebra agate.

Jasper has been found at many localities in the state, and some small-scale mining has taken place. One of the best is the Cave Creek jasper, found in an area about eighteen miles north of Cave Creek, in Maricopa County. There is some orbicular jasper with bright-red bands, about which Sinkankas (1959) reports: "This material ranks among the finest jasper-like quartz gemstones of North America."

Quartz, Variety Petrified Wood

Petrified wood could be considered under chalcedony and agate, but it is such an important lapidary material from Arizona that it deserves a separate discussion. It is the state fossil of Arizona, specifically the petrified wood of

Araucarioxylon arizonicum. Nowhere else in the world has produced the quality and quantity of petrified wood that Arizona has. The wood is bright red, yellow, or multicolored, and the texture and patterns of the wood are preserved in detail. Tons of petrified wood are collected to be used for things such as jewelry, bookends, and tabletops. It is found in several localities around the state, but none can compare with the material from the Triassic Chinle Formation in Apache, Coconino, and Navajo Counties.

The main source of petrified wood is the Holbrook area. Blake (1909) writes:

> The following is clipped from the Sioux Falls (South Dakota) Journal: "The polishing works of this city is now engaged on the stupendous job of getting out $1,000,000 worth of polished chalcedony, or petrified wood, to be taken to the Paris Exposition (1900?). This petrified wood is hauled from its native heath in Arizona, a distance of sixty-five miles, to a railroad, and then shipped to this city to be cut and polished. It shipped here in great logs and stumps weighing many tons each, just as they have lain for many ages during the process required by nature to turn the wood into beautiful stone. After being cut and polished the stone is worked up into every conceivable shape, from cuff buttons to tops for center tables and great columns, which cost a small fortune. All kinds of jewelry are made from it, as well as trinkets and handsome articles for souvenirs."

FIGURE 1.9 Quartz variety petrified wood, Holbrook, Navajo County, polished slab 49 cm in diameter.

The year that this clipping appeared is not known, but to stop the loss of petrified wood, the Petrified Forest National Monument was established in 1906 with 44,000 acres. The monument had 53,000 acres added in 1932, and in 1962 it was declared a national park. Wood is also collected for sale in large areas not in the park.

An unusual petrified wood is found near Winslow, in Navajo County. It is a deep-green color because the chalcedony is colored green by chromium.

Serpentine

In the past, chrysotile asbestos has been mined commercially in the area around the

Salt River Canyon, in Gila County. It was notable because the fibers are up to four inches long. Associated with the asbestos is massive antigorite, a member of the kaolinite-serpentine group of minerals. Sinkankas (1959) notes: "Translucent to opaque serpentine in fair cutting grade is found here in some abundance but though extremely compact, its colors are dull shades of olive-green and yellowish-green." There are similar deposits in the Grand Canyon, but they are inaccessible, and collecting is prohibited.

Shattuckite

Although shattuckite is not a common lapidary material, the interlocking rosettes of radiating crystals from the New Cornelia Mine in Ajo, Pima County, make very fine cabochons. Large quantities in veins up to several inches wide have been found (W. J. Thomas and Gibbs 1983). Shattuckite from Bisbee is also reported as large compact masses (Graeme 1981). There is a cabochon of shattuckite in the American Museum of Natural History from Jerome, in Yavapai County (Sinkankas 1959). The Azurite Mine in Pinal County was mined for lapidary grade shattuckite in white quartz.

FIGURE 1.10 Shattuckite, New Cornelia Mine, Pima County, 30 × 40 mm cabochon and rough shattuckite.

Topaz

Topaz is found in pegmatites throughout the state, but no facet-grade topaz has been found. Several Tertiary rhyolite occurrences have topaz similar to Thomas Range, Utah, specimens. These have been found in the southern Aquarius Mountains and in a few other localities (Burt, Moyer, and Christiansen 1981) but generally are small. The biggest clear crystals in Arizona are from the location referred to as Saddle Mountain, in Pinal County (White 1992). This deposit is on the San Carlos Apache Reservation and is not available for collecting.

Other potential facet-grade topaz localities may exist in Arizona, although none has been very well documented. Arthur Flagg (1958) notes that "A single water-clear crystal of pure topaz was picked up on the alluvium not far west of Bouse." Nothing else is reported about this occurrence.

Tourmaline

The lithium-bearing pegmatites in Arizona have produced some elbaite, a member of the tourmaline group of minerals. The material is generally opaque and often altered. Jahns (1952) describes the elbaite, including some watermelon tourmaline: "A few of the concentrically zoned crystals are typical 'watermelon' tourmaline, in that their green outer rims surround thin colorless layers, which in turn enclose pink cores."

Turquoise

Turquoise, the official gemstone of Arizona, has been collected and mined in Arizona since prehistoric times. Many thousands of tons of turquoise have been recovered from Arizona localities. Some areas are distinctive enough to be readily identifiable by experts. Colors range from green through blue to a fine robin's-egg blue. Early miners found prehistoric diggings where Native Americans mined turquoise for ornamental uses. Since then, most turquoise has come from the major copper mines, where contractors commercially recovered the turquoise. Other properties were mined solely for turquoise. There are not many unclaimed areas where collectors can go and find this gemstone.

One of the major commercial producers was Turquoise Mountain in the Turquoise District near Gleeson, Cochise County. Several claims have been worked almost continuously since their discovery in the 1890s. At one time, Tiffany & Co. of New York is reported to have leased one of the properties and had forty miners working. Other turquoise producers include Bisbee, in Cochise County; the Sleeping Beauty Mine (Copper Cities pit) near Miami, in Gila County; the Pinto Valley Mine (Castle Dome), in Gila County; the Mineral

FIGURE 1.11 Turquoise cabochons: Sleeping Beauty Mine, Miami, Gila County, 31 mm high; Kingman, Mohave County, 45 mm high; and Bisbee, Cochise County, 39 mm high.

Park Mine, in Mohave County; and the Morenci Mine, in Greenlee County. Around the Mineral Park area, extensive prehistoric workings for turquoise have been found.

Today, high-quality untreated turquoise is harder to find, and large amounts of poor-quality turquoise is stabilized and color enhanced by commercial producers for sale to the lapidary market. This material can be difficult to distinguish from high-quality untreated natural turquoise.

Variscite

Good cutting-quality variscite was found in the Cole shaft at Bisbee, in Cochise County, on the 1,200 level, as compact masses up to thirty-three pounds, ranging in color from pale to deep green (Graeme 1981).

Schlegel, in Gem Stones of the United States, *lists variscite from Maricopa County, Alamo Mountain, near Aguila (Schlegel 1957). Present-day information has no Alamo Mountain in Maricopa County, and no other information about this occurrence has been found. Alamo Lake is about thirty miles from Aguila, and the location is probably near there.*

Other Lapidary Materials
Campbellite

Campbellite is a mixture of minerals and is named after the Campbell Mine at Bisbee, in Cochise County (Mueller 2012). Campbellite can have a combination of the following minerals: copper, cuprite, azurite, calcite, chrysocolla,

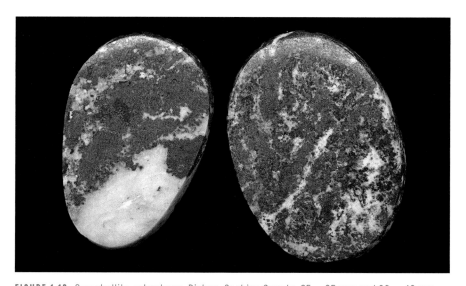

FIGURE 1.12 Campbellite cabochons, Bisbee, Cochise County, 25 × 35 mm and 30 × 40 mm.

malachite, pyrite, quartz, and possibly other minerals. It has been used for cabochons, spheres, and ornamental items. Arizona collectors use the name for the material from Bisbee, but combinations of these minerals can be found at many copper mines in the state. Material containing copper and cuprite with the blue and green copper minerals is very attractive.

Obsidian

Obsidian is not a mineral but a natural glass formed from fast cooling of silica-rich magma. Over time obsidian will hydrate and change to perlite. The remaining areas of unaltered obsidian are called marekanite, named after a river in Russia, or Apache tears in Arizona.

Steve Shackley's website Sources of Archaeological Obsidian in the Greater American Southwest (www.swxrflab.net/swobsrcs.htm) has an interactive map of Apache tear locations in the Southwest. The map has nine locations in Arizona, a couple on the New Mexico–Arizona border, and many others in surrounding states. The text lists a couple of additional Arizona locations. For these localities, the website provides township, range, and section data as well as a detailed description of what was found and, in some cases, photographs and maps. There are a lot of chemical data for the obsidian because the goal was to identify the sources used by prehistoric people. The obsidian from each of the various localities has a unique composition. For example, 220 samples of obsidian were analyzed from Pueblo Grande in Phoenix. Four were from the Sand Tanks location, 67 were from the Sauceda Mountains locality, and so forth.

For collectors, the best-known location is just west of Superior, in Pinal County, where perlite was commercially mined. Thousands of pounds of Apache tears have been collected and used for specimens and jewelry.

An interesting location is Topaz Basin, in Yavapai County. It is close to Interstate 17, by the route 169 turnoff. It seems that the small clear obsidian nodules found there were misidentified as topaz.

Another locality has been described by Sinkankas (1959): "Apache tears are abundant in several places in Maricopa County. Beautifully transparent nodules, frequently cut as faceted gems, are found in a wash a short distance west of Aguila on Highway 60-70."

2 Fluorescent Minerals in Arizona

Arizona is a well-known source of fluorescent minerals due to its varied and widely distributed deposits of lead, zinc, manganese, tungsten, uranium, and other ore commodities. Occurrences include either *self-activated* minerals, where the fluorescence is based purely on intrinsic composition, or minerals containing trace impurities, which serve as *activators*. Given the growing interest in fluorescent mineral collecting, this chapter explores some of Arizona's notable fluorescent occurrences and presents information on unusual fluorescent responses along with remarkable localities.

The mineral district maps in appendix A offer a way of predicting new potential mineral occurrences by examining the relationships of ore commodities, geology, age of mineralization, and existing mineral localities. So, to help in the search for new fluorescent mineral occurrences, the featured mineral examples include notes and references about a location's ore deposits and geologic setting. Discovering new possible sites for fluorescent minerals, however, involves some additional considerations, which are discussed in the background on fluorescence.

Background on Fluorescence

For mineral collectors, fluorescence may be defined as a type of luminescence that involves ultraviolet (UV) light and occurs at room temperature. When a specimen is illuminated by a UV source, the mineral absorbs energy, which stimulates electrons into an excited state. A visible light is emitted as the electrons return to their unexcited state. Usually, when the UV light is removed, the fluorescence decays very quickly, in a microsecond or less. Some minerals, however, will continue to glow for many seconds afterward, a property called *phosphorescence*. A few minerals may produce a visible light when rubbed or scratched. This is called *triboluminescence*.

Not all mineral species are fluorescent. Fluorescence has been observed in only about 15 percent of the currently known minerals. For fluorescence to occur, two conditions must be met. The first condition is a sufficient concentration of luminescence centers, or activators. These localized sources of light emission involve a group of ions and their interactions with a crystal lattice and modes of energy transfer. If the ions are part of a mineral's defined chemical composition, the mineral is said to be intrinsically fluorescent. The number of intrinsic minerals is relatively limited and may involve anions, such as the tungstate (WO_4^{2-}) group. More frequently, luminescence centers are created by the substitution of cations with impurities during the formation of a mineral. These minerals may be referred to as *impurity fluorescent minerals*.

A second key requirement for fluorescence is a low amount of quenching. Quenching involves energy transfer that does not produce light emission and may effectively reduce or eliminate the fluorescence of a mineral. This suppressing effect may be due to the presence of metallic ions, such as iron, cobalt, nickel, or copper; high concentrations of impurities; or temperature-related lattice interactions.

Luminescence centers may produce different fluorescent responses depending on the wavelength and associated energy of the UV light source. Ultraviolet light is divided into bands according to the wavelength, which is measured in nanometers (nm). Shortwave UV has a range of 200 to 280 nm, and lights in this band typically use mercury-vapor bulbs with a peak energy output at 254 nm. As a result of the high-energy photons, most fluorescent minerals (about 90 percent) will fluoresce with a shortwave source.

Longwave UV involves wavelengths from 315 to 400 nm. Lights include conventional mercury-arc bulbs, where the peak output occurs at 365 nm, and more recently low-cost, light-emitting diodes (LEDs), with 365 to 395 nm wavelengths. Approximately 15 percent of fluorescent minerals will respond to the lower energy of a longwave source.

Midrange UV lights lie in between shortwave and longwave sources, with wavelengths between 280 nm and 315 nm. These units represent a relatively new source for illuminating fluorescent minerals and use bulbs with phosphor coatings where the peak output may be at 302 nm or some other wavelength depending on the phosphor. Interest in this band is growing since these lights may reveal different color emissions or possibly fluorescence where no response was observed with either shortwave or longwave UV.

Arizona Fluorescent Occurrences

While Arizona certainly has many outstanding fluorescent occurrences, the number of verified occurrences is somewhat less certain. The literature often does not include information on fluorescence, or if an indication appears in a report, it may not be clear whether the observation was based on samples from Arizona. An additional complication involves potential misidentification due to coatings, mixtures, or backlighting by adjacent minerals.

The third edition of *Mineralogy of Arizona* mentions fluorescence eleven times, and only seven minerals are specifically identified as having fluorescent occurrences (cristobalite/opal, fluorapatite, mimetite, quartz/chalcedony, scheelite, scolecite, and willemite). A review of referenced observations has increased the number of Arizona fluorescent minerals to forty-six species. An annotated mineral list indicating the type and composition of main activators appears in table 2.1. Note that valence states—the number of electrons that an atom or group of atoms may add, lose, or share in bonding with other atoms—of the activators are indicated as superscript numbers. The fluorescent response depends on this state. Table 2.2 summarizes the type of activators and fluorescent responses for these minerals.

TABLE 2.1 Annotated list of Arizona minerals reported to be fluorescent

Bold typeface	indicates intrinsic mineral
Italic typeface	indicates impurity mineral
Regular typeface	indicates no activator has been identified
?	indicates doubtful fluorescence

Andersonite $(UO_2)^{2+}$	*Magnesite* (Mn^{2+})	*Spodumene* (Mn^{2+})
Aragonite (Mn^{2+})	Massicot	**Stolzite** $(WO_4)^{2-}$
Bayleyite $(UO_2)^{2+}$	Matlockite	**Swartzite** $(UO_2)^{2+}$
Becquerelite? $(UO_2)^{2+}$	**Metazeunerite?** $(UO_2)^{2+}$	*Tremolite* (Mn^{2+})
Calcite (Mn^{2+})	Mimetite	Tridymite?
Calomel $(Hg_3)^{2+}$	Minium	**Uranocircite** $(UO_2)^{2+}$
Cerussite (Pb^{2+})	**Natrozippeite** $(UO_2)^{2+}$	**Uranopilite** $(UO_2)^{2+}$
Cristobalite	**Powellite** $(MoO_4)^{2-}$	**Weeksite?** $(UO_2)^{2+}$
Dolomite (Mn^{2+})	*Quartz* $(UO_2)^{2+}$	Wickenburgite
Eucryptite (Fe^{3+})	**Sabugalite** $(UO_2)^{2+}$	*Willemite* (Mn^{2+})
Fluorapatite (Mn^{2+})	**Scheelite** $(WO_4)^{2-}$	*Wollastonite* (Mn^{2+})
Fluorite (Eu^{2+})	**Schröckingerite** $(UO_2)^{2+}$	**Wulfenite** $(MoO_4)^{2-}$
Gypsum (organic)	Scolecite	**Zinczippeite?** $(UO_2)^{2+}$
Hemimorphite	*Smithsonite* (Mn^{2+})	Zunyite
Hydrozincite (Pb^{2+})	**Soddyite?** $(UO_2)^{2+}$	
Leadhillite	*Sphalerite* (Mn^{2+})	

TABLE 2.2 Summary of Arizona minerals reported to be fluorescent

Intrinsic minerals	18
Impurity minerals	17
Minerals with no activator identified	11
Total minerals	**46**
Minerals that respond to shortwave UV	43
Minerals that respond to midrange UV*	14
Minerals that respond to longwave UV	41
Minerals that respond to both shortwave and longwave UV	38
Minerals that respond to both shortwave and midrange UV	14
Minerals that respond to both midrange and longwave UV	14
Minerals that respond to all three UV bands	14

*Information on the midrange responses for Arizona fluorescent minerals was not available. The number is based on the online luminescent database (fluomin.org), which includes responses for various localities involving a 320 nm UV source. The database was accessed on November 5, 2020.

Intrinsic Fluorescent Minerals

Intrinsic minerals are a class of fluorescent minerals where light emission is purely due to a mineral's chemical composition. These minerals are said to be *self-activated* by cations, such as lead (Pb^{2+}) or uranyl (UO_2^{2+}), or by anions, such as the molybdate (MoO_4^{2-}) or tungstate (WO_4^{2-}) groups. Although the number of intrinsic minerals is relatively limited, eighteen of the forty-six reported Arizona fluorescent minerals involve this class, which reflects the widespread lead, molybdenum, uranium, and tungsten mineralization found in the state.

Andersonite, Bayleyite, and Swartzite

Andersonite, bayleyite, and swartzite are secondary uranium carbonate minerals that contain the highly fluorescent uranyl (UO_2^{2+}) ion group. This ion group produces some of the brightest fluorescent responses and is typically associated with a yellow-green fluorescence under shortwave ultraviolet. The emission color, however, can vary depending on mineral's crystal structure and associated energy interactions between uranyl groups.

Anderson, Scholz, and Strobell (1955) reported both shortwave and longwave fluorescence for these uranium carbonates, with andersonite exhibiting a very bright green-white response; bayleyite a weak green-white to indeterminate emission; and swartzite a bright yellowish green. Andersonite, bayleyite, and swartzite were first found at the Hillside Mine in the Eureka supersystem, Hillside District in Yavapai County. The underground lead-gold-zinc-silver-uranium mine, which was discovered in 1887, is located three miles north of Bagdad. The fissure-vein deposit occurs in Proterozoic quartz-mica schist with an associated rhyolite dike in the 76 to 70 Ma (million years old) Bagdad porphyry copper intrusive suite (Barra et al. 2003). The uranyl-bearing minerals occur on the mine walls as intergrown, efflorescent coatings with gypsum.

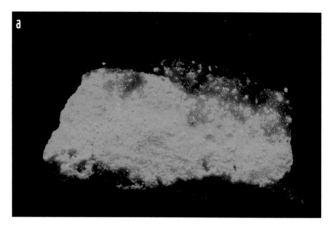

FIGURE 2.1 Andersonite on gypsum, 3.2 × 1.9 cm, Hillside Mine, Yavapai County. Upper image (a) under shortwave UV. Lower image (b) under white light.

Cerussite

Cerussite is a lead carbonate that is found in the oxidized zone of lead ore deposits throughout Arizona. The lead (Pb^{2+}) ion is an intrinsic activator and is associated with a yellow fluorescence that usually appears stronger under longwave ultraviolet. Note that lead has an atomic radius comparable to many metal ions, so it may also occur as an impurity in other minerals.

One notable cerussite occurrence is in the Grand Reef Mine in the Aravaipa supersystem, Grand Reef District, in Graham County, where the mineral may assume various different habits, including large sixling-twinned crystals. Samples may exhibit a yellow fluorescence with shortwave and longwave ultraviolet, but the response may be muted because of the area's extensive iron staining. The presence of only 1 percent or less ferrous iron (Fe^{2+}) may diminish fluorescent emissions (Verbeek 1995). The Grand Reef Mine is a former underground lead-copper-silver-zinc-gold mine discovered in 1890 and operated until 1955. The mineralization is hosted in Precambrian Pinal Schist and silicified limestone breccia. Lead and zinc ore deposits are associated with rhyolite dikes that have been dated at circa 22 Ma (Wrucke et al. 2004).

FIGURE 2.2 Cerussite, 3 cm FOV, Grand Reef Mine, Graham County. Upper image (a) under longwave UV. Lower image (b) under white light.

Cerussite from the Hull Mine in the Castle Dome District, in Yuma County, displays a bright-yellow fluorescence. Many of the cerussites are on fluorite that is also fluorescent (Les Presmyk, pers. comm., 2020). The Hull Mine is part of a lead-copper-silver mine group that was discovered in 1863. Most of the ore was mined prior to 1900, but small-scale production continued through 1974. Mineralization involves altered argentiferous galena with fluorite, baryte, calcite, gypsum, and quartz. The local geology consists of Mesozoic shale, slate, and limestone intruded by diorite and quartz porphyry dikes of probable Laramide age (E. D. Wilson et al. 1951; Stanton B. Keith 1978). At the time of this writing, the Castle Dome Museum in Yuma conducted underground Hull Mine tours, which include a fluorescent mineral cavern about 30 feet wide and over 50 feet tall.

Powellite

Powellite, a calcium molybdate, is an uncommon secondary mineral associated with tungsten ores. The mineral emits a characteristic lemon-yellow color with shortwave UV. It forms a solution series with scheelite, which leads to an interesting caveat relating to powellite fluorescence. During the government-supported tungsten-prospecting programs of the 1940s and 1950s, it was shown that small amounts of molybdate (typically about 10 percent) could be present in the scheelite and would produce a yellowish shortwave fluorescence that mimics the response of powellite. The U.S. Geological Survey subsequently produced fluorescent prospecting cards that would distinguish scheelite-powellite fluorescence gradients as a function of molybdate/tungstate composition (Verbeek 1995; Werner and Nikisher 2016). In this sense, it is quite possible that some reported powellite occurrences may, in fact, be scheelite.

Arizona has more than twenty-seven powellite occurrences (Wilt and Keith 1980). One of these localities is the Flying Saucer Group in Maricopa County's Flying Saucer District. The Flying Saucer District is known for gold-bearing quartz veins in granites. Pyrite, galena, sphalerite, and chalcopyrite are found below the oxidized zone. Mineralization connected with the Wickenburg batholith may be of Laramide age, while quartz veins associated with granite porphyry dikes may have formed in the mid-Tertiary (Welty, Spencer, et al. 1985). The Flying Saucer Group is a tungsten-molybdenum pros-

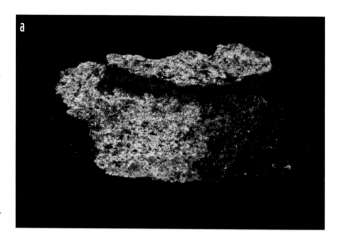

FIGURE 2.3 Powellite, 9.6 × 7.5 cm, Flying Saucer Group, Vulture Mountains, Maricopa County. Upper image (a) under shortwave UV. Lower image (b) under white light.

pect, which is located 2.6 miles west of Wickenburg. The area consists of granite with intrusions and pegmatites in some places and with short, narrow quartz veins in other locations. These pegmatites postdate the copper-related Wickenburg batholith and are considered late Laramide peraluminous intrusions that regionally are related to the tungsten and/or gold mineral systems. Peraluminous refers to a geochemical igneous rock classification where the proportion of alumina is greater than the combined amount of sodium, potassium, and calcium oxides. Powellite occurs as fine to medium grains or flakes disseminated in shear zones or quartz veins. Although production never started at the site, an area near Scorpion Hill was noted where "intense fluorescence is 400 feet long and has a maximum width of nearly 200 feet" (Dale 1959).

Scheelite

Scheelite (calcium tungstate) is the best-known tungstate mineral. It exhibits a distinctive, bright blue-white response under shortwave ultraviolet light. As noted earlier, however, the presence of molybdenum may modify the color. With a molybdenum weight percentage of only 0.1 percent, the color is a light pale blue, and scheelite is nearly white at 0.8 percent and pale yellow at 3 percent (Verbeek 1995). Samples are almost always fluorescent, and scheelite has been used in *ultraviolet prospecting* of tungsten and gold deposits.

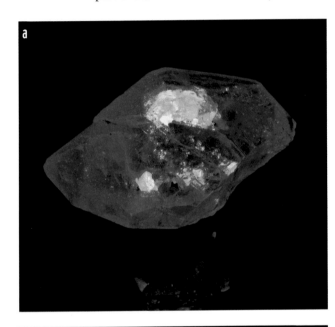

FIGURE 2.4 Scheelite, 2.5 cm, Johnson Camp, Cochise County. Upper image (a) under shortwave UV. Lower image (b) under white light.

Scheelite is widely distributed in Arizona and has been found at most of the state's fourteen tungsten-producing districts (Rubel 1917–18). The Teviston District in Cochise County has several sporadic base-metal sulfide and oxide deposits in irregular quartz veins that cut the Precambrian Pinal Schist. The age of the mineralization is uncertain and may be related to Laramide and mid-Tertiary intrusions (Welty et al. 1985b). One of the district's notable occurrences is the Cohen Tungsten Mine. which is located ten miles east of Wilcox. The small tungsten-lead-silver mine was discovered in 1944 and comprises seven unpatented lode claims that were worked intermittently until 1956. Large scheelite crystals along with small amounts of galena were found in quartz veins near a contact between altered diorite and light-colored trachyte (Dale, Stewart, and McKinney 1960).

The dark-orange scheelite specimens from the Cohen Tungsten Mine are noteworthy not only for their exceptional size, but because they fluoresce an unusual orange color with longwave ultraviolet. Rare-earth element impurities may be a possible cause. Gaft, Reisfeld, and Panczer (2015) reported similar emissions involving trivalent rare-earth impurities in various scheelite samples.

Scheelite has been found as crystals embedded in quartz crystals in Johnson Camp area, near the Republic Mine. The uncommon pale-yellow to white octahedral crystals may be a few millimeters in size and exhibit a bright bluish-white fluorescence under shortwave ultraviolet. The clear to milky-white quartz crystals line small vugs and open fractures in quartz monzonite. Located in the Little Dragoon Mountains, Johnson Camp is part of the Johnson Camp District of the Cochise supersystem. The Precambrian Pinal Schist is the area's oldest formation, which was intruded by quartz monzonite of Late Cretaceous or Early Tertiary age. Copper-zinc deposits are found in extensively metamorphosed Paleozoic limestones, while disseminated scheelite occurs in many contact-metamorphic zones. Mining activity began in 1880s, and intermittent production has continued under various owners. The Republic Mine is a former underground mine that produced from 1882 to 1952. (Wilson 1941; Cooper and Huff 1951; Cooper and Silver 1964).

Impurity Fluorescent Minerals

Impurity minerals represent the largest class of fluorescent minerals because the substitution of cations with impurities is fairly common. These impurities interact with the host mineral's crystal lattice and with each other, resulting in different modes of energy transfer. Energy may be released directly to produce light emission; conveyed to other impurities that will later emit light; or converted by processes that do not emit light.

It is important to note that only trace amounts (less than 1–2 percent) of an impurity may be needed for fluorescence to occur. Higher concentrations can lead to *self-quenching*, where light emissions may be reduced or eliminated because of reabsorption of light energy by unexcited activators. A mineral's fluorescence may also be limited by trace amounts of cobalt (Co^{2+}), copper (Cu^{2+}), iron (Fe^{2+}), or nickel (Ni^{2+}), which introduce non-light-emitting charge-transfer states. For example, nickel may decrease the fluorescence of sphalerite with a level of only one to ten parts per million (Verbeek 1995).

The list of reported Arizona fluorescent minerals includes seventeen impurity minerals, which is a significant underrepresentation of the known impurity minerals. The low number probably reflects a reporting issue and offers an opportunity to add to the list of Arizona fluorescent occurrences.

Calcite

Calcite frequently exhibits an orange-red fluorescence under shortwave and longwave ultraviolet. The main activator for this emission is the manganese (Mn^{2+}) ion, which occurs as an impurity in a wide variety of minerals. Divalent manganese is sensitive to the crystal lattice structure, where the number, size,

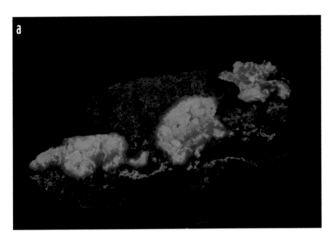

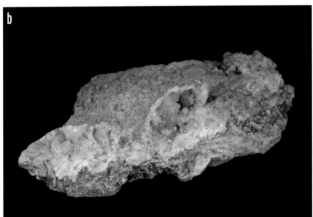

and distance between neighboring atoms may affect the absorption and emission of light energy. It may produce a green, yellow, or orange-red fluorescence, depending on the type of coordination of the host mineral (Tarashchan and Waychunas 1995). Coordination refers to the bonding of transition metal elements where neighboring donor ions contribute electrons to a central metal atom. The number of donor ions involves a specific geometrical arrangement around the central atom. In calcite, a calcium atom is connected with six oxygen ions, forming an octahedron, and this mineral is said to have octahedral coordination.

The manganese (Mn^{2+}) ions in calcite cannot by themselves absorb enough light energy to produce a fluorescent response. A coactivator element is required to add absorption levels and to transfer this extra energy to excite the manganese electrons. Lead (Pb^{2+}) often serves as a coactivator for shortwave fluorescence, while rare-earth elements, such as cerium (Ce^{3+}), may be involved for longwave fluorescence.

FIGURE 2.5 Calcite, Coals of Fire, 15 × 8 × 5 cm, Ruby, Santa Cruz County. Upper image (a) under shortwave UV. Lower image (b) under white light.

With hundreds of calcite occurrences and extensive manganese and lead deposits, it's not surprising that calcite is one of the most prevalent fluorescent minerals found in Arizona. The Montana Mine and surrounding areas near the town of Ruby, in Santa Cruz County, have produced calcite with a brilliant orange-red fluorescence. Specimens have been described as the "most spectacular fluorescing calcite ever found in Arizona" (Flagg 1958) and were sold under the trade name Coals of Fire. In addition to the fluorescent calcite, triboluminescent sphalerite has been found on the mine dumps (Flagg 1958). The Montana Mine was a surface and underground silver-lead-zinc-copper-gold mine that operated from 1877 to 1940. The mineralization is primarily of Laramide age and involves fissure-vein or replacement ore deposits. The host rock for the ores consists of the Cretaceous Oro Blanco Conglomerate, which was intruded by the Ruby, Sidewinder, and Blue Ribbon diorites, which have yielded circa 70 Ma U-Pb dates (E. D. Wilson et al. 1951; Mizer 2018).

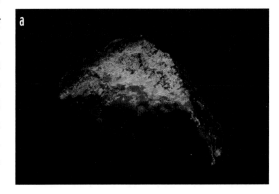

Calomel

Calomel is a mercurous chloride (Hg_2Cl_2) that forms from the alteration of other mercury minerals. It exhibits a bright orange-red shortwave and an orange longwave fluorescence, which originates from clustered heavy metal ion impurities. Trace amounts of calomel decompose into mercuric chloride ($HgCl_2$) and elemental mercury upon exposure to ultraviolet light. The mercury atoms form the cluster cation Hg_3^{2+}, which causes the orange-red emission (Kunkely and Vogler 1995).

A few calomel occurrences have been noted for the Mazatzal Mountains District, in Maricopa County. Mineralization in this district involves Proterozoic felsic metavolcanic activity about 1730 Ma (Ludwig 1974). The mercury-bearing lode deposits are hosted in a phyllitic schist zone between zones of bright-red jasper and brown slate. The deposits consist mainly of veinlets, films, or specks of cinnabar.

FIGURE 2.6 Calomel, 3.5 × 2.2 cm, Saddle Mountain Mercury Mine, Mazatzal Mountains, Maricopa County. Upper image (a) under longwave UV. Middle image (b) under shortwave UV. Lower image (c) under white light.

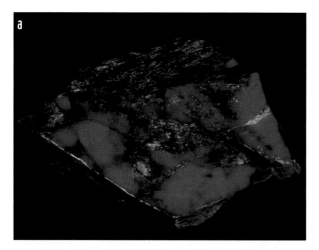

FIGURE 2.7 Eucryptitle, 8 × 5 cm, Midnight Owl Mine, Yavapai County. Upper image (a) under shortwave UV. Lower image (b) under white light.

Eucryptite

Eucryptite is a lithium aluminum silicate that exhibits a vivid crimson response under shortwave ultraviolet. The striking fluorescence is attributed to the substitution of aluminum cations by iron (Fe^{3+}) impurities. Iron usually acts as a quenching element for many fluorescent minerals, but in the trivalent state, it can serve as a substitutional impurity in some silicates, such as feldspars. The emission characteristics are similar to manganese (Mn^{2+}), except the response is shifted to longer wavelengths as a result of energy differences in host mineral coordination sites (Gaft, Reisfeld, and Panczer 2015).

Eucryptite, which is typically dull white to gray, has been found at the Midnight Owl Mine in the White Picacho District, Yavapai County. It is an alteration product of spodumene and occurs with thin rims of fine-grained albite. The Midnight Owl Mine is recognized as one of the best localities for eucryptite. The former feldspar-beryllium-lithium-niobium-tantalum-mica-phosphate mine is located thirteen miles east of Wickenburg and operated mainly from 1947 to 1953. Mineralization involves fracture filling and replacement in the wall, intermediate, and inner zones of irregular pegmatite bodies now dated at circa 1380 Ma (McCauley and Bradley 2014). The pegmatites are hosted in a Precambrian quartz-hornblende-mica gneiss (Jahns 1952).

Fluorite

Fluorite is a common mineral found at hundreds of locations throughout Arizona. Fluorescent specimens typically contain europium (Eu^{2+}) impurities and exhibit a blue to violet response under longwave ultraviolet. The Bluebird Mine and nearby Donna Anna workings in the Bluebird District, Cochise County, however, have produced an unusual type of fluorescent fluorite. Some samples may fluoresce bright pink under shortwave along with the usual blue response under longwave. The Bluebird Mine has also been a source of chlorophane, a rare variety of fluorite that exhibits fluorescence, phosphorescence, and thermoluminescence. The cause of this unique bluish-green shortwave response is not fully understood, but the presence of yttrium and other rare earths along with divalent to trivalent state transitions due to heating has been proposed as a possible mechanism (Robbins 1994). The Bluebird Mine is a former small underground tungsten-copper-bismuth-beryllium mine discovered in 1898 and operated until the early 1950s. The ore deposits contain high concentrations of hübnerite along with scheelite in irregular quartz veins. The fissure-filling mineralization is hosted in the circa 55 Ma Adams Peak Leucogranite that makes up the Bluebird District near Texas Canyon (Cooper and Silver 1964).

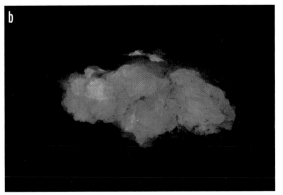

FIGURE 2.8 Fluorite, 6 cm FOV, Donna Anna workings, Dragoon Mountains, Cochise County. Upper image (a) under longwave UV. Middle image (b) under shortwave UV. Lower image (c) under white light.

Quartz/Chalcedony

Chalcedony may contain uranyl impurities since this ion group is fairly stable and soluble. The fluorescence is similar to many intrinsic uranyl minerals, with a bright yellow-green emission under shortwave UV, and may occur with concentrations as low as one part per million (Götze, Gaft, and Möckel 2015).

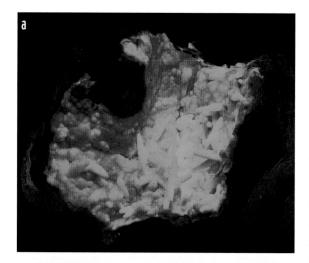

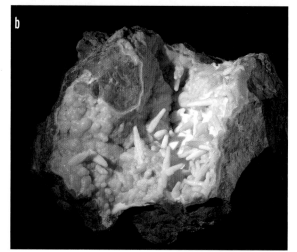

With abundant sources of chalcedony and uranium deposits distributed throughout the state, fluorescent chalcedony is found at many Arizona localities. Potts Canyon in the Reymert supersystem, northeast of Superior in Pinal County, is a notable location where the fluorescent chalcedony occurs as pearly blue-gray linings inside geodes and as roses. Large geodes, more than 12 inches across, have been found (D. Fanger, pers. comm., 2018), while smaller specimens may contain stalactitic perimorphs after calcite. Extensive mid-Tertiary volcanic activity occurred in the Potts Canyon area covering the region with tuff and lava flows. The tuff is part of a major ash-flow tuff sheet that has been redefined as the Apache Leap Tuff and dated 18.6 Ma (Wardwell 1941; Ferguson and Trapp 2001). The chalcedony filled fractures in the tuff.

FIGURE 2.9 Chalcedony epimorph of calcite in geode, 4.8 cm, Potts Canyon, Pinal County. Upper image (a) under shortwave UV. Lower image (b) under white light.

Sphalerite

Sphalerite is one of the five sulfide minerals known to be fluorescent. The zinc sulfide has been synthesized and studied extensively as a phosphor, and numerous activators have been identified. This research has shown that divalent manganese (Mn^{2+}) substituting for zinc ions may result in an orange fluorescence, which is brightest with longwave ultraviolet. The response closely matches natural sphalerite (Robbins 1994). Orange-fluorescing sphalerite has been reported from a few Arizona localities, and the best-known occurrence is the Copper Queen Mine in the Warren District, Cochise County. The Copper Queen Mine is a former large surface and underground copper-gemstone-silver-gold-lead-zinc mine discovered in 1877. The Copper Queen deposits involve two types of mineralization—disseminated copper hosted in circa 200 Ma acidic igneous intrusions (Lang

2001) and limestone replacements. The disseminated copper mineralization consists of primary sulfides, pyrite and bornite, with minor chalcopyrite, hosted in highly altered breccias and including supergene chalcocite found below Cretaceous sediments. Replacement copper mineralization occurs in all of the area's Paleozoic limestones, but high-grade carbonate orebodies appear mainly in the Devonian Martin Limestone and as part of the Mississippian Escabrosa Limestone (Bryant and Metz 1966). Unlike the many spectacular and colorful copper minerals that have been collected at the Copper Queen, sphalerite is found as a fine-grained black massive material. Some specimens may exhibit triboluminescence.

Willemite

Willemite is a zinc silicate that typically exhibits a bright yellowish-green fluorescence under shortwave ultraviolet light. The main activator is divalent manganese (Mn^{2+}), the same impurity involved with calcite's fluorescence. Unlike calcite, however, willemite doesn't require a coactivator since the mineral has tetrahedral coordination, with four oxygen ions surrounding a zinc atom. This tetrahedral arrangement leads to more efficient light absorption and increases the energy of the Mn^{2+} transitions. As a result,

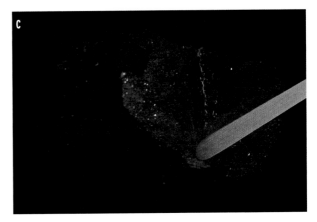

FIGURE 2.10 Sphalerite, 9 × 9 cm, Copper Queen Mine, Bisbee, Cochise County. Upper image (a) under longwave UV. Middle image (b) under white light. Lower image (c) showing triboluminescence, 6.7 × 4.2 cm.

FIGURE 2.11 Willemite, 10 × 10 cm, Nelly James Mine, Miller Canyon, Cochise County. Upper image (a) under shortwave UV. Lower image (b) under white light.

the fluorescent emission shifts to a green color.

Willemite is found as a secondary mineral at many of Arizona's zinc deposits, but most occurrences do not fluoresce. One notable locality with fluorescent willemite is the Nelly James Mine in the Hartford District, Cochise County. The mineral is usually found in white to off-white masses with calcite or hydro-zincite, but specimens may include light-blue vein fillings or rare bluish crystals. The Nelly James Mine is a small zinc-lead-silver-copper-gold prospect located 0.7 miles north of Miller Peak in the Huachuca Mountains. The type of deposit is uncertain but may involve vein or replacement mineralization. The geologic setting involves blocks of Paleozoic sedimentary rocks within Jurassic volcanic rocks and intruded by the Huachuca Quartz Monzonite, all dated between 180 and 167 Ma within the Hartford District (Marvin, Naeser, and Mehnert 1978; Lang 2001).

Fluorescent Minerals with Unknown Activators

A subcategory of impurity fluorescent minerals involves a large number of species in which the main activator is unknown. These minerals may contain several different impurities, and the resulting fluorescence may be a combination of multiple emissions. Determining a specific luminescence center can be difficult given the complex interactions between impurity ions and the host mineral. The list of reported Arizona fluorescent minerals includes eleven minerals with undetermined activators. Some of the fluorescent responses are noteworthy for their color and brightness.

Wickenburgite

Wickenburgite is a rare secondary lead silicate found at several small claims in the Wickenburg area in Maricopa County. The mineral can occur as either small, very lustrous, colorless hexagonal crystals or dull-white fine-grained masses. It may fluoresce pink, orange-pink, orange-red, or crimson under shortwave ultraviolet. The mineral was first discovered at the Lost Spaniard (Potter-Cramer) Mine in the Osborne District, Maricopa County. The lead-zinc-copper-gold-silver mine is located along the northern flank of the Belmont Mountains. It operated in the 1920s–1940s, when activity centered on a vein of oxide minerals in a probable Early Miocene rhyolite dike. The district includes gold-bearing quartz veins in granite along with sulfide minerals below the oxidized zone. Mineralization associated with the quartz veins probably formed in the mid-Tertiary circa 20–15 Ma (G. B. Allen 1985; Stimac et al. 1994; Polman 2021).

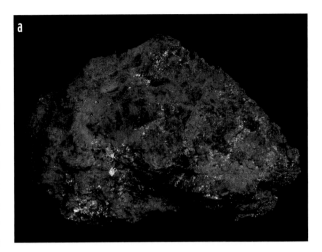

FIGURE 2.12 Wickenburgite, 11 × 8 cm, Potter-Cramer Mine, Maricopa County. Upper image (a) under shortwave UV. Lower image (b) under white light.

Zunyite

Zunyite is a rare aluminosilicate that commonly forms tetrahedral crystals and fluoresces a bright cherry red under shortwave ultraviolet. The Big Bertha Mine in the Sugarloaf Pyrophyllite District, La Paz County, has produced crystals up to 2 cm on an edge. The Sugarloaf Pyrophyllite District is the alunite alteration/vein peripheral zone of the Sugarloaf gold system west of Quartzsite. The mineralization may be sourced in the Diablo high-potassium granite suite, which ranges from 164 to 155 Ma (D. S. Smith 2013; Tosdal and Wooden 2015). The

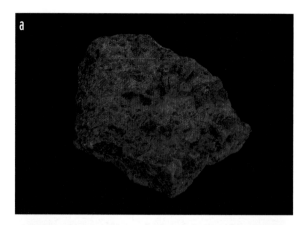

FIGURE 2.13 Zunyite, 4.4 × 4.7 cm, Big Bertha Mine, Dome Rock Mountains, La Paz County. Upper image (a) under shortwave UV. Lower image (b) under white light.

Big Bertha Mine is located eight miles southwest of Quartzsite and consists of an open cut with connecting crystal "caves." Specimen-mining activity occurred from the 1950s to the 1990s and focused on collecting quartz specimens with lustrous tabular crystals and rosettes of specular hematite. The minerals are hosted in highly altered sheared schist that is broadly associated with high-potassium granitic intrusions that have yielded 160 to 155 Ma U-Pb dates (Boettcher and Mosher 1998; Tosdal and Wooden 2015; AZGS 2013a).

Multicolor Fluorescent Specimens

One of the main attractions of fluorescent mineral collecting involves specimens in which different fluorescent species come together in seemingly endless combinations. Under ultraviolet light, these samples often display dazzling, intricate mixtures of brilliant colors. Multicolor fluorescent specimens from some Arizona localities may exhibit up to five colors, and many are comparable to samples from the famous mines in Franklin and Ogdensburg, New Jersey.

Multicolor fluorescent specimens may involve combinations of minerals with intrinsic fluorescence, minerals with impurities, or both. With impurity-activated minerals, it is possible for a single impurity, such as manganese (Mn^{2+}), to produce a variety of different colors. The varicolored response reflects the interaction of the activator with the crystal structure of different host minerals.

Two-Color Fluorescent Specimens

Although scheelite and powellite form a solution series, specimens have been found in which the intrinsic minerals appear as a distinct two-color combination. With shortwave ultraviolet, the scheelite is a bright blue-white, and the powellite fluoresces creamy yellow. Notable samples have been collected from the Three Musketeers Mine in the Three Musketeers District, in the northern Granite Wash Mountains, La Paz County. The Three Musketeers Mine is a former tungsten-gold mine discovered in 1951. Scheelite is the main ore mineral and occurs as small grains and pods or in quartz-filled fractures. The mineralization is hosted in Mesozoic calcareous schist or in Late Cretaceous peraluminous biotite granite and aplopegmatite (Dale 1959). An aplopegmatite is a light-colored pegmatite comprising mainly alkali feldspar and quartz with some biotite.

Manganese (Mn^{2+}) impurities in willemite and calcite may produce a common two-color fluorescent combination. Under shortwave ultraviolet, the willemite fluoresces bright yellow-green, while the calcite response is orange red. This assemblage occurs at numerous locations throughout

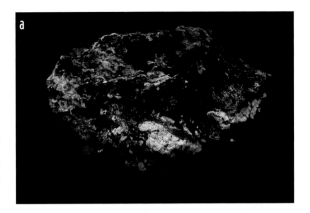

FIGURE 2.14 Two-color fluorescence with intrinsic minerals scheelite and powellite, 13 × 9.8 cm, Three Musketeers Mine, Granite Wash Mountains, La Paz County. Upper image (a) under shortwave UV. Lower image (b) under white light.

the state. The Paul Hinshaw property (sec. 36, T11S, R2E) in the Cimarron Mountains, Pima County, however, has a unique distinction, where tons of this material were quarried and shipped to Los Angeles and Hollywood for ornamental fireplaces (Wilson and Roseveare 1949). The overall geologic setting involves Cretaceous to Late Jurassic sedimentary rocks with minor volcanic rocks (U.S. Geological Survey 1995).

An attractive two-color combination involving an intrinsic mineral and an impurity mineral occurs at the Campo Bonito Group in the Oracle District, Pinal County. Specimens illuminated with shortwave ultraviolet may display swirls of blue-fluorescing scheelite interspersed with orange-red calcite. Campo Bonito is a former tungsten-gold-lead-molybdenum mine that operated from 1908 to 1944. The tungsten-gold mineralization occurs in a silicified breccia hosted in the Escabrosa Limestone, which is surrounded by Precambrian sediments and the Oracle Granite (Dale 1959).

FIGURE 2.15 Two-color fluorescence with impurity minerals calcite and willemite, 7.6 × 7.6 cm, Casa Grande, Pinal County. Upper image (a) under shortwave UV. Lower image (b) under white light.

FIGURE 2.16 Two-color fluorescence with intrinsic and impurity minerals scheelite and calcite, 15 × 14 cm, Campo Bonito Mine, Santa Catalina Mountains, Pinal County. Upper image (a) under shortwave UV. Lower image (b) under white light.

Three-Color Fluorescent Specimens

The oxidized ores found at many of Arizona's lead-zinc deposits may include a classic three-color combination of calcite, fluorite, and willemite. Under shortwave ultraviolet, the calcite fluoresces orange red; the fluorite appears violet; and the willemite is yellow green. Some notable samples of this combination have been collected at the Black Rock Mine in the Silver District, La Paz County. The Black Rock Mine is a former surface and underground lead-silver-zinc-fluorite-manganese-molybdenum mine that started operating in the 1880s. The ore mineralization consists of irregular masses of limonite,

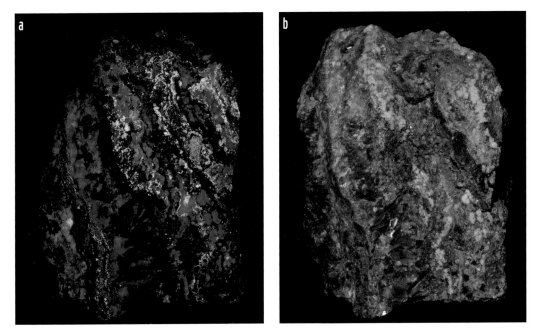

FIGURE 2.17 Three-color fluorescence with calcite, fluorite, and willemite, 10 × 8 cm, Black Rock Mine, La Paz County. Left image (a) under shortwave UV. Right image (b) under white light.

calcite, pyrolusite, smithsonite, willemite, cerussite, anglesite, and minor galena filling vugs and fractures of fine-grained quartz and fluorite. The Tertiary vein deposits are associated with presumed 25 to 20 Ma dioritic intrusions and occur in schist that is correlated with the Orocopia Schist, which in this area was underplated to the base of the Arizona crust during flat subduction 60 to 50 Ma (Haxel et al. 2002).

Four-Color Fluorescent Specimens

Smithsonite may be found at some of the calcite-fluorite-willemite occurrences, which can lead to four-color fluorescent specimens. Due to manganese (Mn^{2+}) impurities, the zinc carbonate may exhibit a blue, violet, red, or pink shortwave response, while longwave emissions vary from yellow to orange. The Purple Passion Mine and nearby Hogan claim in the Red Picacho District, Yavapai County, are known for fluorescent minerals that include some four-color specimens with smithsonite. The former lead-silver-molybdenum-fluorite mine is located about eight miles northeast of Wickenburg. The mineralization appears to be mid-Tertiary and occurs in a brecciated dike intruded in Precambrian granite (Welty et al. 1985b; K. A. Phillips 1987; B. Gardner and Davis 2000).

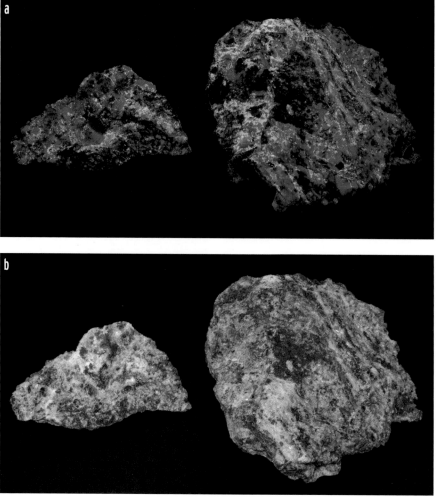

FIGURE 2.18 Four-color fluorescence with fluorite, calcite, smithsonite, and willemite, 17.8 cm wide, Hogan claim, Yavapai County. Upper image (a) under midrange plus shortwave UV. Lower image (b) under white light.

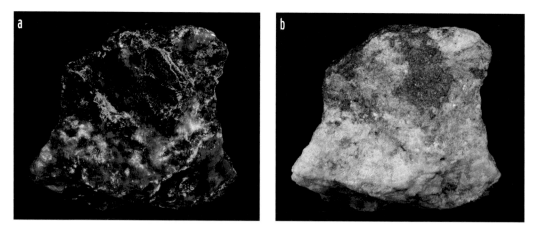

FIGURE 2.19 Five-color fluorescence with calcite, willemite, hydrozincite, sphalerite, and smithsonite, 10.8 × 8.9 cm, Nelly James Mine, Miller Canyon, Cochise County. Left image (a) under shortwave plus longwave UV. Right image (b) under white light.

Five-Color Fluorescent Specimens

Five-color fluorescent specimens, in which each mineral is discernable throughout the sample under a single wavelength, are relatively rare. The Nelly James Mine in the Hartford District, Cochise County, has produced some bright specimens with this combination of qualities. One example involves a mix of calcite, willemite, hydrozincite, powellite, and smithsonite. (See the willemite description under Impurity Fluorescent Minerals for information about the Nelly James Mine.)

Arizona Mineral Occurrences

There are currently 992 mineral species reported from Arizona. This number changes constantly as additional species are found, and species are removed when new research shows they were misidentified. The validity of the data presented here was carefully considered before adding new Arizona species and occurrences. For the past editions of the *Mineralogy of Arizona*, the decision was made to include as much information as possible, knowing that some of it needed better verification. Many of these questionable entries have been retained in this edition, and the reader is warned that there may be errors in the data presented. A reference is included where possible, and the reader can go to the original reference and decide on the validity of the information. Nonetheless, the entries provide a starting point for persons interested in finding different minerals and localities. The RRUFF project at the University of Arizona has been a great source of information and has verified many of the rare Arizona minerals.

Another challenge is to determine the correct locality where the minerals were found and which localities to include. Arizona has 450 mineral districts (Stanley B. Keith et al. 1983) with thousands of mines and prospects. Welty et al. (1985a) lists approximately 5,500 mines in Arizona, and the files at the Arizona Geological Survey have twice that many mines and prospects listed for the state. At each of these locations, there will be some type of mineralization. The vast majority will not have specimens of interest to the collector, but which locality to include becomes problematic. For example, there are more than 250 known wulfenite occurrences in Arizona (Randal Heath, pers. comm., 2014). A separate publication would be necessary to describe all these wulfenite localities, so only the most important ones can be listed here. Additionally, many of the localities have multiple names. For example, the hematite occurrence southwest of Quartzsite has been called the Crystal Cavern claim, Big Bertha

Extension Mine, Veta Grande claim, and Purple Cow Ledge. Preference is given here to the mine name used for the Arizona Geological Survey mine file.

For most locations where the information is available, the mineral district for the location will be given first, followed by the mine name or other location information. The district names have all been updated from the mineral district name listed in Stanley B. Keith et al. (1983), with the addition of nonmetallic deposits and the subdivision of many others. The updated mineral district maps in appendix A also show mineral supersystems, which consist of clusters of districts with a similar composition and age. The maps and tables in appendix A provide additional information about the age, metals present, and locations for each district. This information can be of value to collectors. For example, wulfenite is common in the 30–20 Ma, 75–65 Ma, and 185–155 Ma age groups with Pb-Zn-Ag (lead-zinc-silver) metals. These districts are colored orange on the maps. This category includes the Silver District (Red Cloud Mine) and Castle Dome District (Hull Mine) of southwestern Arizona and the California supersystem (Hilltop District, Hilltop Mine) of Cochise County. Therefore, collectors might want to visit other districts in these age groups with these metals. In earlier editions of the *Mineralogy of Arizona*, the historical mining district names were used. These were based only on geography, and not on age or metals present. Many of the older mining district names were used again for the mineral districts, so there is no name change, but for many other districts, a new name is used in this edition.

Heading Styles and Abbreviations Used in This Chapter

The mineral species names are from the latest available data at the time of this writing from the International Mineralogical Association (IMA). This organization is responsible for the status of valid mineral species. Some minerals from the last edition of the *Mineralogy of Arizona* have been discredited and dropped, and some names have been changed. The chemical formulas given are also from the same IMA List of Minerals.

A letter-and-number code in parentheses, for example (UA 1091), listed under "ACANTHITE, Mohave County," specifies the museum and the catalog number of the specimen. These museum codes are found in table 3.1.

Names in capital letters (e.g., ACANTHITE) indicate mineral species accepted as valid by the IMA. Names in lowercase indicate that the mineral (or substance) is not a recognized species but may be a variety or group name; an explanation typically follows. Minerals added to the Arizona mineral list since the revised third edition of *Mineralogy of Arizona* was published in 1995 are

TABLE 3.1 Museum abbreviations

Office of Legacy Management (formerly Atomic Energy Commission, Grand Junction Operations), Grand Junction, Colorado	AEC
American Museum of Natural History, New York	AM
Arizona-Sonora Desert Museum, Tucson	ASDM
ASDM micromount collection	ASDMx
Harvard University, Cambridge, Massachusetts	H
University of Arizona Mining, Mineral and Natural Resources Education Museum (formerly Arizona Mining and Mineral Museum), Phoenix	MM
Natural History Museum, London	NHM
University of Arizona RRUFF project, Tucson	R
Smithsonian Institution, National Museum of Natural History, Washington D.C.	S
University of Arizona Alfie Norville Gem and Mineral Museum, Tucson	UA
University of Arizona Alfie Norville Gem and Mineral Museum micromount collection	UAX

TABLE 3.2 Dimension abbreviations

millimeter	mm
centimeter	cm
meter	m
inches	in.
feet	ft.

TABLE 3.3 USGS map abbreviations

section	S
township	T
range	R
north	N
south	S
east	E
west	W

preceded by #, as in #BACKITE. Because of frequent citations, abbreviations will be used for the following references:

> (3rd ed.)　Anthony, J. W, S. A. Williams, R. A. Bideaux, and R. W. Grant (1995) *Mineralogy of Arizona*, 3rd ed. (Tucson: University Arizona Press).
>
> (Bull. 181)　Galbraith, F. W., and D. J. Brennan (1970) *Minerals of Arizona*, Arizona Bureau of Mines Bulletin 181 (Tucson: University of Arizona).

Dimensions are given in English or metric units depending on the source of information. Older references using English units have not been converted to metric to preserve the historical nature of the reference. Dimensions used in photo captions are specimen width unless stated otherwise. FOV refers to field of view, which indicates the width of the photograph and is usually used where the entire specimen is not in view. Abbreviations for dimensions used throughout are listed in table 3.2.

Occasionally, locations are noted by township and range as found on U.S. Geological Survey (USGS) topographic maps, and the abbreviations listed in table 3.3. are used. A location in section 23, Township 7 north, and Range 3 east would be shown as S23, T7N R3E.

Alphabetical Listing of
Arizona Mineral Occurrences

#ABERNATHYITE

Potassium uranium arsenate hydrate: $K(UO_2)(AsO_4) \cdot 3H_2O$. A rare secondary mineral found in Colorado Plateau–type, sandstone-hosted, uranium-copper-vanadium deposits.

COCONINO COUNTY: From a locality near Tuba City (J. W. Anthony et al. 2000).

ACANTHITE

Silver sulfide: Ag_2S. Dimorphous with argentite (isometric), which may form in hydrothermal veins at elevated temperatures, inverts upon cooling to the monoclinic dimorph acanthite. An important ore of silver; commonly associated with other silver minerals, such as galena, tetrahedrite, and nickel-cobalt ores. Also formed as a secondary mineral in the zone of sulfide enrichment, with chalcocite, silver, and silver halogen minerals.

COCHISE COUNTY: Warren District, Bisbee, Campbell Mine (Graeme 1993). Tombstone District, in oxidized ores formed from alteration of argentiferous tetrahedrite (Butler, Wilson, and Rasor 1938; Romslo and Ravitz 1947). Pearce District, Commonwealth Mine, with chlorargyrite, bromargyrite, bromian-rich chlorargyrite, and iodargyrite, in quartz veins (Endlich 1897). Chiricahua Mountains, Hilltop District, El Tigre Mine, with cubanite in black-banded quartz (Tsuji 1984).

GILA COUNTY: Richmond Basin District, as the chief primary silver mineral, in masses up to several pounds (Bull. 181) at Stonewall Jackson Mine (Guild 1917).

FIGURE 3.1 Acanthite, Silver King Mine, Pinal County, 4.7 cm high.

GRAHAM COUNTY: Aravaipa supersystem, Grand Reef District, in veins at the Grand Reef Mine (C. P. Ross 1925a).

LA PAZ COUNTY: Silver District, said to have been mined in important quantities in the Princess and other veins (E. D. Wilson 1933); Padre Kino Mine, massive with galena (3rd ed.).

MOHAVE COUNTY: Cerbat Mountains, Mineral Park District, Buckeye Mine (Les Presmyk, pers. comm., 2020), Keystone and Queen Bee Mines; Tennessee District Golden Star Mine (Bastin 1925), Prince George Mine, and veins of the Banner Group; and at various other properties (Schrader 1909; Dings 1951); Rawhide District, Rawhide Mine, altering to chlorargyrite (UA 1091).

PIMA COUNTY: Santa Rita Mountains, Helvetia-Rosemont District, Helvetia portion, Blue Jay Mine (Schrader and Hill 1915). Papago District, Sunshine Mine (Bull. 181), Quijotoa District, Morgan Mine (Bull. 181).

PINAL COUNTY: Pioneer supersystem, Silver King District, Silver King Mine, in large quantities on the upper levels (Romslo and Ravitz 1947); Belmont District, Belmont Mine, as small blebs in galena (Bull. 181); northern Galiuro Mountains, Saddle Mountain District, Little Treasure Mine (Bull. 181); Mammoth District, Mammoth–St. Anthony Mine, as rare, minute, monoclinic crystals on leadhillite and silver; Silver Reef District (Bideaux 1980), Nugget Fraction Mine, with cerussite (David Shannon, pers. comm., 1986); Vekol District, Vekol Mine, as pods with chlorargyrite (3rd ed.); Bunker Hill (Copper Creek) District, Bluebird Mine (Simons 1964); Mineral Hill District, Reymert Mine, with jalpaite, as well as silver in quartz and calcite veins (K. S. Wilson 1984).

SANTA CRUZ COUNTY: Santa Rita Mountains, Tyndall District, Alto, Eureka, Ivanhoe, Montezuma, and Empress of India Mines (Bull. 181); Wrightson District, Augusta, Happy Jack, and Anaconda Mines (Bull. 181); Patagonia Mountains, Red Rock District, La Plata and Meadow Valley Mines (Bull. 181); Flux District, January, Blue Eagle, Flux, and American Mines (Schrader and Hill 1915; Schrader 1917); Oro Blanco District, as 5 mm crystals perched on small quartz crystals (Les Presmyk, pers. comm., 2020).

YAVAPAI COUNTY: Bradshaw Mountains, Hassayampa District, Dos Oris Mine, with silver and chlorargyrite (Bull. 181); northeastern Wickenburg Mountains, Monte Cristo District, Monte Cristo Mine, in primary ores with silver, skutterudite, and proustite (Bastin 1922); Eureka supersystem, Hillside District, Hillside Mine, with arsenopyrite, chalcopyrite, galena, sphalerite, pyrite, and tetrahedrite (Axelrod et al. 1951); Ticonderoga District, Arizona National Mine, in galena with freibergite and in cavities with wire silver (Lindgren 1926).

ACTINOLITE

Calcium magnesium iron silicate hydroxide: $\square Ca_2(Mg_{4.5-2.5}Fe^{2+}_{0.5-2.5})$ $Si_8O_{22}(OH)_2$. An iron-rich member of the amphibole group of rock-forming minerals, actinolite forms a continuous solid-solution series with the iron-free tremolite. Found in thermally metamorphosed gneisses, schists, and marbles and in contact-metamorphosed rocks, particularly limestones and as replacements of ultramafic lenses in schist.

COCHISE COUNTY: Little Dragoon Mountains, Johnson Camp District, in metamorphosed limestones (Bull. 181).

GILA COUNTY: Globe Hills District, Old Dominion Mine, along bedding planes in Mescal Limestone (Bull. 181).

GRAHAM COUNTY: Aravaipa supersystem, Iron Cap and Stanley Districts, as gangue of contact-metamorphosed ores (C. P. Ross 1925a).

GREENLEE COUNTY: Northern part of the Copper Mountain supersystem, Morenci District, where Paleozoic limestones are in contact with the main intrusion, with garnet, diopside, epidote, tremolite, and other calcsilicate minerals (Moolick and Durek 1966).

LA PAZ COUNTY: Granite Wash Mountains, Yuma King District, Yuma Mine; and western Harcuvar Mountains, Cabrolla District, as a replacement of limestone beds (Bull. 181); northern Plomosa Mountains, northern Plomosa District, six miles west of Bouse, as replacements of ultramafic lenses in Orocopia Schist (Stanley B. Keith, pers. comm., 2018); Cemetery Ridge District, very common, and prominent, as veins and pods along contacts between harzburgite, serpentinite, and host Orocopia Schist, as well as a constituent of widespread actinolite-albite gneiss (Haxel et al. 2021).

MOHAVE COUNTY: Black Mountains, Oatman District, Big Jim vein, as thin sheets between layers of quartz (Bull. 181).

PIMA COUNTY: Sierrita Mountains, Pima supersystem, abundant in contact rocks (Bull. 181); Twin Buttes District, Twin Buttes Mine, in skarns and tactites (Stanley B. Keith, pers. comm., 1973); Helvetia-Rosemont District, in contact-metamorphosed sedimentary rocks at several mines (Creasey and Quick 1955).

SANTA CRUZ COUNTY: Patagonia Mountains, Washington Camp District, Pride of the West Mine (Bull. 181).

YAVAPAI COUNTY: Bradshaw Mountains, Big Bug District, Iron Queen Mine, in country rock (Bull. 181), and Boggs Mine, as the fibrous variety, with bournonite (Lindgren 1926).

YUMA COUNTY: Cemetery Ridge District, very common, and prominent, as veins and pods along contacts between harzburgite, serpentinite, and host Orocopia Schist; as a constituent of widespread actinolite-albite gneiss (Haxel et al. 2021); and east of Deadman Tank as bladed green crystals in amphibolite dikes with an asbestiform texture, developed in schist (E. D. Wilson 1933).

ADAMITE

FIGURE 3.2 Adamite, Saginaw Hill, Pima County, FOV 2 mm.

Zinc arsenate hydroxide: $Zn_2(AsO_4)(OH)$. A secondary mineral found in the oxidized portion of some zinc-arsenic deposits.

COCONINO COUNTY: Grandview District, Grand Canyon National Park, Horseshoe Mesa, Grandview (Last Chance) Mine, in very small amount associated with zeunerite, scorodite, and olivenite (Leicht 1971).

PIMA COUNTY: Tucson Mountains, Saginaw Hill District, southern side of Saginaw Hill, about seven miles southwest of Tucson, in association with other oxide-zone minerals (R. Gibbs, pers. comm., 2020).

YAVAPAI COUNTY: Mayer District, S17, T12N R2W, at a small prospect, as pale-green crystalline crusts in vein quartz, associated with alloclasite (3rd ed.).

#ADANITE

FIGURE 3.3 Adanite, Tombstone, Cochise County.

Lead tellurate sulfate: $Pb_2(Te^{4+}O_3)(SO_4)$. A rare mineral found in oxidized deposits containing tellurium. The following locality is a cotype locality with the North Star Mine, Mammoth, Tintic District, in Juab County, Utah (Kampf, Housley, et al. 2020).

COCHISE COUNTY: Tombstone District, as small white crystalline aggregates on euhedral jarosite also associated with frohbergite and rodalquilarite.

AEGIRINE

Sodium iron silicate: $NaFe^{3+}Si_2O_6$. A rock-forming mineral of the pyroxene group, which is primarily produced by late crystallization of alkaline magmas.

APACHE COUNTY: Garnet Ridge SUM District, with diopsidic jadeite and pyrope-almandine garnet in eclogite inclusions from serpentine ultramafic megabreccia pipes (Watson and Morton 1969).

PIMA COUNTY: Northern Sierrita Mountains, Gunsight Mountain, with titanite and scapolite in a pulaskite dike (Bideaux, Williams, and Thomssen 1960). This may be the Late Oligocene lamprophyre dike swarm on Gunsight Mountain mapped by Cooper (1973) that postdates the Eocene-age beryl-bearing aplopegmatite dikes in the overlapping Samaniego District.

AENIGMATITE

Sodium iron titanium oxide silicate: $Na_4[Fe^{2+}_{10}Ti_2]O_4[Si_{12}O_{36}]$. A constituent of syenites; syenitic pegmatites; certain granites; alkali-rich rocks, such as phonolites; and trachytes.

NAVAJO COUNTY: Navajo volcanic field, identified by X-ray diffraction on material from a dike; the mineral was not visible to the naked eye (Laughlin, Charles, and Aldrich 1986).

#AERINITE

Calcium sodium iron magnesium aluminum silicate hydroxide carbonate hydrate: $(Ca,Na)_6(Fe^{3+},Fe^{2+},Mg,Al)_4(Al,Mg)_6Si_{12}O_{36}(OH)_{12}(CO_3)\cdot12H_2O$. A hydrothermal mineral formed at relatively low temperatures, found as coatings on fractures in mafic rocks.

PIMA COUNTY: Gunsight Mountain, found as pale- to deep-blue massive coatings and replacements of iron-bearing minerals, including aegirine and biotite (Bideaux, Williams, and Thomssen 1960).

#AESCHYNITE-(Y)

Yttrium lanthanum calcium thorium titanium niobium oxide hydroxide: $(Y,Ln,Ca,Th)(Ti,Nb)_2(O,OH)_6$. An accessory mineral found in granites and granite pegmatites and in ankerite-dolomitic carbonatites.

MOHAVE COUNTY: Kingman Feldspar District, near Kingman Feldspar Mine, in a small satellite quarry in a pegmatite occurrence located 1.5 km west-southwest of the Kingman Feldspar Mine (S. L. Hanson, Falster, and Simmons 2007).

#AGARDITE-(Nd)

FIGURE 3.4 Agardite-(Nd), Narragansett Mine, Pima County, FOV 1.5 mm.

Neodymium copper arsenate hydroxide hydrate: $NdCu_6$ $(AsO_4)_3(OH)_6 \cdot 3H_2O$. A secondary mineral found in oxidized deposits containing arsenic.

PIMA COUNTY: Helvetia-Rosemont District, Narragansett Mine, as small blue clusters of acicular crystals associated with conichalcite and mimetite found by John Ebner in oxidized vein material in open cuts above the mine (Gibbs, pers. comm., 2020).

AIKINITE

Copper lead bismuth sulfide: $CuPbBiS_3$. A rare mineral found at few localities in the world; may be associated with gold and galena.

COCHISE COUNTY: Signal Hill District, Black Prince claims, as blebs and veinlets between garnet crystals in limestone, about fifteen miles northeast of Tombstone (Bull. 181; UA 3572); Warren District, Bisbee, 2,200 ft. level of the Campbell shaft (Graeme 1993).

GILA COUNTY: Miami-Inspiration District, Miami Mine, in veinlets cutting chalcopyrite, with tennantite and enargite (Legge 1939).

PIMA COUNTY: Roskruge Mountains, Roadside District, Roadside Mine, in small quantities (Bull. 181; S 12049).

AJOITE

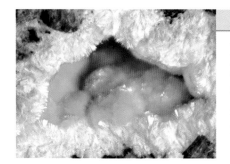

FIGURE 3.5 Ajoite, Magna Mine, Pinal County, FOV 4 mm.

Potassium copper aluminum silicate hydroxide hydrate: $K_3Cu^{2+}_{20}Al_3Si_{29}O_{76}(OH)_{16} \cdot 8H_2O$. A rare mineral intimately associated with shattuckite in the oxidized zone of copper deposits. The New Cornelia Mine at Ajo in Pima County is the type locality.

MARICOPA COUNTY: Belmont Mountains, Osborne District, southwest of Wickenburg, including the Moon Anchor Mine and Potter-Cramer property, associated with wickenburgite, mimetite, willemite, and phoenicochroite (S. A. Williams 1968).

PIMA COUNTY: Ajo District, New Cornelia Mine, as pale aquamarine tufts and crystals filling interstices between radiating spherulitic, dark-blue crystalline shattuckite (Schaller and Vlisidis 1958; Hutton and Vlisidis 1960; Sun 1961; Newberg 1964).

PINAL COUNTY: Bunker Hill (Copper Creek) District, Copper Creek, Magna Mine, as acicular crystals lining and filling vugs with conichalcite and baryte (Joe Ruiz, pers. comm., 2016).

AKAGANEITE

Iron nickel hydroxide oxide chloride hydrate: $(Fe^{3+}Ni^{2+})_8(OH,O)_{16}Cl_{1.25}\cdot nH_2O$. A rare secondary mineral thought to have formed by the alteration of pyrrhotite at the type locality, the Akagané Mine, Iwate Prefecture, Japan.

MOHAVE COUNTY: Wallapai supersystem, Ithaca Peak District, reported as possibly present at the Ithaca Peak Mine (Eidel, Frost, and Clippinger 1968).

SANTA CRUZ COUNTY: Querces District, small prospect north of the Santo Niño Mine, as thin, warty, reddish-brown films on fractured vein quartz containing partly leached pyrite and molybdenite (3rd ed.).

ALABANDITE

Manganese sulfide: MnS. An uncommon primary mineral found in vein deposits, commonly associated with sphalerite, galena, pyrite, rhodochrosite, rhodonite, quartz, and calcite.

COCHISE COUNTY: Tombstone District, Lucky Cuss Mine (Moses and Luquer 1892; Blake 1903; Hewett and Rove 1930; Butler, Wilson, and Rasor 1938) and Oregon-Prompter Mine (Hewett and Fleischer 1960). Warren District, Bisbee, Higgins Mine, as massive material in dolomite with rhodochrosite; Junction shaft massive with rhodochrosite in sphalerite, in veinlets in limestone (Graeme 1981); and Copper Queen Mine (UA 954). Chiricahua Mountains, California supersystem, Humboldt Mine (Hewett and Rove 1930); Hilltop Mine (UA 5417).

SANTA CRUZ COUNTY: Northern Patagonia Mountains, Trench District, Trench Mine, with sphalerite, galena, and rhodochrosite (Hewett and Rove 1930). Flux District, World's Fair Mine, with rhodonite (UA 9924).

ALAMOSITE

Lead silicate: $PbSiO_3$. A rare secondary mineral, associated with wulfenite and leadhillite at the type locality near Alamos, Sonora, Mexico.

COCHISE COUNTY: Tombstone District, Lucky Cuss Mine, as spectacular crystal groupings with queitite and zeolites (3rd ed.).

FIGURE 3.6 Alamosite, Evening Star Mine, Maricopa County, FOV 3 mm.

MARICOPA COUNTY: Big Horn Mountains, Big Horn District, Evening Star Mine, as very small blocky crystals embedded in cerussite at the margin of oxidized pods of galena (Yang et al. 2016).

MOHAVE COUNTY: Rawhide District, Rawhide Mine, as euhedral tabular crystals up to 12 mm across, in quartz druses with luddenite (S. A. Williams 1982).

PINAL COUNTY: Mammoth District, Mammoth–St. Anthony Mine, on a single specimen, as white crystalline sprays and balls up to 5 mm, associated with diaboleite and willemite (3rd ed.).

#ALBITE

Sodium aluminum silicate: $NaAlSi_3O_8$. Albite forms a series with anorthite, $CaAl_2Si_2O_8$, and the series is referred to as the plagioclase feldspars. In the past it was divided into a series of species with sodium-rich ones called albite, oligoclase, and andesine. These are now all the mineral albite (see ANORTHITE for information about calcium-rich plagioclase). Albite is a major constituent of granite and granite pegmatites. It is also found in metamorphic and sedimentary rocks and hydrothermal veins. Albite is an abundant and widespread mineral in Arizona.

COCONINO COUNTY: Canyon Diablo area, as high albite in the Canyon Diablo octahedrite meteorite, associated with kosmochlor, roedderite, richterite, and chromite (Olsen and Fuchs 1968).

LA PAZ, YUMA, AND MARICOPA COUNTIES: Northern Plomosa Mountains, the Cemetery Ridge District, southern Trigo Mountains, Castle Dome Mountains, and Neversweat Ridge District, where outcrops of the Orocopia Schist contain graphite as a characteristic mineral. It coexists with albite feldspar, which reflects high-pressure, low- to moderate-temperature metamorphism of a metagraywacke (Haxel et al. 2002; Strickland, Singleton, and Haxel 2018).

MARICOPA AND YAVAPAI COUNTIES: White Picacho District, albite is a significant constituent of the pegmatites in this district (Jahns 1952).

PIMA COUNTY: Sierrita Mountains, west of Helmet Peak, as coarse perthitic oligoclase in pegmatite dikes (3rd ed.). As abundant, exceptional, andesine-labradorite phenocrysts up to 2 in. long in intrusive and flow porphyritic rocks, locally termed *turkey-track porphyry*, at several localities near Tucson, notably in the Twin Buttes quadrangle, in the lower part of the Helmet fanglomerate (Cooper 1961; Mielke 1964; Percious 1968).

PINAL COUNTY: Galiuro Mountains, as coarse, fresh andesine crystals in the Copper Creek Granodiorite; Copper Creek, as coarse, fresh albite crystals in metasomatized granodiorite (3rd ed.).

#ALLANITE-(Ce)

Calcium cerium aluminum iron silicate oxide hydroxide: $CaCe(Al_2Fe^{2+})$ $[Si_2O_7][SiO_4]O(OH)$. Found as an accessory mineral in granites and in granite pegmatites. The mineral from the Kingman Feldspar Mine and the Rare Metals Mine has been identified as allanite-(Ce); the materials from the other localities listed here have not been analyzed but are likely the cerium species.

COCHISE COUNTY: Tombstone District, as microscopic crystals in granodiorite (Butler, Wilson, and Rasor 1938).

MARICOPA COUNTY: Sierra Estrella Mountains, southwest of Phoenix, small crystals in coarse granite (Bull. 181) and as 1 cm crystals on a specimen in the Flagg Mineral Foundation collection (Les Presmyk, pers. comm., 2020).

MOHAVE COUNTY: Aquarius Mountains, Rare Metals Mine, in pegmatite with monazite-(Ce) and bastnäsite-(Ce) (J. W. Frondel 1964; W. B. Simmons et al. 2012), and Columbite prospect, in pegmatite with monazite (Heinrich 1960). Cerbat Mountains, Kingman Feldspar Mine, in granite pegmatite with allanite-(Nd) and bastnäsite-(Ce) (J. W. Frondel 1964; W. B. Simmons et al. 2012). Greenwood Mountains near Signal (Robert O'Haire, pers. comm., 1972). Garnet Mountain District, as an accessory mineral in granite, a specific location 1 km north of Iron Mountain (Theodore, Blair, and Nash 1987). Near Yucca (UA 8403) and Hualapai Mountains (UA 6477–79).

PIMA COUNTY: Oracle Junction, near Willow Springs Ranch, in pegmatite with schorl (Robert O'Haire, pers. comm., 1972). Cottonwood Ranch, eleven miles southwest of San Xavier Mission, sparingly distributed in black sands over an area several miles across (Adams and Staatz 1969). Southeast of Covered Wells, as spectacular masses in a vein with tourmaline, actinolite, and calcite (S. A. Williams 1960).

YAVAPAI COUNTY: Eureka area, west of Hillside, 7U7 Ranch near Bagdad, in pegmatite knots, with triplite and bermanite (Hurlbut 1936; Leavens 1967). White Picacho District, as a rare accessory mineral in crystals up to 4 in. long (Jahns 1952). Locality three miles west of Congress Junction (UA 6638). Reported near Yarnell (Rich Hill supersystem, Yarnell District) (Bull. 181).

#ALLANITE-(Nd)

Calcium neodymium aluminum iron silicate oxide hydroxide: CaNd$(Al_2Fe^{2+})[Si_2O_7][SiO_4]O(OH)$. Found as a rare accessory mineral in granite pegmatites. This is the first occurrence in the United States.

MOHAVE COUNTY: Cerbat Mountains, Kingman Feldspar Mine, in granite pegmatite with allanite-(Ce), bastnäsite-(Ce), and a mixture of thorite alteration products (S. L. Hanson et al. 2012).

FIGURE 3.7 Allantoin, Rowley Mine, Maricopa County, FOV 0.7 mm.

#ALLANTOIN

A diureide of glyoxylic acid: $C_4H_6N_4O_3$. A widespread compound used in cosmetics and pharmaceutical products, found in some plants and animals, and synthesized in bulk for industry. First found as a naturally occurring mineral thought to have been formed by the evaporation of fluids excreted by bats. The Rowley Mine is the type locality.

MARICOPA COUNTY: Painted Rock District, Rowley Mine, as colorless transparent blades to 0.3 mm long, associated with aphthitalite and urea (Kampf, Celestian, et al. 2020a).

ALLOCLASITE

Cobalt arsenide sulfide: CoAsS. Formed in calcite or quartz veins of apparently low-temperature, late hydrothermal origin.

YAVAPAI COUNTY: Mayer District, small prospect in the Mayer quadrangle, in bull quartz veins, as massive to crudely crystalline coarse-grained material; the same material contains a small amount of adamite (3rd ed.).

ALLOPHANE

Amorphous aluminum oxide silicate hydrate: $Al_2O_3(SiO_2)_{1.3-2.0}\cdot2.5-3.0H_2O$. A widespread constituent of clays; amorphous to X-rays because of minute particle size or disordered structure. Allophane clays are certainly more abundant in Arizona than the few documented localities suggest.

COCHISE COUNTY: Warren District, Bisbee, Sacramento Hill, in a hydrothermally altered granite porphyry stock with sericite, hydromuscovite, kaolinite, and alunite (Schwartz 1947). Courtland District, Maid of Sunshine Mine, as blue glassy material with azurite (3rd ed.).

GILA COUNTY: Miami-Inspiration supersystem, Inspiration District, Inspiration Mine, as a hydrothermal alteration product in granite porphyry (Schwartz 1947, 1956). Pinto Valley District, Castle Dome Mine, a product of hydrothermal alteration of quartz monzonite (N. P. Peterson, Gilbert, and Quick 1951), and Van Dyke claim (AM 30585).

GREENLEE COUNTY: Copper Mountain supersystem, Morenci District, Morenci Mine, as pseudomorphs after plagioclase, with halloysite-10Å, in altered porphyry copper ore and in granite porphyry (Schwartz 1947, 1958), especially around the periphery of the orebody (Moolick and Durek 1966).

MOHAVE COUNTY: Hualapai (Antler) District, Antler Mine, in gangue (Romslo 1948).

PINAL COUNTY: San Manuel District, San Manuel Mine, in hydrothermally altered monzonite and quartz monzonite porphyries (Lovering, Huff, and Almond 1950; Schwartz 1953). Mineral Creek supersystem, Ray District, in a hydrothermally altered porphyry stock and in sericitized veins (Schwartz, 1947, 1952).

ALMANDINE

Iron aluminum silicate: $Fe^{2+}_3Al_2(SiO_4)_3$. A member of the garnet group. Typically formed in regionally metamorphosed argillaceous sediments, but also in the contact-metamorphic environment and in some igneous rocks.

COCONINO COUNTY: Grand Canyon National Park, Inner Gorge, in Archean rocks; Phantom Creek, as crystals over 1 inch in diameter (Bull. 181).

GILA COUNTY: Hope District, Workman Creek, as individual floater crystals to 2 cm in size (Les Presmyk, pers. comm., 2020).

MOHAVE COUNTY: Southern Aquarius Mountains, as crystals up to an inch, in light-colored volcanic rock (Henderson 1941; Burt, Moyer, and Christiansen 1981; Moyer 1982); the manganese content of these garnets is variable, and some are spessartine in composition. Cerbat Mountains, as pink material in migmatite (B. E. Thomas 1953).

PIMA COUNTY: Santa Catalina Mountains, foothills and crest of the main range, in gneiss and granite complexes and in pegmatite (especially garnet schlieren bands near Summerhaven); the garnets contain varying amounts of the end-member components almandine, spessartine, and pyrope (Pilkington 1961). Helvetia-Rosemont District, Peach-Elgin copper deposit, with grossular and diopside in bedded skarn deposits (Heyman 1958).

ALTAITE

Lead telluride: PbTe. Formed in hydrothermal veins with other tellurides, tellurium, pyrite, and other sulfides. The Campbell shaft at Bisbee is the type locality.

COCHISE COUNTY: Warren District, Bisbee, Campbell shaft, in a pod of unusual pyritic ore, as cleavable patches up to 8 mm across; intergrown with canfieldite in an otherwise nearly massive pyrite (Graeme 1981).

R180005　0.5mm

FIGURE 3.8 Alterite, near Cliff Dwellers Lodge, Coconino County.

#ALTERITE

Zinc iron sulfate oxalate hydroxide hydrate: $Zn_2Fe^{3+}_4$ $(SO_4)_4(C_2O_4)_2(OH)_4 \cdot 17H_2O$. A new oxalate mineral found in carbonaceous petrified wood. Cliff Dwellers Lodge is the type locality.

COCONINO COUNTY: Vermilion Cliffs District, unnamed uranium prospect near Cliff Dwellers Lodge, as small yellow-green elongated crystals in fractures in carbonized petrified wood in the Shinarump Conglomerate basal member of the Chinle Formation. Alterite occurs with gypsum, alunogen, natrojarosite, sulphur, and celestine (Yang et al. 2020; R180005).

#ALUM-(K)

Potassium aluminum sulfate hydrate: $KAl(SO_4)_2 \cdot 12H_2O$. Formed with oxidizing pyrite and at fumarole deposits.

COCONINO COUNTY: Sunset Crater National Monument, as crusts on gypsum in the fumarole deposits on Sunset Crater (S. L. Hanson, Falster, and Simmons 2008).

#ALUM-(Na)

Sodium aluminum sulfate hydrate: $NaAl(SO_4)_2 \cdot 12H_2O$. Formed at fumarole deposits and likely a postmining precipitate on mine walls.

COCONINO COUNTY: Sunset Crater National Monument, as pale yellowish-tan masses cementing cinders together in the fumarole deposits on Sunset Crater (S. L. Hanson, Falster, and Simmons 2008).

SANTA CRUZ COUNTY: Flux District, Flux Mine, as colorless ram's horn growths less than 1 inch long (R110204).

ALUNITE

Potassium aluminum sulfate hydroxide: $KAl_3(SO_4)_2(OH)_6$. A common mineral formed in the wall rocks of sulfide orebodies by processes related to hydrothermal activity.

APACHE AND NAVAJO COUNTIES: Monument Valley and Cane Valley Districts, with uranium-vanadium ores in channels at the base of the Shinarump Conglomerate and below channels in the Moenkopi Formation and De Chelly Member of the Cutler Formation (Mitcham and Evensen 1955).

COCHISE COUNTY: Warren District, Bisbee, Sacramento Hill, in the hydrothermally altered granite porphyry stock, as an alteration product of feldspar (Schwartz 1947); Cole Mine, as large light-green masses; Denn Mine shaft as deep-green masses with pyrite; Lavender pit, as large irregular boulders with halloysite-10Å; and Lowell shaft as attractive light to green-banded specimens (Graeme 1981).

COCONINO COUNTY: Cameron District, in uranium ores as white powdery crusts and masses (3rd ed.). Black Peak District, near the Black Peak breccia pipe, as small prismatic crystals in sandstone (Barrington and Kerr 1961), and Black Point–Murphy Mine, in close association with gypsum and secondary uranium minerals in Pleistocene gravels (Austin 1964).

GILA COUNTY: Globe Hills District, Old Dominion Mine, as veins in diabase (Lausen 1923). Dripping Spring Mountains, Banner supersystem, Chilito District, Apex Mine (Bull. 181).

GRAHAM COUNTY: Gila Mountains, Lone Star District, intimately associated with jarosite and turquoise in the oxidized zone of the Safford porphyry copper deposit (R. F. Robinson and Cook 1966).

GREENLEE COUNTY: Copper Mountain supersystem, Copper King District, Ryerson Mine, as grains, irregular masses, and fibrous aggregates in altered porphyry (Lindgren 1905; Reber 1916); Candelaria District, on fractures in the Candelaria breccia pipe with turquoise and wavellite (Bennett 1975).

LA PAZ COUNTY: Sugarloaf Butte, about five miles west of Quartzsite and one mile south of U.S. Highway 60, as branching, irregular veins that constitute a substantial deposit within dacite; unusual in containing up to 4.3 percent Na_2O (Heineman 1935; Thoenen 1941; E. D. Wilson 1944; Omori and Kerr 1963; H 108800).

MARICOPA COUNTY: Vulture District, about three miles west of Morristown on the west side of Hassayampa River, as a major constituent of a hydrothermally altered rhyolite in an area of about 0.5 square mile, associated with kaolinite (Sheridan and Royse 1970).

MOHAVE COUNTY: Wallapai supersystem, Ithaca Peak District, Ithaca Peak Mine, as nodules in a clay-turquoise-sulfide vein that traverses an igneous host rock (Field 1966; Eidel, Frost, and Clippinger 1968).

PIMA COUNTY: Silver Bell District, Silver Bell Mine, Oxide Pit, associated with jarosite in veins (Kerr 1951). Ajo District, in innumerable narrow veins that cut the concentrator volcanics (Gilluly 1937; Hutton and Vlisidis 1960). Pima supersystem, Esperanza District, Esperanza Mine, in veinlets with or without turquoise, in the oxidized capping over the predominantly chalcopyrite-chalcocite orebody (Loghry 1972).

PINAL COUNTY: San Manuel District, San Manuel Mine, abundant in hydrothermally altered monzonite and quartz monzonite porphyries with kaolinite and quartz (Schwartz 1947, 1953, 1958, 1966). Mineral Creek supersystem, rare in the Ray District, associated with the oxidation of sulfides (Ransome 1919).

SANTA CRUZ COUNTY: Patagonia Mountains, Palmetto District, Evening Star prospect; Three-R District, Three-R Mine, disseminated in an altered granite porphyry (Schrader 1913, 1914, 1917; E. D. Wilson 1944). Red Mountain District, Red Mountain copper prospect, where it is thought to be of both primary and secondary origin (Loghry 1972). Kunde Mountain, North Saddle Mountain, and Saddle Mountain, about 2.5 miles northeast of Red Mountain, in mineralized zones (Hall 1978).

ALUNOGEN

Aluminum sulfate hydrate: $Al_2(SO_4)_3(H_2O)_{12} \cdot 5H_2O$. A water-soluble secondary mineral commonly formed by the decomposition of pyrite or under fumarolic conditions. May be associated with various other secondary sulfates. Probably more widespread in the state than suggested by the localities noted below.

COCONINO COUNTY: Sunset Crater National Monument, as coatings in the fumarole deposits on Sunset Crater (S. L. Hanson, Falster, and Simmons 2008). Vermilion Cliffs District, unnamed uranium prospect near Cliff Dwellers Lodge, in fractures in carbonized petrified wood in the Shinarump Conglomerate basal member of the Chinle Formation (Yang et al. 2020; R180005).

YAVAPAI COUNTY: Jerome supersystem, Verde District, United Verde Mine, as a byproduct burning pyritic ore (Lausen 1928).

AMBLYGONITE

Lithium aluminum phosphate fluoride: $LiAl(PO_4)F$. Formed in lithium- and phosphate-rich granite pegmatites. Commonly associated with spodumene, lithiophilite triphylite, lepidolite, and tourmaline.

MARICOPA COUNTY: Mitchell Wash, northeast of Morristown (UA 5623). San Domingo District, near San Domingo Wash, northeast of Wickenburg, White Picacho District, in several pegmatite bodies, as rough crystals up to 5 ft. in diameter (Jahns 1953).

MARICOPA AND YAVAPAI COUNTIES: White Picacho District, in pegmatites, associated with spodumene and zinnwaldite; Midnight Owl Mine (Jahns 1952; UA 2675). London (1981) and London and Burt (1982a) determined that most material from the mines of the White Picacho District formerly identified as amblygonite is actually montebrasite.

AMESITE

Magnesium aluminum silicate hydroxide: $Mg_2Al(AlSiO_5)(OH)_4$. A member of the kaolinite-serpentine group. Produced under conditions of low-grade metamorphism.

PINAL COUNTY: Mammoth District, Mammoth–St. Anthony Mine, as a white powdery matrix on which wulfenite has been deposited in some places (3rd ed.).

#AMMINEITE

Copper chloride ammine: $CuCl_2 \cdot 2NH_3$. Ammineite is the first mineral species found to contain an ammine complex and is formed by the interaction of copper solutions with bat guano.

MARICOPA COUNTY: Painted Rock District, Rowley Mine, occurs as vein fillings and coatings on chrysocolla and as very small equant crystals of beautiful deep-blue color associated with cerussite, mimetite, and phosgenite (R. Gibbs and M. Ascher, pers. comm., 2012; R120154).

FIGURE 3.10 Ammineite with mimetite, Rowley Mine, Maricopa County, FOV 3 mm.

ANALCIME

Sodium aluminum silicate hydrate: $Na(AlSi_2O_6) \cdot H_2O$. A fairly common mineral similar in chemistry and mode of occurrence to members of the

FIGURE 3.11 Analcime, near Santa Maria River, Yavapai County, FOV 3 mm.

zeolite group. Found in some igneous rocks of intermediate and mafic composition; also formed by hydrothermal processes.

APACHE COUNTY: Near Nutrioso, associated with clinoptilolite, as cement in Tertiary sandstone (Wrucke 1961).

COCHISE COUNTY: Along the San Simon River, Bowie District, associated with chabazite, clinoptilolite, erionite, and chabazite-Na in Late Cenozoic tuffs (Sand and Regis 1966; Regis and Sand 1967). Willcox Playa, in Pleistocene mudstone, of authigenic origin (Pipkin 1967).

COCONINO COUNTY: Tuba District, near Tuba City, near Cameron, in the baked border zone of the Tuba monchiquite dike, associated with chlorite and illite (Barrington and Kerr 1962).

MARICOPA COUNTY: Horseshoe Dam District, one mile south of Horseshoe Dam, as small transparent trapezohedral crystals (Shannon 1983b).

MOHAVE COUNTY: Big Sandy District, near Wikieup, where it composes the bulk of a friable green "sandstone" formerly thought to be mostly glauconite; glauconite merely coats the analcime grains (E. D. Wilson 1944; Robert O'Haire, pers. comm., 1972). East of Big Sandy Wash, in the east half of T16N R13W, in Pliocene tuff, associated with chabazite, clinoptilolite, erionite, and phillipsite (C. S. Ross 1928, 1941). Maggie Canyon, in S30, T12N R13W, as cement in sandstone of the Chapin Wash Formation (Lasky and Webber 1949).

NAVAJO COUNTY: Hopi Buttes volcanic field, widespread in the mafic volcanic rocks (H. Williams 1936).

PIMA COUNTY: Santa Rita Mountains, Rosemont area, in cavities in amygdaloidal basalts (Bull. 181).

PINAL COUNTY: Near Eloy, in S25, T7S R8E, in a diamond-drill hole in silty claystone (Sheppard 1969). About 5.5 miles south of Superior in a wash east of State Highway 177, as small crystals in vugs in basalt, with chabazite, mordenite, and thomsonite (Wise and Tschernich 1975).

YAVAPAI COUNTY: Aquarius Cliffs (R. C. Wells 1937), found as nice crystals up to 3 cm in size along the Santa Maria River with other zeolites (Les Presmyk, pers. comm., 2020).

ANATASE

Titanium oxide: TiO_2. Trimorphous with rutile and brookite. An uncommon secondary mineral found in veins and cavities in schists and gneisses; formed from titanium derived through leaching of country rock by hydrothermal solution.

GILA COUNTY: Reported in the Diamond Butte quadrangle, in the suite of authigenic heavy minerals in Precambrian quartzites (Gastil 1958).

GRAHAM COUNTY: Stanley District, Friend Mine (Bull. 181).

PIMA COUNTY: Ajo District, as tiny well-formed pyramidal crystals present in minor amounts, associated with papagoite in altered rock (Hutton and Vlisidis 1960).

SANTA CRUZ COUNTY: Querces District, Patagonia Mountains, around the Santo Niño Mine near Duquesne, with brookite, chalcopyrite, wulfenite, and molybdenite, all as microcrystals in interstices between adularia crystals (collected by William Hunt).

YAVAPAI COUNTY: Eureka supersystem, Bagdad District, Bagdad Mine, as small well-formed crystals with brookite (R. Gibbs, pers. comm., 2020).

FIGURE 3.12 Anatase, Bagdad Mine, Yavapai County, FOV 2 mm.

ANDALUSITE

Aluminum silicate: Al_2SiO_5. Trimorphous with kyanite and sillimanite. Commonly associated with kyanite or sillimanite in regionally metamorphosed rocks such as slates, schists, and gneisses; also typically found in thermally altered rocks, where it may be associated with cordierite. Rarely found in granites.

COCHISE COUNTY: Bluebird District, near the Texas Canyon Adams Peak Leucogranite stock complex, Texas Canyon Quartz Monzonite, as square prismatic porphyroblasts up to 2 in. long in schist, locally as the variety chiastolite (Cooper and Silver 1964). Dos Cabezas Mountains, Apache Pass, as metacrysts in Mesozoic metamorphosed mudstones (3rd ed.). Chiricahua Mountains, as the variety chiastolite (UA 8166).

GILA COUNTY: Southeastern Dripping Spring Mountains, Banner supersystem, Christmas District, Christmas Mine, in small amounts in metamorphosed siltstones of the Naco Formation, associated with muscovite, "biotite," quartz, and orthoclase (Perry 1969).

GILA AND PINAL COUNTIES: Locally abundant in the Pinal Schist near contact of Pinal Schist with the muscovite-bearing, peraluminous Solitude Granite (Ransome 1919; N. P. Peterson 1962).

LA PAZ COUNTY: Near La Paz, in S34, T4N R21W, as small crystalline masses in granite (David Shannon, pers. comm., 1971). Granite Wash

Mountains, west and south flank of Salome Peak, Hot Rock District, as thin prismatic crystals up to 3 cm (Stephen Reynolds, pers. comm., 1989), associated with kyanite, magnetite, and pyrophyllite. K-D District, about three miles southeast of Quartzsite, with kyanite, sillimanite, and dumortierite, as prismatic crystals in schist (E. D. Wilson 1929, 1933; Duke 1960).

MARICOPA COUNTY: Phoenix Mountains District, southwest slope of Piestewa Peak (formerly Squaw Peak) in Phoenix, as a common mineral in quartz-muscovite schist; appears as bright- to dark-green, blocky porphyroblasts up to 1 cm in diameter; chemical composition varies from almost pure andalusite to the manganese-rich end member, kanonaite; most of the andalusite is manganese rich (Thorpe 1980).

MOHAVE COUNTY: Grand Wash Cliffs, in pegmatite (Bull. 181). Hualapai Mountains, Antler District, eleven miles east of Yucca, in quartz veins in schist (Bull. 181).

PIMA COUNTY: Ajo District, in small quantities in the Cardigan Gneiss (Gilluly 1937). Santa Catalina Mountains, locally common in the metamorphic aureole of the peraluminous Wilderness Granite suite along its contact with Paleozoic rocks, Apache Group, and Pinal Schist (Force 1997).

PINAL COUNTY: Gila River Indian Reservation, Sacaton Mountains, about ten miles north of Casa Grande, in metamorphosed sediments encased in granodiorite, as the variety titanandalusite, associated with sillimanite, corundum, and cordierite (Bideaux, Williams, and Thomssen 1960).

YAVAPAI COUNTY: Bradshaw Mountains, as scattered lenses and disseminations in schist; near Middleton, on the Crown King Road, as large pinkish crystals (Bull. 181); also reported from Cleator District, near Cleator and near Granite Mountain in extensive veins (Bull. 181). Santa Maria Mountains, near Camp Wood, as flakes and nodules in schist; also at Bagdad (Bull. 181).

ANDERSONITE

Sodium calcium uranyl carbonate hydrate: $Na_2Ca(UO_2)(CO_3)_3 \cdot 5\text{-}6H_2O$. A very rare water-soluble secondary mineral, which forms as an efflorescence with gypsum and other secondary oxidized minerals. The Hillside Mine deposit is the type locality.

COCONINO COUNTY: Cameron District (Bollin and Kerr 1958).

YAVAPAI COUNTY: Eureka supersystem, Hillside District, Hillside Mine, with gypsum, schröckingerite, bayleyite, swartzite, johannite, and uraninite; as an efflorescence on the walls of mine workings in crusts about 0.625 inch thick on gypsum (Axelrod et al. 1951).

ANDRADITE

Calcium iron silicate: $Ca_3Fe^{3+}_2(SiO_4)_3$. A common member of the garnet group. Typically found in contact-metamorphosed deposits and skarns formed in impure limestones.

COCHISE COUNTY: Dragoon Mountains, Middle Pass District, common in the wall rocks of pyritic ores in the Abrigo Limestone (Perry 1964).

GILA COUNTY: Dripping Spring Mountains, Christmas District, Christmas Mine, in large massive beds (Bull. 181). Banner District, 79 Mine, as an abundant silicate in contact-metamorphosed limestones (Stanley B. Keith 1972). Green Valley District, Harrington claims, near the East Verde River, intergrown with epidote, calcite, and chalcopyrite (C. P. Ross 1925a). Sierra Ancha Mountains, near Workman Creek, as black crystals up to 2 cm, in a contact zone with diabase (3rd ed.).

GRAHAM COUNTY: Aravaipa supersystem, Stanley District, at Stanley Butte, as massive material and as crystals up to 2 in. across; crystals are generally brown, but some display greenish hues because of platy inclusions distributed in layers just beneath their surfaces; almost metallic in luster (Sinkankas 1964, 1966; Bull. 181). Also found as well-formed greenish-brown to black crystals on diopside from the Grey Throne prospect in the Santa Teresa Mountains (Les Presmyk, pers. comm., 2020).

GREENLEE COUNTY: Copper Mountain District, in layered limestone, forming masses from 50 to 100 ft. thick (Lindgren 1905; Guild 1910; Reber 1916; Moolick and Durek 1966).

PIMA COUNTY: Empire Mountains, as zones of massive material at contacts between Paleozoic limestones and quartz monzonite intrusives (Bull. 181). Santa Rita Mountains, Helvetia-Rosemont District, King Mine, in contact-metamorphosed limestones (Michel 1959); as the most common silicate mineral in limestones (Schrader and Hill 1915; Schrader 1917). Sierrita Mountains, Pima supersystem, Twin Buttes District, Twin Buttes Mine, as zones up to 200 ft. wide in limestone (Stanley B. Keith, pers. comm., 1973); Mission Mine, as the primary constituent of tactite formed in Paleozoic limestones (Kinnison 1966).

FIGURE 3.13 Andradite, Grey Throne prospect, Graham County, 7.6 cm.

Patagonia Mountains, Washington Camp District, Pride of the West Mine, as crystals up to 2 in. in diameter, in metamorphosed limestone (Schrader and Hill 1915; Schrader 1917).

FIGURE 3.14 Anglesite, Flux Mine, Santa Cruz County, 5.2 cm.

ANGLESITE

Lead sulfate: $Pb(SO_4)$. Abundant in oxidized lead deposits, most commonly as masses surrounding galena and altering, in turn, to cerussite. Only a few of the many localities in Arizona are listed here.

COCHISE COUNTY: Tombstone District, Tombstone Extension Mine (Butler, Wilson, and Rasor 1938). Warren District, Bisbee, Shattuck shaft and 1,800 ft. level of the Campbell Mine, as glassy spear-shaped crystals with leadhillite (S 114586). Gleeson District, Silver Bill Mine (Bull. 181). Gunnison Hills, Cochise District, Texas-Arizona Mine (Cooper and Silver 1964). California supersystem, Hilltop District, Hilltop Mine (Peter Megaw, UA x4619).

GILA COUNTY: Globe Hills District, Lost Gulch and Apache Mines (E. D. Wilson et al. 1950; N. P. Peterson 1962). Miami-Inspiration supersystem, Pinto Valley District, Castle Dome Mine (N. P. Peterson, Gilbert, and Quick 1951). Banner supersystem, 79 District, 79 Mine, as concentric replacements of galena in the 31 raise (Stanley B. Keith, pers. comm., 2020).

GRAHAM COUNTY: Stanley District (C. P. Ross 1925a). Aravaipa supersystem, Stanley District (Simons 1964), Grand Reef Mine, where a few well-formed translucent crystals up to 1 inch long are associated with linarite in quartz-lined vugs (Richard L. Jones, pers. comm., 1969; R. W. Jones 1980).

MARICOPA COUNTY: Vulture District, in dumps of the Montezuma and Prodigal Mines, west of Morristown, as nodules 1 to 3 in. in diameter (Bull. 181). Painted Rock District, Rowley Mine near Theba (W. E. Wilson and Miller 1974). Belmont Mountains, Osborne District, Tonopah-Belmont Mine (Robert O'Haire, pers. comm., 1972; G. B. Allen and Hunt 1988).

MOHAVE COUNTY: Cerbat Mountains, Wallapai District, Tennessee-Schuylkill Mine (B. E. Thomas 1949).

PIMA COUNTY: Empire District, at several mines of the Hilton Group (Bull. 181). Santa Rita Mountains, Helvetia-Rosemont District, King Mine (UA 7776). Sierrita Mountains, Pima District, abundant at the Paymaster Mine, Olive Camp (Bull. 181). Quijotoa District, Morgan Mine

(Ransome 1922). South Comobabi Mountains, Cababi District, Mildren and Steppe claims (S. A. Williams 1963).

PINAL COUNTY: Galiuro Mountains, Saddle Mountain District, Saddle Mountain Group (Bull. 181). Mammoth District, Mammoth–St. Anthony Mine, as spear-shaped crystals up to about 0.25 inch (Fahey, Daggett, and Gordon 1950; Bideaux 1980; H 98082). Silver Reef District, Nugget Fracture Mine (David Shannon, UA x4345).

SANTA CRUZ COUNTY: Patagonia Mountains, Washington Camp District, Pride of the West Mine (Bull. 181). Patagonia Mountains, Mowry District, Mowry Mine (Schrader and Hill 1915). Santa Rita Mountains, Cottonwood Canyon, Tyndall District, Glove Mine, associated with cerussite and wulfenite in oxidized lead ore (Bideaux, Williams, and Thomssen 1960; H. J. Olson 1966). Flux District, Flux Mine, as clear crystals up to 1 cm long in goethite vugs (W. E. Wilson and Hay 2015).

YAVAPAI COUNTY: Bradshaw Mountains, Castle Creek District, Copperopolis Mine (Lindgren 1926). Eureka supersystem, Hillside District, Hillside Mine (Axelrod et al. 1951). Iron King District, Iron King Mine (C. A. Anderson and Creasey 1958). Jerome supersystem, Verde District, (UA 7711).

YUMA COUNTY: Castle Dome Mountains, Castle Dome District (Foshag 1919; E. D. Wilson 1933); Brush (1873) chemically analyzed a compact variety of anglesite from the district.

ANHYDRITE

Calcium sulfate: $Ca(SO_4)$. Formed in extensive sedimentary deposits with halite, gypsum, and other salts from evaporation of oceanic waters of inland seas. A common gangue mineral in some sulfide mineral deposits.

APACHE AND NAVAJO COUNTIES: In the subsurface of the southern portions of the counties, in the Supai Salt Basin, which embraces about 2,300 square miles; Holbrook Salt District, with halite, dolomite, and clastic red beds (Peirce and Gerrard 1966; Peirce 1969b).

GILA COUNTY: Dripping Spring Mountains, Christmas District, Christmas Mine, between the 1,000 and 1,200 ft. levels, as a fairly abundant constituent of diamond-drill cores (N. P. Peterson and Swanson 1956), and in orebodies replacing dolomite, in veinlets, and interbanded with layers of magnetite (Perry 1969).

PIMA COUNTY: Ajo District, New Cornelia Mine, sparingly as minute crystals of hypogene origin in the orebody (Gilluly 1937) and as large lilac pieces up to 1.5 in. across (Sinkankas 1964; H 107494). Pima supersystem,

Twin Buttes District, Twin Buttes Mine, as a hypothermal mineral (Stanley B. Keith, pers. comm., 1973).

PINAL COUNTY: San Manuel District, San Manuel and Kalamazoo orebodies, in the inner alteration zone of mineralized monzonite and quartz monzonite porphyry, associated with quartz, sericite, and sulfides (J. D. Lowell 1968). A drill hole sunk by the Humble Oil and Refining Company in 1972 penetrated 80 ft. of halite and about 6,000 ft. of anhydrite in the Picacho Basin, just west of the Picacho Mountains, in S5, T8S R8E. The lateral extent of this enormous salt body is unknown (H. Wesley Peirce, pers. comm., 1973).

#ANILITE

Copper sulfide: Cu_7S_4. A primary or supergene copper sulfide found with other copper sulfides.

GREENLEE COUNTY: Copper Mountain supersystem, Morenci District, Morenci Mine, found in the supergene alteration zone in the Morenci Mine with pyrite and chalcopyrite (M. S. Enders, Phelps Dodge Corporation, pers. comm., 2001).

ANKERITE

Calcium iron magnesium carbonate: $Ca(Fe^{2+},Mg)(CO_3)_2$. A member of the dolomite group of minerals in which there is extensive substitution among iron, magnesium, and manganese. Probably widely distributed in metamorphosed limestones and metallic veins in the state, but few localities have been noted.

GILA COUNTY: Globe Hills District, Ramboz (Silver Glance) deposit, as a manganoan variety, with rhodochrosite (N. P. Peterson 1962). Green Valley District, Silver Butte Mine, with tetrahedrite (Lausen and Wilson 1925). Along the Salt River, between Cibecue Creek and Salt River Draw, Tomato Juice and Rock Canyon deposits, as drusy coatings on walls of the central fracture (Granger and Raup 1969). Sierra Ancha District, Sorrel Horse and Horseshoe deposits (3rd ed.).

PIMA COUNTY: Pima supersystem, Esperanza District, Esperanza Mine (UA 9371). Ajo District, New Cornelia orebody, in veins or as druses coated by calcite (Gilluly 1937).

YAVAPAI COUNTY: Jerome supersystem, Verde District, United Verde Mine, in pyritic ore (Bull. 181). Bradshaw Mountains, Big Bug District, Arizona National Mine (Bull. 181); Iron King District, Iron King Mine, where it was formed during the alteration of a massive sulfide deposit, as disseminated grains and in veinlets (Creasey 1952). Black Canyon District, Howard Copper Mine (Bull 181), Kay District, Kay Copper Mine (Bull. 181). Tiger District, M and M veins (Bull. 181). Hassayampa District, Tillie Starbuck Mine, as small rhombs associated with dolomite in cavities (Bull. 181).

YUMA COUNTY: Northern Mohawk Mountains, as thin brownish-gray veins cutting schist and gneiss (E. D. Wilson 1933).

#ANNABERGITE

Nickel arsenate hydrate: $Ni_3(AsO_4)_2 \cdot 8H_2O$. A rare secondary nickel mineral formed from primary nickel arsenide minerals in hydrothermal deposits.

YAVAPAI COUNTY: Northeastern Wickenburg Mountains, Monte Cristo District, Monte Cristo Mine, as an alteration of nickeline and skutterudite (Bastin 1922).

#ANNITE

Potassium iron aluminosilicate hydroxide: $KFe^{2+}_3(AlSi_3 O_{10})(OH)_2$. Annite is likely a very common species in Arizona, where it is found in many igneous rocks and frequently described as "biotite."

PIMA COUNTY: Near Mammoth in a granitic cobble found in alluvium derived from the Santa Catalina Mountains as 0.25 inch crystals of black mica with quartz and feldspars (Gibbs, pers. comm., 2020).

FIGURE 3.15 Annite, near Mammoth, Pinal County, FOV 3 mm.

#ANORTHITE

Calcium aluminum silicate: $Ca(Al_2Si_2O_8)$. Forms a series with albite, $NaAlSi_3O_8$, referred to as the plagioclase feldspars. In the past it was divided into a series of species with calcium-rich ones called labradorite, bytownite, and anorthite. These are now all included as the mineral anorthite. (See ALBITE for information about sodium-rich plagioclase.) Anorthite is the feldspar found in gabbro and basalt, both common rocks in Arizona.

COCHISE, GRAHAM, PIMA AND PINAL COUNTIES: As abundant, exceptional andesine-labradorite (now considered either albite and/or anorthite) phenocrysts up to 2 in. long, in intrusive and flow porphyritic rocks, locally termed *turkey-track porphyry* (Cooper 1961; Mielke 1964; Percious 1968). Labradorite feldspar forms ophitic-textured platy crystals up to 2 in. long, and 60 percent by volume, embedded in augite pyroxenes in pegmatitic differentiates of the 1050–1090 Ma diabase intrusions (Shride 1969). See ALBITE for additional information.

COCHISE COUNTY: Tombstone District, Lucky Cuss Mine, as bytownite, with vesuvianite in metamorphosed shaley limestones (3rd ed.).

LA PAZ AND YUMA COUNTIES: Cemetery Ridge District, common in metasomatic gneiss, with hornblende, epidote, diopside, and spinel; also in calcsilicate schist and granofels derived from calcareous metachert and siliceous marble, Orocopia Schist; with wollastonite, hedenbergite, and quartz (Haxel et al. 2021).

YAVAPAI COUNTY: Bradshaw Mountains, Big Bug District, Blue Bell Mine, as labradorite in panidiomorphic dike rock with augite, brown hornblende, sericite, and chlorite (Lindgren 1926).

ANTHONYITE

Copper hydroxide hydrate: $Cu(OH)_2 \cdot 3H_2O$. A rare water-soluble secondary species that quickly loses water on exposure to air. Most anthonyite in collections has altered to an anhydrous compound.

COCHISE COUNTY: Warren District, Bisbee, 1,300 ft. level of the Cole Mine, as large vividly violet corroded crystals to 5 mm in size. These encrust crumbly pyritic ores and are associated with an unknown copper hydroxide. Chemical analysis shows this anthonyite to be virtually halogen free in contrast to the type locality mineral. Its presence at the Cole Mine may, in some way, be related to the burning of sulfide ores in a nearby stope. After the first few specimens were found, the area was washed before a collecting trip; any remaining anthonyite was inadvertently dissolved (Graeme 1981).

GILA COUNTY: Southeastern Dripping Spring Mountains, Banner supersystem, Christmas District, Christmas Mine, as altered massive blue-green material with probable fresh anthonyite inside (UA 10595).

ANTHOPHYLLITE

Magnesium silicate hydroxide: $\square Mg_2Mg_5Si_8O_{22}(OH)_2$. An orthorhombic member of the amphibole group. A metamorphic mineral common in schists and gneisses; less common in contact-metamorphosed rocks.

COCHISE COUNTY: Cochise supersystem, Johnson Camp District, Johnson Camp, asbestiform, in a narrow vein cutting Horquilla Limestone (Cooper and Silver 1964).

COCONINO COUNTY: Grand Canyon National Park, in anthophyllite-bearing rocks scattered throughout the older Precambrian metamorphic rocks, for example, in cordierite-anthophyllite schists and paragneisses at mile 229, Travertine Canyon (M. D. Clark 1979).

MOHAVE COUNTY: Hualapai (Antler) District, Antler Mine (Romslo 1948). Kingman area, with phlogopite (UA 1193).

PIMA COUNTY: Santa Rita Mountains, Helvetia-Rosemont District, Blue Jay Mine (UA 5233). Santa Catalina Mountains, Catalina District, Kielberg's Iron Mountain claim (UA 5228).

YAVAPAI COUNTY: Reported in the Eureka supersystem (Bull. 181).

ANTIGORITE

Magnesium silicate hydroxide: $Mg_3Si_2O_5(OH)_4$. Antigorite is one of the serpentine minerals. Serpentine specimens typically consist of mixtures of the polymorphs of chrysotile, lizardite, and antigorite. The fibrous mineral is chrysotile (commonly called asbestos, which is a term for any fibrous mineral), and the more massive and platy are antigorite and lizardite. All three polymorphs are found in Arizona. Serpentine is of secondary origin, derived from the alteration of nonaluminous magnesian silicates, particularly olivine, amphibole, or pyroxene. In places, it is formed as large masses derived from peridotite or other mafic igneous rocks. It is a common product of contact metamorphism of magnesian limestones, and it is also found in hydrothermal veins. Antigorite is abundant as a major constituent of the nonasbestiform serpentine rock associated with asbestos ore. Extensive lists of asbestos localities are included in Funnell and Wolfe (1964) and in Arizona Bureau of Mines Bulletins 126 and 180 (E. D. Wilson 1928; Shride 1969). The following are given as antigorite localities, but lizardite may also be present, although exact determinations have not been done.

APACHE COUNTY: Garnet Ridge SUM District constitutes the main matrix of the serpentine ultramafic microbreccia (SUM), abundant in a boulder, pseudomorphous after an orthopyroxene, related to a breccia dike that pierces sedimentary rocks (Gavasci and Kerr 1968).

COCHISE COUNTY: Tombstone District, Lucky Cuss Mine, in altered limestone (Butler, Wilson, and Rasor 1938). Chiricahua Mountains, with chrysotile, in contact-metamorphic ores (Bull. 181). Dos Cabezas Mountains, in metamorphosed limestones (Guild 1910). Little Dragoon Mountains, Cochise District, Johnson Camp area, Moore orebody, as a retrograde product of high-temperature metamorphism of forsterite, tremolite, and diopside (J. R. Cooper and Huff 1951; J. R. Cooper and Silver 1964).

COCONINO COUNTY: Grand Canyon region, Bass and Hance properties, where the Precambrian Bass Limestone has been altered adjacent to diabase sills (Selfridge 1936).

GILA COUNTY: The most extensive and commercially important deposits of chrysotile and associated antigorite in Arizona are north and northeast of Globe along the Salt River near Chrysotile and along Cherry and Ash Creeks (Engineering and Mining Journal 1915; Melhase 1925; L. A. Stewart and Haury 1947; L. A. Stewart 1956); San Carlos Apache Reservation, Emsco and Bear Mines (Bromfield and Schride 1956). The deposits originated through metamorphic action of diabase intrusives upon Precambrian Mescal Limestone. Dark-green antigorite typically occurs closest to the contact with the diabase, with apple-green translucent lizardite forming farther away from the contact. Both are cut by late-stage, fiber-form chrysotile veinlets and seams, which compose the main mined material. Pinal and Mescal Mountains and Pinto Creek region (Bull. 181). Sierra Ancha Mountains, at the head of Pocket Creek (Bateman 1923; Sampson 1924). Southeastern Dripping Spring Mountains, Banner supersystem, Christmas District, Christmas Mine, replacing dolomite and forsterite in the orebodies (Perry 1969). Globe Hills District, Old Dominion Mine, in Mescal Limestone near diabase sills (Bull. 181).

GREENLEE COUNTY: Copper Mountain supersystem, Morenci District, northern part of the Morenci open-pit area, locally formed with other calcsilicate minerals in an extensive contact-metamorphic assemblage (Moolick and Durek 1966); on the ridge just west of Morenci, at the Thompson Mine, as green-banded material associated with magnetite (Lindgren 1905; Reber 1916; Creasey 1959).

PIMA COUNTY: Pima supersystem, Twin Buttes District, Twin Buttes Mine, in metamorphic rocks (Stanley B. Keith, pers. comm., 1973).

PINAL COUNTY: Mammoth District, Mammoth–St. Anthony Mine, as aggregated masses of shining white balls of antigorite that form a matrix for wulfenite and descloizite (Bideaux 1980). Hewett Wash area (L. A. Stewart 1956).

#ANTIMONY

Antimony: Sb. Occurs in Mesozoic ore deposits with gold, silver, copper, lead, zinc, tungsten, and molybdenum.

PIMA COUNTY: Santa Rita Mountains, Cuprite District, Pauline Mine, found in chalcopyrite and pyrite vein material with chalcocite, galena, hematite variety specularite, sphalerite, and molybdenite (Schrader and Hill 1915).

#ANTIPINITE

Potassium sodium copper oxalate: $KNa_3Cu_2(C_2O_4)_4$. Antipinite forms through the interaction of copper-bearing solutions and bat guano.

MARICOPA COUNTY: Painted Rock District, Rowley Mine, as very small equant brilliant deep-blue crystals in bat guano with wheatleyite and rowleyite (Kampf et al. 2016; R160005).

FIGURE 3.16 Antipinite, dark blue, Rowley Mine, Maricopa County, FOV 0.63 mm.

ANTLERITE

Copper sulfate hydroxide: $Cu^{2+}_3(SO_4)(OH)_4$. A rare secondary mineral formed in the oxidized zone of copper deposits; easily mistaken for brochantite, which it resembles. The mineral takes its name from the Antler Mine in Mohave County, which is the type locality.

APACHE COUNTY: Monument Valley area, Cane Valley District, Monument No. 2 Mine (UA 2279).

COCHISE COUNTY: Warren District, Bisbee, unspecified mine, as small crystals of excellent quality, implanted on brochantite (Palache 1939a); Cole shaft, Lavender pit, and Shattuck shaft (Graeme 1981; S 95724).

FIGURE 3.17 Antlerite, Antler Mine, Yavapai County, 2.5 cm high.

COCONINO COUNTY: Grandview District, Grandview (Last Chance) Mine (J. W. Anthony et al. 2003).

GILA COUNTY: Banner supersystem, 79 District, 79 Mine, as pale-green crusts in the 31 stope area (Thomas Trebisky, pers. comm., 1975).

GRAHAM COUNTY: Lone Star District, in small amounts in veins in metamorphosed latites and andesites (Hutton 1959a).

GREENLEE COUNTY: Copper Mountain District, Morenci Mine, Clay orebody (UA 6259).

MARICOPA COUNTY: Painted Rock District, Rowley Mine, as bluish-green mats of small needle-like prismatic crystals (B. Murphy, pers. comm., 2013).

MOHAVE COUNTY: Hualapai (Antler) District, Antler Mine, on material where W. F. Hillebrand (1889b) originally described it as a new species (Romslo 1948), as soft-green lumps (R6075, part of the type material).

PINAL COUNTY: Bunker Hill (Copper Creek) District, Copper Creek, Childs-Aldwinkle Mine, small equant green crystals resembling brochantite from the glory hole above the underground workings (R. Downs, pers. comm., 2012).

YAVAPAI COUNTY: Jerome supersystem, Verde District, United Verde Mine, as perfect crystals up to 3 mm long on fracture surfaces in chlorite schist that contains chalcopyrite; abundant in the lower part of the oxidized zone, with cyanotrichite and brochantite (3rd ed.). Agua Fria District, Copper Queen Mine, as very small stout bladed crystals less than 1 mm long in vugs lined with clinoclase and olivenite (Gibbs, pers. comm., 2014).

FIGURE 3.18 Apachite, Christmas Mine, Gila County, FOV 3 mm.

APACHITE

Copper silicate hydrate: $Cu^{2+}_9Si_{10}O_{29} \cdot 11H_2O$. A retrograde metamorphic mineral. Christmas is the type locality.

GILA COUNTY: Southeastern Dripping Spring Mountains, Banner supersystem, Christmas District, Christmas Mine, as minute blue-matted fibers or blades that resemble shattuckite; fills fractures (with calcite) and has replaced grossular and diopside in some areas; alters to gilalite; first noted by R. A. Jenkins in one fine specimen; later found in relative abundance only along an ill-defined structure that cuts tactites in the southwestern part of the open-pit mine at Christmas (Cesbron and Williams 1980).

apatite

A group name, the most common members of which have this general composition: calcium phosphate fluoride, $Ca_5(PO_4)_3F$, fluorapatite. Members of the group are the most common of the phosphorus-bearing minerals in igneous rocks. Other members of the group are chlorapatite and hydroxylapatite. For the localities listed below, the specific species has not been determined.

APACHE AND NAVAJO COUNTIES: Monument Valley and Cane Valley Districts, as one of the most common heavy minerals in uranium-bearing

Shinarump Conglomerate, associated with baryte, tourmaline, and zircon (R. G. Young 1964).

GILA COUNTY: Miami-Inspiration supersystem, as a common accessory mineral in igneous rocks; Pinto Valley District, Castle Dome area, as veinlets of hydrothermal origin that cut the Scanlan Conglomerate and arkoses of the Pioneer Formation (N. P. Peterson 1962). Dripping Spring Mountains, Christmas District, Christmas Mine, as remnant euhedral grains in metamorphosed diorite (Perry 1969).

FIGURE 3.19 Apatite, Magna Mine, Copper Creek, Pinal County, FOV 2 mm.

GRAHAM COUNTY: Turnbull Mountain, Aravaipa supersystem, Stanley District, Fisher prospect, in micropegmatite (Bull. 181).

GREENLEE COUNTY: Copper Mountain District, Morenci, as a primary mineral in green mica-bearing rock, as small rods embedded in green mica, with pyrite and magnetite (Reber 1916).

MOHAVE COUNTY: Aquarius Mountains, in a pegmatite, with titanite, chevkinite, monazite, and cronstedtite (Kauffman and Jaffe 1946).

PIMA COUNTY: Silver Bell District, as an accessory mineral (Kerr 1951). Pima supersystem, Twin Buttes District, Twin Buttes Mine (UAX 740).

PINAL COUNTY: Bunker Hill (Copper Creek) District, Copper Creek, Childs-Aldwinkle Mine, as a gangue mineral in a breccia pipe; one crystal on the 820 ft. level was 5 in. long (Kuhn 1941; UA 535); Old Reliable Mine, with rutile (UA 9606); Magna Mine (R. Gibbs, pers. comm., 2020). San Manuel District, in the potassium silicate phase of hydrothermal alteration of monzonite and quartz monzonite porphyries, and in the argillic phase of alteration (Creasey 1959). Mineral Hill District, Ajax Mine (William Kurtz, UA x4401).

SANTA CRUZ COUNTY: Palmetto District, as an accessory mineral in granite porphyry (Schrader 1913). Patagonia supersystem, Four Metals District, Four Metals breccia pipe, as an accessory mineral, pale-green hexagonal prisms up to 1 inch long (Stanley B. Keith, pers. comm., 2019).

YAVAPAI COUNTY: Tiger District, Springfield Group, as large crystals in granodiorite (Bull. 181). Eureka supersystem, Copper Ridge District near Bagdad, as an accessory mineral in titaniferous magnetite bodies (Ball and Broderick 1919; Schwartz 1947; C. A. Anderson 1950). White Picacho District, in pegmatites (Jahns 1952). Copper Basin District, as an accessory mineral in several igneous rocks (Johnston and Lowell 1961). Iron King District, Iron King Mine, introduced during hydrothermal alteration of the massive sulfide deposit; as disseminated grains and needle-like crystals (Creasey 1952).

#APHTHITALITE

Potassium sodium sulfate: $K_3Na(SO_4)_2$. An uncommon mineral found around fumaroles, in evaporate deposits, and in guano deposits.

MARICOPA COUNTY: Painted Rock Mountains, Painted Rock District, Rowley Mine. The mineral occurs as tiny dull-white balls and white silky rounded prisms associated with other sulfates and oxalates (A. Kampf, pers. comm., 2019).

FIGURE 3.20 Apophyllite, Christmas Mine, Gila County, FOV 2 mm.

apophyllite

A group name that includes fluorapophyllite-(K), fluorapophyllite-(Na), and hydroxyapophyllite-(K). The specific species in the apophyllite group have not been determined for most Arizona occurrences. Minerals of the apophyllite group primarily form as secondary minerals in amygdules in basalts and are commonly associated with zeolites. They are also found in contact-metamorphosed limestones bordering intrusive rocks.

COCHISE COUNTY: Tombstone District, as subhedral grains with vesuvianite and wollastonite in tactite (Bideaux, Williams, and Thomssen 1960).

GILA COUNTY: Southeastern Dripping Spring Mountains, Banner supersystem, Christmas District, Christmas Mine, enclosing kinoite (S. A. Williams 1976).

PIMA COUNTY: Santa Rita Mountains near Helvetia, Helvetia-Rosemont District, in drill core that penetrated skarn developed in Paleozoic limestones and dolomites; as vuggy crystalline masses in which copper and kinoite are embedded (J. W. Anthony and Laughon 1970).

ARAGONITE

Calcium carbonate: orthorhombic $Ca(CO_3)$. Trimorphous with calcite and vaterite. Metastable under standard conditions, tending to revert to the stable calcite. Formed in spring deposits and from sulfate-bearing saline solutions, in beds with gypsum, in cavities in lavas, and in limestone caverns.

COCHISE COUNTY: Warren District, Bisbee, as magnificent coralloidal groups of the variety flos ferri; also acicular and bladed crystals commonly colored by compounds of copper and iron. Found mainly in limestone caverns in the larger oxidized orebodies (Graeme 1981). Dragoon Mountains,

Gleeson District, especially at the Tom Scott Mine as postmining stalactites and stalagmites lining solution cavities in silver-lead deposits (Bull. 181).

COCONINO COUNTY: Grand Canyon National Park, as part of the travertine common along the canyon where it formed from seeps and streams. Most of the travertine is calcite, but aragonite has been identified at a large travertine cone at mile 62.4 and at Christmas Tree Cave, mile 135.5 along the Colorado River (R. Grant, pers. comm., 1990).

MOHAVE COUNTY: Thirty-five miles southeast of Hackberry, in granite pegmatite, with gadolinite (Palmer 1909).

PIMA COUNTY: Pima supersystem, San Xavier District, as fine crystals from the San Xavier West Mine (Arnold 1964; UA 5782) and Mineral Hill (UA 7630, H 89687); Sierrita Mountains, copper stained in fissures (3rd ed.). Silver Bell District, Silver Bell Mine (UA 6466). Santa Catalina Mountains, Oracle Ridge Mine on Marble Peak (Dewey Wilkins, pers. comm., 1989).

FIGURE 3.21 Aragonite, Southwest Mine, Bisbee, Cochise County, 8 cm high.

SANTA CRUZ COUNTY: Central Santa Rita Mountains, Onyx Cave, as flos ferri (Bull. 181); Cave of the Bells, as large snow-white flos ferri, coral-like groups (3rd ed.), and acicular crystal sprays.

YAVAPAI COUNTY: Jerome supersystem, Verde District, United Verde Extension Mine (UA 7626). Castle Hot Springs, as pseudohexagonal crystals (UA 220). Northeast of Camp Verde, in basalts (E. R. Brenizer, pers. comm., 1974; UA 10185). Copper Canyon District, Camp Verde Salt Mine, as ball-like aggregates of pseudohexagonal crystals and pseudomorphs after glauberite (J. R. Thompson 1983).

YUMA COUNTY: Sheep Tanks District, as part of black-banded calcite veins (Cousins 1972). Castle Dome District, in channels and vugs with smithsonite, hydrozincite, wulfenite, vanadinite, and mimetite (Bull. 181).

ARAVAIPAITE

Lead aluminum fluoride hydrate: $Pb_3AlF_9 \cdot H_2O$. Associated with galena and fluorite from which it is thought to have formed by interaction with supergene fluids. The Grand Reef Mine is the type locality.

GRAHAM COUNTY: Aravaipa supersystem, Grand Reef District, Grand Reef Mine, as thin colorless crystal plates, invariably twinned, in quartz-lined vugs, associated with galena, quartz, fluorite, and anglesite, together

R100004   1 mm

FIGURE 3.22 Aravaipaite, Grand Reef Mine, Graham County.

with the rare species grandreefite, pseudograndreefite, and laurelite (Kampf, Dunn, and Foord 1989).

#ARCANITE

Potassium sulfate: $K_2(SO_4)$. Forms from paloverde (*Parkinsonia microphylla*) tree ash exposed to rain.

MARICOPA COUNTY: Phoenix area, as sparse patches of honey-colored to pale-purple rounded pinnacles in brittle tree ash crust (Garvie 2016).

ARGENTOJAROSITE

Silver iron sulfate hydroxide: $AgFe^{3+}_3(SO_4)_2(OH)_6$. Except for rarity, in most respects similar to the far more common jarosite.

COCHISE COUNTY: Tombstone District, associated with other oxide-zone minerals (S. A. Williams 1980b).

FIGURE 3.23 Arsendescloizite, Tonopah-Belmont Mine, Maricopa County, FOV 0.4 mm.

#ARSENDESCLOIZITE

Lead zinc arsenate hydroxide: $PbZn(AsO_4)(OH)$. An uncommon secondary mineral found in the oxidized portion of arsenic-bearing polymetallic mineral deposits.

MARICOPA COUNTY: Belmont Mountains, Osborne District, Tonopah-Belmont Mine, as minute bladed olive-green crystals in heavily oxidized lead-zinc vein material (Gibbs and Ascher 2012).

FIGURE 3.24 Arsenic, Double Standard Mine, Santa Cruz County, 7 cm.

ARSENIC

Arsenic: As. Found in hydrothermal veins, most commonly associated with cobalt, nickel, and silver ores but also in other sulfide deposits.

SANTA CRUZ COUNTY: Washington Camp District, Double Standard Mine, as reniform masses, some weighing more than 50 pounds, in metamorphosed dolomitic limestone (C. H. Warren 1903; Struthers 1904; Guild 1910; Schrader and Hill 1915).

ARSENIOSIDERITE

Calcium iron oxide arsenate hydrate: $CaFe^{3+}_3O_2(AsO_4)_3 \cdot 3H_2O$. An uncommon oxide product of arsenopyrite in carbonate-hosted ores.

YAVAPAI COUNTY: Kirkland District, Venus prospect, as clusters of deep-red interlocking platy crystals, occupying pseudomorphs after arsenopyrite. The deposit is confined to a narrow shear zone in a large xenolith of Precambrian metavolcanic rocks, which is suspended in a granitic intrusive host (3rd ed.).

ARSENOLITE

Arsenic oxide: As_2O_3. The dimorph of claudetite. A secondary mineral formed by alteration of primary arsenides or arsenic-bearing sulfides or as a product of mine fires. A very toxic substance.

YAVAPAI COUNTY: Jerome supersystem, Verde District, United Verde Mine, as octahedral crystals on burned ore matrix, formed during the mine fires (UA 6708).

ARSENOPYRITE

Iron arsenic sulfide: FeAsS. The most common of the arsenic-bearing minerals. Formed under a wide variety of conditions: in high-temperature gold-quartz veins, in contact-metamorphosed sulfide deposits, and less commonly in pegmatites and low-temperature veins.

GRAHAM COUNTY: Santa Teresa Mountains, Bluebird Cobalt District, Bluebird Mine, near Mt. Trumbull with glaucodot and cobaltite (Bilbrey 1962; USGS 2005a).

MARICOPA COUNTY: White Picacho District, sparingly in pegmatites (Jahns 1952). Unspecified locality near Tempe (3rd ed.)

MOHAVE COUNTY: Cerbat Mountains, in some mines of the Wallapai supersystem, notably the Minnesota-Connor, Windy Point, and Queen Bee properties (Schrader 1909).

PIMA AND SANTA CRUZ COUNTIES: Santa Rita and Patagonia Mountains, especially the Washington Camp District, in several contact-metamorphosed deposits (3rd ed.). Mowry District, Mowry Mine (Schrader 1917).

YAVAPAI COUNTY: Jerome supersystem, Verde District, sparingly at the United Verde Mine (Lausen 1928) and Shea property (Lindgren 1926). Bradshaw Mountains, Big Bug District, as crystals in the Boggs Mine; Iron King District, Iron King Mine, abundant in the massive sulfide ores of en echelon vein deposits, as subhedral grains up to 1.5 mm showing diamond-shaped sections (Creasey 1952). Eureka supersystem, Hillside District, near the Hillside Mine, in a vein with bismuthinite (Axelrod et al. 1951). White Picacho District, sparingly in pegmatites (Jahns 1952). Near Prescott, Old Dick District, Old Dick Mine, cobaltian rich in vein quartz with pyrite; these quartz veins may cut earlier massive arsenopyrite (3rd ed.) as the cobaltian variety (UA 6193).

#ARSENTSUMEBITE

Lead copper arsenate sulfate hydroxide: $Pb_2Cu(AsO_4)(SO_4)(OH)$. A rare secondary mineral found in the oxidized portion of arsenic-bearing polymetallic mineral deposits.

MARICOPA COUNTY: Belmont Mountains, Osborne District, Tonopah-Belmont Mine, as small yellow-green crystals with caledonite and linarite in heavily oxidized lead-zinc vein material (Gibbs and Ascher 2012). Painted Rock District, Rowley Mine, as small yellow-green crystals in oxidized vein material (R. Jenkins, pers. comm., 2013).

PINAL COUNTY: Mammoth District, Mammoth–St. Anthony Mine, as small yellow-green crystals with wulfenite in material acquired by Arthur Flagg (J. Ruiz and R. Jenkins, pers. comm., 2016).

#ARTHURITE

Copper iron arsenate hydroxide hydrate: $CuFe^{3+}_2(AsO_4)_2(OH)_2 \cdot 4H_2O$. A secondary copper arsenate found with other oxide copper minerals in arsenic-rich orebodies.

COCONINO COUNTY: Grand Canyon National Park, Grandview District, Grandview (Last Chance) Mine, as a yellow-green coating with bluish cubes of pharmacosiderite, from a specimen in the Grand Canyon National Park collection (P. Williams, pers. comm., 2008).

ARTROEITE

Lead aluminum fluoride hydroxide: $PbAlF_3(OH)_2$. Associated with galena and fluorite, from which it is thought to have formed by interaction with supergene fluids. The Grand Reef Mine is the type locality.

GRAHAM COUNTY: Aravaipa supersystem, Grand Reef District, Grand Reef Mine, as crystals in a quartz-lined vug, associated with galena, quartz, fluorite, calcioaravaipaite, and anglesite (Kampf and Foord 1995, 1996).

#ASHBURTONITE

Hydrogen copper lead silicate carbonate hydroxide chloride: $HCu_4Pb_4Si_4O_{12}(HCO_3)_4(OH)_4Cl$. A secondary mineral formed in the oxidized portion of lead-zinc-copper deposits.

MARICOPA COUNTY: Belmont Mountains, Osborne District, Tonopah-Belmont Mine, as minute blue prismatic crystals associated with calcite in heavily oxidized lead-zinc vein material (Gibbs and Ascher 2012; R110003). Painted Rock District, Rowley Mine, as sky-blue to dark-blue prismatic crystals in spherical sprays up to 1 mm in diameter (B. Murphy, pers. comm., 2013).

ATACAMITE

Copper chloride hydroxide: $Cu_2Cl(OH)_3$. A secondary mineral formed from the oxidation of other secondary copper minerals, especially under arid, saline conditions. Commonly associated with malachite, cuprite, chrysocolla, brochantite, gypsum, and limonite. Probably more widely represented in Arizona than indicated by the number of localities reported here. Also see CLINOATACAMITE and PARATACAMITE as they are similar and not every specimen has been checked for specific identification.

FIGURE 3.25 Atacamite, Southwest Mine, Bisbee, Cochise County, 8.5 cm.

COCHISE COUNTY: Warren District, Bisbee, associated with connellite (Gene Wright, UA x1201). On the fifth level of the Southwest Mine, as fine crystals up to 2 cm long (Graeme 1993).

GILA COUNTY: Miami-Inspiration supersystem, Pinto Valley District, Castle Dome Mine (H 108296); Inspiration Mine (Olmstead and Johnson 1966).

MARICOPA COUNTY: Painted Rock District, Rowley Mine near Theba, in very small amounts, associated with caledonite and thought to have formed by the alteration of caledonite or linarite (W. E. Wilson and Miller 1974).

PIMA COUNTY: Cerro Colorado District, Cerro Colorado Mine (Bull. 181). Tucson Mountains, Saginaw Hill District, south side of Saginaw Hill, about seven miles southwest of Tucson, associated with other oxidized minerals, including brochantite, pseudomalachite, malachite, libethenite, cornetite, and chrysocolla (Khin 1970). South Comobabi Mountains, Cababi District, Mildren and Steppe claims (S. A. Williams 1963).

PINAL COUNTY: Galiuro Mountains, Bunker Hill (Copper Creek) District, Copper Creek, on the main level of the Old Reliable Mine, as small green crystals, with olivenite (Bull. 181). Mammoth District, Mammoth–St. Anthony Mine, 400 ft. level of the Collins vein, as deep-green coarse granular aggregates (Bull. 181). San Manuel District, San Manuel Mine, as laths in chrysocolla and cornuite (a variety of chrysocolla) and along fractures (Bideaux, Williams, and Thomssen 1960; L. A. Thomas 1966; UA 3180 and others). Lakeshore District, Lakeshore Mine, 900 ft. level, as an isolated occurrence in the brochantite zone (Cook 1988b). Santa Cruz District, Santa Cruz porphyry copper deposit west of Casa Grande, as the major supergene copper mineral. This may be the largest atacamite deposit in the world. The data are from drill-hole studies (Cook 1988a, 1988b); also in the supergene oxide zone at the Sacaton porphyry copper deposit (AZGS 2013b).

YAVAPAI COUNTY: Jerome supersystem, Verde District, in small quantities at the United Verde Extension Mine (Guild 1910).

AUGELITE

Aluminum phosphate hydroxide: $Al_2(PO_4)(OH)_3$. Found in granite pegmatites with other phosphate minerals and with silicates.

MARICOPA AND YAVAPAI COUNTIES: White Picacho District, in pegmatites (London 1981).

AUGITE

Calcium magnesium iron silicate: $(Ca,Mg,Fe)_2Si_2O_6$. A very common rock-forming mineral of the pyroxene group, which is present in a wide variety of mafic igneous rocks such as gabbro, diabase, and basalt. Only a few localities are listed here.

APACHE COUNTY: Garnet Ridge SUM District, Garnet Ridge, in a breccia dike associated with loose garnets and as fragments in sand and soil (Gavasci and Kerr 1968).

COCHISE COUNTY: Tombstone District, in diorite porphyry dikes and basaltic rocks (Butler, Wilson, and Rasor 1938).

GILA AND PINAL COUNTIES: Abundant in diabase sills that intrude rocks of the Apache Group over large areas in central Arizona. Dripping Spring Mountains, with olivine and a mixture of clay minerals and iron oxides in Tertiary basalts (Bull. 181).

NAVAJO COUNTY: North of Bidahochi Butte, as loose crystals in an unnamed diatreme (Bideaux, Williams, and Thomssen 1960).

SANTA CRUZ COUNTY: Patagonia Mountains, Duquesne and Washington Camp, in metamorphosed limestones (Schrader and Hill 1915).

YAVAPAI COUNTY: Copper Basin District, as an accessory mineral in igneous rocks (Johnston and Lowell 1961). Iron King District, Iron King Mine, as phenocrysts in porphyritic basalt flows, with olivine (Creasey 1952).

AURICHALCITE

Zinc copper carbonate hydroxide: $(Zn,Cu)_5(CO_3)_2(OH)_6$. An uncommon secondary mineral associated with other oxidized minerals in the oxidized zones of lead and copper deposits.

COCHISE COUNTY: Warren District, Bisbee, in the upper portions of the Copper Queen Mine (Ransome 1904; Guild 1910), "in beautiful tubes lining cavities" (Kunz 1885); 1200 to 1300 ft. levels of the Cole shaft, with hemimorphite (Graeme 1981). Tombstone District, on the west side of the quarry roll, as plumose aggregates of pale-blue crystals (Butler, Wilson, and Rasor 1938). Gleeson District, as encrustations and drusy linings of cavities in oxidized lead-silver deposits (Bull. 181); Mystery Mine (UA 8942); Silver Bill Mine, as large lathy sky-blue aggregates with hemimorphite, rosasite, and smithsonite (H. Peter Knudsen, UA x852). Little Dragoon Mountains, Cochise supersystem, Johnson Camp District, Johnson Camp, with chrysocolla, malachite, tenorite, copper, and hemimorphite, in the oxidized portions of pyrometasomatic sulfide deposits (J. R. Cooper and Huff 1951; J. R. Cooper and Silver 1964).

COCONINO COUNTY: Grand Canyon National Park, Grandview District, Grandview (Last Chance) Mine, on or near the Redwall Limestone, as greenish-blue lath-like crystals that form tufted encrustations (Leicht 1971).

GILA COUNTY: Banner supersystem, 79 District, 79 Mine, among the finest specimens in the world. The most notable occurrence is on the fourth level of the mine, where the mineral formed delicate sprays of acicular and lath-like crystals that range in color from pale sky blue to deep sea blue green. The sprays typically grew on hemimorphite, smithsonite, and wulfenite and may, in turn, be covered with cerussite, calcite, plattnerite, or murdochite; formed contemporaneously with rosasite (Stanley B. Keith 1972; UA 1201); a specimen described by its collector (W. E.

FIGURE 3.26 Aurichalcite, 79 Mine, Gila County, 5.9 cm.

Wilson) as perhaps the world's finest was recovered from the mine in about 1970 (W. E. Wilson 1987b).

LA PAZ COUNTY: Plomosa Mountains, Ramsey District, Black Mesa Mine, in S16, T3N R17W, as well-formed small crystals with crystalline malachite in limonite gangue with small brilliant hexagonal willemite crystals (David Shannon, pers. comm., 1972). Yuma King District, Yuma Copper Mine, as small equant tablets in oxidized copper ores (Stanley B. Keith, pers. comm., 2019).

PIMA COUNTY: Empire Mountains, Empire District, Lone Mountain Mine, as small radiating fibrous masses and seams with smithsonite and hemimorphite (Bull. 181), as acicular clusters and tufted lathy coatings and rosettes locally with rosasite and hemimorphite at the Chief and 49 Mines (3rd ed.; Stanley B. Keith, pers. comm., 2019). Helvetia-Rosemont District, King in Exile Mine, with malachite, azurite, and smithsonite (Dan Helm, pers. comm., 1978); Omega Mine, with rosasite and calcite (Brian Bond, UA x2511). Sierrita Mountains, Pima supersystem, Twin Buttes District, in fissures in garnet skarn at the Queen Mine at Twin Buttes (UA 5565) and at the San Xavier West Mine (Arnold 1964). Waterman District, Silver Hill Mine, with plattnerite and murdochite (Bideaux, Williams, and Thomssen 1960). South Comobabi Mountains, Cababi District, East silver-lead claim, as small crystals up to 1 mm long on quartz crystals (S. A. Williams 1962).

PINAL COUNTY: Vekol District, Reward Mine (Bull. 181). Mammoth District, Mammoth–St. Anthony Mine, as rare radiating sprays of crystals (3rd ed.). Mineral Mountain District (Robert Mudra, UA x173).

SANTA CRUZ COUNTY: Flux District, Flux Mine, as light-blue clots of thick acicular crystals on limonite, with hemimorphite (Thomas Trebisky, pers. comm., 1972).

AURORITE

Manganese oxide hydrate: $Mn^{2+}Mn^{4+}_3O_7 \cdot 3H_2O$. An associate of black calcite crystals, as small black crystals or massive, with other manganese oxide minerals, including todorokite, pyrolusite, and cryptomelane.

COCHISE COUNTY: Gleeson District, Defiance Mine, as small black crystals with calcite (Hidemichi Hori, pers. comm., 1985).

GREENLEE COUNTY: Copper Mountain District, Morenci, southern part of the open-pit mine, as small black crystals on goethite in the oxidized zone (Hidemichi Hori, pers. comm., 1982). Ash Peak District, Ash Peak Mine near Duncan, as massive material (3rd ed., UA 12294).

YUMA COUNTY: Kofa District, King of Arizona Mine, with todorokite, groutite, chalcophanite, and pyrolusite (Hankins 1984). Sheep Tanks District, as one of several manganese minerals in black calcite veins (Cousins 1972).

#AUROSTIBITE

Gold antimony: $AuSb_2$. Found in hydrothermal gold-quartz veins low in sulfur but including other antimony minerals.

LA PAZ COUNTY: Dome Rock Mountains, Cunningham Pass area, Copper Bottom District, Cunningham Mountain placers, as small gray inclusions in placer gold grains associated with dyscrasite, cuproauride, pyrite, and quartz (Melchiorre 2013).

AUSTINITE

Calcium zinc arsenate hydroxide: $CaZn(AsO_4)(OH)$. A rare secondary mineral found in the oxidized zones of some base-metal deposits.

COCHISE COUNTY: Tombstone District, Oregon-Prompter Mine, as green sprays of crystals less than 1 mm in size and as botryoidal crusts (B. Murphy, pers. comm., 2018). Warren District, Bisbee, Southwest Mine, seventh level; a small open cut on the surface near the Hendricks Gulch portal to this level contained a few specimens that had a crust of 0.1 mm tan to colorless crystals on silica breccia fragments; these crystals were partially overgrown by 0.25 mm gray-green to deep-green crystals of cuprian austinite (Graeme, Graeme, and Graeme 2015).

PINAL COUNTY: Galiuro Mountains, Table Mountain District, Table Mountain Mine, as intergrowths with conichalcite (Bideaux, Williams, and Thomssen 1960).

AUTUNITE

Calcium uranyl phosphate hydrate: $Ca(UO_2)_2(PO_4)_2 \cdot 10-12H_2O$ (water content variable between about 10 and 12 molecules). The most common of the secondary uranium minerals. Widely distributed; commonly formed in the oxidized zones of hydrothermal mineral deposits.

APACHE COUNTY: Cane Valley District, Monument No. 2 Mine, associated with a wide variety of oxidized uranium and vanadium minerals (Witkind and Thaden 1963).

GILA COUNTY: Sierra Ancha District, Red Bluff prospect, with uranophane and uraninite (R. Robinson, pers. comm., 1954).

MARICOPA COUNTY: Lime Creek District, Lucky Find Group, in a mafic dike cutting Precambrian granite (R. L. Robinson 1956).

MOHAVE COUNTY: Chapel District, Chapel prospect, with uranophane and torbernite (Wenrich and Sutphin 1989).

NAVAJO COUNTY: Monument Valley District, Monument No. 1 and Mitten No. 2 Mines (Witkind 1961; Witkind and Thaden 1963).

SANTA CRUZ COUNTY: Near Alamo Spring, with uranophane in a vein with lead ore (Bull. 181). Santa Rita Mountains, Duranium District, Duranium claims, in arkosic sandstone with kasolite and uranophane (R. L. Robinson 1954).

#AWARUITE

Nickel iron: Ni_3Fe. Occurs in serpentinites and meteorites.

LA PAZ AND YUMA COUNTIES: Cemetery Ridge District, as small grains with other sulfides in magnetite in blocks of serpentinite and partially serpentinized harzburgite in the Orocopia Schist at Cemetery Ridge (Haxel et al. 2018).

FIGURE 3.27 Axinite-(Mn), Iron Cap Mine, Graham County, 4 × 4.5 cm.

#AXINITE-(Mn)

Calcium manganese aluminum borosilicate hydroxide: $Ca_4Mn^{2+}_2Al_4[B_2Si_8O_{30}](OH)_2$. Typically formed in contact-metamorphic aureoles where intrusive rocks have invaded sediments, especially limestones; commonly associated with other calcium silicate minerals. This was listed as manganaxinite in the third edition.

GRAHAM COUNTY: Aravaipa supersystem, Iron Cap District, Iron Cap Mine, near Landsman Camp, as yellow-green bladed crystals that commonly have a black coating (Reiter 1980, 1981) and as microcrystals (Marvin Deshler, UA x1850).

PIMA COUNTY: North end of the South Comobabi Mountains, north of the Ko Vaya Hills, as yellow plates being replaced by clinozoisite; in vesicles in andesite flow boulders (Bideaux, Williams, and Thomssen 1960).

AZURITE

Copper carbonate hydroxide: $Cu_3(CO_3)_2(OH)_2$. A widely distributed secondary mineral in the oxidized zones of copper deposits; commonly associated with malachite (to which it alters), cuprite, copper, and limonite.

FIGURE 3.28 Azurite, Bisbee, Cochise County, 8.4 cm.

COCHISE COUNTY: Warren District, Bisbee, as magnificent crystallized specimens with crystals of extraordinary size (up to 4.5 in.) at the Copper Queen Mine, one of the world's premier localities (Kunz 1885; Douglas 1899; Ransome 1903a, 1904; Bideaux 1973; Graeme 1981, 1993; UA 79, 177, 736, 1153); excellent crystallized specimens were noted from the deeper workings of the mine "implanted as a secondary growth in parallel position upon well-formed pseudomorphs of malachite after azurite of longer dimensions" (Palache and Lewis 1927); some crystals are overgrowths on malachite (Schwartz and Park 1932); Junction Mine (Mitchell 1920). Tombstone District, Lucky Cuss and Toughnut Mines (Butler, Wilson, and Rasor 1938). Courtland District, Maid of Sunshine Mine, as large crystallized masses (Bull. 181). Little Dragoon Mountains, Cochise supersystem, Johnson Camp District, Johnson Camp area (Kellogg 1906).

COCONINO COUNTY: Kaibab Plateau, as extensive impregnations in chert beds (Bull. 181); Apex copper property (Tainter 1947b). Grand Canyon National Park, Horseshoe Mesa, Grandview District, Grandview (Last Chance) Mine, as short prismatic crystals on clay and in vugs in sandstone; some crystals of azurite altering to malachite are up to 1 by 3 cm (Leicht 1971). Cameron area, in silica-mineralized plugs near the Cameron District, with malachite, anhydrite, and pyrite; locally, trace uraninite in mineralized silica plugs (Barrington and Kerr 1963).

GILA COUNTY: Miami-Inspiration supersystem, as a common secondary mineral in many deposits (Woodbridge 1906; Schwartz 1921, 1934); Bluebird Mine, as radiating crystal aggregates up to 15 cm (3rd ed.); Pinto Valley District, Castle Dome Mine (N. P. Peterson 1947); Copper Cities District, Copper Cities Mine, associated with malachite and turquoise (N. P. Peterson 1954). Summit District, Blue Ball Mine, as nodular concretions of considerable size (Sinkankas 1964), commonly with a central cavity containing malachite crystals (Grant 1989). Banner supersystem, 79 District, 79 Mine (Kiersch 1949; Stanley B. Keith 1972). Green Valley District, as crystallized masses at the Silver Butte, Golden Wonder, and Bishops Knoll Mines (Bull. 181).

GREENLEE COUNTY: Copper Mountain supersystem, Morenci District, where thousands of specimens of azurite and malachite have been recovered from the Morenci Mine, including azurite stalactites up to 15 cm long (3rd ed.); as large bodies in the Longfellow, Detroit, Manganese Blue, and Shannon Mines; sheaf-like and spherical masses up to 40 pounds were found in kaolinized shale (Kunz 1885; Farrington 1891; Lindgren 1903, 1904, 1905; Reber 1916; Schwartz 1934; Bideaux 1973).

LA PAZ COUNTY: Yuma King District, Yuma Mine, as bright-blue lenticular laminations interlaced with chrysocolla in the oxide zone (Stanley B. Keith, pers. comm., 2019).

MARICOPA AND GILA COUNTIES: Mazatzal Mountains District, at Pine Mountain, near Mt. Ord and Saddle Mountain, on Alder and Slate Creeks, associated with malachite and chalcopyrite in mercury deposits in a schist belt (Lausen 1926).

MOHAVE COUNTY: Cerbat Mountains, Wallapai supersystem, Mineral Park District, as a sparse but widely distributed mineral, associated with sulfide vein deposits and disseminations (B. E. Thomas 1949). Bentley District, Grand Gulch, Bronze L, and Copper King Mines (J. M. Hill 1914b).

PIMA COUNTY: Santa Rita Mountains, widely distributed at mining properties throughout the range, primarily those in the Old Baldy District (Schrader 1917). Copper Mountain prospect of the Anaconda Group, as fine crystalline specimens (Schrader and Hill 1915). Helvetia-Rosemont District, Helvetia, as massive material and crystals and as pseudomorphs of malachite after azurite, up to about 0.5 inch, with rosasite (Peter Megaw, UA x2918). Sierrita Mountains, common in the Pima supersystem, Mission-Pima District (B. M. Williams 2018), Banner Mine, as fine rosettes (UA 6759). Santa Catalina Mountains, Marble Peak District, upper reach of Cañada del Oro (UA 4126). South Comobabi Mountains, Cababi District, Mildren and Steppe claims (S. A. Williams 1963). Ajo District, as granular aggregates that have concentric structures, with malachite and quartz, and as pseudocubic crystals (Schwartz 1934). Waterman District, Silver Hill Mine (UA 1646); Indiana-Arizona Mine, as crystalline material with cerussite in goethite (3rd ed.). Silver Bell District, Silver Bell Mine, with cuprite and malachite in orebodies in limestone (Engineering and Mining Journal 1904).

PINAL COUNTY: Pioneer supersystem, Silver King District, in the open cut of the Silver King Mine, as small but beautifully crystallized groups (Bull. 181). Galiuro Mountains, Bunker Hill (Copper Creek) District, Copper Creek, Childs-Aldwinkle Mine (Kuhn 1941). Mineral Creek District, northeast of Ray Hill and near the Ray deposit, as crystals up to 3 mm long, one of several oxidized minerals, including malachite, cuprite, tenorite, jarosite, and goethite, in Holocene gravels locally replacing a fossil log (C. H. Phillips, Cornwall, and Rubin 1971). San Manuel District, San Manuel Mine, in small amounts in the oxidized portion of the orebody, with chrysocolla, malachite, and cuprite (Schwartz 1949). Mammoth District, Mammoth–St. Anthony Mine, as deep-blue crystals up to 2 in. long, associated with reticulated cerussite, and as pseudomorphs of malachite after azurite (Bideaux 1980; UA 9239, 8549).

SANTA CRUZ COUNTY: Washington Camp District, Duquesne and Washington Camps (Schrader 1917). Tyndall District (Schrader 1917).

YAVAPAI COUNTY: Black Hills, Black Hills District, Yeager Mine, as fine-quality specimens (C. A. Anderson and Creasey 1958). Copper Basin District, in the oxidized zones of several sulfide ore deposits (Johnston and Lowell 1961); in S20 and 21, T13N R3W, as botryoidal specimens (David Shannon, pers. comm., 1971). Jerome supersystem, Verde District, United Verde Extension Mine, as crystals lining vugs and as small radiating spherical aggregates irregularly distributed with malachite in limonitic clay in the replacement orebody (Schwartz 1938).

YUMA COUNTY: Muggins District, Red Knob claims, associated with weeksite, opal, vanadinite, carnotite, gypsum, and calcite (Outerbridge et al. 1960). Gila Mountains, Fortuna District, Blue Butte vein (E. D. Wilson 1933).

#BABINGTONITE

Calcium iron silicate hydroxide: $Ca_2Fe^{2+}Fe^{3+}Si_5O_{14}(OH)$. Found in granitic and volcanic rocks, gneisses, and skarn deposits. Forms a series with manganbabingtonite.

GRAHAM COUNTY: Aravaipa supersystem, Iron Cap District, Iron Cap Mine, as small dark prismatic striated crystals (UA 8742, R060093).

FIGURE 3.29 Babingtonite, Iron Cap Mine, Graham County.

#BACKITE

Lead aluminum telluride chloride: Pb_2AlTeO_6Cl. A rare mineral found in the oxidized portions of tellurium-bearing lead deposits. The Grand Central Mine is the type locality.

COCHISE COUNTY: Tombstone District, Grand Central Mine, as small dark to pale blue-gray rosettes of hexagonal plates associated with schieffelinite, oboyerite, and rodalquilarite (Tait et al. 2014).

FIGURE 3.30 Bairdite, Empire Mine, Tombstone, Cochise County, FOV 3 mm.

#BAIRDITE

Lead copper tellurate hydroxide sulfate hydrate: $Pb_2Cu_4^{2+}Te_2^{6+}O_{10}(OH)_2(SO_4)\cdot H_2O$. A very rare mineral, formed from the partial oxidation of primary sulfides and tellurides.

COCHISE COUNTY: Tombstone District, Empire Mine, as brilliant yellow-green clusters of small bladed crystals (R190040), and from the Grand Central Mine, as very small bladed yellow-green crystals associated with fuettererite, timroseite, and burckhardtite (Carsten Slotta, pers. comm., 2020).

#BANDYLITE

Copper boron hydroxide chloride: $CuB(OH)_4Cl$. A very rare secondary copper mineral.

COCHISE COUNTY: Warren District, Bisbee, Southwest Mine, as a minor part of a complex mineral assemblage occurring at the interface between massive cuprite and claringbullite crystals (Graeme, Graeme, and Graeme 2015).

#BARLOWITE

Copper bromide fluoride hydroxide: $Cu_4BrF(OH)_6$. A rare mineral found in the oxidized portions of copper deposits.

COCHISE COUNTY: Warren District, Bisbee, Southwest Mine, as blue platy crystals associated with cuprite and brochantite (R110007).

BARRINGERITE

Iron nickel phosphide: $(Fe,Ni)_2P$. Found in meteorites.

COCONINO COUNTY: Canyon Diablo meteorite contains small seams and pods of barringerite partly altered to schreibersite, followed in sequence by wüstite and lawrencite (material provided by David Shannon).

#BARROISITE

Sodium calcium magnesium aluminum silicate hydroxide: □(NaCa) $(Mg_3Al_2)(Si_7Al)O_{22}(OH)_2$. A member of the amphibole group.

YAVAPAI COUNTY: Near Prescott, Chino Valley, in the Tertiary Sullivan Buttes latite, occurs in ultramafic xenoliths with other amphiboles and pyroxenes (Schulze and Helmstaedt 1979).

BARYTE

FIGURE 3.31 Baryte, San Manuel Mine, Pinal County, 5.4 cm.

Barium sulfate: $Ba(SO_4)$. A widespread, relatively common mineral that forms under low- to moderate-temperature conditions in hydrothermal veins; also in sedimentary rocks as replacements, veins, and cavity fillings formed by hypogene or meteoric solutions. L. A. Stewart and Pfister (1960) have summarized baryte deposits in the state.

APACHE AND NAVAJO COUNTIES: Monument Valley and Cane Valley Districts, with the uranium-vanadium ores in Shinarump Conglomerate–filled channels at the base of the Chinle Formation (Mitcham and Evensen 1955; R. G. Young 1964).

COCHISE COUNTY: Tombstone District, Ground Hog Mine, as a vein, and near the Lucky Cuss Mine as white crystals (Butler, Wilson, and Rasor 1938; L. A. Stewart and Pfister 1960). Western slope of the Dragoon Mountains, Gleeson District, Johnnie Boy No. 1 claim, as veins and replacements in limestone (L. A. Stewart and Pfister 1960). Warren District, Bisbee, in places associated with manganese ores (Palache and Shannon 1920; Taber and Schaller 1930). Hopeful District, Hopeful claim, north of the Mule Mountains, in NW quarter of S4, T22S R23E, as strong veins of nearly pure baryte that cut Cretaceous sedimentary rocks (Tenney 1936; Funnell and Wolfe 1964); Ramirez District, foothills of the northern Mule Mountains, north side of Gadwell Canyon, Ramirez Group of claims, in veins of nearly pure baryte up to 3 ft. wide that cut Cretaceous sedimentary rocks (L. A. Stewart and Pfister 1960).

COCONINO COUNTY: Grand Canyon National Park, Horseshoe Mesa, Grandview District, Grandview (Last Chance) Mine, in a fault zone in Redwall Limestone, associated with aurichalcite (Leicht 1971); Orphan District, Orphan Mine (UA 7098).

GILA COUNTY: Richmond Basin District, abundant in veins (Bull. 181). Miami-Inspiration supersystem, Pinto Valley District, Castle Dome Mine, in small amounts with fluorite (N. P. Peterson, Gilbert, and Quick

1946, 1951). Reported near Coolidge Dam, in a vein about 500 ft. long, associated with galena (Bull. 181). South of Payson, Green Valley District, "baryte occurs sporadically in more or less parallel fractures for about 15 miles along an east-west zone in granitic rocks" in the Top Hat, Gilmore Spring, and Lone Pine claims (L. A. Stewart and Pfister 1960).

GRAHAM COUNTY: Stanley District, near Stanley Butte, abundant in veins (C. P. Ross 1925a); about five miles east of Turnbull Mountain, Barium King Group of claims, in S19, T4S R20E, and S24, T4S R19E (Funnell and Wolfe 1964). Aravaipa supersystem, Grand Reef District, southeast of Klondyke, Marcotte Group of claims, in veins (L. A. Stewart and Pfister 1960).

GREENLEE COUNTY: Ash Peak District, Luckie Mine, associated with fluorite and psilomelane in a vein in andesite porphyry (Hewett 1964).

LA PAZ COUNTY: Planet District, Planet Mine, as golden-yellow crystals up to 2 cm long on malachite and chrysocolla, and as white sheaves up to 2.5 cm long (3rd ed.). Silver District, Trigo Mountains, Mendevil claims, in veins (Bull. 181); Padre Kino Mine (Peter Megaw, UA x2966). East slope of the Plomosa Mountains, Bouse Hills District, at several deposits, in veins that form an arc around veins of manganese oxides, in layered volcanic host rock (L. A. Stewart and Pfister 1960; Hewett 1964); Red Chief prospect near Bouse, with fluorite (UA 2716). Cottonwood Pass near Salome (E. D. Wilson 1944).

MARICOPA COUNTY: Goldfield Mountains, fourteen miles north of Mesa, in S4 and 5, T2N R7E, as veins at the Granite Reef (Arizona Baryte) Mine (E. D. Wilson 1944; L. A. Stewart and Pfister 1960; UA 5955). Painted Rock District, Rowley Mine near Theba, as colorless crystals up to 2 cm long and 4 mm thick, commonly lining cavities in massive baryte (W. E. Wilson and Miller 1974). Aguila District, Valley View Mine, as layers of crystals with fluorite that alternate with black calcite in a hypogene vein cutting an andesite flow (Hewett 1964). Horseshoe Dam area (Arthur Roe, pers. comm., 1989).

MOHAVE COUNTY: Reported in the Aquarius Mountains District, as veins (Bull. 181). Cerbat Mountains, Wallapai District, as golden tabular crystals up to 7 cm long (Les Presmyk, pers. comm., 2020, UA 4923). Near Alamo Crossing, as veins (Bull. 181). Artillery District, with fluorite in veins with manganese oxides that cut the Artillery Formation (Bull. 181). Rawhide District, Barbee vein, in S1, T11N R14W, associated with chalcedony and mammillary layers of psilomelane in basalt (Hewett and Fleischer 1960). McCracken District, McCracken Mine, about eight miles west of Signal, in large quantities in veins with quartz, carbonates, and galena (H. Bancroft 1911; Funnell and Wolfe 1964). About thirty miles east of Kingman, Rucker Group of claims, in S2, T20N R12W, as sporadic pods, segregations, and stringers, and as veins up to 8 ft. wide, in granitic rocks (L. A. Stewart and Pfister 1960).

NAVAJO COUNTY: Monument Valley District, Starlight No. 3 Mine, as druses of minute brown crystals on fracture surfaces in sandstone (AEC mineral collection).

PIMA COUNTY: Ajo District, sparingly in veins cutting concentrator volcanics and quartz monzonite (Gilluly 1937). South Comobabi Mountains, Cababi District, Mildren and Steppe claims (S. A. Williams 1963). Picacho de Calera Hills. One-half mile east of Colossal Cave, Heavy Boy Mine, as cleavable white masses (L. A. Stewart and Pfister 1960). Silver Bell District, House Canyon, as large bladed crystals (Kerr 1951); Silver Bell Mine, coating fractures (3rd ed.). Santa Rita Mountains, as an abundant gangue mineral in some sulfide deposits (Schrader and Hill 1915; Schrader 1917). Quijotoa Mountains, Quijotoa District, White Prince claims (L. A. Stewart and Pfister 1960), Morgan Mine (UA 7036), and Quijotoa Mine, in S33, T15S R2E (L. A. Stewart and Pfister 1960; Funnell and Wolfe 1964), as bladed crystals that form the matrix for galena crystals; Weldon Mine, as large, pink rosettes of crystals up to 15 cm (D. M. Shannon 1981; MM N 474). Northwest ridge of the Santa Catalina Mountains, as large quartz-encrusted bladed crystals (UA 9601); Santa Catalina foothills, Tucson, north of Campbell Avenue (3rd ed.). Coyote Mountains, as 4 cm off-white to pink bladed crystals with calcite (Presmyk and Hay 2020; UA 1145).

PINAL COUNTY: Pioneer supersystem, Magma District, Magma Mine, as brilliant tabular crystals up to 2 in. in various colors: black, brown, maroon, gray, white, yellow, and golden (Barnes and Hay 1983; Presmyk and Hay 2020; UA 2015). Mammoth District, west side of Tucson Wash, in a vein with calcite and some psilomelane (Schwartz 1953); Mammoth–St. Anthony Mine, as groups of large tabular crystals (N. P. Peterson 1938b). San Manuel District, San Manuel Mine (Hay and Alexander 2020), Bunker Hill (Copper Creek) District, Copper Creek, Old Reliable Mine, as tabular crystals and crystal groups (Bull. 181), Magna Mine, as small individual colorless crystals (Joe Ruiz, pers. comm., 2016). Gonzales Pass deposit, in S16 and 17, T2S R11E, in a vein following a fault in Pinal Schist (L. A. Stewart and Pfister 1960).

SANTA CRUZ COUNTY: Santa Rita Mountains, Tyndall District, and Patagonia Mountains, notably in the Patagonia District, as one of the principal gangue minerals in base- and precious-metal deposits, commonly associated with quartz, fluorite, rhodochrosite, and other carbonates (Schrader 1917).

YAVAPAI COUNTY: Bradshaw Mountains, as a gangue mineral in several properties (Bull. 181); French Creek deposit, in S29, T9N R1W (L. A. Stewart and Pfister 1960). Eureka supersystem, Bagdad District, Bagdad Mine (C. A. Anderson 1950). Bullard District, Hatton Mine (Hewett and Fleischer 1960; Hewett 1964). Northeastern Wickenburg Mountains, Monte Cristo District, Monte Cristo Mine near Wickenburg, with silver, chalcopyrite, and nickel arsenides (Bastin 1922); MGM claims, a few miles northeast of Wickenburg, as filling and replacement in a brecciated zone in volcanic breccia and granite (L. A. Stewart and Pfister 1960).

YUMA COUNTY: Dome District, Castle Dome Mine, in many veins and as large clear crystals with wulfenite and fluorite (Foshag 1919); Hull Mine, as white bladed crystals with cerussite and wulfenite on fluorite (Les Presmyk, pers. comm., 2020). Mohawk District, Baryte Mine, as white to pink radiating crystal aggregates in calcite veins (Bull. 181).

BASSANITE

Calcium sulfate hydrate: $Ca(SO_4) \cdot 0.5H_2O$. An uncommon mineral formed as pseudomorphs after gypsum in cavities in volcanic rocks and as a result of fumarolic activity. Also an alteration product in porphyry copper deposits.

LA PAZ COUNTY: Bouse Hills District, several small prospects near Bouse, in veins in andesite with chrysocolla and tenorite (3rd ed.).

PIMA COUNTY: Sierrita Mountains, Pima District, Sierrita Mine, as tiny subhedral fibrous crystals in veinlets filled with quartz, calcite, and chlorite in hydrothermally altered diorite and quartz diorite porphyry (Roger Lainé, pers. comm., 1972).

BASSETITE

Iron uranyl phosphate hydrate: $Fe^{2+}(UO_2)_2(PO_4)_2(H_2O)_{10}$. A secondary mineral commonly associated with uraninite.

GILA COUNTY: Sierra Ancha District, northwestern part of the county, common in the uranium deposits in Dripping Spring Quartzite; Sue Mine, associated with saléeite and locally with metanováčekite (Granger and Raup 1969); Red Bluff Mine (Granger and Raup 1962).

#BASTNÄSITE-(Ce)

Cerium carbonate fluoride: $Ce(CO_3)F$. The most common rare-earth mineral, typically of hydrothermal origin, found in some granites and alkali syenites and pegmatites; also in carbonatites and contact-metamorphic deposits.

MOHAVE COUNTY: Aquarius Mountains, Rare Metals Mine, pegmatite in the wall zone with allanite-(Ce) and monazite-(Ce) (W. B. Simmons et al. 2012). Cerbat Mountains, Kingman Feldspar District, Kingman Feldspar Mine, as grains replacing allanite in a pegmatite (S. L. Hanson et al. 2011).

BAYLDONITE

Copper lead oxide arsenate hydroxide: $Cu_3PbO(AsO_3OH)_2(OH)_2$. A rare oxide-zone mineral found only in a few localities.

COCHISE COUNTY: Warren District, Bisbee, Southwest Mine, sixth level, as spongy, resinous, greenish-yellow linings of small voids in a quartz breccia (Graeme 1993; AM 21475).

PIMA COUNTY: Santa Rita Mountains, Helvetia-Rosemont District, Frijole prospect, in S24, T18S R15E; common in coarse milky vein quartz; derived from sulfosalts; associated with bindheimite, cerussite, wulfenite, and mimetite (3rd ed.).

BAYLEYITE

Magnesium uranyl carbonate hydrate: $Mg_2(UO_2)(CO_3)_3 \cdot 18H_2O$. A rare water-soluble, secondary mineral associated with other secondary uranium minerals and gypsum as a postmining efflorescence. The Eureka supersystem, Hillside District, Hillside Mine at Bagdad is the type locality.

COCHISE COUNTY: Warren District, Bisbee, Cole Mine (Graeme 1981; UA 10078).

YAVAPAI COUNTY: Hillside District, Hillside Mine, as an efflorescence on the walls, associated with schröckingerite, andersonite, swartzite, gypsum, johannite, and uraninite; locally present as well-formed crystals (Axelrod et al. 1951).

#BAYLISSITE

Potassium magnesium carbonate hydrate: $K_2Mg(CO_3)_2 \cdot 4H_2O$. A secondary mineral found in granitic rocks and as a product of a burned palo verde tree.

MARICOPA COUNTY: Near Phoenix, as a constituent of residual ash from a burned palo verde tree consumed in a wildfire. Found to occur with fairchildite, sylvite, kalicinite, magnesite, and magnesian calcite (Garvie 2016).

#BEAVERITE-(Cu)

FIGURE 3.32 Beaverite-(Cu), 79 Mine, Gila County, FOV 2 mm.

Lead iron copper sulfate hydroxide: $Pb(Fe^{3+}_2Cu)(SO_4)_2(OH)_6$. An uncommon secondary mineral formed in the oxidized portion of lead-copper deposits in arid regions. Commonly associated with plumbojarosite.

COCHISE COUNTY: Tombstone District, Empire and Toughnut Mines, in small quantities with cerussite (B. S. Butler, Wilson, and Rasor 1938). Warren District, Bisbee, Southwest Mine, sixth level (Graeme 1993).

GILA COUNTY: Banner supersystem, 79 District, 79 Mine (R. Gibbs, pers. comm., 2020).

PINAL COUNTY: Mammoth District, Mammoth–St. Anthony Mine, rarely as shining golden-yellow scales around the bases of linarite crystals (Bideaux 1980).

#BECHERERITE

Zinc copper hydroxide silicate sulfate: $Zn_7Cu(OH)_{13}[SiO(OH)_3SO_4]$. A rare secondary mineral found in the oxidized portion of lead-zinc-copper deposits. The Tonopah-Belmont Mine is the type locality.

MARICOPA COUNTY: Belmont Mountains, Osborne District, Tonopah-Belmont Mine, as very small pale-green crystals with a habit resembling spangolite, with willemite, rosasite, hydrozincite, smithsonite, paratacamite, and boleite (Giester and Rieck 1996).

BECQUERELITE

Calcium uranyl oxide hydroxide hydrate: $Ca(UO_2)_6O_4(OH)_6{\cdot}8H_2O$. A secondary mineral typically closely associated with uraninite, from which it is commonly derived; associated with other secondary uranium minerals, such as schoepite, fourmarierite, and curite.

APACHE COUNTY: Cane Valley District, Monument No. 2 and Cato Sells Mines, where it is an important ore mineral; associated with uraninite (C. Frondel 1956; Finnell 1957; Witkind and Thaden 1963).

BEIDELLITE

Sodium calcium aluminum silicate hydroxide hydrate: $(Na,Ca)_{0.3}Al_2(Si,Al)_4O_{10}(OH)_2{\cdot}nH_2O$. Beidellite is a member of the montmorillonite (smectite) group of clay minerals. It is a component of bentonitic clays and is common in hydrothermally altered areas associated with mineral deposits.

COCONINO COUNTY: Along the base of the Echo Cliffs, north of Cameron, as a clay mineral in the Chinle Formation (3rd ed.).

GILA COUNTY: Miami-Inspiration supersystem, Pinto Valley District, Castle Dome Mine, in veinlets and fine aggregates, and as scattered flakes in the altered quartz monzonite (N. P. Peterson, Gilbert, and Quick 1946, 1951).

GREENLEE COUNTY: Copper Mountain supersystem, Morenci District, Morenci, in less altered zones near intense hydrothermal alteration of porphyry copper ores, associated with kaolinite, montmorillonite, and allophane (Schwartz 1947, 1958).

PIMA COUNTY: Ajo District, New Cornelia Mine, commonly formed as pseudomorphic replacement of plagioclase phenocrysts in the outlying parts of the mineralized area (Gilluly 1937).

PINAL COUNTY: San Manuel District, San Manuel Mine, as a minor constituent of hydrothermally altered quartz monzonite and monzonite porphyries, associated with kaolinite, hydromuscovite, and rutile in a quartz-sericite-pyrite-chalcopyrite aggregate (Schwartz 1947).

BEMENTITE

Manganese silicate hydroxide: $Mn_7Si_6O_{15}(OH)_8$. An uncommon mineral formed in contact-metamorphosed manganiferous limestones.

PINAL COUNTY: Mineral Creek supersystem, Ray District, Ray Mine, near the Emperor tunnel, as minute clear platy crystals with hematite in oxidized capping in the Granite Mountain porphyry (Loghry 1972).

BERLINITE

Aluminum phosphate: $Al(PO_4)$. A very rare mineral, first identified at the Westanå Iron Mine, Kristianstad, Sweden, where it is associated with other phosphate minerals.

GILA COUNTY: Miami-Inspiration District, Inspiration Mine, as a single light-brown hollow pod less than 2 mm in size, completely enclosed by hematite in oxidized and leached capping (Loghry 1972).

FIGURE 3.33 Bermanite, 7U7 Ranch, Yavapai County, FOV 3 mm.

BERMANITE

Manganese phosphate hydroxide hydrate: $Mn^{2+}Mn^{3+}_2$ $(PO_4)_2(OH)_2 \cdot 4H_2O$. A rare mineral formed in granite pegmatites. The 7U7 Ranch is the type locality.

YAVAPAI COUNTY: Eureka area, 7U7 Ranch near Bagdad, west of Hillside, in narrow veins cutting a spherical mass of triplite, which formed as a segregation in pegmatite lenses in granite; as single crystals and subparallel aggregates with fan-shaped or rosette-like appearance (Hurlbut 1936); associated with phosphosiderite, leucophosphite, hureaulite, and other phosphate minerals (Leavens 1967; Hurlbut and Aristarain 1968).

BERTRANDITE

Beryllium silicate hydroxide: $Be_4Si_2O_7(OH)_2$. Formed in granite pegmatites, typically as an alteration product of beryl.

MARICOPA COUNTY: White Picacho District, Independence claim (Bull. 181).

FIGURE 3.34 Beryl, Palo Verde claim, Pima County, 6.1 cm high.

BERYL

Beryllium aluminum silicate: $Be_3Al_2Si_6O_{18}$. An important ore mineral of beryllium, typically embedded in granites and pegmatites as a magmatic mineral; also associated with tin ores and mica schists.

COCHISE COUNTY: Swisshelm Mountains, east of Elfrida, as euhedral crystals in vugs with fluorite and as beryl-muscovite masses in veins (Diery 1964; Staatz, Griffitts, and Barnett 1965); Elfrida District, Little Lulu and Silver Drip claims, as aquamarine in aplite dikes in granite at Thompson beryl claims (Meeves 1966). Dragoon Mountains, Middle Pass supersystem, Abril District, Gordon and Abril Mines, in garnet tactite (Balla 1962; Shawe 1966). Dos Cabezas Mountains, Beryl Hill District; Tungsten Bluebird Mine, in scheelite-bearing quartz veins (3rd ed.). Texas Canyon area, Bluebird District, Boericke tungsten property, as small colorless crystals with fluorite (Bull. 181). Tungsten King District, Tungsten King Mine, pale-blue to colorless beryl is a very minor and erratic vein constituent that occurs in irregular grains 0.5 to 5 mm in diameter and in slender prismatic crystals as much as 2 cm long (Cooper and Silver 1964). Texas Canyon stock near Dragoon.

GRAHAM COUNTY: Goodwin Wash (Shawe 1966). Santa Teresa Mountains—a few notable specimens have been recovered from miarolitic cavities (Muntyan 2013).

MARICOPA COUNTY: Near Aguila (UA 6664).

MOHAVE COUNTY: Aquarius Mountains District, Rare Metals Mine and Columbite prospect, as crystals in pegmatite (Heinrich 1960). Beryl Wash near Kingman, as greenish-blue crystals up to 12 by 4 in. (Bull. 181). Groom Peak, G and M pegmatite, fifteen miles southwest of Wikieup, as bluish-green crystals up to 9 ft. long and 18 in. in diameter (Bull. 181). Boriana District, Boriana Mine (Hobbs 1944). Valentine District, about fifteen miles south of Peach Springs, four miles east-southeast of Wright Creek Ranch, in several irregular pegmatite dikes in schist and gneissic granite, with microcline, albite, quartz, fluorite, schorl, titanite, and muscovite; locally, as well-formed prismatic crystals (Schaller, Stevens, and Jahns 1962)—this beryl is quite unusual in its chemical composition and has been assigned the formula $(Na,Cs)Be_3Al(Fe^{2+},Mg)Si_6O_{18}$. Cerbat Mountains, in pegmatite (UA 6309). Kingman Feldspar District, with gadolinite in pegmatite (B. E. Thomas 1953). Virgin Mountains, Hummingbird claims, where it locally constitutes up to 3 percent of quartz-muscovite pegmatite; in crystals up to 15 in. long (J. C. Olson and Hinrichs 1960). Hualapai Mountains (Meeves et al. 1966).

PIMA COUNTY: Sierrita Mountains, Samaniego District, Palo Verde claim, numerous pegmatites with aquamarine crystals, some of faceting quality (Muntyan 2013). Baboquivari Mountains (Shawe 1966). Southern Rincon Mountains, upper reaches of Agua Verde Creek, in a vein in probable muscovite granite (Staatz, Griffitts, and Barnett 1965). Santa Catalina Mountains, near Apache Peak, in quartz veins (Robert O'Haire, pers. comm., 1972); near Summerhaven, as aquamarine beryls to 2 in. long in pegmatite phases of the Lemmon Rock Leucogranite (Keith and Rasmussen 2019).

YAVAPAI COUNTY: Camp Wood District, Black Pearl Mine, in veins that traverse biotite granite (Brownell 1959; Dale 1961; Staatz, Griffitts, and Barnett 1965), and as blue crystals with pyrite and bismuthinite (MM 1062). Bradshaw Mountains, four miles southeast of Wagoner, in pegmatite veins (Bull. 181). Southwestern Bradshaw Mountains, Tussock District, Weatherby Beryl Mine (AZGS 2015). Peck District, three miles east of the Crown King Post Office, in a pegmatite dike (Bull. 181). White Picacho District, as crystals up to 11 in. long in pegmatites; associated with lithium minerals (Jahns 1952; Meeves et al. 1966). Lawler Peak area, in tabular masses of quartz (3rd ed.). Weaver Mountains, Rich Hill District, seven miles northwest of Yarnell, Monte Cristo pegmatite, as green crystals up to 2 ft. long and 14 in. in diameter, associated with albite, muscovite, and quartz; also as honey-brown euhedral and anhedral crystals, and aquamarine crystals up to 6 in. long (Mohon 1975).

YUMA COUNTY: Gila Mountains, Fortuna District, about 1.5 miles east of the Fortuna Mine, as small lavender- and rose-colored crystals in a matrix of yellowish quartz, associated with small masses of an unidentified black mineral containing niobium (Bull. 181).

BEUDANTITE

Lead iron arsenate sulfate hydroxide: $PbFe^{3+}_3(AsO_4)(SO_4)(OH)_6$. A rare secondary mineral formed in the oxidized portions of mineralized veins.

COCHISE COUNTY: Warren District, Bisbee, abundant in the Southwest Mine (Graeme 1993).

GILA COUNTY: Banner supersystem, 79 District, in the northeast workings of the Kullman-McCool (Finch, Barking Spider) Mine, as brown earthy massive material associated with descloizite (Hidemichi Hori, pers. comm., 1984).

MARICOPA COUNTY: Belmont Mountains, Osborne District, Moon Anchor Mine, from a trench on the north side of the hill, with mimetite (William Hunt, pers. comm., 1985).

PIMA COUNTY: Waterman District, Silver Hill Mine (UA 10474). Tucson Mountains, Saginaw Hill District, Saginaw Hill, Tucson, as small dark-green to greenish-yellow crystals with carminite (B. Murphy, pers. comm., 2018).

YAVAPAI COUNTY: Tiger District, Crown King Mine, as greenish crusts in vein matter, with jarosite and iron oxides (3rd ed.).

FIGURE 3.35 Beyerite, Copperopolis, Yavapai County, FOV 3 mm.

BEYERITE

Calcium bismuth oxide carbonate: $CaBi_2O_2(CO_3)_2$. A secondary mineral formed by the alteration of bismutite or other primary bismuth minerals, typically in pegmatites.

MARICOPA AND YAVAPAI COUNTIES: White Picacho District, as grayish-green films on bismutite and bismuth and as dense masses and pearly white flakes in small cavities (Jahns 1952).

BIANCHITE

Zinc sulfate hydrate: $Zn(SO_4) \cdot 6H_2O$. Associated with other soluble secondary sulfates as an efflorescence in mine openings.

COCHISE COUNTY: Warren District, Bisbee (Graeme 1981).

GRAHAM COUNTY: Pinaleño Mountains, at a prospect on the east end of Willow Spring Canyon, as white earthy to botryoidal films on sphalerite; the sphalerite is associated with galena in a quartz vein along a strongly chloritized fault zone (3rd ed.).

BIDEAUXITE

Silver lead fluoride chloride: $AgPb_2F_2Cl_3$. An extremely rare secondary mineral originally discovered at the Mammoth–St. Anthony Mine in the oxide zone with other secondary lead minerals. The Mammoth–St. Anthony Mine is the type locality.

PINAL COUNTY: Mammoth District, Mammoth–St. Anthony Mine, as colorless crystals, up to 7 mm in maximum dimension, that enveloped and replaced boleite in the oxide zone of the Collins vein; associated with cerussite and galena, from which it was ultimately derived, and with leadhillite, matlockite, and anglesite. (S. A. Williams 1970a).

BIEBERITE

Cobalt sulfate hydrate: $Co(SO_4) \cdot 7H_2O$. A moderately soluble uncommon mineral formed through oxidation of cobalt-bearing sulfide and arsenide ores.

COCONINO COUNTY: Cameron District, in most of the mines in the Petrified Forest Member of the Chinle Formation, as pink to rose efflorescences (Austin 1964); Huskon No. 1 Mine, where the powdery mineral stains cross-vein fibrous gypsum or halotrichite (H, unnumbered specimen).

MOHAVE COUNTY: Hack District, Hack Mine (Hack Canyon Uranium Mine), partially coating weathered breccia fractures with chalcanthite, brochantite, erythrite, and yellow uranium minerals (Granger and Raup 1962).

PIMA COUNTY: Ajo District, as a pink efflorescence in a ditch draining the approach to the New Cornelia Mine (Gilluly 1937).

BILINITE

Ferrous ferric iron sulfate hydrate: $Fe^{2+}Fe^{3+}_2(SO_4)_4 \cdot 22H_2O$. An uncommon secondary sulfate mineral of the halotrichite group, formed from the oxidation of iron sulfides.

COCHISE COUNTY: Warren District, Bisbee, Higgins Mine, 100 ft. level, where massive pyritic ore was encrusted and corroded by thick layers of finely fibrous white to tan bilinite (Graeme 1993).

BINDHEIMITE

Lead antimony oxide: $Pb_2Sb^{5+}_2O_7$. An uncommon secondary mineral generally in small quantities in oxidized antimonial lead ores; commonly replaces tetrahedrite. It is considered a questionable species and may be some other mineral.

COCHISE COUNTY: Tombstone District, as yellowish-gray spots in siliceous ores (B. S. Butler, Wilson, and Rasor 1938). Johnny Lyon Hills, Yellowstone District, in a prospect on the bank of Tres Alamos Wash, as a chalky green mixture pseudomorphous after tetrahedrite, associated with partzite, hydroxycalcioroméite, and stibiconite (3rd ed.).

PIMA COUNTY: Santa Rita Mountains, Helvetia-Rosemont District, Frijole prospect (3rd ed.). Tucson Mountains, Saginaw Hill District, Snyder Hill Mine, with cerussite (3rd ed.) and as microcrystals (Bruce Maier, UA x3058).

SANTA CRUZ COUNTY: Patagonia Mountains, Mowry District, Mowry Mine, in small amounts (Schrader and Hill 1915; Schrader 1917).

biotite

The mica group of minerals has been redefined and biotite is no longer recognized as a mineral species. It is currently considered a mica between, or close to, the annite-phlogopite and siderophyllite-eastonite joins; dark micas without lithium. The name is most commonly used for the micas on the Fe-rich end of the series, including annite, fluorannite, tetraferriannite, and siderophyllite (Rieder et al. 1998). Many earlier assignments were based on the micaceous habit and very dark color, and their true species name is not now known in the absence of detailed chemical analysis.

ALL COUNTIES: Dark micas commonly called biotite are found in all Arizona counties and in various rock types.

#BIPHOSPHAMMITE

Ammonium phosphate: $(NH_4)H_2(PO_4)$. Often forms as an alteration product of phosphammite in bat guano.

MARICOPA COUNTY: Painted Rock District, Rowley Mine, forms tiny colorless tetragonal needles with aphthitalite, natrosulfatourea, and urea in an area with bat guano (A. Kampf, pers. comm., 2020).

BIRNESSITE

Sodium calcium potassium manganese oxide hydrate: $(Na,Ca,K)_{0.6}(Mn^{4+}, Mn^{3+})_2O_4 \cdot 1.5H_2O$. An uncommon mineral apparently of secondary origin formed by the breakdown of primary manganese minerals. May be associated with other manganese oxide minerals, rhodonite, and rhodochrosite.

COCHISE COUNTY: Courtland District, several prospects immediately north of Courtland, as a constituent of black scummy crusts on altered quartz monzonite; todorokite is the other major component of these crusts (3rd ed.).

BISMITE

Bismuth oxide: Bi_2O_3. A secondary mineral formed by the oxidation of other bismuth minerals. The validity of the species at some of the localities listed below needs verification.

COCHISE COUNTY: Warren District, Bisbee, Campbell orebody, with argentian wittichenite and bismuth, in a compact silicified hematite matrix (Graeme 1993).

LA PAZ COUNTY: Reported at a locality north of Vicksburg (Bull. 181); Granite Wash Mountains, Three Musketeers District, Three Musketeers Mine (Stanley B. Keith, pers. comm., 2019).

MARICOPA AND YAVAPAI COUNTIES: White Picacho District, as massive material in pegmatites (UA 8970).

YAVAPAI COUNTY: Bradshaw Mountains, Castle Creek District, Swallow Mine, as an alteration product of bismuthinite (Lindgren 1926). Eureka supersystem, Bagdad District, Bagdad Mine (Bull. 181). Bumblebee area (UA 6113).

BISMUTH

Bismuth: Bi. An uncommon primary mineral formed in hydrothermal veins, pegmatites, and topaz-bearing quartz veins.

COCHISE COUNTY: Warren District, Bisbee, Campbell orebody, with wittichenite, bismite, and sphalerite in a matrix of silicified hematite (Graeme 1993).

MARICOPA COUNTY: Vulture District, Cleopatra Mine and at a locality southeast of Granite Reef Dam (Bull. 181).

MOHAVE COUNTY: Aquarius Mountains, thirty miles south of Hackberry, with gadolinite in pegmatite (Bull. 181).

PIMA COUNTY: Sierrita Mountains, Samaniego District, Esmeralda Mine (UA 7163).

YAVAPAI COUNTY: Bradshaw Mountains, in the Humbug Creek placers and on Minnehaha Flats (Bull. 181). Reported in Buckhorn Wash, east of Brooks Hill (Bull. 181). White Picacho District, as a rare constituent of pegmatites, as thin flakes and irregular masses, one of which weighed 2.5 pounds (Jahns 1952).

BISMUTHINITE

Bismuth sulfide: Bi_2S_3. A comparatively rare mineral formed under moderately high-temperature conditions in hydrothermal veins and in tactites, typically with chalcopyrite.

COCHISE COUNTY: Warren District, Bisbee, as bladed crystals up to 1 inch long in chalcopyrite, with pyrophyllite (S 269); Campbell orebody (Graeme 1993).

LA PAZ COUNTY: Granite Wash Mountains, Three Musketeers District, Three Musketeers Mine, with bismutite, kettnerite, and wulfenite (3rd ed.). Dome Rock Mountains, Sugarloaf Gold District, Leadville Mine, where 0.2–0.9 mm needles occur in a late vein cutting highly argillized Middle Camp Quartz Monzonite associated with bismutite, galena, cerussite, silver-rich anglesite, and siderite (Melchiorre 2013).

MOHAVE COUNTY: Aquarius Mountains, in small quantities with bismuth and gadolinite in pegmatite (Bull. 181).

YAVAPAI COUNTY: Bradshaw Mountains, Castle Creek District, Swallow Mine, altering to bismite (Lindgren 1926). Camp Wood District, forty-five miles west of Prescott, in pegmatite (Dale 1961); Camp Wood District, Black Pearl Mine, in quartz veins with "wolframite" (Schmitz 1987);

also with pyrite and beryl (MM 1062). White Picacho District, Midnight Owl Mine, in pegmatite (Jahns 1952).

BISMUTITE

Bismuth oxide carbonate: $Bi_2O_2(CO_3)$. A secondary mineral formed by the alteration of bismuthinite, bismuth, or other bismuth minerals.

COCHISE COUNTY: Warren District, Bisbee, Campbell Mine, as pseudomorphs after aikinite prisms (Graeme 1993). Dos Cabezas Mountains, Teviston District, Comstock Mine, abundant in gossans, as unusually well-crystallized material forming pearly white to pale-tan paper-thin tetragonal platelets that line voids in the rock left by the leaching of pekoite; the mineral superficially resembles sericite in hand specimen, associated with bismuth-rich mottramite (3rd ed.).

GRAHAM COUNTY: Possibly Golondrina District, prospect northwest of Bowie, with kettnerite, malachite, and bismuth-rich mottramite (3rd ed.).

LA PAZ COUNTY: Granite Wash Mountains, Three Musketeers District, Three Musketeers Mine, as earthy masses in quartz, with kettnerite (3rd ed.).

MARICOPA COUNTY: White Picacho District, Outpost pegmatite (Jahns 1952).

MOHAVE COUNTY: Aquarius Mountains District, Rare Metals Mine, as large masses with rusty-brown chalcedony (Heinrich 1960). Hualapai Mountains, east of Yucca (Bull. 181).

YAVAPAI COUNTY: Hillside area, 7U7 Ranch, in prospects at "the granites," as large 3-inch pseudomorphs after bismuthinite (C. Frondel 1943; H 100738). White Picacho District, as an uncommon constituent of pegmatites (Jahns 1952).

BIXBYITE

Manganese oxide: $Mn^{3+}_2O_3$. Commonly formed in cavities in rhyolite with garnet, topaz, beryl, and hematite; also in metamorphosed manganese ores, as at Långban, Sweden.

COCHISE COUNTY: Warren District, Bisbee, Shattuck shaft (Graeme 1981).

PINAL COUNTY: Near Saddle Mountain, San Carlos Apache Reservation, Lookout Mountain District, encrusting and replacing spessartine (UA 8967, 8968); as loose crystals in stream beds (White 1992).

Kofa Gold District, King of Arizona Mine, in the North Star vein with psilomelane and in the No. 2 vein with psilomelane, todorokite, groutite, and manganite (Hankins 1984).

BLÖDITE

Sodium magnesium sulfate hydrate: $Na_2Mg(SO_4)_2\cdot4H_2O$. Fairly widespread in sedimentary sulfate deposits, both lacustrine and oceanic, as large crystals and saline crusts.

COCONINO COUNTY: Paria River at Lee's Ferry (3rd ed.).

FIGURE 3.36 Bobmeyerite, Mammoth–St. Anthony Mine, Pinal County, FOV 2 mm.

#BOBMEYERITE

Lead aluminum copper silicate hydroxide chloride hydrate: $Pb_4(Al_3Cu)(Si_4O_{12})(S_{0.5}Si_{0.5}O_4)(OH)_7Cl(H_2O)_3$. A rare secondary lead mineral; the Mammoth–St. Anthony Mine is the type locality.

PINAL COUNTY: Mammoth District, Mammoth–St. Anthony Mine, as colorless to white or cream-colored needles up to 300 microns in length that taper to sharp points. It occurs with other secondary minerals, including atacamite, caledonite, cerussite, connellite, diaboleite, fluorite, georgerobinsonite, leadhillite, matlockite, murdochite, phosgenite, pinalite, wulfenite, and yedlinite (Kampf et al. 2013).

BOGDANOVITE

Gold tellurium lead iron cuprite: $(Au,Te,Pb)_3(Cu,Fe)$. A rare mineral that, at the type locality in Kamchatka Peninsula, Russia, formed in the supergene zone of oxidation with gold and various tellurides and tellurites.

COCHISE COUNTY: Warren District, Bisbee, Campbell Mine, as tiny bronze-colored grains with pyrite, chalcopyrite, and quartz (Graeme 1993).

BÖHMITE

Aluminum oxide hydroxide: AlO(OH). A major component of bauxites, together with other aluminum oxide minerals, formed under conditions of severe weathering; also found in certain pegmatites.

COCHISE COUNTY: Warren District, Bisbee (Graeme 1981).

BOKITE

Aluminum iron vanadium oxide hydrate: $(Al,Fe)_{1.3}(V^{5+},V^{4+},Fe^{3+})_8O_{20} \cdot 7.5H_2O$. A rare mineral formed as veinlets and crusts in carbonaceous shales.

APACHE COUNTY: Cane Valley District, Monument No. 2 Mine, where it was identified by H. T. Evans Jr. in Triassic fossilized wood (Haynes 1991).

BOLEITE

Potassium silver lead copper chloride hydroxide: $KAg_9 Pb_{26}Cu_{24}Cl_{62}(OH)_{48}$. A rare secondary mineral formed in small amounts in oxidized lead-copper deposits.

GILA COUNTY: Globe Hills District, Apache Mine, with cerussite, brochantite, and matlockite (Bideaux, Williams, and Thomssen 1960).

MARICOPA COUNTY: Painted Rock District, Rowley Mine near Theba, as minute spheres only 0.03 mm in diameter on cerussite associated with linarite, leadhillite, atacamite, diaboleite, and other oxidized-zone minerals (W. E. Wilson and Miller 1974) and as small well-formed crystals with mammothite and diaboleite (Joe Ruiz, pers. comm., 2016). Belmont Mountains, Osborne District, Tonopah-Belmont Mine, as small well-formed equant brilliant blue crystals in oxidized vein material with cerussite (Gibbs and Ascher 2012).

FIGURE 3.37 Boleite, Rowley Mine, Maricopa County, FOV 0.7 mm.

PINAL COUNTY: Mammoth District, Mammoth–St. Anthony Mine, Collins vein, as dark-blue cubes typically on diaboleite; in a leadhillite vug in a small block faulted from the Mammoth vein, associated with cerussite, phosgenite, paralaurionite, quartz, hydrocerussite, diaboleite, matlockite, wherryite, and chrysocolla (Palache 1941b; Fahey, Daggett, and Gordon 1950; Bideaux 1980). Vekol District, Vekol Mine, as small crystals with chlorargyrite (David Shannon, pers. comm., 1980).

BOLTWOODITE

Potassium sodium uranyl silicate hydroxide hydrate: $(K,Na)(UO_2)(SiO_3 OH) \cdot 1.5H_2O$. A moderately common secondary mineral formed by oxidation of black primary uranium ores.

Cameron District, as one of the more common uranium minerals (Austin 1964); Huskon Nos. 17 and 20 Mines, as yellow areas in a blackish sandstone (Daphne Ross, cited in Honea 1961); Ramco Nos. 20 and 22, Jack Daniels No. 1, and Yazzie No. 102 Mines (Austin 1964).

#BONATTITE

Copper sulfate hydrate: $Cu(SO_4) \cdot 3H_2O$. A postmining mineral formed from the decomposition of copper-bearing minerals.

GRAHAM COUNTY: Aravaipa supersystem, Grand Reef District, Grand Reef Mine, as a pale-blue efflorescence with cerussite and malachite (Grant, Bideaux, and Williams 2006; R070424).

#BOOTHITE

Copper sulfate hydrate: $Cu(SO_4) \cdot 7H_2O$. An uncommon postmining mineral formed as the result of the decomposition of copper-bearing minerals in a humid environment.

COCHISE COUNTY: Warren District, Bisbee, Southwest Mine, seventh level, where it is found as 1 cm wide clusters of light-blue radiating acicular crystals. It also occurs as small granular masses composed of crystals up to 2 mm (Graeme, Graeme, and Graeme 2015).

BORNITE

Copper iron sulfide: Cu_5FeS_4. An important ore mineral of copper in many mines in the state. Commonly intimately associated with either chalcocite or chalcopyrite, typically with both. Principally of primary origin, but small amounts of secondary bornite are common in secondarily enriched ores; also in contact-metamorphosed deposits.

APACHE AND NAVAJO COUNTIES: Monument Valley and Cane Valley Districts, with chalcopyrite, chalcocite, and copper in sandstone, associated with uranium-vanadium ores at several properties (Evensen and Gray 1958).

COCHISE COUNTY: Warren District, Bisbee, Campbell Mine and other orebodies, as one of the important ore minerals in hydrothermal deposits in limestone (Trischka, Rove and Barringer 1929; Schwartz and Park 1932; Schwartz 1939; Yagoda 1945; Bain 1952; Bryant 1968); 1,300 to 1,400 ft. levels of the Cole shaft, as 0.25-inch crystals. (For detailed structural information on Bisbee minerals in the bornite-digenite series, see L. Pierce and Buseck 1978). Little Dragoon Mountains, Cochise District (Romslo 1949); Johnson Camp, Black Prince and Peabody Mines area (Kellogg 1906; J. R. Cooper and Huff 1951; J. R. Cooper 1957; Baker 1960; J. R. Cooper and Silver 1964). Courtland District, Leadville, Great Western, Copper Belle, and Tejon Mines (Bull. 181). Whetstone Mountains, Mine Canyon District, in small quantities in quartz veins (DeRuyter 1979). Huachuca Mountains, Hartford District, Barnes Mine (UA 1080).

COCONINO COUNTY: Grand Canyon National Park, reported at several localities in orebodies such as at the Orphan Mine, with chalcocite and cuprite (Gornitz and Kerr 1970).

GILA COUNTY: Globe Hills District, Old Dominion Mine, common as a primary ore mineral; also of secondary origin, forming a distinct blanket beneath the chalcocite zone (Schwartz 1947, 1958; N. P. Peterson 1962). Southeastern Dripping Spring Mountains, Banner supersystem, Christmas District, Christmas Mine, as a common ore mineral (N. P. Peterson and Swanson 1956; Knoerr and Eigo 1963; Perry 1969).

GRAHAM COUNTY: Gila Mountains, Lone Star District, with pyrite and chalcopyrite but much less abundant than the latter; Safford supersystem porphyry copper deposits, in veins and disseminations, especially the early magnetite-bornite zone at Dos Pobres Mine (R. F. Robinson and Cook 1966; Langton and Williams 1982).

LA PAZ COUNTY: Buckskin Mountains, Planet District, Planet Mine (Bull. 181). Harquahala District, with chalcopyrite (UA 9344).

MOHAVE COUNTY: Wallapai supersystem, Mineral Park District, Atlanta and Pinkham Mines (Schrader 1909; Dings 1951). Grand Wash Cliffs, Bentley District, Bronze L Mine (Bull. 181).

PIMA COUNTY: Ajo District, concentrated around pegmatite bodies in the New Cornelia Quartz Monzonite (Joralemon 1914; Gilluly 1937, 1942a, 1942b; Schwartz 1947, 1958). Pima supersystem, Twin Buttes District, as a minor constituent of primary sulfide mineralization at the Twin Buttes Mine (Stanley B. Keith, pers. comm., 1973), and as a minor constituent at the Pima Mine (Himes 1972). Silver Bell Mountains, Silver Bell District, in oxidized ores (Engineering and Mining Journal 1904; Kerr 1951). Santa Catalina Mountains, Marble Peak District, Taylor X claims (Dale, Stewart, and McKinney 1960). Helvetia-Rosemont District, a major mineral in drill cores with djurleite and chalcopyrite; also associated with later kinoite and apophyllite in skarn formed in Paleozoic limestone and dolomite (J. W. Anthony and Laughon

1970). Comobabi Mountains, Cababi District, with silver and chalcocite (UA 1079).

PINAL COUNTY: Pioneer supersystem, Magma District, as exceedingly rich ore found to the deepest levels of the Magma Mine (Ransome 1914; Harcourt 1937, 1942; Short et al. 1943; Brett and Yund 1964; Morimoto and Gyobu 1971; for detailed structural information on Magma Mine minerals in the bornite-digenite series, see L. Pierce and Buseck 1978). Galiuro Mountains, Bunker Hill (Copper Creek) District, Copper Creek, Childs-Aldwinkle Mine (Kuhn 1941; Denton 1947b). Northern Galiuro Mountains, Saddle Mountain District, Adjust Mine (Bull. 181). San Manuel District, San Manuel Mine, sparingly present in sulfide ores (Chapman 1947; Schwartz 1947, 1953; Lovering 1948; J. D. Lowell 1968).

SANTA CRUZ COUNTY: Santa Rita and Patagonia Mountains, at several mining properties (Schrader and Hill 1915; Schrader 1917; Marshall and Joensuu 1961).

YAVAPAI COUNTY: Black Hills District, Yeager Mine, as a shoot of high-grade ore (Lindgren 1926; C. A. Anderson and Creasey 1958). Jerome supersystem, Verde District, United Verde Mine, present in small amounts with chalcopyrite, tennantite, and pyrite as both a primary and a secondary mineral (Lausen 1928; Schwartz 1938).

BOTALLACKITE

Copper chloride hydroxide: $Cu_2Cl(OH)_3$. A rare secondary mineral associated with other copper chlorides.

COCHISE COUNTY: Warren District, Bisbee, Southwest Mine, as very sparse (only two specimens are known) pale-blue tabular crystals, associated with paratacamite and atacamite (Graeme 1993).

MARICOPA COUNTY: Belmont Mountains, Osborne District, Tonopah-Belmont Mine, rarely as small pale-blue flattened sprays of crystals with lead oxides (Gibbs and Ascher 2012).

BOTRYOGEN

Magnesium iron sulfate hydroxide hydrate: $MgFe^{3+}(SO_4)_2(OH)\cdot7H_2O$. Typically formed in arid climates as a result of oxidation of pyritic ores.

COCHISE COUNTY: Warren District, Bisbee, Campbell shaft, postmining with copiapite (Graeme 1981).

BOULANGERITE

Lead antimony sulfide: $Pb_5Sb_4S_{11}$. Formed at low to moderate temperatures in hydrothermal veins.

YUMA COUNTY: Kofa District, North Star Mine, in veinlets with pyrite, rhodochrosite, epidote, and fluorite (Hankins 1984).

BOURNONITE

Copper lead antimony sulfide: $CuPbSbS_3$. One of the most common sulfosalts, formed in hydrothermal veins at moderate temperatures and commonly associated with galena, tetrahedrite, sphalerite, chalcopyrite, and other sulfides and sulfosalts, especially in silver-lead-zinc deposits.

COCHISE COUNTY: Tombstone District, sparingly, with other copper-antimony minerals (B. S. Butler, Wilson, and Rasor 1938).

PIMA COUNTY: Northern Santa Rita Mountains, Cuprite District, Busterville Mine (Bull. 181), in a microveinlet cutting tetrahedrite, associated with galena and chalcopyrite (C. A. Lee and Borland 1935).

SANTA CRUZ COUNTY: Wrightson District, Hosey and Augusta Mines, with tetrahedrite (Bull. 181).

YAVAPAI COUNTY: Northernmost Bradshaw Mountains, Big Bug District, Boggs Mine, as masses in quartz and as crystals with pyrite, chalcopyrite, siderite, and actinolite (Blake 1890; Guild 1910; Lindgren 1926; W. P. Blake's identification of this mineral probably constituted its first recognition in the United States).

BRACKEBUSCHITE

Lead manganese vanadate hydroxide: $Pb_2Mn^{3+}(VO_4)_2(OH)$. A rare mineral formed as a product of oxidation in hydrothermal lead-zinc veins.

COCHISE COUNTY: East side of the Swisshelm Mountains, Swisshelm District, in S29, T20S R28E, at a tiny gold prospect in a quartz vein with coarse oxidized pyrite crystals and traces of cerussite. Distinctive henna-brown crystals up to 1 mm form spongy aggregates in voids left by oxidation and leaching of galena (3rd ed.). Tombstone District, Gallagher vanadium property (Bull. 181).

GILA COUNTY: Troy supersystem, C and B–Grey Horse District, C and B Vanadium Mine (Meyer 2011; R100087).

SANTA CRUZ COUNTY: Palmetto District, Palmetto Mine, as microcrystals with wulfenite (Carl Richardson, UA x3381).

BRANNERITE

Uranium titanium oxide: UTi_2O_6. A rare mineral found in placer gravels, granitic rocks, and quartz veins.

COCHISE COUNTY: Swisshelm Mountains, Elfrida District, with powellite in granite (AEC mineral collection).

BRAUNITE

Manganese oxide silicate: $Mn^{2+}Mn^{3+}_6O_8(SiO_4)$. Formed in veins and lenses as a product of the metamorphism of other manganese minerals, and with pyrolusite, psilomelane, and wad as a secondary mineral formed by weathering.

COCHISE COUNTY: Warren District, Bisbee, Higgins Mine, as radiating masses and compact needles (Palache and Shannon 1920; Hewett and Rove 1930; Taber and Schaller 1930; Hewett and Fleischer 1960); White Tail Deer Mine, as coarse (up to 1 inch long) cleavable crystals in irregular replacement pods in limestone around the mine shaft (3rd ed.).

GILA COUNTY: Miami-Inspiration supersystem, present in supergene oxides, probably derived from alabandite (Hewett and Fleischer 1960).

MARICOPA COUNTY: Southwestern slope of Piestewa Peak (formerly Squaw Peak), Phoenix, as microscopic euhedral rectangular crystals in manganese-rich andalusite (Thorpe 1980).

MOHAVE COUNTY: Rawhide Mountains, Artillery District, associated with a variety of manganese oxide minerals (Head 1941).

PIMA COUNTY: Arivaca area, COD Mine, forming a matrix cut by veinlets of hausmannite (Hewett 1972). Tucson Mountains, Amole District, north end of the Juan Santa Cruz picnic grounds, in some of the piemontite in sandstone (Guild 1935). Coyote and Baboquivari Mountains, with psilomelane and pyrolusite in fractures in andesite (Havens et al. 1954).

SANTA CRUZ COUNTY: Patagonia Mountains, Patagonia supersystem, in supergene oxides that may have been derived from alabandite (Hewett and Fleischer 1960). Harshaw Creek area, associated with manganese-silver ore in the Hardshell-Hermosa District (Havens et al. 1954).

BRAZILIANITE

Sodium aluminum phosphate hydroxide: $NaAl_3(PO_4)_2(OH)_4$. A hydrothermal mineral formed in cavities in pegmatites, typically with other phosphates.

MARICOPA AND YAVAPAI COUNTIES: White Picacho District, in pegmatites (London 1981).

#BREWSTERITE-Ba

Barium aluminum silicate hydrate: $Ba(Al_2Si_6)O_{16}\cdot5H_2O$. An uncommon zeolite reported from few localities in the world. This is the first American occurrence.

LA PAZ COUNTY: Trigo Mountains, Silver District, as small well-formed crystals with stilbite and fluorite in a fault breccia in a road cut near the Hamburg Mine (R. Gibbs, pers. comm., 2018).

FIGURE 3.38 Brewsterite-Ba, Trigo Mountains, La Paz County, FOV 7 mm.

BREZINAITE

Chromium sulfide: Cr_3S_4. A very rare mineral described from the Tucson Ring iron meteorite. The Santa Rita Mountains is the type locality.

PIMA COUNTY: In the Irwin-Ainsa (Tucson) iron meteorite, presumably found near Tucson, in the Santa Rita Mountains, as tiny (5–80 microns) anhedral grains in the metal matrix and contiguous to silicate inclusions (Bunch and Fuchs 1969).

BROCHANTITE

Copper sulfate hydroxide: $Cu_4(SO_4)(OH)_6$. A fairly common secondary mineral formed in oxidized copper deposits especially in arid regions. Associated with various common oxidized copper minerals, including malachite, with which it may be confused on cursory examination.

FIGURE 3.39 Brochantite, Bisbee, Cochise County, 2 cm.

COCHISE COUNTY: Warren District, Bisbee, widely distributed as an intergrowth with malachite; Shattuck Mine, as magnificent coarse crystalline masses; with cuprite at the Copper Queen, and Calumet and Arizona Mines (Ransome 1904; Holden 1922; Palache 1939b; Omori and Kerr 1963). Tombstone District, Toughnut Mine, as needle-like crystals lining vugs in cuprite with connellite and malachite (B. S. Butler, Wilson, and Rasor 1938). Courtland District, Maid of Sunshine Mine, as sharp acicular crystals (3rd ed.).

COCONINO COUNTY: Grand Canyon National Park, Grandview District, Grandview (Last Chance) Mine, associated with cyanotrichite and other sulfate minerals, as radiating groups of acicular crystals and as shorter prismatic crystals (Leicht 1971).

GILA COUNTY: Banner supersystem, 79 District, 79 Mine, where it is intimately associated with oxidizing sulfide minerals (Kiersch 1949; Stanley B. Keith 1972). Globe Hills District, Apache Mine, with cerussite, matlockite, and boleite (Bideaux, Williams, and Thomssen 1960).

GRAHAM COUNTY: Gila Mountains, Lone Star District near Safford, in metasomatized latites and andesites with pseudomalachite, malachite, antlerite, carbonate apatite, chrysocolla, jarosite, and lepidocrocite (Hutton 1959a). Safford supersystem porphyry copper cluster, as an important constituent of oxide mineralization of several porphyry copper deposits, although less so than chrysocolla (R. F. Robinson and Cook 1966).

GREENLEE COUNTY: Copper Mountain District, Morenci, abundant as an intergrowth with malachite, less common as crystals (Lindgren and Hillebrand 1904; Lindgren 1904, 1905; Guild 1910; Moolick and Durek 1966).

LA PAZ COUNTY: Harcuvar Mountains, Cunningham Pass District, Bullard Mine, with djurleite (Roddy 1986). Buckskin Mountains, Planet District, Mineral Hill Mine (Bull. 181). Moon Mountains District, Apache Mine near Salome (UA 7826).

MOHAVE COUNTY: Grand Wash Cliffs, Bentley District, Grand Gulch Mine (J. M. Hill 1914b). Bill Williams Fork, as minute crystals associated with cuprite and chrysocolla (Genth 1868).

PIMA COUNTY: Mission-Pima District, Mission Mine, as well-formed crystals (W. Rhodes, pers. comm., 3rd ed.); Banner Mine (UA 9354); San Xavier West Mine (Arnold 1964); Twin Buttes District with chrysocolla and cuprite at the Twin Buttes Mine (UA 9608). South Comobabi Mountains, Cababi District, Mildren and Steppe claims (S. A. Williams 1963). Ajo District, New Cornelia orebody, uncommon as crusts in the weathered zone (Gilluly 1937). Waterman District, Silver Hill Mine (UA 6799). Tucson Mountains, Saginaw Hill District, on the south side of Saginaw Hill, about seven miles southwest of Tucson, associated with other oxidized-zone minerals, including malachite, pseudomalachite, libethenite, atacamite, and chrysocolla (Khin 1970). Silver Bell District, El Tiro pit as microcrystals (Kenneth Bladh, UA x287).

PINAL COUNTY: Mammoth District, Mammoth–St. Anthony Mine, where it is relatively abundant and typically associated with other sulfate-bearing minerals such as linarite, caledonite, and leadhillite (UA 6114). Galiuro Mountains, Bunker Hill (Copper Creek) District (Bull. 181). Sacaton (Casa Grande) District, Sacaton Mine (ASARCO), in drill core

(Robert O'Haire, pers. comm., 1972); subsequently shown to be common in the oxide zone (3rd ed.). Lakeshore District, Lakeshore Mine, as the principal copper mineral in the highest-grade supergene copper zone (Cook 1988b). Silver Reef District, Nugget Fraction Mine, as large masses with chalcocite (3rd ed.).

YUMA COUNTY: Tule Mountains, Venegas District, Venegas prospect, associated with gypsum (E. D. Wilson 1933).

YAVAPAI COUNTY: Jerome supersystem, Verde District, United Verde Mine, in the lower part of the oxidized zone, in chlorite schist that contains chalcopyrite, with antlerite and cyanotrichite (Phelps Dodge Corp., pers. comm., 1972).

BROCKITE

Calcium thorium cerium phosphate hydrate: $(Ca,Th,Ce)(PO_4) \cdot H_2O$. Formed in veins and altered granitic rocks as earthy coatings and aggregates, presumably under oxidizing conditions.

MOHAVE COUNTY: Rawhide Mountains, Rawhide District, as white fine-grained opaque masses in veins with hematite and geothite (Staatz 1985).

BROMARGYRITE

Silver bromide: AgBr. A complete solid-solution series extends from bromargyrite to chlorargyrite (AgCl). Formed in the oxidized zones of silver deposits from primary silver minerals.

COCHISE COUNTY: Tombstone District, Empire Mine, fourth level of the Skip shaft, as individual gray-green crystals and irregular scales throughout limonitic gangue (B. S. Butler, Wilson, and Rasor 1938; UA 7517, H 97646). Pearce District, Commonwealth Mine, with chlorargyrite, bromian chlorargyrite, iodargyrite, and acanthite in quartz veins (Endlich 1897). Warren District, Bisbee, 700 level of Cole shaft, as sharp lemon-yellow crystals on cuprite; 200 ft. level of the Shattuck Mine, replacing silver (Graeme 1981).

FIGURE 3.40 Bromargyrite, Empire Mine, Tombstone, Cochise County, 6 × 3 cm.

COCONINO COUNTY: Prospect Canyon (Ridenour) District, Ridenour Mine, with vésigniéite, naumannite, tyuyamunite, metatyuyamunite, and tangeite (Wenrich and Sutphin 1989).

GILA COUNTY: Globe Hills District, unspecified locality near Globe, as iodian bromian chlorargyrite (Blake 1905); Hechman Mine, in thin seams and crusts in a quartz vein (Bull. 181).

GRAHAM COUNTY: Aravaipa supersystem, Grand Reef District, Grand Reef Mine, as bright-yellow botryoidal masses with cerussite (George Godas, pers. comm., 1987).

GREENLEE COUNTY: Copper Mountain supersystem, Stargo District, Morenci, Stargo Mine, as microcrystals (Gene Wright, UA x1502).

BROOKITE

FIGURE 3.41 Brookite, Bagdad Mine, Yavapai County, FOV 2 mm.

Titanium oxide: TiO_2. Trimorphous with rutile and anatase. An accessory mineral in igneous and metamorphic rocks; also in hydrothermal veins and as a detrital mineral.

GILA COUNTY: Reported in concentrates derived from the Globe area, with "biotite," quartz, and ilmenite (Bull. 181).

SANTA CRUZ COUNTY: Patagonia Mountains, Querces District, near the Santo Niño Mine close to Duquesne, with anatase, chalcopyrite, wulfenite, and molybdenite, all as microcrystals in interstices between adularia crystals (collected by William Hunt).

BRUCITE

Magnesium hydroxide: $Mg(OH)_2$. Forms as an alteration product of periclase in contact-metamorphosed limestones and dolomites; also as a low-temperature hydrothermal vein mineral.

COCHISE COUNTY: Warren District, Bisbee, Czar shaft (Graeme 1981).

MOHAVE COUNTY: Mag District, three miles northwest of Oatman, in veins with magnesite and serpentine cutting volcanic rocks (E. D. Wilson and Roseveare 1949; Funnell and Wolfe 1964).

#BRUSHITE

Calcium phosphate hydroxide hydrate: $Ca(PO_3OH) \cdot 2H_2O$. Found as a cave mineral forming from organic material such as bat guano.

COCHISE COUNTY: Kartchner Caverns State Park, southwest of Benson, found in the Big Room of the cave as large masses over 2 m long, 0.3 m wide, and 6 cm thick. It is formed from solutions beneath a fresh bat guano pile (C. A. Hill 1999).

#BULTFONTEINITE

Calcium silicate hydroxide fluoride hydrate: $Ca_2SiO_3(OH)F \cdot H_2O$. An uncommon mineral found in the contact zone of thermally metamorphosed limestone and in a kimberlite pipe in South Africa.

GILA COUNTY: Banner supersystem, Christmas District, Christmas Mine, as small pale-blue acicular crystals associated with kinoite and tobermorite. Identified by Dr. Bill Wise (R. Thomssen, pers. comm., 2019).

#BURCKHARDTITE

Lead iron tellurium aluminosilicate: $Pb_2(Fe^{3+}Te^{6+})(AlSi_3O_8)O_6$. A rare secondary mineral found with other arsenates in gold deposits.

COCHISE COUNTY: Huachuca Mountains, Reef District, Reef Mine, as small pale-yellow hemispheres consisting of groups of flat tabular crystals and infrequently as single twinned crystals attached to quartz crystal faces associated with gold. The mineral is often admixed with white kaolinite and with a pale-blue unspecified clay mineral (Walstrom 2013). Tombstone District, as small red to beige masses and irregular aggregates among pale-green rodalquilarite with jarosite (Joy Desor, pers. comm., 2020).

#BUSTAMITE

Calcium manganese silicate: $Mn_2Ca_2MnCa(Si_3O_9)_2$. Formed by the metamorphism of rocks rich in manganese.

YAVAPAI COUNTY: Towers Mountain, near Crown King, as white balls of radiating crystals up to 3 mm across as inclusions in quartz crystals (Kyle Eastman, pers. comm., 2007).

BUTLERITE

Iron sulfate hydroxide hydrate: $Fe^{3+}(SO_4)(OH) \cdot 2H_2O$. Monoclinic, dimorphous with parabutlerite. The United Verde Mine is the type locality.

COCONINO COUNTY: Grandview District, Grand Canyon National Park, Horseshoe Mesa, Grandview (Last Chance) Mine, as bright red-orange microcrystals with brochantite (J. Ruiz, pers. comm., 2015).

FIGURE 3.42 Butlerite, United Verde Mine, Yavapai County.

Jerome supersystem, Verde District, United Verde Mine, as a thin crystalline coating formed as a result of burning pyritic ores (Lausen 1928; Cesbron 1964; Fanfani, Nunzi, and Zanazzi 1971; H 90539; S 95953; UA 52, 7867).

BÜTSCHLIITE

Potassium calcium carbonate: $K_2Ca(CO_3)_2$. Associated with calcite as a product of the hydration of fairchildite, formed in clinkers of fused wood ash in partly burned trees. The type locality is in Grand Canyon National Park.

COCONINO COUNTY: Grand Canyon National Park. Discovered by Ranger William J. Kennedy in an unspecified partly burned tree "at a fire on the north side of Kanabownits Canyon one-quarter mile from the Point Sublime road and one-half mile from the North Entrance road" (Milton 1944; Milton and Axelrod 1947; Mrose, Rose, and Marinenko 1966).

MARICOPA COUNTY: Phoenix area, from palo verde tree ash (Garvie 2016).

BUTTGENBACHITE

Copper nitrate chloride hydroxide hydrate: $Cu_{36}(NO_3)_2Cl_8(OH)_{62}\cdot nH_2O$. McLean and Anthony (1972) specified this formula by analogy to the revised formula of connellite, with which the mineral is isostructural.

COCHISE COUNTY: Warren District, Bisbee, Cole Mine, buttgenbachite was identified in a sample from this mine, which had previously been identified as connellite (Hibbs, Leverett, and Williams 2002, 2003). Czar Mine; a single specimen of massive cuprite from an undocumented level in this mine contained buttgenbachite in very small amounts as tight clusters of radiating crystals (Graeme, Graeme, and Graeme 2015).

PIMA COUNTY: South Comobabi Mountains, Cababi District, Mildren and Steppe claims, as a product of the oxidation of sulfide ores, in quartz veins that cut andesite; identification is based on optical properties and qualitative chemical analysis on very limited material (S. A. Williams 1962, 1963).

CACOXENITE

Iron aluminum oxide phosphate hydroxide hydrate: $Fe^{3+}_{24}AlO_6(PO_4)_{17}(OH)_{12}\cdot 75H_2O$. An uncommon mineral of secondary origin associated with other phosphates and iron oxides in iron deposits and iron-bearing pegmatites.

PIMA COUNTY: Silver Bell District, Silver Bell Mine, Oxide pit, as golden to brownish-yellow acicular crystals up to 1 mm long, clustered in sheaths and forming undulatory mats; associated with turquoise crystals and sharp torbernite crystals in the contact zone of an andesite dike (Kenneth Bladh, pers. comm., 1974).

CALAVERITE

Gold telluride: $AuTe_2$. Typically, in low-temperature hydrothermal deposits, but also in moderate- and high-temperature deposits.

COCHISE COUNTY: Warren District, Bisbee, among the sulfides of the Campbell orebody (Graeme 1993).

#CALCIOARAVAIPAITE

Lead calcium aluminum fluoride: $PbCa_2AlF_9$. A secondary mineral found with other lead and aluminum fluorides, sulfates, and hydroxides. The Grand Reef Mine is the type locality.

GRAHAM COUNTY: Aravaipa supersystem, Grand Reef District, Grand Reef Mine, in a quartz vug with artroeite and anglesite (Kampf and Foord 1996; R100033).

CALCITE

Calcium carbonate: $Ca(CO_3)$. A common mineral formed under a wide range of conditions: as a sedimentary rock–forming mineral composing limestone and chalk, as a metamorphic rock former in marble derived from limestone and dolomite, and as travertine precipitated from thermal springs and other surface waters. Some of the most spectacular calcite specimens are dripstone stalactites and stalagmites formed by the evaporation of underground waters carrying calcium and carbonate ions in aqueous solution. Some limestone caverns in Arizona contain excellent examples of dripstone. Calcite is widely distributed throughout the sedimentary rock sequence in the state, particularly in Paleozoic rocks, for example, in the extensive Redwall and Kaibab Formations and in the Mississippian Escabrosa Limestone and several Pennsylvanian and

FIGURE 3.43 Calcite, Twin Buttes Mine, Pima County, 4.2 × 3.0 cm.

Permian formations in south-central and southeastern Arizona. Calcite is also commonly formed in veins through hydrothermal processes, which allow the mineral to attain its best morphological development. The examples cited here are merely representative of the numerous occurrences of this very abundant mineral.

COCHISE COUNTY: Warren District, Bisbee, as remarkable masses of fine crystals in oxidized ore, as scalenohedra colored bright red by minute cuprite crystals (H 70306) and green by included malachite (Hovey 1900); Copper Queen Mine (Guild 1911), as abundant stalactites in the oxidized-zone workings, locally colored by copper salts; Shattuck Mine, 300 ft. level, in a limestone cavern about 340 ft. in diameter and 80 ft. high, discovered in 1914; some of the magnificent cave-stone material from Bisbee, found during the early days of mining, has fortunately been preserved in collections throughout the United States (Bull. 181). Tombstone District, as coarsely crystalline aggregates along the flanks of the roll deposits, and as snow-white linings of caverns in manganese-bearing orebodies (Bull. 181); Lucky Cuss Mine, as blue crystals (UA 9739); Empire Mine, as black crystals (UA 9747). Chiricahua Mountains, Crystal Cave, as crystal aggregates (Bull. 181); Bowie Marble District, near Fort Bowie in Immigrant Canyon and at the bend of Whitetail Creek, as extensive marble deposits (Bull. 181). Little Dragoon Mountains, northwest of Manzora, as marble; Ligier deposit: marble was quarried a few miles southeast of Dragoon station (Engineering and Mining Journal 1926). Gleeson District, as crystals (UA 1046). Huachuca Mountains, Hartford District, Ramsey Canyon, Hamburg Mine, as well-crystallized scalenohedra with quartz crystals (3rd ed.). Kartchner District, Kartchner Caverns State Park, as numerous speleothem formations, especially cave bacon and stalactites, stalagmites, and columns.

COCONINO COUNTY: Near Cameron, replacing aggregates of lenticular gypsum crystals (AEC mineral collection). Grand Canyon National Park, near Supai, as scalenohedral crystals twinned on the basal pinacoid (S 94238); at Havasupai Falls, as travertine; in Havasu Canyon and at Mooney, Bridal Veil, and other falls (Bull. 181). About twenty miles south of Canyon Diablo, as a brown variety of travertine shading into amber-colored onyx (Engineering and Mining Journal 1922).

GILA COUNTY: Globe Hills District, Old Dominion Mine, as fine groups of scalenohedral crystals in cavities in limestone (Bull. 181); Copper Cities District, at the foot of Sleeping Beauty Mountain, about ten miles west of Globe, quarried as marble; about twenty miles north of Globe in T4N R16E, as large deposits of multicolored and variegated travertine

(Funnell and Wolfe 1964). Banner supersystem, 79 District, 79 Mine, sparingly as "butterfly" habit twins on descloizite, smithsonite, and rosasite; also as 0.625 inch frosty-white plates with well-developed basal pinacoids, perched on aurichalcite (Stanley B. Keith 1972).

GRAHAM COUNTY: Safford area, as "sand crystals," consisting of abundant quartz sand grains incorporated in calcite crystals (UA 4873, 6474).

LA PAZ COUNTY: South of the Harquahala Mountains, about nine miles east of Wenden, as variegated marble (Strong 1962); western Harquahala Mountains, White Marble Mine. Bouse Hills District, Continental Mine, as white scalenohedra up to 1 cm long, on hematite and chrysocolla (MM K404–8).

MARICOPA COUNTY: Cave Creek, as travertine (Engineering and Mining Journal 1892a); Camp Creek, west of Cave Creek, as soft travertine-containing boulders variegated in greens and yellows and veined in browns and reds (Bowles 1940).

MOHAVE COUNTY: Big Sandy Valley (UA 7548) and Kingman area (UA 8921), as "sand crystals." West side of Black Mountains, Pilgrim District, Portland Mine, as aesthetic scalenohedra (J. Callahan, pers. comm., 2018). Oatman District, Moss Mine area, as excellent large clear tabular crystals (UA 880).

PIMA COUNTY: Quijotoa Mountains, as fine groups of large brown scalenohedral crystals associated with pink baryte (UA 7559, 9351). Santa Rita Mountains, Helvetia-Rosemont District, King in Exile Mine (UA 9350); Santa Rita marble quarry, quarried as marble six miles north of Helvetia (Bull. 181). Greaterville District, as brown travertine unusually free of cracks (3rd ed.). Cuprite District, Andrada marble quarries. Mission-Pima supersystem, Twin Buttes District, Twin Buttes Mine, as creamy-white crusts several inches thick, composed of simply terminated beige rhombohedral crystals and white scalenohedra composing crusts up to several square feet wide (Stanley B. Keith, pers. comm., 1973; UA 10163). Colossal Cave, near the southwestern slopes of the Rincon Mountains, as abundant stalactites and stalagmites (Bull. 181); dust layers are typically intercalated with bands of calcite, indicative of dry periods of nondeposition. Ajo District, New Cornelia Mine, as red crystals having minute inclusions of cuprite and as large crystals in vesicles in the volcanic rocks at Well No. 1 (W. J. Thomas and Gibbs 1983).

PINAL COUNTY: Pioneer supersystem, Magma District, Magma Mine, as delicate pink scalenohedral crystal groups (Barnes and Hay 1983), also as clear lustrous twinned crystals over 7 cm across (Les Presmyk, pers. comm., 2020)—the best of these crystals rival English calcites. Bunker

Hill (Copper Creek) District, Copper Creek, Childs-Aldwinkle Mine, with molybdenite (UA 537). Santa Catalina Mountains, Peppersauce Canyon, as dripstone in a cave (3rd ed.); near Condon Mountain, at the northern end of the range, as good quality marble (3rd ed.).

SANTA CRUZ COUNTY: Central Santa Rita Mountains, Sawmill Canyon area, Onyx Hill, Onyx Cave, as magnificent groups of scalenohedral crystals (UA 2025-2027) and as onyx and dripstone; Cave of the Bells, as flos ferri associated with aragonite. Cottonwood Canyon, Tyndall District, Glove Mine, as yellow-stained clusters of scalenohedral crystals (UA 9346) in exceptionally attractive coralloidal material formed in naturally cavernous areas. Patagonia Mountains, Washington Camp District, Holland Mine (UA 6707), as groups of large pink equant and scalenohedral crystals on quartz crystals.

YAVAPAI COUNTY: Bradshaw Mountains, Mount Union District, Cash Mine, as beautiful examples of crystallized calcite, quartz, adularia, and ore minerals (Bull. 181). Mayer Onyx District, on Big Bug Creek area near Mayer, as hot spring deposits of banded travertine that have yielded decorative stone (Bull. 181). Near Cordes, as manganiferous banded travertine (UA 882). Ashfork District, at Ash Fork (De Kalb 1895). Old Dick District, Bruce VMS orebody, as gemmy scalenohedrons and fishtail twins (Muntyan 2015a). Near Montezuma Castle National Monument as travertine (Funnell and Wolfe 1964) and as travertine hot springs in collapsed sinkholes at Montezuma Well as part of the Verde Valley supersystem (Stanley B. Keith, pers. comm., 2019). Verde Valley, Camp Verde District, Verde Salt Mine, as replacements of glauberite crystals from the lake beds (Snyder 1971; UA 9871, 9874). About three miles southeast of Castle Hot Springs, as optical-quality crystals in numerous veins (Bull. 181).

YUMA COUNTY: Reported in the Gila Mountains, south of Dome station, as very pure marble deposits (Bull. 181).

#CALDERÓNITE

Lead iron vanadate hydroxide: $Pb_2Fe^{3+}(VO_4)_2(OH)$. A member of the brackebuschite group found in the oxidized portion of lead deposits.

GILA COUNTY: Dripping Spring Mountains, Troy supersystem, C and B–Grey Horse District, C and B Vanadium Mine, as small blocky orange crystals with descloizite and vanadinite (M. Ascher, pers. comm., 2012).

MARICOPA COUNTY: Big Horn Mountains, Big Horn District, Evening Star Mine, as very small orange tabular and blocky crystals with vanadinite and wickenburgite on the hanging wall of the first level (M. Ascher, pers. comm., 2012).

CALEDONITE

Copper lead sulfate carbonate hydroxide: Cu_2Pb_5 $(SO_4)_3(CO_3)(OH)_6$. An uncommon secondary mineral found in small amounts in some oxidized copper-lead deposits.

GRAHAM COUNTY: Aravaipa supersystem, Grand Reef District, Grand Reef Mine, as microcrystals with brochantite (R. W. Jones 1980; Kenneth Bladh, UA x529).

MARICOPA COUNTY: Painted Rock District, Rowley Mine, as fine clear blue-green crystals up to several millimeters long, in small veins, associated with cerussite, linarite, and leadhillite (W. E. Wilson and Miller 1974; UA 4066). Belmont Mountains, Osborne District, Tonopah-Belmont Mine, as pale-blue crystals up to 2 mm long (G. B. Allen and Hunt 1988).

PIMA COUNTY: South Comobabi Mountains, Cababi District, Mildren Mine, as single anhedral up to 2.5 cm in diameter, surrounded by cerussite altering to malachite (S. A. Williams 1962, 1963). Mission-Pima supersystem, Twin Buttes District, Twin Buttes Mine (William Hefferon, pers. comm., 3rd ed.).

PINAL COUNTY: Mammoth District, Mammoth–St. Anthony Mine, 400 ft. level of the Collins vein, as excellent crystals. An exceptionally large specimen, 3.5 by 3.5 by 2 in., consists of a solid mass of interlocking acicular crystals with a fine deep-blue color (UA 9675).

FIGURE 3.44 Caledonite, Mammoth–St. Anthony Mine, Pinal County, 1 cm crystals.

CALOMEL

Mercury chloride: HgCl. A secondary mineral formed by the alteration of other mercury minerals, principally cinnabar but also eglestonite.

MARICOPA COUNTY: Mazatzal Mountains District, L and N No. 1 claim, with eglestonite, metacinnabar, and mercury in quartz stringers (Lausen and Gardner 1927; UA 217); Mercuria Group, as staining on white sericite schist (Lausen and Gardner 1927); Saddle Mountain Mercury Mine, with cinnabar, corderoite, kenhsuite, and mercury (R. Gibbs, pers. comm., 2019).

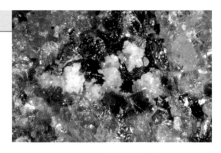

FIGURE 3.45 Calomel, Saddle Mountain Mine, Maricopa County, FOV 5 mm.

CANFIELDITE

Silver tin sulfide: Ag_8SnS_6. Associated with pyrite and other silver sulfo-salts in hydrothermal veins.

COCHISE COUNTY: Warren District, Bisbee, Campbell Mine, intimately associated with stannoidite in pyrite ore. Locally, some ores are reported to have been extremely rich in silver and tin (Graeme 1981).

CANNIZZARITE

Lead bismuth sulfide: $Pb_8Bi_{10}S_{23}$. A rare sulfide mineral.

GRAHAM COUNTY: Aravaipa supersystem, Iron Cap District, Landsman Camp, as tinfoil-like microcrystals on and embedded in quartz veinlets in hedenbergite tactite (3rd ed.).

CARBONATECYANOTRICHITE

Copper aluminum carbonate hydroxide hydrate: $Cu_4Al_2(CO_3)(OH)_{12} \cdot 2H_2O$. A very rare secondary mineral associated with other oxidized copper minerals in some deposits.

COCHISE COUNTY: Warren District, Bisbee, Holbrook shaft, as fibrous radiating needles in tiny spherules on silicified shaley limestones; on azurite and associated with antlerite (Graeme 1993); also from the Cole Mine and Lavender pit (Graeme 1981).

COCONINO COUNTY: Grand Canyon National Park, Horseshoe Mesa, Grandview District, Grandview (Last Chance) Mine, as light-blue sprays of acicular crystals associated with blue cyanotrichite (Colchester et al. 2008).

FIGURE 3.46 Carlosbarbosaite, Belmont Pit, Maricopa County, FOV 3 mm.

#CARLOSBARBOSAITE

Uranyl niobate hydroxide hydrate: $(UO_2)_2Nb_2O_6(OH)_2 \cdot 2H_2O$. A rare mineral formed in the late cavity filling in granitic pegmatites.

MARICOPA COUNTY: Belmont Mountains, Osborne District, Belmont pit, as very small bright-yellow acicular crystals encrusting epidote and as radiating sprays of crystals. The Belmont pit produces landscaping rock from the Belmont Granite, which contains small pegmatitic zones with miarolitic cavities containing unusual minerals (Ron Gibbs, pers. comm., 2016).

CARLSBERGITE

Chromium nitride: CrN. The first nitride identified in iron meteorites; subsequently discovered in more than seventy iron meteorites, as tiny plates in nickel-rich iron; also formed as rings around daubréelite.

COCONINO COUNTY: Meteor Crater, in the Canyon Diablo meteorite, as microscopic oriented laths in nickel-rich iron (E. R. D. Scott, pers. comm., 1988).

#CARMICHAELITE

Titanium chromium iron oxide hydroxide: $(Ti,Cr,Fe)(O,OH)_2$. Found as inclusions in pyrope from an ultramafic SUM diatreme. Garnet Ridge, Arizona, is the type locality.

APACHE COUNTY: Garnet Ridge SUM District, from an ultramafic diatreme on the Navajo Reservation. It occurs as inclusions in pyrope crystals in concentrates collected from the surface. These pyropes also contain rutile, srilankite, spinel, minerals of the crichtonite group, and olivine (Wang et al. 2000).

#CARMINITE

Lead iron arsenate hydroxide: $PbFe^{3+}_2(AsO_4)_2(OH)_2$. Found as an uncommon secondary mineral in the oxidized portions of lead deposits containing arsenopyrite.

PIMA COUNTY: Tucson Mountains, Saginaw Hill District, Amole Group, Papago Queen Mine, rare, as carmine red minute lath-like crystals, less than 0.5 mm in length, associated with beudantite (B. Murphy, pers. comm., 2013).

YAVAPAI COUNTY: Agua Fria District, as very small grains with other arsenates (M. Ascher, pers. comm., 2017).

CARNALLITE

Potassium magnesium chloride hydrate: $KMgCl_3 \cdot 6H_2O$. Found in sedimentary salt deposits formed from oceanic waters, with halite, sylvite, polyhalite, anhydrite, and gypsum.

APACHE AND NAVAJO COUNTIES: Holbrook Basin Potash District, in the subsurface of the southern part of east-central Arizona, in a northeast-trending zone of Permian evaporite deposits (Peirce 1969b); a drill-hole log in S24, T18N R25E listed carnallite, halite, sylvite, polyhalite, anhydrite, and gypsum (H. Wesley Peirce, pers. comm., 1972).

FIGURE 3.47 Carnotite, Anderson Mine, Yavapai County, FOV 3 mm.

CARNOTITE

Potassium uranyl vanadate hydrate: $K_2(UO_2)_2(VO_4)_2 \cdot 3H_2O$. A secondary mineral widely distributed on the Colorado Plateau, where it is an important uranium and vanadium ore. Carnotite can be formed by the action of meteoric waters on preexisting uranium and vanadium minerals, including uraninite and montroseite; disseminated or locally concentrated in sandstone and associated with fossilized tree trunks and other vegetal matter. It is commonly associated with tyuyamunite and metatyuyamunite and other oxidized uranium and vanadium minerals.

APACHE COUNTY: Northeast Carrizo Mountains District, at numerous places in the Salt Wash Member of the Morrison Formation (Isachsen, Mitcham, and Wood 1955; UA 534). Lukachukai Mountains, Lukachukai District (UA 7141). Chuska Mountains (Joralemon 1952; J. D. Lowell 1955; Masters 1955; Wright 1955; Garrels and Larsen 1959). Cane Valley District, Monument No. 2 and Cato Sells Mines, mixed with tyuyamunite (Isachsen, Mitcham, and Wood 1955; Mitcham and Evensen 1955; Wright 1955; Finnell 1957; Evensen and Gray 1958; Witkind and Thaden 1963; R. G. Young 1964).

COCONINO COUNTY: Cameron District, Huskon No. 10 Mine, associated with schröckingerite (Austin 1964). Vermilion Cliffs District, in petrified wood (Bull. 181).

MARICOPA COUNTY: Vulture Mountains, Black Butte District, southeast of Aguila, on minor fractures in tuff (Hewett 1925; H 106863).

NAVAJO COUNTY: Monument Valley District, Monument No. 1 and Mitten No. 2 Mines and other prospects (Evensen and Gray 1958; Holland et al. 1958; Witkind 1961; Witkind and Thaden 1963).

PIMA COUNTY: Cienega Wash near Vail, where the road crosses the Southern Pacific tracks (Robert O'Haire, pers. comm., 1972).

YAVAPAI COUNTY: Date Creek District, Anderson Mine, sparse on sandstone (AEC mineral collection) as small yellow to orange microcrystals (Sherborne et al. 1979). Eureka supersystem, Hillside District, Hillside Mine (Axelrod et al. 1951; S 117681 00-05).

YUMA COUNTY: Muggins U District, Red Knob claims, associated with weeksite, opal, vanadinite, gypsum, calcite, and azurite (Outerbridge et al. 1960).

CASSITERITE

Tin oxide: SnO_2. The most important and widely distributed tin mineral is formed in high-temperature veins and in pyrometasomatic deposits associated with felsic granitic igneous rocks; also in pegmatites and rhyolites.

FIGURE 3.48 Cassiterite, Wood Tin, Apache Tin claims, Graham County, 1.8 cm.

COCHISE COUNTY: Warren District, Bisbee, Campbell Mine, as a very minor accessory mineral in the pyritic ores of the Campbell orebody. Southwest Mine, 5th level, as microscopic anhedral grains in a massive cuprite-hematite mixture. These grains are, most probably, a relic from the primary sulfide ores (Graeme 1993).

GRAHAM COUNTY: Apache Tin claims, twenty-five miles east of Safford on U.S. Highway 70, in spherulitic rhyolite and associated placers, pebbles (UA 7045), and in rhyolite (UA 7028).

MARICOPA COUNTY: White Picacho District, as a rare constituent of pegmatites (Jahns 1952).

PINAL COUNTY: East of Tablelands, as a single small piece of stream tin from rhyolite flows (3rd ed.).

YAVAPAI COUNTY: White Picacho District, in the Outpost pegmatite, as tabular crystals with well-defined faces, honey-yellow to very dark brown, some zoned, up to 1.5 in. in diameter, associated with copper and bismuth minerals (Jahns 1952).

CELADONITE

Potassium magnesium iron silicate hydroxide: $KMgFe^{3+}Si_4O_{10}(OH)_2$. A member of the mica group, with composition and properties like those of glauconite but formed in vesicles in basaltic rocks. It is common throughout the state but seldom mentioned in the literature.

FIGURE 3.49 Celadonite, Malpais Hill, Pinal County, FOV 6 mm.

COCHISE COUNTY: Steele Hills, in seams and vesicles in a basalt within the Threelinks Conglomerate (J. R. Cooper and Silver 1964). Reported in the Warren District, Bisbee (3rd ed.).

NAVAJO COUNTY: Hopi Buttes volcanic field, Seth-La-Kai diatreme, in volcanic sandstone and tuffs, with limonite, laumontite, gypsum, and montmorillonite, associated with weak uranium mineralization (J. D. Lowell 1956).

PINAL COUNTY: Malpais Hill and Midway station, a common lining in vesicle fillings of the andesite of Depression Canyon throughout the northern Galiuro Mountains (Bideaux, Williams, and Thomssen 1960; Thomssen 1983; Stanley B. Keith, pers. comm., 2019).

CELESTINE

Strontium sulfate: $Sr(SO_4)$. Celestine is found in veins, beds, or lenticular masses in sedimentary rocks; also as a gangue mineral with lead-zinc ores.

COCHISE COUNTY: One mile northwest of Portal, as sharp well-formed crystals up to 5 mm long, and as free crystals in vesicles in basalt; associated with prehnite, pyrite, and pumpellyite (3rd ed.).

GILA COUNTY: Banner supersystem, 79 District, 79 Mine, as microcrystals (Lena Marvin, UA x2921).

LA PAZ COUNTY: Central Plomosa Mountains, Southern Cross District, with lead and silver ores (Bull. 181).

MARICOPA COUNTY: About fifteen miles from Gila Bend and three miles east of the Black Rock railroad siding, on the northwest side of a low mountain range with gypsum, sandstone, and conglomerate as beds in sandy tuff (Phalen 1914; B. N. Moore 1935, 1936; UA 907). Southwest pediment of the Vulture Mountains, fifteen miles southeast of Aguila, in NW quarter of S20, T6N R7W, as beds in shaley tuff (B. S. Butler 1929; B. N. Moore 1935, 1936; Hewett et al. 1936; Harness 1942; UA 1154).

MOHAVE COUNTY: Artillery District, Graham prospect, as nodules scattered on the surface (Lasky and Webber 1949). Hack District, Hack Canyon Mine, Pipe No. 1, as massive cleavage pieces up to 4 cm (MM K493, K502).

YAVAPAI COUNTY: Date Creek District, Anderson Mine, as small crystals with gypsum and weeksite (William Hunt, pers. comm., 1985).

CELSIAN

Barium aluminum silicate: $Ba(Al_2Si_2O_8)$. An uncommon monoclinic member of the feldspar group that is commonly associated with manganese deposits.

YAVAPAI COUNTY: Reported in an unspecified locality near Yarnell (Bull. 181).

CERUSSITE

Lead carbonate: $Pb(CO_3)$. Cerussite is a very common secondary mineral found in oxidized lead deposits, formed by reaction between carbonated waters and lead minerals or solutions containing lead. It is commonly formed as concentric layers around the lead sulfate, anglesite, which, in turn, surrounds a core of unaltered galena. Cerussite is so widespread in Arizona that only a few localities can be noted.

COCHISE COUNTY: Tombstone District, as the most abundant lead mineral; Toughnut Mine (B. S. Butler Wilson, and Rasor 1938; Rasor 1938; C. Frondel and Pough 1944; UA 9737). Warren District, Bisbee, Campbell shaft, as sharp pale-gray nearly perfect tabular sixling-twinned crystals up to 2 in. in diameter, on psilomelane (Sinkankas 1964); Hendricks Gulch, as impure "sand carbonate" near a fissure in limestone (Ransome 1904; Guild 1910; Schwartz and Park 1932). Chiricahua Mountains, California supersystem, Hilltop District, Hilltop Mine, as large twinned crystals (Pough 1941). Chiricahua Mountains, Outlook Mine (UA 7642). Gleeson District, as the principal lead ore mineral (3rd ed.); Silver Bill Mine, as microcrystals (Robert Massey, UA x1861). Gunnison Hills, Cochise supersystem, Johnson Camp District, Texas-Arizona Mine (J. R. Cooper and Silver 1964).

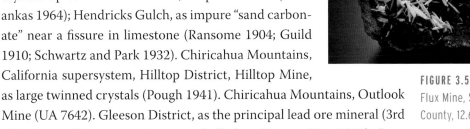

FIGURE 3.50 Cerussite, Flux Mine, Santa Cruz County, 12.6 cm.

COCONINO COUNTY: Grandview District, Grandview (Last Chance) Mine, as twinned and phantom microcrystals (Wolfgang Mueller, UA x119, x121).

GILA COUNTY: Dripping Spring Mountains, Banner supersystem, Chilito District, London-Arizona Mine, as the principal lead ore mineral (Bull. 181); 79 District, 79 Mine, as the most abundant supergene mineral formed as "sand carbonate," and as beautifully crystallized material, including sixling and V-shaped twinned crystals, associated with anglesite, wulfenite, mimetite, and vanadinite (Kiersch 1949; Stanley B. Keith 1972). Dripping Spring Mountains, Troy supersystem, C and B–Grey Horse District, C and B Mine, as Vtype and several other twin types (Crowley 1980). Globe Hills District, Apache Mine, as the principal ore mineral (E. D. Wilson et al. 1950) and in large masses associated with anglesite and galena (E. D. Wilson et al. 1951).

GRAHAM COUNTY: Aravaipa supersystem, Grand Reef District (Denton 1947a; Simons 1964); Grand Reef Mine, as various habits, including large sixling-twinned crystals (some reported up to fist size), reticulated masses, and microcrystals (Kenneth Bladh, George Godas, UA x528). Stanley District (C. P. Ross 1925a).

GREENLEE COUNTY: Copper Mountain District (Reber 1916), Hormeyer Mine, with gold ore (Bull. 181).

LA PAZ COUNTY: Trigo Mountains, Silver District, Red Cloud Mine, as single and twinned crystals and as large reticulated masses; many crystals are arrowhead-shaped twins (Pirsson 1891; Guild 1910; Foshag 1919; Edson 1980).

MARICOPA COUNTY: Painted Rock District, Rowley Mine near Theba, as sharp clear tiny crystals; also massive (W. E. Wilson and Miller 1974). Belmont Mountains, Osborne District, Moon Anchor Mine, Potter-Cramer property and Rat Tail claim, associated with galena, sphalerite, and a variety of lead and chromium oxidized minerals (S. A. Williams 1968; S. A. Williams and Anthony 1970; Williams, McLean, and Anthony 1970).

MOHAVE COUNTY: Wallapai supersystem, Tennessee District, Tennessee-Schuylkill Mine (Schrader 1909; B. E. Thomas 1949). McCracken District, McCracken Mine (Bull. 181). Gold Basin District, at some properties with free gold (Bull. 181).

PIMA COUNTY: Santa Rita Mountains, Helvetia-Rosemont District, Golden Gate and Blue Jay Mines (Bull. 181). Empire District, Total Wreck, Chief, and Hilton Mines (Schrader and Hill 1915); Forty-Nine Mine, as trillings up to 0.75 inch, free in finer "sand carbonate," associated locally with aurichalcite and turquoise (Gene Schlepp, pers. comm., 1974); C and A lease, southwest of Pantano (UA 7680). Sierrita Mountains, Pima supersystem, Mission-Pima District, as "sand carbonate" (Bull. 181); Paymaster Mine, Olive Camp, as massive and crystallized material (Ransome 1922; Nye 1961); Twin Buttes District, Twin Buttes Mine (Stanley B. Keith, pers. comm., 1973). Silver Bell District, as silky crystals and as earthy mixtures with smithsonite; El Tiro and other properties (Bull. 181). Quijotoa District, Morgan Mine, with chlorargyrite (Bull. 181). Tucson Mountains, Old Yuma District, Old Yuma Mine, with wulfenite and vanadinite (Guild 1911). Waterman District, Indiana-Arizona Mine, as white needle-like crystals with azurite (3rd ed.).

PINAL COUNTY: Mammoth District, Mammoth–St. Anthony Mine, Collins vein, as single crystals and magnificent V-twinned and reticulated crystal aggregates (Pogue 1913; N. P. Peterson 1938a, 1938b; Palache 1941b; Fahey, Daggett, and Gordon 1950; Bideaux 1980; UA 850). Galiuro Mountains, Bunker Hill (Copper Creek) District, Copper Creek, Bluebird, and other lead deposits (Bull. 181). Saddle Mountain District (Bull. 181), Saddle Mountain Mine Group. Tortilla Mountains, Ripsey District, Florence Lead-Silver Mine, formed by the alteration of primary ores containing galena, pyrite, sphalerite, and tennantite, and associated with hemihedrite, wulfenite, willemite, vauquelinite, minium, and mimetite (S. A. Williams and Anthony 1970).

SANTA CRUZ COUNTY: Santa Rita Mountains, Salero District, Victor and Rosario properties (Bull. 181); Cottonwood Canyon, Tyndall District, Glove Mine, as large crystalline masses (H. J. Olson 1966); Mansfield District, American Boy Mine (Bull. 181). Patagonia Mountains, Golden

Rose District, Montosa Canyon, Isabella Mine, as "sand carbonate," along a mineralized fault in limestone, associated with specular hematite (J. W. Anthony 1951); Patagonia Mountains, Flux District, Flux Mine, as magnificent groupings of needle-like and pencil-like crystals (jackstraw) and as massive material (Kunz 1885; D. M. Shannon 1981; UA 213, 7134); Ventura District, Domino Mine, and at several other properties, including Washington Camp (Kunz 1885; Schrader 1917).

YAVAPAI COUNTY: Black Hills, Jerome supersystem, Verde District, Copper Chief Mine (Bull. 181). Bradshaw Mountains, Ticonderoga District, Silver Belt Mine, with chlorargyrite, in ancient mine workings (Lindgren 1926). Eureka supersystem, Hillside District, Hillside Mine (Axelrod et al. 1951); Bella Mine (Kunz 1885). Iron King District, Iron King Mine (Creasey 1952; C. A. Anderson and Creasey 1958). Wickenburg Mountains, Black Rock District, S and O claims (also called Amethyst Hill) northeast of Wickenburg, as simple to complex reticulated crystals, often with a dark-brown color (Zimmerman, pers. comm., 2015).

YUMA COUNTY: Castle Dome District, associated with anglesite, and yellow and red lead oxides, and wulfenite (E. D. Wilson 1933; Batty et al. 1947); Hull Mine, as sixling twins with wulfenite and baryte on fluorite (Les Presmyk, pers. comm., 2020; 3rd ed.).

CERVANTITE

Antimony oxide: $Sb^{3+}Sb^{5+}O_4$. A secondary mineral formed through the oxidation of stibnite and other antimony minerals.

LA PAZ COUNTY: Dome Rock Mountains, Cinnabar District, Yellow Devil Mine, in veins, as radiating blades of stibnite partly altered to cervantite and stibiconite (Melchiorre 2013).

#CERVELLEITE

Silver telluride: Ag_4TeS. A rare primary silver mineral found in tellurium-rich orebodies.

COCHISE COUNTY: Tombstone District, Empire Mine (Brent Thorne, pers. comm., 2017).

CESÀROLITE

Lead manganese oxide hydroxide: $PbMn^{4+}_3O_6(OH)_2$. A rare secondary mineral, cesàrolite is sometimes associated with the oxidation of galena.

COCHISE COUNTY: Warren District, Bisbee, in upper Paleozoic lime-stones, probably of supergene origin (Hewett and Fleischer 1960; Hewett, Fleischer, and Conklin 1963).

SANTA CRUZ COUNTY: Patagonia Mountains, Flux District, Flux Mine, in oxide zone associated with cerussite (Stanley B. Keith, pers. comm., via Gunnar Farber, 2017).

CESBRONITE

Copper tellurium oxide hydroxide: $Cu_3Te^{6+}O_4(OH)_4$. Formed in veins with electrum, teineite, and carlfriesite at the type locality in Mexico (S. A. Williams 1974).

COCHISE COUNTY: Tombstone District (3rd ed.).

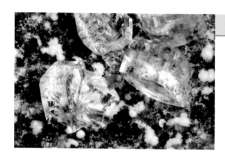

FIGURE 3.51 Chabazite-Ca, Well No. 1, Ajo, Pima County, FOV 3 mm.

#CHABAZITE-Ca

Calcium aluminum silicate hydrate: $Ca_2[Al_4Si_8O_{24}]\cdot13H_2O$. A series in the zeolite group, based on dominant cation: Ca, K, Mg, Na, or Sr. Typically formed in amygdules and fissures in mafic volcanic rocks; also formed through the alteration of volcanic glass in tuffs. Chabazite-Ca is the most common of the chabazite group, and most of these are probably that mineral, but the specific chemistry in these occurrences is not known.

COCHISE COUNTY: San Simon Basin, seven miles northwest of Bowie, in bedded lake deposits with analcime, chabazite-Na, erionite, clinoptilolite, halite, and thénardite (Regis and Sand 1967; Edson 1977). Bowie District, near Bowie, as chalky pale-buff-yellow material (F. A. Mumpton, pers. comm., 3rd ed.; Eyde 1978).

COCONINO COUNTY: Black Peak District, Black Peak, northwest of Tuba City and southeast of the Gap, in the border zones of mafic dikes associated with breccia pipes and carbonate veins and vents cutting Navajo Sandstone; localized in the centers of dolomite replacements of olivine, and as small veins (Barrington and Kerr 1961).

GILA COUNTY: Southeastern Dripping Spring Mountains, Banner super-system, Christmas District, Christmas Mine, as rhombohedral crystals in vugs in hydrothermally altered andesite porphyry, where it appears to be a late-stage mineral (D. Perry, pers. comm., 1967); also as chalky pinkish-white material (F. Mumpton, 3rd ed.). Roosevelt Lake–Tonto

Creek area, in at least five zeolite deposits in altered tuff beds (Eyde 1978). In a road cut twenty-three miles south of Globe on State Highway 77 (Eyde 1978).

GRAHAM COUNTY: Southwest of Pima, in several zeolite deposits in altered tuffs (Eyde 1978).

MARICOPA COUNTY: Seven Springs area, as microcrystals (Robert Mudra, UA x79).

MOHAVE COUNTY: East of Big Sandy Wash, in the east half of T16N R1W, in Pliocene tuff with analcime, clinoptilolite, erionite, and phillipsite (C. S. Ross 1928, 1941).

NAVAJO COUNTY: Coliseum diatreme, near Indian Wells, filling or lining vesicles (3rd ed.).

PIMA COUNTY: Ajo District, Phelps Dodge Well No. 1, as crystals in vugs in altered volcanic rock (William Thomas, pers. comm., 1988).

PINAL COUNTY: In a railway cut at Malpais Hill, north of Mammoth, as crystals intergrown with calcite (Bideaux, Williams, and Thomssen 1960). About 5.5 miles south of Superior, in a wash on the east side of State Highway 177, in basalt with analcime and thomsonite (Tschernich 1992).

YAVAPAI COUNTY: In basalts west of Perkinsville, with phillipsite (McKee and Anderson 1971).

#CHABAZITE-Na

Sodium potassium aluminum silicate hydrate: $(Na_3K)[Al_4Si_8O_{24}]\cdot11H_2O$. Formerly herschelite, it was renamed chabazite-Na in 1997 and is now a member of the chabazite series of the zeolite group.

COCHISE COUNTY: San Simon Basin, Bowie District, seven miles northeast of Bowie, as minute spherules and crystal aggregations formed by alteration of volcanic pyroclastic material, in bedded lake deposits, with analcime, chabazite, erionite, clinoptilolite, halite, and thénardite. The composition of the chabazite varies laterally with its position within the deposit (Regis and Sand 1967).

MARICOPA COUNTY: One mile south of Horseshoe Dam, as transparent to beige-colored hexagonal plates with analcime and phillipsite (D. M. Shannon 1983b).

YAVAPAI COUNTY: Along a dirt road between U.S. Highway 93 and State Highway 96, on the south side of the Santa Maria River, in basalt with phillipsite (William Hunt, pers. comm., 1985).

FIGURE 3.52 Chalcan-thite, Planet Mine, La Paz County, 10.2 cm.

CHALCANTHITE

Copper sulfate pentahydrate: $Cu(SO_4)\cdot5H_2O$. A water-soluble secondary mineral found in the oxidized zone of sulfide copper deposits; commonly forms crusts and stalactites in mine workings. Typically decomposes in a dry atmosphere.

COCHISE COUNTY: Warren District, Bisbee, Copper Queen Mine, as stalactites and irregular porous excrescences several inches thick on mine walls (Merwin and Posnjak 1937); Briggs shaft, as stalactites (Mitchell 1921).

COCONINO COUNTY: Grand Canyon National Park, Grandview District, Grandview (Last Chance) Mine, as cross-fiber veinlets associated with clay, some clearly of postmine origin (Leicht 1971).

GILA COUNTY: Globe Hills District, Old Dominion Mine, as stalactites and as coatings on floors of old openings (Bull. 181). Miami-Inspiration supersystem, Pinto Valley District, Castle Dome Mine, largely as a post-mine mineral (N. P. Peterson, Gilbert, and Quick 1951; N. P. Peterson 1962); Inspiration orebody (Olmstead and Johnson 1966). Dripping Spring Mountains, Banner supersystem, 79 District, 79 Mine, fifth level, with olivenite, sphalerite, galena, siderite, anglesite, and brochantite (Kiersch 1949; Thomas Trebisky, pers. comm., 1972). Sierra Ancha District, First Chance, Little Joe, and Shipp No. 2 Mines, as an efflorescence on mine walls (Granger and Raup 1969); Cherry Creek area, Donna Lee Mine (3rd ed.).

GREENLEE COUNTY: Copper Mountain District, Morenci, as small bodies in the oxidized ores of Copper Mountain, and as stalactites in one of the upper drifts of the Jay shaft (Lindgren 1905; Guild 1910; Moolick and Durek 1966).

LA PAZ COUNTY: Planet District, Planet Mine—superb ram's horn specimens to several inches in length have been recovered from the mine (M. Alter, pers. comm., 2014)

MOHAVE COUNTY: Cerbat Mountains, Wallapai supersystem, as a common secondary mineral (B. E. Thomas, 1949).

NAVAJO COUNTY: Monument Valley District, Mitten No. 2 Mine (Witkind and Thaden 1963).

PIMA COUNTY: Silver Bell Mountains, Silver Bell District, as thick coatings on the walls of old mine workings (Bull. 181). Santa Rita Mountains, Helvetia-Rosemont District, as fibrous veins (Schrader and Hill 1915; Creasey and Quick 1955). Pima supersystem, Mission-Pima District, San Xavier West Mine (Arnold 1964); Twin Buttes Mine (Richardson, UA 13608).

PINAL COUNTY: Galiuro Mountains, Bunker Hill (Copper Creek) District, Copper Creek, Old Reliable, Copper Giant, Glory Hole, and Copper Prince Mines, as coatings on the walls of drifts and in fractures (Simons 1964). Mammoth District, Mammoth–St. Anthony Mine area (UA 4585).

SANTA CRUZ COUNTY: Washington Camp District, Patagonia Mountains, Duquesne and Washington Camp (Schrader 1917).

YAVAPAI COUNTY: Black Hills, Jerome supersystem, Verde District, United Verde Mine, as stalactites up to 2 ft. long (Guild 1910; UA 5569). Tiger District, Springfield Mine, as fine crystals on timbers in the dump (Bull. 181). Eureka supersystem, Bagdad District, Bagdad, abundant in old mine dumps (Therese Murchison, pers. comm., 1972). Mayer District, DeSoto Mine at Cleator, as microcrystals (UA x3771).

CHALCOALUMITE

Copper aluminum sulfate hydroxide hydrate: $CuAl_4(SO_4)$ $(OH)_{12} \cdot 3H_2O$. A very rare secondary mineral discovered in only a few oxidized copper deposits. Warren District, Bisbee, is the type locality.

COCHISE COUNTY: Warren District, Bisbee—Larsen and Vassar (1925) described the mineral, based on material from an unspecified locality in Bisbee. Lavender pit, in vugs in a dense quartz-goethite gossan, associated with cuprite and malachite, as tiny beautifully formed single crystals and twins; commonly altered to pale-blue massive botryoidal gibbsite (S. A. Williams and Khin 1971).

FIGURE 3.53 Chalcoalumite, Bisbee, Cochise County, FOV 5 mm.

COCONINO COUNTY: Grand Canyon National Park, Grandview District, Grandview (Last Chance) Mine, sky blue to greenish in color, as botryoidal crusts on limonite coating earlier zeunerite crystals; also associated with scorodite, olivenite, and brochantite (Leicht 1971).

PIMA COUNTY: Helvetia-Rosemont District, Omega Mine, as small well-formed pale-blue crystals found on the dump (R. Gibbs, pers. comm., 2017).

CHALCOCITE

Copper sulfide: Cu_2S. A widely distributed, important copper-ore mineral. Uncommon as a primary mineral but important as a secondary

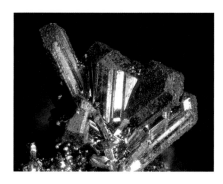

FIGURE 3.54 Chalcocite, Magma Mine, Pinal County, 2 mm crystals.

mineral in the zone of secondary enrichment, where it may replace other sulfides. Of commercial importance in Arizona, where many large low-grade copper orebodies owe much of their value to secondary chalcocite.

APACHE AND NAVAJO COUNTIES: Monument Valley and Cane Valley Districts, associated with other sulfide minerals in uranium-vanadium ores in sandstone (Mitcham and Evensen 1955; Evensen and Gray 1958).

COCHISE COUNTY: Warren District, Bisbee, Campbell and Shattuck Mines, locally abundant as a secondary mineral in limestone replacements; Sacramento Mine, as disseminated ores (Ransome 1904; Guild 1910; Trischka, Rove, and Barringer 1929; Schwartz and Park 1932; Schwartz 1934, 1939; Grout 1946; Hutton 1957; Bain 1952); Lavender pit, as the principal ore mineral (Bryant and Metz 1966). Cochise supersystem, Johnson Camp District, Johnson Camp area, in small amounts in quartz veins cutting granite, schist, and limestone (Kellogg 1906). Tombstone District, with acanthite and stromeyerite (B. S. Butler Wilson, and Rasor 1938); Empire Mine (UA 9731). Courtland District, an important constituent of enriched ores (Bull. 181).

COCONINO COUNTY: Grand Canyon National Park, Horseshoe Mesa, Grandview District, Grandview (Last Chance) Mine (A. F. Rogers 1922; Leicht 1971). Orphan District, breccia pipe, at the Orphan Mine (Gornitz and Kerr 1970; UA 2368).

GILA COUNTY: Globe-Miami area, as an important constituent of disseminated sulfide deposits (Ransome 1903b; Schwartz 1921, 1928, 1934, 1939, 1947, 1958; N. P. Peterson 1962); Globe Hills District, Old Dominion Mine, as compact massive bodies (Ransome 1903b). Miami-Inspiration supersystem, Pinto Valley District, Castle Dome Mine, replacing chalcopyrite in the upper part of the deposit (N. P. Peterson 1947). Copper Cities District, Copper Cities porphyry copper deposit (N. P. Peterson 1954; W. W. Simmons and Fowells 1966). Southeastern Dripping Spring Mountains, Banner supersystem, Christmas District, Christmas Mine, locally associated with bornite in the chalcopyrite-bornite, inner ore zone of the orebodies, replacing dolomite (Perry 1969). Dripping Spring Mountains, Banner supersystem, 79 District, 79 Mine (Kiersch 1949; Stanley B. Keith 1972).

GRAHAM COUNTY: Aravaipa supersystem, Grand Reef District (Denton 1947a; Simons 1964), Ten Strike Group of claims. Gila Mountains, Safford supersystem, Lone Star District, as both steely and sooty varieties, replacing pyrite, chalcopyrite, and bornite in the Safford porphyry copper cluster, where it formed blanket-like deposits on the primary ore zones (R. F. Robinson and Cook 1966).

GREENLEE COUNTY: Copper Mountain District, Morenci Mine, in a thick blanket (Moolick and Durek 1966); as the principal ore mineral of disseminated and vein deposits, in places as solid veins 2 to 3 ft. thick (Lindgren 1903, 1904, 1905; Guild 1910); Ryerson Mine (S 86029; sample collected by W. Lindgren); Montezuma Mine (S 86042).

LA PAZ COUNTY: Cinnabar District, in prospects at Cinnabar, eight miles southwest of Quartzsite (3rd ed.).

MARICOPA COUNTY: Cave Creek District, Red Rover Mine, with argentiferous tetrahedrite (Bull. 181).

MOHAVE COUNTY: Near Bill Williams Fork (Genth 1868). Grand Wash Cliffs, Bentley District, Grand Gulch, Bronze L, and Copper King Mines (J. M. Hill 1914b). Cerbat Mountains, Wallapai supersystem, Mineral Park District, Mineral Park Mine (B. E. Thomas 1949; Field 1966).

PIMA COUNTY: Ajo District, disseminated in a narrow band bordering the southern part of the New Cornelia orebody (Joralemon 1914). Santa Rita Mountains, Helvetia-Rosemont District (Schrader 1917). Sierrita Mountains, Pima supersystem, Esperanza District, Esperanza Mine, in an extensive enriched blanket (Schmitt et al. 1959; D. W. Lynch 1967); Mission-Pima District, Mission-Pima Mine, as a product of secondary enrichment (Journeay 1959); Copper Glance and Queen Mines, as large nearly pure masses (Ransome 1922); Twin Buttes District, Twin Buttes Mine, in secondarily enriched ores (Stanley B. Keith, pers. comm., 1973). Silver Bell Mountains, Silver Bell District, El Tiro and Oxide open-pit mines, as an important secondarily enriched ore mineral (Richard and Courtright 1966). South Comobabi Mountains, Cababi District, Mildren and Steppe claims (S. A. Williams 1963).

PINAL COUNTY: Pioneer supersystem, Magma District, Magma Mine, as large nearly pure secondary bodies and an important constituent of the primary ores of the deeper levels, associated with digenite, bornite, and djurleite; as slender prismatic crystals up to 2 cm long and as V twins (on the 3,260 ft. level; Barnes and Hay 1983); Belmont District, Belmont Mine, as fine-grained, sooty material (Ransome 1914; Guild 1917; Bateman 1929; Harcourt 1942; Short et al. 1943; Morimoto and Gyobu 1971). Mineral Creek supersystem, Ray District, Ray Mine, as the essential mineral of disseminated ores (Ransome 1919; Schwartz 1947; Clarke 1953; D. V. Lewis 1955; Metz and Rose 1966). Galiuro Mountains, Bunker Hill (Copper Creek) District, Copper Creek, at several properties including the Childs-Aldwinkle Mine, where primary chalcocite has formed in the deeper levels (Kuhn 1941; Denton 1947b). Mammoth District, Mammoth–St. Anthony Mine, as thin films and as replacements of chalcopyrite (N. P. Peterson 1938a, 1938b; Fahey, Daggett, and Gordon 1950). San Manuel District, San Manuel Mine, in hydrothermally altered monzonite and quartz monzonite porphyry, replacing chalcopyrite in the secondary sulfide zone; associated with bornite, chalcopyrite, cuprite, chrysocolla, and copper (Chapman 1947; Schwartz 1947, 1949; Lovering 1948; L. A. Thomas 1966; J. D. Lowell 1968).

SANTA CRUZ COUNTY: Santa Rita Mountains, Ivanhoe District, at several properties, including the Ivanhoe Mine; Tyndall District, where it formed large bodies (Schrader and Hill 1915; Schrader 1917). Palmetto District, 3R Mine, as an orebody of secondary origin (Handverger 1963).

YAVAPAI COUNTY: Black Hills, Jerome supersystem, Verde District, in the oxidized zone of the United Verde Mine, and in exceptionally large pure massive bodies in the United Verde Extension Mine (Fearing 1926; Schwartz 1938; C. A. Anderson and Creasey 1958). Eureka supersystem, Bagdad District, Bagdad Mine (C. A. Anderson 1950; Moxham, Foote, and Bunker 1965). Copper Basin District, in the oxidation zone of sulfide deposits in breccia pipes (Johnston and Lowell 1961).

YUMA COUNTY: Castle Dome District, Castle Dome Mine, with malachite (S 65493).

CHALCOMENITE

Copper selenite hydrate: $Cu(Se^{4+}O_3)\cdot 2H_2O$. A secondary mineral formed through the oxidation of copper and lead selenides.

COCHISE COUNTY: Middle Pass District, Middlemarch Mine, Missionary shaft, as isolated pods in a skarn predominantly composed of epidote, quartz, chlorite, calcite, pyrite, sphalerite, and chalcopyrite, plus minor amounts of covellite, azurite, and malachite; orthorhombic crystals are sky blue, arranged in parallel aggregates, bow ties, and radiating clusters on quartz and garnet (Sousa 1980).

#CHALCONATRONITE

Sodium copper carbonate hydrate: $Na_2Cu(CO_3)_2\cdot 3H_2O$. An uncommon mineral found in the oxidized zone of polymetallic deposits and as a component of the patina on ancient bronze artifacts.

MARICOPA COUNTY: Painted Rock District, Rowley Mine, as aggregates of small blue crystals found on a metallic mining artifact (R160004).

CHALCOPHANITE

Zinc manganese oxide hydrate: $ZnMn^{4+}_3O_7 \cdot 3H_2O$. A secondary mineral associated with other manganese and iron oxides.

COCHISE COUNTY: Warren District, Bisbee, Cole shaft, as crystalline material perched on geothite-rich gossan (Graeme 1981; AM 34586).

GILA COUNTY: Banner supersystem, 79 District, southwest workings of the Kullman-McCool (Finch, Barking Spider) Mine (also called D. H. claims), as octahedral microcrystals (3rd ed.); 79 Mine, as brilliant lustrous black microcrystals with hemimorphite and smithsonite (R. Gibbs, pers. comm., 2017).

FIGURE 3.55 Chalcophanite, Cole shaft, Bisbee, Cochise County 1.3 cm.

MARICOPA COUNTY: Belmont Mountains, Osborne District, Tonopah-Belmont Mine, as striated rhombohedra up to 1.5 mm (G. B. Allen and Hunt 1988).

PINAL COUNTY: In a railway cut between San Manuel and Mammoth (UA 9726). Reymert supersystem, Mineral Hill District, Reymert Mine, with psilomelane, hollandite, coronadite, and todorokite (K. S. Wilson 1984).

YUMA COUNTY: Kofa District, King of Arizona Mine, with todorokite, groutite, aurorite, and pyrolusite (Hankins 1984). Sheep Tanks District, Sheep Tanks Mine, vein B, as tabular crystals with hollandite and aragonite in veins of banded black calcite (Cousins 1972).

CHALCOPHYLLITE

Copper aluminum arsenate sulfate hydroxide hydrate: $Cu_{18}Al_2(AsO_4)_4$ $(SO_4)_3(OH)_{24} \cdot 36H_2O$. A rare secondary mineral formed in the oxidized zone of copper deposits with other oxidized copper minerals.

COCHISE COUNTY: Warren District, Bisbee, Calumet and Arizona Mine, associated with cuprite and connellite (Palache and Merwin 1909).

COCONINO COUNTY: Grand Canyon National Park, Grandview District, Grandview (Last Chance) Mine, as small well-formed crystals with chalcoalumite (M. Cline, pers. comm., 2020).

GRAHAM COUNTY: Turtle Mountain area, east of Bonito Creek, as crystals in vesicles in basalt (Bideaux, Williams, and Thomssen 1960).

CHALCOPYRITE

Copper iron sulfide: $CuFeS_2$. Chalcopyrite is the most important ore mineral of copper and a major constituent of nearly all copper sulfide deposits. It is predominantly of primary origin in veins and replacement bodies, as disseminated particles in various rock types, and in contact-metamorphic zones.

FIGURE 3.56 Chalcopyrite, Holland Mine, Santa Cruz County, 8 cm.

APACHE COUNTY: Garnet Ridge SUM District, Garnet Ridge, in vein fillings and in the cement of the Navajo Sandstone, both along its contact with a breccia dike and in nearby areas (Gavasci and Kerr 1968).

APACHE AND NAVAJO COUNTIES: Monument Valley and Cane Valley Districts, associated with chalcocite, bornite, and copper in uranium-vanadium ores in Shinarump Conglomerate, Moenkopi Formation, and De Chelly Member of the Cutler Formation (Mitcham and Evensen 1955; Evensen and Gray 1958).

COCHISE COUNTY: Warren District, Bisbee, as massive orebodies; mined at the Copper Queen, Calumet and Arizona, Junction (Mitchell 1920), and other properties. Tombstone District, as the most abundant copper mineral (Bull. 181). Courtland District, as the principal ore mineral in pyritic bodies (Bull. 181). Little Dragoon Mountains, Cochise supersystem, Johnson Camp District, as an important ore mineral at several properties (J. R. Cooper 1957); Keystone and St. George properties (Romslo 1949). Dragoon Mountains, Dragoon District, Primos Mine (Palache 1941a). Huachuca Mountains, Reef District, Reef Mine (Palache 1941a). Chiricahua Mountains, California supersystem, Leadville District, Humboldt Mine (Hewett and Rove 1930).

GILA COUNTY: Globe Hills District, Old Dominion Mine, as large masses in the Mescal Limestone (Woodbridge 1906). Summit District, Summit Mine, forming the bulk of the ore (Bull. 181). Miami-Inspiration supersystem, Miami District, Miami Mine; Inspiration District, Inspiration Mine, in the protore (Bull. 181); Pinto Valley District, Castle Dome Mine, as the principal ore mineral (Bull. 181); Copper Cities District, Copper Cities porphyry copper deposit (N. P. Peterson 1954). The presence of a chalcopyrite-enriched zone beneath the supergene chalcocite blanket in the Inspiration Mine has led some researchers to believe that the chalcopyrite is of supergene origin; thin chalcopyrite films have formed on the surfaces of pyrite crystals in the mine (Olmstead and Johnson 1966). Green Valley District, as the chief ore mineral of copper-bearing gold veins (Bull. 181). Dripping Spring Mountains, Christmas District, Christmas Mine, as the most abundant ore mineral (Tainter 1948; Knoerr and Eigo 1963; Perry 1969); Banner supersystem, 79 District, 79 Mine, rarely as disphenoidal crystals up to 0.5 in. coated by smithsonite (Stanley B. Keith 1972).

GRAHAM COUNTY: In ores of the Aravaipa supersystem, Stanley District (C. P. Ross 1925a) and Grand Reef District (Denton 1947a; Simons 1964). Safford supersystem, San Juan District, San Juan Mine, in a low-grade sulfide deposit (Rose 1970); the most common copper sulfide in all the districts in the Safford porphyry copper cluster (R. F. Robinson and Cook 1966).

GREENLEE COUNTY: Copper Mountain District, Morenci, in the lower levels of veins and disseminated in limestone near porphyry contacts (Lindgren 1903, 1904; Guild 1910); it is the only important primary copper mineral in the Morenci Mine, where it is associated with pyrite, molybdenite, and sphalerite as disseminations and small veinlets (Moolick and Durek 1966).

LA PAZ COUNTY: Buckskin Mountains, Planet District, Planet Mine (H. Bancroft 1911). Harquahala District, Golden Eagle Mine (Bull. 181); Granite Wash Mountains, Yuma King District, Yuma Copper Mine, main primary copper mineral with magnetite and garnet in replacement skarns in metamorphosed Redwall Limestone (Rasmussen and Hoag 2011).

MOHAVE COUNTY: Cerbat Mountains, Wallapai supersystem, common in nearly all the copper mines and prospects (Bull. 181); Ithaca Peak District, in veins with pyrite, sphalerite, and galena in a quartz monzonite stock (Eidel 1966). Chloride District, Johnny Bull–Silver Knight property (Tainter 1947c). Grand Wash Cliffs, Bentley District, Bronze L and Copper King Mines, as the main ore mineral (Bull. 181). Bill Williams Fork, associated with copper, cuprite, chrysocolla, malachite, brochantite, chalcocite, covellite, and pyrite (Genth 1868). Hualapai (Antler) District, Copper World Mine near Yucca, associated with sphalerite, pyrrhotite, and very minor löllingite (Rasor 1946).

PIMA COUNTY: Santa Rita Mountains, Helvetia-Rosemont District, as the most important copper mineral in the mines and prospects (Schrader and Hill 1915); as the principal ore mineral at several properties, including the Rosemont mines, Copper World Mine, and Leader Mine (Creasey and Quick 1955). Pima supersystem, at several properties (Webber 1929; Eckel 1930; Guild 1934), including the Esperanza Mine (Tainter 1947a; D. W. Lynch 1967); Pima Mine, as a primary ore mineral closely associated with grossular hornfels (Journeay 1959; Himes 1972); Mission Mine, as the principal ore mineral (Richard and Courtright 1959); and Twin Buttes Mine (Stanley B. Keith, pers. comm., 1973). Catalina District, Pontatoc Mine, with epidote and hematite in gneiss associated with limestone (Guild 1934). Marble Peak District, Marble Peak, at Oracle Ridge Mine in contact deposits with garnet, magnetite, and epidote (Dale, Stewart, and McKinney 1960). Ajo District, as scattered grains in the New Cornelia Quartz Monzonite (Joralemon 1914; Gilluly 1937). South Comobabi Mountains, Cababi District, Little Mary Mine (S. A. Williams 1963). Silver Bell District, El Tiro and Oxide open-pit mines (Kerr 1951; Richard and Courtright 1966).

PINAL COUNTY: Pioneer supersystem, Magma District, Magma Mine, as massive replacements of limestone (Short et al. 1943; Sell 1961) and as very fine crystals up to 2.5 cm long (Barnes and Hay 1983); Silver King District, Silver King Mine (Guild 1917); Resolution District, Resolution Mine, as the main ore mineral especially enriched in the diabase where it is associated with biotite and rutile (Hehnke et al. 2012). Galiuro Mountains, Bunker Hill (Copper Creek) District, Copper Creek, Copper Prince Mine, where several thousand tons of nearly pure chalcopyrite were mined (Joralemon 1952); Childs-Aldwinkle Mine (Kuhn 1941); Old Reliable Mine (Denton 1947b). Mineral Creek supersystem, Ray District, Ray Mine, in protore (Bull. 181). San Manuel District, San Manuel Mine, of hypogene origin; the most important ore mineral (Chapman 1947; L. A. Thomas 1966).

SANTA CRUZ COUNTY: Wrightson District, American Boy Mine (Schrader and Hill 1915). Patagonia Mountains, Querces District, Santo Niño Mine, with large bodies of massive molybdenite (Blanchard and Boswell 1930). Washington Camp District, Indiana Mine, with pyrite, and sphalerite (Stanley B. Keith, pers. comm., 1973). Holland Mine (Schrader and Hill 1915). Ruby District, Idaho and Montana Mines; Oro Blanco District, Annie Laurie prospect near Footes Spring (H. V. Warren and Loofbourrow 1932; R. Y. Anderson and Kurtz 1955).

YAVAPAI COUNTY: Black Hills, Jerome supersystem, Verde District, United Verde Mine, as the principal ore mineral in the pyritic orebody (C. A. Anderson and Creasey 1958); also abundant at the Copper Chief and Shea properties (Bull. 181). Bradshaw Mountains, Agua Fria, Black Canyon, and Tiger Districts (Bull. 181). Iron King District, Iron King Mine (Creasey 1952). Northeastern Wickenburg Mountains, Monte Cristo District, Monte Cristo Mine, with silver and nickel arsenides (Bastin 1922).

CHALCOSIDERITE

Copper iron phosphate hydroxide hydrate: $CuFe^{3+}_6$ $(PO_4)_4(OH)_8 \cdot 4H_2O$. This is the iron-bearing analog of turquoise but much less common.

FIGURE 3.57 Chalco-siderite, Silver Bell Mine, Pima County, FOV 7 mm.

COCHISE COUNTY: Reported in the Warren District, Bisbee, in small quantities at the Shattuck Mine (UA 1468) and Cole shaft, green bladed crystals with quartz (Graeme 1981). Courtland District, Turquoise Moun-tain, as small bright-green crystals associated with turquoise (R. Gibbs, pers. comm., 2016).

GILA COUNTY: Globe Hills District, Old Glory Mine (R. Jenkins, pers. comm., 2017).

PIMA COUNTY: Silver Bell District, Silver Bell Mine, Daisy pit, as small green crystals in a quartz vein associated with turquoise and native cop-per (R. Gibbs, pers. comm., 2017).

SANTA CRUZ COUNTY: Patagonia Mountains, Red Mountain District, along a road on the west slope of Red Mountain, as small green microcrystals (R. Jenkins, pers. comm., 2008).

#CHALCOSTIBITE

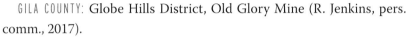

Copper antimony sulfide: $CuSbS_2$. Found with other sulfides in hydro-thermal deposits.

COCHISE COUNTY: Warren District, Bisbee, Junction shaft, in sulfide ore with chalcocite (Schumer 2017).

CHAMOSITE

Iron magnesium aluminum silicate oxide hydroxide: $(Fe^{2+},Mg,Al,Fe^{3+})_6$ $(Si,Al)_4O_{10}(OH,O)_8$. A member of the chlorite sheet-silicate group, formed with clays and iron oxides in lateritic deposits and in sedimen-tary ironstones.

COCHISE COUNTY: Reported in the Warren District, Bisbee (Graeme 1981).

MOHAVE COUNTY: Hualapai (Antler) District, in S3, 4, and 9, T17N R16W, Antler Mine, in schist (Romslo 1948).

PIMA COUNTY: Cababi District, Chicago Mine near Sells, with epidote and pumpellyite as amygdules in altered andesite (3rd ed.). Ajo District, New Cornelia Mine, where it fills fractures and sometimes replaces wall rocks (Gilluly 1946).

SANTA CRUZ COUNTY: Querces District, Santo Niño Mine, in an oligoclase-rutile rock rich in molybdenite; chamosite replaces former dark silicates in the mafic igneous rock that hosts the ore (3rd ed.).

YAVAPAI COUNTY: Old Dick District, as a fine-grained component of schists in the metamorphosed alteration zone associated with the Precambrian Bruce volcanogenic massive sulfide deposit (Larson 1976). Jerome supersystem, United Verde Mine area, where the variety thuringite is present as a constituent of black schist, a chlorite-rich aggregate that has replaced rocks in the region (C. A. Anderson and Creasey 1958).

CHENEVIXITE

FIGURE 3.58 Chenevixite, near Patagonia, Santa Cruz County, FOV 5 mm.

Copper iron arsenate hydroxide: $CuFe^{3+}(AsO_4)(OH)_2$. Chenevixite is a widespread but uncommon mineral of secondary origin in the oxidized portions of copper deposits. It is a relatively common mineral in leached cappings over enargite-pyrite mineralization.

SANTA CRUZ COUNTY: Santa Rita Mountains, Cottonwood Canyon, Tyndall District, Glove Mine (Terry Wallace, pers. comm., 1989; K. Cole, UA 13136). Patagonia Mountains, Red Mountain District, Alum Gulch, as minute green spherules in porous quartzose rock with contact silicates and specular hematite (Loghry 1972; UA 8218). Trench District, Humboldt Mine, as microcrystals (Jessie Hardman, UA 14165). Sunnyside District, Volcano-Sunnyside Mine, as 1.5 cm masses in goethite and pseudomorphs after cuprite (3rd ed.); one mile northeast of Volcano-Sunnyside Mine, with luetheite in highly argillized rhyolite (Joe Ruiz, pers. comm., 2015).

#CHEVKINITE-(Ce)

Cerium titanium iron oxide silicate: $Ce_4(Ti,Fe^{2+},Fe^{3+})_5O_8(Si_2O_7)_2$. A rare mineral found in granite pegmatites, certain granites, and volcanic ash deposits.

MOHAVE COUNTY: Aquarius Mountains, in a vein that traverses a granitic dike; intimately associated with titanite, monazite, cronstedtite, and quartz (Kauffman and Jaffe 1946).

CHLORAPATITE

Calcium phosphate chloride: $Ca_5(PO_4)_3Cl$. A member of the apatite group associated with certain marbles; formed in pegmatites and in veins that cut gabbroic rocks.

COCONINO COUNTY: Winona meteorite (Prinz, Waggoner, and Hamilton 1980).

MARICOPA AND YAVAPAI COUNTIES: White Picacho District, in pegmatites (London 1981).

CHLORARGYRITE

Silver chloride: AgCl. A complete solid-solution series extends from chlorargyrite to bromargyrite (AgBr). It is formed in the oxidized zones of silver deposits from the alteration of primary silver minerals.

COCHISE COUNTY: Tombstone District, in oxidized ores (B. S. Butler, Wilson, and Rasor 1938; Rasor 1938; Romslo and Ravitz 1947); Santa Ana Mine (UA 9162); Bradshaw Mine near Charleston, in granular aggregates (Bull. 181); Contention Mine, as microcrystals, with chrysocolla (Richard Thomssen, UA 29). Pearce Hill, Pearce District, Commonwealth Mine, with bromargyrite, bromian chlorargyrite, iodargyrite, and acanthite in quartz veins (Endlich 1897); with conichalcite (Howell 1977). Dragoon Mountains, Gleeson District, in oxidized lead-silver ores (Bull. 181). Warren District, Bisbee, Cole Mine, in pockets in massive sulfides (Hutton 1957) and on malachite; Campbell Mine (Schwartz and Park 1932); Shattuck Mine, cementing silica breccia (UA 5566).

GILA COUNTY: Globe Hills District, in many surficial ores (N. P. Peterson 1962); Ramboz (Silver Glance) Mine northeast of the Old Dominion Mine, with argentiferous manganese oxides (Bull. 181; N. P. Peterson 1962). Richmond Basin District, Stonewall Jackson Mine, with acanthite, embedded in siderite gangue; formed by alteration of silver with which it is associated (Guild 1917); Richmond Basin District, where massive chlorargyrite formed plates 0.5 in. thick and several inches in diameter (Bull. 181). In the 1950s a local cowboy in the Globe area discovered a mass of pure chlorargyrite that was loose in the soil and weighed more than 50 pounds (3rd ed.). Payson area, Green Valley District, Silver Butte Mine, with silver (Lausen and Wilson 1925).

GRAHAM COUNTY: Aravaipa supersystem, Iron Cap District, Orejana adit, with bromargyrite, and Windsor shaft (C. P. Ross 1925a).

GREENLEE COUNTY: Copper Mountain District, Clifton area, Stargo District, Stargo Mine (UA 5523).

LA PAZ COUNTY: Trigo Mountains, Silver District, Silver Clip and Red Cloud Mines, as the principal silver mineral in oxidized lead ores (E. D. Wilson 1933); Padre Kino Mine, as massive and crudely crystalline material with cerussite and as microcrystals (Kenneth Bladh, UA x3293).

MOHAVE COUNTY: Hualapai Mountains, Maynard District, at several properties (Schrader 1908). Cerbat Mountains, Wallapai supersystem, at several properties, principally in the Mineral Park and Stockton Hill Districts (Engineering and Mining Journal 1892c); Chloride District, Distaff Mine, where, with silver, it was the principal silver mineral (Bastin 1925).

PIMA COUNTY: Santa Rita Mountains, Helvetia-Rosemont District, Blue Jay Mine (Bull. 181); Greaterville District, Conglomerate (Snyder) Mine (Bull. 181; Schrader and Hill 1915). Empire District, Total Wreck Mine (Schrader and Hill 1915; Schrader 1917). Quijotoa District, Morgan Mine (Bull. 181). Cerro Colorado District, Cerro Colorado and other mines (Guild 1910). Papago District, Sunshine-Sunrise Group, with galena, chalcocite, cerussite, and anglesite (Ransome 1922). South Comobabi Mountains, Cababi District, Mildren and Steppe claims (S. A. Williams 1963).

PINAL COUNTY: Pioneer supersystem, Belmont District, Belmont Mine area, as the chief near-surface silver ore mineral (Romslo and Ravitz 1947). Mammoth District, Mammoth–St. Anthony Mine, Collins vein, as tiny yellowish cubo-octahedral crystals implanted on caledonite (Bideaux 1980). Vekol District, Vekol Mine, as 1 mm crystals on jarosite (AM 24967). Tortolita Mountains, Owl Head District, Apache prospect (Bull. 181). Mineral Creek supersystem, Ray District, Ray Mine, as microcrystal pseudomorphs (UA x4440). Bunker Hill (Copper Creek) District, Copper Creek, Bluebird Mine (Kuhn 1951).

SANTA CRUZ COUNTY: Santa Rita Mountains, Tyndall District, Ivanhoe Mine (Guild 1910). Mansfield District, Anaconda Group (Bull. 181). Patagonia Mountains, Red Rock District, La Plata and Meadow Valley Mines (Bull. 181); probably Red Rock District, Prince Rupert Mine, Crittenden, southwest Canelo Hills, with wire silver (Engineering and Mining Journal 1892b). Hardshell District, Hermosa and American Mines (Bull. 181). Palmetto District, Palmetto Mine (Schrader and Hill 1915; Schrader 1917). Ruby District, Noon Camp, as bromian chlorargyrite (S 48704).

Bradshaw Mountains, Hassayampa District, Dos Oris Mine, with acanthite and silver (Bull. 181); Black Canyon District, Thunderbolt Mine, with proustite and silver (Bull. 181); Tuscumbia District, Tuscumbia Mine, with stephanite (Bull. 181). Tip Top District, Tip Top Mine, with ruby silver (Bull. 181); Ticonderoga District, Silver Belt Mine, where the presence of stone hammers and gads in old workings indicates that the deposit was mined in ancient times (Bull. 181); Peck District, Swastika Mine, as fine crystals (Bull. 181). Eureka supersystem, Hillside District, Hillside Mine, associated with anglesite, cerussite, smithsonite, silver, and hemimorphite (Axelrod et al. 1951). Tiger District, Lida Mine (Engineering and Mining Journal 1892d).

CHLORITOID

Iron aluminum oxide silicate hydroxide: $Fe^{2+}Al_2O(SiO_4)(OH)_2$. It is a fairly common product of intermediate-grade thermal metamorphism of aluminum- and iron-rich pelitic sedimentary rocks.

COCHISE COUNTY: Chiricahua Mountains, as crystals in phyllite (Bideaux, Williams, and Thomssen 1960).

MARICOPA COUNTY: Southwest slope of Piestewa Peak (formerly Squaw Peak), Phoenix, as manganese-rich porphyroblasts up to 5 mm in schist, with muscovite and quartz (Thorpe 1980). On a nearby hill in Piestewa Peak Park, in schist with staurolite, "biotite," and garnet (Thorpe and Burt 1980).

CHOLOALITE

Lead calcium copper antimony tellurite chloride: $(Pb,Ca)_3(Cu,Sb)_3$ $Te_6O_{18}Cl$. A very rare mineral associated with other tellurites and tellurates in oxidized ores.

COCHISE COUNTY: Tombstone District, on a dump about halfway between the Joe and Grand Central shafts; associated with granular jarosite and opal in brecciated shale cemented by vuggy quartz; subhedra of choloalite are corroded by crusts of emmonsite (S. A. Williams 1981a).

CHONDRODITE

Magnesium silicate fluoride: $Mg_5(SiO_4)_2F_2$. A member of the humite group; an uncommon mineral formed in contact-metamorphosed limestones and dolomites and in skarns related to ore deposits.

APACHE COUNTY: Buell Park District, Buell Park, found in SUM (serpentine ultramafic microbreccia) diatreme with clinohumite and characterized by high titanium content (Aoki, Fujino, and Akaogi 1976).

COCHISE COUNTY: Cochise supersystem, Johnson Camp District, Johnson Camp area, in tactite formed in the Martin Formation, in bright pink lenses with an unidentified silicate mineral (J. R. Cooper and Silver 1964).

GILA COUNTY: Southeastern Dripping Spring Mountains, Banner supersystem, Christmas District, Christmas Mine, part of the mineral assemblage in the contact-metamorphic rocks (Eastlick 1968).

PINAL COUNTY: Lakeshore District, Lakeshore Mine, locally abundant in forsterite marble (3rd ed.).

CHROMITE

Iron chromium oxide: $Fe^{2+}Cr_2O_4$. A member of the spinel group. Commonly associated with peridotites or serpentines, in which it may form in veins or as segregations. Commonly associated with magnetite, olivine, garnet, vesuvianite, ilmenite, and pyroxene.

APACHE COUNTY: Monument Valley area, Garnet Ridge SUM District, Garnet Ridge, as fragments in a breccia dike that pierces sedimentary rocks (Gavasci and Kerr 1968).

COCONINO COUNTY: Meteor Crater, associated with krinovite, roedderite, high albite, richterite, and kosmochlor in the Canyon Diablo meteorite (Olsen and Fuchs 1968).

LA PAZ COUNTY: Trigo Mountains, Yuma Wash District, as disseminated grains and small masses with mariposite (a variety of muscovite) in Orocopia mica schist associated with albite and graphite (Bull. 181).

CHRYSOBERYL

Beryllium aluminum oxide: $BeAl_2O_4$. An uncommon mineral formed in granite pegmatites, in certain metamorphic rocks, and as detrital grains.

MOHAVE COUNTY: Virgin Mountains District, Hummingbird claims, with beryl crystals up to 15 in. long in pegmatite dikes (J. C. Olson and Hinrichs 1960; Meeves 1966).

YAVAPAI COUNTY: Tungstona District, northeast of Bagdad Mine, as fracture fillings of yellow-green intergrown crystals showing occasional crystal faces associated with small beryl crystals in miarolitic granite (R. Jenkins, pers. comm., 2017).

CHRYSOCOLLA

Copper aluminum hydrogen silicate hydroxide hydrate: $(Cu_{2-x}Al_x)H_{2-x}Si_2O_5(OH)_4 \cdot nH_2O$. A widespread and locally abundant secondary mineral found in practically all oxidized copper deposits of the state.

APACHE COUNTY: Garnet Ridge U District, in veins cutting Navajo Sandstone and as cement along the contacts with a nearby breccia dike; associated with malachite, tangeite, tyuyamunite, limonite, chalcopyrite, and pyrite (Gavasci and Kerr 1968).

FIGURE 3.59 Chrysocolla, overgrown with quartz, Ray Mine, Pinal County, 3.2 × 2.5 cm.

APACHE AND NAVAJO COUNTIES: Monument Valley and Cane Valley Districts, associated with oxidized uranium-vanadium ores in sandstone (Mitcham and Evensen 1955).

COCHISE COUNTY: Warren District, Bisbee, with cuprite, disseminated flakes of copper, azurite, malachite, and small crystals of brochantite, and locally, with tufts of connellite (Ransome 1904; Holden 1922; Schwartz 1934). Tombstone District, widely distributed but not abundant in mine properties (B. S. Butler, Wilson, and Rasor 1938). Cochise supersystem, Johnson Camp District, Johnson Camp area, in copper sulfide orebodies, commonly associated with other oxidized copper minerals (Kellogg 1906; J. R. Cooper and Huff 1951). Sulphur Springs Valley, in small amounts in numerous veinlets cutting porphyritic andesite flows (Lausen 1927b).

COCONINO COUNTY: Vermilion Cliffs District, near Jacob Lake, as a principal constituent of low-grade copper deposits in the Redwall and Kaibab Limestones, associated with malachite and azurite (Fischer 1937). White Mesa District, with malachite-cementing sandstone (J. M. Hill 1914a; Mayo 1955a).

GILA COUNTY: Miami-Inspiration District (Woodbridge 1906), abundant in the Live Oak, Keystone, Black Warrior, and Geneva Mines (R. C. Wells 1937); Bulldog tunnel of the Inspiration Mine, with malachite, chalcedony, and quartz, as aggregates of great beauty (Ransome 1919; N. P. Peterson 1962; Sun 1963); Black Copper portion of the Inspiration orebody, as thin coatings on fractures in Pinal Schist, locally as veins up to 5 cm wide, and as matrix-cementing subangular schist pebbles in an old stream channel composed of White Tail Conglomerate, interlayered with heulandite in some outcrops; analysis showed that the black variety contains 6.7 percent MnO_2, whereas the blue-green variety contains 0.06 percent MnO_2 (Throop 1970; Throop and Buseck 1971; Kemp 1905). Globe Hills District, Old Dominion Mine, as an important ore mineral in the oxidized zone (Schwartz 1921, 1934). Carlota District, Carlota Mine, as the most important copper ore mineral, along with malachite, locally associated with azurite, malachite, and rare powellite and prosopite in a brecciated diabase (Ruiz, pers. comm., 2016; N. P. Peterson 1962). Banner supersystem, 79 District, 79 Mine, as pseudomorphs after hemimorphite and as massive and botryoidal material, some coating and replacing wulfenite (Stanley B. Keith 1972; UA 7587). Southeastern Dripping Spring Mountains, Banner supersystem, Christmas District, Christmas Mine, as a moderately common supergene mineral associated with andradite-bearing skarns in the Naco Formation (David Perry, pers. comm., 1967).

GRAHAM COUNTY: Gila Mountains, Lone Star District near Safford, in metamorphosed latites and andesites, with various other copper minerals (Hutton 1959a); Safford porphyry copper cluster, as the most abundant copper mineral, making up 50 to 85 percent of all the oxide mineralization; as veins, blebs, and coatings, and commonly intimately associated with kaolinite and halloysite-10Å (R. F. Robinson and Cook 1966).

GREENLEE COUNTY: Copper Mountain District, Morenci, as common fine glassy-green specimens (Kunz 1885; Lindgren and Hillebrand 1904; Lindgren 1905; Reber 1916); often found as pseudomorphs after malachite and brochantite (Moolick and Durek 1966).

LA PAZ COUNTY: Bill Williams Fork (Genth 1868), Buckskin Mountains, Planet District, Planet Mine (H. Bancroft 1911; Cummings 1946a, 1946b) as beautifully banded material associated with cuprite in limestone-replacement orebodies (McCarn 1904). Little Harquahala Mountains, Little Harquahala District, south of Salome (NHM 1966, 113); Harquahala Mine, as sky-blue translucent crusts and coatings in fractures in hematitic Bolsa Quartzite, near a microdiorite dike, associated with dioptase, malachite, plancheite, and pseudomalachite (Stanley B. Keith, pers. comm., 2019). New Water District, Eagle Eye Mine, as 2 mm long fibers in vugs (3rd ed.). Granite Wash Mountains, Yuma King District, Yuma Copper Mine, as lapidary-grade material interlaced with azurite (Stanley B. Keith, pers. comm., 2019).

MARICOPA COUNTY: Painted Rock District, Rowley Mine near Theba, where small masses and stringers, locally associated with malachite, have yielded quartz-rich, gem-quality material; most of the deposit, however, is friable to powdery (W. E. Wilson and Miller 1974).

MOHAVE COUNTY: Cerbat Mountains, Emerald Isle District, as excellent specimens at the Emerald Isle Mine, as exotic copper deposits in alluvium, associated with dioptase and "copper pitch," which cements alluvial detritus. Mineral Park District, where chrysocolla has formed in vein and blanket deposits in a granite porphyry (B. E. Thomas 1949; Searls 1950; Newberg 1967).

PIMA COUNTY: Sierrita Mountains, Pima supersystem, Twin Buttes District (Eckel 1930a); Twin Buttes Mine, in large masses with brochantite and cuprite (Henry Worsley, pers. comm., 1972; UA 9608), and as large glassy masses with malachite infilling chrysocolla breccias and diopside skarns (Stanley B. Keith, pers. comm., 2019). Silver Bell District, El Tiro Mine, as clear emerald-green material (UA 8869). Santa Rita Mountains, Helvetia-Rosemont District, Helvetia, with limonite, forming warty masses and encrustation pseudomorphs after gypsum crystals (3rd ed.; UA 3571). South Comobabi Mountains, Cababi District, Mildren and Steppe claims (S. A. Williams 1963). Ajo District, widespread in the ore-body of the New Cornelia Mine (Joralemon 1914; Gilluly 1937). Santa Catalina Mountains, Pusch Ridge (UA 9328), probably Pontatoc Canyon in Catalina District.

PINAL COUNTY: San Manuel District, San Manuel Mine, as the most abundant copper mineral in the oxidized zone of the orebody, locally with atacamite (Chapman 1947; Schwartz 1947, 1949, 1953; L. A. Thomas 1966; J. D. Lowell 1968; Throop and Buseck 1971). Mammoth District, Mammoth–St. Anthony Mine, in places as gem-quality material (N. P. Peterson 1938a, 1938b; Galbraith and Kuhn 1940; Fahey, Daggett, and Gordon 1950). San Tan supersystem, Poston Butte (Florence) District, Florence (R. Grant, pers. comm., 2020, UA 6049). Galiuro Mountains, Virgus Canyon, Table Mountain District, Table Mountain Mine, as the only abundant copper mineral (Simons 1964). Mineral Creek supersystem, Ray District, Ray Mine, as spherulitic aggregations of highly birefringent material, as green and black varieties, with tenorite, malachite, halloysite-10Å, and heulandite (Schwartz 1934; Clarke 1953; Stephens and Metz 1967; Throop and Buseck 1971); 2,200 ft. level of the giant diabase-hosted silica orebody, as showy thumbnail- to cabinet-sized clusters of pseudomorphs, possibly after azurite or gypsum crystals (W. E. Wilson 1977; W. A. Thompson 1980); Copper Butte District, Copper Butte Mine, as infillings with malachite in the matrix of Oligocene-age Pinal Schist–clast hematitic conglomerate (Phelps 1946). Sonora exotic copper deposit in Mineral Creek at Ray, with malachite as fillings in Holocene-age alluvium (C. H. Phillips, Cornwall, and Rubin 1971).

SANTA CRUZ COUNTY: Cottonwood Canyon, Tyndall District, Glove Mine (H. J. Olson 1966). Patagonia Mountains, Washington Camp District, Duquesne and Washington Camp area (Schrader 1917)

YAVAPAI COUNTY: Black Hills, Dundee-Arizona District, Arizona-Dundee property, as mammillary fillings in Tertiary Conglomerate (C. A. Anderson and Creasey 1958). Bradshaw Mountains, Copperopolis District, Whipsaw and Copperopolis properties, as bright-blue material (Lindgren 1926). Eureka supersystem, Bagdad District, Bagdad Mine—large quantities of lapidary material were recovered (Muntyan 2015a).

YUMA COUNTY: Frisco District, Copper Mountain, Betty Lee Group and other properties (E. D. Wilson 1933).

CHRYSOTILE

Magnesium silicate hydroxide: $Mg_3Si_2O_5(OH)_4$. A member of the kaolinite-serpentine group. The fibrous habit of chrysotile is due to the rolled-up tube nature of its kaolinite-like sheet structure. Serpentine is of secondary origin, derived from the alteration of nonaluminous magnesian silicates, particularly olivine, amphibole, or pyroxene. In places, it is formed as large masses derived from peridotite or other mafic igneous rocks. A common product of contact metamorphism of magnesian limestones, it is also found in hydrothermal veins. The most notable serpentine mineral in Arizona is chrysotile, although its exact composition may be unknown. Chrysotile has been extensively mined as asbestos, from late-stage fiber-form seams in veins in lizardite with associated antigorite in contact-metasomatic horizons in Mescal Limestone adjacent to Late Precambrian diabase throughout north-central Gila County. Extensive lists of asbestos localities are included in Funnell and Wolfe (1964) and in Bulletins 126 (Wilson and Butler 1928) and 180 (Shride 1969), published by the Arizona Bureau of Mines (currently the Arizona Geological Survey).

APACHE COUNTY: Garnet Ridge SUM District, Garnet Ridge, abundant in a boulder, as pseudomorphs after an orthopyroxene; related to a breccia dike that pierces sedimentary rocks (Gavasci and Kerr 1968).

COCHISE COUNTY: Tombstone District, Lucky Cuss Mine, in altered limestone (B. S. Butler, Wilson, and Rasor 1938). Chiricahua Mountains, California supersystem, with contact-metamorphic ores (Bull. 181). Dos Cabezas Mountains, in metamorphosed limestones (Guild 1910). Little Dragoon Mountains, Cochise supersystem, Johnson Camp District, Johnson Camp area, Moore orebody, as a retrograde product of high-temperature metamorphism of forsterite, tremolite, and diopside (J. R. Cooper and Huff 1951; J. R. Cooper and Silver 1964). Warren District, Bisbee, Holbrook extension of the Lavender pit, the polytype clinochrysotile, as scaly micaceous intergrowths with stevensite and chrysocolla (3rd ed.); Gardner shaft 700 level, 748 drift, as 1 in. fibers along a 33 ft. zone flanked by tremolite and diopside (Graeme 1981).

COCONINO COUNTY: Grand Canyon region, Bass and Hance asbestos properties, where the Precambrian Bass Limestone has been altered adjacent to diabase sills (Selfridge 1936); cross-fiber asbestos is locally up to 4 in. long but is commonly shorter (Funnell and Wolfe 1964).

GILA COUNTY: The most extensive and commercially important deposits of chrysotile in Arizona are north and northeast of Globe along the Salt River near Chrysotile and along Cherry and Ash Creeks (Engineering and Mining Journal 1915; Melhase 1925; L. A. Stewart and Haury 1947; L. A. Stewart 1956). The deposits originated through metamorphic action of diabase intrusives upon Precambrian Mescal Limestone. Pinal and Mescal Mountains and Pinto Creek region (Bull. 181). Sierra Ancha Mountains, Sierra Ancha supersystem, Asbestos Point, as fiber-form seams up to 4 in. thick; at the head of Pocket Creek (Bateman 1923; Sampson 1924); Salt River supersystem, Salt River Asbestos District, where it was extensively mined at Chrysotile. San Carlos Apache Reservation, Emsco and Bear Mines (Bromfield and Schride 1956). Southeastern Dripping Spring Mountains, Banner supersystem, Christmas District, Christmas Mine, replacing dolomite and forsterite in the orebodies (Perry 1969). Globe Hills District, Old Dominion Mine, in Mescal Limestone near diabase sills (Bull. 181).

GREENLEE COUNTY: Copper Mountain District, Morenci, northern part of the Morenci open-pit area, locally formed with other calcsilicate minerals in an extensive contact-metamorphic assemblage (Moolick and Durek 1966); on the ridge just west of Morenci, at the Thompson Mine, as green-banded material associated with magnetite (Lindgren 1905; Reber 1916; Creasey 1959).

PIMA COUNTY: Pima supersystem, Twin Buttes District, Twin Buttes Mine, in metamorphic rocks (Stanley B. Keith, pers. comm., 1973).

PINAL COUNTY: Mammoth District, Mammoth–St. Anthony Mine, as aggregated masses of shining white balls of antigorite that form a matrix for wulfenite and descloizite (Bideaux 1980). Hewett Wash area (L. A. Stewart 1956).

YAVAPAI COUNTY: Copper Basin District, as an alteration product in ores in breccia pipes (Johnston and Lowell 1961).

CINNABAR

FIGURE 3.61 Cinnabar, San Juan Mine, Cochise County, 3.5 × 2.2 cm.

Mercury sulfide: HgS. Cinnabar is usually of near-surface origin as veins, replacement deposits, or impregnations and is commonly associated with rocks and hot spring deposits of recent volcanic origin.

COCHISE COUNTY: Middle Pass North supersystem, Abril District, San Juan Mine (J. Callahan, pers. comm., 2020).

GILA AND MARICOPA COUNTIES: Mazatzal Mountains District, in several properties, typically as veinlets or thin films on fracture surfaces and as discontinuous, somewhat definite ore shoots in a belt of schist, which consists of various rock types. Cinnabar is the only important ore mineral in the deposits, but calomel, mercury, and metacinnabar are present in small amounts, as are pyrite, chalcopyrite, azurite, and malachite (Lausen 1926); several producers are situated on Alder and Sycamore Creeks (Ransome 1916; Von Bernewitz 1937; Dreyer 1939; Faick 1958; Beckman and Kerns 1965).

LA PAZ COUNTY: Cinnabar District, French, American, and Colonial properties, eight miles southwest of Quartzsite (H. Bancroft 1910; Von Bernewitz 1937; Beckman and Kerns 1965; Melchiorre 2013). Near Ehrenberg, where it is sparsely distributed in veins and mined on a small scale (3rd ed.).

MARICOPA COUNTY: Phoenix Mountains District, Rico, Mercury, and Eureka Groups of claims. Dreamy Draw area, Sam Hughes claims, north of Phoenix (Lausen and Gardner 1927, UA 1391).

MOHAVE COUNTY: Northern Black Mountains (River Range), Gold Basin District, Fry Mine (Bull. 181).

PIMA COUNTY: Roskruge Mountains, Roadside District, Roadside Mine (Beckman and Kerns 1965). Cerro Colorado Mountains, Cerro Colorado District, with malachite (UA 3095), Mary G Mine, as tiny crystals and scales lining cavities in quartz (R. E. Davis 1955).

PINAL COUNTY: Casa Grande Mountains, Mickey Welch claims, south of Casa Grande (Bull. 181). Martinez Canyon District, Martinez Mine, as microcrystals (William Kurtz, UA x4230).

YAVAPAI COUNTY: Copper Basin District (Beckman and Kerns 1965), Black Hills District, Mercury, Cinnabar, Queen, Zero Hour, and Shylock properties (Johnston and Lowell 1961). Near Morristown, White Picacho District, Westerdahl claims (Bull. 181; Beckman and Kerns 1965).

CLARINGBULLITE

Copper fluoride chloride hydroxide: $Cu^{2+}_4FCl(OH)_6$. A very rare oxide-zone mineral.

COCHISE COUNTY: Warren District, Bisbee, in a specimen of cuprite collected by Joseph E. Urban in the late 1960s from the Cole shaft, as minute blue platelets in vugs in the cuprite, with brochantite, malachite, and connellite; Southwest Mine, as crystals up to 6 mm, with atacamite, paratacamite, spangolite, and connellite (Graeme 1993).

CLAUDETITE

Arsenic oxide: As_2O_3. A secondary mineral formed from the oxidation of other arsenic minerals; also formed as a sublimation product of mine fires.

MOHAVE COUNTY: Corkscrew Cave on the Hualapai Indian Reservation (Onac, Hess, and White 2007).

YAVAPAI COUNTY: Jerome supersystem, Verde District, United Verde Mine, as silky crystals filling a small cavity above the burned pyritic orebody (Palache 1934; Buerger 1942) and as a foil, 1 cm by 5 cm (H 92682).

CLAUSTHALITE

Lead selenide: PbSe. A rare lead mineral formed in some complex ores with other sulfides and selenides.

GREENLEE COUNTY: Blue Range Wilderness Area, in a deep drill hole. A sample from a depth of 2,807 ft. showed silvery films of clausthalite lining fractures in a calcite-epidote marble; traces of pyrite and sphalerite were also present (3rd ed.).

#CLINOATACAMITE

Copper chloride hydroxide: $Cu_2Cl(OH)_3$. An uncommon mineral found in the oxidized portion of copper deposits. Polymorphous with atacamite, paratacamite, and botallackite.

COCHISE COUNTY: Warren District, Bisbee, Southwest Mine, fifth level, 14 stope, found as 2 mm prismatic and orthorhombic-appearing crystals in vugs in massive cuprite. They were associated with superb claringbullite and connellite as well as atacamite and what appears to be larger crystals of paratacamite (Graeme, Graeme, and Graeme 2015); Copper Queen Mine, dark-green blocky crystals associated with green acicular atacamite, blue claringbullite, and cuprite. Some crystals are untwinned and tabular (R090062).

MARICOPA COUNTY: Belmont Mountains, Osborne District, Tonopah-Belmont Mine, as very small bluish-green crystals with herbertsmithite in oxidized vein material (J. Ruiz, pers. comm., 2012).

CLINOBISVANITE

Bismuth vanadate: $Bi(VO_4)$. A rare secondary mineral formed under oxidizing conditions.

COCHISE COUNTY: Dos Cabezas Mountains, Teviston District, at a small prospect near the Comstock Mine, as small dull-red tablets resembling wulfenite, up to 1 mm on an edge; the tablets are perched on greasy brown nontronite, which partially fills voids in massive white vein quartz; associated with bismutite and bismuth-rich mottramite (3rd ed.).

CLINOCHLORE

Magnesium aluminum silicate hydroxide: $Mg_5Al(AlSi_3O_{10})(OH)_8$. A member of the chlorite group formed during thermal metamorphism and by hydrothermal processes. The variety penninite is common in the state and is the characteristic chlorite mica in the Orocopia Schist in southwestern Arizona.

COCHISE COUNTY: Warren District, Bisbee, as coarse crystals formed during retrograde metamorphism of garnet-epidote tactites, commonly found in the deeper levels of the Cole and Dallas Mines; the variety penninite as small scaly crystals interlayered with sericite in altered Bolsa Quartzite north of the Dividend fault, and as a matrix of breccia dikes exposed in the upper reaches of Brewery Gulch (3rd ed.).

PIMA COUNTY: East side of the Tortolita Mountains, as the variety penninite, formed as large crystals pseudomorphous after hornblende in abundant syenodiorite dikes (3rd ed.). Pima supersystem, Twin Buttes District, Twin Buttes Mine (William Hefferon, pers. comm., 3rd ed.). Santa Catalina Mountains, in limestone at the contact with the Leatherwood Quartz Diorite (Wood 1963).

PINAL COUNTY: Mammoth District, Mammoth–St. Anthony Mine, as green felted masses in the lower levels of the Collins vein, exact chlorite species not identified (N. P. Peterson 1938a, 1938b).

YAVAPAI COUNTY: Jerome supersystem, Verde District, United Verde Mine area, where the ferroan variety, ripidolite, is a constituent of black schist; associated with other members of the chlorite group (C. A. Anderson and Creasey 1958); Agua Fria District, in the footwall of the copper volcanogenic massive sulfide at Stoddard Mountain (Monte Swan, pers. comm., 2019).

CLINOCLASE

Copper arsenate hydroxide: $Cu_3(AsO_4)(OH)_3$. An uncommon mineral of supergene origin, formed in the oxidized portion of copper deposits.

FIGURE 3.62 Clinoclase, Copper Queen Mine, Yavapai County, FOV 3 mm.

GILA COUNTY: Banner supersystem, 79 District, 31 raise, between sixth and seventh levels of the 79 Mine, associated with ktenasite and olivenite (Stanley B. Keith 1972).

PINAL COUNTY: Mineral Creek supersystem, Ray District, Hull claims, south of Ray, in thin purple crusts and veinlets (AM 35561).

SANTA CRUZ COUNTY: Patagonia Mountains, Ivanhoe District, Temporal Gulch, St. Louis Mine, as druses on baryte, derived from the alteration of tennantite (Bideaux, Williams, and Thomssen 1960).

YAVAPAI COUNTY: Mayer District, Copper (Stoddard) Mountain, Old Robertson claim, four miles east-northeast of Mayer, with olivenite (Robert O'Haire, pers. comm., 1972); also with cornwallite (UA 2211). Agua Fria District, Copper Queen Mine, as brilliant small crystals with olivenite (R. Gibbs, pers. comm., 2016).

CLINOHEDRITE

Calcium zinc silicate hydrate: $CaZn(SiO_4)·H_2O$. An extremely rare mineral.

FIGURE 3.63 Clinohedrite, Christmas Mine, Gila County, FOV 3 mm.

GILA COUNTY: Southeastern Dripping Spring Mountains, Banner supersystem, Christmas District, Christmas Mine, as sprays of delicate pale-lavender prisms on fracture surfaces, with stringhamite, kinoite, and apophyllite (only a few specimens were found; 3rd ed.), and as colorless transparent crystals with kinoite and junitoite (R120105).

CLINOHUMITE

Magnesium silicate fluoride: $Mg_9(SiO_4)_4F_2$. A common member of the humite group, it typically forms in contact-metamorphosed limestones and dolomites and in skarns related to mineral deposits and is a constituent of diatremes in northeastern Arizona.

APACHE COUNTY: Buell Park District, Buell Park SUM diatreme, about fifteen miles north of Fort Defiance, as the variety titanclinohumite, a prominent constituent of the SUM diatreme plug (Balk 1954; Sun 1954; Bideaux, Williams, and Thomssen 1960; McGetchin, Silver, and Chodos 1970).

CLINOPTILOLITE

Calcium sodium potassium aluminum silicate hydrate: (Ca_3,Na_6,K_3) $(Si_{30}Al_6)O_{72} \cdot 20H_2O$. A member of the zeolite group and a very abundant mineral in Arizona formed by the devitrification and alteration of volcanic glass in tuffs. There are three species of clinoptilolite, -Ca, -Na, and -K; the specific chemistry of these occurrences is not known.

APACHE COUNTY: Near Nutrioso, Nutrioso District, with analcime in a Tertiary tuff and sandstone formation (Wrucke 1961; Sheppard 1971).

COCHISE COUNTY: San Simon Basin, Bowie District, seven miles northwest of Bowie, in bedded lake deposits with analcime, chabazite, erionite, thénardite, and halite (Regis and Sand 1967; Edson 1977; Eyde 1978).

GREENLEE COUNTY: About six miles north of Morenci, with mordenite in Tertiary tuff (Sheppard 1969, 1971).

MARICOPA COUNTY: Near Horseshoe Reservoir, S3, T7N R6E, in tuff of the Verde Formation (Sheppard 1971; F. Mumpton, pers. comm., 3rd ed.).

MOHAVE COUNTY: Big Sandy District, near Wikieup, in a Pliocene lacustrine formation (Sheppard 1971); Big Sandy District, east of Big Sandy Wash, in the east half of T16N R13W, in Pliocene tuff with analcime, chabazite, erionite, and phillipsite (C. S. Ross 1928, 1941).

PIMA COUNTY: Ajo District, Phelps Dodge Well No. 1 (William Thomas, pers. comm., 1988).

PINAL COUNTY: Northern Tortilla Mountains, where it has formed by the alteration of feldspar fragments in abundant large pumice masses contained in the middle tuff and tuffaceous sandstone member of the Ripsey Wash beds (Schmidt 1971). Mineral Creek supersystem, Ray District, Ray Mine, as aesthetic material (J. Callahan, pers. comm., 2018).

YUMA COUNTY: Near Dome, with bentonite and tuff in a late Tertiary lacustrine formation (Bramlette and Posnjak 1933; Sheppard 1971; R. C. Wells 1937).

CLINOZOISITE

Calcium aluminum silicate oxide hydroxide: Ca_2 $Al_3[Si_2O_7][SiO_4]O(OH)$. This is a member of the epidote group and is typically a product of both regional and contact metamorphism; much like epidote in mode of occurrence. Much of what has been called zoisite is probably clinozoisite.

FIGURE 3.64 Clinozoisite, Finch Mine, Gila County, FOV 3 mm.

COCHISE COUNTY: Tombstone District, Lucky Cuss Mine, as small vitreous green grains with vesuvianite, monticellite, and thaumasite (B. S. Butler, Wilson, and Rasor 1938). Warren District, Bisbee, as pseudomorphs after "biotite," with chlorite, sericite, titanite, quartz, and muscovite (Schwartz 1958). Dragoon Mountains, Stronghold Canyon, with diopside, vesuvianite, and garnet in metamorphosed Paleozoic rocks (Rushing 1978).

GILA COUNTY: Miami-Inspiration supersystem, Pinto Valley District, Castle Dome Mine, as stringers and scattered grains in sericitized plagioclase of the quartz monzonite (N. P. Peterson, Gilbert, and Quick 1946, 1951; Creasey 1959); common along the north side of the Miami-Inspiration orebody (N. P. Peterson 1962); Copper Cities District, Copper Cities deposit, formed during hydrothermal alteration, associated with chlorite, epidote, pyrite, sericite, and calcite (N. P. Peterson 1954). Banner supersystem, 79 District, 79 Mine area, occurs as crystals and massive vein fillings and rock replacements with epidote in the surrounding limestones (J. Callahan, pers. comm., 2017). Near the Kullman-McCool (Finch, Barking Spider) Mine (also known as the Finch Mine) (K. D. Smith 1995).

PIMA COUNTY: North end of the South Comobabi Mountains and north of the Ko Vaya Hills, with manganaxinite and piemontite, as vesicle fillings in andesite flow boulders (Bideaux, Williams, and Thomssen 1960). Santa Catalina Mountains, with diopside, epidote, actinolite, tremolite, microcline, and plagioclase in arkose at the contact with the Leatherwood Quartz Diorite (Wood 1963).

PINAL COUNTY: San Manuel District, San Manuel Mine, as coarse grains with epidote, as a hydrothermal alteration product (Schwartz 1953).

YAVAPAI COUNTY: Iron King District, Iron King Mine, in diorite, along the northwestern contact with the breccia facies, associated with epidote, chlorite, and hornblende (Creasey 1952).

CLINTONITE

Calcium aluminum magnesium silicate hydroxide: $CaAlMg_2(SiAl_3O_{10})$ $(OH)_2$. A member of the brittle mica group, formed in chlorite schist with talc and in metamorphosed limestones.

GILA COUNTY: Southeastern Dripping Spring Mountains, Banner supersystem, Christmas District, Christmas Mine, north side of 800 ft. level, as the variety xanthophyllite, associated with forsterite, garnet, vesuvianite, and calcite, in contact-metamorphosed diorite (David Perry, pers. comm., 1967, 1969).

COBALTITE

Cobalt arsenic sulfide: CoAsS. Most of the cobaltite localities are from the old literature and not well documented.

APACHE COUNTY: Reported in the White Mountains, exact locality unknown (Bull. 181).

GRAHAM COUNTY: Santa Teresa Mountains, Bluebird Cobalt District, Bluebird Mine, in the center of S5, T5S R21E, as massive and microcrystalline material with pyrite and arsenopyrite (MM K106; Bilbrey 1962).

MARICOPA COUNTY: Mazatzal Mountains, along the Apache Trail between Fish Creek and Roosevelt Dam (Bull. 181).

PIMA COUNTY: Comobabi Mountains, Cababi District, exact locality unknown (Bull. 181).

YAVAPAI COUNTY: Jerome supersystem, Verde District, vicinity of Jerome (Guild 1910), Black Hills, near claims of the old Prudential Copper Mining Company, along the contact between the Bradshaw Granite and greenstone of the Yavapai Schist; altered to erythrite at the surface (Bull. 181; UA 860; UA 861). Black Hills, Grapevine Cobalt District, with associated arsenopyrite and erythrite in shear veins hosted in hornblende gabbro intrusive into the Grapevine Gulch Formation (C. A. Anderson and Creasey 1958). Agua Fria District, Silver Flake Mine, with other sulfide minerals (USGS 1981).

#COBALTPENTLANDITE

Cobalt sulfide: Co_9S_8. A mineral found in hydrothermal veins and in serpentinites.

LA PAZ AND YUMA COUNTY: Cemetery Ridge District, as small grains with other sulfides in blocks of serpentinite and partially serpentinized harzburgite in the Orocopia Schist at Cemetery Ridge (Haxel et al. 2018).

COCONINOITE

Iron aluminum uranyl phosphate sulfate hydroxide hydrate: $Fe^{3+}_2Al_2(UO_2)_2(PO_4)_4(SO_4)$ $(OH)_2 \cdot 20H_2O$. An uncommon secondary mineral formed in the oxidized zones of Colorado Plateau–type, sandstone-hosted uranium-copper-vanadium deposits. The Sun Valley Mine is the type locality.

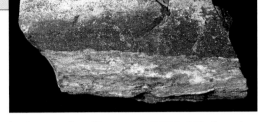

FIGURE 3.65 Coconinoite, Sun Valley Mine, Coconino County, 2.8 × 1.0 cm.

APACHE COUNTY: Cane Valley District, Black Water No. 4 Mine, in the oxidized zone of a uranium deposit in Triassic sandstone, associated with gypsum, jarosite, limonite, and clay; as aggregates of light creamy-yellow microcrystalline grains (E. J. Young, Weeks, and Meyrowitz 1966).

COCONINO COUNTY: Vermilion Cliffs District, Sun Valley Mine, "in seams 1 mm or less thick, predominantly along bedding planes of a light-colored arkosic sandstone that is fine-grained, poorly sorted, and thinly bedded" (E. J. Young, Weeks, and Meyrowitz 1966). Cameron District, Huskon No. 7 Mine (E. J. Young, Weeks, and Meyrowitz 1966).

COESITE

Silicon oxide: SiO_2. First known as a synthetic product created in the laboratory under conditions of extreme pressure by Loring Coes. The natural occurrences are associated with meteoric impacts in high-silica rocks and in high-pressure rocks such as kimberlites and eclogites. The natural occurrence of coesite was predicted by geologic inference before its actual discovery. Meteor Crater is the type locality.

COCONINO COUNTY: Meteor Crater, as an abundant constituent of sheared Coconino Sandstone in debris directly below the crater floor and beyond the crater rim, as well as in breccia deep below the crater floor; also present in lechatelierite in water-lain beds within the crater; as irregular grains, 5 to 50 microns in size (Chao, Shoemaker, and Madsen 1960).

COFFINITE

Uranium silicate hydrate: $U(SiO_4) \cdot nH_2O$. Coffinite is an important primary ore mineral of uranium, presumably of hydrothermal origin, which is associated with uraninite in the deeper unoxidized portions of sandstone uranium deposits.

APACHE COUNTY: Cane Valley District, Monument No. 2 Mine, with uraninite, corvusite, montroseite, and doloresite, as constituents of dark unoxidized uranium-vanadium ores; also in tabular bodies in sandstone-filled paleo-stream channels and scours (R. G. Young 1964). Stinking Spring Mountain (Bull. 181).

COCONINO COUNTY: Cameron District, Yazzie No. 102 Mine (Austin 1964).

GILA COUNTY: Sierra Ancha District, Workman Mine, in Dripping Spring Quartzite (Granger and Raup 1962, 1969).

MOHAVE COUNTY: Hack Canyon District, Hack Canyon Mine (Wenrich and Sutphin 1989).

NAVAJO COUNTY: North of Holbrook, Holbrook District, at the Ruth Group of claims (Bull. 181). Monument Valley District, Monument No. 1 Mine, with uraninite, corvusite, and montroseite (Evensen and Gray 1958).

COHENITE

Carbon iron: CFe_3. Found in iron meteorites, associated with terrestrial iron, and in some altered mafic volcanic rocks.

APACHE COUNTY: Padres Mesa area, as a component of the Navajo meteorite (Buchwald 1975).

COCONINO COUNTY: Meteor Crater, found in meteoritic iron oxide at Meteor Crater with goethite, maghemite, and residual meteoritic taenite and schreibersite (B. A. Hoffman 1992).

COLEMANITE

Calcium borate hydroxide hydrate: $CaB_3O_4(OH)_3 \cdot H_2O$. Colemanite is most commonly formed by the evaporation of inland bodies of saline water.

MARICOPA COUNTY: In gravels on the Gila River, as the well-crystallized colorless mineral with bitumen in a fossil egg 62 mm by 40 mm, enclosed in limestone matrix (Morgan and Tallmon 1904).

#COLORADOITE

Mercury telluride: HgTe. A mineral found in hydrothermal tellurium-bearing precious-metal deposits.

LA PAZ COUNTY: Dome Rock Mountains, Middle Camp District, Oro Fino placer, as mineral inclusions in placer gold associated with tetradymite, calaverite, and petzite (Melchiorre 2010).

#COLUMBITE-(Fe)

Iron niobium oxide: $Fe^{2+}Nb_2O_6$. Formed in granite pegmatites.

MARICOPA COUNTY: Mummy Mountain in Phoenix, as microscopic crystals with niobian rutile in pegmatite (3rd ed.). White Picacho District, as crystals up to 5 in. in diameter in the quartz-rich zone of the Midnight Owl pegmatite (Jahns 1952). Cave Creek District, where ferberite deposits carry 2.19 percent combined niobium and tantalum, probably present as columbite tantalite (Bull. 181).

PIMA COUNTY: Ajo area, San Antonio Group, southwest of the New Cornelia Mine, Valentine 1, 2, and 3 claims, as massive material (MM 7866 is identified as columbite-(Fe); MM 7865 is identified as tantalite-(Fe)).

YAVAPAI COUNTY: White Picacho District, in small quantities in several pegmatites (Jahns 1952). Near Crown King, as a 3 in. twinned crystal in pegmatite (3rd ed.).

COLUSITE

Copper vanadium arsenic sulfide: $Cu_{12}VAs_3S_{16}$. An uncommon mineral associated with bornite, chalcocite, tetrahedrite, tennantite, and enargite in some copper deposits.

COCHISE COUNTY: Warren District, Bisbee, among the ores of the Campbell Mine (Graeme 1993).

PINAL COUNTY: Pioneer supersystem, Magma District, Magma Mine (Hammer and Peterson 1968).

CONICHALCITE

Calcium copper arsenate hydroxide: $CaCu(AsO_4)(OH)$. A member of the adelite group. Formed in the oxidized portions of copper deposits; associated with other secondary minerals, including austinite, olivenite, limonite, brochantite, malachite, azurite, and jarosite.

COCHISE COUNTY: Warren District, Bisbee, Higgins Mine, from which a mineral named "higginsite" (Palache and Shannon 1920) was later shown to be identical with conichalcite; as excellent crystals and small masses in manganese oxides and limonite (Taber and Schaller 1930; Richmond 1940; Radcliffe and Simmons 1971). Tombstone District, Little Joe Mine (UA 43). Pearce District, Commonwealth Mine, with chlorargyrite (Howell 1977).

FIGURE 3.66 Conichalcite, Bagdad Mine, Yavapai County, 11.1 cm.

GILA COUNTY: Miami-Inspiration supersystem, Inspiration District, Thornton pit of the Inspiration Mine, as spheroidal alteration areas in masses of chrysocolla 2 to 3 in. thick, associated with faults (Kenneth Bladh, pers. comm., 1974). Globe Hills District, Copper Hill Mine (UA 6425).

GREENLEE COUNTY: Reported in the Copper Mountain supersystem, Candelaria District, in the Candelaria breccia pipe (3rd ed.).

LA PAZ COUNTY: New Water District, Eagle Eye Mine, with chrysocolla (3rd ed.).

PIMA COUNTY: Ajo District, New Cornelia Mine, as small green crystals in vugs associated with shattuckite (Schaller and Vlisidis 1958; UA 8987). Santa Rita Mountains, Helvetia-Rosemont District, near Rosemont (S 112807).

PINAL COUNTY: Galiuro Mountains, Table Mountain District, Table Mountain Mine, as good quality crystals in vugs with willemite, plancheite, and malachite, and as massive material, with gold (Bideaux, Williams, and Thomssen 1960). Sawtooth Mountains in manganese deposits (Robert O'Haire, pers. comm., 1972). Suizo Mountains, Durham-Suizo District, Azurite Mine, with malachite, plancheite, and shattuckite (Bideaux, Williams, and Thomssen 1960).

SANTA CRUZ COUNTY: North end of the Tumacacori Mountains (UA 9265).

YAVAPAI COUNTY: In a shallow pit thirteen miles west of Congress, just off U.S. Highway 93 (Bull. 181). Eureka supersystem, Bagdad District, Bagdad Mine, lining vugs with malachite (Muntyan 2015a). Date Creek (Brian Sage, UA 16446). Agua Fria District, Old Robinson claim at Copper Mountain as microcrystals (Brian Sage, UA x6914).

CONNELLITE

Copper sulfate hydroxide chloride hydrate: $Cu_{36}(SO_4)(OH)_{62}Cl_8 \cdot 6H_2O$. A very rare secondary mineral from oxidized copper deposits; commonly formed in small cavities in cuprite, with brochantite and malachite.

COCHISE COUNTY: Warren District, Bisbee, Copper Queen (NHM 55921), Calumet and Arizona and Czar Mines, as small radiating aggregates of slender crystals (Ransome 1904; Palache and Merwin 1909; Holden 1922, 1924; C. Frondel 1941); crystals of connellite on type specimens of paramelaconite from the Copper Queen Mine are as large as matchsticks (Ford and Bradley 1915); Southwest Mine, acicular crystals in cuprite with paratacamite (Graeme 1993). Tombstone District,

Toughnut Mine, as slender needle-like crystals and aggregates in cavities in cuprite (McLean and J. W. Anthony 1972; UA 26).

PIMA COUNTY: Ajo District, New Cornelia Mine, in vugs with cuprite (NHM 1965,257). Pima supersystem, Mission-Pima District, Mineral Hill Mine, Daisy shaft, with pseudoboleite, gerhardtite, and atacamite in cuprite (S. A. Williams 1961).

PINAL COUNTY: Mammoth District, Mammoth–St. Anthony Mine, as rare microcrystals associated with caledonite (3rd ed.). Bunker Hill (Copper Creek) District, Copper Creek, Childs-Aldwinkle Mine, very small elongated blue crystals with cuprite from the glory hole above the workings (M. Ascher, pers. comm., 2012).

YAVAPAI COUNTY: Eureka supersystem, Bagdad District, Bagdad Mine, as beautiful sprays of elongated acicular crystals to 1 cm nearly filling a vug in cuprite from the open pit (R. Jenkins, pers. comm., 2014).

FIGURE 3.67 Connellite, Bagdad Mine, Yavapai County, 3.5 × 5.0 cm.

COOKEITE

Aluminum lithium silicate hydroxide: $(Al,Li)_3Al_2(Si,Al)_4O_{10}(OH)_8$. A rare member of the chlorite group. Formed in granite pegmatite; associated with beryl, lepidolite, and spodumene.

MARICOPA COUNTY: White Picacho District, as rare coatings of pale-pink flakes and foils commonly grouped in felt-like aggregates on lepidolite and spodumene, filling fractures in the interior parts of several lithium-bearing pegmatites (Jahns 1952).

COPIAPITE

Iron sulfate hydroxide hydrate: $Fe^{2+}Fe^{3+}_4(SO_4)_6(OH)_2\cdot20H_2O$. A fairly common secondary mineral formed with other sulfates such as melanterite, alunogen, and halotrichite, from the alteration of sulfide minerals, principally pyrite.

COCHISE COUNTY: Warren District, Bisbee, Copper Queen Mine, in crusts several inches thick, with coquimbite and voltaite (Merwin and Posnjak 1937); Campbell Mine, 2,100 ft. level, as bright-yellow crystals (Graeme 1981).

COCONINO COUNTY: Cameron District, a fairly common mineral associated with jarosite and other sulfates in uranium deposits (Austin 1964); Yazzie No. 102 Mine, in large pure masses of yellow scales (3rd ed.).

GILA COUNTY: Sierra Ancha District, First Chance Mine (Granger and Raup 1969).

PIMA COUNTY: Sierrita Mountains, Pima supersystem, as silky fibers and foliated masses (Bull. 181).

YAVAPAI COUNTY: Jerome supersystem, Verde District, United Verde Mine, as encrustations nearly 1 cm thick, crystals, and crystalline masses, formed by burning of sulfide ore (C. A. Anderson 1927; Lausen 1928; H 90538).

FIGURE 3.68 Copper, New Cornelia Mine, Ajo, Pima County, 16 × 7 cm.

COPPER

Copper: Cu. Most often of secondary origin, copper is widely distributed in the oxidized zones of many sulfide copper deposits, where it may be accompanied by cuprite, malachite, azurite, tenorite, and limonite. Also in sedimentary rocks and in cavities in volcanic rocks. May replace other minerals such as malachite, cuprite, azurite, and chalcopyrite, even wood from mine timbers. Very abundant in Arizona.

APACHE AND NAVAJO COUNTIES: Monument Valley and Cane Valley Districts, as a primary mineral associated with copper sulfides in uranium-vanadium deposits (Evensen and Gray 1958).

COCHISE COUNTY: Warren District, Bisbee, abundant at several mines, locally as pseudomorphs after cuprite crystals (3rd ed.); Copper Queen Mine above the third level, in oxidized ore as masses weighing several hundred pounds (Petereit 1907; Guild 1910); Campbell Mine, as fine specimens of crystallized material, some coated with silver, recovered from a single pocket (Schwartz and Park 1932; Schwartz 1934); Calumet and Arizona Mine, as small crystals and irregular networks throughout cuprite and in earthy mixtures of cuprite, limonite, and kaolinite (Ransome 1904; Holden 1922; Papish 1928; Schwartz 1934; C. Frondel 1941); Junction Mine (Mitchell 1920). Tombstone District, as microscopic particles in cuprite (B. S. Butler, Wilson, and Rasor 1938). Courtland District, as large arborescent masses (Bull. 181). Cochise supersystem, Johnson Camp District, Johnson Camp area, Mammoth, Republic, and Copper Chief Mines, in trace amounts in oxidized ores (Kellogg 1906; J. R. Cooper and Silver 1964). Huachuca Mountains, Reef District, in dark basaltic rock on claims of the Jack Wakefield Mining Company (R. Davis, pers. comm., 3rd ed.).

GILA COUNTY: Globe Hills District, Old Dominion Mine, in quartzite as small hackly particles (Ransome 1903b; N. P. Peterson 1962) and as fine pseudomorphs after cuprite (Les Presmyk, pers. comm., 2020). Miami-Inspiration supersystem, Pinto Valley District, Castle Dome Mine, as an uncommon secondary mineral (N. P. Peterson, Gilbert, and Quick 1946, 1951). Banner supersystem, 79 District, 79 Mine, associated with tenorite, chrysocolla, and cerussite (Stanley B. Keith, pers. comm., 1973).

GRAHAM COUNTY: Gila Mountains, Safford supersystem, Lone Star District, rare along thin veinlets or fractures, typically associated with cuprite in the oxidized zone of the Safford porphyry copper cluster (R. F. Robinson and Cook 1966).

GREENLEE COUNTY: Copper Mountain supersystem, Morenci, common in the upper parts of veins, partly as branching coralloid forms and groups of indistinct crystals (Kunz 1885), mostly with cuprite, at the upper limits of the chalcocite zone (Bull. 181); Arizona Central Mine, Williams vein, as solid copper up to 8 in. thick, with fibrous structure and probably pseudomorphs after chalcocite (Lindgren 1904, 1905; Reber 1916).

MARICOPA COUNTY: Cave Creek District, Red Rover Mine, as tiny scales impregnating schist (A. S. Lewis 1920).

MOHAVE COUNTY: Cerbat Mountains, Wallapai supersystem, in small amounts as a secondary mineral (B. E. Thomas 1949); Mineral Park District, King claim, in small amounts in thin plate-like masses, associated with chalcocite (Bastin 1925; Eidel, Frost, and Clippinger 1968). Bill Williams Fork (Genth 1868).

PIMA COUNTY: Santa Rita Mountains, at several properties in the Helvetia-Rosemont District (Schrader and Hill 1915); in exploratory drill core between Helvetia and Rosemont, associated with kinoite, apophyllite, djurleite, bornite, and chalcopyrite (J. W. Anthony and Laughon 1970). Ajo District, New Cornelia Mine, abundant as an oxidation product (Schwartz 1934; Gilluly 1937; W. J. Thomas and Gibbs 1983). Pima supersystem, Mission-Pima District, Pima Mine, as a secondary mineral (Journeay 1959); Mission Mine, as dendritic crystals in gypsum (UA 5560) and with cuprite (Stanley B. Keith, pers. comm., 1973; B. M. Williams 2018). Pima District—W. P. Blake (1855) saw "the pure metal" at a locality he described as "near Altar," in a vein containing "the red *oxyd* of copper . . . and green crusts of carbonate." Sierrita Mountains, Esperanza District, New Year's Eve Mine, with malachite and cuprite (Stanley

B. Keith, pers. comm., 1973; Kuck 1978). South Comobabi Mountains, Cababi District, Mildren and Steppe claims (S. A. Williams 1963). Silver Bell District, El Tiro pit (Kerr 1951).

PINAL COUNTY: Mineral Creek supersystem, Ray District, Ray Central Mine, as large masses (Ransome 1919) and as unusually large sawtooth-shaped single crystals and aggregates up to 5.5 by 1 in., loose in clay (White 1974; R. W. Jones and Wilson 1983); Pearl Handle pit, associated with cuprite and malachite (Stanley B. Keith, pers. comm., 1973). Galiuro Mountains, Bunker Hill (Copper Creek) District, Copper Creek, Copper Prince Mine, as twisted and wire-like masses in the oxidized ore (UA 536). San Manuel District, San Manuel Mine, generally in the lower part of the supergene sulfide zone (Schwartz 1949, 1953; UA 6769, a specimen with a tetrahexahedral crystal) and as spinel twins up to 3 in. long (Les Presmyk, pers. comm., 2020). Pioneer supersystem, Magma District, Magma Mine, in small amounts (Short et al. 1943) and as crystal groups (Barnes and Hay 1983).

SANTA CRUZ COUNTY: Patagonia Mountains, Three-R District, Three-R Mine, as thin sheets and films apparently derived from chalcocite (Schrader and Hill 1915; Schrader 1917).

YAVAPAI COUNTY: Jerome supersystem, Verde District, United Verde Extension Mine, locally abundant with cuprite and near Walker, as fine specimens (Lindgren 1926; Schwartz 1938; C. A. Anderson and Creasey 1958). Copper Basin District (Johnston and Lowell 1961). Eureka District, Bagdad Mine (C. A. Anderson 1950).

COQUIMBITE

Aluminum iron sulfate hydrate: $AlFe^{3+}_3(SO_4)_6(H_2O)_{12} \cdot 6H_2O$. A common efflorescent mineral formed in the oxidized portions of base-metal deposits with other secondary sulfates.

COCHISE COUNTY: Warren District, Bisbee, Copper Queen Mine, as porous crusts several inches thick (Merwin and Posnjak 1937).

PIMA COUNTY: Pima supersystem, Mission-Pima District (UA 5905), San Xavier West Mine (Arnold 1964).

PINAL COUNTY: Pioneer supersystem, Magma District, Magma Mine, with other sulfates from the 1,000 ft. level (3rd ed.).

YAVAPAI COUNTY: Jerome supersystem, Verde District, United Verde Mine, as an aluminous variety, with copiapite and other sulfate minerals, formed by the burning of pyritic ore (C. A. Anderson 1927; Lausen 1928; UA 54; H 91845; H 90623). Near Yarnell, upper reaches of Antelope

Creek, as striking arborescent intergrown spinel twins in a fracture zone in granite (Stanley B. Keith, pers. comm., 2019).

#CORDEROITE

Mercury sulfide chloride: $Hg_3S_2Cl_2$. A rare mercury mineral found as a secondary mineral in hydrothermal deposits associated with other mercury minerals.

MARICOPA COUNTY: Mazatzal Mountains District, Saddle Mountain Mercury Mine, as small colorless well-formed dodecahedral crystals associated with calomel, mercury, and cinnabar (R. Gibbs, pers. comm., 2019).

FIGURE 3.69 Corderoite, Saddle Mountain Mine, Maricopa County, FOV 3 mm.

CORDIERITE

Magnesium aluminum silicate: $Mg_2Al_4Si_5O_{18}$. An early formed product of the metamorphism of argillaceous sedimentary rocks, which persists into conditions of high-grade metamorphism.

COCONINO COUNTY: Grand Canyon National Park, as a common constituent of anthophyllite-bearing rocks, for example, at mile 229, Travertine Canyon (M. D. Clark 1979).

GRAHAM COUNTY: Graham Mountains, as dark-gray slightly greenish large (up to 0.5 inch in diameter) crystals, which have been completely replaced by muscovite in graphic granite (John S. White Jr., pers. comm., 3rd ed.; locality noted by John W. Donowick).

MOHAVE COUNTY: Unspecified locality, as sharp 1 in. crystals (H 91900).

PINAL COUNTY: Sacaton Mountains, Gila River Indian Reservation, ten miles north of Casa Grande, as cyclic twins, with sillimanite, corundum, and titanian andalusite (Bideaux, Williams, and Thomssen 1960).

CORKITE

Lead iron sulfate phosphate hydroxide: $PbFe^{3+}_3(SO_4)(PO_4)(OH)_6$. An uncommon secondary mineral formed in the oxidized cappings of certain base-metal deposits.

GILA COUNTY: Banner supersystem, 79 District, 79 Mine, as very small pale-tan rhombic crystal clusters from the fourth level (R. Gibbs, pers. comm., 2017).

YAVAPAI COUNTY: Old Dick District, prospect near the Bruce Mine, as drusy crusts of brilliant yellow-green crystals on vein quartz (3rd ed.).

FIGURE 3.70 Cornetite, Silver Bell Mine, Pima County, FOV 1.5 mm.

CORNETITE

Copper phosphate hydroxide: $Cu_3(PO_4)(OH)_3$. A rare mineral formed in the oxidized portions of base-metal deposits at only a few localities in the world.

PIMA COUNTY: Tucson Mountains, Saginaw Hill District, southern side of Saginaw Hill, about seven miles southwest of Tucson, in association with other oxide-zone minerals, including brochantite, pseudomalachite, malachite, libethenite, atacamite, and chrysocolla; in well-crystallized fine-grained soft clusters plastering the fracture surfaces in chert or chert gangue; as clear light peacock-blue crystal aggregates and as darker deep-blue or deep greenish-blue crystalline clots, which may alter to pseudomalachite. Crystals are typically between 0.05 mm and 0.20 mm in size (Khin 1970). Silver Bell District, Silver Bell Mine (R. Gibbs, pers. comm., 2020).

CORNUBITE

Copper arsenate hydroxide: $Cu_5(AsO_4)_2(OH)_4$. An exceedingly rare mineral formed in the oxide zone of some base-metal deposits.

SANTA CRUZ COUNTY: Patagonia Mountains, Trench District, prospect near the Humboldt Mine, as lovely pistachio-green crystals associated with luetheite and chenevixite. The crystals are measurable and clearly triclinic, verifying the symmetry proposed in the original description (3rd ed.). Marshall Mine, as microcrystals (Jessie Hardman, UA x3010).

CORNWALLITE

Copper arsenate hydroxide: $Cu_5(AsO_4)_2(OH)_4$. A rare secondary mineral first recognized in Cornwall, England, and recently in several localities in the western United States. Associated with other copper arsenates such as clinoclase, chenevixite, and duftite.

COCHISE COUNTY: Tombstone District, Oregon-Prompter Mine, as small light greenish-blue balls with olivenite (B. Murphy, pers. comm., 2018).

YAVAPAI COUNTY: Mayer District, Old Robertson claim, four miles east-northeast of Mayer, as films and coatings (S 117492), with clinoclase (UA 5758).

CORONADITE

Lead manganese oxide: $Pb(Mn^{4+}_6Mn^{3+}_2)O_{16}$. Related to cryptomelane. A rare mineral formed in the oxidized portion of mineralized veins. First described from the Coronado vein in the Copper Mountain supersystem, Morenci District (Lindgren and Hillebrand 1904). The Morenci District is the type locality.

COCHISE COUNTY: Warren District, Bisbee, near the White Tail Deer Mine, associated with neltnerite, hübnerite, and tilasite (3rd ed.). Gleeson District, Silver Bill Mine (Knudsen 1983).

GREENLEE COUNTY: Copper Mountain supersystem, Morenci District, west end of the Coronado vein, in fairly large amounts (Lindgren and Hillebrand 1904; Lindgren 1905; Guild 1910; Fairbanks 1923; C. Frondel and Heinrich 1942; Fleischer and Richmond 1943; Hewett and Fleischer 1960; H 83825).

MARICOPA COUNTY: Belmont Mountains, Osborne District, Tonopah-Belmont Mine, as botryoidal encrustations (G. B. Allen and Hunt 1988).

MOHAVE COUNTY: Artillery District, in numerous veinlets of manganese oxides along fractures, joints, bedding planes, and breccia zones in Tertiary rocks; associated with cryptomelane, hollandite, psilomelane, pyrolusite, ramsdellite, and lithiophorite (Mouat 1962).

PINAL COUNTY: Reymert supersystem, Mineral Hill District, Reymert Mine, with psilomelane, hollandite, chalcophanite, and todorokite (K. S. Wilson 1984). Pioneer supersystem, Magma District, Magma Mine, in small amounts near the lower limits of the oxidized zone, with sauconite (Fleischer and Richmond 1943; Short et al. 1943; Hewett and Fleischer 1960; Hewett, Fleischer, and Conklin 1963).

SANTA CRUZ COUNTY: Santa Rita Mountains, Cottonwood Canyon, Tyndall District, Glove Mine, coating wulfenite (David Shannon, UA 12793).

CORRENSITE

Calcium sodium potassium magnesium iron aluminum silicate hydroxide hydrate: $(Ca,Na,K)_{1-x}(Mg,Fe,Al)_9(Si,Al)_8O_{20}(OH)_{10}\cdot nH_2O$. A silicate clay mineral that consists of repeating layers of chlorite and vermiculite-smectite-like components, common in some sedimentary rocks.

COCONINO COUNTY: Grand Canyon region, abundant in various units of the widespread Supai Formation.

CORUNDUM

Aluminum oxide: Al_2O_3. A relatively common mineral formed in pegmatites and in contact-metamorphic zones related to intrusions of silica-undersaturated rocks such as nepheline syenites, also with andalusite in the contact zone of aluminum-rich granitic rocks.

MOHAVE COUNTY: Grand Wash Cliffs, as blue, red, and white material with andalusite in an aluminous pegmatite dike (Bull. 181); the exact location of this pegmatite is not well known. McConnico District, Ruby No. 1 claim, about five miles southwest of Kingman, as single crystals up to 7.5 cm long in mica schist (R. Grant, pers. comm., 1990).

PINAL COUNTY: Sacaton Mountains, Gila River Indian Reservation, with sillimanite, cordierite, and titanian andalusite (Bideaux, Williams, and Thomssen 1960); in S12, T5S R5E, with rutile and quartz in felsite (E. D. Wilson 1969).

YAVAPAI COUNTY: Mingus Mountain quadrangle, as an accessory mineral in trachyandesite (McKee and Anderson 1971).

CORVUSITE

Sodium calcium potassium vanadium iron oxide hydrate: $(Na,Ca,K)_{1-x}$ $(V^{5+},V^{4+},Fe^{2+})_8O_{20}\cdot4H_2O$. A widespread mineral in the sandstone uranium-vanadium deposits of the Colorado Plateau; locally an abundant ore mineral of vanadium, probably of primary origin.

APACHE COUNTY: Cane Valley District, abundant at the Monument No. 2 and Cato Sells Mines, where it is associated with coffinite, montroseite, uraninite, and doloresite (Weeks, Thompson, and Sherwood 1955; Finnell 1957; Witkind and Thaden 1963; R. G. Young 1964). Lukachukai Mountains, Lukachukai District, in the Salt Wash Member of the Morrison Formation (Chenoweth 1967).

NAVAJO COUNTY: Monument Valley District, Monument No. 1 and Mitten No. 2 Mines, in the dark uranium-vanadium ores, associated with a variety of uranium and vanadium minerals (Evensen and Gray 1958; Holland et al. 1958; Witkind 1961; Witkind and Thaden 1963).

COSALITE

Lead bismuth sulfide: $Pb_2Bi_2S_5$. A rare sulfosalt formed in moderate-temperature veins, in contact-metamorphic deposits, and in pegmatites.

COCHISE COUNTY: Warren District, Bisbee, Campbell shaft (Graeme 1981).

GRAHAM COUNTY: Aravaipa supersystem, Iron Cap District, Landsman claim, with calcite and diopside (Bull. 181).

COTUNNITE

Lead chloride: $PbCl_2$. A rare mineral associated with cerussite and anglesite in oxidized mineral deposits as an alteration product of galena.

LA PAZ COUNTY: Dome Rock area, La Paz Placer District, as encrustations on galena associated with gypsum, silver-rich anglesite, and phosgenite, as galena-rich nodules in the placer deposits (Melchiorre 2013).

MOHAVE COUNTY: Grand Wash Cliffs, Bentley District, Grand Gulch Mine, reported as small veinlets replacing chalcocite (J. M. Hill 1914b).

PIMA COUNTY: Santa Catalina Mountains, western end of Pusch Ridge. The mineral occurs as very small white individual crystals and as clusters of crystals less than 0.5 mm long in fractures where groundwater has seeped into the rock. It was found several feet below the surface in mylonitized Wilderness Granite (Yang, pers. comm., 2020).

FIGURE 3.71 Cotunnite, Pusch Ridge, Santa Catalina Mountains, Pima County, FOV 1.5 mm.

COVELLITE

Copper sulfide: CuS. Commonly formed as a secondary mineral in the zone of oxidation and secondary enrichment of copper deposits associated with chalcocite and other copper sulfides. Covellite can also form as a primary mineral. It is widespread in small amounts in most copper deposits, typically as coatings and iridescent tarnish on other sulfides.

COCHISE COUNTY: Tombstone District, lining boxwork structures formed by the removal of primary sulfide minerals (B. S. Butler, Wilson, and Rasor 1938). Warren District, Bisbee, Campbell Mine, in massive sulfide orebodies in limestone, believed to be of supergene origin; Cole Mine, in veins in limestone with chalcocite (Schwartz and Park 1932; Bain 1952; Hutton 1957) and as platy crystals to 4 mm (Les Presmyk, pers. comm., 2020); Junction Mine (Mitchell 1920). Little Dragoon Mountains, Cochise

FIGURE 3.72 Covellite, Campbell Mine, Bisbee, Cochise County, 7 cm.

supersystem, Johnson Camp District, Johnson Camp area, as a common mineral in the upper portions of cupriferous veins and pyrometasomatic deposits (J. R. Cooper and Silver 1964). Dragoon District, Primos Mine near Dragoon (Palache 1941a).

GILA COUNTY: Miami-Inspiration supersystem, as a widespread but minor constituent of the disseminated copper deposits (N. P. Peterson 1962); Roseboom (1966) suggested that its association with djurleite in the district indicates an unstable assemblage; Pinto Valley District, Castle Dome Mine (N. P. Peterson 1947; N. P. Peterson, Gilbert, and Quick 1951); Copper Cities District, Copper Cities Mine (N. P. Peterson 1954). Green Valley District, in small amounts replacing chalcopyrite and bornite (Bull. 181). Dripping Spring Mountains, Banner supersystem, Christmas District, Christmas Mine, as a replacement of bornite, with secondary chalcocite (N. P. Peterson and Swanson 1956; Perry 1969); 79 District, 79 Mine, in minor amounts, associated with oxidation products of galena (Kiersch 1949; Stanley B. Keith 1972).

GRAHAM COUNTY: Aravaipa supersystem, as films and blebs in enriched ores (C. P. Ross 1925a; Denton 1947a).

GREENLEE COUNTY: Copper Mountain supersystem, Morenci District, Ryerson and Montezuma Mines (Lindgren 1905; Guild 1910).

MOHAVE COUNTY: Cerbat Mountains, Wallapai supersystem, common as an accessory mineral in sulfide deposits (B. E. Thomas 1949).

PIMA COUNTY: Ajo District, New Cornelia Mine, as minute blebs and coatings on other copper minerals (Gilluly 1937). Sierrita Mountains, La Coronado Mine (3rd ed.).

PINAL COUNTY: San Manuel District, widespread but sparse at the San Manuel Mine (Chapman 1947; Schwartz 1949, 1953). Mammoth District, Mammoth–St. Anthony Mine, replacing chalcopyrite (Bull. 181). Pioneer supersystem, Magma District, sparingly distributed in the Magma Mine (Bateman 1929; Short et al. 1943).

SANTA CRUZ COUNTY: Patagonia Mountains, Three-R District, Three-R Mine and other properties, as films on other sulfides (Schrader and Hill 1915). Ruby District, Idaho and Montana Mines, near Ruby (Warren and Loofbourrow 1932).

YAVAPAI COUNTY: Jerome supersystem, Verde District, United Verde Extension Mine, as fine specimens (Lindgren 1926; Schwartz 1938). Eureka supersystem, Bagdad District, Bagdad Mine, as films on chalcopyrite in the chalcocite zone (C. A. Anderson 1950).

COWLESITE

Calcium aluminum silicate hydrate: $Ca(Al_2Si_3)O_{10} \cdot 5\text{-}6H_2O$. An uncommon member of the zeolite group. Data from the Superior occurrence were used in the original description of cowlesite.

PINAL COUNTY: About 5.5 miles south of Superior, at Arnett Creek, as tiny white bladed crystals up to 1 mm long in vesicles in olivine bombs and scoria from middle Tertiary cinder cones (H 116448); associated with calcite and the zeolites thomsonite, chabazite, analcime, and mordenite (Wise and Tschernich 1975).

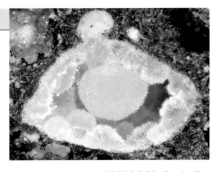

FIGURE 3.73 Cowlesite, Arnett Creek, Pinal County, FOV 5 mm.

CRANDALLITE

Calcium aluminum phosphate hydroxide: $CaAl_3(PO_4)(PO_3OH)(OH)_6$. As fibrous aggregates in phosphate deposits, associated with other phosphate minerals and as an alteration of some primary phosphate minerals.

MARICOPA AND YAVAPAI COUNTIES: White Picacho District, in pegmatites (London 1981).

CREASEYITE

Copper lead iron silicate hydrate: $Cu_2Pb_2Fe^{3+}_2Si_5O_{17} \cdot 6H_2O$. A rare mineral known from only a few localities. Formed in partially oxidized lead-copper ores. The Mammoth–St. Anthony Mine is the type locality.

MARICOPA COUNTY: Belmont Mountains, Osborne District, Potter-Cramer Mine, in vuggy, leached fluorite gangue with willemite, mimetite, wickenburgite, and ajoite (S. A. Williams and Bideaux 1975); Osborne District, Tonopah-Belmont Mine, as pale-green acicular crystals up to 0.8 mm long (G. B. Allen and Hunt 1988). Big Horn Mountains, Big Horn District, Evening Star Mine and nearby prospects, as small aggregates of brilliant acicular crystals (R. Gibbs, pers. comm., 2015).

FIGURE 3.74 Creaseyite, Tonopah-Belmont Mine, Maricopa County, FOV 6 mm.

PINAL COUNTY: Mammoth District, Mammoth–St. Anthony Mine, as pale-green fibrous tufts in or on cerussite, wulfenite, and fluorite (S. A. Williams and Bideaux 1975). Tortolita District, small prospect south of Big Mine as small spherical aggregates of bright-green acicular crystals (Gibbs 2003).

CREDNERITE

Copper manganese oxide: $CuMnO_2$. A secondary mineral formed by the oxidation of copper and manganese mineral deposits.

COCHISE COUNTY: Warren District, Bisbee, identified on an old specimen in the collections of the University of Arizona (UA 6993); Dallas shaft (Graeme 1993).

LA PAZ COUNTY: Copper Bottom District, at the Green Glory Mine, in a fractured quartzite bed with oxidized copper minerals as a component of chrysocolla, cuprite, malachite and silver-rich copper wad, which is a mixture of tenorite, crednerite, bixbyite, and silver-rich jarosite (Melchiorre 2013).

CREEDITE

Calcium aluminum sulfate hydroxide fluoride hydrate: $Ca_3Al_2(SO_4)$ $(OH)_2F_8 \cdot 2H_2O$. A rare mineral formed in hydrothermal veins.

GRAHAM COUNTY: Aravaipa supersystem, Grand Reef District, Grand Reef Mine, with linarite and anglesite in quartz-lined veins (R. W. Jones 1980); Iron Cap District, Landsman Camp, Iron Cap Mine, as a rare gangue mineral in the ore deposit (W. E. Wilson 1988).

#CRICHTONITE

Strontium manganese yttrium uranium iron titanium chromium vanadium oxide hydroxide: $Sr(Mn,Y,U)Fe_2(Ti,Fe,Cr,V)_{18}(O,OH)_{38}$. Usually found in alpine fissure veins.

APACHE COUNTY: Garnet Ridge SUM District, found in an ultramafic diatreme on the Navajo Reservation as inclusions in pyrope with srilankite, carmichaelite, rutile, and many other minerals (Wang, Essene, and Zhang 1999).

FIGURE 3.75 Cristobalite, Big Lue Mountains, Greenlee County, FOV 5 mm.

CRISTOBALITE

Silicon oxide: SiO_2. A high-temperature polymorph of SiO_2. Typically associated with volcanic rocks as a metastable mineral.

GREENLEE COUNTY: Big Lue Mountains, Big Lue District, as well-formed crystal vesicles with tridymite (R. Gibbs, pers. comm., 2020).

PIMA COUNTY: Roskruge Mountains, in cavities in andesite, associated with tridymite and clay (Bull. 181).

CROCOITE

Lead chromate: $Pb(CrO_4)$. An unusual secondary mineral formed in oxidized, chromium-bearing lead deposits. It is commonly associated with galena, sphalerite, cerussite, and secondary lead chromate minerals.

GILA COUNTY: Sierra Ancha District, Workman Creek, as smears in the upper Dripping Spring Quartzite, with metatorbernite (Robert O'Haire, pers. comm., 1972).

GREENLEE COUNTY: Copper Mountain supersystem, Granville District, Lime Cap Mine, as blocky crystals to 1 cm with cerussite on a specimen obtained by Richard Flagg. The location has not been verified (R. Gibbs, pers. comm., 2016).

MARICOPA COUNTY: Reported in the "Vulture region" at the Collateral, Chromate, Blue Jay, and Phoenix properties (Silliman 1881). The exact location of these mines is not known, and an examination of Silliman's specimens at Yale has not found crocoite. South of Wickenburg, Belmont Mountains, Osborne District, Moon Anchor Mine, Potter-Cramer property, and Rat Tail claim, as a minor oxide-zone mineral derived from lead-zinc ores; associated with various secondary lead, zinc, and chromium minerals (S. A. Williams 1968; S. A. Williams, McLean, and Anthony 1970).

PIMA COUNTY: South Comobabi Mountains, Cababi District, Mildren and Steppe claims (S. A. Williams 1963).

YAVAPAI COUNTY: Castle supersystem, Black Rock District, Black Butte claim, with vanadinite (R. Grant, pers comm. 2010; specimen from A. L. Flagg Collection) and with fornacite (John S. White, pers. comm., 3rd ed.).

CRONSTEDTITE

Iron silicate hydroxide: $(Fe^{2+},Fe^{3+})_3(Si,Fe^{3+})_2O_5(OH)_4$. An uncommon member of the kaolinite-serpentine group. Formed under conditions similar to those of the chlorites.

MOHAVE COUNTY: Aquarius Mountains, Aquarius Mountains District, in a pegmatite at the Rare Metals Mine, with titanite, monazite, apatite, and chevkinite (Kauffman and Jaffe 1946; Heinrich 1960).

CRYPTOMELANE

Potassium manganese oxide: $K(Mn^{4+}_7Mn^{3+})O_{16}$. Typically associated with the psilomelane manganese ores; probably most commonly of secondary origin.

COCHISE COUNTY: Tombstone District—material collected by A. E. Granger exhibited fine-grained, cleavable, and fibrous varieties in the same specimen (Richmond and Fleischer 1942).

COCONINO COUNTY: Prospect Canyon (Ridenour) District, as fine-grained massive material intimately associated with hollandite; contains 0.23 percent thallium and 5.5 percent BaO (Crittenden et al. 1962); Adams-Woodie prospect, along the Aubrey Cliffs, about twenty-two miles northeast of Peach Springs, cementing rock fragments in veins cutting Kaibab Limestone; associated with psilomelane (Hewett, Fleischer, and Conklin 1963).

GILA COUNTY: Ramsdell District, Apache Mine, as botryoidal material, intimately associated with hollandite; contains 0.34 percent thallium, about 9 percent BaO, and about 2 percent K_2O (Crittenden et al. 1962; Hewett, Fleischer, and Conklin 1963).

MOHAVE COUNTY: Artillery District, Black Jack, Price, and Priceless veins, and the Plancha and Maggie Canyon bedded deposits (Mouat 1962).

PINAL COUNTY: Galiuro Mountains, Bunker Hill (Copper Creek) District, Copper Creek, Blake Place (UA 7037). Pioneer supersystem, Magma District, Magma Mine (Hewett and Fleischer 1960).

SANTA CRUZ COUNTY: Patagonia Mountains, Mowry District, Mowry Mine (Fleischer and Richmond 1943); argentian cryptomelane present in large quantities (S. R. Davis 1975). Santa Rita Mountains, Tyndall District, Glove Mine, on wulfenite, with calcite (3rd ed.).

YAVAPAI COUNTY: Burmister District, Burmister Mine near Mayer, in S17, T11N R3W, with other manganese oxides interlayered with volcanic ash, clastic sediments, and a basalt flow; also found in and near mounds of opalized dolomite deposited after travertine and silica by an extinct spring (Hewett and Fleischer 1960; Hewett, Fleischer, and Conklin 1963).

CUBANITE

Copper iron sulfide: $CuFe_2S_3$. Found typically as a high-temperature sulfide commonly associated with chalcopyrite (with which it may be intimately intergrown), pyrrhotite, and pentlandite.

COCHISE COUNTY: Chiricahua Mountains, California supersystem, Hilltop District, El Tigre Mine, as microscopic blebs with acanthite in black-banded quartz (Tsuji 1984).

GILA COUNTY: Southeastern Dripping Spring Mountains, Banner supersystem, Christmas District, Christmas Mine, as lamellae of probable exsolution origin in chalcopyrite, in the pyrrhotite-chalcopyrite zone of the lower Martin orebody (Perry 1969; McCurry 1971). Sierra Ancha District, Workman Creek area, Workman adit No. 1, in chalcopyrite (Granger and Raup 1969).

PINAL COUNTY: Pioneer supersystem, Magma District, Magma Mine (Barnes and Hay 1983).

YAVAPAI COUNTY: Old Dick District, in the Precambrian, volcanogenic, Bruce massive sulfide deposit, with mackinawite and chalcopyrite (Larson 1976).

#CUMENGEITE

Lead copper chloride hydroxide hydrate: $Pb_{21}Cu_{20}Cl_{42}$ $(OH)_{40} \cdot 6H_2O$. An uncommon mineral formed in the oxidized portions of lead copper deposits.

MARICOPA COUNTY: Belmont Mountains, Osborne District, Tonopah-Belmont Mine, as small equant brilliant blue crystals with herbertsmithite in oxidized vein material (Gibbs and Ascher 2012). Painted Rock District, Rowley Mine, as very small blue flattened tetragonal crystals on boleite (J. Ruiz, pers. comm., 2012).

FIGURE 3.76 Cumengeite, Tonopah-Belmont Mine, Maricopa County, FOV 0.8 mm.

CUMMINGTONITE

Magnesium silicate hydroxide: $\square Mg_2Mg_5Si_8O_{22}(OH)_2$. A monoclinic amphibole, which is typically a product of regional metamorphism, although it is known to form as a primary igneous mineral in certain mafic rocks.

LA PAZ COUNTY: Northern Plomosa Mountains, possibly northern Plomosa District, where it may be a component of mafic igneous rocks interlayered with the Orocopia Schist. Bouse Hills District, BBC claims, as yellow-green fibers in hematite matrix (David Shannon, pers. comm., 1988).

MOHAVE COUNTY: Hualapai (Antler) District, Antler Mine, in mafic schist (Bull. 181), in a matrix of cordierite, anthophyllite, biotite phlogopite, actinolite tremolite, muscovite, chlorite, garnet sillimanite, carbonate, epidote, quartz, and trace cummingtonite and grunerite (More 1980).

FIGURE 3.77 Cuprite, Morenci Mine, Greenlee County, 1.2 cm high.

CUPRITE

Copper oxide: Cu_2O. A widespread secondary copper mineral; formed in many of the oxidized copper mines and prospects of Arizona, where it may be associated with malachite, azurite, tenorite, limonite, or copper. The variety called *chalcotrichite* consists of intergrown clusters of thin elongated crystals, often showing parallel growth. Chalcotrichite is widespread in Arizona copper deposits.

COCHISE COUNTY: Warren District, Bisbee, as an important ore constituent; magnificent specimens from Bisbee grace mineral collections throughout the world; Copper Queen Mine, mainly as earthy material mixed with limonite, but also as crystals and the variety chalcotrichite (H 97648); Calumet and Arizona Mine, as large crystalline masses associated with copper and in beautiful druses of ruby-red crystals, which are mostly simple cubes (Douglas 1899; Ransome 1903a, 1904; Mitchell 1920; Holden 1922; Schwartz and Park 1932; Schwartz 1934; C. Frondel 1941); the remarkable crystal habit of the variety chalcotrichite is nicely illustrated by scanning electron microscope photography (Dunn 1978); Southwest Mine, as transparent modified cubic crystals up to 5 cm (Graeme 1993). Tombstone District, as bright-red cubic crystals associated with malachite, brochantite, and locally, connellite, which lines small cavities in cuprite at the Toughnut Mine (B. S. Butler, Wilson, and Rasor 1938). Cochise supersystem, Johnson Camp District, as deep-red splendent crystals lining pockets in quartz veins in the Texas Canyon Quartz Monzonite (J. R. Cooper and Silver 1964).

GILA COUNTY: Miami-Inspiration supersystem, Pinto Valley District, Castle Dome Mine (Schwartz 1921, 1934; N. P. Peterson 1947). Aravaipa supersystem, Iron Cap Mine, sparse throughout the leached and chalcocite zones of porphyry copper deposits; associated with malachite, azurite, copper, and turquoise (Schwartz 1921, 1934; N. P. Peterson 1947). Globe Hills District, Buffalo and Continental Mines, as massive material and as chalcotrichite (Bull. 181). Globe Hills District, Old Dominion Mine, as large octahedra (Ransome 1903b; N. P. Peterson 1962). Dripping Spring Mountains, Banner supersystem, 79 District, 79 Mine (Stanley B. Keith 1972).

GREENLEE COUNTY: Copper Mountain supersystem, Morenci District, in several mines at the upper limit of the chalcocite zone, as cubic crystals and as chalcotrichite (Kunz 1885; Lindgren 1903, 1904, 1905; Reber 1916).

LA PAZ COUNTY: Planet District, Planet Mine, with chrysocolla (McCarn 1904).

MARICOPA COUNTY: White Picacho District, as a supergene mineral in some pegmatites (Bull. 181).

MOHAVE COUNTY: Bill Williams Fork (Genth 1868). Cerbat Mountains, Wallapai supersystem, Tennessee District, Clyde Mine, where it occurs with copper in an area of rich chalcopyrite ore (B. E. Thomas 1949); Mineral Park District, Altata Mine (Dings 1951), as "rich and beautifully crystalline" material (3rd ed.).

PIMA COUNTY: Santa Rita Mountains, Helvetia-Rosemont District, at Rosemont, as crystal aggregates lining cavities (Schrader and Hill 1915). Silver Bell District, as cubic crystals and as chalcotrichite in small fractures (Engineering and Mining Journal 1904; Kerr 1951). Waterman District, Silver Hill Mine, with azurite and malachite (UA 6719). Pima supersystem, Twin

FIGURE 3.78 Cuprite, variety chalcotrichite, Bisbee, Cochise County, 8.2 cm.

Buttes District, Twin Buttes Mine, associated with copper that has replaced an iron pipe in a sump (UA 6386); Twin Buttes Mine, in oxidized ores (Stanley B. Keith, pers. comm., 1972). Tucson Mountains, Saginaw Hill District, Saginaw and Arizona Tucson properties, in porphyry that is disseminated over a wide area (Bull. 181). Ajo District, New Cornelia Mine, so abundant that miners made baseballs from the variety chalcotrichite (Schwartz 1934; Gilluly 1937; W. J. Thomas and Gibbs 1983). South Comobabi Mountains, Cababi District, Mildren and Steppe claims (S. A. Williams 1963).

PINAL COUNTY: Bunker Hill (Copper Creek) District, Copper Creek, Childs-Aldwinkle Mine (Kuhn 1941). San Manuel District, San Manuel Mine, near the base of the oxide zone (Schwartz 1949, 1953; L. A. Thomas 1966). Mammoth District, Mammoth–St. Anthony Mine, as rare crystalline masses associated with chalcotrichite (3rd ed.). Mineral Creek supersystem, Ray District, Mineral Creek property, as slender ruby-red crystals up to 1 cm long, in a Holocene gravel deposit, with jarosite and goethite (C. H. Phillips, Cornwall, and Rubin 1971); as sparkling ruby-red aggregates commonly on copper in a stope worked from the old Ray shaft (Ransome 1919), and as crystalline material and chalcotrichite in the oxidized zone of the Ray Mine (Metz and Rose 1966), in the Pearl Handle Pit (Les Presmyk, pers. comm., 2020).

SANTA CRUZ COUNTY: Patagonia Mountains, Patagonia District, Washington Camp District, Pride of the West Mine (Bull. 181). Oro Blanco District, Montana Mine, as fine bright crystals in vugs (Schrader 1917).

YAVAPAI COUNTY: Black Hills, Jerome supersystem, Verde District, United Verde Extension Mine, where it is locally abundant and commonly accompanied by copper; as beautiful druses of crystallized material and as chalcotrichite (Lindgren 1926; Schwartz 1938; C. A. Anderson and Creasey 1958). White Picacho District, as a supergene mineral in some pegmatites (Jahns 1952). Copper Basin District, in S20 and 21, T13N R3W, as massive crystalline material (David Shannon, pers. comm., 1971). Zonia District, Zonia Mine, with malachite and chrysocolla (Kumke 1947).

YUMA COUNTY: Muggins District, Red Knob Mine, associated with wulfenite, vanadinite, chalcedony, and limonite (Honea 1959).

#CUPROAURIDE

Copper gold: Cu_3Au. Found in hydrothermal gold-quartz veins. Cuproauride has a grandfathered approval by the IMA but is considered a questionable species.

LA PAZ COUNTY: Dome Rock Mountains, Cunningham Pass area, Copper Bottom District, Cunningham Mountain placers, as small pinkish-gold inclusions in placer gold grains associated with dyscrasite, aurostibite, pyrite, and quartz (Melchiorre 2013).

#CUPROCOPIAPITE

Copper iron sulfate hydroxide hydrate: $Cu^{2+}Fe^{3+}_4(SO_4)_6(OH)_2 \cdot 20H_2O$. A secondary mineral formed from the oxidation of sulfides.

COCHISE COUNTY: Warren District, Bisbee, 1800 level of the Junction Mine, where it was found as a minor constituent of postmining crusts with rhomboclase, copiapite, and other sulfates (Graeme, Graeme, and Graeme 2015).

CUPROPAVONITE

Copper silver lead bismuth sulfide: $Cu_{0.9}Ag_{0.5}Pb_{0.6}Bi_{2.5}S_5$. Found in zoned mesothermal base- and precious-metal deposits.

COCHISE COUNTY: Warren District, Bisbee, Campbell sulfide orebody, with chalcopyrite, bornite, sphalerite, chalcocite, galena, minor hessite, silver, and wittichenite (Graeme 1993).

#CUPRORIVAITE

Calcium copper silicate: $CaCuSi_4O_{10}$. A rare mineral found in volcanic and furnace environments. It is a natural analog of the Egyptian Blue pigment.

PIMA COUNTY: Ajo District, Ajo, as small blue crystal fragment from bricks of a copper-smelting furnace (Mazzi and Pabst 1962).

CUPROTUNGSTITE

Copper tungstate hydroxide: $Cu^{2+}_3(WO_4)_2(OH)_2$. A rare secondary mineral formed by the alteration of scheelite; usually found in concentric layers around a scheelite core. The Cave Creek occurrence was the first place the mineral was described, but the description was not adequate, and the location is not listed as the type locality.

COCHISE COUNTY: Little Dragoon Mountains, Bluebird District, Burrell claim, 1.5 miles west of Dragoon, altering from scheelite (Dale, Stewart, and McKinney 1960).

LA PAZ COUNTY: Reported in the Bright Star District, Livingston claims (Bull. 181).

MARICOPA COUNTY: Cave Creek District, associated with ferberite (UA 1192; Schaller 1932).

MOHAVE COUNTY: Boriana District, Boriana Mine (Chatman 1988).

PIMA COUNTY: Santa Rita Mountains, Helvetia-Rosemont District, 200 ft. level of the Helvetia Mine (Bull. 181); Omega Mine, found as small green octahedrons in the mine dump (R. Gibbs, pers. comm., 2015).

CYANOTRICHITE

Copper aluminum sulfate hydroxide hydrate: $Cu_4Al_2(SO_4)(OH)_{12}(H_2O)_2$. A rare secondary mineral associated with other oxidized copper minerals in some deposits.

COCHISE COUNTY: Warren District, Bisbee, with azurite and malachite (Graeme 1981; UA 3107). Courtland District, Maid of Sunshine Mine, as abundant acicular crystal groups, with azurite, malachite, and spangolite (MM 9791).

COCONINO COUNTY: Grandview District, Grand Canyon National Park, Grandview (Last Chance) Mine, as radiating crystals of exceptional beauty and massive nodules and veins, closely associated with brochantite (A. F. Rogers 1922; Gordon 1923; Leicht 1971).

FIGURE 3.79 Cyanotrichite, Grandview (Last Chance) Mine, Coconino County, FOV 5 mm.

Copper Mountain District, Morenci Mine; Copper Mountain Mine near Morenci, as narrow seams in siliceous gangue, coated with earthy hematite, as encrustations up to 2 mm thick, and as thin fibers and small tufts in cavities (Genth 1890).

PIMA COUNTY: Helvetia-Rosemont District, Omega Mine, as small acicular clusters of pale-blue crystals (Gibbs, pers. comm., 2015).

YAVAPAI COUNTY: Jerome supersystem, Verde District, in the lower part of the oxidized zone, in a chlorite schist that contains chalcopyrite, with antlerite and brochantite (3rd ed.).

DANALITE

Beryllium iron silicate sulfide: $Be_3Fe^{2+}_4(SiO_4)_3S$. A rare accessory mineral found in granite and greisen.

YAVAPAI COUNTY: Black Hills, southwest of Jerome (UA 7970).

DANBURITE

Calcium borosilicate: $CaB_2Si_2O_8$. Formed in contact-metamorphosed limestones and dolomites (skarns and tactites); may be associated with feldspar, axinite, scapolite, titanite, and tourmaline.

COCHISE COUNTY: Gleeson District, Maud Hill, as granular patches in drill core taken from the garnet-epidote tactite (3rd ed.).

DARAPSKITE

Sodium sulfate nitrate hydrate: $Na_3(SO_4)(NO_3)\cdot H_2O$. An uncommon water-soluble mineral typically formed in very arid desert regions, associated with other soluble sulfates and with nitrates in some places, as in northern Chile.

GREENLEE COUNTY: Peloncillo Mountains, northeast of Bowie, where it forms in caves during wet, cold weather, together with niter; these minerals disappear during hot, dry weather (3rd ed.).

DATOLITE

Calcium borosilicate hydroxide: $CaB(SiO_4)(OH)$. An uncommon mineral found in metamorphosed limestones or basaltic rocks. It may form in amygdules in basalts associated with prehnite, diopside, and grossularite.

PIMA COUNTY: Pima supersystem, Mission-Pima District, Pima Mine, in thin veins in diopside-garnet tactites, as microscopic material (3rd ed.). Southern half of the Helvetia-Rosemont District, with prehnite in skarn or tactite (McNew 1981).

DAUBRÉELITE

Iron chromium sulfide: $FeCr_2S_4$. A fairly common constituent of meteorites, in small amounts, commonly associated with troilite.

COCONINO COUNTY: Meteor Crater near Winslow, in the Canyon Diablo meteorite, with troilite (C. Frondel and Marvin 1967b).

#DAVIDBROWNITE-(NH₄)

Ammonium vanadium phosphate oxalate hydroxide hydrate: $(NH_4)_5(V^{4+}O)_2(C_2O_4)[PO_{2.75}(OH)_{1.25}]_4 \cdot 3H_2O$. A new and very rare secondary organic mineral formed in an old mine working with bat guano. The Rowley Mine is the type locality.

MARICOPA COUNTY: Painted Rock Mountains, Painted Rock District, Rowley Mine. The mineral is found in a hot and humid area of the mine with bat guano and growing on a baryte-quartz-rich matrix. Associated minerals include antipinite, fluorite, mimetite, mottramite, quartz, rowleyite, salammoniac, struvite, vanadinite, willemite, and wulfenite (Kampf et al. 2018; IMA 2018-129).

FIGURE 3.80 David-brownite-(NH4), Rowley Mine, Maricopa County, FOV 1.14 mm.

#DAVIDITE-(Ce)

Cerium yttrium uranium iron titanium chromium vanadium oxide hydroxide fluoride: $Ce(Y,U)Fe_2(Ti,Fe,Cr,V)_{18}(O,OH,F)_{38}$. A primary mineral found in high-temperature hydrothermal veins in norite and anorthosite, alkali, and granitic rocks.

YAVAPAI COUNTY: Eureka supersystem, Bagdad District, Bagdad Mine, as small glassy brown grains in drill core (R. Jenkins, pers. comm., 2017).

DAVIDITE-(La)

Lanthanum yttrium uranium iron titanium chromium vanadium oxide hydroxide fluoride: $La(Y,U)Fe_2(Ti,Fe,Cr,V)_{18}(O,OH,F)_{38}$. A rare primary mineral found in hydrothermal veins.

PIMA COUNTY: Quijotoa District, Pandora prospect, about five miles west of Covered Wells, as dark-brown pitchy-lustered masses in a matrix of titanite, epidote, and feldspar, in a transition zone between a metaspessartine dike and a quartz-monzonite intrusive; nearly entirely metamict (noncrystalline due to α-particle bombardment; Pabst and Thomssen 1959; Pabst 1961; R070698). Gatehouse et al. (1979) established the atomic structure of davidite from material from the Quijotoa Mountains; they also provided a detailed electron-microprobe chemical analysis.

DELAFOSSITE

FIGURE 3.81 Delafossite, Bisbee, Cochise County, 2.3 cm.

Copper iron oxide: $Cu^{1+}Fe^{3+}O_2$. A rare secondary mineral formed in the oxidized portions of copper deposits with hematite, cuprite, tenorite, and copper.

COCHISE COUNTY: Warren District, Bisbee, Cole shaft, with cuprite (UA 7124); Hoatson shaft, in "kaolin and ferruginous clay at about the lowest zone of oxidation" (A. F. Rogers 1913; Pabst 1938; see also C. Frondel 1935).

PINAL COUNTY: Mineral Creek supersystem, Ray District, Ray Mine, with chalcotrichite (UA 3925).

YAVAPAI COUNTY: Jerome supersystem, Verde District, United Verde Mine, as crusts of black tabular crystals, up to 8 mm on an edge, perched on milky quartz (3rd ed.).

#DENDORAITE-(NH₄)

Sodium aluminum ammonium oxalate phosphate hydrate: $(NH_4)_2NaAl(C_2O_4)(PO_3OH)_2(H_2O)_2$. A new organic mineral, the Rowley Mine is the type location.

MARICOPA COUNTY: Painted Rock District, Rowley Mine, as very small clear bladed crystals associated with relianceite-(K), antipinite, fluorite, rowleyite, salammoniac, and others on a baryte-quartz matrix. The minerals are found in a postmining assemblage associated with bat guano (Kampf, Cooper, Celestian, et al. 2021a, 2021b).

DESCLOIZITE

Lead zinc vanadate hydroxide: $PbZn(VO_4)(OH)$. A secondary mineral formed in small amounts in some oxidized lead-zinc or copper deposits;

associated with cerussite, vanadinite, and other secondary minerals. See also the related species MOTTRAMITE.

COCHISE COUNTY: Warren District, Bisbee, found in Shattuck shaft, Higgins Mine, and Dallas shaft (Graeme 1981). Tombstone District, Lucky Cuss, Toughnut, and Tombstone Extension Mines (B. S. Butler, Wilson, and Rasor 1938).

COCONINO COUNTY: Havasu Canyon District, as fine stalactitic crystal groups (Bull. 181).

GILA COUNTY: Green Valley District, in small amounts at the Oxbow and Zulu Mines (Lausen and Wilson 1925). Globe Hills District; a locality two miles north of the Old Dominion Mine; the 400 ft. level of the Comstock Extension Mine (N. P. Peterson 1962); Apache Mine, as brown to reddish poorly crystallized masses and coatings, commonly associated with vanadinite (W. E. Wilson 1971). Banner supersystem, 79 District, 79 Mine, fourth level, as tiny wedge-shaped crystals on hemimorphite; locally, crystals up to 0.25 in. in crust with wulfenite and lining cavities (Stanley B. Keith 1972); northeast workings of the Kullman-McCool (Finch, Barking Spider) Mine (also known as the Finch Mine), as common crystalline masses (K. D. Smith 1995). Dripping Spring Mountains, C and B–Grey Horse District, C and B Vanadium Mine, ten miles northwest of Christmas, as crystal aggregates and sharp brown to brown-black crystals up to 5 mm, associated with vanadinite and mimetite (Trebisky and Keith 1975; S R16420).

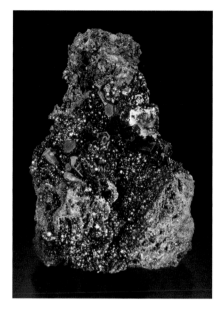

FIGURE 3.82 Descloizite, with vanadinite, Ben Hur Mine, Pinal County, 4 cm high.

MARICOPA COUNTY: Belmont Mountains, Osborne District, south of Wickenburg, at several localities, including the Moon Anchor Mine, Potter-Cramer property, and Rat Tail claim; associated with a suite of secondary minerals (S. A. Williams 1968). Vulture Mountains, Vulture District, Black Hawk property, one mile south of the Vulture Mine, as a velvety coating of fine crystals (Bull. 181). White Picacho District, in pegmatites (Bull. 181). Painted Rock District, Rowley Mine near Theba, as sparse black coatings of minute crystals associated with vanadinite (W. E. Wilson and Miller 1974).

MOHAVE COUNTY: Grand Wash Cliffs, Bentley District, Grand Gulch Mine (J. M. Hill 1914b).

PIMA COUNTY: Tucson Mountains, Old Yuma District, Old Yuma Mine, as rare brownish-orange crystals up to 5 mm long (3rd ed.). Pima supersystem, Twin Buttes District, Twin Buttes Mine (William Hefferon, pers. comm., 3rd ed.).

PINAL COUNTY: Mammoth District, Mammoth–St. Anthony Mine, with mottramite, forming crusts of small pointed crystals (Galbraith and Kuhn 1940; Bideaux 1980; R120097); Genth (1887) first noted the mineral at the mine. Ben Hur Mine with vanadinite (W. E. Wilson 2020a). Bunker Hill (Copper Creek) District, Copper Creek, Bluebird Mine (Kuhn 1951). Mineral Creek supersystem, Ray District, Ray Mine (S 93866). Slate District, Turning Point Mine, as tiny red crystals coating vanadinite (Hammer 1961). Dripping Spring Mountains, C and B–Grey Horse District, Grey Horse Mine, as single crystals and crystal aggregates (A. Clark and Fleck 1980).

YUMA COUNTY: Near Radium Hot Springs (UA 7894). Reported in the Castle Dome District (Guild 1910).

#DESTINEZITE

Iron phosphate sulfate hydroxide hydrate: $Fe^{3+}_2(PO_4)(SO_4)(OH) \cdot 6H_2O$. A secondary mineral that is the visibly crystalline counterpart of amorphous diadochite.

SANTA CRUZ COUNTY: Patagonia Mountains, as rounded masses of colorless acicular crystals (R050631).

DEVILLINE

Calcium copper sulfate hydroxide hydrate: $CaCu_4(SO_4)_2(OH)_6 \cdot 3H_2O$. A rare species formed in the oxidized portion of copper ore deposits.

COCHISE COUNTY: Warren District, Bisbee, as blue microcrystals in Southwest Mine (Graeme, Graeme, and Graeme 2015). Middle Pass District, San Juan Mine, as microscopic crystals with gypsum (Bruce Maier, pers. comm., 1990).

COCONINO COUNTY: Grand Canyon National Park, Horseshoe Mesa, Grandview District, Grandview (Last Chance) Mine, as bluish-green crusts of lath-like crystals with gypsum (Leicht 1971).

GRAHAM COUNTY: Aravaipa supersystem, Grand Reef District, Grand Reef Mine (George Godas, pers. comm., 1987).

PIMA COUNTY: South Comobabi Mountains, Cababi District, Little Mary Mine, as a zincian variety associated with brochantite, anglesite, zincian dolomite, and gypsum; in quartz veins cutting andesite (S. A. Williams 1962, 1963). Ajo District, New Cornelia Mine as sky-blue hemispheres up to 2 cm with malachite (Les Presmyk, pers. comm., 2020; William Thomas, pers. comm., 1988).

PINAL COUNTY: Mammoth District, Mammoth–St. Anthony Mine, as a rare, scaly alteration of powdery djurleite (Bideaux 1980).

DIABOLEITE

Copper lead chloride hydroxide: $CuPb_2Cl_2(OH)_4$. A rare secondary mineral formed in oxidized lead ores with boleite, cerussite, and hemimorphite. From a study of diaboleite and its associated minerals, H. E. Wenden (Winchell and Wenden 1968) concluded that diaboleite formed at low temperatures consistent with data from their experiments on synthesizing the species under hydrothermal conditions (between 100° and 170°C).

COCHISE COUNTY: Warren District, Bisbee, Campbell shaft, small blue crystals with leadhillite and silver on chalcocite (Graeme, Graeme, and Graeme 2015). Tombstone District, Gallagher Mine, tentatively identified as tiny blue crystals with anglesite, leadhillite, and linarite (Bowell, Luetcke, and Luetcke 2015).

GILA COUNTY: Dripping Spring Mountains, C and B–Grey Horse District, C and B Vanadium Mine, as small tabular blue crystals with leadhillite and cerussite (R. Gibbs, pers. comm., 2015).

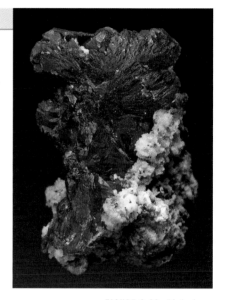

FIGURE 3.83 Diaboleite, Mammoth–St. Anthony Mine, Pinal County, 2.6 cm high.

MARICOPA COUNTY: Painted Rock District, Rowley Mine near Theba, in sparse amounts in association with linarite and anglesite, as sandy aggregates, not as distinct crystals (W. E. Wilson and Miller 1974). Belmont Mountains, Osborne District, Tonopah-Belmont Mine, as small dark-blue acicular crystals with other secondary lead minerals (Gibbs and Ascher 2012).

PINAL COUNTY: Mammoth District, Mammoth–St. Anthony Mine, 400 ft. level of the Collins vein, as crystals of superb quality, associated with cerussite, wulfenite, phosgenite, and boleite (Palache 1941b; Fahey, Daggett, and Gordon 1950; UA 5168, 6202, 6473; NHM 1947, 91).

DIADOCHITE

Iron phosphate sulfate hydroxide hydrate: $Fe^{3+}_2(PO_4)(SO_4)(OH) \cdot 6H_2O$. A secondary mineral found in gossans and as a secondary mineral formed in mine openings.

Copper Mountain supersystem, Morenci District, Morenci Mine, found with yellow clay in drill core sample (Enders 2000).

SANTA CRUZ COUNTY: Santa Rita Mountains, Ivanhoe District, Gringo Gulch, about three miles north-northwest of Patagonia, in fist-sized masses of fine-grained crystalline material (sample collected by James A. Yanez).

DIAMOND

Carbon: C. In Arizona, diamond has been found only in meteorites. Evidence has been put forward to support the concept that meteoric diamonds were formed by shock on impact with the Earth (N. L. Carter and Kennedy 1964).

COCONINO COUNTY: In 1891, a forty-pound mass of the Canyon Diablo meteorite was dissolved and found to contain tiny black diamonds. Subsequently, small diamonds embedded in graphite, and in places associated with lonsdaleite, have been found in other fragments from the same fall (Foote 1891; Kunz and Huntington 1893; Ksanda and Henderson 1939; C. Frondel and Marvin 1967b).

DIASPORE

Aluminum oxide hydroxide: $AlO(OH)$. The dimorph of böhmite, diaspore is a constituent of bauxite ores formed by extensive weathering and leaching of aluminous rocks and common in solfatarically altered volcanic rocks with alunite.

COCHISE COUNTY: Warren District, Bisbee, as a microscopic constituent of sericitized quartz monzonite (Graeme 1981).

GREENLEE COUNTY: Steeple Rock District, northwest of Duncan, associated with widespread nacrite, which formed by solfataric alteration of dacites and quartz latites (3rd ed.).

MARICOPA COUNTY: Unspecified locality near Tempe (UA 7451).

PIMA COUNTY: Ajo District, New Cornelia Mine, found in the contact-metamorphous zone on the east side of the pit (W. J. Thomas and Gibbs 1983).

SANTA CRUZ COUNTY: Red Mountain District, Patagonia, abundant in hydrothermally altered andesites; associated with quartz, pyrite, and alunite (3rd ed.).

#DICKINSONITE-(KMnNa)

Potassium sodium manganese calcium aluminum phosphate hydroxide: $K(Na,Mn)CaNa_3AlMn_{13}(PO_4)_{12}(OH)_2$. Found in granite pegmatites.

MARICOPA AND YAVAPAI COUNTIES: White Picacho District, in pegmatites (London 1981).

DICKITE

Aluminum silicate hydroxide: $Al_2Si_2O_5(OH)_4$. A member of the kaolinite group and of hydrothermal origin, it is commonly found in mineral deposits with sulfides.

COCHISE COUNTY: Warren District, Bisbee, Lavender pit, as dense, white, and earthy material cementing massive pyrite and euhedral pyrite to 2 cm in size (Graeme 1981).

PINAL COUNTY: San Manuel District, San Manuel Mine, associated with alunite in the most intensely altered rocks (Schwartz 1953).

DIGENITE

Copper sulfide: $Cu_{1.8}S$. Common in copper deposits, where it is typically associated with chalcocite; forms in both hypogene and supergene environments. A detailed discussion of crystallographic aspects of the bornite-digenite series, based partly on some Arizona minerals, is provided in L. Pierce and Buseck (1978).

COCHISE COUNTY: Warren District, Bisbee, associated with djurleite (Roseboom 1966; L. Pierce and Buseck 1978). Cochise supersystem, Johnson Camp District, Johnson Camp, Black Prince adit, intergrown with chalcopyrite (J. R. Cooper and Silver 1964; Roseboom 1966).

MOHAVE COUNTY: Wallapai supersystem, Gross Peak District, Alum prospect, about one mile northwest of Mineral Park, where it is the dominant supergene mineral replacing pyrite (Vega 1984).

PIMA COUNTY: Pima supersystem, Twin Buttes District, Twin Buttes Mine, in copper ores (William Hefferon, pers. comm., 3rd ed.).

PINAL COUNTY: Pioneer supersystem, Magma District, forming a part of all chalcocite-bornite intergrowths on and below the 3,400 ft. level of the

FIGURE 3.84 Digenite, Campbell Mine, Bisbee, Cochise County, 2 cm.

Magma Mine (Short et al. 1943; Morimoto and Gyobu 1971; L. Pierce and Buseck 1978); Resolution District, Resolution Mine, in the later phyllic argillic stage loosely associated with chalcocite, molybdenite, alunite, dickite, and topaz (Hehnke et al. 2012).

SANTA CRUZ COUNTY: Washington Camp District, Simplot Mine, replacing chalcopyrite (Lehman 1978).

YAVAPAI COUNTY: Black Mountains, Jerome supersystem, Verde District, United Verde Mine, in the fire zone as distinct crystals (Harcourt 1942).

DIOPSIDE

Calcium magnesium silicate: $CaMgSi_2O_6$. Diopside is a member of the pyroxene group and generally of metamorphic origin; most abundant in crystalline limestones with other prograde proximal contact-metamorphic calcium silicate minerals.

APACHE COUNTY: Buell Park District, Buell Park, a chromian variety found as detrital grains (Bideaux, Williams, and Thomssen 1960). Garnet Ridge SUM District, Garnet Ridge, as an emerald-green mineral formed in a breccia dike, associated with garnet (Gavasci and Kerr 1968).

COCHISE COUNTY: Warren District, Bisbee, a common alteration mineral in the contact zone around Sacramento Hill (Graeme 1981). Tombstone District, Comstock Hill, as small pale-green crystals in the contact zone (B. S. Butler, Wilson, and Rasor 1938). Little Dragoon Mountains, Cochise supersystem, Johnson Camp District, Johnson area, Republic and Moore Mines, as one of the abundant contact-metamorphic silicate minerals in limestone (J. R. Cooper and Huff 1951; A. Baker 1960). Middle Pass supersystem, Abril District, Abril Mine, abundant in contact-metamorphosed limestones (Perry 1964).

GILA COUNTY: Southeastern Dripping Spring Mountains, Banner supersystem, Christmas District, Christmas Mine, as a common constituent of hornfels and an important skarn mineral, with tremolite, a prominent gangue mineral (N. P. Peterson and Swanson 1956; Perry 1969). Banner supersystem, 79 District, 79 Mine, in contact-metamorphosed limestones with andradite, tremolite, and epidote (Stanley B. Keith 1972).

GRAHAM COUNTY: Stanley District, Grey Throne prospect, as single crystals and massive with andradite garnet (3rd ed.).

GREENLEE COUNTY: Copper Mountain supersystem, Morenci District, abundant in contact-metamorphosed limestones of the Longfellow Formation (Lindgren 1905; Moolick and Durek 1966).

LA PAZ COUNTY: Harquahala Mountains, in S14, T5N R10W, in coarse masses with sparse crystals of very light-colored material, along contact zones (Funnell and Wolfe 1964; David Shannon, pers. comm., 1972).

PIMA COUNTY: Santa Rita Mountains, Helvetia-Rosemont District, in the wall rocks of orebodies in metamorphic limestones (Schrader and Hill 1915; Schrader 1917); Peach-Elgin copper deposit, with almandine, and grossular, as bedded replacements (Heyman 1958). Sierrita Mountains, Pima supersystem, Mission-Pima District, in contact-metamorphosed limestones; Pima supersystem, Twin Buttes District, Twin Buttes Mine, in early prograde skarns (Stanley B. Keith, pers. comm., 1973). Mission-Pima District, Pima Mine, abundant in limestone hornfels with grossular and tremolite (Journeay 1959); Mission Mine, as the most abundant mineral in the hornfels host rock (Richard and Courtright 1959; Kinnison 1966).

SANTA CRUZ COUNTY: Patagonia Mountains, Washington Camp District, Pride of the West Mine, Duquesne area (Schrader and Hill 1915).

YAVAPAI COUNTY: Bradshaw Mountains, Ticonderoga District, with magnetite at the Henrietta Mine (Bull. 181).

DIOPTASE

Copper silicate hydrate: $CuSiO_3 \cdot H_2O$. A mineral of secondary origin found with other oxidized copper minerals.

GILA COUNTY: Miami-Inspiration District, Inspiration Mine, Live Oak Pit, as clusters of radiating crystals and sheaves partly coated by later chalcotrichite, with cubic pseudomorphs after fluorite (3rd ed.). Green Valley District, Oxbow and Summit Mines, as small prismatic crystals (Lausen and Wilson 1925). Southeastern Dripping Spring Mountains, Banner supersystem, Christmas District, Christmas Mine (UA 2549).

GREENLEE COUNTY: Copper Mountain supersystem, Morenci District, Bon Ton Group, near the head of Chase Creek, about nine miles from Clifton, "as brilliant crystals lining cavities in what is called locally 'mahogany ore'" (Hills 1882; see also Kunz 1885; Lindgren and Hillebrand 1904; Lindgren 1905; Guild 1910), Morenci Mine, Northwest Extension Pit, as brilliant green crystals to 3 mm on chrysocolla (Les Presmyk, pers. comm., 2020).

LA PAZ COUNTY: Salome area, Little Harquahala District, Harquahala Mine, as microcrystals (Marvin Deshler, UA x2547), and with crystalline

FIGURE 3.85 Dioptase, Mammoth–St. Anthony Mine, Pinal County, 2.1 cm high.

hematite and malachite on cherty-appearing brownish hematitic quartz-ite (John S. White Jr., pers. comm., 1972); also as microcrystals from the Pride Mine (Francis Sousa, UA x3090). Buckskin Mountains, Midway District, in small quantities at the Chicago prospect (Bull. 181). Cienega District, as microcrystals from a locality near Parker (Michael Rausch-kolb, UA x2688).

MOHAVE COUNTY: Cerbat Mountains, Wallapai supersystem, in minor amounts with chrysocolla in both blanket and vein deposits (B. E. Thomas 1949).

PIMA COUNTY: Santa Rita Mountains, Helvetia-Rosemont District, Narragansett Mine (UA 8075).

PINAL COUNTY: Mammoth District, Mammoth–St. Anthony Mine, as aggregates of small deep-emerald-green crystals and as druses of stout crystals in chrysocolla (Galbraith and Kuhn 1940; Palache 1941b; Bideaux 1980; UA 7921; NHM 1966,112; H 104226, 104227). Pioneer supersystem, Magma District, Magma Mine, as deep-green crystal encrustations, partly coated with small olivenite crystals, from the upper levels, particularly from the outcrop of the No. 1 glory hole (Short et al. 1943). At an unspecified locality near Riverside on the Gila River (W. B. Smith 1887). Galiuro Mountains, Virgus Canyon, Table Mountain District, Table Mountain Mine, where it is fairly common in vugs in jasperoid of the Escabrosa Limestone (Bideaux, Williams, and Thomssen 1960; Simons 1964; UA 548, 673). Mineral Creek supersystem, Ray District, Ray area (UA 1276). Silver Reef District, Nugget Fraction Mine (David Shannon, pers. comm., 1993). Lakeshore District, Lakeshore copper deposit, associated with chrysocolla (Romslo 1950).

YAVAPAI COUNTY: Amazon Wash, Black Rock District, near the Gold Bar Mine, fifteen miles northwest of Wickenburg (Bull. 181).

DJURLEITE

Copper sulfide: $Cu_{31}S_{16}$. Djurleite was described by E. H. Roseboom in 1966, and his prediction that it would prove to be a relatively common supergene mineral has been borne out. It forms in intimate association with the more common chalcocite, and the distinction between the two usually rests on X-ray diffraction analysis. It is highly probable that as details of the sulfide mineralogy of the copper mines of Arizona are worked out, many more djurleite occurrences will be discovered.

COCHISE COUNTY: Warren District, Bisbee, in association with digenite (Roseboom 1966).

GILA COUNTY: Miami-Inspiration District, in association with covellite (Roseboom 1966; AM 34772).

LA PAZ COUNTY: Dome Rock Mountains, La Cholla District, as large pure masses in vein quartz; partial oxidation has produced graemite, teineite, and brochantite (S. A. Williams and Matter 1975).

PIMA COUNTY: Near Helvetia-Rosemont District, in drill core that penetrated contact-metamorphosed Paleozoic limestones and dolomites, with chalcopyrite, bornite, apophyllite, copper, and kinoite (J. W. Anthony and Laughon 1970).

PINAL COUNTY: Pioneer supersystem, Magma District, Magma Mine, with digenite and bornite (Morimoto and Gyobu 1971). Bunker Hill (Copper Creek) District, at a small prospect in the Copper Creek area (UA 9607). Mammoth District, Mammoth–St. Anthony Mine, replacing galena (3rd ed.).

DOLOMITE

Calcium magnesium carbonate: $CaMg(CO_3)_2$. Most commonly found as a sedimentary rock-forming mineral in a rock of the same name and abundant in dolomitic limestones; also as a gangue mineral in hydrothermal mineral deposits.

FIGURE 3.86 Dolomite, Vekol Mine, Pinal County, 3.8 × 3.0 cm.

COCHISE COUNTY: Tombstone District, as massive beds interbedded with limestones and shales of the Naco Formation (B. S. Butler, Wilson, and Rasor 1938). Cochise supersystem, Johnson Camp District, Johnson Camp area (J. R. Cooper and Huff 1951). Warren District, Bisbee, in several beds of the Martin Formation (Hewett and Rove 1930).

COCONINO COUNTY: Portions of the Kaibab Limestone in the Flagstaff and Grand Canyon areas are highly dolomitic, especially beds west of El Tovar (Weitz 1942).

GILA COUNTY: Southeastern Dripping Spring Mountains, Banner supersystem, Christmas District, Christmas Mine area, in many units in the Martin and Escabrosa Formations (Perry 1969).

GREENLEE COUNTY: Copper Mountain supersystem, Morenci District, in beds in the lowest part of the Modoc Formation and in the upper part of the Morenci shale; sparingly in the Longfellow Limestone (Lindgren 1905).

LA PAZ COUNTY: Little Harquahala District, as a bed several feet thick near the Bonanza Mine (Bull. 181).

MARICOPA COUNTY: Agua Fria–Humbug area, thirty-eight miles north of Marinette; also east of Agua Fria, near Castle Hot Springs (Weitz 1942).

MOHAVE COUNTY: As thick beds on Tassai Ridge, below Pearce Ferry, Lake Mead area (3rd ed.); Peach Springs area, in flat-lying beds (Weitz 1942).

PIMA COUNTY: Sierrita Mountains, Pima supersystem, as coarsely crystallized material in fissures (Bull. 181); Mission-Pima District, Mission Mine, as microcrystals with fluorite (UA x3550); Twin Buttes District, Twin Buttes Mine (William Hefferon, pers. comm., 3rd ed.). South Comobabi Mountains, Cababi District, Mildren and Steppe claims (S. A. Williams 1963).

PINAL COUNTY: Vekol District, Vekol Mine, as lustrous chocolate-brown rhombohedral crystals; the color is present as a very thin coating of iron-rich dolomite (D. M. Shannon 1983a; UA 7661).

SANTA CRUZ AND PIMA COUNTIES: Santa Rita and Patagonia Mountains, widely distributed in metamorphosed areas in limestones near contacts with intrusive rocks (Schrader 1917).

YAVAPAI COUNTY: Black Hills, Jerome supersystem, Verde District, United Verde Mine, as a fairly abundant gangue mineral (Bull. 181). Bradshaw Mountains, Mount Union District, Tillie Starbuck Mine (Bull. 181). Hassayampa District, as small rhombohedral crystals coating walls of cavities (Bull. 181). Mayer Onyx District, Burmister Mine near Mayer, in mounds that are partly replaced by opal and chalcedony, deposited by an extinct spring (Hewett and Fleischer 1960; Hewett, Fleischer, and Conklin 1963) in beds up to 80 ft. thick south and southwest of Seligman (Weitz 1942).

DOLORESITE

Vanadium oxide hydroxide: $V^{4+}_3O_4(OH)_4$. Formed intimately mixed with oxide minerals, especially paramontroseite, which it replaces, in slightly oxidized uranium-vanadium ores in sandstone.

APACHE COUNTY: Cane Valley District, Monument No. 2 Mine, as crystals and chocolate-brown bladed masses with satin-like cleavage surfaces; associated with various other vanadium minerals (Stern et al. 1957; Jensen 1958; Witkind and Thaden 1963; R. G. Young 1964).

DOMEYKITE

Copper arsenide: Cu_3As. A rare mineral, probably of primary origin; locally associated with nickeline.

COCHISE COUNTY: Massive material with copper (UA 130; Guild 1910).

DRAVITE

Sodium magnesium aluminum silicate borate hydroxide: $NaMg_3Al_6$ $(Si_6O_{18})(BO_3)_3(OH)_3(OH)$. A member of the tourmaline group, mainly formed as a product of metamorphism involving boron. Schorl is the most abundant tourmaline variety, but some of the localities listed under it could be dravite, as they have not been checked for exact identification.

PINAL COUNTY: Copper Creek Canyon, Bunker Hill (Copper Creek) District, Copper Creek, at the western end of the Aldwinkle and Longstreet claims, as gray to green crystals in masses and small fan-shaped aggregates enclosed in quartz crystals, some of which are Japan law twins (William and Mildred Schupp, pers. comm., 2001; Gary M. Edson, pers. comm., 1972).

DUFRÉNOYSITE

Lead arsenic sulfide: $Pb_2As_2S_5$. A rare mineral known from only a few localities; associated with sphalerite and other sulfides and sulfosalts.

MOHAVE COUNTY: Reported in the Mineral Park District, but the exact locality is unknown (Bull. 181).

DUFTITE

Lead copper arsenate hydroxide: $PbCu(AsO_4)(OH)$. A rare mineral formed in the oxidized zones of base-metal deposits.

GILA COUNTY: Southeastern Dripping Spring Mountains, Banner supersystem, Christmas District, Christmas Mine, as brilliant pea-green crusts of microcrystals hidden beneath drusy quartz on which diopside spherules are perched (3rd ed.).

MARICOPA COUNTY: Osborne District, at several localities south of Wickenburg, including the Moon Anchor Mine and Potter-Cramer property, as a secondary mineral in the oxidized portion of galena-sphalerite veins, associated with various secondary minerals of lead and chromium (S. A. Williams 1968).

PIMA COUNTY: Pima supersystem, Twin Buttes District, Twin Buttes Mine (William Hefferon, pers. comm., 3rd ed.).

PINAL COUNTY: Table Mountain District, Table Mountain Mine, as microcrystals (H. Peter Knudsen, UA x266).

YAVAPAI COUNTY: Jerome supersystem, Verde District, as microcrystals from an unspecified locality at Jerome (Thomas Trebisky, UA x1084).

FIGURE 3.87 Dugganite, Empire Mine, Tombstone, Cochise County.

DUGGANITE

Lead zinc tellurate arsenate: $Pb_3Zn_3(TeO_6)(AsO_4)_2$. An extremely rare secondary mineral formed by oxidation of other tellurium minerals in acid waters. Tombstone is the type locality.

COCHISE COUNTY: Tombstone District, in the dumps at three properties, the Emerald and Old Guard Mines and the Joe shaft. It is found in either quartz or manganese oxide gangue associated with khinite and khinite-3T, from which it may form by alteration. Crystals are curved stubby prisms, varying in color from yellow green to water green, and are up to 0.3 mm long (S. A. Williams 1978); Empire Mine, 400 ft. level, as druse of prismatic green crystals (R060498).

DUMONTITE

Lead uranyl oxide phosphate hydrate: $Pb_2(UO_2)_3O_2(PO_4)_2 \cdot 5H_2O$. A rare secondary mineral associated with kasolite and other secondary uranium minerals.

SANTA CRUZ COUNTY: Pajarito District, White Oak Mine, with kasolite, uranophane, and autunite (Granger and Raup 1962).

FIGURE 3.88 Dumortierite, Dome Rock Mountains, La Paz County, FOV 3 mm.

DUMORTIERITE

Aluminum borosilicate: $AlAl_6BSi_3O_{18}$. An uncommon mineral typically found in the same environments as kyanite, andalusite, staurolite, and sillimanite, with which it may be associated in schists and gneisses and in aluminum metasomatic zones associated with peraluminous intrusions.

COCHISE COUNTY: About four miles north-northeast of Willcox, just east of U.S. Highway 191, as a purple vein (Duke 1960; Funnell and Wolfe 1964).

LA PAZ COUNTY: Dome Rock Mountains near Quartzsite, as microcrystals (Marvin Deshler, UA x584); K-D District, about three miles south of Quartzsite, as rich masses in quartzite, some used as lapidary material, with kyanite, sillimanite, lazulite, pyrophyllite, and andalusite (E. D. Wilson 1929; Bideaux, Williams, and Thomssen 1960; Reynolds et al. 1988).

In boulders along the Colorado River, between Yuma and Ehrenberg, as fine fibrous dumortierite altering to pyrophyllite and associated with kyanite (E. D. Wilson 1929). Near Clip on Colorado River (Diller and Whitfield 1889; Ford 1902; Schaller 1905; Bowen and Wyckoff 1926; E. D. Wilson 1933).

SANTA CRUZ COUNTY: Patagonia Mountains, twelve miles northeast of Nogales (UA 6268).

DYSCRASITE

Silver antimonide: $Ag_{3+x}Sb_{1-x}$ ($x \approx 0.2$). A rare mineral found in silver vein deposits with galena, silver, and silver sulfosalts.

LA PAZ COUNTY: Dome Rock Mountains, Copper Bottom District, Copper Bottom Pass area near the Copper Bottom Mine, as inclusions in placer gold grains (Melchiorre 2013).

PIMA AND SANTA CRUZ COUNTIES: Santa Rita and Patagonia Mountains, mainly the Tyndall and Old Baldy Districts, as a secondary mineral in quartz diorite (Schrader 1917).

ECLARITE

Copper iron lead bismuth sulfide: $(Cu,Fe)Pb_9Bi_{12}S_{28}$. A rare mineral formed in hydrothermal systems.

COCHISE COUNTY: Middle Pass supersystem, Abril District, Abril Mine, as 0.1 mm prisms embedded in sphalerite in veins cutting massive garnet-hematite skarn (3rd ed.); Gordon Camp, as acicular prisms up to 5 mm long embedded in dense massive skarn composed of andradite, hedenbergite, magnetite, and hematite (3rd ed.).

#EDDAVIDITE

Lead copper bromine oxide: $Pb_2Cu_{12}O_{15}Br_2$. Found in the oxidized portion of copper deposits, it is the bromine analog of murdochite. The Southwest Mine is the type locality.

COCHISE COUNTY: Warren District, Bisbee, Southwest Mine, occurs as small lustrous black cubic crystals with malachite and cuprite (R050381) (Yang and Downs 2018).

FIGURE 3.89 Eddavidite, Southwest Mine, Bisbee, Cochise County.

EDENITE

Sodium calcium magnesium aluminosilicate hydroxide: $NaCa_2Mg_5(Si_7Al)$ $O_{22}(OH)_2$. A member of the amphibole group of silicates and often found in metamorphic rocks.

COCHISE COUNTY: Warren District, Bisbee, Lowell shaft, a white to gray fibrous mineral, associated with tremolite and kaolinite (Graeme 1981).

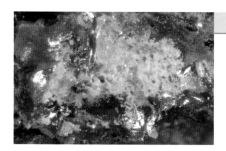

FIGURE 3.90 Eglestonite, Saddle Mountain Mine, Maricopa County, FOV 2 mm.

EGLESTONITE

Mercury oxychloride hydroxide: $([Hg^{1+}]_2)_3OCl_3(OH)$. Commonly associated with other mercury minerals in ore deposits of that metal; closely associated with calomel, from which it may be derived by oxidation.

MARICOPA COUNTY: Mazatzal Mountains District, near Sunflower, L and N No. 1 claim, with calomel and metacinnabar (UA 217; specimen collected by Carl Lausen); Saddle Mountain Mine (R. Gibbs, pers. comm., 2020).

ELBAITE

Sodium aluminum lithium borosilicate hydroxide: $Na(Al_{1.5}Li_{1.5})Al_6$ $(Si_6O_{18})(BO_3)_3(OH)_3(OH)$. A member of the tourmaline group, found in lithium-bearing granite pegmatites.

MARICOPA COUNTY: White Picacho District, in pegmatites, some being of the colorful "watermelon" variety (Jahns 1952).

YAVAPAI COUNTY: Southwestern Bradshaw Mountains, Tussock District, Weatherby Beryl Mine, occurs with lepidolite and muscovite (MM 9461, AZGS 2015).

EMMONSITE

Iron tellurium oxide hydrate: $Fe^{3+}_2(Te^{4+}O_3)_3 \cdot 2H_2O$. A rare secondary mineral formed by the oxidation of tellurides and tellurium. The occurrence near Tombstone is the type locality. It seems likely, however, that the type description was based on rodalquilarite.

COCHISE COUNTY: Tombstone District, unknown locality near Tombstone, as "yellowish green, translucent, crystalline scales and patches throughout a rather hard brownish gangue composed of lead carbonate, quartz, and a brown substance containing oxidized iron and tellurium plus water" (Hillebrand 1885). Toughnut-Empire Mine, with mackayite and tellurium (Bideaux, Williams, and Thomssen 1960). C. Frondel and Pough (1944) noted that the mineral from Tombstone formed in a hard brownish gangue composed of an intimate mixture of cerussite, quartz, and a brownish oxygenated compound of iron and tellurium. The mineral is associated with cerussite and rodalquilarite on a specimen found at the surface between the dumps of the Joe and Grand Central shafts (S. A. Williams 1981b).

EMPLECTITE

Copper bismuth sulfide: $CuBiS_2$. In hydrothermal veins with other sulfides and sulfosalts formed at moderate temperatures.

COCHISE COUNTY: Warren District, Bisbee, from the sulfide ores of the Campbell Mine (Graeme 1993).

EMPRESSITE

Silver telluride: AgTe. A rare mineral found in low-temperature gold-poor hydrothermal deposits.

COCHISE COUNTY: Tombstone District, Joe Mine, as large (1 in.) masses of tin-white crystalline material in flinty quartz-opal gangue, partly altered to rickardite, then to anglesite and rodalquilarite (3rd ed.).

ENARGITE

Copper arsenic sulfide: Cu_3AsS_4. Found in veins and replacement deposits formed under moderate-temperature conditions; associated with other sulfides.

COCHISE COUNTY: Warren District, Bisbee, Campbell Mine, as rounded grains and blades, primarily in chalcocite but in bornite as well; also associated with tetrahedrite, tennantite, and famatinite (Schwartz and Park 1932).

GILA COUNTY: Miami-Inspiration District, Miami Mine, with tennantite and aikinite in veins cutting chalcopyrite (Legge 1939).

PINAL COUNTY: Pioneer supersystem, Magma District, where it is the most important ore mineral of copper in the lowest levels of the Magma Mine (Short et al. 1943). Galiuro Mountains, Bunker Hill (Copper Creek) District, Copper Creek, Childs-Aldwinkle Mine, sparingly with tennantite (Kuhn 1941).

SANTA CRUZ COUNTY: Sunnyside District, Volcano Mine (UA 5858), as a rare vein mineral with pyrite, beneath a chalcocite enrichment blanket (Kistner 1984).

YAVAPAI COUNTY: Wickenburg Mountains, Monte Cristo District, Monte Cristo Mine, associated with tennantite, nickeline, and silver (Bastin 1922).

ENSTATITE

Magnesium silicate: $Mg_2Si_2O_6$. A member of the pyroxene group of rock-forming minerals commonly found in mafic igneous rocks such as gabbro, norite, and peridotite and in their extrusive equivalents.

GILA COUNTY: Dripping Spring Mountains, Banner supersystem, 79 District, on the Reagan claims near the 79 Mine (Bull. 181).

LA PAZ AND YUMA COUNTIES: Common in partially serpentinized harzburgite (Haxel et al. 2018).

PIMA COUNTY: Ajo District, in the basal facies of the Batamonte andesite series (Gilluly 1937).

SANTA CRUZ COUNTY: Santa Rita and Patagonia Mountains, sparingly in some andesites (Bull. 181).

EOSPHORITE

Manganese aluminum phosphate hydroxide hydrate: $Mn^{2+}Al(PO_4)(OH)_2 \cdot H_2O$. Found in granite pegmatite with other phosphate minerals.

MARICOPA AND YAVAPAI COUNTIES: White Picacho District, in granite pegmatites (London 1981).

FIGURE 3.91 Epidote, near Mammoth, Pinal County, FOV 4 mm.

EPIDOTE

Calcium aluminum iron silicate oxide hydroxide: $Ca_2(Al_2Fe^{3+})[Si_2O_7][SiO_4]O(OH)$. Forms in a wide variety of rock types; characteristically a product of low- to medium-grade thermal metamorphism of igneous and sedimentary rocks. In the southwestern United States, it is a product of propylitic alteration of country rock, associated with base-metal mineralization. Also commonly found in contact-metamorphosed limestones with other calcium silicates. It is widespread in the southern part of the state, where it is related to propylitic alteration in early dioritic-stage preporphyry copper hydrothermalism.

APACHE COUNTY: Garnet Ridge SUM District, Garnet Ridge, in the matrix of ejected boulders of basement garnet gneiss (Gavasci and Kerr 1968).

COCHISE COUNTY: Tombstone District, in shale and quartzite (B. S. Butler, Wilson, and Rasor 1938). Chiricahua Mountains, common in the California supersystem, where a copper-bearing epidote vein up to 5 ft. wide extends for over one mile (Dale, Stewart, and McKinney 1960). Courtland and Gleeson Districts, in the wall rocks of pyritic deposits in Abrigo Limestone (Bull. 181). Warren District, Bisbee, Sacramento Hill, in hydrothermally altered porphyry dikes in limestone (Schwartz 1947, 1958, 1959). Cochise supersystem, abundant in metamorphosed limestone skarns and shales adjacent to the Texas Canyon stock (Kellogg 1906; J. R. Cooper and Silver 1964).

GILA COUNTY: Miami-Inspiration supersystem, Pinto Valley District, Castle Dome Mine, in the marginal parts (especially in diabase host rocks) of the mineralized area (N. P. Peterson, Gilbert, and Quick 1946; Creasey 1959); Copper Cities District, Copper Cities porphyry copper deposit (N. P. Peterson 1954). Green Valley District, Harrington claims, with chalcopyrite (Bull. 181). Dripping Spring Mountains, Christmas District, Christmas Mine (C. P. Ross 1925a; Perry 1969). Banner supersystem, 79 District, 79 Mine, in contact-metamorphosed limestones with diopside, andradite, and tremolite (Stanley B. Keith 1972). Banner supersystem, widely distributed with calcite and chlorite in veins and splotches throughout the Williamson Canyon volcanics in propylitic alteration zones peripheral to and earlier than the porphyry copper deposits.

GRAHAM COUNTY: Aravaipa supersystem, Iron Cap District (Simons 1964) and Stanley District (C. P. Ross 1925a), widely distributed in contact-metamorphic base-metal deposits. Santa Teresa Mountains—in 2011, fine-quality epidote crystals associated with crystallized smoky quartz and microcline were found in remote areas in the north and south ends of the mountains. The locations are accessible only by foot and require a long hike to the area (Evan Jones, pers. comm., 2018). San Juan District, San Juan property, widely distributed in the volcanics throughout the Safford supersystem as distinctive splotches in early, strong propylitic alteration zones that predate the porphyry copper mineralization; locally peripheral to a chlorite-pyrite zone in andesite, associated with chlorite and carbonates (Rose 1970).

GREENLEE COUNTY: Copper Mountain supersystem, Morenci District, widespread in contact-metamorphosed rocks, rarely as well-defined crystals (Lindgren 1905; Reber 1916).

LA PAZ COUNTY: Abundant in metamorphosed limestones at several localities. Granite Wash Mountains, Yuma King District, associated with actinolite, magnetite, chalcopyrite, diopside and late sillimanite (Stanley B. Keith, pers. comm., 2019). Dome Rock Mountains, Cinnabar District, in wall rocks of cinnabar veins (Bull. 181).

MOHAVE COUNTY: Cerbat Mountains, Wallapai supersystem, common as an alteration product of wall rocks in sulfide vein deposits in gneisses marginal to the Mineral Park porphyry copper system (B. E. Thomas 1949).

PIMA COUNTY: Santa Rita Mountains, Helvetia-Rosemont and Greaterville Districts, widespread in metamorphosed limestones in the wall rocks of copper deposits (Schrader and Hill 1915; Schrader 1917). Sierrita Mountains, common as a metamorphic mineral and as an alteration product in igneous dikes; Sierrita Mine, a product of the hydrothermal alteration of dioritic rocks (Roger Lainé, pers. comm., 1973); the results of many chemical analyses of epidotes in the Sierrita Mountains are given in Fellows (1976). Pima supersystem, in contact-metamorphic deposits in limestones, in considerable amounts with magnetite, garnet, wollastonite, and hedenbergite (Webber 1929; Eckel 1930a); Twin Buttes District, in the tactites and skarns of the Twin Buttes Mine (Stanley B. Keith, pers. comm., 1973). Santa Catalina Mountains, Marble Peak District, in contact-metamorphic copper deposits near Marble Peak, in places as splendid crystals (Dale, Stewart, and McKinney 1960). Catalina District, Pontatoc Mine (Guild 1934). Summerhaven area, as a metamorphic mineral that is widespread in the Leatherwood Quartz Diorite pluton, where it formed during tectonic burial of the Santa Catalina crystalline complex (Anderson, Barth, and Young 1988). Ajo District, as a widespread but sparse mineral formed as an alteration product of dark silicate minerals (Gilluly 1937; Schwartz 1947, 1958; Hutton and Vlisidis 1960). Tucson Mountains, Amole District, in contact-metamorphic zones adjacent to the Amole intrusive complex (UA 7139). Silver Bell District, as an alteration product in dacite porphyry (Kerr 1951).

PINAL COUNTY: San Manuel District, San Manuel and Kalamazoo orebodies, in hydrothermally altered quartz monzonite and monzonite porphyries, associated with the less intensely altered areas; associated with zoisite, chlorite, hydrobiotite, and secondary biotite (Schwartz 1947; Creasey 1959; J. D. Lowell 1968). Mammoth District, near Ben Hur Mine, lining vesicles in volcanics filled with calcite (R. Gibbs, pers.

comm., 2019). Saddle Mountain District, widely distributed with calcite and chlorite in veins and splotches throughout the Williamson Canyon volcanics in propylitic alteration zones peripheral to and earlier than the porphyry copper mineralization.

SANTA CRUZ COUNTY: Santa Rita and Patagonia Mountains, abundant in metamorphosed limestones (Schrader and Hill 1915); Tyndall District, Glove Mine (H. J. Olson 1966).

YAVAPAI COUNTY: Bradshaw Mountains, in lenses in schist; in dikes at Rich Hill (Bull. 181). Reported in Pylan Creek, twelve miles southeast of Wagoner, as crystals with a prism diameter of 5 in. (Bull. 181) White Picacho District, as small widely dispersed crystals (Jahns 1952). Iron King District, Iron King Mine, in gabbro and diorite, associated with clinozoisite (Creasey 1952).

EPSOMITE

Magnesium sulfate hydrate: $Mg(SO_4) \cdot 7H_2O$. A secondary mineral commonly formed as efflorescences in old mine workings or caves; also an accessory mineral in saline deposits.

COCHISE COUNTY: Warren District, Bisbee, as a late secondary mineral in some mines (Graeme 1981).

GREENLEE COUNTY: Copper Mountain supersystem, Morenci District, as a delicate efflorescence on the wall of mine openings (Lindgren 1905; Guild 1910).

PIMA COUNTY: Silver Bell District, El Tiro Mine, as capillary hair-like crystals. Pima supersystem (Bull. 181). Ajo District, in the oxidized portion of the New Cornelia orebody, possibly of postmine origin (Gilluly 1937).

PINAL COUNTY: San Manuel District, San Manuel Mine, with other secondary sulfate minerals coating mine openings (3rd ed.).

SANTA CRUZ COUNTY: Unspecified locality south of Patagonia (UA 6732).

FIGURE 3.92 Epsomite, Southwest Mine, Bisbee, Cochise County, 18 cm.

#ERIOCHALCITE

Copper chloride hydrate: $CuCl_2 \cdot 2H_2O$. A rare mineral found around volcanic fumaroles and as a product of weathering of copper sulfides in an arid climate.

PINAL COUNTY: Sacaton (Casa Grande) District, ASARCO Sacaton Mine, as light blue-green crusts with brochantite on pyrite partially altered to chalcocite (P. A. Williams, pers. comm., 2005) and is associated with atacamite (Stanley B. Keith, pers. comm., 2019).

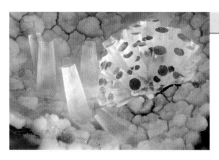

FIGURE 3.93 Erionite, Well No. 1, Ajo, Pima County, FOV 1.8 mm.

ERIONITE

Calcium potassium sodium aluminosilicate hydrate: $(Ca_5,K_{10},Na_{10})[Si_{26}Al_{10}O_{72}]\cdot 30H_2O$. A member of the zeolite group. Erionite is a series name, and three separate species are recognized based on the most abundant cation (Coombs et al. 1997). The specific species have not been determined for localities listed below. See also ERIONITE-K.

COCHISE COUNTY: San Simon Basin, Bowie District, seven miles northeast of Bowie, in bedded lake deposits associated with analcime, chabazite-Na, chabazite, clinoptilolite, thénardite, and halite (Regis and Sand 1967; Edson 1977).

PIMA COUNTY: Well No. 1, seven miles north of Ajo, in volcanics excavated from the well shaft (William Thomas, pers. comm., 1988).

YAVAPAI COUNTY: On U.S. Highway 89, 4.2 miles south of Kirkland Junction, in altered tuff with clinoptilolite (Eyde 1978).

#ERIONITE-K

Potassium aluminosilicate hydrate: $K_{10}[Si_{26}Al_{10}O_{72}]\cdot 30H_2O$. A member of the zeolite group often found in extrusive volcanic rocks.

GRAHAM COUNTY: Ash Peak District, two miles north of Thumb Butte (Tschernich 1992).

GREENLEE COUNTY: Road cut 0.1 mile north of Clifton (Wise and Tschernich 1976).

MOHAVE COUNTY: East of Big Sandy Wash, in the east half of T16N R13W, in Pliocene tuff (C. S. Ross 1928, 1941).

PINAL COUNTY: In a railway cut one mile north of Malpais Hill, on the west side of the San Pedro River, with chabazite, phillipsite, heulandite, and calcite, on celadonite (Thomssen 1983; UA 10547).

SANTA CRUZ COUNTY: Red Mountain District, in a porphyry copper prospect, as a rare mineral, with pyrite (Kistner 1984).

ERYTHRITE

Cobalt arsenate hydrate: $Co_3(AsO_4)_2 \cdot 8H_2O$. Commonly contains nickel; a complete solid-solution series extends to annabergite, $Ni_3(AsO_4)_2 \cdot 8H_2O$. A secondary mineral typically formed by the oxidation of cobalt and nickel arsenides.

APACHE COUNTY: Reported in the White Mountains, exact locality unknown, with cobaltite (Bull. 181).

GILA COUNTY: Found at a locality 0.5 mile northeast of Mule Shoe Bend of the Salt River (Bull. 181). Between Canyon Creek and Carrizo Canyon (UA 1996).

YAVAPAI COUNTY: Black Hills, Grapevine District, near claims of the Old Prudential Copper Company, as powdery encrustations from the alteration of cobaltite and arsenopyrite in a northeast-trending vein hosted in gabbro intruding the Grapevine Gulch metasediments (Guild 1910; C. A. Anderson and Creasey 1958).

ETTRINGITE

Calcium aluminum sulfate hydroxide hydrate: $Ca_6Al_2(SO_4)_3(OH)_{12} \cdot 26H_2O$. A mineral typically formed by the alteration of contact-metamorphosed limestones.

COCHISE COUNTY: Tombstone District, Lucky Cuss Mine, as an alteration product of calcium and aluminum silicates (B. S. Butler, Wilson, and Rasor 1938; H 68800).

EUCRYPTITE

Lithium aluminum silicate: $LiAlSiO_4$. A mineral formed from the alteration of spodumene and commonly intergrown with albite in pegmatites.

YAVAPAI COUNTY: White Picacho District, Midnight Owl and Independence pegmatites (London and Burt 1978, 1982b; London 1981).

EUGENITE

A silver mercury intermetallic compound: $Ag_{11}Hg_2$.

COCHISE COUNTY: Warren District, Bisbee, Southwest Mine, as grains measuring one micron or less, in massive cuprite (Graeme 1993).

#EULYTINE

FIGURE 3.94 Eulytine, Flying Saucer Group, Vulture Mountains, Maricopa County.

Bismuth silicate: $Bi_4(SiO_4)_3$. An uncommon mineral found in bismuth-rich hydrothermal veins.

MARICOPA COUNTY: Vulture Mountains, Flying Saucer District, Flying Saucer Group, Scorpion Hill, as very small white blocky lustrous crystals with kettnerite and laumontite (M. Ascher, pers. comm., 2013; R130114; Dale 1959).

#EUXENITE-(Y)

Yttrium calcium cerium uranium thorium niobium tantalum titanium oxide: $(Y,Ca,Ce,U,Th)(Nb,Ta,Ti)_2O_6$. A member of the euxenite-polycrase series, a rare mineral formed in granite pegmatites and heavy mineral sands.

MARICOPA COUNTY: White Tank Mountains, Caterpillar tractor testing grounds, as massive to poorly crystalline material with zircon crystals in a pegmatite (Phil Hooker, pers. comm., 1985). Buckeye Hills, Buckeye Hills District, in granite at two small prospects, in S16, T1S R3W (Parker 1963).

MOHAVE COUNTY: Near the Aquarius Mountains, Aquarius Mountains District, east of the Big Sandy Wash and south of Burro Creek, in pegmatites as small pockets or kidneys (Shaw 1959); Rare Metals pegmatite (W. B. Simmons et al. 2012).

NAVAJO COUNTY: Holbrook District, Hugh Baron claim (Bull. 181).

PIMA COUNTY: Sierrita Mountains, Pima supersystem, Esperanza District, New Year's Eve Mine, associated with molybdenite and chalcopyrite (Bull. 181).

FAIRBANKITE

FIGURE 3.95 Fairbankite, Tombstone, Cochise County.

Lead tellurite sulfate: $Pb^{2+}_{12}(Te^{4+}O_3)_{11}(SO_4)$. An extremely rare mineral formed by the oxidation of other tellurium-bearing minerals. The Tombstone occurrence is the type locality.

COCHISE COUNTY: Tombstone District, restricted to one fist-sized specimen found on the waste dump of the Grand Central Mine. As tiny clear colorless crystals less than 0.5 mm in size in a thin crust plastered on the walls of voids left by the leaching of galena. Closely associated with oboyerite (S. A. Williams 1979).

FAIRCHILDITE

Potassium calcium carbonate: $K_2Ca(CO_3)_2$. Found in clinkers formed in partly burned trees, with bütschliite and calcite. Coconino County is a cotype locality.

COCONINO COUNTY: Grand Canyon National Park. Discovered by Ranger William J. Kennedy in an unspecified partly burned tree "at a fire on the north side of Kanabownits Canyon one-quarter mile from the Point Sublime road and one-half mile from the North Entrance road" (Milton 1944; Milton and Axelrod 1947; Mrose, Rose, and Marinenko 1966).

FAIRFIELDITE

Calcium manganese phosphate hydrate: $Ca_2Mn^{2+}(PO_4)_2 \cdot 2H_2O$. Usually found in granite pegmatites.

MARICOPA AND YAVAPAI COUNTIES: White Picacho District, in pegmatites (London 1981).

FAMATINITE

Copper antimony sulfide: Cu_3SbS_4. Less common than enargite, famatinite is typically associated with it and is similar to it in structure. It is typically formed in moderate-temperature replacement deposits and veins with other sulfides.

COCHISE COUNTY: Tombstone District, in small amounts in the Ingersoll and Toughnut Mines (B. S. Butler, Wilson, and Rasor 1938). Warren District, Bisbee, Campbell Mine, as rounded grains and blades, largely confined to chalcocite and bornite but also associated with tetrahedrite, tennantite, and enargite; may form coarse graphic intergrowths with chalcocite (Schwartz and Park 1932).

PINAL COUNTY: Reported in the Pioneer supersystem, Magma District, Magma Mine, with the hypogene ores (Hammer and Peterson 1968).

FAYALITE (see also Olivine)

Iron silicate: $Fe^{2+}_2(SiO_4)$. Present in small amounts in certain felsic and alkaline volcanic and plutonic igneous rocks; also formed during the regional metamorphism of iron-rich sedimentary rocks.

GILA COUNTY: Peridot District, near Peridot and Tolklai, as a minor constituent of volcanic bombs and stream gravels (Mason 1968).

#FELSÖBÁNYAITE

Aluminum sulfate hydroxide hydrate: $Al_4(SO_4)(OH)_{10} \cdot 4H_2O$. A rare secondary mineral formed by the oxidation of ores rich in marcasite. Basaluminite, named in 1948, was found to be identical to felsőbányaite, which was named earlier and has priority.

COCHISE COUNTY: Warren District, Bisbee, Lavender pit and Southwest Mine, as white earthy masses with hydrobasaluminite and other aluminum sulfates (Graeme 1981; UA x4504).

FERBERITE

Iron tungstate: $Fe^{2+}(WO_4)$. The end members, iron-rich ferberite ($\geq 50\%$ Fe) and manganese-rich hübnerite ($\geq 50\%$ Mn), form a solid-solution series. Most abundant in quartz veins in granite and in schist associated with pegmatites. Known ferberite localities are listed here, while samples of unknown iron/manganese ratio appear under HÜBNERITE.

MARICOPA COUNTY: Cave Creek District, containing 2.19 percent combined niobium tantalum as oxides (Wherry 1915); Gold Cliff Group, in S11, T6N R4E, about twenty-seven miles north of Phoenix, as ferberite, associated with tungstite, cuprotungstite, fluorite, molybdenite, pyrite, and chalcopyrite (Dale 1959).

MOHAVE COUNTY: Hualapai Mountains, Boriana District, in quartz veins of the Boriana Mine (Hobbs 1944); Myers (1983) reported ferberite in veins in phyllite from this property.

SANTA CRUZ COUNTY: Cox Gulch, Ventura District, as ferberite in the Ventura breccia pipe (Kuck 1978).

YAVAPAI COUNTY: Tungstona District, Boulder Creek, as small brown prismatic crystals ranging from one-eighth to one-fourth of an inch in length with beryl and quartz (C. A. Anderson, Scholz, and Strobell et al. 1955).

FERGUSONITE

Cerium neodymium yttrium niobium oxide: $(Ce,Nd,Y)NbO_4$. A rare mineral found in granite pegmatites with other rare-earth minerals, niobates, and tantalates. Fergusonite is a member of the fergusonite-formanite series, in which the rare-earth element content varies (J. R. Butler and Hall 1960). The exact fergusonite species listed here have not been determined.

MOHAVE COUNTY: Aquarius Mountains District, Rare Metals Mine, in pegmatite; the sample showed an X-ray diffraction pattern that resembled fergusonite upon ignition (Heinrich 1960).

YAVAPAI COUNTY: Rich Hill District, near Yarnell, mica-feldspar quarry northwest of the highway maintenance camp on White Spar Road (Bull. 181).

FERNANDINITE

Calcium sodium potassium vanadium iron titanium oxide hydrate: $(Ca, Na,K)_{0.9}(V^{5+},V^{4+},Fe^{2+},Ti)_8O_{20}\cdot4H_2O$. This rare vanadium mineral is found in only a few locations in the world.

APACHE COUNTY: Cane Valley District, Monument No. 2 Mine, with doloresite and other oxidized vanadium minerals (Witkind and Thaden 1963).

FERRICOPIAPITE

Iron sulfate hydroxide hydrate: $Fe^{3+}_{0.67}Fe^{3+}_4(SO_4)_6(OH_2)\cdot20H_2O$. One of several yellowish secondary iron sulfates formed through the breakdown of pyrite, probably the result of oxidation of part of the ferrous iron of copiapite.

COCHISE COUNTY: Warren District, Bisbee, with the sulfide minerals of the Campbell orebody, as silky-yellow to greenish-yellow rounded crystals (Graeme 1993).

FERRIERITE

Sodium potassium magnesium calcium aluminosilicate hydrate: $\{(K,Na)_5,[Mg_2(K,Na)_2Ca_{0.5}],(Na,K)_5\}[(Si_{31}Al_5),(Si_{29}Al_7)]O_{72}\cdot18H_2O$. An uncommon member of the zeolite group of silicates. Ferrierite is a series, in which the most predominant cations are used to identify a specific species (Coombs et al. 1997). Which of the three ferrierite species appear in the occurrences below is unknown.

MARICOPA COUNTY: In road cuts on the new Lake Pleasant Road, in vugs in basalt (William Hunt, pers. comm., 1992).

PIMA COUNTY: Ajo District, New Cornelia Mine, in a late postmineral mafic dike, with heulandite and calcite (W. J. Thomas and Gibbs 1983; UA 12291); as microcrystals (UA x3495).

FIGURE 3.96 Ferrierite, New Cornelia Mine, Ajo, Pima County, FOV 3 mm.

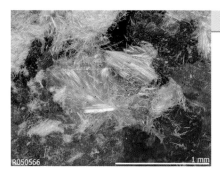

FIGURE 3.97 Ferrimolybdite, Kingman, Mohave County.

FERRIMOLYBDITE

Iron molybdate hydrate: $Fe^{3+}_2(Mo^{6+}O_4)_3 \cdot 7H_2O$. A secondary mineral typically formed by alteration of molybdenite.

COCHISE COUNTY: Warren District, Bisbee, along the Dividend fault, in the Lavender pit (Graeme 1993).

GILA COUNTY: Miami-Inspiration supersystem, Pinto Valley District, Castle Dome Mine; and Copper Cities District, Copper Cities Mine (N. P. Peterson, Gilbert, and Quick 1951; N. P. Peterson 1962).

MOHAVE COUNTY: Kingman area, as short fibrous material encrusting quartz (H 95978; UA 8207–11). Wallapai supersystem, Mineral Park District, in the oxide zone at Ithaca Peak as a common replacement product of molybdenite (Eidel, Frost, and Clippinger 1968).

PIMA COUNTY: South Comobabi Mountains, Cababi District, Mildren and Steppe claims (S. A. Williams 1963). Santa Rita Mountains, near Madera Canyon, as hair-like crystals and tufts (Guild 1907). Twin Buttes District, Twin Buttes Mine, skarns of the Northeast pit (Kuck 1978).

PINAL COUNTY: Galiuro Mountains, Bunker Hill (Copper Creek) District, Copper Creek, Childs-Aldwinkle Mine, as yellow powder and radiating crystal aggregates (Kuhn 1941). Troy supersystem, Troy Ranch District and Rattler District, in oxidized veins as yellow silky fiber-form encrustations on molybdenite (Stanley B. Keith, pers. comm., 2019). Tortolita District, Rare Metals Mine (UA 9488).

SANTA CRUZ COUNTY: Patagonia Mountains, Red Mountain District, Red Mountain Mine (Schrader and Hill 1915).

YAVAPAI COUNTY: Copper Basin District, in the oxidized portions of sulfide deposits (Johnston and Lowell 1961). Northern Bradshaw Mountains, Cleator District, Crazy Basin property (St. Louis claim) near Cleator, as yellow fibrous masses in vugs in quartz, with molybdenite (3rd ed.).

FERRO-ACTINOLITE

Calcium magnesium iron silicate hydroxide: $\square Ca_2(Mg_{2.5-0.0}Fe^{2+}_{2.5-5.0})Si_8O_{22}(OH)_2$. Typically formed in regionally or contact-metamorphosed rocks, especially those with a high amount of iron.

GILA COUNTY: Banner supersystem, 79 District, as compact radiating aggregates in the southwestern workings of the Kullman-McCool Group; identified by electron-microprobe analysis (Jeff Roberts, pers. comm., 1991).

#FERROHEXAHYDRITE

Iron sulfate hydrate: $Fe^{2+}(SO_4)·6H_2O$. A rare mineral formed at volcanic fumaroles.

COCONINO COUNTY: Sunset Crater District, found at Sunset Crater as a fumarolic deposit. It occurs as a white powder with pickeringite, locally coating the cinders (S. L. Hanson, Falster, and Simmons 2001).

#FERRO-HORNBLENDE

Calcium, iron, aluminum silicate hydroxide: $\square Ca_2(Fe^{2+}_4Al)(Si_7Al)O_{22}$ $(OH)_2$. A member of the amphibole group found in metamorphic rocks.

YAVAPAI COUNTY: Wabayuma Peak Wilderness Study Area, southeast of Wabayuma Peak, where a black rather than more common dark-green amphibole occurs. Some of the hornblende occurs in rocks with magnetite. This suggests ferro-hornblende rather than the more common magnesio-hornblende (Conway et al. 1990).

#FERROSELITE

Iron selenide: $FeSe_2$. Found in sedimentary uranium deposits in the Colorado Plateau.

NAVAJO COUNTY: Monument Valley District, at the Mitten No. 2 Mine (S. A. Williams, pers. comm., 2002).

PIMA COUNTY: Well No. 1, seven miles north of Ajo, in volcanics excavated from the Well shaft (R. Gibbs, pers. comm., 2020).

FIGURE 3.98 Ferroselite, Well No. 1, Ajo, Pima County, FOV 2 mm.

FERVANITE

Iron vanadate oxide hydrate: $Fe^{3+}_4V^{5+}_4O_{16}·5H_2O$. Found among the uranium-vanadium ores of the Colorado Plateau.

APACHE COUNTY: Cane Valley District, Monument No. 2 Mine, where H. T. Evans identified it in a Triassic fossilized tree (Haynes 1991).

FIBROFERRITE

Iron sulfate hydroxide hydrate: $Fe^{3+}(SO_4)(OH)·5H_2O$. A secondary mineral formed through the oxidation of pyrite, typically associated with other iron-bearing sulfates such as jarosite, copiapite, and melanterite.

COCHISE COUNTY: Warren District, Bisbee, 700 ft. level of the Shattuck Mine, abundant as crusts on pyritic ores. The mineral is readily recognizable by its peculiar gummy or sectile behavior (Graeme 1981).

FILLOWITE

Sodium calcium manganese phosphate: $Na_3CaMn^{2+}_{11}(PO_4)_9$. Found in granite pegmatites.

MARICOPA AND YAVAPAI COUNTIES: White Picacho District, in pegmatites (London 1981).

#FLAGGITE

Lead copper tellurium sulfate oxide hydroxide hydrate: $Pb_4Cu^{2+}_4Te^{6+}_2$ $(SO_4)_2O_{11}(OH)_2(H_2O)$. A new secondary tellurium mineral named to honor the late Arthur Flagg, a noted Arizona mineralogist and educator.

COCHISE COUNTY: Tombstone District, Tombstone, Grand Central Mine, as small bright lime-green to yellow-green tabular crystals up to 0.5 mm across associated with alunite, backite, cerussite, jarosite, and rodalquilarite (Kampf, Mills, Celestian, et al. 2021).

FLAGSTAFFITE

Terpin hydrate: $C_{10}H_{22}O_3$. A rare mineral formed in fossil logs. The locality north of Flagstaff is the type locality.

COCONINO COUNTY: In debris washed down from the San Francisco Mountains, a few miles north of Flagstaff, as a filling in radial cracks of certain buried tree trunks; a yellowish resinous material probably derived from natural resins of the tree through hydration or oxidation; as orthorhombic crystals up to 1 mm long in drusy cavities (Guild 1920, 1921, 1922; Strunz and Contag 1965).

FLUORAPATITE

Calcium phosphate fluoride: $Ca_5(PO_4)_3F$. A widespread member of the apatite group.

APACHE COUNTY: Garnet Ridge SUM District, as an accessory mineral in ejection boulders of garnet gneiss, associated with zircon and muscovite (Gavasci and Kerr 1968). Cane Valley District, Monument No. 2 Mine, as carbonate fluorapatite, probably related to bone material (Witkind and Thaden 1963).

GRAHAM COUNTY: Lone Star District near Safford, as a minor constituent in metasomatized volcanic rocks (Hutton 1959a).

MARICOPA COUNTY: Mazatzal Mountains, Four Peaks District, Four Peaks Amethyst Mine, as 1.5 mm tabular crystals (J. Lowell and Rybicki 1976).

MOHAVE COUNTY: Hualapai Mountains, as white crystals, which exhibit golden-yellow fluorescence under ultraviolet light, associated with muscovite (Bull. 181). Cleopatra District, Red Top Mine (T. Kampf, pers. comm., 2001).

PIMA COUNTY: Ajo District, New Cornelia Mine (W. J. Thomas and Gibbs 1983).

SANTA CRUZ COUNTY: Patagonia supersystem, Four Metals District, Four Metals breccia pipe, as an accessory mineral, pale-pink hexagonal prisms up to 1 in. long, occurring with minor chalcopyrite and interstitial pyrite, filling open spaces between breccia clasts of chloritic granodiorite (Stanley B. Keith, pers. comm., 2019).

YAVAPAI COUNTY: Eureka supersystem, Bagdad District near Bagdad, as an accessory mineral in titaniferous magnetite bodies (Ball and Broderick 1919; Schwartz 1947; C. A. Anderson 1950). White Picacho District, in pegmatites (London and Burt 1982a).

#FLUORAPOPHYLLITE-(K)

Potassium calcium silicate fluoride hydrate: $KCa_4Si_8O_{20}F \cdot 8H_2O$. A member of the apophyllite group, which was renamed in 2013 to include a prefix and a suffix to indicate specific anion and cation (Hatert et al. 2013). Fluorapophyllite-(K) primarily forms as a secondary mineral in amygdules in basalts and is commonly associated with zeolites. It is also found in contact-metamorphosed limestones bordering intrusive rocks.

COCHISE COUNTY: Tombstone District, as subhedral grains with vesuvianite and wollastonite in tactite (Bideaux, Williams, and Thomssen 1960). The exact species determination is not known for this locality.

GILA COUNTY: Southeastern Dripping Spring Mountains, Banner supersystem, Christmas District, Christmas Mine, coating kinoite (S. A. Williams 1976).

PIMA COUNTY: Santa Rita Mountains, Helvetia-Rosemont District, near Helvetia, in a drill core that penetrated skarn developed in Paleozoic limestones and dolomites; as vuggy, crystalline masses in which copper and kinoite are embedded (J. W. Anthony and Laughon 1970). The exact apophyllite species is not known.

FIGURE 3.99 Fluorite, Homestake Mine, Mohave County, 5.6 cm high.

FLUORITE

Calcium fluoride: CaF_2. Fluorite is a widespread and common mineral of hydrothermal origin. Commonly formed as a primary mineral in veins, of which it is the chief constituent, or in the gangue of lead, zinc, and silver ores. It is also found in sedimentary rocks, such as limestones and dolomites, and in plutonic igneous rocks, such as granites and monzonites.

COCHISE COUNTY: Tombstone District, locally abundant in some silicified areas, particularly at the Empire Mine (B. S. Butler, Wilson, and Rasor 1938). Southeast Tombstone Hills, Fluorita District, as pink translucent to transparent cubo-octahedral crystal aggregates filling fractures in dolomite; at the La Fluorita Dulcita claim as pale to dark purple octahedrons to 1 in. on quartz casts after calcite crystals (Les Presmyk, pers. comm., 2020). Reportedly found near Government Draw, as purple crystals with quartz (Bull. 181). Chiricahua Mountains, California supersystem, as small quantities mined from quartz veins near Paradise (Bull. 181). Little Dragoon Mountains, Bluebird District, in a granite pegmatite, with hübnerite (Guild 1910; Palache 1941a; J. R. Cooper and Huff 1951; J. R. Cooper and Silver 1964). Northeastern Whetstone Mountains, Lone Star District, Lone Star Mine, as clusters of cubic crystals in veins west of San Juan siding (Hewett 1964; Stanton B. Keith 1973).

GILA COUNTY: Green Valley District, Oxbow Mine, with epidote (Lausen and Wilson 1925). Eastern Tonto Basin, Sierra Ancha District, Packard claims, as white, light-blue, and purple masses in veins (Batty et al. 1947). Miami-Inspiration supersystem, Pinto Valley District, Castle Dome Mine, in small amounts in open fractures, associated with baryte (N. P. Peterson, Gilbert, and Quick 1951).

GRAHAM COUNTY: Aravaipa supersystem, Grand Reef District, Grand Reef and Ten Strike Mines, as purple fluorite associated with quartz and hematite (J. Callahan, pers. comm., 2018); in the Santa Teresa Mountains as purple octahedrons up to 2 in. (Les Presmyk, pers. comm., 2020); Iron Cap District, in veins of the Landsman Group (Simons 1964); Stanley District, as crystals in baryte veins (C. P. Ross 1925a).

GREENLEE COUNTY: Ash Peak District, in gouge in veins in andesite at several properties near Duncan (M. A. Allen and Butler 1921a; Ladoo 1923), especially the Luckie and Fourth of July Mines, where fluorite is covered by mammillary layers of psilomelane (Hewett, Fleischer, and Conklin 1963; Hewett 1964); Ellis Mine (Batty et al. 1947).

LA PAZ COUNTY: Trigo Mountains, Silver District, as crystalline masses, dense varicolored bands coating quartz, and as vein material with quartz and baryte (Bull. 181); Padre Kino Mine, as microcrystals with willemite and wulfenite (Peter Megaw, UA x2978); Silver King Mine, as microcrystals (Peter Megaw, UA x1829). Eastern slopes of the Plomosa Mountains, Bouse Hills District, in most baryte veins, which form in association with manganese oxide veins (Hewett 1964); McWilliams Group, as purple octahedra up to 1 cm on an edge (MM L530). Buckskin Mountains, Midway District, Chicago and Mammoth properties, with baryte (3rd ed.).

MARICOPA COUNTY: Harquahala District, Snowball property, in veins in Precambrian rocks, with baryte (Denton and Kumke 1949; Hewett 1964). Vulture District, west of Morristown, in veins (Bull. 181). Reported in the White Tank Mountains. In a quartz quarry at Pinnacle Peak, east of Paradise Valley (Bull. 181). Painted Rock District, Rowley Mine near Theba, in small amounts as colorless to violet masses in quartz veins; commonly as cubic or dodecahedral microcrystals, some of which are elongated (W. E. Wilson and Miller 1974). Crest of the Belmont Mountains, Osborne District, in the Belmont Granite, as small purple crystals in miarolitic cavities with quartz crystals, muscovite, "biotite," and epidote (Reynolds, Scot, and O'Haire 1985).

MOHAVE COUNTY: Oatman District, as white to pale-green bands or as linings of cavities (Schrader 1909); green octahedrons in veins from Oregon, Homestake, Moss, and Hardy Mines (Hay and Morris 2015); Skinner lode, south of Silver Creek, "as beautiful octahedral crystals of green, white, and purple . . . in a quartzose and feldspathic gangue with occasional gray spots of minutely diffused sulphide of silver" (Silliman 1866); at other smaller properties of the district, including the Caledonia, Dayton, Quackenbush, and Knickerbocker properties (Silliman 1866). Cerbat Mountains, Wallapai supersystem, Mineral Park District, Altata and Tintic Mines, as a late gangue mineral known from two veins in a belt of sulfide-containing fissure-vein deposits, associated with a granite porphyry stock (B. E. Thomas 1949). Artillery Peak District, with baryte in veins that contain manganese oxides and cut the Artillery Formation (Hewett and Fleischer 1960; Hewett 1964). Potts Mountain, Owens

District, as purple material, cementing breccia (Robert O'Haire, pers. comm., 1972). Boriana District, Boriana Mine, of a rich purple color, in quartz veins (Hobbs 1944). Valentine District, near Wright Creek Ranch on the railroad (about thirteen miles west-southwest of Peach Springs; Moore 1936 in Hewett et al. 1936), Bluebird or Bountiful Beryl prospect, in several irregular pegmatite dikes in schist and granite, with titanite, garnet, schorl, and beryl, and in microcline-albite-quartz pegmatite (Schaller, Stevens, and Jahns 1962).

PIMA COUNTY: Santa Rita Mountains, Helvetia-Rosemont District, New York Mine (Hewett 1964); Gunsight Mine, as microcrystals (William Kurtz, UA x2677). Silver Bell District, at several properties (Guild 1910). Sierrita Mountains, Samaniego District, Neptune property, as veins from a few inches to 2 ft. wide (M. A. Allen and Butler 1921a); Mission-Pima District, Mission pit, as microcrystals, with tetrahedrite and dolomite (UA x3548). As geodes in the southern Santa Catalina Mountains (UA 1571) and as microcrystals at the Old Spanish Mine (William Kurtz, UA x436).

PINAL COUNTY: Mammoth District, Mammoth–St. Anthony Mine, abundant as microscopic crystals in the lower levels; as crystals up to 1 in. on an edge (N. P. Peterson 1938a, 1938b; Fahey 1955; Bideaux 1980).

SANTA CRUZ COUNTY: Patagonia Mountains, Trench District, Alta Mine, as red material with bromian chlorargyrite and chlorargyrite (Schrader and Hill 1915; Schrader 1917).

YAVAPAI COUNTY: South of Bagdad Mine, in small pockets in pegmatite, associated mainly with triplite and a green mica (Hurlbut 1936). Bradshaw Mountains, Tiger District, Springfield Group (Bull. 181). Castle Creek District, Swallow Mine (Bull. 181). McCloud Mountains, Rich Hill District, at a prospect near the Leviathan Mine from which fluorite has been shipped (Bull. 181). Northeast of Congress Junction, as dodecahedral crystals 3 to 4 in. in diameter, in pegmatite dikes (Jahns 1953). Black Rock District, Monarch Copper Mine, common as white and green octahedral crystals up to 5 cm (3rd ed.).

YUMA COUNTY: Castle Dome Mountains, Castle Dome District, as greenish, purple, and rose-colored crystals and clear cleavable masses up to several inches in diameter, locally associated with galena, baryte, and wulfenite; dark-blue to almost black fluorite from this district, particularly from the Hull Mine, is photosensitive and changes to light gray or pink on exposure to sunlight (Ladoo 1923; E. D. Wilson 1933; Batty et al. 1947; N. P. Peterson 1947); Big Dome claim (M. A. Allen and Butler 1921a). Kofa District, in deposits in layered volcanic rocks (Hewett 1964).

FORNACITE

Copper lead chromate arsenate hydroxide: $CuPb_2(CrO_4)$ $(AsO_4)(OH)$. A rare secondary mineral found in lead deposits.

FIGURE 3.100 Fornacite, Evening Star Mine, Maricopa County, FOV 5 mm.

COCHISE COUNTY: Warren District, Bisbee, Shattuck shaft (Graeme 1981).

GILA COUNTY: Dripping Spring Mountains, Banner supersystem, 79 District, 79 Mine (UA 12173), as microcrystals associated with mimetite and wulfenite (UA x2319).

LA PAZ COUNTY: New Water District, Eagle Eye Mine, as small crystalline rosettes (3rd ed.).

MARICOPA COUNTY: Big Horn Mountains, Big Horn District, Evening Star Mine (R. Gibbs, pers. comm., 2019).

PIMA COUNTY: Tucson Mountains, Old Yuma District, Old Yuma Mine, on hematite and quartz (Bideaux, Williams, and Thomssen 1960), as microcrystals (William Kurtz, UA x3978); Gila Monster Mine, with cerussite and wulfenite (Richard L. Jones, pers. comm., 3rd ed.). Ajo Cornelia District, 0.5 mile south of the New Cornelia pit, with cerussite and chlorargyrite, as films coating fracture surfaces in altered andesite (3rd ed.). Empire Mountains, Empire District, Copper Point prospect, as microcrystals (UA x3361). Roadside District, in dumps near the Roadside Mine, as sharply formed crystals with shattuckite and cerussite (3rd ed.).

PINAL COUNTY: Mammoth District, Mammoth–St. Anthony Mine, with wulfenite and fluorite, as crystals up to several millimeters long (Bideaux, Williams, and Thomssen 1960; Bideaux 1980; UA 3401), and as fiber-like microcrystals with creaseyite (Bruce Maier, UA x1933). Vekol District, Vekol Mine, with wulfenite and chlorargyrite (3rd ed.).

YAVAPAI COUNTY: Black Rock District, near Constellation, with crocoite (John S. White Jr., pers. comm., 3rd ed.).

FORSTERITE

Magnesium silicate: $Mg_2(SiO_4)$. The magnesium-rich portion of the olivine series, of which fayalite is the iron-rich end member. Forsterite is associated with dunites, peridotites, some basalts, and olivine-rich layered intrusive rocks. It is also formed during thermal and contact metamorphism of dolomites and magnesium-bearing limestones.

APACHE COUNTY: Buell Park District, Buell Park, ten miles north of Fort Defiance. Garnet Ridge SUM District, at Garnet Ridge, as clear green to brown gem-quality material (Gregory 1916).

COCHISE COUNTY: Little Dragoon Mountains, Cochise supersystem, Johnson Camp District, near Johnson Camp, with epidote, zoisite, garnet, and other silicates, as a product of contact metamorphism of dolomitic rocks around the Texas Canyon stock (J. R. Cooper 1957). Tombstone District, in the contact zone at Comstock Hill; Lucky Cuss Mine, as an important constituent of skarn (B. S. Butler, Wilson, and Rasor 1938). Reported in the Warren District, Bisbee (3rd ed.). San Bernardino (Geronimo) volcanic field, as the principal constituent of ellipsoidal lherzolite inclusions and nodules in alkali basalts with chrome diopside and spinel (D. J. Lynch 1972; Kempton et al. 1984).

COCONINO COUNTY: San Francisco Mountains, in basalts (Gregory 1917).

GILA COUNTY: Southeastern Dripping Spring Mountains, Banner supersystem, Christmas District, Christmas Mine, as an important constituent of skarns formed in the Martin Formation and Escabrosa Limestone; associated with garnet, vesuvianite, calcite, anhydrite, and clintonite (Perry 1969). Peridot District, San Carlos Apache Reservation, near Peridot and San Carlos, as the main mineral in Type I lherzolite inclusions with diopside and spinel in alkali basanitic basalts (Frey and Prinz 1978; originally characterized as volcanic bombs by Lausen 1927a) and stream gravels (H 89313); a cut stone from a quarry on the north side of the diatreme at the south end of Peridot Mesa weighed 25.75 carats; also occurs as occasional large euhedral crystals up to 1.5 in. diameter. Sierra Ancha, abundant in a differentiated diabase sill complex (D. Smith 1970) and is generally a major constituent of the 1050 to 1090 Ma diabase sills throughout southeastern Arizona (J. R. Cooper and Silver 1964).

LA PAZ AND YUMA COUNTIES: Cemetery Ridge District, common in partially serpentinized harzburgite; and in metasomatic schist and granofels, with clinochlore, tremolite, olivine, and spinel (Haxel et al. 2021).

MARICOPA COUNTY: North end of the Sierra Estrella Mountains, at the contact between marble and pelitic schists, with zoisite, diopside, and tremolite (Sommer 1982).

MOHAVE COUNTY: Cerbat Mountains, in Quaternary basalts (B. E. Thomas 1953).

PIMA COUNTY: Pima supersystem, Twin Buttes District, Twin Buttes Mine, in garnetite (Stanley B. Keith, pers. comm., 1973); in skarns or tactites where the altered limestones were relatively pure (McNew 1981).

PINAL COUNTY: Lakeshore District, Lakeshore Mine, abundant in some portions of the metadolomite (3rd ed.). As an accessory mineral in diabase sills that intrude rocks of the Apache Group in this and adjacent counties (Bull. 181). Dripping Spring Mountains, with augite and a mixture of clay minerals and iron oxides in Tertiary basalts (Bull. 181). Galiuro Mountains, in the contact zone between granodiorite and Cretaceous sedimentary rocks (Bull. 181).

SANTA CRUZ COUNTY: Santa Rita and Patagonia Mountains, sparingly in gabbro, diabase, and andesites (Bull. 181).

YAVAPAI COUNTY: Iron King District, Iron King Mine, as phenocrysts in porphyritic basalt flows associated with mineralized volcanic breccias, with augite (Creasey 1952).

FOURMARIERITE

Lead oxide uranyl hydroxide hydrate: $Pb_{1-x}O_{3-2x}(UO_2)_4(OH)_{4+2x} \cdot 4H_2O$. A secondary mineral formed from lead and uranium derived from the alteration of uraninite; also formed as a constituent of a mixture of secondary uranium minerals through alteration of uraninite.

APACHE COUNTY: Cane Valley District, Monument No. 2 Mine, as small reddish grains in uraninite, with schoepite and becquerelite (C. Frondel 1956; Witkind and Thaden 1963).

FRAIPONTITE

Zinc aluminum silicate hydroxide: $(Zn,Al)_3(Si,Al)_2O_5(OH)_4$. An uncommon clay mineral and a member of the kaolinite-serpentine group.

COCHISE COUNTY: Gleeson District, on the dumps of the Defiance and Silver Bill Mines (Foord, Taggert, and Conklin 1983); Silver Bill Mine, as microcrystals (UA x4357).

FREIBERGITE

Silver copper iron antimony sulfide: $Ag_6[Cu_4Fe_2]Sb_4S_{12}$. A member of the tetrahedrite solid-solution series (see also TETRAHEDRITE); some deposits of tetrahedrite group minerals in mines yielding silver-rich ores are probably freibergite, although only the localities listed here are known to be authenticated.

COCHISE COUNTY: Warren District, Bisbee, in the sulfide minerals of the Campbell orebody (Graeme 1993).

PIMA COUNTY: Cerro Colorado District, Cerro Colorado Mine, occurs with chlorargyrite, cerussite, anglesite, chrysocolla, copper carbonates, galena, and sphalerite; New Colorado Mine, found with argentiferous galena, copper carbonates and oxides, sphalerite, and rich silver mineralization (Stanton B. Keith, 1974).

PINAL COUNTY: Pioneer supersystem, Silver King District, Silver King Mine, as microcrystals (David Shannon, UA x4344).

YAVAPAI COUNTY: Big Bug District, Arizona National Mine, in galena with acanthite and in cavities with wire silver (Lindgren 1926).

FREIESLEBENITE

Silver lead antimony sulfide: $AgPbSbS_3$. A rare medium- to low-temperature mineral formed in veins with acanthite, galena, siderite, and ruby silvers.

YUMA COUNTY: Castle Dome District, from which a small amount was mined and shipped with other argentiferous ores (Bull. 181).

FRIEDRICHITE

Copper lead bismuth sulfide: $Cu_5Pb_5Bi_7S_{18}$. Found in vein quartz with other copper sulfides.

COCHISE COUNTY: Northeast flank of the Johnny Lyon Hills, Yellowstone District, as microscopic beads in auriferous vein quartz in Pinal Schist (3rd ed.).

FROHBERGITE

Iron telluride: $FeTe_2$. An uncommon mineral formed in hydrothermal veins with gold, copper sulfides, and other tellurides.

COCHISE COUNTY: Tombstone District, in one rock from the dump of the Joe shaft. The host rock is brecciated Bisbee Group shale cemented by granular pyrite (S. A. Williams 1980b). Gold Hill District, Gold Hill, southeastern Mule Mountains (J. W. Anthony et al. 1990).

#FUETTERERITE

Lead copper telluride hydroxide chloride: $Pb_3Cu_6Te^{6+}O_6(OH)_7Cl_5$. A rare mineral formed from the partial oxidation of primary sulfides and tellurides.

COCHISE COUNTY: Tombstone District, Grand Central Mine, as light-blue clusters of very small crystals associated with bairdite, timroseite, and burckhardtite (Carsten Slotta, pers. comm., 2020).

#FURUTOBEITE

Copper silver lead sulfide: $(Cu,Ag)_6PbS_4$. Found with stromeyerite and bornite in hydrothermal sulfide deposits.

COCHISE COUNTY: Warren District, Bisbee, Campbell Mine, 1600 level in bornite sulfide ore (Schumer 2017).

GADOLINITE

Cerium neodymium yttrium iron beryllium oxide silicate: $(Ce_2,Nd_2,Y_2)Fe^{2+}Be_2O_2(SiO_4)_2$. A rare mineral principally formed in granites and granite pegmatites. Gadolinite has been designated as a supergroup of minerals, which is divided into two groups—gadolinite and herderite. The gadolinite group is defined by the presence of the tetrahedral cation Si^{4+}; which of the gadolinite species occur in the localities below have not been verified.

MOHAVE COUNTY: Aquarius Mountains District, Rare Metals Mine, thirty miles south of Hackberry, in pegmatites from which several tons of the mineral have been mined (Heinrich 1960); W. B. Simmons et al. (2012) could not verify gadolinite from their studies at the Rare Metals Mine. Near Kingman and in the northern part of the county, as fragments in sand dunes (Bull. 181). Cerbat Mountains, Kingman Feldspar District, Kingman Feldspar Mine, as black, vitreous, rough prismatic crystals a few inches long, with beryl in pegmatite (B. E. Thomas 1953; Heinrich 1960).

PIMA COUNTY: North side of the Rincon Mountains, with xenotime-(Y) in biotite gneiss (Bideaux, Williams, and Thomssen 1960).

GAHNITE

Zinc aluminum oxide: $ZnAl_2O_4$. An uncommon member of the spinel group, gahnite is found in schists, high-temperature replacement bodies in metamorphic rocks and in granite pegmatites.

MARICOPA AND YAVAPAI COUNTIES: White Picacho District, as small blue-gray to deep-green crystals in pegmatites that also contain greenish-black crystals of the ferroan variety of spinel, pleonaste (Jahns 1952).

PIMA COUNTY: Santa Catalina Mountains, with spessartine garnet and scheelite in heavy mineral placers in Lower Sabino Canyon (Hinkle and Ryan 1982; Force 1997).

YAVAPAI COUNTY: Old Dick District, Bruce Mine, as dark-green equant octahedral crystals in mica schist, up to several millimeters in size (R. Jenkins, pers. comm., 2017).

FIGURE 3.101 Galena, Iron Cap Mine, Graham County, 5.4 cm high.

GALENA

Lead sulfide: PbS. A widely distributed primary mineral typically associated with zinc and copper sulfides and silver minerals. Galena is the most important ore mineral of lead and often silver bearing.

APACHE COUNTY: Lukachukai District, associated with secondary uranium and vanadium minerals in bedded deposits in the Salt Wash Member of the Morrison Formation (Joralemon 1952).

COCHISE COUNTY: Warren District, Bisbee, Campbell Mine, from which considerable quantities were mined from 1945 to 1949 (E. D. Wilson et al. 1950); Junction and Shattuck shafts (Graeme 1981). Tombstone District, in the orebodies of both the roll deposits and in the fissure veins; extensively replaced by cerussite and, to a lesser extent, by anglesite (Hewett and Rove 1930; B. S. Butler, Wilson, and Rasor 1938; Rasor 1939). Southern Dragoon Mountains, Gleeson District, as scattered bunches in copper sulfide ores; Mystery Mine (UA 7188). Little Dragoon Mountains, Cochise supersystem, Johnson Camp District, Primos Mine (E. D. Wilson et al. 1950; J. R. Cooper and Huff 1951; J. R. Cooper and Silver 1964). Huachuca Mountains, Reef District, Reef Mine (Palache 1941a). Chiricahua Mountains, California supersystem, Leadville District, Humboldt Mine (Hewett and Rove 1930); State of Texas and Panama Mines (E. D. Wilson et al. 1951). Swisshelm Mountains, as replacements in Naco Limestone (E. D. Wilson et al. 1951).

COCONINO COUNTY: Grand Canyon National Park, Orphan District, Orphan Mine breccia pipe, on the South Rim 1 km west of Bright Angel Lodge, associated with the uranium component of copper-uranium-lead ores in Coconino Sandstone (Isachsen, Mitcham, and Wood 1955).

GILA COUNTY: Miami-Inspiration supersystem, in sparse amounts in all deposits (N. P. Peterson 1962); Pinto Valley District, Castle Dome Mine (Brush 1873; N. P. Peterson, Gilbert, and Quick 1946; N. P. Peterson 1947); Globe Hills District, Apache Mine, as remnants enclosed in shells of anglesite and cerussite (E. D. Wilson et al. 1950). Dripping Spring Mountains, Banner supersystem, 79 District, 79 Mine, as the most abundant hypogene ore mineral, commonly coated with smithsonite, and as relics in the upper oxide zone (Kiersch 1949; E. D. Wilson et al. 1951; D. V. Lewis 1955; Stanley B. Keith 1972); Christmas District, Christmas Mine, in small amounts in orebodies replacing dolomite (Perry 1969).

GRAHAM COUNTY: Aravaipa supersystem, Grand Reef District (Denton 1947a), as the main ore mineral at the Grand Reef Mine; Iron Cap District, Head Center, Iron Cap, and Grand Central Mines (C. P. Ross 1925a; Simons and Munson 1963; Simons 1964; W. E. Wilson 1988).

GREENLEE COUNTY: Copper Mountain supersystem, Morenci District, gold bearing in the ores of the King Mine; Stevens Group of claims (3rd ed.).

LA PAZ COUNTY: Trigo Mountains, Silver District, Black Rock, Chloride, Silver King, and Silver Glance properties (Bull. 181); Red Cloud Mine, highly argentiferous, probably contains acanthite (Hamilton 1884; Foshag 1919). Harquahala District, Bonanza Mine (H. Bancroft 1911; E. D. Wilson 1933; E. D. Wilson et al. 1951). Plomosa Mountains, New Water District, in baryte veins with manganese oxides in layered volcanic rocks (Hewett 1964). Southern Cross District, Southern Cross Mine, as the main hypogene lead mineral (Stanton B. Keith 1978).

MARICOPA COUNTY: Painted Rock District, Rowley Mine, as nodular masses as cores enveloped by anglesite, then cerussite (W. E. Wilson and Miller 1974; UA 4005). Belmont Mountains, Osborne District, south of Wickenburg, Moon Anchor Mine, Potter-Cramer property, and Rat Tail claim, where oxidization has produced a variety of secondary lead and chromium minerals (S. A. Williams, McLean, and Anthony 1970).

MOHAVE COUNTY: Common at many properties of the Cerbat Mountains, Wallapai supersystem, especially in the Golconda and Tennessee lead-zinc-silver districts (Haury 1947; Tainter 1947c) and the Grand Wash Cliffs area (Schrader 1909; B. E. Thomas 1949); Mineral Park District, Ithaca Peak area, with pyrite, chalcopyrite, and sphalerite in veins in a quartz-monzonite stock (Eidel 1966).

PIMA COUNTY: Santa Rita Mountains, Helvetia-Rosemont District, abundant at several properties, including the Ridley Mine near Helvetia.

Empire District, Chief, Prince, and other properties of the Hilton Group (Schrader and Hill 1915; Schrader 1917; E. D. Wilson et al. 1950, 1951). Cerro Colorado District, Cerro Colorado Mine, with stromeyerite, tetrahedrite, and silver (Bull. 181). Sierrita Mountains, Pima supersystem, Mission-Pima District, Daisy Mine, Mineral Hill Mine, Eagle-Picher Mine (B. M. Williams 2018). Papago District, Sunshine Mine, as the main ore mineral (Ransome 1922). South Comobabi Mountains, Cababi District, Mildren and Steppe claims (S. A. Williams 1963). Tucson Mountains, Old Yuma District, Old Yuma Mine, with sparse tetrahedrite and supergene anglesite (Guild 1917; Stanley B. Keith, pers. comm., 1973).

PINAL COUNTY: Pioneer supersystem, Silver King District, Silver King Mine (Guild 1917); Belmont District, in fairly large bodies in the Belmont Mine; Magma District in the ores of the Magma Mine (Short et al. 1943). Mammoth District, Mammoth–St. Anthony Mine, in the sulfide zone and altered to anglesite and cerussite in the oxidized zone (N. P. Peterson 1938a, 1938b; E. D. Wilson et al. 1950; Fahey 1955). Bunker Hill (Copper Creek) District, Copper Creek, Bluebird Mine, as the principal ore mineral (Simons 1964). Saddle Mountain District, Adjust Mine, silver bearing; Saddle Mountain and Little Treasure properties (Bull. 181). Central Dripping Spring Mountains, Troy supersystem, C and B–Grey Horse District, in several vanadium prospects near Kearny (Stanley B. Keith, pers. comm., 1973).

SANTA CRUZ COUNTY: Santa Rita Mountains, Santa Rita supersystem, in many districts and mines, including the Tyndall District, Glove Mine (H. J. Olson 1966). Patagonia Mountains, abundant in nearly all districts: Mowry District, Mowry Mine, in ore having silver values up to 3,800 oz/ton (Bull. 181); Flux District, Flux Mine, as excellent specimens of cubo-octahedral crystals (Bull. 181); Washington Camp District, Holland Mine (Marshall and Joensuu 1961); Stella Mine near Duquesne, with chalcopyrite, diopside, and quartz (Stanley B. Keith, pers. comm., 1973); Trench District, Trench Mine, Haggin shaft, with sphalerite, alabandite, and rhodochrosite (Hewett and Rove 1930). Cobre Ridge area, Ruby District, Montana Mine, with sphalerite (E. D. Wilson et al. 1951).

YAVAPAI COUNTY: Bradshaw Mountains, at several properties in the Walker, Hassayampa, Tiger, Tip Top, and Castle Creek Districts, in many places with tetrahedrite (Bull. 181). Iron King District, Iron King Mine, as the most abundant ore mineral (Creasey 1952; C. A. Anderson and Creasey 1958). Eureka supersystem, Hillside District, Hillside Mine (Axelrod et al. 1951) and Bagdad District, Bagdad Mine (C. A. Anderson 1950). Black Hills, Verde District, United Verde Mine; Shea District, Shea property; Black Hills, Shylock District, Shylock Mine (Bull. 181).

YUMA COUNTY: Castle Dome District, Flora Temple, Señora, Little Dome, Hull, Lincoln, and Adams properties (Bull. 181).

#GANOMALITE

Lead calcium silicate oxide: $Pb_9Ca_6(Si_2O_7)_4(SiO_4)O$. A mineral found in skarn assemblages.

PIMA COUNTY: Sells area, as yellowish-red crystalline masses (R050387).

GARRONITE

Calcium sodium aluminosilicate hydrate: $(Ca_3,Na_6)(Al_6Si_{10}O_{32})\cdot(14,8.5)$ H_2O. A silica-poor zeolite found in low-silica rocks accompanied by other silica-poor zeolites. Garronite-(Ca) is more common, but which of the two species occurs in the locality below has not been determined.

PINAL COUNTY: About 5.5 miles south of Superior, in a middle Tertiary cinder cone, in olivine bombs and scoria, with cowlesite, lévyne, and phillipsite (Tschernich 1992).

GEARKSUTITE

Calcium aluminum fluoride hydroxide hydrate: $CaAl\ F_4(OH)\cdot H_2O$. An uncommon mineral, formed in pegmatites, hydrothermal veins, sedimentary rocks, and as a result of hot spring activity.

GRAHAM COUNTY: Aravaipa supersystem, Grand Reef District, Grand Reef Mine, as porcelaneous masses and powder, associated with baryte, linarite, and other minerals in quartz-lined vugs (R. W. Jones 1980).

GEDRITE

Magnesium aluminosilicate hydroxide: $\square Mg_2(Mg_3Al_2)(Si_6Al_2)O_{22}(OH)_2$. A member of the amphibole group, typically found in metamorphic rocks; it is similar to the species anthophyllite but contains aluminum (Deer, Howie, Zussman 1962).

SANTA CRUZ COUNTY: Patagonia Mountains, Washington Camp District, Pride of the West Mine, in contact-metamorphosed limestone (Schrader and Hill 1915).

#GEERITE

Copper sulfide: Cu_8S_5. An intermediate secondary sulfide in the chalcocite-covellite replacement series (Sikka et al. 1991).

GREENLEE COUNTY: Copper Mountain supersystem, Metcalf District, Metcalf Mine, as replacement of secondary copper sulfides (Enders 2000).

GEIKIELITE

Magnesium titanium oxide: $MgTiO_3$. A mineral of the ilmenite group, commonly formed in magnesium-rich metamorphic rocks and serpentines.

COCHISE COUNTY: Cottonwood Basin District, Skeleton Canyon, near the New Mexico border, as lustrous black crystals up to 1 in. across in silicified tuff, found as float (David Garske, pers. comm., 3rd ed.).

FIGURE 3.102 Georgerobinsonite, Mammoth–St. Anthony Mine, Pinal County.

#GEORGEROBINSONITE

Lead chromate hydroxide fluoride chloride: $Pb_4(CrO_4)_2$ $(OH)_2FCl$. A rare secondary mineral found in the oxidized portions of polymetallic ore deposits. The Mammoth–St. Anthony Mine is the type locality.

PINAL COUNTY: Mammoth District, Mammoth–St. Anthony Mine, as minute intergrowths of thin tabular orange crystals less than 0.1 mm across, associated with caledonite, cerussite, diaboleite, leadhillite, matlockite, murdochite, pinalite, wulfenite, and yedlinite in vugs in a silicified matrix (M. A. Cooper et al. 2011; R110148). This is the type and only known locality.

GERHARDTITE

Copper nitrate hydroxide: $Cu_2(NO_3)(OH)_3$. A rare secondary mineral formed under oxidizing conditions in copper deposits in arid and semi-arid regions and associated with minerals such as atacamite, brochantite, malachite, and azurite. The United Verde Mine is the type locality.

GREENLEE COUNTY: Copper Mountain supersystem, Morenci District, Chase Creek Canyon, on cliffs of granite porphyry, as bright-green coating of small rough mammillary forms (Lindgren and Hillebrand 1904; Lindgren 1905; Guild 1910); Morenci, as microcrystals (Carl Richardson, UA x5963).

MARICOPA COUNTY: Belmont Mountains, Osborne District, Tonopah-Belmont Mine, as very small pale-blue clusters of bladed crystals (Joe Ruiz, pers. comm., 2012; R120082).

PIMA COUNTY: Pima supersystem, Mission-Pima District, Mineral Hill Mine, Daisy shaft, as thin seams of granular material in cuprite, which is coated by aurichalcite (S. A. Williams 1961).

YAVAPAI COUNTY: Jerome supersystem, Verde District, United Verde Mine, as crystals up to 0.25 in., on fractures in massive cuprite (H. L. Wells and Penfield 1885).

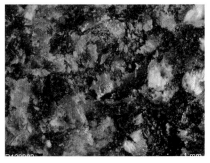

FIGURE 3.103 Gerhard-tite, Tonopah-Belmont Mine, Maricopa County.

GERSDORFFITE

Nickel arsenic sulfide: NiAsS. An uncommon mineral found in a few hydrothermal vein deposits formed at moderate temperatures. Gersdorffite is a member of the cobaltite group of minerals. There are three polytypes: $-P2_13$, $-Pa3$, and $-Pca2_1$. The specific species of the occurrence below is unknown.

MOHAVE COUNTY: Hack District, Hack No. 2 Mine (Casebolt, Faurote, and Pillmore 1986).

GIBBSITE

Aluminum hydroxide: $Al(OH)_3$. A secondary mineral derived from the alteration of aluminous minerals; locally the chief constituent of bauxite deposits formed by weathering of aluminous rocks. It can also be formed as a low-temperature hydrothermal mineral in veins or cavities in igneous rocks.

COCHISE COUNTY: Warren District, Bisbee, Sacramento pit, as pale-blue massive botryoidal material, formed as an alteration product of chalcoalumite (S. A. Williams and Khin 1971).

GREENLEE COUNTY: Copper Mountain supersystem, Morenci District, reported as white massive granular material (3rd ed.).

FIGURE 3.104 Gibbsite, Copper Queen Mine, Bisbee, Cochise County, 9.6 cm high.

FIGURE 3.105 Gilalite, Christmas Mine, Gila County, FOV 7 mm.

GILALITE

Copper silicate hydrate: $Cu_5Si_6O_{17} \cdot 7H_2O$. A silicate mineral found in late retrograde skarns. The Christmas Mine is the type locality.

GILA COUNTY: Southeastern Dripping Spring Mountains, Banner supersystem, Christmas District, Christmas Mine, abundant as green to blue-green coatings or thick botryoidal crusts on fracture surfaces or embedded in the rock, replacing diopside; as microscopic green spheres (Gene Wright, UA 15952); common as a matrix mineral with kinoite, junitoite, and ruizite (Cesbron and Williams 1980).

GISMONDINE

Calcium aluminosilicate hydrate: $Ca_2(Si_4Al_4)O_{16} \cdot 8H_2O$. A member of the zeolite group usually found in basaltic lavas.

LA PAZ COUNTY: Near Salome, Little Harquahala District, Harquahala Mine, with dioptase (William Panczner, pers. comm., 1972).

GLADITE

Copper lead bismuth sulfide: $CuPbBi_5S_9$. A primary mineral of hydrothermal origin.

COCHISE COUNTY: Dos Cabezas Mountains, Teviston District, Comstock Mine, as small beads strung along the margins of pekoite prisms, which gladite appears to be replacing (3rd ed.).

GLAUBERITE

Sodium calcium sulfate: $Na_2Ca(SO_4)_2$. A constituent of sedimentary saline deposits; also formed as isolated crystals in sedimentary rocks and associated with fumarolic activity.

MOHAVE COUNTY: Detrital Valley, in S12 and 13, T29N R21W, as crystals associated with halite and anhydrite in fine-grained sediments in the subsurface (H. Wesley Peirce, pers. comm., 1973).

YAVAPAI COUNTY: Camp Verde District, Verde Salt Mine, about two miles southwest of Camp Verde and in nearby Copper and Lucky Canyons, abundant as tabular crystals up to 3 in. long, as loose fragments in silts and clays and in masses of thénardite in the Verde Formation. Commonly replaced by gypsum, calcite, and aragonite to form pseudomorphs (Snyder 1971; J. R. Thompson 1983; see also Blake 1890; Guild 1910).

#GLAUCOCERINITE

Zinc aluminum sulfate hydroxide hydrate: $(Zn_{1-x}Al_x)(SO_4)_{x/2}(OH)_2 \cdot nH_2O$ ($x < 0.5$, $n > 3x/2$). A rare secondary mineral found in some oxidized copper and zinc deposits.

COCHISE COUNTY: Courtland District, Maid of Sunshine Mine, as fibrous masses (Frost et al. 2004).

PINAL COUNTY: Riverside District, Black Copper Wash, south of Kelvin, where it occurs as a cement in recent conglomerate with opal on the floor of the wash (Tim Hayes, pers comm., 2008).

#GLAUCODOT

Cobalt iron arsenic sulfide: $(Co_{0.5}Fe_{0.5})AsS$. Found in hydrothermal veins.

GRAHAM COUNTY: Santa Teresa Mountains, Bluebird Cobalt District, Bluebird Mine, near Mt. Trumbull with cobaltite (Bilbrey 1962; USGS 2005a).

#GLUSHINSKITE

Magnesium oxalate hydrate: $Mg(C_2O_4) \cdot 2H_2O$. Found associated with decaying organic material.

MARICOPA COUNTY: Formed from the weathering of weddellite in decaying saguaro cacti in the Gila Bend Mountains (Garvie 2003).

GMELINITE

Sodium calcium aluminum silicate hydrate: $(Ca_2K_4Na_4)(Si_8Al_4)O_{24} \cdot 11H_2O$. A member of the zeolite group, commonly formed in cavities in basaltic and related igneous rocks. Gmelinite is a series name with three species based on the dominant cation (Coombs et al. 1997). The specific species has not been determined for this location.

COCHISE COUNTY: Bowie District, reported near Bowie in zeolite-replaced tuffs (Ted H. Eyde, pers. comm., 3rd ed.).

GOETHITE

FIGURE 3.106 Goethite, Bisbee, Cochise County, 11 cm.

Iron oxide hydroxide: FeO(OH). Goethite is the most common and abundant of the iron oxides after hematite. It is trimorphous with lepidocrocite and akaganeite and forms under a wide range of oxidizing conditions but most typically as a weathering product of iron-bearing minerals. It is commonly referred to as limonite, which is so abundant near mineralized areas in Arizona as to be practically ubiquitous.

COCHISE COUNTY: Warren District, Bisbee, Shattuck and Copper Queen Mines, as thick botryoidal crusts having fibrous structure; may be mixed with cuprite, copper, or tenorite (C. Frondel 1941); Junction Mine, as attractive stalactitic specimens (Les Presmyk, pers. comm., 2020).

GREENLEE COUNTY: Copper Mountain supersystem, Morenci District, Morenci Mine, as the most abundant oxidation product, with hematite (Moolick and Durek 1966).

LA PAZ COUNTY: Near Parker, as well-crystallized sheaves and fan-like aggregates, with malachite and hematite (Robert O'Haire, pers. comm., 1972).

MOHAVE COUNTY: Cerbat Mountains, Wallapai supersystem, in widespread limonites in the oxidized gossans of vein sulfide deposits; may be associated with hematite, jarosite, and plumbojarosite (B. E. Thomas 1949).

PIMA COUNTY: Cerro Colorado District, Princess claim (UA 7472). South Comobabi Mountains, Cababi District, Mildren and Steppe claims (S. A. Williams 1963). New Cornelia District, New Cornelia Mine, as fine earthy powder in matrix of Locomotive fanglomerates (Gilluly 1937). Pima supersystem, Twin Buttes District, Twin Buttes Mine, abundant in the oxidized capping (Stanley B. Keith, pers. comm., 1973).

PINAL COUNTY: Mineral Creek supersystem, Ray District, Mineral Creek deposit, northeast of former Ray Hill, as part of the iron oxide gossans above the porphyry copper deposit; also in copper-rich Holocene gravels, associated with hematite, azurite, malachite, cuprite, and tenorite (C. H. Phillips, Cornwall, and Rubin 1971). San Manuel District, San Manuel Mine, abundant in surface rocks that have been stained red by hematite gossan above the San Manuel porphyry copper deposit (Schwartz 1953).

GOLD

Gold: Au. Gold is widely distributed but commonly only in very small amounts. Rarely combines with other elements to form minerals, with the important exception of the gold tellurides. Gold is typically found in hydrothermal quartz veins and, because of its superior chemical inertness and mechanical resistivity, is found concentrated in placer deposits derived from them. It is also associated with copper sulfide deposits, especially in the oxidized zones.

In early times, American Indians and Spanish explorers of the Southwest had worked placer gold deposits. The mineral wealth of Arizona was not substantially exploited, however, until the middle and latter half of the nineteenth century, after the rediscovery of these deposits, especially those in Mohave, Yuma, and Cochise Counties. As the limited placer and lode deposits were rapidly depleted, and modern copper-mining techniques were developed, especially for the huge low-grade sulfide orebodies, more and more gold was produced as a byproduct of the smelting of base-metal ores. Most placer gold in the state had been mined out by 1885 (R. T. Moore 1969a), and the total output slightly exceeded only 400,000 ounces (Elsing and Heineman 1936). Since the late 1930s, almost all the gold produced in the state has come from secondary sources as a byproduct of copper mining. Shortly before World War II, gold production in Arizona reached an all-time-high annual rate of about 300,000 ounces. Arizona ranks among the top gold producers in the United States.

The number of lode and placer gold mines and prospects in the state is large, and only a few are listed here. For more detailed information, see Bulletin 137 (lists seventy-two lode gold districts and hundreds of mines; E. D. Wilson, Cunningham, and Butler 1934), Bulletin 168 (lists sixty placer gold districts; E. D. Wilson et al. 1961), and Bulletin 180 (lists ninety-seven lode gold districts and fifty-one placer gold districts: R. T. Moore 1969a), all published by the Arizona Bureau of Mines (now called the Arizona Geological Survey).

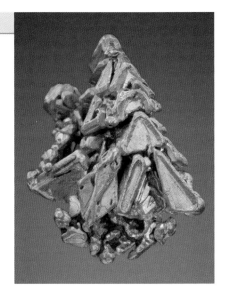

FIGURE 3.107 Gold, Quartzsite, La Paz County, 2.2 cm high.

LODE DEPOSITS

COCHISE COUNTY: Warren District, Bisbee, as a byproduct of copper, lead, and silver ores (UA 9525), on bornite; Shattuck Mine, as rich, spongy gold matte in breccia (Ransome 1904; Bain 1952; S 95730). Tombstone

District, with lead-silver ores (B. S. Butler, Wilson, and Rasor 1938). Pearce District, Commonwealth Mine, in broad splotches and leaf form (Endlich 1897). Dos Cabezas Mountains, Dos Cabezas and Teviston Districts (Bull. 181). Huachuca (Reef Mine) and Swisshelm Mountains (Bull. 181).

GILA COUNTY: Globe Hills District, with ores of copper and silver (Ransome 1903b; N. P. Peterson 1962), as fine specimens from the Old Dominion Mine (3rd ed.). Dripping Spring Mountains, C and B–Grey Horse District, in a prospect one-eighth mile east of the Cowboy Mine, about eight miles northwest of Christmas (Stanley B. Keith, pers. comm., 1973).

GRAHAM COUNTY: Galiuro Mountains, Rattlesnake District (Blake 1902). Aravaipa supersystem, Grand Reef and Stanley Districts (Bull. 181).

GREENLEE COUNTY: Copper Mountain supersystem, Morenci District, with copper and silver ores (Lindgren 1905).

LA PAZ COUNTY: Harquahala District, with ores of copper and lead, rich in quartz, locally as large masses (Wilson, Cunningham, and Butler 1934; H 100665). Williams Mountains, Cienega District, as a byproduct of copper ores (Bull. 181). Northern Plomosa Mountains, Dutchman District, Dutchman Mine, as flecks on red earthy massive hematite (S R15089). Dome Rock and Plomosa Mountains, placer and lode gold deposits in several different districts (Melchiorre 2011, 2013, 2017). Gila Bend Mountains (Bull. 181).

MARICOPA COUNTY: Vulture District (Metzger 1938), with lead and silver ores, with wulfenite (H 100312). Cave Creek District and Big Horn Mountains (Bull. 181). Mystic District, Mystic Mine, near Twin Buttes at the south end of the Hieroglyphic Mountains; in 1991 and 1992, many outstanding specimens were recovered (Tony Potucek, pers. comm., 2018).

MOHAVE COUNTY: Black Mountains, Oatman District (Silliman 1866; Ransome 1923; E. D. Gardner 1936); Ruth vein, one-quarter mile south of the Moss Mine, and Tom Reed vein, massive in quartz (Lausen 1931a, 1931b; H 81355). Cerbat Mountains, Wallapai District (B. E. Thomas 1949). Hualapai Mountains, Maynard District, with silver ores (Bull. 181). Gold Basin District (Schrader 1908). Williams Fork of the Colorado River (Blake 1865).

PIMA COUNTY: Ajo District, as a major byproduct of copper and silver ores (Gilluly 1937); New Cornelia Mine, cementing some fractures in bornite (3rd ed.). South Comobabi Mountains, Cababi District, Mildren and Steppe claims (S. A. Williams 1963); Cahuabi Mine, as flecks in quartz, with limonite, galena, and malachite (Raphael Pumpelly collection, H 81380); Sun-Gold Mine, irregular quartz vein with pockets of high-grade gold values (Stanton B. Keith 1974). Santa Catalina Mountains, Molino Basin, in quartz stained with chrysocolla and shattuckite (3rd ed.). Tucson Mountains, on limonite that pseudomorphed magnetite (UA 9724). Santa Rita Mountains, Greaterville District, as rusty wire gold (Schrader 1917; UA 6734); Yuba Mine, Gold Ledge claim, as excellent small specimens of wire gold collected by Dewey Keith in about 1955 (W. E. Wilson 1987a). Helvetia-Rosemont District, Golden Gate Mine (UA 344).

PINAL COUNTY: Pioneer and Mineral Creek supersystems, as a byproduct of copper, silver, and lead ores (Bull. 181); Magma Mine, as flakes and small wires on bornite and hematite (Les Presmyk, pers. comm., 2020). Mammoth District, Mammoth–St. Anthony Mine, as flecks in rhyolite porphyry (N. P. Peterson 1938a, 1938b; H 99916). Oracle District, from John's Ranch, as hackly gold in quartz (S R221); in rusty pockets in quartz, with brochantite (H 108579). Sacaton (Casa Grande) District, Sacaton Mine, as byproduct of copper and silver ores (Bull. 181). Goldfield Mountains, Goldfield District. Pinto Creek, as stout one-inch wires (H 100386).

SANTA CRUZ COUNTY: Santa Rita and Patagonia Mountains (Schrader 1917). Oro Blanco District, with ores of lead, silver, and copper; as crystalline flakes of "white gold," the color of which was reportedly due to small amounts of gallium (Engineering and Mining Journal 1900). Nogales District, Little Annie Mine (UA 2180).

YAVAPAI COUNTY: Verde District, as byproduct of copper and silver ores (Bull. 181). Bradshaw Mountains, Big Bug, Peck, Walker, and Tiger Districts, as byproduct of ores of copper, silver, and lead (C. A. Anderson and Creasey 1958). Hassayampa and Black Canyon Districts (Guiteras 1936), with lead-silver ores. Agua Fria District, with copper-silver ores (Bull. 181). Groom Creek, Turkey Creek, and Tip Top Districts, with silver ores (Bull. 181); Copper Basin District (Johnston and Lowell 1961). Santa Maria Mountains, Eureka District, as byproduct of copper-silver-lead ores (E. D. Wilson, Cunningham, and Butler 1934). Date Creek Mountains, Martinez (Congress) District (Metzger 1938). Wickenburg Mountains, Black Rock District (Lindgren 1926). Rich Hill District, Octave Mine and widespread in the Rich Hill placers (Melchiorre 2008).

YUMA COUNTY: Kofa District (E. L. Jones 1915) and Gila Mountains, Dome District, in paleoplacer deposits (Bull. 181). Castle Dome District, with lead-silver ores (Bull. 181). Gila Mountains, Fortuna District, Fortuna Mine (Blake 1897). Gila Bend Mountains (Bull. 181).

PLACER DEPOSITS

COCHISE COUNTY: Dos Cabezas, Teviston, Huachuca, Gleeson, Pearce, Gold Gulch (Bisbee), and Hartford (Huachuca) placers, produced in small amounts (E. D. Wilson et al. 1961). Ash Canyon, ten miles south of Hereford, as coarse placer gold (Robert O'Haire, pers. comm., 1972).

COCONINO COUNTY: Lee's Ferry and Paria Creek, in the lower shales of the Chinle Formation, as flakes associated with small amounts of mercury (Lausen 1936).

GILA COUNTY: Central Dripping Spring Mountains, Barbarossa paleoplacer, Green Valley, Globe-Miami, and Payson placers, produced in small amounts (E. D. Wilson et al. 1961).

GREENLEE COUNTY: Clifton-Morenci placers, along San Francisco and Chase Creeks (E. D. Wilson et al. 1961).

LA PAZ COUNTY: La Paz, Plomosa, Trigo, and Harquahala Districts (E. D. Wilson et al. 1961).

MARICOPA COUNTY: Cave Creek, Agua Fria, Pikes Peak, Big Horn, Vulture, San Domingo, and Hassayampa (T. L. Carter 1911) placers (E. D. Wilson et al. 1961). Near Phoenix, in a gravel pit near the Salt River, as a waterworn nugget weighing 3.5 ounces (Scott Williams collection, H 106179).

MOHAVE COUNTY: Gold Basin, Chemehuevi, Lost Basin, Lewis, Wright Creek, Lookout, and Silver Creek placers (E. D. Wilson et al. 1961).

PIMA COUNTY: Greaterville, Quijotoa, Horseshoe Basin, Arivaca, and Papago placers; a nugget valued at $228 was found at Greaterville in 1924. Las Guijas, Old Baldy, Old Hat, Baboquivari, Armagosa, and Alder Canyon placers, as less commercially important deposits (E. D. Wilson et al. 1961). Tohono O'odham (Papago) Reservation, Golden Green Mine, where William Coplen discovered a placer nugget that weighed slightly more than 8 ounces (*Arizona Daily Star*, March 1967). Northwest flank of the Santa Catalina Mountains, Cañada del Oro placers (E. D. Wilson et al. 1961).

SANTA CRUZ COUNTY: In many small placer deposits, including the Oro Blanco, Mowry, Harshaw, Patagonia, Tyndall, Nogales, and Palmetto placers (E. D. Wilson et al. 1961).

YAVAPAI COUNTY: Weaver Creek, Rich Hill (Melchiorre 2009), Lynx Creek, Big Bug, Minnehaha, Hassayampa, Groom Creek, Copper Basin, Placerita, and Black Canyon (Guiteras 1936) areas, as the most important localities; a nugget weighing 271 grams was found on Weaver Creek in 1930, and in 1932–33 several nuggets weighing 3 ounces or more were recovered from that area. Weaver District, Red Bank, where coarse goldbearing quartz veins yielded a nugget shaped like a human molar that measured 53 mm by 47 mm and weighed 270.90 grams (Heineman 1931). Placers near Congress Junction, where a nugget weighing 4.81 ounces was produced (UA 3470).

YUMA COUNTY: Dome, Castle Dome, Muggins, Kofa, and Fortuna Districts, as less important deposits (E. D. Wilson et al. 1961). Near Yuma, Myers claim on the Colorado River, where a water-worn nugget, 0.5 in. across, was produced in July 1863 by dry washing (S 55749).

GOLDFIELDITE

Copper tellurium sulfide: $(Cu_4\square 2)Cu_6Te_4S_{13}$. Found in epithermal precious-metal veins.

COCHISE COUNTY: Warren District, Bisbee, with the sulfide ores of the Campbell Mine (Graeme 1993).

GONNARDITE

Sodium calcium aluminosilicate hydrate: $(Na,Ca)_2(Si,Al)_5O_{10}\cdot 3H_2O$. A member of the zeolite group formed in vesicular basalts as well as in metamorphic rocks; commonly associated with other zeolites.

PIMA COUNTY: Santa Catalina Mountains, Golder Dam spillway, collected by Douglas Shakel (3rd ed.).

#GORDAITE

Sodium zinc sulfate hydroxide chloride hydrate: $NaZn_4(SO_4)(OH)_6Cl\cdot 6H_2O$. Found in the oxidized zone of hydrothermal deposits.

YAVAPAI COUNTY: Found at an unspecified locality in San Miguel Wash. It forms a white crust with thénardite in a calcite matrix (Vajdak 2002).

GORMANITE

Iron aluminum phosphate hydroxide hydrate: $Fe^{2+}_3Al_4(PO_4)_4(OH)_6\cdot 2H_2O$. Found as low-temperature fracture fillings.

COCHISE COUNTY: Warren District, Bisbee, in drill core as dark-green rosettes in open fractures cutting an equigranular tonalite, as isolated radial sprays or prisms up to 1 cm long; the green color belies the spectacular pleochroism seen in thin section; associated with chlorite, calcite, and quartz (Graeme 1993).

GOSLARITE

Zinc sulfate hydrate: $Zn(SO_4) \cdot 7H_2O$. A water-soluble secondary mineral formed through the alteration of sphalerite; common as efflorescences on the walls of old mine openings.

COCHISE COUNTY: Warren District, Bisbee, locally abundant as a post-mining mineral (Graeme 1981).

GILA COUNTY: Globe Hills District, Continental and Old Dominion Mines, as efflorescences (Bull. 181). Pinto Valley District, Castle Dome Mine (N. P. Peterson, Gilbert, and Quick 1951; N. P. Peterson 1962).

GREENLEE COUNTY: Copper Mountain supersystem, Morenci District, as efflorescences; Arizona Central Mine (Lindgren 1905; Guild 1910).

MOHAVE COUNTY: Cerbat Mountains, Wallapai supersystem, Stockton Hill District, De la Fontaine property (Bull. 181).

PIMA COUNTY: Silver Bell District, as copper-bearing variety in old mine workings (Bull. 181). South Comobabi Mountains, Cababi District, Mildren and Steppe claims (S. A. Williams 1963).

FIGURE 3.108 Graemite, Cole Mine, Bisbee, Cochise County, FOV 3 cm.

GRAEMITE

Copper tellurite hydrate: $Cu^{2+}(Te^{4+}O_3) \cdot H_2O$. A very rare secondary mineral found in hydrothermal deposits. The Cole Mine is the type locality.

COCHISE COUNTY: Warren District, Bisbee, Cole Mine, in a specimen from the 1,200 ft. level; the graemite is attached to the surface of and has partially replaced teineite crystals embedded in malachite associated with a loose spongy aggregate of cuprite crystals (S. A. Williams and Matter 1975; R100207). Shattuck shaft, replacing and overgrown by teineite (Graeme 1993).

LA PAZ COUNTY: Dome Rock Mountains, in a small prospect, associated with goethite, gypsum, and teineite in small cavities in djurleite in quartz-tourmaline gangue; as at Bisbee, the graemite appears to replace teineite (S. A. Williams and Matter 1975).

GRANDREEFITE

Lead sulfate fluoride: $Pb_2SO_4F_2$. Associated with galena and fluorite, from which it is thought to have formed by interaction with supergene fluids. The Grand Reef Mine is the type locality.

GRAHAM COUNTY: Aravaipa supersystem, Grand Reef District, Grand Reef Mine, as colorless striated prismatic orthorhombic crystals in quartz-lined vugs, associated with galena, fluorite, and anglesite (Kampf, Dunn, and Foord 1989; R100012).

#GRANDVIEWITE

Copper aluminum sulfate hydroxide: $Cu_3Al_9(SO_4)_2(OH)_{29}$. A secondary copper mineral found with other oxidized copper minerals. The Arizona locality is the type locality.

COCONINO COUNTY: Grand Canyon National Park, Grandview District, Grandview (Last Chance) Mine, as blue acicular sprays in goethite-rich gossan with chalcoalumite and cyanotrichite (Colchester et al. 2008).

FIGURE 3.109 Grandviewite, Grandview (Last Chance) Mine, Coconino County, 1.5 cm.

GRAPHITE

Carbon: C. Formed by the reduction of carbon-containing compounds during metamorphism or hydrothermal activity; possibly also formed as a primary constituent in igneous rocks.

COCHISE COUNTY: Dos Cabezas Mountains, in thin veins or streaks in gold-quartz veins (Bull. 181). Near Benson, as graphitic clay (Guild 1910). Warren District, Bisbee (Graeme 1981).

COCONINO COUNTY: Canyon Diablo and Elden meteorites, as small nodules (Ksanda and Henderson 1939).

GILA COUNTY: Sierra Ancha District, Rainbow deposit, as uraniferous graphite in the Dripping Spring Quartzite (Granger and Raup 1969).

LA PAZ COUNTY: Yellowbird graphite prospects as black laminated layers intercalated with white illitic muscovite; extensive Raman spectroscopy indicates that the main carbon component in many specimens is graphene (Stanley B. Keith, pers. comm., 2019).

MOHAVE COUNTY: Cerbat Mountains, Canyon Station Wash, disseminated in Precambrian schist (Schrader 1908).

YUMA, LA PAZ, AND MARICOPA COUNTIES: Northern Plomosa Mountains, Cemetery Ridge, southern Trigo Mountains, Castle Dome Mountains, and Neversweat Ridge, where numerous windows into the Orocopia Schist contain graphite as a characteristic mineral where it ubiquitously coexists with albite feldspar, which reflects high-pressure, low- to moderate-temperature metamorphism of a metagraywacke protolith (Haxel et al. 2002; Strickland, Singleton, and Haxel 2018).

GREENOCKITE

FIGURE 3.110 Greenockite, San Juan Mine, Cochise County, 4.8 × 3.0 cm.

Cadmium sulfide: the hexagonal dimorph of CdS. A rare mineral most commonly formed as coatings on sphalerite; rarely in cavities in mafic igneous rocks.

COCHISE COUNTY: Reported in the Warren District, Bisbee, as a yellow coating on sphalerite (Graeme 1981). Middle Pass North supersystem, Abril District, San Juan Mine, as minute needles (Bruce Maier, UA x3364).

COCONINO COUNTY: Cameron District, Huskon Mine, with uraninite, marcasite, pyrite, calcite, and siliceous gangue, replacing wood structures as well as forming cement in sandstones of the Chinle Formation (Bollin and Kerr 1958; Maucher and Rehwald 1961).

GROSSULAR

Calcium aluminum silicate: $Ca_3Al_2(SiO_4)_3$. Typically a product of thermal and contact metamorphism of calcareous and aluminous rocks; a member of the garnet group.

COCHISE COUNTY: Warren District, Bisbee, as rounded crystals in unoxidized pyritic ores (Graeme 1981). Tombstone District, cinnamon brown, in contact-metamorphic zones; forms massive beds at Comstock Hill (B. S. Butler, Wilson, and Rasor1938). Little Dragoon Mountains, Cochise supersystem, Johnson Camp District, Johnson area and I-10 deposit, as a gangue mineral with copper zinc ores (Kellogg 1906; J. R. Cooper and Huff 1951; J. R. Cooper 1957; J. R. Cooper and Silver 1964).

PIMA COUNTY: Pima supersystem, Mission-Pima District, Pima Mine, abundant in limestone hornfels, with diopside and tremolite (Journeay 1959; Himes 1972); Twin Buttes District, in the skarns of the Twin Buttes Mine with magnetite and chalcopyrite (Stanley B. Keith, pers. comm., 1973). Silver Bell District, Atlas Mine area, abundant in tactites as a product of pyrometasomatic alteration of limestone (Agenbroad 1962). Helvetia-Rosemont District, Peach-Elgin copper deposit, with garnet, diopside, chalcopyrite, and locally bornite, as bedded replacements (Heyman 1958).

SANTA CRUZ COUNTY: Santa Rita and Patagonia Mountains, in the latter range at Washington Camp District, Duquesne and Washington Camp, in limestone in contact-metamorphic deposits associated with other contact-silicate minerals and copper-zinc sulfides (Schrader 1917).

YAVAPAI COUNTY: Eureka supersystem, Bagdad District, Bagdad, with minor fluorite (UA 8039).

GROUTITE

Manganese oxide hydroxide: $Mn^{3+}O(OH)$. Groutite was originally described from the iron ranges of Minnesota, where it formed in the iron ores with manganite in vuggy cavities.

COCHISE COUNTY: Warren District, Bisbee, Campbell Mine, as tiny dark brownish-black crystals coating sooty manganese oxides (Graeme 1981).

NAVAJO COUNTY: Petrified Forest Wood District, near Holbrook, as needles in petrified wood (3rd ed.).

PINAL COUNTY: Pioneer supersystem, Magma District, Magma Mine, as coatings of small prismatic crystals (ASDM x1331).

YUMA COUNTY: Kofa District, No. 2 vein of the North Star Mine, with psilomelane, bixbyite, todorokite, and manganite (Hankins 1984); King of Arizona Mine, with todorokite, chalcophanite, aurorite, and pyrolusite in the King of Arizona vein, and with nsutite in the No. 5 vein (Hankins 1984). Sheep Tanks District, as the chief manganese mineral in one vein (Cousins 1972).

GRUNERITE

Iron silicate hydroxide: $\square Fe^{2+}_2 Fe^{2+}_5 Si_8 O_{22}(OH)_2$. A member of the amphibole group; common in iron-rich, regionally metamorphosed rocks; also formed in contact-metamorphic rocks with fayalite, hedenbergite, and almandine.

GRAHAM COUNTY: Santa Teresa Mountains, Stanley Butte District, in contact-metamorphosed limestone (Bull. 181).

YAVAPAI COUNTY: Tip Top District, north of the Foy Mine (Bull. 181).

GUILDITE

Copper iron sulfate hydroxide hydrate: $CuFe^{3+}(SO_4)_2(OH) \cdot 4H_2O$. A very rare mineral formed during a mine fire in the United Verde Mine at Jerome, which is the type locality.

YAVAPAI COUNTY: Black Hills, Jerome supersystem, Verde District, United Verde Mine, formed under fumarolic conditions as a result of burning pyritic ores; as relatively rare crystals up to 5 mm (Lausen 1928; Laughon 1970).

GYPSUM

Calcium sulfate hydrate: $Ca(SO_4) \cdot 2H_2O$. A common mineral of widespread and abundant distribution that often formed by the evaporation of inland seas and salt lakes and by the hydration of anhydrite. It can also

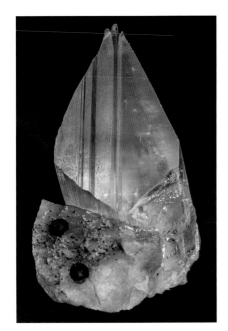

FIGURE 3.111 Gypsum, Duval Mine, Santa Cruz County, 4 × 6 cm.

form as a saline mineral precipitated from deep-sourced saline brine injections into shallow lacustrine or marine environments. It is a common constituent of the oxidized zones of sulfide ores.

APACHE AND NAVAJO COUNTIES: Monument Valley and Cane Valley Districts, in the copper-uranium-vanadium ores in channels at the base of the Shinarump Conglomerate (Mitcham and Evensen 1955).

COCHISE COUNTY: Tombstone District, widespread as small crystals and scales lining small fissures in shale and as coatings in mine stopes (B. S. Butler, Wilson, and Rasor 1938). Near Douglas, where it was quarried (Santmyers 1929). Sulphur Springs Valley, as beds in recent lake sediments (Bull. 181). San Pedro Valley, Benson and St. David Gypsum Districts, both north and south of Benson as roses and spherical clusters at an oxidation/reduction boundary (Guild 1910; Wenrich 2013; Muntyan 2015b). Warren District, Bisbee, as excellent examples of the curved ram's horn variety (ASDM 03533). Courtland District, Shannon Mine (UA 6790).

COCONINO COUNTY: Along the eastern bank of the Little Colorado River, between Cameron and Leupp, as beds of varying thickness alternating with mudstones, in the Moenkopi Formation (J. W. Anthony, DuBois, and Krumlauf 1955; Baldwin 1971).

GILA COUNTY: Miami-Inspiration supersystem, Copper Cities District, Copper Cities Mine, as large crystals in the lower levels (N. P. Peterson 1962), Castle Dome Mine, as crystals up to 2 in. on azurite (Les Presmyk, pers. comm., 2020). San Carlos Gypsum District, as thick beds of massive transparent gypsum in Pliocene-Pleistocene sediments (Stanley B. Keith, pers. comm., 2019).

GREENLEE COUNTY: Copper Mountain supersystem, Morenci District, common in oxidized deposits in limestone (Lindgren 1905; Guild 1910).

LA PAZ COUNTY: Plomosa Mountains, Mudersbach Camp, as a bed several feet thick and as beds at the eastern base of the Harquahala Mountains (Bull. 181). Silver District, massive and crystalline in the upper portions of veins (E. D. Wilson 1933); Red Cloud Mine, as gangue in veins in andesite, with fluorite, calcite, and baryte (Foshag 1919).

MARICOPA COUNTY: About fifteen miles south of Gila Bend, as beds in sandstone and conglomerate, with celestine (Phalen 1914).

MOHAVE COUNTY: Abundant in the Virgin Valley badlands, as thick beds in South Mountain and Quail Canyon. Bill Williams River, in beds northeast of the Planet Mine. Mammoth claim, sixty miles southeast of Kingman, as satin spar (all Mohave occurrences from Bull. 181).

NAVAJO COUNTY: Near Winslow, mined as large plates of selenite in the upper member of the Permian Supai Formation (Peirce and Gerrard 1966; R. T. Moore 1968).

PIMA COUNTY: Empire and Santa Rita Mountains, as Permian beds up to 50 ft. thick (Schrader and Hill 1915). Santa Catalina Mountains foothills, north of Tucson in Rillito I beds (Bull. 181). In the Pantano Formation interbedded with clay near Vail (Guild 1910). Pima supersystem, Mission-Pima District, Minnie Mine (UA 7162) and near San Xavier Mine, as clear crystals up to 4 in. (3rd ed.); Mission Mine (B. M. Williams 2018), enclosing copper and, rarely, chalcotrichite (UA 9198); also interbedded with dolomite in Epitaph dolomite. South Comobabi Mountains, Cababi District, Mildren and Steppe claims (S. A. Williams 1963).

PINAL COUNTY: Winkelman Gypsum District, east side of the San Pedro River near Feldman, as thick beds in lake deposits, where it has been extensively mined (Hardas 1966). San Pedro Valley northeast of San Manuel as layers in Quiburis Formation (Bull. 181).

SANTA CRUZ COUNTY: Santa Rita Mountains, Montosa Canyon area, as thick Permian beds (J. W. Anthony 1951); Cottonwood Canyon, Tyndall District, Glove Mine, as a secondary mineral (H. J. Olson 1966).

YAVAPAI COUNTY: Eureka supersystem, Hillside District, Hillside Mine, in the oxidized portion of the vein, associated with secondary uranium minerals (Axelrod et al. 1951). Jerome supersystem, Verde District, United Verde Mine, abundant in decomposed dikes (Lindgren 1926). Verde Valley, Camp Verde District, in saline deposits (Guild 1910), as pseudomorphs after glauberite (UA 9872).

YUMA COUNTY: Castle Dome District, massive and crystalline in the upper portions of veins (E. D. Wilson 1933).

GYROLITE

Sodium calcium aluminosilicate hydroxide hydrate: $NaCa_{16}(Si_{23}Al)O_{60}(OH)_8 \cdot 14H_2O$. A secondary mineral formed through the alteration of calcium-bearing silicate minerals.

COCHISE COUNTY: Tombstone District, Lucky Cuss Mine, as light-green spherules (resembling prehnite) composed of thin interlocking tabular crystals. The spherules may attain 7 mm in diameter. The gangue minerals are mainly quartz natrolite hosting alamosite and queitite (3rd ed.).

MARICOPA COUNTY: Horseshoe Dam District, one mile south of Horseshoe Dam, as rare 3 mm rose-like crystal clusters, with apophyllite (D. M. Shannon 1983b).

FIGURE 3.112 Halite, Verde Salt Mine, Yavapai County, 4.9 cm.

HALITE

Sodium chloride: NaCl. The source of common salt often found in sedimentary beds formed by the evaporation of inland seas and salt lakes commonly associated with gypsum, anhydrite, and other salts of sodium and potassium. It may also form as a saline mineral precipitated from deep-sourced saline brine injections into shallow lacustrine or marine environments.

APACHE AND NAVAJO COUNTIES: Southern regions, in the subsurface of the Supai Salt Basin, which embraces about 2,300 square miles; Holbrook Salt District interbedded with anhydrite, dolomite, sylvite, and red and gray siliceous clastics (Peirce and Gerrard 1966; Peirce 1969b).

COCHISE COUNTY: San Simon Basin, seven miles northeast of Bowie, in bedded lake deposits, associated with thénardite and a variety of zeolite minerals (Regis and Sand 1967).

GILA AND MARICOPA COUNTIES: Salt River Valley, as encrustations derived from evaporation of saline springs (Guild 1910; E. D. Wilson 1944).

MARICOPA COUNTY: A well drilled in S19, T2N R1W, revealed substantial thicknesses of halite at a depth of 2,350 ft. in Cenozoic rocks; the Luke Salt deposit is currently being mined by solution methods (Peirce 1969b; Richardson et al. 2019).

MOHAVE COUNTY: In the badlands of the Virgin River Valley, near the Nevada border, with gypsum (E. D. Wilson 1944). Detrital Valley District, in the subsurface of Detrital and Hualapai Valleys, in S12 and 13, T29N R21W, with anhydrite and glauberite. Red Lake District—the deposits south of Red Lake flat are at least 1,200 ft. thick (Peirce 1969b; H. Wesley Peirce, pers. comm., 1973), and those northwest of the flat are 500 to 700 ft. thick (W. G. Pierce and Rich 1962).

PIMA COUNTY: Tohono O'odham (Papago) Reservation (Bull. 181).

PINAL COUNTY: Picacho Basin District, west of the Picacho Mountains, in S5, T8S R8E—a drill hole sunk by the Humble Oil and Refining Company in 1972 penetrated 80 ft. of halite and about 6,000 ft. of anhydrite (H. Wesley Peirce, pers. comm., 1973).

Verde Valley, Camp Verde District, in the salt deposits, some of which are deep purple, associated with glauberite, gypsum, mirabilite, and thénardite (Blake 1890; Guild 1910; J. R. Thompson 1983).

#HALLOYSITE-10Å

Aluminum silicate hydroxide hydrate: $Al_2Si_2O_5(OH)_4 \cdot 2H_2O$. A hydrated member of the kaolinite group that can be formed by the action of sulfate-containing waters on kaolinite but may also form independently of it. A hydrothermal alteration mineral associated with porphyry copper deposits. In the third edition, halloysite-10Å was called endellite, but this mineral name was discredited in 2006 (Burke 2006). No distinction was made with the unhydrated species, which is now known as halloysite-7Å, for the reported occurrences.

COCHISE COUNTY: Warren District, Bisbee, Southwest Mine, as a product of hydrothermal alteration (Schwartz 1956).

GILA COUNTY: Miami-Inspiration supersystem, as a product of hydrothermal alteration in granite porphyry, near orebodies in schist (Schwartz 1947); Pinto Valley District, Castle Dome Mine, in small amounts in the capping and in the chalcocite zone in quartz monzonite (N. P. Peterson, Gilbert, and Quick 1946).

GRAHAM COUNTY: Gila Mountains, Lone Star District (3rd ed.).

GREENLEE COUNTY: Copper Mountain supersystem, Morenci District, with allophane, as a product of intense hydrothermal activity (Schwartz 1947, 1958).

PIMA COUNTY: Silver Bell District, Silver Bell Mine (3rd ed.).

PINAL COUNTY: San Manuel District, San Manuel Mine, in small veinlets in alunite-kaolinite rock (Lovering, Huff, and Almond 1950; Schwartz 1953). Mineral Creek supersystem, Ray District, Ray orebody, in the hydrothermally altered porphyry stock, associated with sericite, kaolinite, and illite; as veinlets and coating fracture surfaces; commonly exhibits colloform texture and may be intergrown with chrysocolla (Schwartz 1934, 1947; Stephens and Metz 1967).

HALOTRICHITE

Iron aluminum sulfate hydrate: $Fe^{2+}Al_2(SO_4)_4 \cdot 22H_2O$. Water soluble, formed from the weathering of pyritic and aluminous rocks as efflorescence in old mine workings. It is commonly associated with gypsum and other secondary sulfate minerals.

COCHISE COUNTY: Tombstone District (Bull. 181).

COCONINO COUNTY: Cameron District, as cross-fiber veins in carbonaceous fossil wood and in the surrounding sediments, some stained inky blue by ilsemannite (Austin 1964). Grand Canyon National Park, Orphan District, Orphan Mine, as loose aggregates of pale-blue silky fibers (R070673).

GILA COUNTY: Banner supersystem, 79 District, 79 Mine, as an abundant secondary mineral throughout the oxide-sulfide transition on the fifth level, commonly as arching whiskers up to 12 in. long growing from the walls of the workings and twisting in all directions; growth rates of up to 3 in. per year have been measured for the currently forming mineral (Stanley B. Keith 1972).

PINAL COUNTY: Mineral Creek supersystem, Ray District, Ray Mine, as microcrystals (David Shannon, UA x4319).

SANTA CRUZ COUNTY: Patagonia Mountains (UA 4992); Trench District, Blue Nose (Big Chief) Mine (UA 9592); Palmetto District, Denver Mine, as microcrystals (David Shannon, UA x4320).

HARMOTOME

Barium aluminosilicate hydrate: $Ba_2(Si_{12}Al_4)O_{32} \cdot 12H_2O$. A member of the zeolite group, it is typically formed in cavities and veins in igneous rocks.

GREENLEE COUNTY: Unspecified locality near Duncan (UA 5991).

PINAL COUNTY: Table Mountain District, Table Mountain Mine, as small clear elongated crystals, occasionally twinned, with fornacite and mimetite, (Joe Ruiz, pers. comm., 2015; ID by S. A. Williams).

HAUSMANNITE

Manganese oxide: $Mn^{2+}Mn^{3+}_2O_4$. Typically found in high-temperature veins. Also a contact-metamorphic mineral formed as a recrystallization product of preexisting manganese minerals under conditions of regional metamorphism.

COCHISE COUNTY: Warren District, Bisbee, White Tail Deer Mine, replacing limestone that contains braunite (3rd ed.).

PIMA COUNTY: Arivaca District, COD Mine, in veinlets cutting a matrix of braunite (Hewett 1972).

SANTA CRUZ COUNTY: Cottonwood Canyon, Tyndall District, Glove Mine, as a constituent of fine-grained brownish manganese oxides (Bideaux, Williams, and Thomssen 1960; H. J. Olson 1966). Hardshell-Hermosa District, Hardshell Mine, as a constituent of manganese ores (S. R. Davis 1975).

HAWLEYITE

Cadmium sulfide: CdS. The cubic dimorph of greenockite, commonly formed as an earthy coating on sphalerite.

COCHISE COUNTY: Cochise supersystem, Dragoon District, Buena Vista Mine, as a yellow powder derived from sphalerite, which coats fractures in a quartz-rich garnet tactite (3rd ed.).

HAXONITE

Iron nickel carbide: $(Fe, Ni)_{23}C_6$. Found in meteorites associated with nickel-rich iron and taenite, in particles up to 40 microns in diameter. Meteor Crater is the type locality.

COCONINO COUNTY: Meteor Crater near Winslow, in the Canyon Diablo meteorite (Scott 1971).

#HEAZLEWOODITE

Nickel sulfide: Ni_3S_2. A mineral found in serpentinites.

LA PAZ AND YUMA COUNTY: Cemetery Ridge District, as small grains with other sulfides in magnetite in blocks of serpentinite and partially serpentinized harzburgite in the Orocopia Schist at Cemetery Ridge (Haxel et al. 2018).

#HECHTSBERGITE

Bismuth oxide vanadate hydroxide: $Bi_2O(VO_4)(OH)$. A rare secondary mineral.

LA PAZ COUNTY: La Cholla District, La Cholla gold placers, as inclusions in placer gold (Melchiorre 2013).

HECTORITE

Sodium magnesium lithium silicate fluoride hydroxide hydrate: $Na_{0.3}$ $(Mg,Li)_3Si_4O_{10}(F,OH)_2 \cdot nH_2O$. A member of the smectite group of clay minerals.

YAVAPAI COUNTY: Lyles District, mined at the Lyles hectorite deposit northeast of Hillside (Eyde 1986).

FIGURE 3.113 Hedenbergite, Iron Cap Mine, Graham County, 5.2 × 4.1 cm.

HEDENBERGITE

Calcium iron silicate: $CaFe^{2+}Si_2O_6$. A member of the pyroxene group that is typically formed in contact-metamorphosed limestones, associated with other calcium silicates.

GRAHAM COUNTY: Stanley District, Stanley Butte (UA 6987). Aravaipa supersystem, Iron Cap District, Iron Cap Mine, as a manganoan variety associated with johannsenite and other manganese-rich skarn minerals (Reiter 1980, 1981).

LA PAZ AND YUMA COUNTIES: Cemetery Ridge District, in calcsilicate schist and granofels derived from calcareous metachert and siliceous marble of the Orocopia Schist; with wollastonite, anorthite or bytownite, and quartz (Haxel et al. 2021).

PIMA COUNTY: Silver Bell District, Atlas Mine area (Agenbroad 1962). Sierrita Mountains, Pima supersystem, common as gangue in contact-metamorphosed limestones (Eckel 1930a; Irvin 1959).

SANTA CRUZ COUNTY: Patagonia Mountains, Washington Camp District, Pride of the West Mine, with diopside and other contact silicates (Schrader and Hill 1915; Schrader 1917).

#HEDYPHANE

Calcium lead arsenate chloride: $Ca_2Pb_3(AsO_4)_3Cl$. A rare secondary mineral found in manganese and zinc orebodies.

MARICOPA COUNTY: Painted Rock District, Rowley Mine, as pale-yellow hexagonal crystals, also found in spheroidal clusters (W. E. Wilson 2020b).

PIMA COUNTY: Empire District, Total Wreck Mine, as clusters of small yellow elongated crystals about 1 mm in length associated with wulfenite (J. Callahan, pers. comm., 2020).

HELVINE

Beryllium manganese silicate sulfide: $Be_3Mn^{2+}_4(SiO_4)_3S$. Found in granites and granite pegmatites and in contact-metamorphic rocks, where it may be associated with other beryllium-bearing minerals.

COCHISE COUNTY: Dragoon Mountains, Middle Pass su-persystem, Abril District, Abril and Gordon (San Juan) Mines, with beryllium-bearing epidote, in tactite in a limestone replacement deposit (Warner et al. 1959; Meeves 1966, UA x1851).

FIGURE 3.114 Helvine, San Juan Mine, Cochise County, FOV 5 mm.

HEMATITE

Iron oxide: Fe_2O_3. A mineral formed in many ways, in igneous, metamorphic, and sedimentary rocks; in the latter, as bedded deposits of commercial significance. It is also in hydrothermal veins and in the gossans or leached cappings of base-metal deposits. Only a few typical localities are listed here.

COCHISE COUNTY: Near Willcox, as fine-grained massive specularite (Bull. 181). Tombstone District, near the Lucky Cuss Mine, as radiating crystal aggregates in a vein in granodiorite (B. S. Butler, Wilson, and Rasor 1938). Warren District, Bisbee, Lavender pit, as the iridescent variety (turgite) (UA 7827). Dragoon Mountains, Middle Pass supersystem, Black Diamond District, Black Diamond claim (UA 7357).

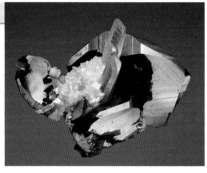

FIGURE 3.115 Hematite, Big Bertha Extension, Quartzsite, La Paz County, 4.4 cm.

GILA COUNTY: Globe Hills District, as massive bodies of specularite along the veins of the Old Dominion fault system, as replacements of limestone or diabase; Buckeye, Black Oxide, Big Johnnie, Stonewall, and other veins (Sanders 1911; N. P. Peterson 1962). Cibecue District, Fort Apache Reservation, Bear Spring Canyon, as extensive deposits with chert beds in the Mescal Limestone (Burchard 1930, 1943). Aztec Peak area, reniform masses (Sinkankas 1964). Banner supersystem, 79 District, 79 Mine, as colorful iridescent coatings on quartz (R. Gibbs, pers. comm., 2020). Apache Iron District, Fort Apache Reservation, Canyon Creek, as a large deposit of siliceous hematite, estimated at 10 million tons, ranging from soft pulverulent bright-red material to hard dark-blue oxide (L. A. Stewart 1947); extensive deposits in the middle member of the Precambrian Mescal Limestone, interbedded with ferruginous cherts, sandstones, and shales (R. T. Moore 1968).

GRAHAM COUNTY: Aravaipa supersystem, Grand Reef District, Grand Reef area, as specularite, very abundant in the Lead King and Cobre Grande Mines (Simons 1964).

GREENLEE COUNTY: Copper Mountain supersystem, Morenci District, Manganese Blue Mine (Lindgren 1905).

LA PAZ COUNTY: Buckskin Mountains, Planet District, Planet Mine, as extensive replacement deposits with carbonate and silicate copper ores in limestone (Cummings 1946a). Dome Rock Mountains, Middle Camp District, Big Bertha Extension Mine (also called Crystal Caverns claim and Veta Grande claim), as high-quality crystals up to 6.5 cm across; these crystals, some of which are mirror bright and twinned, with quartz crystals, may be the finest U.S. specimens (Sprunger 1980). Locality southeast of Quartzsite, as well-formed "iron roses" up to 3.5 cm across (3rd ed.). Bouse supersystem, Black Mountain District, BBC Mine, as grouped crystals, some of which are up to 4 cm, with some quartz and rarely actinolite tufts; Burro Hill prospect, 2 miles north of the BBC Mine, as large lustrous masses of crystals (MM 9797); west of Bouse as iridescent globular masses (turgite); Dutchman District, abundant at Dutchman claim, where it hosts gold flakes, and at the Heart's Desire property, where it occurs with chrysocolla and malachite. Little Butte District, widespread coarsely crystalline vein mineral in veins and fractures with local chrysocolla (Stanley B. Keith, pers. comm., 2019).

MARICOPA COUNTY: Pikes Peak District, northeast of Beardsley, as syngenetic banded iron Formation (BIF) in schist (Farnham and Havens 1957). Painted Rock District, Rowley Mine near Theba, where the powdery mineral is responsible for the ubiquitous red coloration (W. E. Wilson and Miller 1974).

MOHAVE COUNTY: Cerbat and Aquarius Mountains, widely distributed, chiefly as the specular variety (Blake 1865); Wallapai supersystem, Chloride District, Chloride, as red ocher in the oxidized gossan (UA 1836).

PIMA COUNTY: Santa Rita Mountains, Helvetia-Rosemont District, Cuprite Mine area (3rd ed.). Empire District, Hilton Mines, as red ocher (UA 7388). Pima supersystem, Mission-Pima District, Copper Glance Mine (UA 7389); Twin Buttes District, Twin Buttes Mine (Stanley B. Keith, pers. comm., 1973). Ajo District, New Cornelia Mine, as specularite in brilliant adamantine crystals up to 2 in. across, of hypogene origin; as the earthy variety, as supergene gossan (Gilluly 1937).

PINAL COUNTY: Mammoth District, Mammoth–St. Anthony Mine, as glistening black masses in the lower levels of the Collins vein (N. P. Peterson 1938a, 1938b). Near Winkelman, as an iridescent variety (Bull. 181).

Pioneer supersystem, Pioneer District, Magma Mine, as bladed crystals up to 2 cm (Les Presmyk, pers. comm., 2020).

SANTA CRUZ COUNTY: Patagonia Mountains, widely distributed in mines and properties (3rd ed.); Mowry District, Mowry Mine (Schrader 1917); Washington Camp District, at Duquesne and Washington Camp, in limestone in contact-metamorphic deposits; as a primary mineral in numerous siliceous veins filling fault fissures (Schrader 1917). Santa Rita Mountains, Tyndall District, Montosa Canyon, Isabella Mine, as massive material of the specular variety with cerussite ("sand carbonate"), replacing Permian limestones (Schrader and Hill 1915; J. W. Anthony 1951).

YAVAPAI COUNTY: Deposits near Camp Wood (Bull. 181). Little Copper Creek District, McBride claims, seventeen miles south of Seligman, as large deposits of earthy material that form irregular lenses in limestone near the contact with diorite (Bull. 181). Near Townsend Butte and the Kay District, Howard copper property, in high concentrations with magnetite, in schist (Bull. 181). Jerome supersystem, Verde District, United Verde Mine, as the specular variety, in late-state veinlets that cut massive sulfides; south of Jerome, common in Precambrian rocks (C. A. Anderson and Creasey 1958). Black Canyon area, at Cortes, as a major constituent of the BIFs with silica and magnetite; also in the Rich Hill area at the Stanton BIF (Stanley B. Keith, pers. comm., 2019).

HEMIHEDRITE

Zinc lead chromate silicate hydroxide: $ZnPb_{10}(CrO_4)_6$ $(SiO_4)_2(OH)_2$. A rare secondary mineral formed in the oxidized portions of galena-sphalerite veins. The Florence Lead-Silver Mine is the type locality.

MARICOPA COUNTY: Belmont Mountains, Osborne District, at several localities, including the Rat Tail claim, Moon Anchor Mine, and Potter-Cramer property; formed by alteration of galena in the oxidized portion of galena-bearing quartz veins that cut andesite agglomerate; associated with phoenicochroite, vauquelinite, willemite, and mimetite; crystals are orange to nearly black (McLean and Anthony 1970; S. A. Williams and Anthony 1970; S. A. Williams, McLean, and Anthony 1970; R141184).

FIGURE 3.116 Hemihedrite, Rat Tail claim, Maricopa County.

PINAL COUNTY: Tortilla Mountains, Florence Lead District, Florence Lead-Silver Mine, in the oxidized zone of lead-bearing veins that cut strongly brecciated Paleozoic limestone intruded by an altered latite porphyry, and Precambrian quartzite containing a diabase dike. Associated primary minerals include galena, sphalerite, pyrite, and tennantite; oxidation products include cerussite, phoenicochroite, vauquelinite, wulfenite, and willemite. Hemihedrite forms contemporaneously with wulfenite, after the formation of cerussite, and may be replaced by wulfenite. Crystals are bright orange to henna brown to nearly black (S. A. Williams and Anthony 1970; R140133).

YAVAPAI COUNTY: Black Rock District, near the confluence of Amazon Wash and the Hassayampa River, with phoenicochroite, mimetite, and descloizite (William Hunt, pers. comm., 1985).

HEMIMORPHITE

Zinc silicate hydroxide hydrate: $Zn_4(Si_2O_7)(OH)_2 \cdot H_2O$. A secondary mineral formed in the oxidized portion of zinc deposits; commonly associated with smithsonite, cerussite, anglesite, galena, and sphalerite.

COCHISE COUNTY: Tombstone District, Empire and Toughnut Mines, as sparse radiating aggregates in oxidized ore (B. S. Butler, Wilson, and Rasor 1938). Warren District, Bisbee, with aurichalcite (UA 9343) and

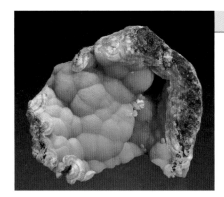

FIGURE 3.117 Hemimorphite, 79 Mine, Gila County, 7 cm.

rosasite (UA 8685), also on the 1300 level of the Cole shaft as fine large crystals up to 4 in. (Graeme 1981). Gleeson District, Mystery and Silver Bill Mines, as encrustations and druses (Bull. 181). Little Dragoon Mountains, Cochise supersystem, Johnson Camp District, Johnson Camp (J. R. Cooper and Huff 1951); Gunnison Hills, as small colorless crystals; Texas-Arizona Mine, as the most abundant oxidized ore mineral (J. R. Cooper 1957; J. R. Cooper and Silver 1964).

GILA COUNTY: Globe Hills District, Irene, Albert Lea, and Apache Mines, common in the oxidized zones of veins (N. P. Peterson 1962). Miami-Inspiration supersystem, Pinto Valley District, Castle Dome Mine, as tiny rounded grains in clay (N. P. Peterson, Gilbert, and Quick 1951). Cherry Creek area, Sierra Ancha District, Horseshoe deposit, sparingly coating fracture surfaces (Granger and Raup 1969). Banner supersystem, 79 District, 79 Mine, as very fine specimens in various habits, associated with rosasite, chrysocolla, and cerussite (UA 1164, 8307), and altering to chrysocolla, as pseudomorphs of chrysocolla after hemimorphite on wad (UA 7587); as odd, blue but in some places white, hollow eggshell-like balls on matrix (Stanley B. Keith 1972); northeast workings of the Kullman-McCool (Finch, Barking Spider) Mine, as light-green to blue botryoidal masses, associated with wulfenite (K. D. Smith 1995).

GREENLEE COUNTY: Copper Mountain supersystem, Shannon District, Shannon Mine, as small transparent crystals in decomposed garnet rock (Lindgren and Hillebrand 1904; Lindgren 1905).

LA PAZ COUNTY: Silver District, Red Cloud Mine, as small crystal groups (Edson 1980).

MARICOPA COUNTY: White Picacho District, as a supergene mineral after sphalerite; also associated with hydrozincite (Jahns 1952).

MOHAVE COUNTY: McCracken District, McCracken Mine near Signal, with fluorite (UA 9271).

PIMA COUNTY: Empire District, Hilton Mines, as small colorless crystals in smithsonite (Bull. 181); Total Wreck Mine, with plattnerite on limonite (Bideaux, Williams, and Thomssen 1960); also with rosasite at the 49 Mine (Stanley B. Keith, pers. comm., 2019). Pima supersystem, Mission–Pima District, as light green-blue mammillary crusts on mine dumps (Therese Murchison, pers. comm., 1972); San Xavier West Mine (Arnold 1964); Queen Mine (UA 5634). Waterman District, common at the Silver Hill Mine (UA 6005). Tucson Mountains, Amole District, north base of Amole Peak, as botryoidal crusts (Robert O'Haire, pers. comm., 1972).

PINAL COUNTY: Mammoth District, Mammoth–St. Anthony Mine, as porous to compact masses and as slender needles bristling from quartz crystals on the walls of open cavities (N. P. Peterson 1938a, 1938b; Palache 1941b; Fahey 1955).

SANTA CRUZ COUNTY: Flux District, Flux Mine, on limonite with aurichalcite (Thomas Trebisky, pers. comm., 1972).

YAVAPAI COUNTY: Eureka supersystem, Hillside District, Hillside Mine, in the oxidation zone of a sulfide-bearing vein in mica schist, associated with silver, anglesite, cerussite, chlorargyrite, and smithsonite (Axelrod et al. 1951).

HENRYITE

Copper silver telluride: $(Cu,Ag)_{3+x}Te_2$ (x ~ 0.4). A rare species associated with primary sulfide and telluride minerals. The Campbell orebody at Bisbee is the type locality.

COCHISE COUNTY: Warren District, Bisbee, Campbell orebody, very sparingly in sulfide ores, associated with hessite, petzite, sylvanite, altaite, rickardite, and pyrite (Criddle et al. 1983).

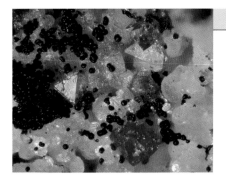

FIGURE 3.118 Herbertsmithite, Tonopah-Belmont Mine, Maricopa County, FOV 0.5 mm.

#HERBERTSMITHITE

Copper zinc hydroxide chloride: $Cu_3Zn(OH)_6Cl_2$. A rare secondary mineral found in the oxidized portions of copper-zinc deposits.

MARICOPA COUNTY: Belmont Mountains, Osborne District, Tonopah-Belmont Mine, as very small blue to blue-green equant crystals with a pseudo-octahedral habit occurring with cumengeite (Gibbs and Ascher 2012).

HERCYNITE

Iron aluminum oxide: $Fe^{2+}Al_2O_4$. A member of the spinel group most commonly found in metamorphosed argillaceous sediments. It also can be a constituent of mantle-sourced peridotitic inclusions in alkali basalts.

GILA COUNTY: Peridot District, unspecified locality near Rice (now San Carlos), in olivine bombs (UA 3701). San Carlos Apache Reservation, Peridot Mesa, as black opaque crystals in xenoliths (Uchida et al. 2005).

HESSITE

Silver telluride: Ag_2Te. Found in hydrothermal veins with other tellurides, gold, and tellurium.

COCHISE COUNTY: Tombstone District (Romslo and Ravitz 1947); West Side Mine, as bands and disseminations in quartz, with chlorargyrite and gold (Genth 1887); Flora Morrison Mine, altering to silver (B. S. Butler, Wilson, and Rasor 1938; Rasor 1938). Warren District, Bisbee, Campbell orebody, sparingly in the sulfide ores (Graeme 1993).

YAVAPAI COUNTY: Jerome supersystem, Verde District, United Verde Mine, as small blebs with probable zincian tennantite in chlorite matrix (3rd ed.).

HETAEROLITE

Zinc manganese oxide: $ZnMn^{3+}_2O_4$. A rare secondary mineral associated with other oxidized manganese and zinc minerals.

COCHISE COUNTY: Tombstone District, Lucky Cuss Mine, as tiny veinlets in manganite (B. S. Butler, Wilson, and Rasor 1938; Rasor 1939; Hewett and Fleischer 1960). Warren District, Bisbee, Campbell (UA 7717) and

Junction Mines, as splendent botryoidal and stalactitic masses and coatings; 1,300 ft. level of the Cole shaft (Graeme 1981); White Tail Deer Mine, as crystals in calcite (W. E. Wilson 1987b).

GILA COUNTY: Banner supersystem, 79 District, 79 Mine, as massive encrustations on hemimorphite in the stope system above the 470 ft. level (Stanley B. Keith 1972).

MARICOPA COUNTY: Painted Rock Mountains, Painted Rock District, Rowley Mine, as black pseudocubic crystals resembling murdochite (B. Murphy, pers. comm., 2018).

PINAL COUNTY: Pioneer supersystem, Silver King District, Domeroy property, four miles north of Superior, with psilomelane and pyrolusite (Dean, Snedden, and Agey 1952).

FIGURE 3.119 Hetaerolite, Campbell shaft, Bisbee, Cochise County, 4×3 cm.

HEULANDITE

Barium calcium potassium sodium strontium aluminosilicate hydrate: $[(Ba,Ca,K)_5,(Ca, Na,K)_5,(K,Ca,Na)_5,(Na,-Ca,K)_6,(Sr,Ca,Na)_5][(Si_{27}Al_9),(Si,Al)_{36}]O_{72}\cdot22-26H_2O$. A member of the zeolite group and commonly found in vesicles in basaltic rocks. It may also be formed by late low-temperature alteration of porphyry copper deposits. Five specific species are now recognized: heulandite-Ba, heulandite-Ca, heulandite-K, heulandite-Na, and heulandite-Sr. The specific species is not known for the occurrences listed below.

FIGURE 3.120 Heulandite, Malpais Hill, Pinal County, FOV 3 mm.

GILA COUNTY: Miami-Inspiration supersystem, Inspiration District, Black Copper portion of the Inspiration orebody, interlayered with chrysocolla in some schist outcrops (Throop and Buseck 1971).

GREENLEE COUNTY: Along road cuts on U.S. Highway 191, five miles south of Hannagan Meadow, in vesicles in basalt, with phillipsite (3rd ed.). Big Lue Mountains, Big Lue District, in road cuts along State Highway 78, three to six miles west of the New Mexico state line with other zeolites (Gibbs 1992).

MOHAVE COUNTY: Unspecified locality near Oatman, with thomsonite in andesite (UA 8476).

PIMA COUNTY: Ajo area, New Cornelia Mine, as small clear crystals with calcite and ferrierite in fractures in postcopper mineralization, Hospital porphyry dike. Also found with orange stilbite in leached chalcopyrite veins in the north end of the pit (W. J. Thomas and Gibbs 1983). Well No. 1, seven miles north of Ajo, as small clear crystals with calcite and other zeolites in vesicles in the Batamonte volcanic rock excavated from the well (R. Gibbs, pers. comm., 2017).

PINAL COUNTY: Mineral Creek supersystem, Ray District, Ray Mine, mixed with green and black chrysocolla (Throop and Buseck 1971). In a railway cut in Malpais Hill, north of Mammoth, as crystals with calcite in vesicles (Bideaux, Williams, and Thomssen 1960). Mammoth District, Mammoth–St. Anthony Mine, as rare twinned microcrystals, with wulfenite (Bideaux 1980).

SANTA CRUZ COUNTY: Sierrita Mountains, as microcrystals, with epidote (George Sanders, UA x1306).

YAVAPAI COUNTY: U.S. Highway 17, Badger Springs exit, as small crystals in vesicles exposed in road cuts of the highway interchange (R. Gibbs, pers. comm., 2018).

HEWETTITE

Calcium vanadium oxide hydrate: $CaV^{5+}_6O_{16}\cdot 9H_2O$. A rare secondary mineral formed in the oxidized portions of vanadium or uranium-vanadium deposits.

APACHE COUNTY: Cane Valley District, Monument No. 2 Mine, moderately abundant as individual crystals up to 15 mm long and about 0.5 mm thick, as cross-fibrous seams, and in crusts up to 10 cm thick (Weeks, Thompson, and Sherwood 1955; Witkind and Thaden 1963). Carrizo Mountains, as hair-like crystals and as fibrous encrustations in sandstone, with carnotite (Bull. 181). Lukachukai Mountains, Lukachukai District, in the Salt Wash Member of the Morrison Formation (Chenoweth 1967).

NAVAJO COUNTY: Monument Valley District, Monument No. 1 and Mitten No. 2 Mines (Evensen and Gray 1958; Holland et al. 1958; Witkind 1961; Witkind and Thaden 1963).

HEXAHYDRITE

Magnesium sulfate hydrate: $Mg(SO_4)\cdot 6H_2O$. An uncommon secondary mineral found sparingly in mine workings with epsomite, from which it may form by dehydration.

COCHISE COUNTY: Warren District, Bisbee, Campbell shaft (Graeme 1981).

COCONINO COUNTY: In breccia pipes mined for uranium (Wenrich and Sutphin 1988).

PINAL COUNTY: San Manuel District, 2,015 and 2,700 ft. levels of the San Manuel Mine, with epsomite, starkeyite, and an unknown selenium-bearing aluminum sulfate hydrate mineral (material collected by Joseph Urban; J. W. Anthony and McLean 1976).

HIDALGOITE

Lead aluminum sulfate arsenate hydroxide: $PbAl_3(SO_4)(AsO_4)(OH)_6$. A rare secondary mineral found in the oxidized portions of lead deposits and a member of the beudantite group.

YAVAPAI COUNTY: Tip Top District, Silver Crown Mine, in the oxidized zone of massive sulfide ores, as a white crust on silicified schist (3rd ed.).

HILLEBRANDITE

Calcium silicate hydroxide: $Ca_2SiO_3(OH)_2$. A rare mineral formed during contact metamorphism of limestone.

COCHISE COUNTY: Tombstone District, Lucky Cuss Mine, with monticellite and vesuvianite (B. S. Butler, Wilson, and Rasor 1938).

HINSDALITE

Lead aluminum sulfate phosphate hydroxide: hexagonal $PbAl_3(SO_4)(PO_4)$ $(OH)_6$. A rare primary or early supergene mineral, formed in sulfide veins.

COCHISE COUNTY: Huachuca Mountains, Reef District, Big Dipper Mine, as small pale-green prismatic, tapering crystals, lining pockets in quartz. The mineral is closely associated with galena, pyromorphite, and plumbogummite (Walstrom 2013).

MOHAVE COUNTY: Wallapai supersystem, Ithaca Peak District, Ithaca Peak Mine, as spherical clusters of pearly-white scaly tablets, in voids on corroded, sooty sulfides; abundant in a vein with pyrite, sphalerite, bornite, minor amounts of galena, and abundant supergene chalcocite (3rd ed.); Mineral Park District, Mineral Park Mine, as hexagonal microcrystals (Wilkinson, Roe, and Williams 1980; UA x629).

HISINGERITE

Iron silicate hydroxide hydrate: $Fe_2Si_2O_5(OH)_4 \cdot 2H_2O$. A secondary mineral formed by hydrothermal alteration or weathering of iron-bearing silicates or sulfides.

GILA COUNTY: Miami-Inspiration supersystem, Pinto Valley District, Castle Dome Mine, a supergene mineral associated with malachite, limonite, jarosite, and wulfenite (N. P. Peterson 1947; N. P. Peterson, Gilbert, and Quick 1951).

PIMA COUNTY: South Comobabi Mountains, Cababi District, Mildren and Steppe claims, with various secondary minerals, which are products of the oxidation of sulfide ores in quartz veins cutting andesite (S. A. Williams 1963).

PINAL COUNTY: Mammoth District, Mammoth–St. Anthony Mine, in the central parts of radiating fibrous creaseyite spherules (Bideaux 1980).

SANTA CRUZ COUNTY: Washington Camp District, North Belmont Mine, as replacements of pyrite crystals up to 3 in. (Muntyan 2019).

HOCARTITE

Silver iron tin sulfide: Ag_2FeSnS_4. An uncommon tin-bearing mineral, commonly found as small grains in other sulfides and typically associated with other tin minerals such as stannite.

COCHISE COUNTY: Warren District, Bisbee, in the sulfide ores of the Campbell Mine (Graeme 1993).

HODRUŠITE

Copper bismuth sulfide: $Cu_8Bi_{12}S_{22}$. In polymetallic ore deposits of subvolcanic type, in quartz veins developed in propylitized pyroxenite andesite.

COCHISE COUNTY: Warren District, Bisbee, with the sulfide ores of the Campbell Mine, as steel-gray laths up to 0.5 mm long, with pyrite and chalcopyrite (Graeme 1993).

FIGURE 3.121 Hoganite, Holbrook Mine, Bisbee, Cochise County, 6 × 4 cm.

#HOGANITE

Copper acetate: $Cu(CH_3COO)_2 \cdot H_2O$. A very rare naturally formed copper acetate. The specimens from the Warren District need additional study to verify their natural origin.

COCHISE COUNTY: Courtland District, Courtland, from an unspecified location, as small blue-green crystals and fragments up to 1 mm in size found loose in a clay-rich gossan (M. Origlieri, pers. comm.,

2017). Warren District, Bisbee, Holbrook Mine, as blue-green crystals up to 2 mm on cuprite (Les Presmyk, pers. comm., 2020).

HOLLANDITE

Barium manganese oxide: $Ba(Mn^{4+}_6Mn^{3+}_2)O_{16}$. Hollandite is one of several manganese oxide minerals that can form in contact-metamorphic deposits or as a weathering product of other manganese minerals.

COCONINO COUNTY: Peach Springs area; fine-grained, massive, and intimately associated with cryptomelane; contains 0.23 percent thallium and 5.5 percent BaO (Crittenden et al. 1962).

GILA COUNTY: Roosevelt District, Apache claims, one of several manganese oxides concentrated in breccia zones of a sandstone conglomerate; botryoidal, intimately associated with cryptomelane; contains 0.34 percent thallium, about 9 percent BaO, and about 2 percent K_2O (Crittenden et al. 1962; Hewett, Fleischer, and Conklin 1963). Banner supersystem, 79 District, as a constituent of wad in the 79 Mine (Stanley B. Keith 1972).

GREENLEE COUNTY: Big Lue Mountains, Big Lue District, in vesicles in andesite exposed in road cuts along Highway 78 east of the New Mexico state line (Gibbs 1992).

MOHAVE COUNTY: Artillery District, as one of the most abundant minerals in the Black Jack and Priceless deposits, where it formed very fine anhedral grains and needles and is associated with cryptomelane, psilomelane, pyrolusite, coronadite, ramsdellite, and lithiophorite; Price vein; Maggie Canyon bedded deposit (Mouat 1962).

PINAL COUNTY: Mammoth District, Mammoth–St. Anthony Mine, as the most widespread manganese oxide mineral (3rd ed.). Reymert supersystem, Mineral Hill District, Reymert Mine, with the other manganese oxide minerals, psilomelane, coronadite, chalcophanite, and todorokite (K. S. Wilson 1984).

YUMA COUNTY: Sheep Tanks District, Sheep Tanks Mine, Booth Bonanza deposit, as black fibers in calcite as part of banded black calcite veins (Cousins 1972).

#HOPEITE

Zinc phosphate hydrate: $Zn_3(PO_4)_2 \cdot 4H_2O$. A secondary mineral found in hydrothermal zinc deposits.

GRAHAM COUNTY: Aravaipa supersystem, Iron Cap District, found at the Iron Cap Mine as white botryoidal masses with axinite-(Mn), hedenbergite, and clinochrysotile (Vajdak 1995).

HÖRNESITE

Magnesium arsenate hydrate: $Mg_3(AsO_4)_2 \cdot 8H_2O$. A secondary mineral formed in blocks of metamorphosed limestone in volcanic tuff at the type locality in Banat, Hungary.

MOHAVE COUNTY: West side of Peach Springs Canyon, in a cave as radiating white crystals, with talmessite, powellite, carnotite, conichalcite, calcite, and aragonite, among bright-white encrustations with green bands and spotty yellow patches (Wenrich and Sutphin 1989). Corkscrew Cave on the Hualapai Indian Reservation (Wenrich and Sutphin 1994; Onac, Hess, and White 2007).

#HOUSLEYITE

Lead copper tellurate hydroxide: $Pb_6CuTe_4O_{18}(OH)_2$. A very rare mineral found in oxidized lead copper tellurium veins.

COCHISE COUNTY: Tombstone District, Tombstone, Empire Mine, as very small greenish, sea-blue prismatic crystals (B. Murphy, pers. comm., 2013).

HÜBNERITE

Manganese tungstate: $Mn^{2+}(WO_4)$. The end members, iron-rich ferberite (>50% Fe) and manganese-rich hübnerite (>50% Mn), form a solid-solution series. In the past, minerals of intermediate or unspecified composition were called *wolframite*, but this is no longer a valid mineral species. It is found in quartz veins in granite and in schist associated with pegmatites; also found as an accessory mineral in gold-bearing polymetallic veins. Some of the following are identified as hübnerite, but the composition has not been determined for all the localities listed here; some are possibly ferberite (see also FERBERITE).

COCHISE COUNTY: Warren District, Bisbee, in the sulfide ores of the Cole shaft (Graeme 1993). Northeastern slope of the Whetstone Mountains, Whetstone District, in veins and as replacements with scheelite and associated pyrite and bornite (Hess 1909; Dale, Stewart, and McKinney 1960). Bluebird District, near the Southern Pacific Railroad station at Dragoon (Guild 1910), as crystals of hübnerite 2 in. long (see also Kuck

1978); Russellville ghost town (near former Peabody Copper-Zinc Mine at Johnson Camp), eight miles north of Dragoon, as a highly manganiferous hübnerite in quartz veins that cut porphyritic granite; as prismatic and blade-like masses, associated with small quantities of scheelite and gold (Blake 1898); Little Dragoon Mountains, in quartz veins and placer gravels (Kellogg 1906; Palache 1941a; E. D. Wilson 1941; Dale, Stewart, and McKinney 1960), and as hübnerite, the principal ore mineral in tungsten veins in the Texas Canyon Quartz Monzonite; as platy crystals up to 10 cm long (Palache 1941a; E. D. Wilson 1941; J. R. Cooper and Silver 1964). Tungsten King District, Tungsten King Mine, occurs with scheelite in narrow parallel quartz veins (Dale, Stewart, and McKinney 1960).

GILA COUNTY: Pinal Mountains District, at several places south of Globe; Samsell Mine, where the quartz locally contains scattered platy and prismatic crystals of "wolframite" partly altered to scheelite (N. P. Peterson 1963); Bobtail Mine, west of Miami on Spring Creek (Bull. 181). Spring Creek District, southwest of Young, as hübnerite, with galena and sphalerite (Ransome 1903b; Dale 1961). Mazatzal Mountains, El Oso District (Dale 1961).

LA PAZ COUNTY: Livingston Hills, southern Plomosa Mountains, Livingston District, Colorado Mine (Livingston claims), as large black platy crystals over 5 cm long in a quartz-muscovite vein with scheelite and powellite in a specimen collected by J. R. Livingston in 1943 (MM K128).

MARICOPA COUNTY: White Picacho District, in pegmatites (Bull. 181).

MOHAVE COUNTY: Maynard District, Telluride Chief, Laxton, and Moon properties, in silver-rich veins with scheelite, minor copper oxides, and molybdenite (Bull. 181). Aquarius Mountains District (Engineering and Mining Journal 1910; Dale 1961); and in quartz veins in the Williams and other mines near Boner Canyon (Hobbs 1944). Greenwood District, west of Cottonwood Cliffs and west of Greenwood Peak (Bull. 181). Gold Basin District, O.K. Mine (Schrader 1909; E. D. Wilson 1941; Dale 1961).

PIMA COUNTY: Baboquivari District (Dale, Stewart, and McKinney 1960) and Comobabi District (Guild 1930; E. D. Wilson 1941; H 89695). Las Guijas District, as striated crystals up to 1.5 in. across, in quartz veins that cut granite (Blake 1896). Cerro Colorado District, in quartz veins (H 89698).

PINAL COUNTY: Mammoth District, Tarr property, northwest of the Mammoth–St. Anthony Mine, with schorl (Dale 1959). Galiuro Mountains, Bunker Hill (Copper Creek) District, Bluebird Mine (E. D. Wilson 1941). Pioneer area, Antelope Peak area. Summit supersystem, Bronx District, Swede Mine, as platy crystals partly altered to scheelite. At several properties in the Crook National Forest, northeast of Superior, with scheelite in quartz veins (Dale 1959).

SANTA CRUZ COUNTY: Patagonia Mountains, Nogales District, Mt. Benedict and Reagan Camp, with scheelite in narrow quartz veins in lamprophyre dikes that cut quartz monzonite (Schrader 1917). San Cayetano District, southeast of Calabasas (Pellegrin 1911; Dale, Stewart, and McKinney 1960).

YAVAPAI COUNTY: Bradshaw Mountains, Tip Top District, Tip Top Mine, Thule Creek area (De Wolf 1916; Dale 1961); Camp Wood District, in quartz veins at the Black Pearl and Joy properties, south of Camp Wood (Dale 1961); Silver Mountain District (E. D. Wilson 1941).

HUMMERITE

Potassium magnesium vanadium oxide hydrate: $KMgV^{5+}_5O_{14} \cdot 8H_2O$. A rare water-soluble secondary mineral found in veins and efflorescences in sedimentary rocks.

APACHE COUNTY: Lukachukai District, Mesa No. 1 Mine (Bull. 181), 4B Mine (Dodd 1955).

HUNTITE

Calcium magnesium carbonate: $CaMg_3(CO_3)_4$. Typically, a weathering product of magnesian rocks such as dolomites and some limestones, precipitated at low temperatures from aqueous solutions in cavities and vugs.

PIMA COUNTY: Cimarron Mountains, Tohono O'odham (Papago) Reservation, as nodular masses in altered limestone with magnesite (James Vacek, pers. comm., 1986) and as microcrystals (David Shannon, UA x4333).

FIGURE 3.122 Hureaulite, 7U7 Ranch, Yavapai County, FOV 3 mm.

HUREAULITE

Manganese phosphate hydroxide hydrate: $Mn^{2+}_5(PO_3 OH)_2(PO_4)_2 \cdot 4H_2O$. Formed by alteration of primary phosphate minerals such as triphylite and associated with other secondary phosphates in granite pegmatites.

YAVAPAI COUNTY: White Picacho District, locally formed between crystals of lithiophilite and triphylite and as a coating on sicklerite; also as crystalline aggregates and in fractures in these crystals; amber to beige colored (Jahns 1952). Hillside area, 7U7 Ranch west of Hillside, associated with bermanite in triplite seams and with phosphosiderite and leucophosphite (Leavens 1967).

#HYALOTEKITE

Barium lead potassium calcium yttrium boron beryllium silicate fluoride: $(Ba,Pb,K)_4(Ca,Y)_2(B,Be)_2(Si,B)_2Si_8O_{28}F$. A very rare secondary silicate mineral.

MOHAVE COUNTY: Rawhide District, Rawhide Mine, as a minor component of fine-grained melanotekite veins cutting alamosite (S. A. Williams 1982).

HYDROBASALUMINITE

Aluminum sulfate hydroxide hydrate: $Al_4(SO_4)(OH)_{10} \cdot 15H_2O$. A rare secondary mineral associated with felsőbányaite, to which it alters on dehydration.

COCHISE COUNTY: Warren District, Bisbee, Lavender pit, Holbrook Extension, as pearly-white flakes with other aluminum sulfates encrusting silicified limestones (3rd ed.); Southwest Mine, seventh level, as white to colorless blades (Graeme 1981).

HYDROBIOTITE

Potassium magnesium iron aluminosilicate hydroxide hydrate: $K(Mg, Fe^{2+})_6(Si,Al)_8O_{20}(OH)_4 \cdot nH_2O$. A mixed-layer mica composed of interstratified sequences of vermiculite and "biotite."

COCHISE COUNTY: Warren District, Bisbee, as an alteration product associated with the intrusion of a quartz monzonite (3rd ed.).

PINAL COUNTY: San Manuel District, San Manuel Mine, as a minor constituent of hydrothermal alteration of monzonite and quartz-monzonite porphyries in the marginal, less intense alteration areas (Schwartz 1958; Creasey 1959; J. D. Lowell 1968). Mineral Mountain District, southeast of Apache Junction, abundant in selvages of ore veins containing lead and copper (3rd ed.).

HYDROCERUSSITE

Lead carbonate hydroxide: $Pb_3(CO_3)_2(OH)_2$. A rare mineral of secondary origin commonly associated with other products of the alteration of galena.

COCHISE COUNTY: Warren District, Bisbee, as a coating on cerussite in a quartz-galena vein west of the Lavender pit (Graeme 1993).

GILA COUNTY: Central Dripping Spring Mountains, C and B–Grey Horse District, C and B Vanadium Mine, with cerussite (3rd ed.).

LA PAZ COUNTY: Silver District, Red Cloud Mine, associated with galena (3rd ed.).

MARICOPA COUNTY: Belmont Mountains, Osborne District, Potter-Cramer Mine, as thin milky films on cerussite (3rd ed.).

PINAL COUNTY: Mammoth District, Mammoth–St. Anthony Mine, Collins vein, as snow-white hexagonal pyramidal crystals, accompanied by diaboleite and leadhillite. These steep-sided crystals are up to nearly 1 in. long (Fahey, Daggett, and Gordon 1950; Bideaux 1980; R080104).

#HYDROGLAUBERITE

Sodium calcium sulfate hydrate: $Na_{10}Ca_3(SO_4)_8 \cdot 6H_2O$. Found as a weathering product of glauberite.

YAVAPAI COUNTY: Camp Verde District, Camp Verde Salt Mine. Crystals of glauberite are found altering to hydroglauberite in the walls of the pit (T. Nikischer, pers. comm., 2006). Pseudomorphs of aragonite, calcite, gypsum, and hydroglauberite after glauberites are found in the area.

#HYDROHONESSITE

Nickel iron sulfate hydroxide hydrate: $(Ni_{x-1}Fe^{3+}_x)(SO_4)_{x/2}(OH)_2 \cdot nH_2O$ ($x <$ 0.5, $n > 3x/2$). Found as a secondary encrustation on primary nickel ores.

COCONINO COUNTY: Cameron District, from a shallow open cut eight miles east southeast of Gray Mountain. It is found as a postmine encrustation on the walls of a uranium prospect (S. A. Williams and Cesbron 1995).

#HYDROKENORALSTONITE

Aluminum fluoride hydrate: $\square_2Al_2F_6(H_2O)$. An uncommon mineral that was previously known as ralstonite until it was renamed in 2017 (Atencio et al. 2017). It is associated with thomsenolite in cryolite in Greenland and with pachnolite and thomsenolite in altered rhyolite near Pikes Peak, Colorado.

COCHISE COUNTY: Warren District, Bisbee, Southwest Mine; one specimen was found by Richard Graeme in a void, associated with later paratacamite (Graeme 1993).

COCONINO COUNTY: Sunset Crater National Monument, Sunset Crater Fumaroles District, as coatings on gypsum in the fumarole deposits on Sunset Crater (S. L. Hanson, Falster, and Simmons 2008).

HYDROMAGNESITE

Magnesium carbonate hydroxide hydrate: $Mg_5(CO_3)_4(OH)_2 \cdot 4H_2O$. A secondary mineral formed by the alteration of magnesian rocks such as dolomites and serpentines.

COCONINO COUNTY: Grand Canyon National Park, as white crusts in a cave in Nautiloid Canyon, a few hundred meters from the Colorado River (3rd ed.).

HYDRONIUMJAROSITE

Hydronium iron sulfate hydroxide: $(H_3O)Fe^{3+}_3(SO_4)_2(OH)_6$. At an occurrence in Poland, the origin of the mineral is ascribed to the low alkali content of mine waters due to the more rapid breakdown of sulfide than of rock-forming minerals. Moss (1957) and Van Tassel (1958) showed that the mineral previously termed *carphosiderite* was actually hydroniumjarosite (Brophy and Sheridan 1965).

PINAL COUNTY: Cherry Creek area, Sierra Ancha District, Black Brush deposit (Granger and Raup 1969).

#HYDROXYAPOPHYLLITE-(K)

Potassium calcium silicate hydroxide fluoride hydrate: $KCa_4Si_8O_{20}$ $(OH,F) \cdot 8H_2O$. This is a member of the apophyllite group, which was renamed in 2013 to include a prefix and suffix to indicate specific anion and cation (Hatert et al. 2013). Hydroxyapophyllite-(K) primarily forms as a secondary mineral in amygdules in basalts and is commonly associated with zeolites. It is also found in contact-metamorphosed limestones bordering intrusive rocks.

Southeastern Dripping Spring Mountains, Banner supersystem, Christmas District, Christmas Mine, coating kinoite (R050169).

MARICOPA COUNTY: Horseshoe Dam District, one mile south of Horseshoe Dam, as clear tabular crystals up to 10 mm, with natrolite (D. M. Shannon 1983b).

#HYDROXYCALCIOROMÉITE

Calcium antimony titanium oxide hydroxide: $(Ca,Sb^{3+})_2(Sb^{5+},Ti)_2O_6(OH)$. A secondary mineral formed from the alteration of tetrahedrite. It was originally known as lewisite until revised in 2010.

COCHISE COUNTY: Johnny Lyon Hills, Yellowstone District, at a prospect along the banks of Tres Alamos Wash, abundant as an alteration product of tetrahedrite. It typically forms cell walls in the pseudomorphs, whereas the interiors are filled with partzite (discredited in 2016 as a mixture of plumboroméite-like and chrysocolla-like phases). The tetrahedrite crystals must have been spectacular, as euhedra 1 in. or larger, before oxidation occurred. Mineralization is confined to milky quartz veins that cut limestone (3rd ed.).

#HYDROXYKENOELSMOREITE

Lead tungsten iron aluminum hydroxide: $(\square,Pb)_2(W,Fe^{+3},Al)_2(O,OH)_6(OH)$. A rare secondary mineral found in gold deposits.

MARICOPA COUNTY: Vulture District, in a small prospect 1.5 miles northwest of the Vulture Mine, Vulture Mountains, as very small octahedral dark red crystals, 10 to 50 microns in size, associated with powellite and wulfenite, in a quartz vein (Robert Jenkins, pers. comm., 2020).

HYDROXYLAPATITE

Calcium phosphate hydroxide: $Ca_5(PO_4)_3OH$, a member of the apatite group.

COCHISE COUNTY: Whetstone Mountains, Kartchner District, Kartchner Caverns (C. A. Hill 1999).

YAVAPAI COUNTY: White Picacho District, Midnight Owl Mine (London and Burt 1982a).

#HYDROXYLPYROMORPHITE

Lead phosphate hydroxide: $Pb_5(PO_4)_3(OH)$. An uncommon mineral of the apatite group, the OH analog of pyromorphite, was first approved in 2017.

PIMA COUNTY: Santa Catalina Mountains, western end of Pusch Ridge. The mineral occurs as small beige-tan elongated crystals less than 0.5 mm long, associated with chrysocolla and with lazaraskeite in mylonitized Wilderness Granite (Yang, pers. comm., 2020).

FIGURE 3.123 Hydroxylpyromorphite, Pusch Ridge, Santa Catalina Mountains, Pima County, FOV 2 mm.

HYDROZINCITE

Zinc carbonate hydroxide: $Zn_5(CO_3)_2(OH)_6$. A secondary mineral formed in the oxidized portions of mineral deposits by the alteration of sphalerite; commonly associated with smithsonite, hemimorphite, and cerussite.

COCHISE COUNTY: Tombstone District, west side of the quarry roll, in a small seam with aurichalcite and hemimorphite (B. S. Butler, Wilson, and Rasor 1938); Empire Mine (UA 12423). Gunnison Hills, Cochise supersystem, Johnson Camp District, Texas-Arizona Mine, in metamorphosed limestone as white chalky masses up to 6 × 24 in. (J. R. Cooper 1957; J. R. Cooper and Silver 1964).

COCONINO COUNTY: Havasu Canyon District, with other secondary lead and zinc minerals (Bull. 181).

GILA COUNTY: Banner supersystem, 79 District, 79 Mine, as fine-grained white acicular crystal clusters with hemimorphite and aurichalcite (R. Gibbs, pers. comm., 2018).

MARICOPA COUNTY: White Picacho District, as a supergene mineral formed from the alteration of sphalerite and associated hemimorphite, in pegmatites (Jahns 1952).

PINAL COUNTY: Pioneer supersystem, Magma District, 1,600 ft. level of the Magma Mine, as a white film on sphalerite (Short et al. 1943); Hancock property near Superior (Bull. 181). Slate District, Jackrabbit Mine, intimately associated with pulverulent yellow limonite and formed as fracture fillings (Hammer 1961). Mammoth District, Mammoth–St. Anthony Mine, with wulfenite (UA 12993).

YUMA COUNTY: Castle Dome District, in vugs and channels, associated with smithsonite, wulfenite, vanadinite, and mimetite (E. D. Wilson 1933); Señora claims, in fissures with gypsum (Bull. 181).

#ICE

Hydrogen dioxide: H_2O. A common mineral found throughout Arizona when temperatures are below freezing.

ALL COUNTIES: Found as ice, snow, sometimes as perfect hexagonal crystals under favorable temperature conditions, especially in winter.

COCONINO COUNTY: Ice River Cave, fourteen miles north of Flagstaff—a partially collapsed lava tube often has ice inside even during the summer months (Coconino National Forest, Flagstaff).

#IDAITE

Copper iron sulfide: Cu_3FeS_4. A mineral that forms from the weathering of bornite (Rice et al. 1979), currently considered an inadequately described species.

GREENLEE COUNTY: Copper Mountain supersystem, Morenci District, Morenci, Morenci pit, occurs with bornite replacing chalcopyrite (Enders 2000).

#IIMORIITE-(Y)

Yttrium silicate carbonate: $Y_2(SiO_4)(CO_3)$. A rare mineral found in quartz-microcline pegmatite or in veins in riebeckite-aegirine granite.

MOHAVE COUNTY: Probably McConnico District, southwest of Kingman, in a prospect formerly known as the Guy Hazen Group, as a replacement of thalénite-(Y) (Fitzpatrick and Pabst 1986; Škoda et al. 2015).

#IKAITE

Calcium carbonate hydrate: $CaCO_3 \cdot 6H_2O$. Ikaite forms as a precipitate in cold water. Above 8°C, it is unstable and will change to calcite.

PINAL COUNTY: Superstition Mountains, as microscopic single crystals on the weeping wounds of the Arizona ash tree when the temperature is near freezing. Specimens converted to monohydrocalcite after warming to room temperature (Garvie, pers. comm., 2020).

illite

Potassium aluminum silicate hydroxide: $K_{0.65}Al_{2.0}\square Al_{0.65}Si_{3.35}O_{10}(OH)_2$. Illite is usually considered to be a variety of muscovite as it is structurally similar. It has been redefined as a series name for dioctahedral interlayer-deficient micas with the preceding generalized formula (see Rieder et al. 1998). Structurally similar to muscovite but a partial alteration product, as it contains more hydroxide and less potassium and aluminum. A widespread and common mica component of clays, illites are especially abundant on the Colorado Plateau. They are a major constituent of the Salt Wash Member of the Morrison formation (Keller 1962), as well as the dominant clay mineral in most of the Moenkopi formation. In the latter unit, illite is typically associated with kaolinite, chlorite, or montmorillonite, in places forming mixed-layer assemblages of illite montmorillonite and chlorite illite (Schultz 1963). Illite is a common constituent of many clay-bearing sedimentary rocks throughout the state and is a widespread product of hydrothermal alteration of some mineral deposits, especially certain porphyry copper orebodies. Only a few localities are listed.

COCHISE COUNTY: Warren District, Bisbee, Sacramento Hill, in a hydrothermally altered granite porphyry stock with sericite, kaolinite, allophane, and alunite, as an alteration of feldspar; in dikes that cut limestones peripheral to the stock (Schwartz 1947).

GILA COUNTY: Miami-Inspiration District, Miami Mine, associated with kaolinite, halloysite-10Å, and sericite as a product of hydrothermal alteration of granite porphyry and schist (Schwartz 1947).

GREENLEE COUNTY: Copper Mountain supersystem, Morenci District, as pseudomorphs after plagioclase, with sericite, in the intensely altered porphyry orebody, with a large suite of phyllic and argillic hydrothermal alteration products, including kaolinite and allophane (Schwartz 1947, 1958).

PINAL COUNTY: Mineral Creek supersystem, Ray District, Ray area, in a hydrothermally altered stock within the orebody, as pseudomorphs after plagioclase phenocrysts, associated with sericite, allophane, and kaolinite, and as pseudomorphs after biotite (Schwartz 1947, 1952). San Manuel District, San Manuel Mine, as an abundant major constituent, in phyllically and hydrothermally altered monzonite and quartz-monzonite porphyries (Schwartz 1947, 1958; Creasey 1959); Kalamazoo orebody (J. D. Lowell 1968).

ILMENITE

Iron titanium oxide: $Fe^{2+}Ti^{4+}O_3$. Commonly found as an accessory mineral in certain igneous rocks and, locally, as an important detrital constituent of black sands. An important early crystallizing accessory mineral found in low-oxidation-state ilmenite series granitic rocks (Ishihara 1977; Stanley B. Keith et al. 1991).

APACHE COUNTY: Garnet Ridge SUM District, Garnet Ridge, as fragments in a breccia dike that pierces Mesozoic sediments; as needles, the most common inclusions in garnets (Gavasci and Kerr 1968).

GILA AND PINAL COUNTIES: As a constituent of the Pinal Schist and diabase (N. P. Peterson 1962).

GILA COUNTY: Miami-Inspiration supersystem, Pinto Valley District, near Castle Dome Mine, as tabular pieces in quartz (N. P. Peterson, Gilbert, and Quick 1946, 1951). Sierra Ancha Mountains, intimately associated with hematite in Precambrian quartzite (E. C. Peterson 1966). East slope of Mazatzal Mountains, Three Bar Wildlife Management Area, in quartz veins with chalcopyrite (MM 7383).

GRAHAM COUNTY: Galiuro Mountains, in disseminated form in the northern part of the range (Bull. 181). Santa Teresa Mountains, east of Klondyke, in miarolitic cavities as individual crystals to 1.8 cm (Les Presmyk, pers. comm., 2020).

MARICOPA COUNTY: White Tank District, as grains and intergrowths with magnetite in Precambrian schist and in pegmatites (Harrer 1964). Big Horn District, with titaniferous magnetite constituting about 3 to 7 percent of extensive placer deposits (Harrer 1964).

PIMA COUNTY: Ajo District, formed both as a primary constituent of the Cornelia Quartz Monzonite and by alteration of titanite (Gilluly 1937).

PINAL COUNTY: Florence Junction area, with titaniferous magnetite in widespread alluvial deposits (Harrer 1964).

SANTA CRUZ COUNTY: Patagonia Mountains, Washington Camp District, especially at Duquesne and Washington Camp, in several contact-metamorphic deposits (Schrader 1917).

YAVAPAI COUNTY: Eureka supersystem, Bagdad District, with magnetite, as dikes and irregular bodies in gabbro (Ball and Broderick 1919); reported near the Bagdad Mine, as a large low-grade deposit. Bradshaw Mountains, in low-oxidation-state granite pegmatite near Cleator (Bull. 181). Chino Valley, Paulden quadrangle, as minute lamellae intimately intergrown with pyroxene-garnet assemblages, in xenoliths in the Sullivan Buttes latite (Krieger 1965; Schulze, Helmstaedt, and Cassie 1978).

ILSEMANNITE

Molybdenum oxide hydrate: $Mo_3O_8 \cdot nH_2O$. Listed by IMA as a questionable mineral, ilsemannite is mainly amorphous blue coatings found as a secondary mineral formed by oxidation of molybdenite and other molybdenum minerals. In places, it is of postmine origin.

APACHE COUNTY: Cane Valley District, Monument No. 2 Mine, as dispersed powdery fine-grained material associated with corvusite, navajoite, hewettite, uraninite, and gypsum (Witkind and Thaden 1963).

COCHISE COUNTY: Warren District, in a prospect pit near Warren, as blue stains on intensely silicified, brecciated limestone, with fluorite and scheelite (Graeme 1981).

COCONINO COUNTY: Cameron District, Huskon No. 11 Mine, with marcasite in sandstone, as ink-blue masses and stains; Huskon No. 10 Mine, as ink-blue stain on halotrichite (AEC mineral collection). East of Jacob Lake, Vermilion Cliffs District, Sun Valley Mine, abundant on the walls of older mine workings, in uranium deposits in a paleo-stream channel filled with Shinarump Conglomerate (Petersen, Hamilton, and Myers 1959).

ILVAITE

Calcium iron oxide silicate hydroxide: $CaFe^{3+}Fe^{2+}_2O(Si_2O_7)(OH)$. It is found in contact-metamorphic deposits, typically in limestones and dolomites, with other calcium silicate minerals.

COCHISE COUNTY: Dragoon Mountains, Middlemarch Pass, Middle Pass supersystem, Black Diamond District, abundant in hedenbergite hornfels in several mines, including Black Diamond Mine, as poorly developed crystals up to 0.5 in. long, associated with sphalerite and fluorite (3rd ed.).

GRAHAM COUNTY: Aravaipa supersystem, Iron Cap District, Iron Cap Mine, as vitreous fine-grained material associated with fluorite, sphalerite, and quartz (W. E. Wilson 1988).

INGODITE

Bismuth tellurium sulfide: Bi_2TeS. A rare primary mineral in hydrothermal deposits.

COCHISE COUNTY: Cochise supersystem, Johnson Camp District, Bluebird Mine (S C7073), as small tabular cleavable silver-gray crystals in scheelite-bearing garnetite (3rd ed.).

IODARGYRITE

Silver iodide: AgI. A secondary mineral formed in the oxidized zone of silver deposits; alters to silver.

COCHISE COUNTY: Pearce District, Commonwealth Mine, with chlorargyrite, bromargyrite, bromian chlorargyrite, and acanthite (Endlich 1897; Guild 1910). Warren District, Bisbee, Campbell shaft, with mottramite, argentite, silver, and kettnerite (Graeme 1993).

MARICOPA COUNTY: Painted Rock District, Rowley Mine, as euhedral pale-yellow to clear crystals (R. Gibbs, pers. comm., 2016).

MOHAVE COUNTY: Silliman (1866) reported the presence of iodyrite in quartz veins in "ash-colored feldspathic porphyry" in Mohave County, associated with "fluorspar, green carbonate of copper, free gold, and abundant iron gossan in cellular quartz." Oatman District, in several lodes in the Caledonia, Dayton, Quackenbush, and Knickerbocker properties (3rd ed.). Cerbat Mountains, Wallapai supersystem, as the chief near-surface ore mineral in oxidized sulfide vein deposits, especially the Stockton Hill and Chloride Districts (B. E. Thomas 1949).

PIMA COUNTY: Cerro Colorado District, Cerro Colorado Mine (Guild 1910). South Comobabi Mountains, Cababi District, Mildren and Steppe claims (S. A. Williams 1960).

PINAL COUNTY: Mammoth District, Mammoth–St. Anthony Mine, as pale-green droplets, with caledonite and boleite (Bideaux 1980).

SANTA CRUZ COUNTY: Flux District, World's Fair Mine, as small isolated crystals, globules, and specks in "contour boxwork" limonite deposits, which were derived from tetrahedrite (Blanchard and Boswell 1930).

IRANITE

Copper lead chromate silicate hydroxide: $CuPb_{10}(CrO_4)_6$ $(SiO_4)_2(OH)_2$. Closely related to hemihedrite chemically and structurally but differs by having a cation site principally occupied by copper rather than zinc. Formed in the oxidized zone of lead- and chromium-bearing veins.

MARICOPA COUNTY: Big Horn Mountains, Big Horn District, Evening Star Mine, as brown crystals with fornacite, phoenicochroite, cerussite, and other lead secondary minerals (R. Gibbs, pers. comm., 2015). At a small prospect near the Evening Star Mine, as small well-developed crystals with fornacite, wickenburgite, cerussite, mimetite, and creaseyite (R. Gibbs, pers. comm., 2015).

PINAL COUNTY: Mammoth District, Mammoth–St. Anthony Mine, as small crystals on older specimens (Richard W. Thomssen, pers. comm.,

1987). Vekol District, Vekol and Reward Mines, in the oxide zone of sulfide deposits in dolomite arsenical sulfide-sulfosalt veins broadly related to the Vekol porphyry copper deposit (Stanley B. Keith, pers. comm., 2016).

IRIGINITE

Uranyl molybdenum oxide hydrate: $(UO_2)Mo^{6+}_2O_7 \cdot 3H_2O$. An uncommon secondary mineral found in uranium deposits.

COCONINO COUNTY: Cameron District, in some uranium prospects (3rd ed.).

#IRON

Iron: Fe. Terrestrial iron is extremely rare, and iron occurrences in Arizona involve meteorite falls. The variety alpha nickel iron, α-(Ni,Fe), contains about 5.5 percent nickel and was previously known as kamacite until this mineral name was discredited in 2006 (Burke 2006). It is a major constituent of iron meteorites, including those found in Arizona, especially the Canyon Diablo meteorite fall around and northwest of Meteor Crater in southeastern Coconino County. It is commonly associated with taenite.

#JACOBSITE

Manganese iron oxide: $Mn^{2+}Fe^{3+}_2O_4$. A rare hypogene mineral found with other manganese oxides.

COCHISE COUNTY: Warren District, Bisbee, No. 4 claim. Found in a small manganese deposit that was largely braunite with cryptomelane veins in the surrounding limestone. The jacobsite is included as 5 mm blebs in the cryptomelane (Graeme, Graeme, and Graeme 2015).

JADEITE

Sodium aluminum silicate: $NaAlSi_2O_6$. A member of the pyroxene group found in high-pressure metamorphic rocks.

APACHE COUNTY: Garnet Ridge SUM District, Garnet Ridge, as diopsidic jadeite with an appreciable proportion of aegirine, coexisting with pyrope-almandine garnet in eclogite inclusions from kimberlite pipes (Watson and Morton 1969).

JAHNSITE

Calcium iron magnesium manganese zinc sodium iron phosphate hydroxide hydrate: $[CaFe^{2+}Mg_2, CaMn^{2+}Fe^{2+}_2, CaMn^{2+}Mg_2, CaMn^{2+}Mn^{2+}_2, CaMn^{2+}Zn_2, Mn^{2+}Mn^{2+}Fe^{2+}_2, Mn^{2+}Mn^{2+}Mg_2, Mn^{2+}Mn^{2+}Mn^{2+}_2, Mn^{2+}Mn^{2+}Zn_2, NaFe^{3+}Mg_2, (Na,Ca)Mn^{2+}(Mg,Fe^{3+})_2, NaMn^{2+}(Mn^{2+}Fe^{3+})_2]Fe^{3+}_2(PO_4)_4(OH)_2 \cdot 8H_2O$. Formed in iron-manganese-phosphate zones of pegmatites, where it is a supergene oxidation product of earlier-formed phosphates. Jahnsite is a member of the whiteite-jahnsite group where species are recognized based on the dominant elements in four nontetrahedral cation sites (P. B. Moore and Ito 1978; Kampf, Steele, and Loomis 2008). Because the chemistry of the Arizona material is unknown, the exact species is unknown.

MARICOPA COUNTY: White Picacho District, Midnight Owl Mine near Wickenburg, with hureaulite in lithiophilite (UA 6916).

JALPAITE

Silver copper sulfide: Ag_3CuS_2. A primary mineral formed under low-temperature hydrothermal conditions.

COCHISE COUNTY: Warren District, Bisbee, with the sulfide ores of the Campbell Mine (Graeme 1993).

PINAL COUNTY: Mineral Hill District, Reymert Mine, with acanthite and silver in quartz veins (K. S. Wilson 1984).

JAMESONITE

Lead iron antimony sulfide: $Pb_4FeSb_6S_{14}$. Formed in hydrothermal veins under low to moderate temperatures; typically associated with galena, sphalerite, stibnite, tetrahedrite, and other sulfosalts.

YAVAPAI COUNTY: Reported in the Bradshaw Mountains, in some of the ores, with free gold (Bull. 181).

JAROSITE

Potassium iron sulfate hydroxide: $KFe^{3+}_3(SO_4)_2(OH)_6$. A member of the alunite group and a secondary mineral widely distributed throughout the Southwest. Formed in the oxidized cappings of base-metal deposits; may be a major component of the limonites of gossans, especially those in the tops and fringes of porphyry copper deposits, and, to a lesser extent, volcanogenic massive sulfide deposits.

APACHE COUNTY: Cane Valley District, Monument No. 2 Mine, in Shinarump Conglomerate (Mitcham and Evensen 1955; Witkind and Thaden 1963).

COCHISE COUNTY: Tombstone District, abundant in the Toughnut and Empire Mines (B. S. Butler, Wilson, and Rasor 1938). Gleeson District, as small flaky bunches (Bull. 181). Pearce District, Commonwealth Mine, with chlorargyrite and silver (Bull. 181). Warren District, Bisbee, Sacramento Hill, in the hydrothermally altered granite porphyry stock (Schwartz 1947), possibly the most common secondary mineral in the district found in the gossan zones of all the producing mines (Graeme 1981). Dragoon Mountains, Red Cloud Peak, as microcrystals (Peter Megaw, UA x3365).

FIGURE 3.126 Jarosite, Morenci Mine, Greenlee County, FOV 3 mm.

COCONINO COUNTY: Cameron District, very abundant in the uranium deposits (Austin 1964).

GILA COUNTY: Miami-Inspiration supersystem, Pinto Valley District, Castle Dome Mine; Copper Cities District, Copper Cities Mine, as an abundant supergene mineral (N. P. Peterson 1947, 1962; N. P. Peterson, Gilbert, and Quick 1951). Sierra Ancha District, Little Joe deposit; Cherry Creek, Shepp 2 deposit (Granger and Raup 1969). Southern Dripping Spring Mountains, Banner supersystem, Chilito District, Arizona Apex Mine, as sharp brown crystals up to 0.5 mm (David Shannon, pers. comm., 1988).

GRAHAM COUNTY: Gila Mountains, Safford supersystem, Lone Star District, with pseudomalachite, malachite, brochantite, antlerite, and other secondary oxidized minerals (Hutton 1959a); Safford porphyry copper deposit cluster, in fractures and veins as a replacement of primary minerals, fault gouge, and clay, and intimately associated with turquoise and alunite (R. F. Robinson and Cook 1966).

GREENLEE COUNTY: Copper Mountain supersystem, Morenci District, Morenci Mine, where it is "widespread but most dramatic as an oxidation product of pyritic veinlets in areas of weak (copper) mineralization" (Moolick and Durek 1966).

MARICOPA COUNTY: Vulture District, Vulture Mine, as "fine, transparent, yellow and dark-brown rhombic crystals" filling cavities formed from the oxidation of pyrite; associated with gold particles (Silliman 1879). Black Rock District, Black Rock Mine (Bull. 181).

MOHAVE COUNTY: Cerbat Mountains, Wallapai supersystem, as a widespread secondary mineral (B. E. Thomas 1949; Field 1966).

NAVAJO COUNTY: Monument Valley District, Monument No. 1 Mine, in the largely oxidized uranium-vanadium ores, associated with torbernite, carnotite, corvusite, and tyuyamunite (Holland et al. 1958).

PIMA COUNTY: Empire District, Total Wreck Mine, an oxide zone mineral with wulfenite, vanadinite, chlorargyrite, cerussite, and smithsonite (W. E. Wilson 2015); Jerome No. 2 Mine, Hilton Group of claims, in lead-zinc veins (Schrader and Hill 1915). Silver Bell District, Silver Bell Mine, with alunite (Kerr 1951). Sierrita Mountains, Pima supersystem, Mission-Pima District, Mineral Hill area (UA 7857); Twin Buttes District, Twin Buttes Mine (Stanley B. Keith, pers. comm., 1973); San Xavier District, San Xavier North Mine, in drill cores as fracture coatings and botryoidal masses in open spaces (B. A. Koenig 1978). South Comobabi Mountains, Cababi District, Mildren and Steppe claims (S. A. Williams 1963), as lovely color-zoned tablets up to 5 mm across in vugs in milky quartz. Ajo District, where it is largely confined to weathered capping of areas of notably pyritic ore (Gilluly 1937). Helvetia-Rosemont District (Schrader 1917).

PINAL COUNTY: Mineral Creek supersystem, Ray District near Ray, originally as one of the oxidized iron minerals in the supergene blanket above the Ray porphyry copper deposit, and at the Sonora copper deposit in a copper-rich Holocene gravel deposit leached from a nearby secondary copper sulfide blanket, variously associated with goethite, hematite, malachite, azurite, cuprite, and tenorite (C. H. Phillips, Cornwall, and Rubin 1971). Bunker Hill (Copper Creek) District, Copper Creek, as a minor oxidation product of pyrite (Simons 1964). San Manuel District, San Manuel deposit, in small amounts, formed from the breakdown of pyrite in the oxidized cap of the porphyry copper deposit (Schwartz 1953).

SANTA CRUZ COUNTY: Red Mountain District, near Patagonia, as small honey-brown, transparent crystals less than 0.5 mm in size coating breccia and altered porphyritic rock (Kenneth Bladh, pers. comm., 1974). Patagonia Mountains, Mowry District, Mowry Mine (3rd ed.). Flux District, Flux Mine (3rd ed.).

YAVAPAI COUNTY: Copper Basin District, in the oxidized zone of mineralized breccia pipes, with hematite, limonite, chrysocolla, and other secondary minerals (Johnston and Lowell 1961). Eureka supersystem, Bagdad District, Bagdad, in leached cappings of copper sulfide deposits, with goethite (C. A. Anderson 1950). Jerome supersystem, Verde District, United Verde Mine, in minor amounts dusted over yavapaiite and other secondary sulfates that were formed from the burning of pyritic ores (Hutton 1959b).

Copper Mountains, La Posa District, Engesser prospect (E. D. Wilson 1933).

JOHANNITE

Copper uranyl sulfate hydroxide hydrate: $Cu(UO_2)_2(SO_4)_2(OH)_2 \cdot 8H_2O$. A secondary mineral typically formed by the alteration of uraninite in sulfide veins.

COCHISE COUNTY: Warren District, Bisbee, Cole Mine—this species has been recognized on only a small number of specimens collected in the late 1940s from an unspecified level in this mine. On these, it was found as an efflorescence on decomposing sulfides as a green to blue-green crust with gypsum and minor zippeite as well as uranopilite (Graeme, Graeme, and Graeme 2015).

YAVAPAI COUNTY: Eureka supersystem, Hillside District, Hillside Mine, with several other uranium minerals (Axelrod et al. 1951).

JOHANNSENITE

Calcium manganese silicate: $CaMnSi_2O_6$. A member of the pyroxene group that is similar to diopside-hedenbergite in properties and paragenesis, but much less common.

GRAHAM COUNTY: Aravaipa supersystem, Iron Cap District, Black Hole prospect, in NW quarter of S36, T5S R19E, in tabular bodies and irregular masses replacing limestone, as radiating or spherulitic aggregates of prisms or needles a few centimeters in diameter and larger masses a few feet thick and several tens of feet long, associated with neotocite, chalcopyrite, galena, and sphalerite (Simons and Munson 1963); Iron Cap Mine, as the major pyroxene, both as massive material and as crystals (Reiter 1980, 1981).

SANTA CRUZ COUNTY: Washington Camp District, near the Dudley Standard Mine, as radiating clusters (Lehman 1978).

#JÔKOKUITE

Manganese sulfate hydrate: $Mn^{2+}(SO_4) \cdot 5H_2O$. A secondary mineral formed by the alteration of sulfide minerals.

COCHISE COUNTY: Warren District, Bisbee, Junction Mine, 2300 level—minor amounts of postmining jôkokuite were found in broken alabandite left in the stope during mining. It occurs as a soft granular pink to white material to 5 mm thick filling fractures (Graeme, Graeme, and Graeme 2015).

JORDISITE

Amorphous molybdenum sulfide: MoS_2. A poorly characterized amorphous material of hydrothermal origin.

COCONINO COUNTY: East of Jacob Lake, Vermilion Cliffs District, Sun Valley Mine, in a uranium deposit located in a bend in a paleo-stream channel filled with Shinarump Conglomerate, associated with uraninite, ilsemannite, pyrite, sphalerite, and hematite (Petersen 1960).

FIGURE 3.127 Junitoite, Christmas Mine, Gila County, FOV 5 mm.

JUNITOITE

Calcium zinc silicate hydrate: $CaZn_2Si_2O_7 \cdot H_2O$. Probably of late-stage hydrothermal origin; the type locality is the Christmas Mine.

GILA COUNTY: Southeastern Dripping Spring Mountains, Banner supersystem, Christmas District, in the open-pit mine at Christmas, in tactite ores with kinoite and apophyllite, as clear platy crystals up to 5 mm across, typically associated with pink smectite (clay) in rings around nuggets of unaltered sphalerite (S. A. Williams 1976; R050154).

FIGURE 3.128 Jurbanite, San Manuel Mine, Pinal County.

JURBANITE

Aluminum sulfate hydroxide hydrate: $Al(SO_4)(OH) \cdot 5H_2O$. A sparse secondary mineral formed at the San Manuel Mine under conditions of high humidity. Associated with other water-soluble secondary sulfates, including epsomite and pickeringite. The San Manuel Mine is the type locality.

PINAL COUNTY: San Manuel District, 2,075 ft. level of the San Manuel Mine, as sparse minute clear colorless crystals, intimately associated with epsomite, hexahydrite, pickeringite, and starkeyite deposited on lagging and overhead pipes (J. W. Anthony and McLean 1976; R130817).

KAERSUTITE

Sodium calcium magnesium aluminum titanium aluminosilicate oxide: $NaCa_2(Mg_3AlTi^{4+})(Si_6Al_2)O_{22}O_2$. May be regarded as a titanian oxo-amphibole; a member of the amphibole supergroup. Formed in under-saturated nepheline-alkalic plutonic and volcanic rocks.

COCHISE COUNTY: San Bernardino Valley, San Bernardino (Geronimo) volcanic field, as porphyrocrysts in xenoliths formed in alkali-olivine basalts and basanites of the Quaternary Geronimo lavas (S. H. Evans and Nash 1979; D. J. Lynch 1976).

COCONINO COUNTY: San Francisco Mountain, as microcrystals (Kenneth Bladh, UA x1672).

GILA COUNTY: Peridot Mesa, Peridot District, on the San Carlos Apache Reservation, associated with spinel and augite in Group II ultramafic inclusions (Frey and Prinz 1978). Mason (1968) thoroughly studied the mineral from xenocrysts in a basalt flow (see also Cross and Holloway 1974); Garcia, Muenow, and Liu (1980) discussed details of the chemistry of its volatile content.

MOHAVE COUNTY: Hoover District, about eight miles south of Hoover Dam on U.S. Highway 93, as large porphyrocrysts up to 15 cm long in a camptonite dike (Campbell and Schenk 1950; Garcia, Muenow, and Liu 1980). At the southern end of the Uinkaret Plateau, as large poikilitic crystals in ultramafic inclusions in geologically young flows and associated cinder cones of basanitic composition (Best 1970).

#KALICINITE

Potassium bicarbonate: $KH(CO_3)$. A mineral found in decomposing dead trees.

MARICOPA COUNTY: Phoenix area, found in brittle ash crust of palo verde (*Parkinsonia microphylla*) trees (Garvie 2016).

#KAMITUGAITE

Lead aluminum uranyl phosphate oxide hydroxide hydrate: $PbAl(UO_2)_5(PO_4)_3O_2(OH)_2(H_2O)_{11.5}$. A very rare secondary uranium mineral.

MARICOPA COUNTY: Big Horn Mountains, Osborne District, Evening Star Mine, associated with wickenburgite, mimetite, and quartz (Vajdak 2002).

KANONAITE

Manganese aluminum oxide silicate: $Mn^{3+}AlOSiO_4$. Found in manganese-rich metamorphic schists.

MARICOPA COUNTY: Phoenix, southwestern slope of Piestewa Peak (formerly Squaw Peak), as green porphyroblasts in quartz-muscovite schist; kanonaite from this locality has a range of compositions and may be distinguished from the local manganese-rich andalusite only through chemical analysis (Thorpe 1980).

KAOLINITE

Aluminum silicate hydroxide: $Al_2Si_2O_5(OH)_4$. This is the most important member of the kaolinite group of clay minerals, which also includes dickite, nacrite, halloysite-7Å, and halloysite-10Å. Members of the group are formed by hydrothermal processes during alteration accompanying mineral deposit formation; they are also extensive in sedimentary formations, to which they have been introduced either by transportation from previously weathered rocks or by in situ weathering of feldspathic rocks. Common beneath coal beds as underclay that is produced by downward leaching of feldspathic component of footwall rocks by humic acid derived from the overlying coal. Only a few Arizona localities can be listed.

APACHE COUNTY: Cane Valley District, Monument No. 2 Mine, associated with the uranium-vanadium ores (Witkind and Thaden 1963).

COCHISE COUNTY: Warren District, Bisbee, second level of the Copper Queen Mine, as nearly pure kaolinite in white waxy masses (Bull. 181); Sacramento Hill, in hydrothermally altered granite porphyry stock, with sericite, illite, allophane, and alunite, as an alteration product of feldspar (Schwartz 1947, 1958). Tombstone District, Toughnut Mine (B. S. Butler, Wilson, and Rasor 1938). Turquoise District, Silver Bill Mine (Bull. 181). Willcox Playa, very sparse in clays near the surface (Pipkin 1968).

COLORADO PLATEAU: Prevalent in certain sedimentary formations, including the sandstone of the Shinarump Member of the Chinle Formation (Schultz 1963) and the Salt Wash Member of the Morrison Formation (Keller 1962).

GILA COUNTY: Globe Hills District, Old Dominion Mine, with chalcocite in the oxidized zones (Bull. 181). Miami-Inspiration supersystem, Pinto Valley District, Castle Dome Mine, associated with halloysite-10Å, as small masses filling open fractures in quartz monzonite (Schwartz 1947);

Miami and Inspiration orebodies, as a product of hydrothermal alteration (Schwartz 1947).

GREENLEE COUNTY: Copper Mountain supersystem, Morenci District, widespread as a product of pervasive hydrothermal alteration (Schwartz 1947, 1958; Moolick and Durek 1966); in large masses at the Longfellow Mine, and in snow-white mammillary masses with azurite and malachite at the Copper Mountain and Mammoth Mines (Lindgren 1905); Humboldt, Ryerson, and other properties (3rd ed.).

MARICOPA COUNTY: Vulture Mountains, Newsboy District, on the west side of the Hassayampa River, associated with alunite, making up a major phase within a hydrothermally altered rhyolite (Sheridan and Royse 1970).

NAVAJO COUNTY: Along the eastern and southeastern edge of Black Mesa, as probable underclay beneath coal beds or as cement or matrix of sandstones in the Westwater Canyon Member of the Morrison Formation and in the Cow Springs and Mesa Verde Formations. Near Coal Mine Canyon and the Hopi Mesas (Kiersch 1955). Pinedale coalfield, as kaolinite underclay beneath the coal beds (Eyde and Wilt 1989).

PIMA COUNTY: Silver Bell District, upper levels of the El Tiro Mine, as large masses in wall rock (Kerr 1951). Pima supersystem, Twin Buttes District, Twin Buttes Mine, as a product of hydrothermal alteration (Stanley B. Keith, pers. comm., 1973). Ajo, Phelps Dodge Well No. 1, as the variety anauxite (William Thomas, pers. comm., 1988).

PINAL COUNTY: San Manuel District, San Manuel Mine, as a common product of hydrothermal alteration of igneous rocks; with alunite, it makes up most of the rock in the intensely altered phyllic and argillic zones (Schwartz 1947, 1953, 1958; Creasey 1959). Mineral Creek supersystem, Ray District, Ray area, with sericite, illite, and alunite, in hydrothermally altered granitic rocks (Ransome 1919; Schwartz 1947, 1952).

KASOLITE

Lead uranyl silicate hydrate: $Pb(UO_2)SiO_4 \cdot H_2O$. A secondary mineral probably formed by the reaction of silica-bearing meteoric waters with earlier secondary uranium minerals.

MARICOPA COUNTY: South of Buckeye, in a pegmatite, associated with polycrase-(Y) (Bull. 181).

NAVAJO COUNTY: Southeast of Fort Apache, Seven Mile Canyon, Shinarump 1B Mine, as yellow specks on gray sandstone (AEC mineral collection).

SANTA CRUZ COUNTY: Santa Rita Mountains, Tyndall District, Kinsley property, east of Amado (UA 2704); Duranium claims, disseminated in arkosic sandstone with uranophane and autunite (R. L. Robinson 1954; Shawe 1966). Pajarito District, White Oak property, as good crystalline material, with uranophane, dumontite, and autunite, associated with oxidized lead ore in a shear zone in rhyolite (Granger and Raup 1962; Stanton B. Keith 1975). Walnut Canyon, as encrustations along fractures in felsite porphyry (Robert O'Haire, pers. comm., 1972).

YAVAPAI COUNTY: Steel City Mine, with soddyite and uranium-rich thorite (C. Frondel 1958).

FIGURE 3.129 Kenhsuite, Saddle Mountain Mine, Maricopa County, FOV 3 mm.

#KENHSUITE

Mercury sulfide chloride: $Hg_3S_2Cl_2$. A rare secondary mercury mineral found in hydrothermal deposits associated with other mercury minerals. Kenhsuite is a polymorph of corderoite.

MARICOPA COUNTY: Mazatzal Mountains District, Saddle Mountain Mercury Mine, as small colorless well-formed bladed crystals associated with calomel, mercury, and cinnabar (R. Gibbs, pers. comm., 2019).

KËSTERITE

Copper zinc tin sulfide: Cu_2ZnSnS_4. Found in quartz-sulfide hydrothermal veins in tin deposits.

COCHISE COUNTY: Warren District, Bisbee, Campbell Mine, among the sulfide minerals (Graeme 1993).

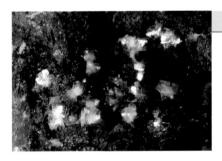

FIGURE 3.130 Kettnerite, Scorpion Hill, Vulture Mountains, Maricopa County, FOV 3 mm.

KETTNERITE

Calcium bismuth oxide carbonate fluoride: CaBiO $(CO_3)F$. Found in quartz veins with other bismuth minerals and fluorite.

COCHISE COUNTY: Warren District, Bisbee, Campbell orebody, with aikinite, argentite, and silver (Graeme 1993).

GRAHAM COUNTY: Golondrina District, prospect northwest of Bowie, as abundant yellowish-tan tabular crystals up to 5 mm (3rd ed.). The kettnerite is associated with bismutite, malachite, and bismuth-rich mottramite as pseudomorphs after very large (up to 3 in. long) prisms of a bismuth sulfosalt, which cannot be characterized because of its weathered condition.

LA PAZ COUNTY: Northwest Granite Wash Mountains, Three Musketeers District, Three Musketeers Mine, as masses in quartz veins, with other secondary bismuth minerals (collected by David Shannon).

MARICOPA COUNTY: Central Vulture Mountains, Flying Saucer District, Scorpion Hill, small white crystals with eulytine and bismuth-rich mottramite in a thin quartz-chrysocolla vein (Gibbs, pers. comm., 2012; R130113).

KHINITE

Copper lead tellurate hydroxide: $Cu^{2+}_3PbTe^{6+}O_6(OH)_2$. An extremely rare secondary mineral formed by oxidation of other tellurium minerals in acid waters. Tombstone is the type locality. There are two polytypes, khinite-3T and khinite-4O. Khinite-3T was a separate species named *parakhinite* but was discredited in 2009.

FIGURE 3.131 Khinite-3T, Empire Mine, Tombstone, Cochise County.

COCHISE COUNTY: Tombstone District, in dump material of the Old Guard Mine, as corroded deep-green minute crystals up to 1.15 mm that form rings on fracture surfaces of silicified granodiorite. The centers of the rings are composed of massive chlorargyrite, whereas the outer portions have been replaced by sparkling druses of minute dugganite crystals. Associated with xocomecatlite, bromargyrite, and several other tellurates and tellurites (S. A. Williams 1978); Empire Mine (R080124). Reef District, Reef Mine (Walstrom 2012).

KIDDCREEKITE

Copper tungsten tin sulfide: Cu_6WSnS_8. A rare primary sulfide mineral.

COCHISE COUNTY: Warren District, Bisbee, Campbell Mine, as minute grains in the sulfide orebody (Harris et al. 1984).

#KIESERITE

Magnesium sulfate hydrate: $Mg(SO_4).H_2O$. Found commonly in evaporite deposits.

COCONINO COUNTY: Found as an efflorescent mineral in Grand Canyon National Park (Quick, Corbett, and Manner 1989).

FIGURE 3.132 Kinoite, Christmas Mine, Gila County, 6 cm.

KINOITE

Calcium copper silicate hydrate: $Ca_2Cu_2Si_3O_{10} \cdot 2H_2O$. Found in skarns formed late in the paragenetic sequence, with apophyllite and copper. The Santa Rita Mountains occurrence is the type locality.

GILA COUNTY: Southeastern Dripping Spring Mountains, Banner supersystem, Christmas District, Christmas Mine, as well-formed bright-blue crystals with gilalite, apachite, ruizite, and often covered by apophyllite (S. A. Williams 1976).

PIMA COUNTY: Northern Santa Rita Mountains, Helvetia-Rosemont District, on claims formerly owned by ASARCO, in a core from diamond-drill holes, which penetrated contact-metamorphosed Paleozoic limestones and dolomites; with apophyllite, copper, djurleite, bornite, and chalcopyrite, as small tabular euhedral crystals and in veinlets (J. W. Anthony and Laughon 1970; Laughon 1971). Pima supersystem, Twin Buttes District, Twin Buttes Mine, as microcrystals, with apophyllite (Gene Wright, pers. comm., 1980; UA x1448); as crystals up to 0.5 cm, with stringhamite, copper, and apophyllite in wollastonite (William Hefferon, pers. comm., 3rd ed.).

#KINOSHITALITE

Barium magnesium aluminosilicate hydroxide: $BaMg_3(Si_2Al_2O_{10})(OH)_2$. Found in manganese-bearing rocks.

APACHE COUNTY: Garnet Ridge SUM District, found in an ultramafic diatreme at Garnet Ridge on the Navajo Reservation as inclusions in pyrope with srilankite, carmichaelite, rutile, and several other minerals (Wang, Essene, and Zhang 1999).

KOECHLINITE

Bismuth molybdenum oxide: Bi_2MoO_6. Found with other bismuth minerals in quartz veins; also formed by the alteration of tetradymite.

GILA COUNTY: Miami-Inspiration supersystem, Inspiration District, reported in the Live Oak Pit of the Inspiration Mine, as yellowish tabular crystals on chrysocolla (William Roberts, pers. comm., 1986).

#KOLOVRATITE

Hydrous nickel zinc vanadate: $(Ni,Zn)_x(VO_4)\cdot nH_2O$. A secondary mineral found in quartz schists and carbonaceous slates. Although an approved mineral by the IMA, it is considered questionable and would benefit from further work.

APACHE COUNTY: Cane Valley District, Monument No. 2 Mine, as a yellow crust with nolanite and pascoite (Jim McGlasson, pers. comm., 2020).

KORNELITE

Iron sulfate hydrate: $Fe^{3+}_2(SO_4)_3\cdot 7H_2O$. An uncommon secondary mineral formed by the oxidation of pyrite; found as efflorescences in old mine workings.

COCHISE COUNTY: Warren District, Bisbee, Copper Queen Mine, as irregular porous crusts (Merwin and Posnjak 1937); Campbell shaft, as pale-pink fibers (Graeme 1981).

KOSMOCHLOR

Sodium chromium silicate: $NaCr^{3+}Si_2O_6$. A rare member of the pyroxene group, found in iron meteorites.

COCONINO COUNTY: Canyon Diablo area, near Meteor Crater, with krinovite in graphite nodules in the Canyon Diablo octahedrite meteorite; also associated with roedderite, high albite, richterite, and chromite (Olsen and Fuchs 1968).

KOSTOVITE

Gold copper telluride: $AuCuTe_4$. An uncommon primary mineral formed with sulfides and other tellurium minerals.

COCHISE COUNTY: Warren District, Bisbee, as minute grains in the sulfide ores of the Campbell Mine, with pyrite, chalcopyrite, altaite, goldfieldite, and melonite (Graeme 1993).

KRENNERITE

Gold silver telluride: Au_3AgTe_8. An uncommon mineral found in hydrothermal gold ores.

COCHISE COUNTY: Warren District, Bisbee, in the ores of the Campbell Mine, associated with other tellurides (Graeme 1993). Tombstone District, Joe Mine, in an ore specimen, as a small patch of granular material oxidizing to gold and paratellurite; probably a common original constituent of the primary ores but readily oxidized, even to the lowest mine levels (3rd ed.).

KRINOVITE

Sodium magnesium chromium oxide silicate: $Na_4[Mg_8Cr^{3+}_4]O_4[Si_{12}O_{36}]$. Found in iron meteorites. Meteor Crater is the type locality.

COCONINO COUNTY: Meteor Crater, Canyon Diablo octahedrite meteorite, as minute deep emerald-green subhedral grains disseminated within graphite nodules, associated with roedderite, high albite, richterite, kosmochlor, and chromite (Olsen and Fuchs 1968).

#KRUPKAITE

Lead copper bismuth sulfide: $PbCuBi_3S_6$. Found with other sulfides in hydrothermal deposits.

COCHISE COUNTY: Warren District, Bisbee, Campbell shaft, in chalcocite sulfide ore (Schumer 2017).

FIGURE 3.133 Ktenasite, DH Mine, Gila County.

KTENASITE

Zinc copper sulfate hydroxide hydrate: $ZnCu_4(SO_4)_2$ $(OH)_6 \cdot 6H_2O$. A rare secondary mineral found in copper zinc deposits.

COCHISE COUNTY: Warren District, Bisbee, Campbell Mine, as thin crystalline crusts on pyrite-chalcopyrite ores (Graeme 1993).

GILA COUNTY: Banner supersystem, 79 District, 79 Mine, abundant as blue crusts in the 31 raise area (Thomas Trebisky, pers. comm., 1975); Kullman-McCool (Finch, Barking Spider) Mine (also known as the D. H. claims), in the southwestern workings, as very small tabular crystals and rosettes of crystals, with serpierite and gypsum (3rd ed.).

#KUKSITE

Lead zinc tellurate phosphate: $Pb_3Zn_3TeO_6(PO_4)_2$. A rare mineral found in the oxidized portions of lead-zinc deposits rich in tellurium.

COCHISE COUNTY: Huachuca Mountains, Reef District, Big Dipper and Reef Mines, as very small pale golden-yellow long prismatic barrel-shaped 1 mm hexagonal crystals on quartz prism faces and goethite. SEM photography reveals a few of the double-terminated crystals exhibit a hollow cavity along the c-axis (Walstrom 2013). Kuksite is widely distributed but nowhere locally abundant.

KULANITE

Barium iron aluminum phosphate hydroxide: $BaFe^{2+}_2Al_2(PO_4)_3(OH)_3$. Found in pegmatites with other phosphate minerals.

MARICOPA AND YAVAPAI COUNTIES: White Picacho District, in pegmatites (London 1981).

KURAMITE

Copper tin sulfide: Cu_3SnS_4. An uncommon sulfide mineral found in hydrothermal gold quartz veins.

COCHISE COUNTY: Warren District, Bisbee, in the sulfide ores of the Campbell Mine, associated with pyrite, altaite, and melonite (Graeme 1993).

KUTNOHORITE

Calcium manganese carbonate: $CaMn^{2+}(CO_3)_2$. A member of the dolomite group, it is often found in some mineral deposits as a gangue mineral.

GRAHAM COUNTY: Aravaipa supersystem, Iron Cap District, Iron Cap Mine, as pale-pink to white botryoidal coatings on quartz (Reiter 1980).

KYANITE

Aluminum oxide silicate: Al_2OSiO_4. Kyanite is typically formed in schist and gneiss as a product of the metamorphism of aluminous rocks.

COCHISE COUNTY: Northwesternmost Dos Cabezas Mountains, Drury prospect in Precambrian granitic rocks near Willcox, in S15, T13S R25E (Elevatorski 1978).

LA PAZ COUNTY: K-D District, locality three miles southeast of Quartzsite, with dumortierite (E. D. Wilson 1929; Bideaux, Williams, and Thomssen 1960; Reynolds et al. 1988). Granite Wash Mountains, west flank of Salome Peak, Hot Rock District, as large blades up to 6 cm long (Stephen Reynolds, pers. comm., 1989), and at several other localities where it occurs with magnetite, rutile, and pyrophyllite (Reynolds et al. 1988). Dome Rock Mountains, at several kyanite districts (Stray Elephant kyanite, Kyanite 1–3, Sugarloaf Gold system, and Dome Rock Kyanite Districts) with muscovite and one or more of the following: rutile, hypersthene, tourmaline, dumortierite, pyrophyllite, and sillimanite (Reynolds et al. 1988). Edge of the Trigo Mountains, one mile west of the lower Yuma Wash, as blades up 5 cm, with pyrophyllite and tourmaline (Gordon Haxel, pers. comm., 1989). Clip Wash, near Clip on the Colorado River, as long-bladed crystals in quartzose schist in Colorado River terrace gravels, associated with dumortierite (E. D. Wilson 1933).

MARICOPA COUNTY: Phoenix area, Phoenix Mountains District, Piestewa Peak (formerly Squaw Peak), in a quartz-rich metarhyolite, where 38 tons of kyanite were reportedly mined (E. D. Wilson and Roseveare 1949).

MOHAVE COUNTY: Found in Precambrian metamorphic rocks at several areas in the Lost Basin District (see Theodore, Blair, and Nash 1987 for exact locations).

YAVAPAI COUNTY: Tire claims in S½ S6, T15N R4W, a commercial kyanite prospect with sillimanite and some rare-earth minerals (AZGS files).

YUMA COUNTY: Gila Mountains, about eight miles west of Wellton, on the east side of the range (Bull. 181).

LANARKITE

Lead oxide sulfate: $Pb_2O(SO_4)$. A secondary mineral formed in lead deposits.

COCHISE COUNTY: Tombstone District, Gallagher Mine, microcrystals in galena (Bowell, Luetcke, and Luetcke 2015).

GILA COUNTY: Central Dripping Spring Mountains, Troy supersystem, C and B–Grey Horse District, C and B Vanadium Mine (Crowley 1980).

MARICOPA COUNTY: Big Horn Mountains, Big Horn District, Evening Star Mine, as very small crystals in minute vugs in galena (Yang et al. 2016).

LANGITE

Copper sulfate hydroxide hydrate: $Cu_4(SO_4)(OH)_6 \cdot 2H_2O$. A secondary mineral formed by the oxidation of copper sulfides; commonly associated with gypsum.

COCHISE COUNTY: Warren District, Bisbee, uncommon but widely distributed in the deeper workings of mines, as thin sky-blue crusts of small tabular crystals on fractures in or near chalcopyrite, associated with greenish films of brochantite (3rd ed.).

COCONINO COUNTY: Grand Canyon National Park, Horseshoe Mesa, Grandview District, Grandview (Last Chance) Mine, as silky greenish crusts lining cavities (Leicht 1971).

PINAL COUNTY: Mammoth District, Mammoth–St. Anthony Mine, as small blue crystal clusters (Brent Thorne, pers. comm., 2017).

#LANSFORDITE

Magnesium carbonate hydrate: $Mg(CO_3)·5H_2O$. An alteration mineral usually formed in coal mines and in serpentine masses.

MARICOPA COUNTY: Formed from the weathering of weddellite in decaying saguaro cacti in the Gila Bend Mountains (Garvie 2003).

LAUMONTITE

Calcium aluminum silicate hydrate: $CaAl_2Si_4O_{12}·4H_2O$. A member of the zeolite group, formed in veins and cavities in igneous rocks, where it is of hydrothermal origin, in skarns, and as a product of incipient metamorphism in sedimentary rocks.

FIGURE 3.134 Laumontite, Christmas Mine, Gila County, FOV 4 mm.

COCHISE COUNTY: Huachuca Mountains, probably Van Horn District, 1.5 miles east of Sunnyside, formed as pinkish crystals up to 1 in. long, with calcite crystals (H 107581). Warren District, Bisbee, as a product of hydrothermal alteration (3rd ed.).

GILA COUNTY: Southeastern Dripping Spring Mountains, Banner supersystem, Christmas District, Christmas Mine, in vugs and veinlets in diorite and replacing garnet in skarn (Perry 1969).

MARICOPA COUNTY: Belmont Mountains, Osborne District, at several localities, including the Moon Anchor Mine and Potter-Cramer property (S. A. Williams 1968). Flying Saucer District, Flying Saucer Group, Scorpion Hill, as elongated semitransparent white crystals with powellite in open fractures in granitic rocks (Gibbs, pers. comm., 2013).

NAVAJO COUNTY: Hopi Buttes volcanic field, Seth-La-Kai diatreme, in volcanic sandstone and tuffs, with limonite, gypsum, celadonite, and montmorillonite, with weak uranium mineralization (J. D. Lowell 1956).

PIMA COUNTY: Sierrita Mountains, Pima supersystem, Twin Buttes District, Twin Buttes Mine, with epidote in calcite veinlets (Fellows 1976). Ajo District, New Cornelia Mine, as elongate pink crystals with calcite in andesite dikes (W. J. Thomas and Gibbs 1983).

PINAL COUNTY: Suizo Mountains, Durham-Suizo District, Azurite Mine, with stilbite in quartz veins with copper silicates (Bideaux, Williams, and Thomssen 1960).

FIGURE 3.135 Laurelite, Grand Reef Mine, Graham County, 5 cm.

LAURELITE

Lead fluoride chloride: $Pb_7F_{12}Cl_2$. A secondary mineral found in oxidized lead ores and that may have formed by the interaction of supergene fluids with galena and fluorite. The Grand Reef Mine is the type locality.

GRAHAM COUNTY: Aravaipa supersystem, Grand Reef District, Grand Reef Mine, as colorless slightly flexible fibers up to 1 cm long, in quartz-lined vugs, associated with galena, quartz, fluorite, and anglesite (Kampf, Dunn, and Foord 1989; R100011).

FIGURE 3.136 Lausenite, United Verde Mine, Yavapai County.

LAUSENITE

Iron sulfate hydrate: $Fe^{3+}_2(SO_4)_3 \cdot 5H_2O$. A rare mineral originally named rogersite; found only at Jerome, where it formed as a result of a mine fire. The United Verde Mine is the type locality.

YAVAPAI COUNTY: Jerome supersystem, Verde District, United Verde Mine, formed as a result of the burning of a pyritic orebody (G. M. Butler 1928; Lausen 1928; X070004).

LAUTITE

Copper arsenic sulfide: CuAsS. Found in hydrothermal deposits formed at moderate temperatures.

COCONINO COUNTY: As a relatively late-stage primary sulfide mineral in uranium-bearing breccia pipes (Wenrich and Sutphin 1988).

LAWRENCITE

Iron chloride: $FeCl_2$. A rare mineral found in meteoric and terrestrial iron and associated with volcanic fumarolic action. It has been reported as a weathering product from some iron meteorites of Arizona.

COCONINO COUNTY: Small seams and pods of barringerite in the Canyon Diablo meteorite have altered in part to schreibersite, followed by wüstite, and then lawrencite (David Shannon specimen). The intergrowth of nickel iron and graphite in the Elden meteorite contains lawrencite (Bull. 181).

LAWSONITE

Calcium aluminum silicate hydroxide hydrate: $CaAl_2(Si_2O_7)(OH)_2 \cdot H_2O$. A mineral found in low-temperature metamorphic rocks.

APACHE COUNTY: Garnet Ridge SUM District, with garnet and pyroxene in eclogite inclusions in kimberlite pipes (Watson and Morton 1969).

#LAZARASKEITE

Copper glycolate: $Cu(C_2H_3O_3)_2$. A new mineral formed through the interactions of fluids containing glycolic acid (C2H4O3) with copper produced by the oxidation of primary and secondary minerals. This is the first organic mineral that contains the glycolate anionic group $(C_2H_3O_3)^-$. There are currently two structure forms: lazaraskeite form-1 and lazaraskeite form-2. The two forms have the same structure topology, but the CuO_6 octahedron in lazaraskeite form-1 is more distorted and elongated than that in lazaraskeite form-2. In addition, there is a relative change in the molecular orientation between the two structures. Pusch Ridge is the type locality for both forms.

FIGURE 3.137 Lazaraskeite, Pusch Ridge, Santa Catalina Mountains, Pima County, FOV 3 mm.

PIMA COUNTY: Santa Catalina Mountains, western end of Pusch Ridge. The mineral occurs as small lustrous bright-blue or pale-blue bladed or acicular crystals elongated on the c-axis and associated with chrysocolla, phosphohedyphane, wulfenite, and mimetite in mylonitized Wilderness Granite (Yang et al. 2019).

LAZULITE

Magnesium aluminum phosphate hydroxide: $MgAl_2(PO_4)_2(OH)_2$. A member of the lazulite-scorzalite [$Fe^{2+}Al_2(PO_4)_2(OH)_2$] solid-solution series found in granite pegmatites, quartz veins, and aluminous high-grade thermally metamorphosed rocks.

LA PAZ COUNTY: Several localities in the Dome Rock Mountains near Quartzsite, in quartzite, with kyanite, andalusite, pyrophyllite, and dumortierite (Bideaux, Williams, and Thomssen 1960). Northern Plomosa Mountains, Dutchman District, with pyrophyllite and rutile in highly argilized exotic blocks hosted in trachyandesite (Spencer and Pearthree 2015).

MARICOPA COUNTY: Reported in a locality on the Phoenix Cave Creek road, 1.2 miles north of Hyatt's Camp (Bull. 181).

PIMA COUNTY: Santa Rita Mountains, on a ridge between Stone Cabin and Madera Canyons (Robert O'Haire, pers. comm., 1972).

LEAD

Lead: Pb. Uncommon as the metallic element in the oxidized zone of lead-bearing vein deposits. All localities noted below, except Tubac, require confirmation.

MARICOPA COUNTY: Reported in the benches of Oxbow Creek, Old Woman Gulch, and Little San Domingo Creek, in red sands with magnetite (Bull. 181).

FIGURE 3.138 Lead, Tubac, Santa Cruz County, 9.8 × 3.6 cm.

SANTA CRUZ COUNTY: Tubac, replacing tree roots (UA 8577). Sonoita Creek–Alum Canyon placers in gravel with gold (Stanton B. Keith 1975).

YAVAPAI COUNTY: Reported in Gold Crater, fifteen miles west of Congress, in fist-sized masses; also reported at La Paz, in red-stained quartz (Bull. 181). Humbug District, Columbia Mine, small masses of lead recovered in placer mining (Brian Beck, pers. comm., 2016), possibly the result of contamination from the historic fire assay laboratories at the localities.

LEADHILLITE

Lead sulfate carbonate hydroxide: $Pb_4(SO_4)(CO_3)_2(OH)_2$. A rare secondary mineral formed in oxidized lead deposits.

COCHISE COUNTY: Warren District, Bisbee, Campbell Mine, 1,800 ft. level, and Cole Mine, with malachite and silver on chalcocite (H 94731, 94733; Graeme 1981). Charleston District—a fine specimen was found in 1942 on the dump of the old Manila Mine on the Tombstone Charleston Road, adjoining the Gallagher properties, about one mile from the Charleston railroad crossing (Bull. 181).

GILA COUNTY: Central Dripping Spring Mountains, C and B–Grey Horse District, C and B Vanadium Mine, as small well-formed crystals with cerussite and diaboleite (Gibbs, pers. comm., 2015).

GRAHAM COUNTY: Aravaipa supersystem, Grand Reef District, Grand Reef Mine, as tabular and pseudohexagonal twinned crystals up to 1 cm across (R. W. Jones 1980).

MARICOPA COUNTY: Painted Rock District, Rowley Mine near Theba, sparingly as fine water-clear, highly modified plate-like crystals up to 6 mm across; associated with caledonite and anglesite (W. E. Wilson and Miller 1974). Belmont Mountains, Osborne District, Tonopah-Belmont Mine, as crystal crusts and coatings on cerussite, with caledonite and linarite (G. B. Allen and Hunt 1988). Big Horn Mountains, Big Horn District, Evening Star Mine, as very small crystals in vugs in galena (Yang et al. 2016).

PIMA COUNTY: South Comobabi Mountains, Cababi District, Mildren and Steppe claims, with a wide variety of other oxidized minerals (S. A. Williams 1963).

PINAL COUNTY: Mammoth District, Mammoth–St. Anthony Mine, 400 ft. level of the Collins vein, with other rare oxide-zone minerals; as crystals up to 2.5 cm, with basal cleavage having brilliant luster. Some of the crystals are prismatic, composed of sectors of monoclinic symmetry; others are pseudo-rhombohedral or tabular, composed of two, three, or six individuals, twinned after the Artini law (Palache 1941b; Fahey, Daggett, and Gordon 1950; Bideaux 1980; H 104513).

FIGURE 3.139 Leadhillite, C and B Vanadium Mine, Gila County, 3 mm crystal.

#LECHATELIERITE

Fused silica (glass): SiO_2. When formed by the application of intense heat from a lightning stroke to rock and soil containing quartz, lechatelierites are called fulgurites. They can also be formed by the intense heat generated by meteoric impact. Lechatelierite has also been found beneath power lines, where it has formed by electrical discharge between the earth and the conductors during electrical storms.

COCONINO COUNTY: Meteor Crater, west of Winslow, where it was formed by fusion of fine-grained Coconino Sandstone by meteorite impact (A. F. Rogers 1930). Several fulgurites, as black to green obsidian-like coatings and tubes, have been found on the summit of Humphreys Peak and other peaks of the San Francisco Mountains near Flagstaff; the composition of these fulgurites has not been established (D. G. Davis and Breed 1968).

PIMA COUNTY: Numerous fulgurites, formed by lightning activity, have been noted under power lines along Ajo Road, near the road to Tucson Mountain Park. Such electrical discharges have, in some instances, followed the roots of desert plants downward into the soil, carbonizing the plant material and forming a sheaf of fused soil or rock (which may include lechatelierite) in and near the root (3rd ed.).

LEPIDOCROCITE

Gamma iron oxide hydroxide: $\gamma\text{-}Fe^{3+}O(OH)$. Formed under essentially the same conditions as goethite, with which it is commonly associated. Probably far more abundant in Arizona than indicated by the few localities noted.

COCHISE COUNTY: Warren District, Bisbee, with goethite (UA 8969).

GILA COUNTY: Banner supersystem, 79 District, 79 Mine, as a constituent of the limonites formed in the oxidized portions of the deposit (Stanley B. Keith 1972).

GRAHAM COUNTY: Lone Star District, near Safford, as a minor constituent of metasomatized volcanic rocks (Hutton 1959a).

PIMA COUNTY: South Comobabi Mountains, Cababi District, Mildren and Steppe claims (S. A. Williams 1963).

PINAL COUNTY: Mammoth District, Mammoth–St. Anthony Mine (Bull. 181). Pioneer supersystem, Magma District, Magma Mine (3rd ed.).

lepidolite

Potassium lithium aluminum silicate fluoride hydroxide: $K(Li,Al)_3(Si,Al)_4 O_{10}(F,OH)_2$. A lithium-rich member of the mica group and a typical constituent of lithium-bearing granite pegmatites. Lepidolite has been discredited as a species and is listed in *Fleischer's Glossary* (Back 2018) as an incompletely investigated trioctahedral mica.

MARICOPA COUNTY: White Picacho District, as light pinkish-gray to deep lilac or lavender compact aggregates and small books, in lithium-bearing pegmatites (Jahns 1952). Central Vulture Mountains, Flying Saucer District, six miles southwest of Wickenburg, Boyd-Fortner claims on the Lucky Mica dike (Bull. 181; AGS file data). Garcia Mountains, west of Morristown (Bull. 181). Eastern Vulture Mountains. Harquahala Mountains, in masses that are nearly identical in appearance to muscovite (Bull. 181).

YAVAPAI COUNTY: In Precambrian pegmatites of the Bagdad area (Bull. 181). Southwestern Bradshaw Mountains, Tussock District, Weatherby Beryl Mine, as pink crystals with elbaite and muscovite (MM 9461; AGS file data).

LEUCITE

Potassium aluminum silicate: $K(AlSi_2O_6)$. Leucite typically forms in potassium-rich, silica-poor volcanic flow rocks, as well as in chemically equivalent hypabyssal rocks.

NAVAJO COUNTY: Navajo and Hopi Reservations, Hopi Buttes volcanic field, in certain olivine-rich dikes (H. Williams 1936).

LEUCOPHOSPHITE

Potassium iron phosphate hydroxide hydrate: $KFe^{3+}_2(PO_4)_2(OH)\cdot2H_2O$. An uncommon secondary mineral found in pegmatites; also of sedimentary origin.

YAVAPAI COUNTY: Hillside area, 7U7 Ranch, with bermanite in seams in triplite, also associated with phosphosiderite (metastrengite) and hureaulite (Leavens 1967).

LEUCOSPHENITE

Sodium barium titanium boron silicate: $Na_4BaTi_2B_2Si_{10}O_{30}$. Found in pegmatite veins and as an authigenic mineral in some shales.

PIMA COUNTY: Silver Bell District, Silver Bell Mine, as microcrystals (A. Roe 1980).

LÉVYNE

Calcium sodium aluminum silicate hydrate: $(Ca_3, Na_6)(Si_{12}Al_6)O_{36}\cdot18H_2O$. A zeolite group mineral, typically formed in cavities in basalt. The

FIGURE 3.140 Lévyne, Horseshoe Dam, Maricopa County, FOV 3 mm.

chemistry has not been determined, so it is not known if occurrences are lévyne-Ca or lévyne-Na.

GREENLEE COUNTY: As microcrystals in a road cut three miles north of Clifton (A. Roe 1980).

MARICOPA COUNTY: Horseshoe Dam District, one mile south of Horseshoe Dam, in a road cut and in boulders along the Verde River (R. Gibbs, pers. comm., 2020).

PINAL COUNTY: South of Superior, as microcrystals in volcanics in Arnett Creek (Robert Mudra, UA x167).

FIGURE 3.141 Libethenite, Old Reliable Mine, Pinal County, FOV 4 mm.

LIBETHENITE

Copper phosphate hydroxide: $Cu_2(PO_4)(OH)$. An uncommon secondary mineral formed in the oxidized zones of copper deposits, especially porphyry copper deposits.

APACHE COUNTY: Fort Defiance area, as microcrystals (Jessie Hardman, UA x3011).

COCHISE COUNTY: Unspecified locality in the Little Dragoon Mountains (Robert O'Haire, pers. comm., 1972) (possibly in the Cochise supersystem).

GILA COUNTY: Miami-Inspiration supersystem, Pinto Valley District, Castle Dome Mine, as crusts composed of small emerald-green prismatic crystals or as drusy mats of acicular crystals along open fractures (N. P. Peterson, Gilbert, and Quick 1946, 1951); Inspiration District, Inspiration Mine (Reed and Simmons 1962). Banner supersystem, 79 District, 79 Mine (Robert O'Haire, pers. comm., 1970).

GREENLEE COUNTY: Copper Mountain supersystem, Coronado District, Coronado Mine, as small crystals in cavities (identified by S. L. Penfield). This was the first discovery of the mineral in the United States (Lindgren and Hillebrand 1904).

PIMA COUNTY: Tucson Mountains, Saginaw Hill District, southern side of Saginaw Hill, about seven miles southwest of Tucson, associated with other oxidized minerals, including cornetite, pseudomalachite, malachite, atacamite, and chrysocolla, coating fractures in chert; as pale-green greasy masses surrounding cornetite crystals, in places apparently formed from partially corroded, fairly large cornetite crystalline masses; also formed as clusters of yellowish-green prismatic crystals (Khin 1970). Waterman District, Silver Hill Mine, as microcrystals (ASDM x503). North end of the South Comobabi Mountains, on fractures in quartz monzonite, with pseudomalachite (3rd ed.). Santa Rita Mountains, Helvetia-Rosemont District, near Rosemont (Bideaux, Williams, and Thomssen 1960). Silver Bell District, Silver Bell Mine (Robert O'Haire, pers. comm., 1972), as microcrystals (A. Roe 1980; Kenneth Bladh, UA x285; MM 958-959).

PINAL COUNTY: Bunker Hill (Copper Creek) District, Copper Creek, Old Reliable Mine (UA 529, 9609). Mineral Creek supersystem, Ray District, Ray Mine (UA 9611); many specimens with small single prismatic crystals and crusts of crystals with dioptase and chrysocolla have come from the open-pit copper mine at Ray (R. W. Jones and Wilson 1983).

YAVAPAI COUNTY: Agua Fria District, of volcanogenic massive sulfide deposits; Binghampton Mine (Grant, Bideaux, and Williams 2006).

LIEBIGITE

Calcium uranyl carbonate hydrate: $Ca_2(UO_2)(CO_3)_3 \cdot 11H_2O$. An uncommon secondary mineral formed from alkaline carbonate solutions; may be associated with calcite, schröckingerite, bayleyite, and gypsum.

COLORADO PLATEAU: Reported in a diatreme on the Hopi and Navajo Reservations, but the exact location is unknown (Bull. 181), probably in the Hopi Buttes volcanic field.

YAVAPAI COUNTY: Martinez (Congress) District, Congress Mine, occurs in hanging wall fault where gold in quartz is found (Cannaday 1977).

#LIKASITE

Copper nitrate hydroxide hydrate: $Cu_3NO_3(OH)_5 \cdot 2H_2O$. A rare secondary mineral found in the oxidized zone of copper deposits.

GILA COUNTY: Globe-Miami area, Copper Cities District, Copper Cities deposit (J. W. Anthony et al. 2003).

LIME

Calcium oxide: CaO. Found in thermally metamorphosed calcareous rocks and as the result of burning coal beds.

COCHISE COUNTY: Warren District, Bisbee, Campbell Mine, in calcined limestone walls in the main vent for the Campbell fire (Graeme 1981).

MARICOPA COUNTY: In burned palo verde tree ash in the Phoenix area (Garvie 2016).

limonite

Limonite is used as a general term for mixtures of cryptocrystalline minerals, predominantly goethite and hematite. Most of the brownish material seen in the oxidized outcrops of copper and other base-metal deposits in Arizona is limonite. Note, however, that jarosite, alunite, and other secondary minerals may be common constituents of the so-called limonites that form in oxidized cappings (gossans). Limonite is abundant in the southwestern United States and found in all mineral deposits in the region that contain iron-bearing minerals subjected to oxidation.

FIGURE 3.142 Linarite, Grand Reef Mine, Graham County, 3 cm.

LINARITE

Copper lead sulfate hydroxide: $CuPb(SO_4)(OH)_2$. A sparse but widely distributed secondary mineral found in the oxidized zone of copper-lead deposits, which may be easily mistaken for azurite.

COCHISE COUNTY: Cochise supersystem, Johnson Camp District, Texas-Arizona Mine (J. R. Cooper and Silver 1964). Huachuca Mountains, Reef District, Reef Mine, as small blue crystals and coatings on fractures (Walstrom 2012). Charleston District, Gallagher Mine, with leadhillite, anglesite and diaboleite in dump material (Bowell, Luetcke, and Luetcke 2015). Tombstone District, Tranquility Mine (H 101593). Courtland District, as microcrystals at the Maid of Sunshine Mine (Brian Bond, UA x2595). Warren District, Bisbee, rare patches up to 5 mm from Cole Mine (Graeme, Graeme, and Graeme 2015).

GILA COUNTY: Banner supersystem, 79 District, 79 Mine, with caledonite in altered galena (Stanley B. Keith 1972)

GRAHAM COUNTY: Aravaipa supersystem, Grand Reef District, Ten Strike Group of claims (C. P. Ross 1925a); Grand Reef Mine, as brilliant druses and splendent groups of crystals up to 1 in. long, associated with cerussite, anglesite, and leadhillite in quartz-lined cavities (R. W. Jones 1980); Iron Cap District, Sinn Fein Mine, as electric blue crystals lining fractures in rhyolite (Flagg Collection).

LA PAZ COUNTY: Silver District, Red Cloud Mine (P. Bancroft and Bricker 1990).

MARICOPA COUNTY: Painted Rock District, Rowley Mine near Theba, associated with atacamite and diaboleite (W. E. Wilson and Miller 1974), and as microcrystals (Carl Richardson, UA x4675). Belmont Mountains, Osborne District, Tonopah-Belmont Mine (G. B. Allen and Hunt 1988).

PIMA COUNTY: South Comobabi Mountains, Cababi District, Mildren Mine, in anglesite-cerussite aggregates with paratacamite, chlorargyrite, leadhillite, and matlockite, as excellent crystals (S. A. Williams 1962). Mission-Pima District, Banner Mine, as microcrystals with brochantite (William Hunt, pers. comm., 1985).

PINAL COUNTY: Mammoth District, Mammoth–St. Anthony Mine, in places as excellent crystals, as thin films filling crevices in brecciated rock, and as small to large euhedral crystals up to 4 in. long; invariably associated with brochantite (Guild 1911; Bideaux 1980; Palache 1941b; Sinkankas 1964; Omori and Kerr 1963).

LINDGRENITE

Copper molybdate hydroxide: $Cu_3(Mo^{6+}O_4)_2(OH)_2$. An uncommon secondary mineral found in the oxidized portions of copper deposits.

GILA COUNTY: Miami-Inspiration supersystem, Inspiration District, Inspiration Mine, Live Oak pit, as platy aggregates in hydrothermally altered schist, in seams with molybdenite, and rarely, associated with powellite (N. P. Peterson 1962).

MARICOPA COUNTY: Cave Creek District, with cuprotungstite (Schaller 1932).

PIMA COUNTY: Pima supersystem, Esperanza District, very sparse at the Esperanza Mine (UA 6445).

PINAL COUNTY: Bunker Hill (Copper Creek) District, Copper Creek, Childs-Aldwinkle Mine (Richard W. Thomssen, pers. comm., 3rd ed.; UA 4880). Mineral Creek supersystem, Ray District, Hull claims, south of Ray (H 108666). Pioneer District, Superior (H 105628).

FIGURE 3.143 Lindgrenite, Childs-Aldwinkle Mine, Pinal County, FOV 3 mm.

LINNAEITE

Cobalt sulfide: $Co^{2+}Co_2^{3+}S_4$. A primary mineral found in hydrothermal veins with various copper, iron, cobalt, and nickel sulfides and sulfosalts.

COCONINO AND MOHAVE COUNTIES: In breccia pipes mined for uranium (Wenrich and Sutphin 1988).

#LIROCONITE

Copper aluminum arsenate hydroxide hydrate: $Cu_2Al(AsO_4)(OH)_4 \cdot 4H_2O$. A rare secondary mineral found in the oxide zone of copper deposits.

COCHISE COUNTY: Warren District, Bisbee, Lavender pit, Holbrook Extension. Found as tabular bright-blue crystals up to 2.2 mm on chrysocolla (Graeme, Graeme, and Graeme 2015).

LITHARGE

Alpha lead oxide: α-PbO. One of four lead oxides, an uncommon mineral formed under highly alkaline and oxidizing conditions in deposits containing lead sulfide. Litharge is probably a more common mineral in oxidized environments containing lead than is generally supposed.

GRAHAM COUNTY: Aravaipa supersystem, Grand Reef District, Grand Reef Mine (Roberts et al. 1995).

LA PAZ COUNTY: Silver District, Hamburg Mine (P. Bancroft and Bricker 1990).

MARICOPA COUNTY: Belmont Mountains, Osborne District, Tonopah-Belmont Mine, as pink replacements of cerussite and as microcrystals (David Shannon, UA x4316).

PIMA COUNTY: South Comobabi Mountains, Cababi District, Mildren and Steppe claims, as an alteration product of wulfenite, associated with a large variety of secondary minerals formed during the oxidation of sulfide-bearing veins that cut andesite (S. A. Williams 1962, 1963).

LITHIOPHILITE

Lithium manganese phosphate: $LiMn^{2+}(PO_4)$. A primary mineral formed in granite pegmatites. Probably forms a solid-solution series with triphylite $[LiFe^{2+}(PO_4)]$.

MOHAVE COUNTY: Aquarius Mountains District, northeast of Wikieup (UA 6437, 7889), in pegmatites.

YAVAPAI COUNTY: White Picacho District, with triphylite, at the Midnight Owl and other pegmatites (Jahns 1952; London and Burt 1982a; UA 5880).

LITHIOPHORITE

Aluminum lithium manganese oxide hydroxide: $(Al,Li)(Mn^{4+},Mn^{3+})O_2(OH)_2$. Found in vein and bedded deposits with other, more common manganese oxide minerals.

MOHAVE COUNTY: Artillery District, Priceless vein, associated with cryptomelane, hollandite, pyrolusite, psilomelane, and coronadite; at the Plancha bedded manganese deposits it is superimposed on hard silvery manganese ore, and locally it may replace ramsdellite-pyrolusite grains (Mouat 1962).

YAVAPAI COUNTY: About five miles north-northeast of Ash Fork, as blue-black coatings in cavities in limestones of the Kaibab Formation (UA 9152) and replacing calcite cement in red beds of the Toroweap Formation, as minor amounts of spongy to colloform blebs in siltstone and limestones (Mullens 1967; UA 9153).

#LIUDONGSHENGITE

Zinc chromium hydroxide carbonate hydrate: Zn_4Cr_2 $(OH)_{12}(CO_3) \cdot 3H_2O$. Liudongshengite is a rare member of the hydrotalcite group and the chromium analog of zaccagnaite-3R. The 79 Mine is the type locality.

GILA COUNTY: Banner supersystem, 79 District, 79 Mine, where it occurs as very small thin pink hexagonal plates on the fourth level with cerussite, rosasite, hemimorphite, and malachite in corroded galena crystals embedded on a small quartz vein with scheelite and stolzite (Miyawaki et al. 2019a; R180016).

FIGURE 3.144 Liudongshengite, 79 Mine, Gila County.

LIZARDITE

Magnesium silicate hydroxide: $Mg_3Si_2O_5(OH)_4$. A member of the serpentine group of silicate minerals most typically found with chrysotile in serpentines. Most Arizona lizardite occurrences in widespread asbestos deposits in Gila County occur as a contact-metasomatic mineral adjacent to diabase sills.

GILA COUNTY: Salt River Asbestos District, at the Chrysotile deposits in the Salt River Canyon. Sierra Ancha Asbestos District, Seneca District, Mystery District, and Chiricahua District as translucent apple-green pearly masses that postdate gray-green fiberform antigorite, crosscut by late strongly fiberform satiny green chrysotile veins and seams. This paragenesis describes most of the lizardite occurrences in the districts cited above (Stanley B, Keith, pers. comm., 2019).

LA PAZ COUNTY: Northern Granite Wash Mountains, Yuma King District, as green to greenish-yellow translucent bands intermixed with tremolite, forming swirl-like patterns (*verde antique*) in skarns that have replaced lower Paleozoic carbonates (Stanley B. Keith, pers. comm., 2019).

PIMA COUNTY: Silver Bell District, Silver Bell Mine, El Tiro pit (3rd ed.).

LÖLLINGITE

Iron arsenide: $FeAs_2$. An uncommon mineral formed in mesothermal veins with iron and copper sulfides in calcite gangue.

MARICOPA COUNTY: Disseminated in pegmatites throughout the White Picacho District (Jahns 1952).

MOHAVE COUNTY: Boriana District, Copper World Mine, with sphalerite, chalcopyrite, and pyrrhotite (Rasor 1946; H 104101).

LONSDALEITE

Carbon: C. This is the hexagonal dimorph of diamond. Meteor Crater is the type locality.

COCONINO COUNTY: Canyon Diablo area, near Meteor Crater, in the Canyon Diablo meteorite, as black cubes and cubo-octahedra coated with graphite, up to 0.7 mm (C. Frondel and Marvin 1967a, 1967b).

#LOVERINGITE

Calcium cerium lanthanum zirconium iron magnesium titanium chromium aluminum oxide: $(Ca,Ce,La)(Zr,Fe)(Mg,Fe)_2(Ti,Fe,Cr,Al)_{18}O_{38}$. Occurs as a late-stage mineral in mafic igneous rocks.

APACHE COUNTY: Garnet Ridge SUM District, found in an ultramafic diatreme at Garnet Ridge on the Navajo Reservation as inclusions in pyrope with srilankite, carmichaelite, rutile, and many other minerals (Wang, Essene, and Zhang 1999).

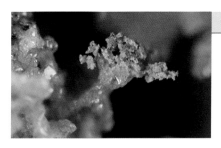

FIGURE 3.145 Luddenite, Evening Star Mine, Maricopa County, FOV 5 mm.

LUDDENITE

Copper lead silicate hydrate: $Cu_2Pb_2Si_5O_{14}·14H_2O$. A rare secondary mineral found in a strongly oxidized fault breccia in a base-metal deposit. The Rawhide Mine occurrence is the type locality.

MARICOPA COUNTY: Big Horn Mountains, Big Horn District, Evening Star Mine, and a nearby unnamed prospect, as apple-green aggregates of very small crystals in vugs with cerussite in galena pods (Jenkins, pers. comm., 2012).

MOHAVE COUNTY: Rawhide District, Rawhide Mine, in the oxidized copper-lead ores associated with fluorite, alamosite, mimetite, wulfenite, wickenburgite, and cerussite (S. A. Williams 1982).

LUDWIGITE

Magnesium iron oxide borate: $Mg_2Fe^{3+}O_2(BO_3)$. An uncommon mineral apparently formed only in high-temperature contact-metamorphic environments.

PIMA COUNTY: Santa Catalina Mountains, Marble Peak District, Leatherwood Mine Group, near and west of the Control Mine, in the contact zone with the Leatherwood Granodiorite (James Post, pers. comm., 3rd ed.).

LUETHEITE

Copper aluminum arsenate hydroxide: $CuAl(AsO_4)_2(OH)_2$. A secondary mineral formed as an oxidation product of arsenic sulfides. The Sunnyside District is the type locality.

SANTA CRUZ COUNTY: Sunnyside District, near the Humboldt Mine about one mile southwest of the Volcano-Sunnyside Mine, in veinlets and vugs in rhyolite breccia, as small tabular crystals of distinctive Indian blue color; associated with chenevixite, cornubite, and alunite (S. A. Williams 1977).

FIGURE 3.146 Luetheite, near Humboldt Mine, Santa Cruz County, FOV 3 mm.

LUZONITE

Copper arsenic sulfide: Cu_3AsS_4. A hydrothermal sulfosalt mineral formed at moderate to low temperatures.

COCHISE COUNTY: Warren District, Bisbee, Junction shaft, as massive fine-grained material (Graeme 1993).

MACKAYITE

Iron tellurium oxide hydroxide: $Fe^{3+}Te^{4+}_2O_5(OH)$. A rare secondary mineral formed in the oxidized portions of tellurium-bearing base-metal deposits, with emmonsite and other secondary minerals.

COCHISE COUNTY: Tombstone District, Toughnut-Empire Mine, with tellurium and emmonsite (Bideaux, Williams, and Thomssen 1960).

MACKINAWITE

Iron nickel sulfide: $(Fe,Ni)_{1+x}S$ ($x = 0–0.07$). A primary mineral formed by hydrothermal activity and in the reducing environment found in river bottom muds.

COCONINO COUNTY: Meteor Crater near Winslow, in the Canyon Diablo meteorite, with troilite (Bunch and Keil 1969).

YAVAPAI COUNTY: Old Dick District, about 5 km south of Bagdad, with cubanite, as fine-grained fracture fillings in chalcopyrite in a fine-grained schist in the metamorphosed alteration zone associated with a massive volcanogenic sulfide deposit (Larson 1976).

MACPHERSONITE

Lead sulfate carbonate hydroxide: $Pb_4(SO_4)(CO_3)_2(OH)_2$. A secondary mineral formed under oxidizing conditions in hydrothermal lead deposits.

MARICOPA COUNTY: Belmont Mountains, Osborne District, found on one specimen from the Moon Anchor Mine, where tabular crystals up to 5 mm on an edge occupy voids in a granular cerussite matrix (3rd ed.).

MACQUARTITE

Copper lead chromate silicate hydroxide: $Cu_2Pb_7(CrO_4)_4$ $(SiO_4)_2(OH)_2$. An extremely rare secondary mineral formed under oxidizing conditions at somewhat elevated temperatures. The Mammoth–St. Anthony Mine is the type locality.

PINAL COUNTY: Mammoth District, Mammoth–St. Anthony Mine, found on old specimens in quartz-rich matrix with dioptase and reddish wulfenite (S. A. Williams and Duggan 1980).

FIGURE 3.147 Macquartite, Mammoth–St. Anthony Mine, Pinal County, FOV 3 mm.

MAGHEMITE

Iron oxide: $(Fe^{3+}_{0.67}\square_{0.33})Fe^{3+}_2O_4$. Maghemite is formed from magnetite or lepidocrocite by slow oxidation at low temperatures and from the oxidation of meteoric iron.

COCONINO COUNTY: Meteor Crater area, in elongate isotropic grains, alternating with a goethite-like iron oxide as a minor component of metallic spheroids formed by the impact of the meteorite (Mead, Littler, and Chao 1965).

MOHAVE COUNTY: Lost Basin area in Precambrian iron formation at unnamed shaft 303 in NE quarter of S17, T29N R17W (Theodore, Blair, and Nash 1987).

PIMA COUNTY: Tucson Mountains, as a thin brownish surface-alteration product on magnetite pebbles and boulders (3rd ed.).

#MAGNESIOALTERITE

Magnesium iron sulfate oxalate hydroxide hydrate: $Mg_2Fe^{3+}_4(SO_4)_4(C_2O_4)_2(OH)_4 \cdot 17H_2O$. A new oxalate mineral that is the magnesium analog of alterite. The Cliff Dwellers Lodge is the type locality.

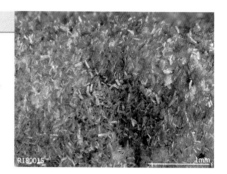

FIGURE 3.148 Magnesioalterite, near Cliff Dwellers Lodge, Coconino County.

COCONINO COUNTY: Vermilion Cliffs District, unnamed uranium prospect near Cliff Dwellers Lodge, as small yellow-green elongated crystals in fractures in carbonized petrified wood in the Shinarump Conglomerate basal member of the Chinle Formation. Magnesioalterite occurs with alterite, gypsum, alunogen, natrojarosite, sulphur, and celestine (Yang et al. 2020).

MAGNESIOCHROMITE

Magnesium chromium oxide: $MgCr_2O_4$. A member of the spinel group, found as a primary mineral in peridotites and other ultramafic rocks.

APACHE COUNTY: Garnet Ridge SUM District, Garnet Ridge, as mineral fragments in a breccia dike that pierces sedimentary rocks (Gavasci and Kerr 1968).

MAGNESIOCOPIAPITE

Magnesium iron sulfate hydroxide hydrate: $MgFe^{3+}_4(SO_4)_6(OH)_2 \cdot 20H_2O$. A secondary mineral formed by the oxidation of pyrite. Similar in most respects to the more common copiapite.

COCONINO COUNTY: Cameron District, in a uranium prospect with wupatkiite and other sulfates (S. A. Williams and Cesbron 1995).

SANTA CRUZ COUNTY: Palmetto District, Denver Mine, as microcrystals with halotrichite (identified by Bayliss and Atencia, 1985; UA x4321; R060155).

YAVAPAI COUNTY: Jerome supersystem, Verde District, United Verde Mine, with guildite or ransomite (J. W. Anthony et al. 2003).

#MAGNESIO-FOITITE

Magnesium aluminum silicate borate hydroxide: $\square(Mg_2Al)Al_6(Si_6O_{18})$ $(BO_3)_3(OH)_3(OH)$. An uncommon member of the tourmaline group found in igneous rocks.

PINAL COUNTY: Bunker Hill (Copper Creek) District, Copper Creek, Old Reliable Mine, as small gray to black elongated crystals in the groundmass of the breccias (J. Ebner, pers. comm., 2018).

MAGNESIO-HORNBLENDE

Calcium magnesium iron aluminum silicate hydroxide: $\square Ca_2(Mg_4Al)$ $(Si_7Al)O_{22}(OH)_2$. A member of the amphibole group, found in a wide variety of igneous and metamorphic rocks. Some hornblende in igneous rocks is a product of the hydration of primary pyroxenes during supercritical hydrous metasomatism/implosion under magmatic conditions.

COCHISE COUNTY: Tombstone District, as long prismatic crystals in the Schieffelin Granodiorite (B. S. Butler, Wilson, and Rasor 1938).

GREENLEE COUNTY: Copper Mountain supersystem, Morenci District, abundant in diorite porphyry (Bull. 181).

LA PAZ COUNTY: Harcuvar Mountains, as crystals up to 1 in. long, near dikes in Precambrian granite (Bull. 181).

PIMA COUNTY: Ajo District, Ajo, as bodies of hornblendite, the largest of which is about 2,000 by 1,000 ft. in planar dimensions, in the Cardigan Gneiss (Gilluly 1937). Empire District, Prince Mine, as phenocrysts in a diorite porphyry dike (Bull. 181). Near Vail (UA 5556). Quinlan Mountains, forty miles southwest of Tucson (UA 5459).

PINAL COUNTY: In intrusive bodies near Picketpost Mountain near Superior, as the principal constituent of the greenschist facies of the Pinal Schist (Bull. 181). North of Saddle Mountain, as large porphyrocrysts to 3 cm long in basaltic dikes that cut the Williamson Canyon volcanics (Stanley B. Keith, pers. comm., 2019).

YAVAPAI COUNT: Lenticular bodies composed largely of hornblende are found at many places in the Yavapai Schist (Bull. 181).

MAGNESITE

Magnesium carbonate: $Mg(CO_3)$. Magnesite is a member of the calcite group. It forms as a product of the metamorphism of magnesian rocks, through the alteration of calcite by magnesium-bearing waters, as sedimentary deposits, and occasionally, as a hydrothermal gangue mineral in magnesium-rich hot springs. It can also form as a product of serpentinization of ultramafic peridotites.

GREENLEE COUNTY: Sparingly in horizons of magnesium skarns in the Longfellow Limestone (Bull. 181).

MOHAVE COUNTY: Mag District, three miles northwest of Oatman, in veins with brucite and serpentine cutting volcanic rocks (E. D. Wilson and Roseveare 1949; Funnell and Wolfe 1964; E. D. Wilson 1944). Lower Burro Creek Wilderness Area, as white cryptocrystalline material in mound-like masses of up to 90 percent (volume %) magnesite and 10 to 40 ft. thick (R. J. Miller et al. 1987).

PIMA COUNTY: Cimarron Mountains, Tohono O'odham (Papago) Reservation, with huntite (3rd ed.).

MAGNETITE

Iron oxide: $Fe^{2+}Fe^{3+}_2O_4$. This is the most abundant and widespread member of the spinel group. Found as an accessory mineral in many igneous rocks, abundant in metamorphic rocks, and widely distributed in sedimentary rocks and in sands as a detrital mineral. Only a few representative localities can be noted. An important early crystallizing accessory mineral, along with titanite (sphene), in high-oxidation-state magnetite or magnetite-sphene series granitic rocks (Ishihara 1977; Stanley B. Keith et al. 1991). Magnetite-titanite (sphene) assemblage is an important oxidation-state indicator in plutons associated with porphyry copper deposits.

COCHISE COUNTY: Dragoon Mountains, Middle Pass supersystem, Black Diamond District, Black Diamond Mine, as granular masses with chalcopyrite and ilvaite (3rd ed.).

COCONINO COUNTY: Grand Canyon National Park, as octahedral crystals up to 1 in. across, in pegmatite (Bull. 181).

GILA COUNTY: Sierra Ancha District, Asbestos Peak, as sharp dodecahedral crystals up to 0.5 in. diameter, in calcite with tremolite and serpentine (Sinkankas 1964), and as a widespread accessory mineral in the asbestos deposits. Southeastern Dripping Spring Mountains, Banner supersystem, Christmas District, Christmas Mine, a very abundant mineral in the lower Martin Formation orebody, locally abundant in the skarn rocks (N. P. Peterson and Swanson 1956; David Perry, pers. comm., 1967). Peridot District, near Peridot, as large, titanomagnetite crystals associated with spinel, kaersutite, olivine, and clinopyroxene, in spinel inclusions in Type II inclusions dominated by spinel (Bull. 181; Frey and Prinz 1978).

GREENLEE COUNTY: Copper Mountain supersystem, Morenci District, abundant in metamorphosed limestone, with garnet, amphibole, pyroxene, and sulfides; mined as flux at the Manganese Blue and Arizona Central Mines (Lindgren 1905). Gila Hot Springs District, pyrolusite claims, twelve miles south of Morenci, with pyrolusite and psilomelane (Potter, Ipsen, and Wells 1946).

LA PAZ COUNTY: Bouse Hills District, as microcrystals near Bouse (Robert Mudra, UA x951). Northwest Granite Wash Mountains, Yuma King District, as large lenticular magnetite skarn bodies (Harrer 1964) replacing overturned Kaibab Limestone near monzodiorite sills associated with actinolite, epidote, chalcopyrite, garnet, and lizardite-tremolite (verde antique).

MARICOPA COUNTY: Big Horn District, with ilmenite in extensive placer deposits up to 100 ft. thick (Harrer 1964).

PIMA COUNTY: Santa Rita Mountains, Helvetia-Rosemont District, at Rosemont, abundant in contact-metamorphic copper ores (Schrader and Hill 1915). Sierrita Mountains, Pima supersystem, in contact-metamorphic copper orebodies (Guild 1934). In large amounts on the surface, five miles from Tule Wells, near pegmatite bodies containing copper sulfides (Gilluly 1937). Tucson Mountains, as rounded transported blocks, in many places pitted like meteorites (Guild 1910), probably the Amole District, near the Sus picnic area. Catalina District, Pontatoc Mine, in an ore deposit in mylonitic diorite associated with metamorphosed Mescal Limestone (Guild 1934; Shakel 1974).

SANTA CRUZ COUNTY: Patagonia Mountains, Querces District, Line Boy Mine, south of Duquesne near the Mexican border, as lodestone in considerable quantities (Schrader 1917).

YAVAPAI COUNTY: Bradshaw Mountains, Big Bug Creek, as large stream-worn masses up to 18 in. in a stream bed (David Shannon, pers. comm., 1971); Black Canyon District, near Townsend Butte and the Howard property, with hematite and silica as banded iron Formation horizons in schist (Bull. 181). Tiger District, Springfield Group, as large crystals with apatite and titanite in granodiorite (Bull. 181); near Stoddard, as large pieces of lodestone (Lindgren 1926). Eureka supersystem, south of Centipede Mesa and along Boulder Creek, as titaniferous magnetite in dikes and irregular bodies in gabbro intergrown with ilmenite and minor hematite (Harrer 1964; Anderson 1955). Old Dick District, Copper King Mine, as the main ore mineral along with chalcopyrite (Anderson Scholz, and Strobell 1955). Seligman Iron District, McBride claims, seventeen miles south of Seligman, as segregations of titaniferous magnetite in gabbro (Bull. 181).

MALACHITE

Copper carbonate hydroxide: $Cu_2(CO_3)(OH)_2$. Commonly found as a secondary mineral in oxidized copper deposits. Generally associated with other secondary copper minerals, particularly azurite, cuprite, and tenorite, and commonly with limonite. One of the most common secondary copper minerals and found in every county in Arizona; only a few localities can be listed.

FIGURE 3.149 Malachite, Morenci Mine, Greenlee County, 8.1 cm.

APACHE COUNTY: Garnet Ridge U District, in vein fillings and in cement in the Navajo Sandstone, associated with chrysocolla, tangeite, tyuyamunite, limonite, chalcopyrite, and pyrite (Gavasci and Kerr 1968). Cane Valley District, associated with uranium-vanadium ores in the Moenkopi Formation, the Shinarump Member of the Chinle Formation, and the De Chelly Member of the Cutler Formation (Mitcham and Evensen 1955; Evensen and Gray 1958).

COCHISE COUNTY: The Warren District, Bisbee, has produced some of the world's most remarkable malachite, and specimens hold places of honor in many major collections. Large masses of malachite were early noted in several mines in the district: Copper Queen Mine, on Queen Hill, south of Bisbee (Kunz 1885; Douglas 1899; Ransome 1903a; Lindgren 1904; Palache and Lewis 1927; C. Frondel 1941); Campbell Mine (Schwartz and Park 1932), including fine pseudomorphs after azurite (Les Presmyk, pers. comm., 2020); Cole Mine (Trischka, Rove, and Barringer 1929; Schwartz 1934; Hutton 1957); Sacramento Mine, pseudomorphs after azurite to 5 cm (Les Presmyk, pers. comm., 2020); and Junction Mine (Mitchell 1920). Courtland District, in large masses at the Maid of Sunshine Mine, in some places as small but superb crystals (Bull. 181). Gleeson District, Silver Bill Mine, as pseudomorphs after azurite up to 3 cm (3rd ed.). Tombstone District, where it is widespread but not abundant (B. S. Butler, Wilson, and Rasor 1938). Sulphur Springs Valley, Pat Hills,

Pat District, in numerous calcite veins cutting porphyritic andesite flow rocks, associated with chrysocolla (Lausen 1927b). Cochise supersystem, Johnson Camp District, Johnson Camp area, as an abundant mineral in tabular bodies with sulfides in contact-metamorphosed limestone (Kellogg 1906; J. R. Cooper and Huff 1951).

COCONINO COUNTY: Kaibab Plateau, with azurite and chrysocolla as impregnations in extensive chert beds (Bull. 181). Grand Canyon National Park, Grandview District, Grandview (Last Chance) Mine, as botryoidal crusts coating limestone and in fissures in clay (Leicht 1971); Orphan District, Orphan Mine, with azurite (3rd ed.). White Mesa District, with chrysocolla, cementing Navajo Sandstone (J. M. Hill 1914a; Mayo 1955a). Warm Springs District, Jacob Lake, Apex Mine, with azurite and minor amounts of chalcopyrite and chalcocite in the Kaibab Limestone (Tainter 1947b). Cameron District, in mineralized silica plugs (Barrington and Kerr 1963).

GILA COUNTY: Globe Hills District, where it constitutes a considerable part of the ores at the Buffalo, Big Johnnie, Buckeye, and other mines, but nowhere found in large masses (Ransome 1903b); Old Dominion Mine (Schwartz 1921, 1934). Miami-Inspiration supersystem, Pinto Valley District, Castle Dome Mine. Copper Cities District, Copper Cities Mine (N. P. Peterson 1947); Miami-Inspiration supersystem, Inspiration District, Inspiration Mine, in the Live Oak Pit as beautiful aggregates of malachite, chrysocolla, and chalcedony, including malachite pseudomorphs after azurite (Ransome 1919; N. P. Peterson 1962; Sun 1963; Les Presmyk, pers. comm., 2020); northern Pinal Mountains, Summit District, Blue Ball Mine, as nodules of pure malachite, mixed azurite-malachite nodules, and fibrous malachite in hollow azurite geodes (Grant 1989). Green Valley District, Silver Butte Mine, as stout prismatic crystals in porous quartz (Lausen and Wilson 1925). Southeastern Dripping Spring Mountains, Banner supersystem, Christmas District, Christmas Mine, as a common supergene mineral in skarn developed in the Naco Formation (David Perry, pers. comm., 1967). Banner supersystem, 79 District, 79 Mine, where it is scattered throughout the supergene zone as veinlets and small spheres up to 0.25 in. diameter (Stanley B. Keith 1972).

GRAHAM COUNTY: Aravaipa supersystem, Grand Reef District, Klondyke area, as a widespread but sparse mineral (Simons 1964).

GREENLEE COUNTY: Copper Mountain supersystem, Morenci District, common in irregular deposits in limestone; mined at the Detroit, Manganese Blue, and Longfellow Mines; Metcalf District, as fine radiating groups of crystals at the Standard Mine near Metcalf (Lindgren 1903,

1904, 1905; Reber 1916; Schwartz 1934; Grout 1946). The Detroit Mine produced chatoyant malachite pseudomorphs after azurite (Les Presmyk, pers. comm., 2020).

LA PAZ COUNTY: Buckskin Mountains, Planet District, Planet Mine, associated with azurite and chrysocolla (H. Bancroft 1911). Near Bouse (UA 3966). Little Harquahala Mountains, Little Harquahala District, Harquahala Mine, as green sprays and crystals associated with dioptase, chrysocolla, and pseudomalachite (Stanley B. Keith, pers. comm., 2019).

MARICOPA AND GILA COUNTIES: Mazatzal Mountains District, in small quantities associated with mercury deposits, in schist (Lausen 1926).

MOHAVE COUNTY: Bill Williams Fork (Genth 1868). Cerbat Mountains, Wallapai supersystem, widely distributed in the oxide zones of porphyry copper deposits and as an oxidation product of copper sulfide vein deposits, where it occurs with cuprite and chrysocolla; Emerald Isle District, locally, as cement in alluvium, with cuprite, chrysocolla, and copper (B. E. Thomas 1949); Mineral Park District, Mineral Park Mine, in granite gneiss (3rd ed.). Bentley District, Grand Gulch, Bronzell, and Copper King Mines (J. M. Hill 1914b).

NAVAJO COUNTY: Monument Valley District, associated with uranium-vanadium deposits in sandstone as a probable oxidation product of chalcopyrite (Holland et al. 1958).

PIMA COUNTY: Santa Rita Mountains, Helvetia-Rosemont District, as globular masses and in veinlets; also widely distributed in copper deposits throughout the range in small amounts (Bull. 181). Sierrita Mountains, abundant in the Pima supersystem, Mission-Pima District, and the most important ore mineral at Mineral Hill Mine (Schrader and Hill 1915; Schrader 1917); Twin Buttes District, Twin Buttes Mine, in the oxidized portion, especially as a component of the chrysocolla breccias (Stanley B. Keith, pers. comm., 1973). Ajo District, New Cornelia Mine, the most common product of the weathering of copper-bearing minerals and the dominant mineral in leaching ores (Joralemon 1914; Gilluly 1937); also, as pseudomorphs of blocky azurite crystals to 1 in. (Thomas and Gibbs 1983) and in concentric encrustations with azurite and quartz in vugs, as well as covering the outside of porous nodules of cuprite that, in turn, surround copper (Schwartz 1934). Ajo District, Copper Giant exotic copper deposits, S10, S11, and S15, T13S R6W, in fanglomerate and volcanic rocks (Romslo and Robinson 1952). Waterman District, Silver Hill Mine, with rosasite (UA 6709). Silver Bell District, Silver Bell Mine, with cuprite and azurite in oxidized limestone replacements (Engineering and Mining Journal 1904; C. A. Stewart 1912).

PINAL COUNTY: Mineral Creek supersystem, Ray District, Ray Mine area, with chrysocolla and tenorite cementing part of the White Tail Conglomerate east of Ray (Clarke 1953); also cementing Holocene gravels with other oxidized copper and iron minerals in the Pearl Handle open-pit mine at Ray, in the Sonora exotic copper deposit. This mineralization was shown to be not older than about 7,000 years, based on a radiocarbon date obtained from a fossil log incorporated in the mineralized area (C. H. Phillips, Cornwall, and Rubin 1971). Pioneer supersystem, Magma District, Magma Mine, with chrysocolla as the principal oxidized copper mineral (Short et al. 1943). San Manuel District, San Manuel Mine, where it is present in the oxidized zone of the orebody but accounts for little of the copper values; associated with chrysocolla, cuprite, and azurite (Schwartz 1949). Mammoth District, Mammoth–St. Anthony Mine, common as powdery masses and microcrystals and as pseudomorphs after azurite crystals, up to 4 in. long (Bideaux 1980).

SANTA CRUZ COUNTY: Santa Rita and Patagonia Mountains, widespread as a product of the oxidation of sulfide copper minerals at numerous mines and prospects in many of the districts in the area (Schrader and Hill 1915; Schrader 1917).

YAVAPAI COUNTY: Black Hills District, Yeager Mine, as fine specimen material with crystallized azurite; also common in the oxidized portions of all copper deposits in the region (Schwartz 1934, 1938; C. A. Anderson and Creasey 1958). Zonia District, Zonia Mine, with abundant chrysocolla (Kumke 1947). Bloody Basin District, Piedmont Mine, as fine malachite pseudomorphs after azurite, coated with clear to white quartz (Muntyan 2020b).

FIGURE 3.150 Mammothite, Rowley Mine, Maricopa County, FOV 0.6 mm.

MAMMOTHITE

Lead copper aluminum antimony oxide sulfate chloride hydroxide: $Pb_6Cu_4AlSb^{5+}O_2(SO_4)_2Cl_4(OH)_{16}$. A rare secondary mineral found in the oxidized portion of a base-metal deposit with other secondary copper and lead minerals. The Mammoth–St. Anthony Mine at Tiger is the type locality.

MARICOPA COUNTY: Painted Rock District, Rowley Mine, as small green to teal-colored striated bladed crystals with cerussite, boleite, diaboleite, phosgenite, and other secondary lead minerals. Originally found on the dump in 2013 and later on the 60 ft. level of the Jobs shaft (Keith Wentz, pers. comm., 2018).

PINAL COUNTY: PINAL COUNTY: Mammoth District, Mammoth–St. Anthony Mine, as bright-blue sprays of radial acicular crystals and flat plates embedded in anglesite and associated with phosgenite, wulfenite, leadhillite, and caledonite (Peacor et al. 1985).

MANANDONITE

Lithium aluminum boron silicate hydroxide: $Li_2Al_4(Si_2AlB)O_{10}(OH)_8$. A member of the chlorite group of micas.

LA PAZ COUNTY: Near Quartzsite, as minute tablets in quartzite, with kyanite, andalusite, and dumortierite (Bideaux, Williams, and Thomssen 1960).

MANGANBABINGTONITE

Calcium manganese iron silicate hydroxide: $Ca_2Mn^{2+}Fe^{3+}Si_5O_{14}(OH)$. Similar to babingtonite, but manganese is substituted for iron in the structure; very similar in provenance, that is, in skarns, granite pegmatites, and hydrothermal veins; probably of low-temperature origin.

GRAHAM COUNTY: Santa Teresa Mountains, Aravaipa supersystem, Iron Cap District, Landsman claims, as exceptional dark-brown bladed euhedral crystals up to 7 mm long, from a vein containing galena and sphalerite, in a contact-metamorphic zone developed in limestone and shale; associated with euhedral diopside-hedenbergite and yellow garnet (material collected by Raymond Rhodes and Richard L. Schick in 1972); also associated with axinite and polylithionite (3rd ed.); and Iron Cap Mine, as black bladed crystals up to 1 cm long (Reiter 1980, 1981).

MANGANITE

Manganese oxide hydroxide: $Mn^{3+}O(OH)$. Typically found with other manganese and iron oxide minerals in low-temperature veins and in deposits of secondary origin.

FIGURE 3.151 Manganite, Stovall Mine, Pinal County, 9 cm.

COCHISE COUNTY: Tombstone District, as needle-like crystals in parallel groups and as soft fibers lining cavities (B. S. Butler, Wilson, and Rasor 1938; Rasor 1939; Romslo and Ravitz 1947; Needham and Storms 1956).

GILA COUNTY: Globe Hills District, where, with pyrolusite and psilomelane, it forms the bulk of the gangue in the manganese-zinc-lead-silver deposits (N. P. Peterson 1962). Banner supersystem, 79 District, 79 Mine, as irregular masses with oxidized copper ore (Kiersch 1949).

LA PAZ COUNTY: Planet District, Planet Mine (Bull. 181). Artillery District, Kaiserdoom claims (Hewett, Fleischer, and Conklin 1963). Cienega District, War Eagle claims, twenty-six miles north of Bouse (Havens et al. 1947).

MARICOPA COUNTY: Big Horn Mountains, Aguila District, with pyrolusite and wad (Bull. 181). White Picacho District, as crusts coating lithium phosphate minerals, in pegmatites (Jahns 1952).

MOHAVE COUNTY: Artillery District, associated with a variety of manganese oxide minerals in the Artillery Peak deposit (Head 1941).

PINAL COUNTY: Pioneer supersystem, Silver King District, Silver King Mine, as crystals in baryte (UA 7437). South of Casa Grande, Stovall Mine (Arthur Flagg Collection).

SANTA CRUZ COUNTY: Patagonia Mountains, Mowry District, Mowry Mine (Havens et al. 1954). Santa Rita Mountains, Cottonwood Canyon, Tyndall District, Glove Mine (H. J. Olson 1966).

YAVAPAI COUNTY: Black Rock District, Castle Creek, with pyrolusite and wad (Bull. 181). Box Canyon District, on the property of the Mohave Mining and Milling Company, east of Wickenburg, as shiny black crystals (Bull. 181). White Picacho District, as thick crusts coating lithium phosphate minerals, in pegmatites (Jahns 1952). Burmister District, Burmister Mine, twelve miles southeast of Mayer, with pyrolusite and psilomelane (Long, Batty, and Dean 1948).

MANJIROITE

Sodium manganese oxide: $Na(Mn^{4+}_7Mn^{3+})O_{16}$. The sodium analog of cryptomelane and probably most commonly formed as a secondary mineral.

COCHISE COUNTY: Tombstone District, Prompter Mine, as large half-inch cleavable prisms, some of which have hollow fibrous centers; also as crystals that form large black to dark-brown masses in a white crystalline limestone (3rd ed.).

GILA COUNTY: Banner supersystem, 79 District, 79 Mine, found as massive material mixed with chalcophanite (John Callahan, pers. comm., 2007).

MARCASITE

Iron sulfide: FeS_2. A low-temperature mineral formed under near-surface or surface conditions. Less common and less stable than pyrite, with which it is dimorphous. Generally found in replacement deposits or as concretions in sedimentary rocks.

APACHE COUNTY: Near Sanders (UA 7285).

COCHISE COUNTY: Cochise supersystem, Johnson Camp District, Johnson Camp area, in narrow seams and as crystals along faults (J. R. Cooper and Silver 1964). Warren District, Bisbee, Cole shaft, as concretions in limestone (Graeme 1981).

COCONINO COUNTY: Cameron District, as cyclically twinned inclusions in amethystine quartz crystals in petrified wood (Bull. 181); Alyce Tolino Mine, enclosed by small cubes of cobalt-rich pyrite, associated with umohoite in sooty masses and in carbonaceous trash replacements (Hamilton and Kerr 1959); Huskon Mines, with uraninite, greenockite, pyrite, calcite, and siliceous gangue replacing wood of a Triassic conifer in the Chinle Formation (Maucher and Rehwald 1961); Huskon No. 11 Mine, in sandstone with ink-blue masses and stains of ilsemannite (AEC mineral collection).

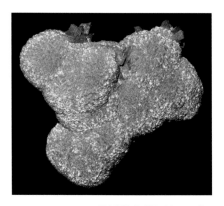

FIGURE 3.152 Marcasite, Antler Mine, Mohave County, 4.5 × 3.5 cm.

GILA COUNTY: Sierra Ancha Mountains, Sierra Ancha District, common in veins in uranium deposits in the Dripping Spring Quartzite (Granger and Raup 1969).

MOHAVE COUNTY: Hualapai (Antler) District, Antler Mine (J. Callahan, pers. comm., 2020). Cerbat Mountains, Wallapai supersystem, Golconda and Tennessee Districts, as a primary mineral deposited in a belt of sulfide-bearing fissure veins, associated with arsenopyrite (B. E. Thomas 1949). Black Mountains, Oatman District, Moss Mine, as thin plates in quartz (Bull. 181).

PIMA COUNTY: Santa Rita Mountains, Pima supersystem, Mission-Pima District (Schrader 1917); Copper Glance and Queen Mines, as an alteration product of pyrrhotite (Webber 1929); Mineral Hill and San Xavier Mines, as thin coatings and veinlets (Bull. 181).

SANTA CRUZ COUNTY: Santa Rita Mountains, Tyndall District (Schrader 1917).

YAVAPAI COUNTY: Northern Bradshaw Mountains, Big Bug District, Iron Queen Mine, as small colloform masses in partly oxidized ore (Lindgren 1926). White Picacho District, in pegmatites (Jahns 1952). Martinez (Congress) District, Congress Mine, in gold-bearing quartz veins with pyrite, associated with intrusive mafic dikes in Precambrian granite (Engineering and Mining Journal 1904).

MARIALITE

Sodium aluminum silicate chloride: $Na_4Al_3Si_9O_{24}Cl$. A member of the scapolite group, the two end members are sodium-rich marialite and calcium-rich meionite. The Dragoon Mountains material has been identified as marialite, the other two have not had chemistry reported. Mainly confined to regionally or contact-metamorphosed rocks such as schists, gneisses, and crystalline limestones.

COCHISE COUNTY: Dragoon Mountains, Middle Pass District, at the northern end of Stronghold Canyon, as short curved fibrous masses that replace vesuvianite in skarns (Rushing 1978).

GILA COUNTY: Reported in the Diamond Butte quadrangle, in metamorphosed Precambrian rocks (Gastil 1958).

YAVAPAI COUNTY: Bradshaw Mountains, Black Canyon District, six miles south of Cleator, in Yavapai schist (Bull. 181).

FIGURE 3.153 Maricopaite, Moon Anchor Mine, Maricopa County.

MARICOPAITE

Calcium lead aluminosilicate hydrate hydroxide: Ca_2Pb_7 $(Si_{36}Al_{12})O_{99} \cdot n(H_2O,OH)$. A secondary mineral and a member of the zeolite group. The Moon Anchor Mine is the type locality.

MARICOPA COUNTY: Belmont Mountains, Osborne District, Moon Anchor Mine, as sprays of translucent white acicular crystals, associated with mimetite; discovered by William Hunt; characterized by Peacor et al. (1988; R120147).

FIGURE 3.154 Markascherite, Childs-Aldwinkle Mine, Pinal County, FOV 3 mm.

#MARKASCHERITE

Copper molybdate hydroxide: $Cu_3(MoO_4)(OH)_4$. Very rare mineral found in the oxidized portion of copper-molybdenum deposits. The Childs-Aldwinkle Mine is the type locality.

PINAL COUNTY: Bunker Hill (Copper Creek) District, Copper Creek, Childs-Aldwinkle Mine, a secondary mineral found as small green transparent elongated crystals associated with brochantite, antlerite, lindgrenite, wulfenite, natrojarosite, and chalcocite (Yang et al. 2012; R100030).

MASSICOT

Lead oxide: PbO. An uncommon secondary mineral that is dimorphous with litharge, formed by the alteration of galena or secondary lead-bearing minerals.

GILA COUNTY: Globe Hills District, Albert Lea Mine, as a yellow powdery deposit on cerussite (N. P. Peterson 1962). Green Valley District, Silver Butte Mine, as an earthy yellow powder, associated with anglesite and galena (Lausen and Wilson 1925).

GRAHAM COUNTY: Aravaipa supersystem, Grand Reef District, Dog-water Mine, as yellow earthy masses with wulfenite (David Shannon, pers. comm., 1988); Grand Reef Mine with shannonite (Roberts et al. 1995).

LA PAZ COUNTY: Trigo Mountains, Silver District, at many mines in the district, as an earthy powder associated with cerussite and smithsonite (E. D. Wilson 1933).

MARICOPA COUNTY: Belmont Mountains, Osborne District, Tonopah-Belmont Mine, with minium (Bull. 181); Belmont Mountains, Osborne District, south of Wickenburg, at the Moon Anchor Mine, Potter-Cramer property, and Rat Tail claim (S. A. Williams 1968).

PIMA COUNTY: South Comobabi Mountains, Cababi District, Little Mary Mine and silver-lead claim, as an earthy yellow alteration product of galena (S. A. Williams 1962).

SANTA CRUZ COUNTY: Patagonia Mountains, Flux District, Flux Mine, with cerussite (Schrader and Hill 1915; E. D. Wilson 1933).

MATILDITE

Silver bismuth sulfide: $AgBiS_2$. A primary sulfide found in hydrothermal veins at moderate to high temperatures, and in pegmatites.

COCHISE COUNTY: Warren District, Bisbee, with sulfide ores of the Campbell Mine (Graeme 1993).

MATLOCKITE

Lead chloride fluoride: PbClF. This rare secondary mineral is known from only a few localities in the world.

GILA COUNTY: Globe Hills District, Apache Mine, with cerussite, brochantite, and boleite (Bideaux, Williams, and Thomssen 1960).

PIMA COUNTY: South Comobabi Mountains, Cababi District, Mildren Mine, as a single crystal in parallel growth with a prismatic anglesite crystal (S. A. Williams 1962).

PINAL COUNTY: Mammoth District, Mammoth–St. Anthony Mine, Collins vein, generally as minute crystals and as massive nodules coated with cerussite (Fahey, Daggett, and Gordon 1950; Bideaux 1980; R140538).

#MATTHEDDLEITE

Lead silicate sulfate chloride: $Pb_5(SiO_4)_{1.5}(SO_4)_{1.5}Cl$. A rare secondary mineral found in oxidized lead deposits.

MARICOPA COUNTY: Big Horn Mountains, Big Horn District, Evening Star Mine, as very small elongated hexagonal crystals in clusters with anglesite in pods of galena (R. Jenkins, pers. comm., 2012).

PINAL COUNTY: Mammoth District, Mammoth–St. Anthony Mine, as small clear hexagonal crystals (R. Jenkins, pers. comm., 2017).

#MAUCHERITE

Nickel arsenide: $Ni_{11}As_8$. Found with other nickel minerals associated with peridotite and serpentinite.

LA PAZ AND YUMA COUNTIES: Cemetery Ridge District, as small grains with other nickel minerals in magnetite in blocks of serpentinite and partially serpentinized harzburgite in the Orocopia Schist at Cemetery Ridge (Haxel et al. 2018).

MAWSONITE

Copper iron tin sulfide: $Cu_6Fe_2SnS_8$. In massive to disseminated copper ores within highly altered volcanic rocks at the Australian type locality.

COCHISE COUNTY: Warren District, Bisbee, Campbell orebody (Graeme 1993).

#MCALPINEITE

Copper tellurate: $Cu_3Te^{6+}O_6$. A very rare secondary mineral formed by the alteration of tellurides and tellurium-bearing sulfides.

COCHISE COUNTY: Huachuca Mountains, Reef District, Reef Mine, as small pale-green crystals and coatings spread over fractures in close proximity to galena and other sulfides. The mineral is associated with khinite, timroseite, dugganite, malachite and aurichalcite (Walstrom 2013).

MCKINSTRYITE

Silver copper sulfide: $Ag_5Cu_3S_4$. Of hydrothermal origin, associated with chalcopyrite, stromeyerite, silver, and arsenopyrite.

COCHISE COUNTY: Tombstone District; a specimen in the Smithsonian Institution (S 105199) labeled stromeyerite was shown to be mckinstryite (John S. White Jr., pers. comm., 1989).

#MEIONITE

Calcium aluminum silicate carbonate: $Ca_4Al_6Si_6O_{24}(CO_3)$. A member of the scapolite group found mainly in metamorphic rocks and as hydrothermal alteration mineral.

COCHISE COUNTY: Warren District, Bisbee, 500 level of the Holbrook Mine, found in altered Abrigo Limestone with tremolite and wollastonite (Tenney 1913). It is almost certain that the species noted is meionite (Graeme, Graeme, and Graeme 2015).

MELANOTEKITE

Lead iron oxide silicate: $Pb_2Fe^{3+}_2O_2(Si_2O_7)$. A very rare lead silicate mineral found in oxidized lead-copper deposits.

MOHAVE COUNTY: Rawhide District, Rawhide Mine, as fine-grain veins cutting alamosite (S. A. Williams 1982)

PINAL COUNTY: Mammoth District, Mammoth–St. Anthony Mine, as minute brownish spherules on diaboleite (Bideaux 1980). Mineral Hill District, Reymert Mine, as small spheres of black platy crystals (Ray Demark, pers. comm., 1984).

MELANOVANADITE

Calcium vanadium oxide hydrate: $Ca(V^{5+},V^{4+})_4O_{10}\cdot 5H_2O$. A rare mineral found in Colorado Plateau–type uranium deposits.

APACHE COUNTY: Lukachukai District, Mesa No. 1 and No. 5 Mines (Bull. 181); Kerr-McGee 4-1 Mine (Bull. 181); 4B Mine (Dodd 1955).

MELANTERITE

Iron sulfate hydrate: $Fe(SO_4) \cdot 7H_2O$. Water soluble; commonly formed in old mine workings as a product of the oxidation of pyritic ores. Probably more common in abandoned mine workings than the localities mentioned below would suggest.

COCHISE COUNTY: Warren District, Bisbee, as a widely distributed post-mine mineral in stalactites and coatings of walls, in Campbell shaft as large pseudo-octahedral crystals to 18 mm in stagnant waters; commonly in pyrite-rich areas (Graeme 1981).

GILA COUNTY: Banner supersystem, 79 District, 79 Mine, sparingly as white encrustations and efflorescences in the mine workings, altering from pyrite; melanterite blankets formed on the floor of the fifth level contain short acicular prismatic hair-like crystals, also in a crosscut on the sixth level, sparingly as the cuprian variety (Stanley B. Keith 1972). Globe Hills District, fifteenth level of the Old Dominion Mine, as the cuprian variety (3rd ed.).

GREENLEE COUNTY: Copper Mountain supersystem, Morenci District, sparingly in the upper levels of mines (Lindgren 1905) and locally in the Morenci Mine, formed by the oxidation of sulfide minerals (Moolick and Durek 1966).

PIMA COUNTY: Sierrita Mountains, Pima supersystem, as efflorescences on the walls of old mine workings (Bull. 181); San Xavier West Mine, associated with chalcanthite (Arnold 1964). Silver Bell District, El Tiro Mine, in small amounts as the cuprian variety (3rd ed.). Ajo District, New Cornelia Mine, occurs as yellowish and green concretionary crusts on exposed surfaces of seeps and drainage channels (W. J. Thomas and Gibbs 1983).

SANTA CRUZ COUNTY: Mansfield District, Temporal Gulch, St. Louis Mine, as the cuprian variety, formed as efflorescences on tunnel walls (Bideaux, Williams, and Thomssen 1960).

MELILITE

A group name for a family of sorosilicates with the general formula $Ca_2M(XSiO_7)$, where M is usually either Mg or Al and X usually Si or Al, but both can rarely include other elements. They are rock-forming minerals, typically found in rocks with relatively low silica content. The specific mineral species within this group is not known.

APACHE COUNTY: Garnet Ridge SUM District, as a constituent of the antigorite-dominated matrix in the diatreme (Witkind and Thaden 1963).

COCHISE COUNTY: Tombstone District, Comstock Hill, in skarn with vesuvianite, andradite and wollastonite (Michael Rubenach, pers comm. 2012).

NAVAJO COUNTY: Navajo volcanic field, as a constituent of lamprophyre dikes (Laughlin, Charles, and Aldrich 1986).

MELONITE

Nickel telluride: $NiTe_2$. A primary mineral often formed in hydrothermal vein deposits with other tellurides and sulfides.

COCHISE COUNTY: Warren District, Bisbee, among the sulfide ores of the Campbell Mine, with altaite in quartz (Graeme 1993).

MERCURY

Mercury: Hg. This is a comparatively rare secondary mineral that is formed from the alteration of cinnabar.

COCONINO COUNTY: Near Lee's Ferry, in minute quantities in the Chinle Formation, associated with gold (Lawson 1913; Lausen 1936).

GILA COUNTY: Mazatzal Mountains District, Pine Mountain Mine (Lausen 1926); Ord Mine, in small amounts in the Slate Creek deposits (Faick 1958).

FIGURE 3.155 Mercury, Saddle Mountain Mine, Maricopa County, FOV 3 mm.

MARICOPA COUNTY: Mazatzal Mountains District (Beckman and Kerns 1965), Saddle Mountain Mine, (R. Gibbs, pers. comm., 2020).

MOHAVE COUNTY: Hualapai Mountains, Maynard District, with cerussite (Bull. 181).

PINAL COUNTY: About nine miles east of Apache Junction, as fine globules in schist (Beckman and Kerns 1965).

YAVAPAI COUNTY: Lower Copper Basin Wash, Kirkland District, in appreciable quantities, probably derived from the low-grade cinnabar deposits in the area (Bull. 181). Walnut Grove District, associated with cinnabar (Engineering and Mining Journal 1897).

#MERRILLITE

Calcium sodium magnesium phosphate: $Ca_9NaMg(PO_4)_7$. An anhydrous phosphate found in stony meteorites, but reported occurrences also include lunar samples from the Fra Mauro Highlands and Taurus-Littrow Valley. It should be noted that a controversy exists involving the distinction between merrillite and whitlockite. X-ray diffraction and spectrographic results for *meteoric merrillite* and *terrestrial whitlockite* appear to match, and the two mineral names have been used interchangeably in the literature. Further analysis, however, revealed significant differences in crystal structure due to either the absence of hydrogen (merrillite) or the presence of hydrogen in the form of PO_4OH (whitlockite). In 1976, the IMA revalidated merrillite as a phosphate species deficient in hydrogen.

GILA COUNTY: Clover Springs, Clover Springs meteorite (Mittlefehldt et al. 1998).

NAVAJO COUNTY: Holbrook, Holbrook meteorite (Fuchs 1969).

FIGURE 3.156 Mesolite, near Clifton, Greenlee County, FOV 3 mm.

MESOLITE

Sodium calcium aluminum silicate hydrate: $Na_2Ca_2(Si_9Al_6)O_{30}·8H_2O$. Found in cavities in volcanic rocks, associated with other zeolites.

GILA COUNTY: Southeastern Dripping Spring Mountains, Banner supersystem, Christmas District, Christmas Mine, as fine hair-like crystals in hydrothermally altered andesite (David Perry, pers. comm., 1967).

GRAHAM COUNTY: Thumb Butte area, S4 and 5, T7S R29E, in vesicular basalt with other zeolites (William Hunt, pers. comm., 1985).

GREENLEE COUNTY: Road cuts north of Clifton along San Francisco River (R. Gibbs, pers. comm., 2020).

SANTA CRUZ COUNTY: Southwest of Patagonia, in drill core, as silky white fibers, with calcite in veinlets that cut fresh granodiorite (3rd ed.).

META-ALUNOGEN

Aluminum sulfate hydrate: $Al_2(SO_4)_3·14H_2O$. A secondary mineral formed in veins with pickeringite in altered andesite. Readily dehydrates to a white powder.

PIMA COUNTY: Ajo District, New Cornelia Mine, as crusts forming from a seep, with gypsum (William Thomas, pers. comm., 1988).

META-AUTUNITE

Calcium uranyl phosphate hydrate: $Ca(UO_2)_2(PO_4)_2·6H_2O$. A secondary uranium mineral formed by the oxidation of other uranium minerals, notably uraninite, and closely associated with the higher hydrate autunite.

COCHISE COUNTY: Courtland District, as scaly yellow crusts on pyrite ores (3rd ed.).

COCONINO COUNTY: Cameron District, Black Point–Murphy Mine; Jackpot No. 24 Mine (Austin 1964).

GILA COUNTY: Sierra Ancha District, Sue, Red Bluff, and Little Joe deposits, in weathered deposits in the Dripping Spring Quartzite (Granger and Raup 1962, 1969).

METACINNABAR

Mercury sulfide: HgS. The dimorph of cinnabar and an uncommon mineral of secondary origin formed in the upper portion of cinnabar deposits.

GILA AND MARICOPA COUNTIES: Mazatzal Mountains District, Alder and Slate Creeks area in small quantities with cinnabar (Lausen 1926). Found at the Mt. Ord deposit (Les Presmyk, pers. comm., 2020).

LA PAZ COUNTY: Dome Rock Mountains, La Cholla District, Colonial property, as thin coatings on cinnabar (Bull. 181); Cinnabar District, Cinnabar Mine, minor gray metacinnabar with cinnabar (Melchiorre 2013).

METAHEWETTITE

Calcium vanadium oxide hydrate: $CaV^{5+}_6O_{16} \cdot 3H_2O$. A secondary mineral formed as impregnations in sandstone.

APACHE COUNTY: Cane Valley District, Monument No. 2 Mine, as one of the principal ore minerals in a channel in the De Chelly sandstone that is filled by Shinarump Conglomerate; as an impregnation of the sandstone and a replacement of fossil plant matter; associated with tyuyamunite, carnotite, becquerelite, corvusite, hewettite, rauvite, navajoite, and uraninite (Finnell 1957). Black Mountain District in the Toreva Formation at unspecified mines (Chenoweth and Malan 1973).

COCONINO COUNTY: Vermilion Cliffs District, Sandy No. 1–3 claims, with uranium minerals (Peirce, Keith, and Wilt 1970).

METANOVÁČEKITE

Magnesium uranyl arsenate hydrate: $Mg(UO_2)_2(AsO_4)_2 \cdot 8H_2O$. A rare secondary mineral derived from nováčekite.

GILA COUNTY: Cherry Creek area, Sierra Ancha District, Sue Mine, in Dripping Spring Quartzite, as individual flakes, and as alteration rims to 0.5 mm wide on bassetite and saléeite (Granger and Raup 1969).

METAROSSITE

Calcium vanadium oxide hydrate: $CaV^{5+}_2O_6 \cdot 2H_2O$. A rare mineral found in sandstone uranium-vanadium deposits on the Colorado Plateau. Formed from the alteration of rossite, with which it is intimately associated.

APACHE COUNTY: Lukachukai District, as an alteration product of rossite in the Salt Wash Member of the Morrison Formation (Chenoweth 1967); 4B Mine (Dodd 1956). Also reported in the Cane Valley District, Monument No. 2 Mine (Evensen and Gray 1958).

METASIDERONATRITE

Sodium iron sulfate hydroxide hydrate: $Na_2Fe^{3+}(SO_4)_2(OH) \cdot H_2O$. A rare secondary mineral associated with other secondary sulfates.

COCONINO COUNTY: Cameron District, as a deep-yellow cleavable fibrous material from the Yazzie No. 101 Mine (Austin 1957, 1964).

FIGURE 3.157 Metatorbernite, Silver Bell Mine, Pima County, 5.2 cm high.

METATORBERNITE

Copper uranyl phosphate hydrate: $Cu(UO_2)_2(PO_4)_2 \cdot 8H_2O$. A secondary mineral commonly formed by dehydration of torbernite but may crystallize directly from solution; much like torbernite in mode of occurrence and association.

APACHE AND NAVAJO COUNTIES: Monument Valley and Cane Valley Districts, in the oxidized "yellow ore" of uranium-vanadium deposits in the Shinarump Conglomerate and De Chelly sandstone, in conglomerate-filled paleo-stream channels; associated with carnotite, tyuyamunite, metarossite, tangeite, and hewettite (Evensen and Gray 1958).

COCONINO COUNTY: Cameron District, in uranium deposits in the Shinarump Conglomerate, associated with uraninite and meta-autunite (Holland et al. 1958); Huskon No. 7 Mine and Riverside No. 1 claims (Austin 1964); Arrowhead claim southeast of Cameron, with uraninite in the Riverview collapse feature (Barrington and Kerr 1963). Reported in the breccia pipes, Riverview Group of claims, with uranophane and torbernite in a breccia pipe (Wenrich and Sutphin 1989).

GILA COUNTY: Sierra Ancha District, northwestern region, associated with nearly all the uranium deposits in the Dripping Spring Quartzite

(Granger and Raup 1969); Workman Creek, in the upper Dripping Spring Quartzite, with crocoite (Robert O'Haire, pers. comm., 1972); and Wilson Creek area, in Dripping Spring Quartzite (R. L. Wells and Rambosek 1954). Miami-Inspiration supersystem, Copper Cities District, Copper Cities deposit, as small amounts in disseminated copper ore in quartz monzonite and as tiny rosettes on the walls of minute fractures along the Coronado fault (N. P. Peterson 1954, 1962); Melinda Mine (Bull. 181). Pinto Valley District, Castle Dome Mine, commonly on wavellite crusts (N. P. Peterson 1947).

MOHAVE COUNTY: Hack District, Hack Canyon Mine (Granger and Raup 1962).

NAVAJO COUNTY: Monument Valley District, Monument No. 1 and Mitten No. 2 Mines (Witkind 1961; Witkind and Thaden 1963). Petrified Forest Wood District, Ruth Group of claims near Holbrook (Bull. 181).

PIMA COUNTY: Silver Bell District, Silver Bell Mine, as microcrystals (ASDM x1093).

PINAL COUNTY: Bunker Hill (Copper Creek) District, Old Reliable Mine, as small clusters of transparent green tablets with libethenite (Gibbs, pers. comm., 2016)

YAVAPAI COUNTY: Tiger District, Crown King area, as microcrystals (Richard Thomssen, UA x598).

METATYUYAMUNITE

Calcium uranyl vanadate hydrate: $Ca(UO_2)_2(VO_4)_2 \cdot 3H_2O$. A secondary mineral commonly associated with tyuyamunite in the oxidized portions of uranium-vanadium deposits.

APACHE COUNTY: Northeast Carrizo Mountains District, Cove Mesa, Sycamore Group of claims, and King Tut Mine (Bull. 181). Lukachukai District, Mesa No.1 Mine (Bull. 181). Cane Valley District, Monument No. 2 Mine, associated with tyuyamunite; forms a thick zone around the dark resinous material that is the matrix for small grains of uraninite (Rosenzweig, Gruner, and Gardiner 1954).

COCONINO COUNTY: Cameron District, Montezuma Group of claims, in Shinarump Conglomerate; Huskon No. 12 Mine (Austin 1964). Prospect Canyon (Ridenour) District, Ridenour Mine, with vésigniéite, naumannite, bromargyrite, tangeite, and tyuyamunite (Wenrich and Sutphin 1989).

NAVAJO COUNTY: Petrified Forest Wood District, Ruth Group of claims (Bull. 181). Monument Valley District, Monument No. 1 and Mitten No. 2 Mines (Witkind 1961; Witkind and Thaden 1963).

#METAURANOCIRCITE-I

Barium uranyl phosphate hydrate: $Ba(UO_2)_2(PO_4)_2 \cdot 6H_2O$. An uncommon secondary mineral and a member of the meta-autunite group.

COCONINO COUNTY: Cameron District, as an abundant ore mineral that forms fine-grained yellow masses, associated with fossil logs in the Petrified Forest Member of the Chinle Formation (Austin 1964).

GILA COUNTY: Sierra Ancha District, reported in uranium deposits in the Dripping Spring Quartzite (Granger 1955).

METAVOLTINE

Potassium sodium iron oxide sulfate hydrate: $K_2Na_6Fe^{2+}Fe^{3+}_6O_2(SO_4)_{12} \cdot 18H_2O$. A rare mineral formed in arid climates, commonly by the oxidation of pyritic ores.

COCHISE COUNTY: Warren District, Bisbee, 2,100 ft. level of the Campbell Mine, as vermicular stacks of minute greenish-yellow hexagonal platelets; associated with copiapite, coquimbite, voltaite, and römerite (Graeme 1981).

METAZEUNERITE

Copper uranyl arsenate hydrate: $Cu(UO_2)_2(AsO_4)_2 \cdot 8H_2O$. A secondary mineral formed in the oxidized portions of uranium deposits.

APACHE COUNTY: Cane Valley District, Monument No. 2 Mine (Witkind and Thaden 1963).

COCONINO COUNTY: Cameron District, in minor amounts in uranium ore in the Shinarump Conglomerate, paleo-stream channels, and the sandy portions of the Chinle Formation; associated with uraninite, metatorbernite, and meta-autunite (Holland et al. 1958). Grand Canyon National Park, Grandview District, Grandview (Last Chance) Mine, as transparent emerald-green to leek-green tabular crystals, associated with scorodite and olivenite (X-ray data suggest that the leek-green variety is close to metazeunerite, the emerald-green variety to zeunerite); some crystals, up to 2 cm across, are among the largest known for the species (Leicht 1971).

GILA COUNTY: Sierra Ancha District, Easy deposit, where it coats limonite on fracture surfaces and is locally coated by hyalite (Granger and Raup 1969).

NAVAJO COUNTY: Petrified Forest Wood District, Ruth Group of claims near Holbrook (Bull. 181).

MIARGYRITE

Silver antimony sulfide: $AgSbS_2$. It forms in low-temperature hydrothermal veins with galena and other silver minerals.

MOHAVE COUNTY: Cerbat Mountains, Wallapai supersystem, probably Stockton Hill or Chloride Districts, as a primary mineral in minor amounts, with pyrargyrite and polybasite; with proustite in veinlets that cut galena and chalcopyrite (B. E. Thomas 1949).

YAVAPAI COUNTY: Peck District, Black Warrior and Peck Mines, as coatings on other minerals (Beck 2011).

MICROCLINE

Potassium aluminum silicate: $K(AlSi_3O_8)$. A widespread rock-forming feldspar that is present in all the granite and pegmatites in the state. Much of the potash feldspar commonly identified as orthoclase is microcline. Microcline is the main potassium feldspar mineral in peraluminous pegmatites, which are widespread in Arizona. Only a few localities are listed below.

FIGURE 3.158 Microcline, Santo Niño Mine, Santa Cruz County, 6.1 cm.

COCHISE COUNTY: Reported in the Warren District, Bisbee (3rd ed.).

MARICOPA AND YAVAPAI COUNTIES: White Picacho District, as the variety perthite, the most abundant mineral in the pegmatites; as crystals up to 13 ft. (Jahns 1952).

MOHAVE COUNTY: Cerbat Mountains, Kingman Feldspar District, Kingman Feldspar Mine, commercially mined from pegmatites since 1924 (Heinrich 1960). Aquarius Mountains District, Rare Metals Mine, with euxenite-(Y), polycrase-(Y), fluorite, quartz, muscovite, garnet (K. A. Phillips 1987).

PIMA COUNTY: Ajo District, south-central part of the New Cornelia orebody, as coarse crystals in the copper-rich pegmatite masses (Gilluly 1937); all of the mineral has been removed by mining. Pima supersystem, Twin Buttes District, Twin Buttes Mine (William Hefferon, pers. comm., 3rd ed.). Santa Catalina Mountains, south of Summerhaven as large crystals up to 2 ft. across in pegmatites associated with spessartine garnet schlieren, quartz, and muscovite in the Lemmon Rock Leucogranite

(Force 1997); Santa Catalina foothills, as distinctive rolled augen in deformed Oracle Granite, especially well-displayed in many roadcuts along the Catalina Highway.

PINAL COUNTY: Oracle area, as large, perthitic, "boxcar" porphyrocrysts up to 2 in. long in porphyritic phases of undeformed Oracle Granite (Force 1997; Banerjee 1957).

SANTA CRUZ COUNTY: Querces District, Santo Niño Mine (B. Muntyan, pers. comm., 2020).

MICROLITE

Generalized formula: $A_{2-m}Ta_2X_{6-w}Y_{1-n}$, where A may be Na, Ca, Sn^{2+}, Sr, Pb^{2+}, Sb^{3+}, Y, U^{4+}, H_2O, □; X may be O,OH,F; and Y may be vacancy, H_2O, very large >> 1.0 Å monovalent cation. The subscript values represent incomplete occupancy of A, X, and Y sites and may assume $m = 0-2$, $w = 0-0.7$, and $n = 0-1$ (see Atencio et al. 2017). Microlite is now a group name under the pyrochlore supergroup. The exact species for the Arizona occurrences has not been determined.

MOHAVE COUNTY: Valentine District, Bountiful Beryl prospect, as a minor constituent in a pegmatite fifteen miles southwest of Peach Springs near the railroad tracks just northeast of Valentine (Schaller, Stevens, and Jahns 1962).

YAVAPAI COUNTY: White Picacho District, Outpost, Midnight Owl, and Picacho View pegmatites, as tiny olive-green to dark-brown and black crystals with sharply defined octahedral and dodecahedral faces; associated with the related species pyrochlore (Jahns 1952).

FIGURE 3.159 Miersite, on atacamite, Southwest Mine, Bisbee, Cochise County, FOV 1 cm.

MIERSITE

Silver iodide: AgI. A rare secondary mineral formed in the oxidized zones of base-metal deposits.

COCHISE COUNTY: Warren District, Bisbee, Southwest Mine, as small lemon-yellow crystals, with atacamite, malachite, and cuprite (Graeme 1993).

PIMA COUNTY: South Comobabi Mountains, Cababi District, Mildren and Steppe claims, as overgrowths on iodargyrite; a product of the oxidation of sulfide ores in quartz veins that cut andesite (S. A. Williams 1962, 1963).

#MILARITE

Potassium calcium beryllium aluminosilicate hydrate: $KCa_2(Be_2AlSi_{12})O_{30}·H_2O$. An accessory mineral found in low-temperature hydrothermal veins, aplites, syenites, and granite pegmatites.

MARICOPA COUNTY: Belmont Mountains, Osborne District, Belmont pit (a small landscape rock mine), as very small elongated clear hexagonal crystals on orthoclase and albite in miarolitic cavities in the Belmont Granite associated with fluorite and epidote (Gibbs and Turzi 2007).

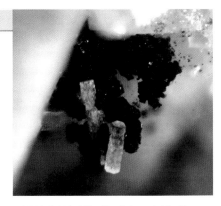

FIGURE 3.160 Milarite, Belmont Pit, Maricopa County, FOV 3 mm.

MILLERITE

Nickel sulfide: NiS. Typically, a low-temperature mineral formed in cavities in veins, where it is commonly associated with carbonates; also formed in geodes in limestone.

COCONINO COUNTY: Orphan District, Orphan Mine, in Grand Canyon National Park, in cavities in baryte (3rd ed.).

MIMETITE

Lead arsenate chloride: $Pb_5(AsO_4)_3Cl$. Structurally, a member of the apatite group; an end member of the pyromorphite-mimetite series. A secondary mineral formed in the oxidized portions of arsenical lead deposits; typically associated with cerussite, plattnerite, wulfenite, smithsonite, hemimorphite, anglesite, limonite, and other oxide-zone minerals.

COCHISE COUNTY: Cochise supersystem, Johnson Camp District, Empire Mine, as microcrystals (Peter Megaw, UA x4620). Tombstone District, Gallagher vanadium property near Charleston (Bull. 181). Warren District, Bisbee, Campbell Mine, with wulfenite; Cole Mine, with malachite in a fault zone on the 600 ft. level; Shattuck Mine, with shattuckite and malachite; Southwest Mine, with plattnerite and calcite (Graeme 1981).

GILA COUNTY: Banner supersystem, 79 District, 79 Mine, as fine specimens, notably as brilliant-orange, orange-yellow, and bright-yellow crusts and reniform masses, some of which are stalactitic; some of the

FIGURE 3.161 Mimetite, Rowley Mine, Maricopa County, 2.2 cm.

finest mimetites are associated with thick clear-orange wulfenite crystals, many of which are 0.5 in. on an edge (Stanley B. Keith 1972).

GRAHAM COUNTY: Aravaipa supersystem, Grand Reef District, Grand Reef Mine, as microcrystals (Frank Valenzuela, UA x713).

LA PAZ COUNTY: Silver District, Red Cloud Mine, as red crystalline masses with small doubly terminated crystals 1 mm in size (Edson 1980).

MARICOPA COUNTY: Harquahala District, east of the Alaska Mine, southwest of Aguila (Bull. 181). Painted Rock District, Rowley Mine near Theba, as microcrystals with a great variety of habits: as minute perfectly hexagonal crystals less than 1 mm long and as nearly perfectly spherical aggregates. W. E. Wilson and Miller (1974) described the latter as an extreme example of "wheat-sheaf"-like crystal growth; the crystal aggregates are curled back on themselves, producing the spherical shape. Colors range from red to yellow. Some crystals are color zoned: the centers are orange, and the ends are yellow (W. E. Wilson 2020b). Vulture Mountains, Domingo Mine (Bull. 181). Belmont Mountains, Osborne District, Moon Anchor Mine, as acicular yellow crystals and as pseudomorphs of galena cubes (Peacor et al. 1988). Amazon Wash, Broken Finger Mine, as red-fluorescing microcrystals (Arthur Roe, pers. comm., 3rd ed.).

MOHAVE COUNTY: Reported in the Cerbat Mountains, Wallapai supersystem, (Bull. 181). Buckskin Mountains, Rawhide District, Rawhide Mine, as botryoidal crusts (3rd ed.).

PIMA COUNTY: Waterman District, Indiana-Arizona Mine, as microcrystals with wulfenite (Francis Sousa, UA x1034). Empire District, Total Wreck and Verde Queen Mines, as microcrystals with wulfenite (Francis Sousa and Brian Bond, UA x181, x2710). Silver Bell District, Ironwood Mine (Sammy Dog, God's claim), as bright red to yellow crystals to several millimeters, often bicolored with mushroom terminations (Andy Anderson, pers. comm., 2015).

PINAL COUNTY: Mammoth District, Mammoth–St. Anthony Mine, as bright-orange and canary-yellow crusts and as coatings of tiny prismatic to tabular crystals, with wulfenite (Bideaux 1980). Mineral Hill District, Reymert Mine, with anglesite, cerussite, wulfenite, and vanadinite (K. S. Wilson 1984). Table Mountain District, Table Mountain Mine, as microcrystals, associated with wulfenite (Kenneth Bladh, UA x526). Mineral Creek supersystem, Ray District, Ray Mine, as microcrystals (UA x3494).

SANTA CRUZ COUNTY: Trench District, Hardshell Mine, as microcrystals, with descloizite (Fleetwood Koutz, UA x339). Tyndall District, Glove Mine, with wulfenite (UA x4693).

Bradshaw Mountains, Black Canyon District, at a prospect on the Slate Creek property of Kalium Chemicals, as botryoidal crusts, with pyromorphite and mottramite (William Berridge, pers. comm., 1973).

MINIUM

Lead oxide: $Pb^{2+}_2Pb^{4+}O_4$. A secondary mineral formed in lead deposits under extreme oxidizing conditions.

FIGURE 3.162 Minium, Rowley Mine, Maricopa County, 5.8 cm.

LA PAZ COUNTY: Silver District, Silver King Mine, as microcrystals (Brian Bond, UA x3542).

MARICOPA COUNTY: Belmont Mountains, Osborne District, Tonopah-Belmont Mine, with massicot (G. B. Allen and Hunt 1988); Osborne District, south of Wickenburg, at the Moon Anchor Mine, Potter-Cramer property, and Rat Tail claim, with various secondary minerals formed by the oxidation of galena-sphalerite ores (S. A. Williams 1968); Painted Rock District, Rowley Mine, as reddish-orange masses to 6 cm long (Les Presmyk, pers. comm., 2020).

MOHAVE COUNTY: Black Mountains, Oatman District, Big Jim vein, as pulverulent material in cavities (Bull. 181).

PIMA COUNTY: South Comobabi Mountains, Cababi District, Silver-Lead and Beacon claims, Mildren Mine, as coatings on wulfenite and other secondary minerals (S. A. Williams 1962).

PINAL COUNTY: Tortilla Mountains, Ripsey District, Florence Lead-Silver Mine (S. A. Williams and Anthony 1970). Silver Reef District, Casa Grande, (UA 7324). Mammoth District, Mammoth–St. Anthony Mine, as a rare powdery coating on wulfenite crystals (Bideaux 1980).

SANTA CRUZ COUNTY: Patagonia Mountains, Flux District, Flux Mine, with cerussite (Schrader and Hill 1915).

YUMA COUNTY: Castle Dome Mountains, Castle Dome District, with cerussite and wulfenite (Bull. 181).

MIRABILITE

Sodium sulfate hydrate: $Na_2(SO_4) \cdot 10H_2O$. A water-soluble mineral formed in playas, saline lakes, and clay soils of the desert. It is also found as an efflorescence in old mine workings.

YAVAPAI COUNTY: Verde Valley, in salt deposits, associated with halite, glauberite, and thénardite (Silliman 1881; Blake 1890; Guild 1910; Peirce 1969b; J. R. Thompson 1983).

FIGURE 3.163 Mixite, Table Mountain Mine, Pinal County, FOV 3 mm.

MIXITE

Copper bismuth arsenate hydroxide hydrate: Cu_6Bi $(AsO_4)_3(OH)_6 \cdot 3H_2O$. A rare secondary mineral formed in some base-metal deposits.

COCHISE COUNTY: Tombstone District, Little Joe Mine (Sid Williams, pers. comm., 2006).

GILA COUNTY: Southeastern Dripping Spring Mountains, Banner supersystem, Christmas District, Christmas Mine, as spherules of delicate twisted or matted fibers in cavities in gangue, associated with fibrous malachite; the peculiar yellow-green color of this mixite strongly resembles creaseyite (3rd ed.).

PINAL COUNTY: Mammoth District, Mammoth–St. Anthony Mine, as rare pale-green radiating sprays associated with wulfenite and mimetite on baryte matrix (Bideaux 1980). Table Mountain District, Table Mountain Mine, as microcrystals (Arthur Roe, UA x4591).

YAVAPAI COUNTY: Cherry Creek District, as microcrystals (Brian Sage, UA x1473).

#MOGÁNITE

Silicon dioxide hydrate: $SiO_2 \cdot nH_2O$. A polymorph of quartz found as a minor component of finely crystalline chalcedony, chert, and flint.

GRAHAM COUNTY: Deer Creek Fire Agate District, Deer Creek Fire Agate Mine, mixed with chalcedony, quartz, and opal forming a colloform texture in fire agate (Natkaniec-Nowak et al. 2020).

MOISSANITE

Silicon carbide: SiC. Material originally thought to be diamond in the Canyon Diablo meteorite was found to be naturally occurring silicon carbide. The material was named in 1905 after Ferdinand Henri Moissan, who discovered the mineral. Although there have been some suggestions that the material was a contaminant of silicon carbide (carborundum) introduced during the cutting of the meteorite, it has been found in many locations throughout the world. Moissan used no silicon carbide to prepare the sample. Moissanite is also known from a few upper mantle rocks such as kimberlites and lamproites. Meteor Crater is the type locality.

COCONINO COUNTY: Meteor Crater, in samples of the Canyon Diablo meteorite (Kunz 1905).

MOLYBDENITE

Molybdenum sulfide: MoS_2. The most common molybdenum mineral, it is widely distributed but in small amounts. A primary sulfide mineral formed in granitic rocks, quartz-orthoclase veins with chalcopyrite, and tin and tungsten ores. It may also be formed in contact-metamorphic deposits. An important associated mineral in porphyry copper deposits of the Southwest, where it is an important byproduct of copper ores.

FIGURE 3.164 Molybdenite, Childs-Aldwinkle Mine, Pinal County, FOV 5 mm.

COCHISE COUNTY: Little Dragoon Mountains, Cochise supersystem, Johnson Camp District, Johnson Camp, in copper ores (Ransome 1919; J. R. Cooper and Huff 1951; J. R. Cooper 1957; J. R. Cooper and Silver 1964). Warren District, Bisbee, Campbell shaft, as film on quartz and schist, Lavender pit with bornite (Graeme 1981); Mine Canyon District, as disseminations and veinlets in quartz monzonite porphyry (DeRuyter 1979).

GILA COUNTY: Miami-Inspiration supersystem, Pinto Valley District, in small quantities in ores, particularly at the Castle Dome Mine (N. P. Peterson 1947, 1962; N. P. Peterson, Gilbert, and Quick 1946, 1951); Copper Cities District, Copper Cities deposit (N. P. Peterson 1954). Banner supersystem, 79 District, with chalcopyrite and pyrite, locally coating the walls of veinlets in the 79 Mine (Stanley B. Keith 1972). Sierra Ancha District, Workman Creek area, in hornfels near diabase; Suckerite deposit, associated with uraninite (Granger and Raup 1969).

GREENLEE COUNTY: Copper Mountain supersystem, Morenci District, in veins with pyrite, chalcopyrite, and sphalerite (Lindgren 1905; Guild 1910; Schwartz 1947), and as thin films in fractures devoid of other sulfides (Moolick and Durek 1966).

LA PAZ COUNTY: Yuma King District, Yuma Copper Mine area, as veinlet fillings with chalcopyrite and quartz in a several-hundred-foot drill-hole intercept and as veins in chalcopyrite-bearing skarn with associated fluorite, magnetite, and andradite garnet (Rasmussen and Hoag 2011).

MARICOPA COUNTY: White Picacho District, sparsely scattered in pegmatites, but locally abundant (Jahns 1952).

MOHAVE COUNTY: White Hills, Gold Basin District, O.K. Mine, with galena and "wolframite" (Bull. 181). Cerbat Mountains, Wallapai supersystem, Mineral Park District, Mineral Park Mine (Garrison 1907; B. E. Thomas 1949; Field 1966); Tennessee District, Samoa Mine (Schrader 1909; Blanchard and Boswell 1930); Ithaca Peak District, Ithaca Peak, with pyrite in the core of a quartz-monzonite stock (Eidel 1966). Hualapai Mountains, Maynard District, Leviathan and American Mines (Bull. 181). Deluge Wash area, in small quantities at several properties (J. W. Frondel and Wickman 1970).

PIMA COUNTY: Santa Rita Mountains, Helvetia-Rosemont District (J. W. Frondel and Wickman 1970), at the Leader, Ridley, and Pauline Mines and in many prospects in Madera and Providencia Canyons (Guild 1907; Schrader and Hill 1910; Creasey and Quick 1955). Pima supersystem, Mission-Pima District, Pima Mine (Himes 1972). Mission Mine as rare well-formed pseudomorphs after andradite up to 1 in. in diameter (B. M. Williams 2018); Mineral Hill Mine (Eckel 1930a; Guild 1934); Twin Buttes District, a primary ore mineral at the Twin Buttes Mine (Eckel 1930a). Sierrita-Esperanza District, Sierrita Mine, as a byproduct mineral (up 133 million pounds Mo recovered through 1979; highest Arizona molybdenum producer) notable for its high rhenium content (Wilt and Keith 1980); Esperanza Mine, as fine specimens as fillings in the Esperanza breccia. Northern Santa Rita Mountains, Cuprite District, Cuprite Mine, as small masses in chalcopyrite ore (J. F. Browne 1958). Silver Bell District, at a small prospect north of the Kurtz shaft (Guild 1910; C. A. Stewart 1912). Baboquivari Mountains, Mildred Peak District, Gold Bullion Mine, in quartz veins (Bull. 181). Ajo District, in sparse amounts (Gilluly 1937; Schwartz 1947), locally common in the eastern part of the New Cornelia pit. Near Redington, in limestone (UA 7288). South Comobabi Mountains, Cababi District, Mildren and Steppe claims (S. A. Williams 1963).

PINAL COUNTY: Galiuro Mountains, Bunker Hill (Copper Creek) District, Copper Creek, as the most important ore mineral at the Childs-Aldwinkle Mine, from which 70 million pounds were produced from 1933 to 1938 (C. A. Anderson 1969); as fine crystallized specimens (Kuhn 1941; Fleischer 1959). The rhenium content of the molybdenite concentrates from the Childs-Aldwinkle Mine contained from 320 to 580 parts per million, among the highest known; also at Copper Prince Mine, Old Reliable Mine, and other properties in lesser quantities (Simons 1964). Mineral Creek supersystem, Ray District, at Ray (UA 7129). San Manuel District, San Manuel Mine (Lovering, Huff, and Almond 1950; Schwartz 1953); Kalamazoo orebody (J. D. Lowell 1968). Pioneer supersystem, Resolution District, Resolution Mine, an important rhenium-bearing mineral occurring in an early potassic stage with chalcopyrite and a later phyllic argillic stage loosely associated with digenite, alunite, dickite, and topaz (Hehnke et al. 2012).

SANTA CRUZ COUNTY: Patagonia Mountains, Santo Niño District, Santo Niño Mine, 2.5 miles southwest of Duquesne, in a breccia pipe as large bodies of fine-grained massive material and as good crystals in quartz veins with pyrite (Blanchard and Boswell 1930; J. W. Frondel and Wickman 1970); Washington Camp District, Bonanza Mine at Duquesne, where small quantities were mined; Querces District, Benton and Line Boy Mines (Schrader and Hill 1915; Schrader 1917).

YAVAPAI COUNTY: Copper Basin District, Copper Hill Mine, Loma Prieta Mine, and other properties, as a common widespread mineral (Johnston and Lowell 1961). Bradshaw Mountains, in several districts at the Black Hawk, Bluebird, and Squaw Peak Mines (Bull. 181). Eureka supersystem, Bagdad District, in thin veins at the Bagdad Mine (Lindgren 1926). White Picacho District, in pegmatites (Jahns 1952). Black Canyon District, Crazy Basin property (St. Louis claim) near Cleator, as fine single crystals up to 4 cm across, in the quartz core of a pegmatite (3rd ed.).

#MOLYBDITE

Molybdenum oxide: MoO_3. A secondary mineral formed from the oxidation of molybdenite.

GILA COUNTY: Pinto Valley, Miami-Inspiration supersystem, Pinto Valley District, Castle Dome Mine, with other secondary minerals (N. P. Peterson, Gilbert, and Quick 1951).

YAVAPAI COUNTY: White Picacho District, Picacho View Mine, in E half of S10, T7N R3W, in fractures in quartz in a pegmatite (Jahns 1952).

MOLYBDOFORNACITE

Copper lead molybdate arsenate hydroxide: $CuPb_2(MoO_4)(AsO_4)(OH)$. An uncommon mineral formed during severe oxidation of sulfide deposits, related to fornacite and vauquelinite.

MARICOPA COUNTY: Belmont Mountains, Osborne District, Tonopah-Belmont Mine, with wulfenite and pyromorphite (G. B. Allen and Hunt 1988).

MOHAVE COUNTY: Buckskin Mountains, Rawhide District, Rawhide Mine, as bladed crystals with orange mimetite (3rd ed.).

PIMA COUNTY: South Comobabi Mountains, Cababi District, Mildren claim, as very small yellow-green crystals in cavities on quartz crystals (Gibbs, pers. comm., 2016).

#MOLYBDOPHYLLITE

Lead magnesium silicate hydroxide carbonate: $Pb_8Mg_9[Si_{10}O_{28}(OH)_8$ $O_2(CO_3)_3]\cdot H_2O$. A lead mineral found in skarn deposits in limestone.

PINAL COUNTY: Mammoth District, Mammoth–St. Anthony Mine, as transparent platy crystals with leadhillite and susannite (R050642).

#MONAZITE-(Ce)

Cerium phosphate: $Ce(PO_4)$. Monazite is an important ore mineral of the rare-earth elements and thorium. It may contain various rare-earth and other elements, in addition to the typically preponderant cerium: for example, lanthanum, neodymium, yttrium, thorium, uranium, calcium, and silicon. Formed as an accessory mineral in peraluminous granites and pegmatites; also found in thermally metamorphosed rocks and as an abundant constituent of certain sands. The monazite group contains several species of monazite. The chemistry of most of the following has not been determined, but the two Mohave County occurrences are monazite-(Ce) (see S. L. Hanson, Falster, and Simmons 2007).

GRAHAM COUNTY: Santa Teresa Mountains, as small crystals in pegmatite (Bull. 181).

MARICOPA AND YAVAPAI COUNTIES: White Picacho District, as a minor accessory mineral, associated with tantalum-niobium minerals in several pegmatites (Jahns 1952).

MOHAVE COUNTY: Chemehuevi District, about twenty miles southeast of Topock, sparingly in stream gravels (Heineman 1930; Overstreet 1967). Virgin Mountains, with xenotime-(Y), in Precambrian gneiss (E. J. Young and Sims 1961). Aquarius Mountains, Rare Metals Mine, in pegmatites, associated with titanite, chevkinite, apatite, and cronstedtite (Kauffman and Jaffe 1946; Heinrich 1960); Wagon Bow No. 3 pegmatite (S. L. Hanson, Falster, and Simmons 2007). Near the Nevada border, opposite Mesquite (Clark County, Nevada), in granite augen gneiss, with xenotime-(Y) (Overstreet 1967). Near Hoover Dam (H 102368). Wallapai District, small prospect near the Kingman Feldspar Mine (S. L. Hanson, Falster, and Simmons 2007)

PIMA COUNTY: Two miles east of Papago Wells, as massive material (MM 1008). Santa Catalina Mountains, common as an accessory mineral along with zircon in muscovite-bearing sills of the Wilderness Granite intrusive suite (Shakel, Silver, and Damon 1977; Stanley B. Keith et al. 1980; Force 1997).

Black Canyon Creek, sparse in sands, with magnetite, hematite, garnet, and gold (Day and Richards 1906). Squaw Peak District, Squaw Peak Mine, as rosette-shaped aggregates in quartz-chalcopyrite veins (R. R. Roe 1976).

#MONOHYDROCALCITE

Calcium carbonate hydrate: $Ca(CO_3) \cdot H_2O$. An unstable mineral formed in sediments and from decaying vegetation.

MARICOPA COUNTY: Gila Bend Mountains, found as a weathering product of weddellite in decaying saguaro cacti (Garvie 2003).

#MONTANITE

Bismuth tellurium oxide hydrate: $Bi^{3+}_2 Te^{6+} O_6 \cdot 2H_2O$. A secondary mineral formed from the alteration of earlier tellurium minerals.

COCHISE COUNTY: Tombstone District, Little Joe shaft, as brown massive fine-grained material that appears to be pseudomorph after a prismatic mineral, where it occurs with mixite in altered rhyodacite (J. W. Anthony et al. 2003).

MONTEBRASITE

Lithium aluminum phosphate hydroxide: $LiAl(PO_4)(OH)$. Found in pegmatites as crystals, locally of large size, and as massive material that may be easily mistaken for the closely related amblygonite.

MARICOPA COUNTY: White Picacho District, North Morning Star Mine (London 1981); much of what has been identified as amblygonite in the district is probably montebrasite (London and Burt 1982b).

YAVAPAI COUNTY: White Picacho District, common in the Independence, Lone Giant, Midnight Owl, Picacho View, and White Ridge pegmatites (London 1981).

MONTICELLITE

Calcium magnesium silicate: $CaMg(SiO_4)$. An uncommon member of the olivine group that is usually found in high-temperature skarns.

COCHISE COUNTY: Tombstone District, fourth level of the Toughnut Mine, as narrow bands in a contact-metamorphic zone, with calcite, thaumasite, clinozoisite, and vesuvianite (B. S. Butler, Wilson, and Rasor 1938).

MONTMORILLONITE

Sodium calcium aluminum magnesium silicate hydroxide hydrate: $(Na,Ca)_{0.3}(Al,Mg)_2Si_4O_{10}(OH)_2 \cdot nH_2O$. This is the most common and widespread member of the smectite group of clay minerals. It is the principal constituent of the bentonite clays, which result from the alteration of volcanic ash and tuffs; also formed by hydrothermal activity. Montmorillonites are characterized by high ion-exchange capacities and by an ability to swell markedly when wetted.

APACHE COUNTY: Sanders–Defiance Plateau area, Cheto Bentonite District, at Cheto, Allentown, Barnwater Wash, and Ganado Mesa, in linear channels and lens-like bodies underlying the upper member of the Bidahochi Formation, derived from volcanic tuff; as a calcian variety (Kiersch and Keller 1955). Cane Valley District, Monument No. 2 Mine, associated with uranium-vanadium ores (Witkind and Thaden 1963).

COCHISE COUNTY: Reported in an area east of Elgin and about two miles south of Benson (Bull. 181). Willcox Playa, as the most abundant clay mineral (after illite) in sediments (Pipkin 1967).

COCONINO COUNTY: Black Peak District, Black Peak, northwest of Tuba City and southeast of the Gap, as an abundant alteration product of mafic dike rocks associated with the Black Peak breccia pipe (Barrington and Kerr 1961).

COLORADO PLATEAU REGION: As an abundant constituent of various sedimentary rock units, notably the Chinle Formation (Schultz 1963), and in various units of the Morrison Formation, with illite and chlorite (Keller 1962).

GILA COUNTY: Miami-Inspiration supersystem, Pinto Valley District, abundant in the host rocks of the Castle Dome Mine; Copper Cities District, Copper Cities Mine, where it formed by the hydrothermal alteration of rock-forming silicate minerals (N. P. Peterson 1962).

GREENLEE COUNTY: Copper Mountain supersystem, Morenci District, as yellowish-brown material in small amounts in the less intensely altered rock below the supergene zone, formed by hydrothermal alteration; associated with kaolinite, allophane, and beidellite (Schwartz 1947; Moolick and Durek 1966).

LA PAZ COUNTY: Near Bouse, in bentonite (Bull. 181).

MARICOPA COUNTY: Reported about two miles northeast of Wickenburg, in bentonites, also reported near Phoenix and near Carl Pleasant Dam, in poor-quality bentonites (Bull. 181).

MOHAVE COUNTY: Reported in the southern part of the county and east of the Big Sandy Wash, as bentonite (Bull. 181). Near Kingman, in altered tuff (UA 8859).

PIMA COUNTY: Silver Bell District, as an alteration mineral that replaces feldspar in dacite (Kerr 1951). Pima supersystem, Twin Buttes District, Twin Buttes Mine, as a product of hydrothermal alteration (Stanley B. Keith, pers. comm., 1973).

PINAL COUNTY: Mineral Creek supersystem, Ray District, near Ray and Superior, as bentonite (Bull. 181).

YAVAPAI COUNTY: Thompson Valley, Lyles District, Lyles deposit, between Kirkland and Yava, in W half of S12, T12N R6W, as bentonite; the clay is characterized as being intermediate between normal montmorillonite and hectorite and contains 0.3 to 0.5 percent Li_2O (Norton 1965). Reported near Wagoner (Bull. 181).

YUMA COUNTY: Muggins Clinoptilolite District, reported near Wellton, in zeolitized bentonite tuff (Bull. 181).

MONTROSEITE

Vanadium iron oxide hydroxide: $(V^{3+},Fe^{2+},V^{4+})O(OH)$. An essentially unoxidized vanadium mineral formed with uraninite and sulfide minerals and believed to be of primary origin. It will alter to paramontroseite, corvusite, and melanovanadite.

APACHE COUNTY: Cane Valley District, Monument No. 2 Mine (Johnson 1963). Black Rock Point District, Martin Mine (Bull. 181). Lukachukai District, Mesa 4½ Mine (Bull. 181). Cove Mesa District, Cove Mine (Bull. 181).

NAVAJO COUNTY: Monument Valley District, Monument No. 1 Mine (Evensen and Gray 1958).

#MONTROYDITE

Mercury oxide: HgO. A hydrothermal mineral found in mercury deposits.

YAVAPAI COUNTY: Castle Creek District, found as small bright-red patches at the Tres Amigos prospect near Copperopolis (R. Jenkins, pers. comm., 2007).

MOORHOUSEITE

Cobalt sulfate hydrate: $Co(SO_4)\cdot6H_2O$. A rare water-soluble secondary mineral found as an efflorescence.

COCONINO COUNTY: Cameron District, as crusts on sandstone (3rd ed.).

MORDENITE

Sodium calcium potassium aluminosilicate hydrate: $(Na_2,Ca,K_2)_4(Al_8Si_{40})O_{96} \cdot 28H_2O$. A member of the zeolite group that formed as an alteration product of volcanic glass in tuffs; also formed in fissures and as vesicle fillings in mafic volcanic rocks.

GREENLEE COUNTY: About six miles north of Morenci, in Tertiary volcanic tuffs (Sheppard 1969).

LA PAZ COUNTY: New Water District, near the Eagle Eye Mine, as fine acicular crystals (David Shannon, pers. comm., 1988).

MOHAVE COUNTY: Black Mountains Mordenite District, north side of Union Pass, in tuff and lapilli tuff in the Golden Deer volcanics (Sheppard 1969).

PIMA COUNTY: Sikort Chuapo Mountains, fifteen miles northeast of Ajo, in altered tuff with clinoptilolite (Eyde 1978).

PINAL COUNTY: Midway station, as vesicle fillings in flow rock (Bideaux, Williams, and Thomssen 1960). On Route 177, 5.5 miles south of Superior, with other zeolites (Wise and Tschernich 1975).

YAVAPAI COUNTY: Cottonwood Basin in an altered tuff bed (Eyde 1978).

MORENOSITE

Nickel sulfate hydrate: $Ni(SO_4) \cdot 7H_2O$. A rare secondary mineral formed by the oxidation of nickel-bearing sulfides.

MARICOPA COUNTY: Reported near Wickenburg (possibly the Monte Cristo District) (Bull. 181).

MOTTRAMITE

Lead copper vanadate hydroxide: $PbCu(VO_4)(OH)$. An uncommon secondary mineral formed in the oxidized portions of base-metal deposits and commonly associated with vanadinite and cerussite. The mineral "duhamelite" (3rd ed.) has been shown to be bismuth- and calcium-bearing mottramite, and those localities are included here.

COCHISE COUNTY: Warren District, Bisbee, as crystals in the Higgins Mine (Taber and Schaller 1930; Schaller 1934) and as reniform masses in the Shattuck Mine (Taber and Schaller 1930; R. C. Wells 1937). Tombstone District, Lucky Cuss and Toughnut Mines, as brilliant black crystals (Guild 1911; Hillebrand 1889a); Charleston District, Gallagher vanadium property (UA 6061). Reported in the Pat Hills (Bull. 181). Cochise

District, Texas Canyon Quartz Monzonite, in quartz-tungsten veins (J. R. Cooper and Silver 1964). Prospect northwest of Bowie, "duhamelite" with kettnerite and malachite (3rd ed.). Dos Cabezas Mountains, Teviston District, Comstock Mine, "duhamelite" with bismutite (3rd ed.).

GILA COUNTY: Globe Hills District, Apache Mine, as rich black druses associated with vanadinite, in a few places as freestanding arborescent-botryoidal forms (W. E. Wilson 1971). Banner supersystem, 79 District, 79 Mine, on the fourth level, as crystal druses encrusting wulfenite and mimetite in breccia of the main fault zone (Stanley B. Keith 1972). Green Valley District, at an old mine 5 km southwest of Payson, on the banks of Lousy Gulch; formed in small gold-bearing veins that cut Precambrian greenstones; associated with chrysocolla and hematite as bundles of fibers on corroded chrysocolla or directly on gangue. (This was the type locality for "duhamelite"; S. A. Williams 1981b.)

FIGURE 3.165 Mottramite, 79 Mine, Gila County, 6.1 cm.

MARICOPA COUNTY: Painted Rock District, Rowley Mine, as microcrystals (Bruce Maier, UA x3063).

PIMA COUNTY: Tucson Mountains, Old Yuma District, Old Yuma Mine (Guild 1910, 1911). Empire District, Total Wreck Mine, as microcrystalline masses and as complete pseudomorphs of wulfenite crystals (H. Peter Knudsen, UA x320). South Comobabi Mountains, Cababi District, Mildren and Steppe claims (S. A. Williams 1963). Ajo District, southwest of the New Cornelia Mine, as a chromian variety formed in small quantities as crusts on weathered keratophyre (Gilluly 1937). Pima supersystem, Twin Buttes District, Twin Buttes Mine (Stanley B. Keith, pers. comm., 1973).

PINAL COUNTY: Mammoth District, Mammoth–St. Anthony Mine, as crusts of small pointed crystals (Guild 1911; Galbraith and Kuhn 1940; Bideaux 1980).

SANTA CRUZ COUNTY: Santa Rita supersystem, J. C. Holmes District, J. C. Holmes claim, as microcrystals with vanadinite (Arthur Roe, UA x2942). Unspecified locality near Nogales, as "cuprodescloizite," in fibrous reddish chestnut-brown layers up to 0.5 in. thick, enclosed in crystallized calcite, which may, in turn, be covered by a layer of amber-colored indistinct crystals (Headden 1903; Guild 1911). Tyndall District, Glove Mine, as 3 mm thick khaki-green crusts coating wulfenite (Les Presmyk, pers. comm., 2020).

YAVAPAI COUNTY: Bradshaw Mountains, Mount Union District, at a prospect on the Slate Creek property of Kalium Chemicals, associated with pyromorphite and mimetite (William C. Berridge, pers. comm., 1973).

MROSEITE

Calcium tellurium oxide carbonate: $CaTe^{4+}O_2(CO_3)$. A very rare mineral found in a hydrothermal gold-tellurium deposit.

COCHISE COUNTY: Tombstone District, T. E. I. pit, in brecciated vein quartz, mixed with massive oboyerite; associated with gold, choloalite, and an unknown copper tellurite (3rd ed.).

FIGURE 3.166 Munaka-taite, Tonopah-Belmont Mine, Maricopa County.

#MUNAKATAITE

Lead copper selenium oxide sulfate hydroxide: Pb_2Cu_2 $(Se^{4+}O_3)(SO_4)(OH)_4$. A rare mineral found in the oxidized portions of selenium-bearing lead-copper deposits.

MARICOPA COUNTY: Belmont Mountains, Osborne District, Tonopah-Belmont Mine, as small acicular to stout bladed dark-blue crystals associated with linarite, cerussite, and caledonite (Gibbs and Ascher 2012; R110005). This material was originally identified from this locality as schmiederite, but samples retested after the new species was described confirmed its true identity.

MARICOPA COUNTY: Mammoth District, Mammoth–St. Anthony Mine, as powder-blue acicular crystals, averaging 0.5 mm long, with altered boleite and leadhillite (Meyer, pers. comm., 2009).

FIGURE 3.167 Murdochite, Mammoth Mine, Pinal County.

MURDOCHITE

Copper lead oxide chloride: $Cu_{12}Pb_2O_{15}Cl_2$. A rare secondary mineral formed in the oxidized portion of copper-lead deposits. Associated with wulfenite, in which it may be embedded, and with hemimorphite, willemite, and quartz. The Mammoth–St. Anthony Mine is the type locality.

COCHISE COUNTY: Warren District, Bisbee (UA 7715); Southwest Mine; Shattuck Mine (Graeme 1993).

GILA COUNTY: Banner supersystem, 79 District, 79 Mine, as rare tiny black cubic crystals, up to 0.1 mm in size, associated with plattnerite, aurichalcite, and rosasite (Stanley B. Keith 1972).

MARICOPA COUNTY: Big Horn Mountains, Big Horn District, prospect near the Evening Star Mine as small black cubes on chrysocolla; at the Evening Star Mine as small black crystals with mimetite (Gibbs, pers. comm., 2015). Belmont Mountains, Osborne District, Tonopah-Belmont Mine, as small cubo-octahedrons with willemite (G. B. Allen and Hunt 1988).

PIMA COUNTY: Waterman District, Silver Hill Mine, with plattnerite and aurichalcite (Bideaux, Williams, and Thomssen 1960).

PINAL COUNTY: Mammoth District, Mammoth–St. Anthony Mine, as tiny black octahedra on the surface of and embedded within plates of wulfenite, and on the surfaces of fluorite crystals, with hemimorphite and willemite (Fahey 1955; R180001).

MUSCOVITE

FIGURE 3.168 Muscovite, Santo Niño Mine, Santa Cruz County, 6.8 cm.

Potassium aluminum silicate hydroxide: $KAl_2(Si_3Al)O_{10}(OH)_2$. An important rock-forming member of the mica group that is most abundant in schists and gneisses and in granite pegmatites, where it may form large "books." Many of the references use the term sericite, which is a finely divided shredded variety of muscovite primarily formed by hydrothermal alteration. It is common in porphyry copper deposits. It is an abundant alteration product in the wall rocks of many mineral deposits, especially low-pH phyllic alteration marginal to chalcopyrite mineralization in porphyry copper systems. It is also widely distributed in sediments and sedimentary rocks. Muscovite is an abundant and widely distributed mineral in Arizona; only a few relatively unusual occurrences are listed below.

COCHISE COUNTY: Johnny Lyon Hills, Yellowstone District, at a prospect on the banks of Tres Alamos Wash, as the chromian variety fuchsite, which forms thick folia in altered and silicified limestone. Its color is so vivid that it is likely that the prospect was mistakenly opened up for copper! (3rd ed.) Warren District, Bisbee, in the Sacramento Hill stock (Schwartz 1947, 1958).

GILA COUNTY: Miami-Inspiration supersystem, Pinto Valley District, Castle Dome Mine (N. P. Peterson, Gilbert, and Quick 1946; Schwartz 1947; Creasey 1959); Copper Cities District, Copper Cities deposit (N. P.

Peterson 1954), as a widespread phyllic alteration mineral; Miami and Inspiration orebodies (Schwartz 1947; Olmstead and Johnson 1966).

GRAHAM COUNTY: Gila Mountains, Lone Star District, Safford area, at and near the Safford area in the Safford porphyry copper cluster (R. F. Robinson and Cook 1966; Rose 1970), as a widespread alteration mineral.

GREENLEE COUNTY: Copper Mountain supersystem, Morenci District (Reber 1916; Schwartz 1947, 1958; Creasey 1959).

LA PAZ COUNTY: Trigo Mountains, Yuma Wash District, disseminated in Orocopia Schist and accompanied by chromite (Bull. 181).

MARICOPA COUNTY: White Picacho District, as an abundant mineral in pegmatites (Jahns 1952). Crusher mica quarry, as well-formed single crystals in pegmatite (Ken Phillips, pers. comm., 1989). White Tank Mountains, Caterpillar testing grounds, as crystals over 12 in. across (3rd ed.).

MOHAVE COUNTY: Unspecified locality north of the Colorado River, in clear transparent sheets up to 6 by 10 in. (Engineering and Mining Journal 1892a). Cerbat Mountains, Wallapai supersystem, associated with sulfide-bearing veins (B. E. Thomas 1949).

PIMA COUNTY: Ko Vaya Hills, north of Sells, as the variety fuchsite (UA 4174). San Antonio Mica Mine near Ajo, as large masses (3rd ed.). Ajo District, New Cornelia Mine area (Gilluly 1937; Schwartz 1958; Creasey 1959). Silver Bell District, as the most widespread phyllic alteration product (Kerr 1951). Santa Catalina Mountains, Summerhaven, in garnet schlieren outcrops (Force 1997), as large euhedral "books" up to 2.5 in. across in microcline-garnet pegmatites, locally displaying stacked aggregates decreasing in size toward the direction of crystallization (Stanley B. Keith, pers. comm., 2019).

PINAL COUNTY: Willow Springs Ranch in Oracle, as pseudomorphs after tourmaline (UA 8617). San Manuel District, San Manuel orebody (Schwartz 1947; Creasey 1959); Kalamazoo orebody (J. D. Lowell 1968). Mineral Creek supersystem, Ray District, at Ray Mine, as extensive phyllic alteration with pyrite and chalcopyrite (Schwartz 1947, 1952, 1959; Rose 1970).

SANTA CRUZ COUNTY: Querces District, Santo Niño Mine (B. Muntyan, pers. comm., 2020).

YAVAPAI COUNTY: Weaver Mountains, near Peeples Valley, as segregations in Yavapai Schist (Bull. 181). Bradshaw Mountains, as segregations in a pegmatite dike that extends about five miles, from Middleton to Horsethief Basin (Bull. 181). White Picacho District, as an accessory mineral in pegmatites (Jahns 1952). Copper Basin District (Johnston and Lowell 1961). Iron King District, Iron King Mine (Creasey 1952;

Moxham, Foote, and Bunker 1965). Eureka supersystem, Bagdad District, Bagdad area (Schwartz 1947; C. A. Anderson 1950; Creasey 1959; Moxham, Foote, and Bunker 1965); as coarse crystals with quartz, pyrite, and molybdenite in the northeast part of the open pit (Barra et al. 2003). Jerome supersystem, Verde District, United Verde Mine (Moxham, Foote, and Bunker 1965).

NACRITE

Aluminum silicate hydroxide: $Al_2Si_2O_5(OH)_4$. An uncommon member of the kaolinite group.

GREENLEE COUNTY: Steeple Rock District, northeast of Duncan, as a widespread constituent of solfatarically altered dacites and quartz latites; associated with diaspore in some places (3rd ed.).

#NAMIBITE

Copper bismuth oxide vanadate hydroxide: $Cu(BiO)_2$ $(VO_4)(OH)$. A rare secondary mineral found in bismuth-bearing hydrothermal deposits.

COCHISE COUNTY: Warren District, as very small deep grassy-green elongated crystals with tangeite and chrysocolla (J. McGlasson, pers. comm., 2021).

GILA COUNTY: Central Dripping Spring Mountains, Troy supersystem, C and B–Grey Horse District, C and B Vanadium Mine, as small dark-green crystals associated with descloizite, cerussite, and other secondary lead minerals (J. Ruiz, pers. comm., 2014).

FIGURE 3.169 Namibite, Copperopolis, Yavapai County.

MARICOPA COUNTY: Flying Saucer District, Scorpion Hill, as very small dark-green elongated crystals with kettnerite, chrysocolla, powellite, laumontite, and bismuth-rich mottramite (Gibbs, pers. comm., 2013).

YAVAPAI COUNTY: Castle Creek District, Copperopolis, as small lustrous dark-green, equant crystals with bismuth-rich mottramite, beyerite, and alunite (J. Ruiz, pers. comm., 2012).

NANTOKITE

Copper chloride: CuCl. A secondary mineral found in the oxide portions of copper deposits, associated with copper, cuprite, atacamite, and hematite.

COCHISE COUNTY: Warren District, Bisbee, Southwest Mine, as clear paraffin-like masses impregnating the cores of vuggy cuprite nodules; under humid conditions, it changes to paratacamite powder (Graeme 1993).

NATROALUNITE

Sodium aluminum sulfate hydroxide: $NaAl_3(SO_4)_2(OH)_6$. A relatively rare member of the alunite group formed during alunitization, a process of solfataric action, commonly accompanied by kaolinization and silicification.

LA PAZ COUNTY: Sugarloaf Peak, Sugarloaf Pyrophyllite District, five miles west-southwest of Quartzsite, as large porcelanic masses in highly argillized, sheared metaplutonic rocks (Omori and Kerr 1963; UA 9348).

#NATROCHALCITE

Sodium copper sulfate hydroxide hydrate: $NaCu_2(SO_4)_2(OH)\cdot H_2O$. A rare secondary mineral formed in oxidized copper deposits in arid climates.

COCHISE COUNTY: Tombstone District, Empire Mine, as emerald-green crystalline crusts (B. Murphy, pers. comm., 2018).

#NATRODUFRÉNITE

Sodium iron phosphate hydroxide hydrate: $NaFe^{2+}Fe^{3+}_5(PO_4)_4(OH)_6\cdot 2H_2O$. A rare secondary mineral related to dufrenite.

GRAHAM COUNTY: Safford supersystem, Lone Star District. Found in drill core as warty deep-green to black crusts on open fracture surfaces with apatite, calcite, and stilbite (Sid Williams, pers. comm., 2005).

NATROJAROSITE

Sodium iron sulfate hydroxide: $NaFe^{3+}_3(SO_4)_2(OH)_6$. Isostructural with jarosite and found with other sulfates, as an oxidation product of pyrite.

GILA COUNTY: Reported in the Globe area, exact locality unknown (Bull. 181).

MOHAVE COUNTY: McConnico District, Georgia Sunset claim, four miles south of Kingman, as compact to earthy golden-brown to yellow masses made up of tabular crystals (E. V. Shannon and Gonyer 1927).

NATROLITE

Sodium aluminosilicate hydrate: $Na_2(Si_3Al_2)O_{10}\cdot2H_2O$. A member of the zeolite group found in vesicular volcanic rocks.

COCHISE COUNTY: Warren District, Bisbee, Lowell shaft, in an andesite dike (Graeme 1981). Tombstone District, Lucky Cuss Mine (3rd ed.)

GRAHAM COUNTY: Aravaipa supersystem, Grand Reef District, Grand Reef Mine, with linarite (R. W. Jones 1980).

MARICOPA COUNTY: Horseshoe Dam District, one mile south of Horseshoe Dam, in a road cut and in boulders along the Verde River, abundant in basalt with crystals up to 1 cm long in radiating hemispheres (D. M. Shannon 1983b).

MOHAVE COUNTY: On U.S. Highway 93, near mile marker 446, seven miles southeast of Hoover Dam, in a kaersutite-camptonite dike (Bideaux, Williams, and Thomssen 1960).

YAVAPAI COUNTY: Eureka area, south of Hillside Mine, in vesicular basalt about ten miles south of Bagdad, in S14 and 15, T13N R8W (William Hunt, pers. comm., 1984).

FIGURE 3.170 Natrolite, Horseshoe Dam, Maricopa County, 7.2 × 5.3 cm.

#NATROSULFATOUREA

Sodium sulfate urea: $Na_2(SO_4)[CO(NH_2)_2]$. A secondary mineral formed in deposits of bat guano; the Rowley Mine is the type locality.

MARICOPA COUNTY: Painted Rock District, Rowley Mine, as colorless prisms up to about 0.3 mm in length with aphthitalite, biphosphammite, and urea in an area with bat guano (Kampf, Celestian, et al. 2020b).

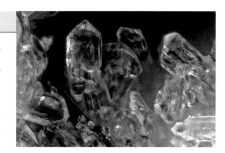

FIGURE 3.171 Natrosulfatourea, Rowley Mine, Maricopa County, FOV 0.29 mm.

#NATROZIPPEITE

Sodium uranyl sulfate oxide hydroxide hydrate: $Na_5(UO_2)_8(SO_4)_4O_5$ $(OH)_3\cdot12H_2O$. The sodium analog of zippeite; its properties, mode of occurrence, and associated minerals are very similar to those of that mineral. The name has been changed since it appeared in the third edition as sodium zippeite.

GILA COUNTY: Sierra Ancha District, reported in the Sue Mine; Granger and Raup (1969) originally reported it as the sodium analog of zippeite;

C. Frondel et al. (1976) subsequently established the mineral as a distinct species.

YAVAPAI COUNTY: Eureka supersystem, Hillside District, Hillside Mine, with other secondary uranium minerals (C. Frondel et al. 1976).

NAUMANNITE

Beta silver selenide: ß-Ag_2Se. An uncommon mineral associated with sulfides and other selenium minerals in hydrothermal mineral deposits.

COCONINO COUNTY: Prospect Canyon (Ridenour) District, Ridenour Mine, with vésigniéite, bromargyrite, tyuyamunite, metatyuyamunite, and tangeite (Wenrich and Sutphin 1989).

LA PAZ COUNTY: Near Quartzsite, as small beads within massive djurleite in quartz veins (S. A. Williams and Matter 1975).

NAVAJOITE

Vanadium iron oxide hydrate: $(V^{5+},Fe^{3+})_{10}O_{24} \cdot 12H_2O$. An uncommon secondary mineral formed as impregnations of sandstone, associated with oxidized vanadium-uranium minerals. The Monument No. 2 Mine is the type locality.

APACHE COUNTY: Cane Valley District, Monument No. 2 Mine, as dark-brown silky to fibrous minute columns normal to fracture surfaces in channel fillings in Shinarump Conglomerate, with a host of other secondary uranium and vanadium minerals (Weeks, Thompson, and Sherwood 1954, 1955; Finnell 1957; M. Ross 1959; Witkind and Thaden 1963; R. G. Young 1964).

FIGURE 3.172 Navajoite, Monument No. 2 Mine, Apache County, 4.0 × 3.3 cm.

NEKOITE

Calcium silicate hydrate: $Ca_3Si_6O_{15} \cdot 7H_2O$. A dimorph of okenite found in contact-metamorphosed limestone deposits.

GRAHAM COUNTY: Aravaipa supersystem, Iron Cap District, Iron Cap Mine, near Landsman Camp, in contact-metamorphosed limestone as white radiating fibers filling fractures in quartz-epidote hornfels (3rd ed.).

FIGURE 3.173 Nekoite, Iron Cap Mine, Graham County, FOV 3 mm.

NEKRASOVITE

Copper vanadium tin sulfide: $Cu_{13}VSn_3S_{16}$. An unusual mineral found with other sulfides in propylitic andesites and dacites.

COCHISE COUNTY: Warren District, Bisbee, Campbell shaft, as blebs up to 0.7 mm in quartz, with pyrite and calcite (Graeme 1993).

NELTNERITE

Calcium manganese oxide silicate: $CaMn^{3+}_6O_8(SiO_4)$. A member of the braunite mineral group. Found in veins and lenses as a secondary mineral associated with braunite, marokite, and crednerite.

COCHISE COUNTY: Warren District, Bisbee, White Tail Deer Mine, as small 1 mm oval crystals embedded in braunite II or at the contact between braunite II cores and braunite rims. Associated minerals are hübnerite, coronadite, and tilasite (Graeme 1993).

NEOTOCITE

Manganese iron silicate hydrate: $(Mn,Fe)SiO_3 \cdot H_2O$. This is a poorly defined species that may form partly through alteration of manganese-bearing silicate minerals.

GRAHAM COUNTY: Aravaipa supersystem, Iron Cap District, Black Hole prospect, in drusy cavities associated with johannsenite, which has replaced limestone containing galena, sphalerite, chalcopyrite, quartz, and calcite (Simons and Munson 1963).

GREENLEE COUNTY: Copper Mountain supersystem, Morenci District, Morenci Mine, (Enders 2000).

PIMA COUNTY: Ajo District, New Cornelia Mine, as black dendrites (W. J. Thomas and Gibbs 1983).

PINAL COUNTY: Mineral Creek supersystem, Ray District, Ray Mine (R. W. Jones and Wilson 1983).

NEPHELINE

Sodium potassium aluminum silicate: $Na_3K(Al_4Si_4O_{16})$. A feldspathoid mineral commonly formed in alkaline igneous rocks and similar rocks thought to be of metasomatic origin.

COCONINO COUNTY: Black Peak District, Black Peak, as small anhedral crystals filling interstices in the augite matrix of mafic dike rocks associated with a breccia pipe (Barrington and Kerr 1961); Tuba District, as anhedral crystals between blades of augite in the monchiquite Tuba dike (Barrington and Kerr 1962).

PIMA COUNTY: Sierrita Mountains, Gunsight Mountain, sparse in a pulaskite dike (Bideaux, Williams, and Thomssen 1960). Esperanza Mine area, in nepheline syenite encountered in an exploratory drill hole (Schmitt et al. 1959).

YAVAPAI COUNTY: House Mountain, just west of the Village of Oak Creek, in nephelinite dikes (Wittke and Holm 1996).

#NESQUEHONITE

Magnesium carbonate hydrate: $Mg(CO_3) \cdot 3H_2O$. An alteration mineral usually formed in coal mines and in serpentine masses.

MARICOPA COUNTY: Gila Bend Mountains, formed from the weathering of weddellite in decaying saguaro cacti (Garvie 2003).

NEYITE

Silver copper lead bismuth sulfide: $Ag_2Cu_6Pb_{25}Bi_{26}S_{68}$. Found in hydrothermal quartz veins.

MARICOPA COUNTY: Gila Bend Mountains, Cortez Peak District, Cortez prospect, as minute beads in vein quartz, with bismuthinite (3rd ed.).

#NICKELBOUSSINGAULTITE

Ammonium nickel sulfate hydrate: $(NH_4)_2Ni(SO_4)_2 \cdot 6H_2O$. A rare secondary nickel mineral often found as efflorescence.

COCONINO COUNTY: Cameron District, from a shallow open cut eight miles east-southeast of Gray Mountain. It is found as a postmine encrustation with hydrohonessite (S. A. Williams and Cesbron 1995).

NICKELINE

Nickel arsenide: NiAs. An uncommon mineral of primary origin found in significant quantities at only a few localities. It is typically associated with cobalt and silver-arsenic minerals and with primary silver.

COCONINO COUNTY: Typically formed with rammelsbergite and pararammelsbergite as an early formed primary phase in breccia pipes mined for uranium (Wenrich and Sutphin 1988).

Cristo Mine, with nickel skutterudite and silver (Bastin 1922).

#NICKELPHOSPHIDE

Nickel phosphide: Ni_3P. Present in numerous iron meteorites; the nickel end member of a solid-solution series with schreibersite.

COCONINO COUNTY: Canyon Diablo area, Canyon Diablo meteorite (Britvin et al. 1999; Jambor, Kovalenker, and Roberts 2000).

NICKELSKUTTERUDITE

Nickel cobalt iron arsenide: $(Ni,Co,Fe)As_3$. Found in moderate-temperature hydrothermal veins.

YAVAPAI COUNTY: Wickenburg Mountains, Monte Cristo District, Monte Cristo Mine, with silver (Bastin 1922).

NICKELZIPPEITE

Nickel uranyl sulfate hydroxide hydrate: $Ni_2(UO_2)_6(SO_4)_3(OH)_{10}\cdot16H_2O$. Probably a secondary mineral partly formed under oxidizing conditions from primary uranium minerals and sulfides.

YAVAPAI COUNTY: Eureka supersystem, Hillside District, Hillside Mine, in very small amounts, associated with other minerals of the zippeite group, especially zippeite proper, natrozippeite, and probably cobaltzippeite, as well as schröckingerite, johannite, bayleyite, and gypsum (C. Frondel et al. 1976).

NITER

Potassium nitrate: $K(NO_3)$. A product of evaporation; formed from guano or by bacterial action on other animal remains. Found in caves or old mine workings in small amounts.

COCONINO COUNTY: Walnut Canyon, as a thin white covering on limestone shelves in ancient cliff dwellings (Guild 1910).

GRAHAM COUNTY: Peloncillo Mountains, in caves (Bull. 181).

PINAL COUNTY: Galiuro Mountains, Aravaipa Canyon, as thin crusts and in cracks in rock below caves (Bull. 181).

NITRATINE

Sodium nitrate: $Na(NO_3)$. A water-soluble mineral formed in arid environments as coatings and efflorescences on rocks and soil.

MARICOPA COUNTY: Superstition Mountains (UA 7838).

MOHAVE COUNTY: Rawhide Mountains, three miles south of Artillery Peak (Bull. 181).

PINAL COUNTY: Gila River Indian Reservation, as a minor constituent of sodium nitrate– and sodium chloride–bearing crusts in salt flats of the Santa Cruz Wash (E. D. Wilson 1969).

NITROCALCITE

Calcium nitrate hydrate: $Ca(NO_3)_2 \cdot 4H_2O$. A water-soluble mineral formed as efflorescences in limestone caverns. Locally, the nitrate is derived from guano.

COCHISE COUNTY: Kartchner Caverns (C. A. Hill 1999).

GILA AND PINAL COUNTIES: Along the Gila River, two miles above Winkelman, in fissures up to 6 to 8 in. wide in Mississippian limestone (Bull. 181). Banner supersystem, Christmas District, Christmas Mine, as white to colorless needles (R 120047).

PINAL COUNTY: Casa Grande area (UA 7837).

NOLANITE

Vanadium iron oxide hydroxide: $V^{3+}_8Fe^{3+}_2O_{14}(OH)_2$. Found in hydrothermal uranium deposits.

COCHISE COUNTY: Warren District, Bisbee, Campbell orebody (Graeme 1993).

NONTRONITE

Sodium iron aluminosilicate hydroxide hydrate: $Na_{0.3}Fe^{3+}_2(Si,Al)_4O_{10}$ $(OH)_2 \cdot nH_2O$. The iron-rich end member of the montmorillonite group of clay minerals. A secondary mineral of uncertain origin, known to form in hydrothermally altered copper deposits.

COCHISE COUNTY: Dos Cabezas Mountains, Teviston District, at a small prospect near the Comstock Mine, partially filling voids in massive white vein quartz; associated with clinobisvanite, bismutite, and bismuth-rich mottramite (3rd ed.). Kartchner Caverns (C. A. Hill 1999).

GILA COUNTY: Miami-Inspiration supersystem, Miami District, Miami Mine, abundant as pseudomorphs after plagioclase phenocrysts, associated with sericite, kaolinite, and illite (Schwartz 1947). Sierra Ancha District, Workman Creek area, associated with sulfide minerals in Dripping Spring Quartzite and Red Bluff Mine (Granger and Raup 1969).

GRAHAM COUNTY: Stanley District, Friend Mine (3rd ed.).

GREENLEE COUNTY: Copper Mountain supersystem, Morenci District, as silky seams in limy shale in the Arizona Central Mine (Schwartz 1947, 1958) and on the periphery of the Morenci orebody (Moolick and Durek 1966).

PIMA COUNTY: Santa Rita Mountains, Helvetia-Rosemont District, Pauline Mine, in metamorphosed wall rock (Schrader and Hill 1915). Ajo District, uncommon as a secondary mineral (Gilluly 1937). Pima supersystem, Twin Buttes District, Twin Buttes Mine, in the supergene zone in collapsed chrysocolla-filled breccias (Stanley B. Keith, pers. comm., 1973).

PINAL COUNTY: Mineral Creek supersystem, Ray District, Ray deposit, in hydrothermally altered quartz monzonite with sericite, illite, and quartz (Schwartz 1952). Mammoth District, impregnating younger sediments on Copper Creek, where it emerges from the box canyon (Herbert E. Hawkes, pers. comm., 1973).

YAVAPAI COUNTY: Black Hills, Jerome supersystem, Verde District, United Verde Mine, with limonite in gossans (Bull. 181).

#NORDSTRANDITE

Aluminum hydroxide: $Al(OH)_3$. This secondary aluminum mineral is found in some soils, in sedimentary rocks, and in miarolitic cavities of igneous rocks.

COCHISE COUNTY: Warren District, Bisbee, with gibbsite and malachite (Graeme, Graeme, and Graeme 2015).

FIGURE 3.174 Nordstrandite, on azurite, Southwest Mine, Bisbee, Cochise County, 21.5 cm.

NORDSTRÖMITE

Lead copper bismuth sulfide selenide: $Pb_3CuBi_7(S,Se)_{14}$. An uncommon sulfosalt of hydrothermal origin.

COCHISE COUNTY: Northeast side of the Johnny Lyon Hills, in vein quartz (3rd ed.).

NSUTITE

Manganese oxide hydroxide: $Mn^{2+}_x Mn^{4+}_{1-x} O_{2-2x}(OH)_{2x}$, with the value of x being small. A fairly widespread secondary manganese mineral.

SANTA CRUZ COUNTY: Patagonia Mountains, Trench District, Hardshell Mine, with several other manganese oxide minerals (Koutz 1984).

YUMA COUNTY: Kofa Mountains, Kofa District, King of Arizona Mine, No. 5 vein, with groutite (Hankins 1984).

NUKUNDAMITE

Copper iron sulfide: $Cu_{3.4}Fe_{0.6}S_4$. Commonly a primary mineral of hypogene origin, also formed as an alteration product of primary chalcopyrite.

PINAL COUNTY: Pioneer supersystem, Superior area, as polygonal orange grains that resemble stannoidite, in bornite; reported by Ben Leonard, U.S. Geological Survey (John S. White Jr., pers. comm., 1981).

OFFRETITE

Potassium calcium magnesium aluminosilicate hydrate: $KCaMg(Si_{13}Al_5)$ $O_{36} \cdot 15H_2O$. An uncommon mineral of the zeolite group, found in amygdaloidal cavities in basaltic flow rocks, associated with other zeolites.

GREENLEE COUNTY: East side of the San Francisco River, about one mile north of Clifton, as tiny elongate prismatic crystals in amygdules in basalt flows exposed in a road cut (William Hunt and Dan Caudle, pers. comm., 1976).

PINAL COUNTY: On Queen Creek six miles west of Superior, on lévyne, and at Malpais Hill with other zeolites (Wise and Tschernich 1976).

YAVAPAI COUNTY: South side of the Santa Maria River, between U.S. Highway 93 and State Highway 96, in a basalt flow (William Hunt, pers. comm., 1985).

OKENITE

Calcium silicate hydrate: $Ca_{10}Si_{18}O_{46} \cdot 18H_2O$. Formed as amygdule fillings in eruptive volcanic rocks and often found with zeolites.

MARICOPA COUNTY: Horseshoe Dam District, near Horseshoe Dam, as bladed fibrous white microcrystals that line vug walls, associated with zeolites (D. M. Shannon 1983b).

OLIVENITE

Copper arsenate hydroxide: $Cu_2(AsO_4)(OH)$. A secondary mineral formed in the oxidized zones of base-metal mineral deposits, where it is associated with other secondary lead and copper minerals.

COCHISE COUNTY: Tombstone District (UA 6439, 8240). Oregon-Prompter Mine Group (Jenkins, pers. comm., 2014).

COCONINO COUNTY: Grand Canyon National Park, Horseshoe Mesa, Grandview District, Grandview (Last Chance) Mine, as short olive-green prismatic crystals and lighter-colored acicular groups, on altered metazeunerite (Leicht 1971).

GILA COUNTY: Banner supersystem, 79 District, 79 Mine, as clusters of tiny green crystals on sphalerite and galena, associated with brochantite, siderite, chalcanthite, and anglesite (Stanley B. Keith 1972; Thomas Trebisky, pers. comm., 1972).

GRAHAM COUNTY: Aravaipa supersystem, Grand Reef District, Grand Reef Mine, as acicular green tufts on azurite (R. W. Jones 1980).

GREENLEE COUNTY: Copper Mountain supersystem, Morenci District, Morenci Mine, (Enders 2000).

PIMA COUNTY: Ajo District, New Cornelia Mine, as small crystals with ajoite (W. J. Thomas and Gibbs 1983).

PINAL COUNTY: Galiuro Mountains, Bunker Hill (Copper Creek) District, Copper Creek, Old Reliable Mine, as small olive-green crystals (Bull. 181). Pioneer supersystem, Magma District, Magma Mine, in the outcrop at the No. 1 glory hole, as small crystals with dioptase (Barnes and Hay 1983).

YAVAPAI COUNTY: Copper Mountain, Agua Fria District, Old Robertson claims, with clinoclase (Robert O'Haire, pers. comm., 1972); Copper Queen Mine, as small well-formed crystals with clinoclase (Gibbs, pers. comm., 2017). Eureka supersystem, Bagdad District, Bagdad Mine, as microcrystals on chrysocolla (William Hunt, pers. comm., 1985).

FIGURE 3.175 Olivenite, Copper Queen Mine, Yavapai County, FOV 3 mm.

OMPHACITE

Calcium sodium magnesium iron aluminum silicate: $(Ca,Na)(Mg,Fe,Al)Si_2O_6$. A member of the pyroxene group of rock-forming silicate minerals, whose presence indicates high-pressure origins.

APACHE COUNTY: Garnet Ridge SUM District, associated with garnet and lawsonite in eclogite inclusions in kimberlite pipes (Watson and Morton 1969).

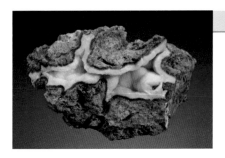

FIGURE 3.176 Opal, Bagdad Mine, Yavapai County, 8.8 cm.

#OPAL

Silicon oxide hydrate: $SiO_2 \cdot nH_2O$. Most opal is deposited from hot spring, underground, and surface waters. Common opal is also produced as a late low-temperature mineral in hydrothermal mineral deposits, or by the weathering of silicate minerals, especially in the oxidized caps of porphyry copper deposits. Opal may also be produced by the accumulation of the tests of silica-secreting freshwater or marine animals, such as diatoms. Opal is not a mineral in the strict definition as it is composed of either cristobalite or amorphous silica but is recognized by the IMA as a mineral species. Only a few localities are listed below, as opal is very common as a late mineral in hydrothermal deposits and silica-rich volcanic rocks.

COCHISE COUNTY: Tombstone District, Lucky Cuss Mine, as small seams of common opal in ore, presumed to be of late-stage hydrothermal origin (B. S. Butler, Wilson, and Rasor 1938).

GILA COUNTY: Miami-Inspiration supersystem, Miami-Inspiration District, Inspiration Mine, as common opal, in the oxidized zone as thin layers and in vugs, with chrysocolla and malachite (Sun 1963). Sierra Ancha District, as fluorescent hyalite, at several uranium deposits in the Dripping Spring Quartzite (Granger and Raup 1969).

GREENLEE COUNTY: Copper Mountain supersystem, Morenci District, in monzonite porphyry, where it forms in cavities left after the leaching of pyrite (Reber 1916), especially on the periphery of the Morenci orebody (Moolick and Durek 1966).

MARICOPA COUNTY: North slope of Saddle Mountain, south-southwest of Tonopah, as common opal (Bull. 181). Near Castle Hot Springs, in pink rhyolite, as precious opal (Bull. 181). Colorful precious opal from Morgan City Wash, near Lake Pleasant, specimen from A. L. Flagg Collection.

MOHAVE COUNTY: Union Pass, west of Kingman, as tiny specks of fire opal. Eastern slopes of the Black Mountains, northwest of Kingman, as hyalite (Bull. 181).

PIMA COUNTY: Ajo District, associated with shattuckite, ajoite, and quartz (Sun 1961). Silver Bell District, Silver Bell Mine, as common opal veinlets in the oxidized zone, with jarosite (Kerr 1951). Pima District, Copper Glance Mine, as common opal formed in the old dump (Therese Murchison, pers. comm., 1972).

PINAL COUNTY: Picketpost Mountain, near Superior, as hyalite in cavities in dacite (UA 6879).

SANTA CRUZ COUNTY: Grosvenor Hills, near Santa Cruz, as hyalite (Bull. 181). Jay R claim and Arizona blue fire opal claim, in S25, T22S R11E, as commercially mined blue and precious opal (Lurie 2012).

YAVAPAI COUNTY: Bradshaw Mountains, fourteen miles from Mayer on the Agua Fria River, as common opal with chalcedony (Hewett, Fleischer, and Conklin 1963). Eureka supersystem, Bagdad District, Bagdad Mine, as common opal in quartz veins (Bull. 181). East of the Black Canyon Road, near Moore Wash, as common opal (Bull. 181).

YUMA COUNTY: Muggins District, Red Knob claims, associated with secondary uranium minerals (Outerbridge et al. 1960).

#ORCELITE

Nickel arsenide: $Ni_{5-x}As_2$ ($x \approx 0.25$), An uncommon mineral found in serpentinites with sulfides and sulfosalts.

LA PAZ AND YUMA COUNTIES: Cemetery Ridge District, as small grains with other sulfides in magnetite in blocks of serpentinite and partially serpentinized harzburgite in the Orocopia Schist at Cemetery Ridge (Haxel et al. 2018).

ORPIMENT

Arsenic sulfide: As_2S_3. A secondary mineral formed by the alteration of other arsenic-bearing minerals. It is typically associated with realgar, but less common.

PINAL COUNTY: In 1915, several pounds of orpiment and realgar were discovered at an unidentified locality near the junction of the Gila River and Hackberry Wash (Bull. 181).

ORTHOCLASE

Potassium aluminum silicate: $K(AlSi_3O_8)$. An important rock-forming alkali feldspar. An abundant constituent of felsic igneous rocks and detrital sedimentary rocks derived from them. It may also be formed by the

action of hydrothermal solutions on wall rocks of mineral deposits. Widely distributed in the felsic igneous rocks of the state, especially in metaluminous granites, where phenocrysts up to several inches long have been found. An important alteration mineral in potassic alteration zones with phlogopite, chalcopyrite, and molybdenite in porphyry copper deposits.

COCHISE COUNTY: Cochise supersystem, Texas Canyon Quartz Monzonite stock, as phenocrysts up to several inches long and as abundant excellent examples of Carlsbad twins (3rd ed.).

MARICOPA COUNTY: Cave Creek District, in pipes up to 175 ft. in diameter in porphyritic granite southeast of the town of Cave Creek (Bull. 181). Gila Bend Mountains, Quail Wash (S. Celestian, pers. comm., 2020).

MOHAVE COUNTY: Reported near Hackberry, as thick veins (Bull. 181).

PIMA COUNTY: Ajo District, New Cornelia orebody, as crystals up to several inches across, as part of a massive replacement of quartz monzonite (Gilluly 1937). Silver Bell District, as a late-stage alteration mineral (Kerr 1951). Northwest Santa Catalina Mountains, Catalina Quartz Monzonite, as large porphyrocrysts up to 1.5 in. long (Hoelle 1974; Force 1997).

PINAL COUNTY: Mineral Creek supersystem, Teapot District, Crystal Pass, northwest of Ray, as small euhedra, commonly twinned after the Carlsbad and other laws (3rd ed.) as porphyrocrysts in Teapot Mountain quartz-feldspar porphyry (Creasey, Peterson, and Gambell 1983); reported near Kearny, as twinned crystals (3rd ed.); North Butte area, Box O District, as single and twinned crystals in a quartz-feldspar porphyry dike (Ray Grant, pers. comm., 2019). Northern Santa Catalina Mountains, as Carlsbad twins (UA 4034, 4035), in Rice Peak Granodiorite. Bunker Hill (Copper Creek) District, Childs-Aldwinkle Mine, as crystals up to 6 in. long, in a pegmatite zone at the base of the Childs-Aldwinkle breccia pipe (Kuhn 1941).

YAVAPAI COUNTY: Bradshaw Mountains, near the old townsite of Middleton on the Crown King Road, as relatively pure material (Bull. 181).

#ORTHOSERPIERITE

Calcium copper sulfate hydroxide hydrate: $CaCu_4(SO_4)_2(OH)_6 \cdot 3H_2O$. Found as postmining encrustations on mine workings from the alteration of polymetallic sulfide deposits.

PINAL COUNTY: Copper Creek District, Childs-Aldwinkle Mine, forms small clusters of small blue crystals and blue crusts on rocks outside an adit where mine waters have seeped out (D. M. Shannon 1996). Mammoth District, Mammoth–St. Anthony Mine, as small blue crystal laths with cerussite (P. Haynes, pers. comm., 2003). Riverside District, Black Copper Wash, west of Kelvin breccia pipe, as bright-blue encrustations at a seep near the head of the wash (Barton et al. 2007).

FIGURE 3.178 Orthoserpierite, Childs-Aldwinkle Mine, Pinal County, FOV 5 mm.

OSARIZAWAITE

Lead aluminum copper sulfate hydroxide: $Pb(Al_2Cu^{2+})(SO_4)_2(OH)_6$. An uncommon mineral formed in the oxidized zone of some base-metal veins.

COCHISE COUNTY: Warren District, Bisbee, as massive pistachio-green material in goethite gossans at several localities; particularly common at the Shattuck Mine (3rd ed.); on the dumps at the Copper Prince Mine (Graeme 1981). Tombstone District, Emerald Mine, as bluish-green botryoidal crystalline masses (Murphy, pers. comm., 2018).

FIGURE 3.179 Osarizawaite, Silver Hill Mine, Pima County.

PIMA COUNTY: Helvetia-Rosemont District, Omega Mine, as microcrystalline yellow-green crusts (David Shannon, UA x4351). Waterman District, Silver Hill Mine, as attractive blue massive to microcrystalline material (Stolburg 1988; Hal Frazier, UA x2146).

YAVAPAI COUNTY: Silver Mountain District, Fat Jack Mine, with stolzite, cerussite, pyromorphite, and mottramite (Modreski and Scovil 1990; Scovil and Wagner 1991).

#OTAVITE

Cadmium carbonate: $Cd(CO_3)$. A secondary mineral formed in the oxidized portions of orebodies rich in cadmium.

COCHISE COUNTY: Huachuca Mountains, Reef District, Big Dipper and Reef Mines, as milk-white masses and small translucent rhombohedral crystals associated with galena, sphalerite, corkite, cerussite, and phosphates in quartz veins (Walstrom 2013).

FIGURE 3.180 Ottoite, Grand Central Mine, Tombstone, Cochise County.

R130126   1 mm

#OTTOITE

Lead tellurate: Pb_2TeO_5. A rare mineral found in oxidized lead deposits rich in tellurium.

COCHISE COUNTY: Tombstone District, Tombstone, Grand Central Mine, as crème white wheat-sheave aggregates of crystals to 3 mm associated with tellurium raspite (Murphy, pers. comm., 2013; S. A. Williams 1979).

#PACEITE

Calcium copper acetate hydrate: $CaCu(CH_3COO)_4 \cdot 6H_2O$. A very rare naturally formed acetate mineral.

COCHISE COUNTY: Warren District, Bisbee, Holbrook Mine, found as small euhedral crystals on native copper and on gypsum with malachite associated with hoganite (Jim McGlasson, pers. comm., 2019). The specific mode of formation for this occurrence is not known.

PALYGORSKITE

Magnesium aluminum silicate hydroxide hydrate: $(Mg,Al)_2Si_4O_{10}(OH) \cdot 4H_2O$. A secondary mineral sometimes known by the synonym *attapulgite*, it is commonly classified as a clay mineral because of its finely divided habit in soils. It may also be formed by hydrothermal activity. A variety consisting of intertwined fibers is termed *mountain leather.*

COLORADO PLATEAU: In certain zones of the Petrified Forest and Owl Rock Members of the Chinle Formation (Schultz 1963).

GILA COUNTY: Globe Hills District, Black Copper pit, Inspiration Consolidated Copper Company, along foliation in gouge near the dacite hanging wall of the Keystone fault system, as irregular white to pale-lavender slickenside seams up to 0.5 in. thick that apparently cement gouge fragments (Bladh 1973); Apache Mine (Brian Bond, UA x3325).

PIMA COUNTY: Ajo District, New Cornelia Mine (W. J. Thomas and Gibbs 1983). Pima supersystem, Twin Buttes District, Twin Buttes Mine (William Hefferon, pers. comm., 3rd ed.).

PINAL COUNTY: Mammoth District, Mammoth–St. Anthony Mine, in some fault gouges (Bideaux 1980).

SANTA CRUZ COUNTY: Patagonia Mountains, Washington Camp District, South Empire and Silver Bell Mine areas, as tough matted fibrous masses on fault surfaces (Lehman 1978).

PAPAGOITE

Calcium copper aluminum silicate hydroxide: CaCu AlSi$_2$O$_6$(OH)$_3$. A rare secondary mineral found in the oxide zone of copper deposits. The New Cornelia Mine is the type locality.

PIMA COUNTY: Ajo District, New Cornelia Mine, as elongated, somewhat flattened cerulean-blue crystals in narrow veinlets and in veneers on slip surfaces in altered granodiorite porphyry, associated with aurichalcite, shattuckite, ajoite, baryte, and iron oxides (Hutton and Vlisidis 1960; Guillebert and Le Bihan 1965; UA 168; NHM 1959, 192).

FIGURE 3.181 Papagoite, New Cornelia Mine, Pima County, FOV 2 mm.

#PARACOQUIMBITE

Iron sulfate hydrate: Fe^{3+}$_4$(SO$_4$)$_6$(H$_2$O)$_{12}$·6H$_2$O. Found as a rare postmining mineral.

COCHISE COUNTY: Warren District, Bisbee, in the Campbell Mine, as small bright pale-violet crystals with coquimbite, voltaite, rhomboclase, and copiapite; Higgins Mine, 100 level, as violet scepter growths to 0.5 mm in size with coquimbite and halotrichite (Graeme, Graeme, and Graeme 2015).

PARAGONITE

Sodium aluminum silicate hydroxide: NaAl$_2$(Si$_3$Al)O$_{10}$(OH)$_2$. A member of the mica group that resembles muscovite and is mostly found in metamorphic rocks. There would likely be more localities of this mineral if chemical analyses were made.

YAVAPAI COUNTY: Governors Peak quadrangle, in schist (Shafiqullah et al. 1980).

PARALAURIONITE

Lead chloride hydroxide: PbCl(OH). This uncommon secondary mineral is the monoclinic dimorph of laurionite. It is formed in the oxidized zones of lead-copper deposits.

MARICOPA COUNTY: Belmont Mountains, Osborne District, Tonopah-Belmont Mine, found sparingly as minute clear bladed crystals and as clear glassy encrustations with cerussite and linarite (Gibbs and Ascher 2012).

FIGURE 3.182 Paralaurionite, Mammoth–St. Anthony Mine, Pinal County, 4.1 cm.

PINAL COUNTY: Mammoth District, Mammoth–St. Anthony Mine, Collins vein, as small slender isolated yellowish-white crystals in cavities with cerussite, and as coarser crystal aggregates with leadhillite and diaboleite; characterized by extremely good cleavage and some flexibility; many crystals are bent (Fahey, Daggett, and Gordon 1950; Palache 1950; Bideaux 1980).

PARAMELACONITE

Copper oxide: $Cu^{1+}_2Cu^{2+}_2O_3$. A rare secondary mineral found in hydrothermal copper deposits. The Copper Queen Mine at Bisbee is the type locality.

COCHISE COUNTY: Warren District, Bisbee, Copper Queen Mine, in a matrix of goethite, associated with cuprite, copper, tenorite, malachite, and connellite, as unusually large crystals (G. A. Koenig 1891; C. Frondel 1941; O'Keefe and Bovin 1978).

PARAMONTROSEITE

FIGURE 3.183 Paramelaconite, Copper Queen Mine, Bisbee, Cochise County, 1.9 cm high.

Vanadium oxide: VO_2. A rare secondary mineral formed by the oxidation of montroseite.

APACHE COUNTY: Cane Valley District, Monument No. 2 Mine, as shiny black crystals up to 0.5 mm (Evensen and Gray 1958; Witkind and Thaden 1963; R. G. Young 1964).

PARARAMMELSBERGITE

Nickel arsenide: $NiAs_2$. Formed in nickel- and cobalt-bearing hydrothermal veins, it is dimorphous with rammelsbergite.

COCONINO COUNTY: In breccia pipes mined for uranium; may be associated with its dimorph rammelsbergite and nickeline (Wenrich and Sutphin 1988). Grand Canyon National Park, Orphan District, Orphan Mine (Gornitz 2004).

PARATACAMITE

Copper zinc hydroxide chloride: $Cu_3(Cu,Zn)Cl_2(OH)_6$. A secondary mineral found at numerous localities; the rhombohedral dimorph of atacamite.

COCHISE COUNTY: Warren District, Bisbee, Holbrook and Cole Mines, where it is associated with cuprite, malachite, and azurite but is rare (Graeme 1981); Southwest Mine, as crystals up to 1.5 cm long, probably the finest examples of the species (Graeme 1993).

MARICOPA COUNTY: Belmont Mountains, Osborne District, Tonopah-Belmont Mine, found with bechererite, rosasite, hydrozincite, willemite and boleite (Giester and Rieck 1996)

PIMA COUNTY: South Comobabi Mountains, Cababi District, Mildren Mine, as fibrous and granular masses embaying linarite and as pseudomorphs after atacamite in quartz veins cutting andesite (S. A. Williams 1962, 1963).

PINAL COUNTY: Mammoth District, Mammoth–St. Anthony Mine, with cerussite, boleite, and diaboleite (Bideaux 1980; UA 5154).

PARATELLURITE

Tellurium oxide: TeO_2. A fairly rare oxide mineral formed in acid waters from tellurides or tellurium. It may be easily mistaken for anglesite in appearance.

COCHISE COUNTY: Warren District, Bisbee, in the ores of the Campbell Mine (Graeme 1993). Tombstone District, Joe Mine, in numerous samples of partly oxidized pyritic telluride ore, as crystals up to 3 mm long in cavities, with anglesite, rodalquilarite, and emmonsite (3rd ed.).

PARGASITE

Sodium calcium magnesium aluminum silicate hydroxide: $NaCa_2(Mg_4Al)$ $(Si_6Al_2)O_{22}(OH)_2$. A member of the amphibole group of rock-forming silicate minerals mostly found in metamorphosed rocks.

COCHISE COUNTY: San Bernardino Valley, San Bernardino (Geronimo) volcanic field, associated with ultramafic inclusions (Kempton 1983).

GILA COUNTY: San Carlos Apache Reservation, Peridot District, Peridot Mesa, associated with Group I forsterite-rich ultramafic inclusions (Frey and Prinz 1978).

PARISITE

Calcium cerium lanthanum carbonate fluoride: $(CaCe_2,CaLa_2)(CO_3)_3F_2$. Used here as a group name. The relative amounts of the rare-earth elements may be unknown; thus, the suffix -(Ce) or -(La) cannot always be assigned to local species. A fairly widespread mineral formed in several geological environments: pegmatites, certain high-sodium granites, and veinlets in shales.

MOHAVE COUNTY: White Hills, Gold Basin District, in an unnamed prospect, as small crystals in miarolitic cavities in small leucosyenite intrusive bodies, with fluorite and gold (Blacet 1969).

#PARKINSONITE

Lead molybdate chloride: $Pb_7MoO_9Cl_2$. A rare mineral found in oxidized ores.

MARICOPA COUNTY: Painted Rock District, Rowley Mine, as cherry red pyramidal needle-like crystals (Bruce Murphy, pers. comm., 2018).

FIGURE 3.184 Parnauite, Grandview (Last Chance) Mine, Coconino County, FOV 2 mm.

PARNAUITE

Copper arsenate sulfate hydroxide hydrate: $Cu_9(AsO_4)_2$ $(SO_4)(OH)_{10}\cdot 7H_2O$. A rare secondary mineral formed by oxidation of primary copper-bearing ores at low temperatures.

COCONINO COUNTY: Grandview District, Grandview (Last Chance) Mine on the Grand Canyon's South Rim, as a rare member of an extensive suite of secondary sulfates, arsenates, and carbonates formed by the oxidation of primary copper ores. The mineral, designated as "unknown no. 6" in Leicht's (1971) description of the Grandview (Last Chance) Mine, was subsequently identified by William Wise (pers. comm., 1980).

FIGURE 3.185 Parsonsite, near Fat Jack Mine, Yavapai County, FOV 5 mm.

PARSONSITE

Lead uranyl phosphate: $Pb_2(UO_2)(PO_4)_2$. A rare secondary mineral found in hydrothermal uranium deposits.

COCHISE COUNTY: Huachuca Mountains, Hartford District, as pale-yellow crusts on quartz in a small lead prospect (3rd ed.).

YAVAPAI COUNTY: Bradshaw Mountains, from a small prospect near the Fat Jack Mine, where it is found with wulfenite and vanadinite (Gibbs, pers. comm., 2019).

PASCOITE

Calcium vanadium oxide hydrate: $Ca_3V^{5+}_{10}O_{28} \cdot 17H_2O$. A water-soluble secondary mineral formed under surficial oxidizing conditions and locally associated with carnotite.

APACHE COUNTY: Cane Valley District, Monument No. 2 Mine, as water-soluble coatings on the mine walls (Witkind and Thaden 1963). Lukachukai District, Mesa No. 1, No. 5, and No. 6 Mines (Chenoweth 1967). Carrizo Mountains, Black Rock Point District, Zona No. 1 claim (Bull. 181).

PAULKERRITE

Potassium magnesium titanium phosphate hydroxide hydrate: $KMg_2TiFe^{3+}_2(PO_4)_4(OH)_3 \cdot 15H_2O$. A rare mineral formed in granitic pegmatites as small yellow-brown to nearly colorless minute vitreous crystals. The 7U7 Ranch is the type locality.

YAVAPAI COUNTY: 7U7 Ranch, west of Hillside, associated with altered triplite and other secondary phosphate minerals (Peacor, Dunn, and Simmons 1984).

PEARCEITE

Silver copper arsenic antimony sulfide: $[Ag_9CuS_4][(Ag,Cu)_6(As,Sb)_2S_7]$. An uncommon mineral formed under low to moderate temperatures in veins, typically associated with lead and silver sulfosalts, galena, and acanthite.

COCHISE COUNTY: Warren District, Bisbee, Campbell shaft (Graeme 1993; S 100455).

MOHAVE COUNTY: Cerbat Mountains, Wallapai supersystem, Chloride and/or Stockston Hill Districts, reported in small amounts from some of the high-grade silver ores (Bull. 181); Chloride District, Distaff Mine (Bastin 1925).

SANTA CRUZ COUNTY: Patagonia Mountains, Trench District, Trench Mine, 500 ft. level; Flux District, World's Fair Mine (Kartchner 1944).

PECORAITE

Nickel silicate hydroxide: $Ni_3Si_2O_5(OH)_4$. An unusual mineral formed by the weathering of nickel-iron meteorites, in ultramafic rocks and in geodes.

COCHISE COUNTY: Southeastern Dragoon Mountains, Barrett Camp District, at the ghost town of Barrett Camp, in crystalline dolomite; the Colina-Epitaph formations, as bright-green, microcrystalline material that apparently replaced another mineral, which, in turn, had replaced rock fragments of possible fossil hash; the dolomite, despite its degree of crystallinity, exudes a pronounced fetid odor when cut or broken (3rd ed.).

PECTOLITE

Sodium calcium silicate hydroxide: $NaCa_2Si_3O_8(OH)$. A widespread mineral formed in various environments: in cavities in basaltic rocks, commonly with zeolites, and in metamorphosed rocks high in calcium.

COCHISE COUNTY: Warren District, Bisbee, Lowell Mine, 1200 level, 3 drift, found as radiating needle-like crystals to 4 mm with wollastonite, epidote, chlorite, quartz, and pyrite (Tenney 1913).

NAVAJO COUNTY: Navajo volcanic field, in lamprophyre dike rocks (Laughlin, Charles, and Aldrich 1986).

PEKOITE

Copper lead bismuth sulfide: $CuPbBi_{11}S_{18}$. At the type locality in Australia, formed in magnetite-bearing pipes of hydrothermal origin that cut felsic sediments and volcanic rocks; the pipes are probably hydrothermal replacements associated with volcanism.

COCHISE COUNTY: Dos Cabezas Mountains, Teviston District, near the Comstock Mine, as slender prisms up to 8 mm long in vein quartz, associated with pyrite and bismutite; the pekoite is corroded by gladite along crystal margins (3rd ed.).

#PENTAHYDRITE

Magnesium sulfate hydrate: $Mg(SO_4)\cdot5H_2O$. A member of the chalcanthite group, this secondary mineral often occurs with other efflorescent salts.

COCONINO COUNTY: Found as an efflorescent mineral in Grand Canyon National Park (Quick, Corbett, and Manner 1989).

PENTLANDITE

Nickel iron sulfide: $(Ni,Fe)_9S_8$. A primary mineral commonly associated with pyrrhotite; in places, thought to be a product of basic magmatic segregation. It is an important nickel ore.

LA PAZ AND YUMA COUNTIES: Cemetery Ridge District, as small grains with other nickel minerals in magnetite in blocks of serpentinite and partially serpentinized harzburgite in the Orocopia Schist at Cemetery Ridge (Haxel et al. 2018).

MOHAVE COUNTY: Reported near Littlefield, in mafic dikes with pyrrhotite, chalcopyrite, and pentlandite (Bull. 181).

PERICLASE

Magnesium oxide: MgO. A high-temperature metamorphic mineral found in dolomitic marbles.

COCHISE COUNTY: Tombstone District (3rd ed.).

MOHAVE COUNTY: Black Mountains, Mag District, about three miles northwest of Oatman, as corroded relicts in marble, with brucite (3rd ed.).

PERITE

Lead bismuth oxide chloride: $PbBiO_2Cl$. A rare secondary mineral found in the oxidized portions of base-metal deposits rich in bismuth.

COCHISE COUNTY: Johnny Lyon Hills, Yellowstone District, at a prospect on Tres Alamos Wash northeast of Benson, as bright-yellow scaly masses in vugs in contorted quartz veins that cut epidote-plagioclase gneiss, associated with minor amounts of chrysocolla, pyromorphite, and vauquelinite (Sid Williams, pers. comm., 3rd ed.).

PEROVSKITE

Calcium titanium oxide: $CaTiO_3$. Found as a widespread accessory mineral in alkaline and mafic igneous rocks and as a mineral of deuteric origin.

PIMA COUNTY: Sierrita Mountains, Gunsight Mountain, as polysynthetically twinned cores of titanite crystals in a pulaskite dike (Bideaux, Williams, and Thomssen 1960).

#PERROUDITE

Silver mercury sulfide halide: $Ag_4Hg_5S_5(I,Br)_2Cl_2$. An uncommon mineral created by the oxidization of sulfides.

LA PAZ COUNTY: Trigo Mountains, Silver District, Red Cloud Mine, very small red acicular crystals of perroudite were found in small vugs in masses of altered galena and anglesite in the main vein above the open cut of the mine. The mineral identification is based on Raman spectroscopy and physical characteristics (R. Gibbs, pers. comm., 2016).

FIGURE 3.186 Perroudite, Red Cloud Mine, La Paz County, FOV 4 mm.

#PETERSITE-(Ce)

Copper cerium phosphate hydroxide hydrate: $Cu_6Ce(PO_4)_3(OH)_6\cdot3H_2O$. A rare secondary mineral found in oxidized copper deposits. The Cherry Creek location is the type locality.

YAVAPAI COUNTY: Black Hills, Cherry Creek District, as sprays of very small yellowish-green acicular crystals with malachite, chlorite, a biotite phase, quartz, albite, orthoclase, hematite, chalcopyrite, and a hisingerite-like mineral. The type specimen is a micromount specimen donated to the Mineral Museum of the University of Arizona by Dr. Arthur Roe (Morrison et al. 2016; R050541).

PETZITE

Silver gold telluride: Ag_3AuTe_2. Commonly found in vein deposits of gold associated with other tellurides.

COCHISE COUNTY: Warren District, Bisbee, Campbell Mine, in the massive sulfide orebody (Criddle et al. 1983).

#PHARMACOLITE

Calcium arsenate hydroxide hydrate: $Ca(AsO_3OH)\cdot2H_2O$. A secondary mineral formed from other arsenic minerals.

MOHAVE COUNTY: Found at Corkscrew Cave on the Hualapai Indian Reservation as green nodular speleothems with hörnesite and talmessite (Onac, Hess, and White 2007).

PHARMACOSIDERITE

Potassium iron arsenate hydroxide hydrate: $KFe^{3+}_4(AsO_4)_3(OH)_4 \cdot 6\text{-}7H_2O$. A widespread but uncommon mineral formed from hydrothermal solutions, as well as from alteration of arsenopyrite and other primary arsenic minerals.

COCHISE COUNTY: Warren District, Bisbee, Lavender pit, octahedrons on psilomelane (Graeme 1981).

COCONINO COUNTY: Grand Canyon National Park, Grandview District, Grandview (Last Chance) Mine (Michael Cline, pers. comm., 2018).

SANTA CRUZ COUNTY: Santa Rita Mountains, Tyndall District, St. Louis Mine, Temporal Gulch, on baryte, with azurite, formed by the alteration of tennantite (Bideaux, Williams, and Thomssen 1960).

YAVAPAI COUNTY: Eureka supersystem, Hillside District, Hillside Mine, emerald-green crystals tentatively identified as pharmacosiderite (C. A. Anderson, Scholz, and Strobell 1955).

#PHILIPSBORNITE

Lead aluminum arsenate hydroxide: $PbAl_3(AsO_4)(AsO_3OH)(OH)_6$. A rare secondary mineral found in the oxide zone of hydrothermal polymetallic deposits.

COCONINO COUNTY: Grand Canyon National Park, Grandview District, Grandview (Last Chance) Mine, as yellowish clusters of minute crystals with osarizawaite and brochantite (Michael Cline, pers. comm., 2018).

PHILLIPSITE

Calcium potassium sodium aluminum silicate hydrate: $(Ca_3,K_6,Na_6)(Si_{10}Al_6)O_{32} \cdot 12H_2O$. A member of the zeolite group that includes three species based on dominant cation. The specific species has not been determined for the following occurrences.

GILA COUNTY: North side of Roosevelt Lake, in an altered volcanic ash (3rd ed.).

GREENLEE COUNTY: About five miles south of Hannagan Meadow, along road cuts on U.S. Route 491, as limpid clear fourlings with heulandite in vesicles in basalt (3rd ed.).

LA PAZ COUNTY: Harquahala Mountains, as microcrystals (Brian Bond, UA x3311).

FIGURE 3.187 Phillipsite, Malpais Hill, Pinal County, FOV 3 mm.

MARICOPA COUNTY: Horseshoe Dam District, in altered tuff about 2.25 miles south of Horseshoe Dam (Eyde 1978), and in basalt about one mile south of the dam (D. M. Shannon 1983b).

MOHAVE COUNTY: Big Sandy District, east of Big Sandy Wash, in the east half of T16N R13W, in Pliocene tuff with analcime, chabazite, erionite, and clinoptilolite (C. S. Ross 1928, 1941).

NAVAJO COUNTY: Wood Chop Mesa, as twinned fiveling crystals (A. Hampson, UA x4523).

PIMA COUNTY: Ajo District, Phelps Dodge Well No. 1 (William Thomas, pers. comm., 1988).

PINAL COUNTY: San Pedro Valley, Malpais Hill, in a railway cuts, with chabazite, heulandite, and calcite on celadonite (Thomssen 1983). Silver Reef Mountains, Silver Reef District, Nugget Fracture property, as microcrystals (David Shannon, UA x4346).

YAVAPAI COUNTY: Cottonwood Basin, in an altered tuff bed, with mordenite (Eyde 1978). Lyles District, James Stewart Company lithium claims, nine miles west of Kirkland, in altered tuff (Eyde 1978). At two localities in the Verde Valley, in altered tuffs (Eyde 1978).

PHLOGOPITE

Potassium magnesium aluminum silicate hydroxide: $KMg_3(AlSi_3O_{10})(OH)_2$. A member of the mica group that is commonly formed in metamorphosed limestones and dolomites and in ultramafic igneous rocks and an important alteration mineral in potassic alteration zones with orthoclase, chalcopyrite, and molybdenite in porphyry copper deposits.

FIGURE 3.188 Phlogopite, Big Lue Mountains, Greenlee County, FOV 2 mm.

GILA COUNTY: Northwestern region, associated with the uranium ores in the Dripping Spring Quartzite; Sierra Ancha District, Little Joe, Lucky Boy, Big Six, and Last Chance properties (Granger and Raup 1969). Southeastern Dripping Spring Mountains, Banner supersystem, Christmas District, Christmas Mine, associated with talc and tremolite adjacent to anhydrite sulfide veinlets in diopside hornfels, forming most of the footwall of the lower Martin orebody; also in skarn rocks and some of the hornfels (David Perry, pers. comm., 1967).

GREENLEE COUNTY: Big Lue Mountains, Big Lue District, in vesicles in volcanics in roadcuts along Highway 78 near the New Mexico border with tridymite, hematite, pseudobrookite, and titanite (R. Gibbs, pers. comm., 2005).

PIMA COUNTY: Pima supersystem, Twin Buttes District, Twin Buttes Mine (William Hefferon, pers. comm., 3rd ed.).

PINAL COUNTY: Slate Mountains, Lakeshore District, Lakeshore Mine, in metadolomite magnesian skarns, with forsterite and chondrodite (3rd ed.).

SANTA CRUZ COUNTY: Patagonia Mountains, Washington Camp District, Simplot Mine, as a product of the alteration of volcanic rocks (Lehman 1978).

YAVAPAI COUNTY: Eureka supersystem, Bagdad District, Bagdad open pit, as an important potassic zone alteration mineral with chalcopyrite, molybdenite, and orthoclase (McCandless and Ruiz 2002).

PHOENICOCHROITE

FIGURE 3.189 Phoenicochroite, Rat Tail claim, Maricopa County, FOV 6 mm.

Lead oxide chromate: $Pb_2O(CrO_4)$. A rare secondary mineral associated with crocoite, vauquelinite, and cerussite in oxidized galena-bearing veins.

MARICOPA COUNTY: Wickenburg Mountains, possibly San Domingo District, Collateral, Chromate, Blue Jay, and Phoenix claims, east of the present location of Trilby Wash. In 1881, Benjamin Silliman noted the presence of phoenicochroite, vauquelinite, and crocoite, and the probable presence of volborthite and descloizite, from the "Vulture region." Belmont Mountains, Osborne District, at several localities, including the Moon Anchor Mine, Potter-Cramer property, and Rat Tail claim, as dark cochineal-red cleavable and polycrystalline masses formed by the oxidation of lead-zinc ores and associated with wickenburgite, willemite, mimetite, hemihedrite, and vauquelinite (S. A. Williams 1968; S. A. Williams, McLean, and Anthony 1970; S. A. Williams and Anthony 1970). Big Horn Mountains, Big Horn District, Evening Star Mine, with fornacite, iranite, and cerussite in the oxidized portion of the vein (Gibbs, pers. comm., 2012).

YAVAPAI COUNTY: Black Rock District, near the confluence of Amazon Wash and the Hassayampa River, from several unnamed properties, with mimetite, hemihedrite, and descloizite (William Hunt, pers. comm., 1985).

PHOSGENITE

Lead carbonate chloride: $Pb_2(CO_3)Cl_2$. A rare secondary mineral formed by the oxidation of galena and other lead minerals and commonly associated with cerussite and anglesite and other distinctive chlorine-bearing minerals.

COCHISE COUNTY: Warren District, Bisbee, Cole Mine, 1100 level; phosgenite was recognized on specimens from here as 5 mm colorless prismatic crystals on calcite with copper. The phosgenite from here could easily be confused with calcite; however, the calcite from this locality is highly fluorescent, while the phosgenite is not (Graeme, Graeme, and Graeme 2015).

MARICOPA COUNTY: Painted Rock Mountains, Painted Rock District, Rowley Mine, as clear bladed striated crystals with diaboleite and cerussite (Gibbs, pers. comm., 2013). Belmont Mountains, Osborne District, Tonopah-Belmont Mine, as small blocky lustrous brilliant-yellow crystals with brochantite (J. Ruiz, pers. comm., 2012).

PINAL COUNTY: Mammoth District, Mammoth–St. Anthony Mine, 400 ft. level of the Collins vein, as slender prismatic crystals with diaboleite, paralaurionite, boleite, and cerussite (Palache 1941b; Fahey, Daggett, and Gordon 1950; Bideaux 1980; H 104522).

FIGURE 3.190 Phosphohedyphane, Tonopah-Belmont Mine, Maricopa County, FOV 3 mm.

#PHOSPHOHEDYPHANE

Calcium lead phosphate chloride: $Ca_2Pb_3(PO_4)_3Cl$. It occurs in the oxidized zone of lead-bearing deposits. Many specimens previously thought to be pyromorphite have been checked and found to be phosphohedyphane.

COCHISE COUNTY: Found at the Great Eastern Mine, as colorless steep pyramids (Kampf et al. 2006). Huachuca Mountains, Reef District, Reef Mine, as small crystals (Walstrom 2012). Charleston District, Gallagher and Manila Mines, as pale yellowish-green needles (Bowell, Luetcke, and Luetcke 2015). Johnny Lyon Hills, Yellowstone District, at an unnamed property in Tres Alamos Wash (Robert Jenkins, pers. comm., 2018).

GILA COUNTY: Banner supersystem, 79 District, 79 Mine, as clusters of small pale to bright-yellow crystals with cerussite and wulfenite (Gibbs, pers. comm., 2012).

MARICOPA COUNTY: Belmont Mountains, Osborne District, Tonopah-Belmont Mine, as colorless tapering prisms (Kampf et al. 2006).

PINAL COUNTY: Mammoth District, Mammoth–St. Anthony Mine, as yellow acicular crystals (Kampf et al. 2006).

SANTA CRUZ COUNTY: Patagonia Mountains, Hardshell-Hermosa District, Hardshell Mine, as clear-yellow stout prisms (Kampf et al. 2006).

PHOSPHOSIDERITE

Iron phosphate hydrate: $Fe^{3+}(PO_4)\cdot2H_2O$. A rare mineral found in pegmatites with other phosphate minerals and a dimorph of strengite.

YAVAPAI COUNTY: Hillside area, 7U7 Ranch, with bermanite in seams in triplite, associated with leucophosphite and hureaulite (Leavens 1967).

PHOSPHURANYLITE

Potassium calcium hydronium uranyl phosphate oxide hydrate: $KCa(H_3O)_3(UO_2)_7(PO_4)_4O_4\cdot8H_2O$. A secondary mineral commonly associated with torbernite and other uranium minerals.

COCONINO COUNTY: Cameron District, Huskon No. 17 Mine, where it replaces part of a fossilized log; Jack Daniels No. 1 Mine (Austin 1964; Finch 1967).

#PHOXITE

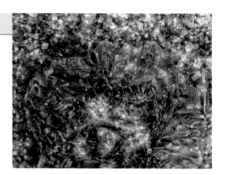

Ammonium magnesium oxalate hydroxyphosphate hydrate: $(NH_4)_2Mg_2(C_2O_4)(PO_3OH)_2\cdot4(H_2O)$. A very rare secondary organic mineral formed in an old mine working with bat guano. The Rowley Mine is the type locality.

MARICOPA COUNTY: Painted Rock District, Theba, Rowley Mine, found growing on a baryte-quartz–rich matrix and is associated with antipinite, aphthitalite, bassanite, struvite, thénardite, and weddellite (Kampf et al. 2018).

FIGURE 3.191 Phoxite, Rowley Mine, Maricopa County, FOV 0.68 mm.

PICKERINGITE

Magnesium aluminum sulfate hydrate: $MgAl_2(SO_4)_4\cdot22H_2O$. A secondary mineral found in the gossans of oxidized pyrite-bearing bodies with other hydrated secondary sulfates.

COCHISE COUNTY: Reported as abundant in an unspecified locality about thirty miles east of Douglas (Bull. 181). Warren District, Bisbee, Shattuck shaft (Graeme 1981).

COCONINO COUNTY: Cameron District, at an unnamed uranium prospect with wupatkiite (S. A. Williams, and Cesbron 1995). Sunset Crater National Monument in the fumarole deposits on Sunset Crater (S. L. Hanson, Falster, and Simmons 2008).

FIGURE 3.192 Pickeringite, 79 Mine, Gila County, 4 cm.

GILA COUNTY: Sierra Ancha District, Pueblo Canyon, Ancient deposit, in Mescal Limestone (Granger and Raup 1969); near the Rock Canyon deposit, as efflorescences identified only as members of the pickeringite-halotrichite group (Granger and Raup 1969); Banner supersystem, 79 District, 79 Mine, 470 level, as masses of parallel fibers over an inch in length (Gibbs, pers. comm., 2019).

PINAL COUNTY: San Manuel District, 2075 ft. level of San Manuel Mine (J. W. Anthony and McLean 1976).

PIEMONTITE

Calcium aluminum manganese silicate oxide hydroxide: $Ca_2(Al_2Mn^{3+})$ $[Si_2O_7][SiO_4]O(OH)$. A member of the epidote group, formed in low-grade metamorphic rocks and by hydrothermal activity.

COCHISE COUNTY: Pat Hills, Pat District, north of Pearce, as small rosettes of minute crystals in andesite (Lausen 1927b).

MARICOPA COUNTY: Phoenix Mountains District, southwestern slope of Piestawa Peak in Phoenix, as acicular to bladed crystals up to 3 mm long in quartz-muscovite schist (Thorpe 1980).

PIMA COUNTY: Tucson Mountains, Tucson Mountain Park, Amole District, Brown Mountain, southeast of Juan Santa Cruz picnic area, in rhyolitic intrusion and adjacent sandstone (Guild 1935). Santa Rita Mountains, near Madera Canyon (Guild 1935). Northern Comobabi Mountains, with epidote and axinite-(Mn), as vesicle fillings in andesite flow boulders (Bideaux, Williams, and Thomssen 1960). About six miles from Vail (UA 2256).

PINAL COUNTY: Near Casa Grande (UA 2218).

YAVAPAI COUNTY: Northwest of Prescott, in the Government Canyon and Prescott Granodiorites, as small veinlets and disseminated grains (Krieger 1965).

PIGEONITE

Magnesium iron calcium silicate: $(Mg,Fe,Ca)_2Si_2O_6$. A member of the pyroxene group and a rock-forming mineral largely confined to quickly cooled mafic flow rocks.

COCHISE COUNTY: Tombstone District, Lucky Boy Mine, as dark-green grains with vesuvianite in the contact-silicate zone (B. S. Butler, Wilson, and Rasor 1938).

PINALITE

Lead tungstate oxychloride: $Pb_3(WO_4)OCl_2$. A very rare secondary mineral formed by late oxidation processes. The Mammoth–St. Anthony Mine is the type locality.

PINAL COUNTY: Mammoth District, Mammoth–St. Anthony Mine, typically as isolated crystals and as divergent sprays in irregular cavities that are pseudomorphic molds of an uncertain precursor, perhaps calcite; in some places, enclosed by chromian leadhillite (Dunn, Grice, and Bideaux 1989).

FIGURE 3.193 Pinalite, Mammoth–St. Anthony Mine, Pinal County, FOV 1.5 mm.

PINTADOITE

Calcium vanadium oxide hydrate: $Ca_2V^{5+}_2O_7 \cdot 9H_2O$. A secondary mineral formed as water-soluble efflorescences.

APACHE COUNTY: Lukachukai District, Mesa No. 5 and No. 6 properties, in the Salt Wash Member of the Morrison Formation (Bull. 181).

PLANCHEITE

Copper silicate hydroxide hydrate: $Cu_8(Si_4O_{11})_2(OH)_4 \cdot H_2O$. A rare secondary mineral formed in the oxidized portions of copper deposits.

COCHISE COUNTY: Warren District, Bisbee, Shattuck shaft, 400 level (Graeme 1981).

LA PAZ COUNTY: New Water District, Eagle Eye Mine, in quartz (David Shannon, pers. comm., 1990). Little Harquahala Mountains, Little Harquahala District, Harquahala Mine, as sky-blue radial sprays coating fractures in hematitic Bolsa Quartzite, near a microdiorite dike, associated with dioptase, chrysocolla, malachite, and pseudomalachite (Stanley B. Keith, pers. comm., 2019).

FIGURE 3.194 Plancheite, Nugget Fraction Mine, Pinal County, 3 cm.

MARICOPA COUNTY: Big Horn District, about twenty miles south of Aguila, as deep-blue fibrous veinlets and masses in quartzite, associated with chrysocolla and "copper pitch" (Montoya 1967).

PIMA COUNTY: Ajo District, New Cornelia Mine, as massive material in the first oxide zone (W. J. Thomas and Gibbs 1983).

PINAL COUNTY: Galiuro Mountains, Table Mountain District, Table Mountain Mine, as blue crystalline grains and as masses disseminated in compact green conichalcite (Bull. 181). Suizo Mountains, Durham-Suizo District, Azurite Mine, with shattuckite (Bideaux, Williams, and Thomssen 1960). San Manuel District, San Manuel Mine, as blue tablets in chrysocolla (L. A. Thomas 1966; UA 5390, 6491); Mammoth District, Mammoth–St. Anthony Mine, as veinlets and as coatings on chrysocolla (Bideaux 1980). Silver Reef District, Nugget Fraction Mine, with brochantite (David Shannon, pers. comm., 1988).

YAVAPAI COUNTY: Buckhorn Mountains, Castle Creek District, Whipsaw Mine (H 90834).

#PLANERITE

Aluminum phosphate hydroxyphosphate hydrate: $Al_6(PO_4)_2(PO_3OH)_2$ $(OH)_8 \cdot 4H_2O$. An unusual mineral found in the oxidized portion of deposits rich in phosphate.

PIMA COUNTY: Ajo District, Ajo, New Cornelia Mine, identified as a weathering crust from Camelback Mountain and other nearby localities noted by James Gilluly in 1946. (W. J. Thomas and Gibbs 1983).

PLATINUM

Platinum: Pt. A rare mineral mostly restricted to ultramafic rocks, their metamorphic equivalents, and placers derived from them. The localities listed below have not been verified and the occurrence in Arizona is doubtful.

MARICOPA COUNTY: Reported in the Santo Domingo placers, along the Gila River, opposite the old Riverside stage station (Bull. 181).

YAVAPAI COUNTY: Reported in black sands near Columbia and Prescott (Guild 1910).

PLATTNERITE

Lead oxide: PbO_2. An uncommon mineral formed in lead deposits by extreme oxidation. Associated with cerussite, smithsonite, pyromorphite, wulfenite, and other oxidized minerals.

COCHISE COUNTY: Warren District, Bisbee, with murdochite and malachite on goethite (Bideaux, Williams, and Thomssen 1960; Graeme 1993). Gleeson District, Silver Bill, Defiance, and Tom Scott Mines, with wulfenite (Bideaux, Williams, and Thomssen 1960).

GILA COUNTY: Banner supersystem, 79 District, 79 Mine, as tiny needle-like crystals up to 1.5 mm long, associated with aurichalcite, rosasite, and, in a few places, murdochite (Stanley B. Keith 1972).

LA PAZ COUNTY: Trigo Mountains, Silver District, Red Cloud Mine, as minute black needles with vanadinite (Edson 1980).

MARICOPA COUNTY: Belmont Mountains, Osborne District, Tonopah-Belmont Mine (G. B. Allen and Hunt 1988).

FIGURE 3.195 Plattnerite, 79 Mine, Gila County, FOV 3 mm.

PIMA COUNTY: Tucson Mountains, Old Yuma District, Old Yuma Mine, as tiny crystals on vanadinite (Bideaux, Williams, and Thomssen 1960). Waterman Mountains, Waterman District, Silver Hill Mine, with murdochite, malachite, and aurichalcite (Bideaux, Williams, and Thomssen 1960). Empire Mountains, Empire District, Total Wreck Mine, with hemimorphite on limonite (Bideaux, Williams, and Thomssen 1960); Lone Mountain Mine (UA 6364); Chief Mine (Scott Richardson, UA x1510); Hilton Mine, as microcrystals (Gene Wright, UA x4498). Gunsight District, Gunsight Mine (Scott Richardson, UA x2676).

SANTA CRUZ COUNTY: Santa Rita Mountains, Cottonwood Canyon, Tyndall District, Glove Mine, as crystals on wulfenite (Bull. 181).

PLUMBOGUMMITE

Lead aluminum phosphate hydroxide hydrate: $PbAl_3(PO_4)(PO_3OH)(OH)_6$. A secondary mineral formed by oxidation of lead ores.

COCHISE COUNTY: Reef District, Reef Mine (Walstrom 2013).

GILA COUNTY: Miami-Inspiration supersystem, Copper Cities District, Coronado fault area, Copper Cities deposit, with chalcopyrite in quartz (H 106829).

PLUMBOJAROSITE

Lead iron sulfate hydroxide: $Pb_{0.5}Fe^{3+}{}_3(SO_4)_2(OH)_6$. A member of the alunite group found in oxidized lead deposits typically in arid regions. It is less common than jarosite, to which it is similar in origin and mode of occurrence.

COCHISE COUNTY: Tombstone District, abundant in the brown oxide ore of the Holderness roll; Empire Mine, as material that assayed 58.92 oz/ton in silver; Toughnut Mine, as material that assayed up to 200 oz/ton in silver (B. S. Butler, Wilson, and Rasor 1938). Warren District, Bisbee, Southwest and Shattuck Mines, abundant in lead ore adjacent to quartz breccia (Graeme 1993).

GILA COUNTY: Banner supersystem, 79 District, 79 Mine, as ocherous brown massive material in the surface and near-surface workings (Stanley B. Keith 1972).

GRAHAM COUNTY: Aravaipa supersystem, Grand Reef District, Dogwater Mine (Simons 1964); Grand Reef Mine (Roberts et al. 1995).

MARICOPA COUNTY: Belmont Mountains, Osborne District, Tonopah-Belmont Mine, with creaseyite (G. B. Allen and Hunt 1988).

MOHAVE COUNTY: Cerbat Mountains, Wallapai supersystem, Tennessee District, as a secondary mineral (B. E. Thomas 1949); Tennessee-Schuylkill Mine (Bull. 181).

PIMA COUNTY: Empire District, Hilton Mines, as earthy masses (Bull. 181). Pima supersystem, Twin Buttes District, Twin Buttes Mine (Stanley B. Keith, pers. comm., 1972).

YAVAPAI COUNTY: Castle Creek District, Humbug Mine, Castle Hot Springs area (3rd ed.).

PLUMBONACRITE

Lead carbonate oxide hydroxide: $Pb_5(CO_3)_3O(OH)_2$. An inadequately characterized mineral, long confused with hydrocerussite and probably formed under similar conditions during oxidation of lead-rich sulfide ores.

PINAL COUNTY: Mammoth District, Mammoth–St. Anthony Mine, 1,000 ft. level of the Collins vein, on a specimen collected by S. C. Creasey (S 71), as small pearly-white scales, associated with anglesite and linarite on galena (3rd ed.).

#PLUMBOTELLURITE

Lead tellurite: $PbTe^{4+}O_3$. A rare mineral formed by the oxidation of other tellurium minerals.

COCHISE COUNTY: Tombstone District, in the dumps of the Grand Central Mine, associated with many tellurium oxysalt minerals. Material originally described and identified as oboyerite has subsequently been found to be a mixture of at least two species: ottoite and plumbotellurite (Miyawaki et al. 2019b).

PLUMBOTSUMITE

Lead silicate hydroxide: $Pb_5Si_4O_8(OH)_{10}$. A secondary mineral in oxidized lead deposits.

PINAL COUNTY: Mammoth District, Mammoth–St. Anthony Mine, as massive white material intergrown with wulfenite (Pete J. Dunn, pers. comm., 1988).

POLLUCITE

Cesium aluminum silicate hydrate: $Cs(Si_2Al)O_6 \cdot nH_2O$. This rare member of the zeolite group is usually found in lithium-rich granite pegmatites.

MARICOPA COUNTY: White Picacho District, Independence Mine, a reported but unconfirmed locality (Bull. 181).

POLYBASITE

Silver copper antimony arsenic sulfide: $[Ag_9CuS_4][(Ag,Cu)_6(Sb,As)_2S_7]$. Found in silver veins formed at low to moderate temperatures associated with other silver sulfosalts.

COCHISE COUNTY: Warren District, Bisbee, in massive galena from an unspecified mine (Graeme 1981).

MOHAVE COUNTY: Cerbat Mountains, Wallapai supersystem, Chloride and Stockton Hill Districts, common in the silver ores of the area (Bastin 1925; B. E. Thomas 1949).

PINAL COUNTY: Pioneer supersystem, Silver King District, Silver King Mine, as fine specimens (Guild 1910).

SANTA CRUZ COUNTY: Santa Rita Mountains, Tyndall District, Ivanhoe Mine, near Squaw Gulch, with stromeyerite, proustite, and silver (Engineering and Mining Journal 1912).

YAVAPAI COUNTY: Northwestern Bradshaw Mountains, Groom Creek District, Davis Mine, with proustite (Lindgren 1926). Tip Top District, Tip Top Mine, with pyrargyrite and tetrahedrite in quartz veins (C. P. Kortemeier 1984).

YUMA COUNTY: Kofa Mountains, Kofa District, King of Arizona Mine (Hankins 1984).

#POLYCRASE-(Y)

Yttrium titanium niobium oxide hydroxide: $Y(Ti,Nb)_2(O,OH)_6$. A rare mineral formed in granite pegmatites and a member of the euxenite-polycrase series.

MARICOPA COUNTY: Buckeye Hills District, south of Buckeye, in a pegmatite dike with kasolite (Robert O'Haire, pers. comm., 1972).

MOHAVE COUNTY: Aquarius Mountains, Aquarius Mountains District, Rare Metals pegmatite with other rare-earth minerals (W. B. Simmons et al. 2012).

#POLYDYMITE

Nickel sulfide: $Ni^{2+}Ni^{3+}_2S_4$. An uncommon mineral found in hydrothermal ore deposits.

COCONINO COUNTY: South Rim of the Grand Canyon, Orphan District, Orphan Mine, as discrete zones in small euhedral zoned but complex Ni-Co-Fe-As-S minerals. Associated minerals include siegenite, violarite, pyrite, and vaesite (Dr. Karen Wenrich, unpublished microprobe analysis, 1994).

POLYHALITE

Potassium calcium magnesium sulfate hydrate: $K_2Ca_2Mg(SO_4)_4 \cdot 2H_2O$. Primarily formed as a precipitate from oceanic waters, it is widely distributed and commonly associated with halite and anhydrite.

APACHE AND NAVAJO COUNTIES: Holbrook Potash District, east-central Arizona, encountered in drill holes that delineated a northeast-trending potash zone underlying an area of about three hundred square miles, in Permian saline deposits (Peirce 1969b); the log of a hole drilled in S24, T18N R24E, also listed carnallite, sylvite, halite, anhydrite, and gypsum (H. Wesley Peirce, pers. comm., 1972).

POLYLITHIONITE

Potassium lithium aluminum silicate fluoride: $KLi_2AlSi_4O_{10}F_2$. Found in nepheline syenites associated with microcline, natrolite, and other minerals, and in lithium-cesium-tantalum (LCT) pegmatites.

GRAHAM COUNTY: Aravaipa supersystem, Iron Cap District, near Landsman Camp, in axinite- and babingtonite-rich metamorphosed limestone (3rd ed.).

POSNJAKITE

Copper sulfate hydroxide hydrate: $Cu_4(SO_4)(OH)_6 \cdot H_2O$. A rare secondary mineral formed under acidic oxidizing conditions by the alteration of copper sulfides.

LA PAZ COUNTY: Bouse Hills District, a specimen in the Natural History Museum in London (NHM 1972,201) is labeled as coming from Bouse (Peter G. Embrey, pers. comm., 1973).

PIMA COUNTY: Ajo District, New Cornelia Mine, with gypsum, associated with partially altered pyrite (Hidemichi Hori, pers. comm., 1984).

FIGURE 3.196 Posnjakite, Childs-Aldwinkle Mine, Pinal County, FOV 2 mm.

PINAL COUNTY: Childs-Aldwinkle District, Childs-Aldwinkle Mine, as small bright-blue lustrous crystals and crystal aggregates to 1 mm long and as crusts cementing soil outside the main adit of the mine. It is associated with orthoserpierite and chalcanthite (Ron Gibbs, pers. comm., 2020).

YAVAPAI COUNTY: Bloody Basin District, prospect in the Turret Peak quadrangle in S30, T13N R5E, as a powder-blue film on chalcopyrite (3rd ed.).

#POUGHITE

Iron tellurate sulfate hydrate: $Fe_2^{3+}(Te^{4+}O_3)_2(SO_4) \cdot 3H_2O$. A secondary mineral usually formed from the alteration of pyrite in gold-tellurium deposits.

COCHISE COUNTY: Tombstone District, Grand Central Mine, as small crystals with fairbankite, cerussite, chlorargyrite, jarosite, and rodalquilarite in boxwork cavities in quartz gossan. The mineral was found during the reexamination of the type specimen of fairbankite (Missen et al. 2021).

POWELLITE

Calcium molybdate: $Ca(MoO_4)$. Commonly contains tungsten. An uncommon secondary mineral found in tungsten ores, especially those in skarns, and rarely with zeolites in basalts. Also found with scheelite in peraluminous pegmatites and aplogranites.

COCHISE COUNTY: Little Dragoon Mountains, Bluebird District, near Adams Peak, as pseudomorphs after molybdenite in quartz-tungsten veins (J. R. Cooper and Silver 1964). Reported in the Warren District; Bisbee—one verified locality is the Bisbee Queen shaft east of Warren (3rd ed.).

GILA COUNTY: Miami-Inspiration supersystem, Inspiration District, Inspiration Mine, as crusts of tiny crystals in a seam adjacent to veins containing molybdenite and lindgrenite; thought to be a product of late hydrothermal solutions, which attacked molybdenite (Fredrick E. Pough, pers. comm., 3rd ed.). Carlota District, Carlota Mine, as elongated yellow crystals to 2 mm in azurite with szenicsite and prosopite (W. E. Wilson and Jones 2015).

MARICOPA COUNTY: White Picacho District, northeast of Morristown, on the upper Santo Domingo Wash, with scheelite in pegmatites (Bull. 181).

FIGURE 3.197 Powellite, Carlota Mine, Gila County, FOV 5 mm.

Vulture Mountains, Flying Saucer District, Flying Saucer Group, with scheelite as disseminations in peraluminous granitic rocks that fluoresce bright yellowish green (Dale 1959).

MOHAVE COUNTY: Reported in the Cerbat Mountains (Bull. 181). Corkscrew District, west side of Peach Springs Canyon, in the Green Room of Corkscrew Cave with talmessite, carnotite, conichalcite, calcite, and aragonite (Wenrich and Sutphin 1989, 1994).

PIMA COUNTY: Northern Santa Rita Mountains, Helvetia-Rosemont District, disseminated with scheelite in the skarn contact zone near the Black Horse shaft (Dale, Stewart, and McKinney 1960). Pima supersystem, Twin Buttes District, Twin Buttes Mine in skarns (Stanley B. Keith, pers. comm., 1973; Kuck 1978); Senator Morgan Mine, with scheelite in quartz veins (Dale, Stewart, and McKinney 1960). Eastern flank of the Santa Catalina Mountains, Redington District, at the Korn Kob Mine, as reaction rims around molybdenite (J. R. Wilson 1977).

PINAL COUNTY: Southern Tortilla Mountains, Gold Circle District, near Antelope Peak, Upshaw Tungsten Mines Group, with "wolframite" and scheelite in quartz veins (Dale 1959). Central Galiuro Mountains, Bunker Hill (Copper Creek) District, Copper Creek, Childs-Aldwinkle Mine, as pseudomorphs after molybdenite (Ascher, pers. comm., 2016).

YAVAPAI COUNTY: White Picacho District, as a rare mineral in pegmatites (Jahns 1952).

PREHNITE

Calcium aluminum silicate hydroxide: $Ca_2Al(Si_3Al)O_{10}(OH)_2$. Primarily formed in cavities and in veins in mafic lavas, where it is commonly associated with zeolite minerals. It is also formed in veins in granitic rocks and in contact-metamorphosed limestones.

COCHISE COUNTY: Warren District, Bisbee, Cole shaft, 800 and 1000 levels with epidote and grossular (Graeme 1981). One mile northwest of Portal, in vesicles in basalt with celestine, pyrite, and pumpellyite (3rd ed.). Central Dragoon Mountains, Middlemarch Canyon, Middle Pass supersystem, Abril District, Middlemarch Mine, as sparse veinlets in garnet-wollastonite schist (Sousa 1980).

GILA COUNTY: Coolidge Dam area, as large pale-green crystalline masses (Les Presmyk, pers. comm., 1988); Haystack Butte area, four miles south-southwest of Haystack Butte, as sheaf-like crystal masses and clusters in east-northeast-trending veins in chloritic Precambrian diabase associated with tremolite (Muntyan 2020a). Sierra Ancha District and Salt River District, as a widespread component of chlorite alteration that is marginal to uranium veins associated with granophyre phase of the Precambrian diabase (Granger and Raup 1969).

PIMA COUNTY: Northern Santa Rita Mountains, Cuprite District, near Mt. Fagan, as crystalline masses with copper, epidote, and quartz in andesite (Bideaux, Williams, and Thomssen 1960). Cerro Colorado District, Cerro Colorado Mine area, on the road to Arivaca (Robert O'Haire, pers. comm., 1972).

FIGURE 3.198 Prehnite, Haystack Butte, Gila County, 13 × 10 cm.

PROSOPITE

Calcium aluminum fluoride hydroxide: $CaAl_2F_4(OH)_4$. An uncommon mineral found in greisen in tin veins, associated with other fluorine-bearing minerals; also in pegmatites.

GILA COUNTY: Miami-Inspiration District, Carlota Mine, as small light-green crystals and aggregates with azurite and powellite (W. E. Wilson and Jones 2015, R100181).

GRAHAM COUNTY: Aravaipa supersystem, Grand Reef District, Grand Reef Mine, as light-green masses (George Godas, pers. comm., 1988).

LA PAZ COUNTY: Bouse Hills District, at a copper prospect about ten miles north of Bouse, in hematite veins in andesite with chrysocolla, malachite, and tenorite, as complex limpid-sea-green crystals up to 6 mm long in voids in hematite gangue (3rd ed.).

FIGURE 3.199 Prosopite, Carlota Mine, Gila County, FOV 7 mm.

PROUSTITE

Silver arsenic sulfide: Ag_3AsS_3. A late-stage mineral formed in hydrothermal veins at low temperatures with other silver sulfosalts.

COCHISE COUNTY: Pearce District, Commonwealth Mine, associated with tetrahedrite and silver halides (Bull. 181).

MOHAVE COUNTY: Cerbat Mountains, Wallapai supersystem, Tennessee District, Minnesota-Connor; Chloride District, Distaff Mine and Merrimac Mines (Bastin 1925); Tennessee District, Gold Star Mine (Bull. 181); Stockton Hill District, Paymaster Mine (B. E. Thomas 1949); Cupel Mine, in relatively large quantities (Schrader 1908).

PINAL COUNTY: Pioneer supersystem, Belmont District, Belmont Mine, as minute blebs in galena (Bull. 181).

SANTA CRUZ COUNTY: Nogales District, Mt. Benedict area, with gold in quartz monzonite (Schrader 1917). Santa Rita Mountains, Ivanhoe District, Ivanhoe Mine, with stromeyerite, polybasite, and silver (Engineering and Mining Journal 1912).

YAVAPAI COUNTY: Wickenburg Mountains, Monte Cristo District, Monte Cristo Mine, with silver and acanthite (Bastin 1922). Bradshaw Mountains, Hassayampa District, Davis and Catoctin Mines, with polybasite (Bull. 181). Northeast Bradshaw Mountains, Turkey Creek District, near the Thunderbolt Mine, with silver and chlorargyrite in a low-angle vein (Bull. 181). Southern Bradshaw Mountains, Tip Top District, Tip Top Mine, with silver and chlorargyrite in high-angle northeast-trending veins associated with felsic dikes in the Crazy Basin pluton (Bull. 181).

PSEUDOBOLEITE

Lead copper chloride hydroxide: $Pb_{31}Cu_{24}Cl_{62}(OH)_{48}$. A rare secondary mineral formed in small amounts in oxidized lead-copper deposits.

PIMA COUNTY: Pima supersystem, Mission-Pima District, Mineral Hill Mine, Daisy shaft, with connellite, gerhardtite, and atacamite in cuprite (S. A. Williams 1961).

PINAL COUNTY: Mammoth District, Mammoth−St. Anthony Mine, as overgrowths on boleite, associated with other secondary lead minerals (Palache 1941b; Bideaux 1980).

PSEUDOBROOKITE

Iron titanium oxide: $(Fe^{3+}_2Ti)O_5$. An uncommon mineral formed in volcanic rocks by pneumatolytic or fumarolic action.

GREENLEE COUNTY: Big Lue Mountains, Big Lue District, road cuts on Arizona State Route 78, about six miles from New Mexico border (William Hunt, pers. comm., 1989).

PIMA COUNTY: Near Ajo, Phelps Dodge Well No. 1 (William Thomas, pers. comm., 1988).

PINAL COUNTY: Ash Creek, San Carlos Apache Reservation, Lookout Mountain District, as a few needle-like crystals, with topaz, spessartine, and bixbyite, in rhyolite (White 1992). Copper Creek District, in float boulder near Blue Bird Spring, as small dark-red bladed crystals (Gibbs, pers. comm., 2018).

FIGURE 3.200 Pseudo-brookite, Big Lue Mountains, Greenlee County, FOV 3 mm.

PSEUDOGRANDREEFITE

Lead sulfate fluoride: $Pb_6(SO_4)F_{10}$. Associated with galena and fluorite, from which it is thought to have formed by interaction with supergene fluids. The Grand Reef Mine is the type locality.

GRAHAM COUNTY: Aravaipa supersystem, Grand Reef District, Grand Reef Mine, as colorless tabular crystals and as a subparallel crystal aggregate, in a quartz-lined vug, associated with galena, fluorite, and anglesite (Kampf, Dunn, and Foord 1989).

PSEUDOMALACHITE

Copper phosphate hydroxide: $Cu_5(PO_4)_2(OH)_4$. A rare secondary mineral found in the oxidized zone of copper deposits with other oxidized copper minerals.

COCHISE COUNTY: Warren District, Bisbee, Lavender pit, as a crust on malachite (Graeme 1981).

GILA COUNTY: Miami-Inspiration supersystem, Pinto Valley District, Castle Dome Mine, as small dark emerald-green crystals (Bull. 181); Live Oak Pit at the Inspiration Mine, as 2 mm hemispheres on chrysocolla (Les Presmyk, pers. comm., 2020).

GRAHAM COUNTY: Lone Star District, Safford area, as prismatic crystals, many of which are almost hair-like, in metasomatized volcanic rocks,

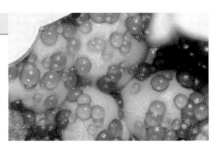

FIGURE 3.201 Pseudo-malachite, Old Reliable Mine, Pinal County, FOV 5 mm.

associated with malachite, brochantite, antlerite, carbonate-rich apatite, chrysocolla, jarosite, lepidocrocite, and sulfide minerals (Hutton 1959a).

LA PAZ COUNTY: Little Harquahala Mountains, Little Harquahala District, Harquahala Mine, as clusters of sharp dark-green crystal spheres in vugs, partially coated by chrysocolla; (John S. White Jr., pers. comm., 1972), associated with dioptase, plancheite, malachite, and chrysocolla in hematitic quartzite near a microdiorite dike (Stanley B. Keith, pers. comm., 2019). Cunningham Pass District, near Wenden, at the Critic Mine (3rd ed.).

PIMA COUNTY: East end of the South Comobabi Mountains, as green crystals with slightly curved faces, up to 3 mm long, and as films on the walls of fractures in severely deformed quartz monzonite, associated with libethenite (3rd ed.). Tucson Mountains, Saginaw Hill District, southern side of Saginaw Hill, about seven miles southwest of Tucson, associated with other oxidized minerals, including cornetite, brochantite, malachite, libethenite, atacamite, and chrysocolla, in fractures in chert (Khin 1970). Silver Bell District, Silver Bell Mine, Oxide pit, as microcrystals (Kenneth Bladh, UA, x3720). Sacaton Hill, found in drill core (3rd ed.).

PINAL COUNTY: Galiuro Mountains, Bunker Hill (Copper Creek) District, Copper Creek, with botryoidal malachite on quartz crystals, some of which are Japan twins (William and Mildred Schupp, pers. comm., 3rd ed.); Old Reliable Mine, as dark-green lustrous balls with libethenite (Gibbs, pers. comm., 2016).

psilomelane

Psilomelane is a general term for massive undifferentiated hard manganese oxides of secondary origin, formed under surface conditions from the alteration of manganous carbonates or silicates and associated with materials of similar origin, such as pyrolusite, goethite, limonite, and wad. Psilomelane has been reported widely in Arizona, but specific species identification is largely absent.

APACHE COUNTY: Sonsela Buttes area, Sonsela Buttes District, south side of a minette diatreme, filling pore spaces between pebbles or fragments in Chuska sandstones and conglomerates, with pyrolusite (Mayo 1955b).

COCHISE COUNTY: Tombstone District, as the most abundant manganese oxide mined in the ore deposits (Rasor 1939; Romslo and Ravitz 1947;

Hewett and Fleischer 1960). Warren District, Bisbee, Higgins Mine, with pyrolusite and braunite (Palache and Shannon 1920; Taber and Schaller 1930).

COCONINO COUNTY: Aubrey Cliffs, Johnson and Hayden District, with braunite at the Adams-Woodie prospect and with other manganese oxides cementing rock fragments in veins that cut and partly replace Kaibab Limestone; associated with braunite, pyrolusite, and cryptomelane (Potter and Havens 1949; Hewett, Fleischer, and Conklin 1963); same as Johnson and Hayden deposit, NW quarter of S2, T26N R7W, in a steeply dipping fracture or brecciated zone in Kaibab Limestone (Farnham and Stewart 1958). Long Valley District, as numerous small irregular disconnected lenticular masses in Kaibab Limestone, with pyrolusite in S19, 20, 29, and 30, T14N R10E; Denison, Shoup, Blue Ridge, and Lost Apache claims (Farnham and Stewart 1958). Heber District, in a small deposit in parts of S17, 18, 19, and 20, T11N R15E (Farnham and Stewart 1958).

GILA COUNTY: Miami-Inspiration supersystem, where, with manganite and pyrolusite, it forms the bulk of the gangue in the manganese-zinc-lead-silver deposits (N. P. Peterson 1962). Banner supersystem, 79 District, 79 Mine, as a common supergene mineral in replacement deposits in limestone and rhyolite porphyry (Kiersch 1949). Medicine Butte area, Ramsdell District, Apache and Accord manganese deposits, with wad, as fracture fillings and cement in Cenozoic Conglomerate (R. T. Moore 1968). Sierra Ancha Mountains, Sunset District, Armer Wash (Sunset) Mine, as fine-grained mammillary material in breccia zones in quartzite of the Apache Group; contains 0.02 percent thallium, approximately 12 percent BaO, and 0.2 percent K_2O (Crittenden et al. 1962); about six miles north of Roosevelt Lake, in S29, T5N R13E, in Apache Group rocks (Hewett and Fleischer 1960).

GREENLEE COUNTY: Ash Peak District, Fourth of July Mine, as mammillary layers on fluorite in veins in andesite porphyry (Hewett and Fleischer 1960; Hewett, Fleischer, and Conklin 1963; Hewett 1964). Gila Hot Springs District, pyrolusite claims, twelve miles southeast of Morenci, with pyrolusite and manganite (Potter, Ipsen, and Wells 1946).

LA PAZ COUNTY: Artillery District, Kaiserdoom claims, in S22, T11N R11W, in bedded manganese oxides underlying a bed of volcanic ash in the Artillery Formation, with associated soft manganite (Hewett, Fleischer, and Conklin 1963); Spring Mine, forty miles west of Congress Junction, as irregular nodular fragments cemented by clay (Long, Batty, and Dean 1948). Lincoln Ranch District, Doyle-Smith claims, fifteen miles from Artillery Peak (Ipsen and Gibbs 1952).

MARICOPA COUNTY: Black Vulture District, Black Vulture Mine, thirty-two miles south of Wickenburg, with pyrolusite as replacements in limestone (Long, Batty, and Dean 1948). Big Horn Mountains, Aguila District, Black Queen and Black Nugget Mines, with pyrolusite and manganite in small fissure veins in volcanic flows and breccia (Sandell and Holmes 1948; Farnham and Stewart 1958).

MOHAVE COUNTY: Artillery Mountains, Artillery District, Black Jack, Price, and Priceless veins and Maggie Canyon bedded deposit, in numerous veinlets and fractures in Tertiary volcanic rocks, associated with cryptomelane, hollandite, coronadite, pyrolusite, ramsdellite, and lithiophorite (Hewett and Fleischer 1960; Mouat 1962); Black Warrior Mine, six miles northeast of Alamo Crossing, with pyrolusite (Long, Batty, and Dean 1948). Pilot Rock District, Arizona manganese claims, thirty miles north of Parker Dam (Havens et al. 1947).

PIMA COUNTY: Reported near Tucson, but the exact locality is unknown (R. C. Wells 1937; Palache, Berman, and Frondel 1944). Northern end of the Coyote Mountains, as massive hard coatings up to 3 in. thick on porphyritic rocks (Blake 1910).

PINAL COUNTY: Cochise Group of claims, as fine-grained mammillary material, in veins and breccia zones in Gila Conglomerate; contains less than 0.01 percent thallium and about 12 percent BaO (Crittenden et al. 1962). Bunker Hill (Copper Creek) District, Copper Creek, Bluebird Mine (Kuhn 1951). Mammoth District, in a vein with baryte and calcite on the west side of Tucson Wash (Schwartz 1953). Riverside District, Almino and Cochise Mines, cementing fragments in breccia zones (Hewett and Fleischer 1960; Hewett, Fleischer, and Conklin 1963). Steamboat Mountain District, Geronimo claims, eight miles northwest of Winkelman, with manganite cementing brecciated zones of conglomerate (Dean, Snedden, and Agey 1952). Central Tortilla Mountains, Crescent District, Benningfield property, 11.5 miles south of Winkelman, with pyrolusite, as nodules and lenses replacing limestone in capping over an igneous dike (Dean, Snedden, and Agey 1952). North Star Group and Crescent District, Orsen Branch claim near Winkelman (Dean, Snedden, and Agey 1952). Pioneer supersystem, Silver King District, Pomeroy property, four miles north of Superior, with pyrolusite and hetaerolite (Dean, Snedden, and Agey 1952).

SANTA CRUZ COUNTY: Cottonwood Canyon, Tyndall District, Glove Mine (H. J. Olson 1966). Patagonia Mountains, Mowry District, Mowry Mine, in replacement deposits in limestone, with pyrolusite and hematite (Schrader 1917).

YAVAPAI COUNTY: Black Rock District, near Constellation, Black Rock deposit, with pyrolusite in Precambrian granite gneiss (Fleischer and Richmond 1943; Hewett and Fleischer 1960; Hewett, Fleischer, and Conklin 1963). Burmister District, Burmister Mine, near Sycamore Creek above its junction with the Agua Fria River, with cryptomelane and other manganese oxides interlayered with volcanic ash, clastic sediments, and a basalt flow (Hewett and Fleischer 1960; Hewett, Fleischer, and Conklin 1963). Big Bug District, thirteen miles from Mayer, at the confluence of Big Bug and Sycamore Creeks, as nodules and irregular veins, with minor amounts of pyrolusite (Engineering and Mining Journal 1918).

YUMA COUNTY: Little Horn Mountains, Sheep Tanks District, near Kofa Butte, Sheep Tanks Mine (Romslo and Ravitz 1947). Kofa District, North Star Mine, in the North Star vein, with bixbyite, and in the No. 2 vein, with bixbyite, todorokite, groutite, and manganite (Hankins 1984).

PUMPELLYITE

$Ca_2XY_2(Si_2O_7)(SiO_4)A_2 \cdot H_2O$, where X = Mg, Fe^{2+}, Fe^{3+}, Al, Mn^{2+}, V^{3+}; Y = Al, Fe^{3+}, Mn^{3+}, V^{3+}, Cr^{3+}; and A = OH, O. Associated with prehnite and zeolites in vesicles in volcanic rocks. Pumpellyite is a mineral group comprising eleven individual species, combining pumpellyites, julgoldites, and three other minerals. The root name *pumpellyite* is used where Y = Al, and the dominant ion in the X site is appended in parentheses. None of the Arizona localities has been assigned a specific pumpellyite species, of which five are currently recognized.

COCHISE COUNTY: About one mile northwest of Portal, in vesicles in basalt with prehnite, celestine, and adularia; as amygdule fillings up to 1 in. across, in radiating fibrous masses (3rd ed.). Reported in the Warren District (3rd ed.).

GILA COUNTY: Sierra Ancha Mountains, Aztec Creek, Workman Creek Falls, as amygdule fillings in a 34 m thick basalt flow above the Mescal Limestone of the Precambrian Apache Group, with quartz, chlorite, pyrite, and chalcopyrite (Heidecker 1978).

PIMA COUNTY: Santa Rita Mountains, Mt. Fagan, with thomsonite, prehnite, epidote, and copper in andesite (Bideaux, Williams, and Thomssen 1960).

PURPURITE

FIGURE 3.202 Purpurite, Little Giant Mine, Yavapai County, FOV 3 mm.

Manganese phosphate: $Mn^{3+}(PO_4)$. A rare mineral formed in granite pegmatites as an alteration product of triphylite and lithiophilite; also associated with sicklerite.

YAVAPAI COUNTY: White Picacho District, as tiny needles and plates forming crusts and cavity fillings, associated with strengite (Jahns 1952); Little Giant Mine (R. Gibbs, pers. comm., 2020).

PYRARGYRITE

Silver antimony sulfide: Ag_3SbS_3. An important ore of silver formed in veins and commonly associated with galena, tetrahedrite, pyrite, and other silver-bearing sulfosalt minerals as a product of late-stage mineralization.

COCHISE COUNTY: Warren District, Bisbee, Campbell Mine, inclusions in bornite (Graeme, Graeme, and Graeme 2015).

MOHAVE COUNTY: Cerbat Mountains, Wallapai supersystem, mined at many localities in the Cerbat and Stockton Hill Districts (B. E. Thomas 1949).

PINAL COUNTY: Vekol District, Vekol Mine, as massive material in quartz (AM 24969). Galiuro Mountains, Saddle Mountain District, Little Treasure Mine, associated with acanthite, baryte, galena, pyrite, silver, quartz, and calcite (Bull. 181).

SANTA CRUZ COUNTY: Patagonia Mountains, between Hardshell and Trench Districts, Alta Mine, with bromian chlorargyrite and fluorite (Bull. 181). Palmetto District, Sonoita Mine, associated with pyrite, galena, and chalcocite (Schrader 1917).

YAVAPAI COUNTY: Bradshaw Mountains, Mount Union District, Tillie Starbuck Mine (Bull. 181). Black Canyon District, Thunderbolt Mine (3rd ed.). Tip Top District, Tip Top and Davis Mines (C. P. Kortemeier 1984). Reported in the Tiger District (Lindgren 1926).

PYRITE

Iron sulfide: FeS_2. The most common of all sulfide minerals, pyrite forms under a wide range of conditions. In Arizona it is typically an abundant associate of most metallic mineral deposits. The localities noted here are only representative of a very widespread mineral.

APACHE AND NAVAJO COUNTIES: Monument Valley and Cane Valley Districts, in the uranium-vanadium deposits, in vein fillings and as cement in sandstone (Joralemon 1952; Rosenzweig, Gruner, and Gardiner 1954; Coleman and Delevaux 1957; Jensen 1958).

COCHISE COUNTY: Warren District, Bisbee, Copper Queen and other mines, in large massive bodies (Ransome 1904; Mitchell 1920; Trischka, Rove, and Barringer 1929; Schwartz and Park 1932; Bain 1952; Bryant 1968).

COCONINO COUNTY: South Rim of the Grand Canyon, Orphan District, Orphan Mine, associated with copper-uranium-lead ores in Coconino Sandstone (Isachsen, Mitcham, and Wood 1955). Cameron District, commonly associated with uranium minerals in sedimentary rocks; Alyce Tolino Mine, where it is cobalt bearing (Rosenzweig, Gruner, and Gardiner 1954; Hamilton and Kerr 1959).

FIGURE 3.203 Pyrite, Trench Mine, Santa Cruz County, 6.4 cm.

GILA COUNTY: Banner supersystem, 79 District, 79 Mine, as a massive replacement body on the sixth level containing some highly modified crystals (Stanley B. Keith 1972). Southeastern Dripping Spring Mountains, Banner supersystem, Christmas District, Christmas Mine (C. P. Ross 1925b; N. P. Peterson and Swanson 1956; Perry 1969). Miami-Inspiration supersystem, Copper Cities District, Copper Cities Mine (N. P. Peterson 1954); Pinto Valley District, Castle Dome Mine (N. P. Peterson 1947); Miami and Inspiration Mines (Schwartz 1947). Globe Hills District, Old Dominion Mine (Ransome 1903b).

GRAHAM COUNTY: Lone Star District, as a primary sulfide mineral in the Safford porphyry copper deposit cluster (R. F. Robinson and Cook 1966), especially in the phyllic alteration zones. Aravaipa supersystem, Iron Cap District, Iron Cap Mine, with sphalerite and galena in calcsilicate skarns (Simons 1964).

GREENLEE COUNTY: Copper Mountain supersystem, Morenci District, Hudson and Fairplay veins, as large crystals (Lindgren 1905; Reber 1916; Schwartz 1947, 1958; Creasey 1959), especially in the phyllic alteration zones.

LA PAZ COUNTY: Dome Rock Mountains, Moon Mountains District, Don Welsh prospect, as crystals more than 2.5 cm in diameter (Bull. 181) in an east-west-trending quartz vein.

MOHAVE COUNTY: Cerbat Mountains, Wallapai supersystem, widespread in a belt of sulfide-bearing fissure-vein deposits (B. E. Thomas 1949; Field

1966); Ithaca Peak, with chalcopyrite, galena, sphalerite, and molybdenite in quartz-monzonite porphyry (Eidel 1966).

PIMA COUNTY: Tucson Mountains, Amole District, Arizona Tucson Mine, as crystals up to 0.25 in. diameter, with remarkably abundant crystal faces (Ayres 1924). Ajo District, New Cornelia Mine, particularly abundant in dioritic border facies; also in the Concentrator volcanics (Gilluly 1937), as a few fine crystal groups. Pima supersystem, Twin Buttes District, Twin Buttes Mine, with other primary sulfide minerals (Stanley B. Keith, pers. comm., 1973); Mission-Pima District, Mission Mine, as lustrous crystals to 10 cm (B. M. Williams 2018). Santa Rita Mountains, Santa Rita supersystem, in most of the mines and prospects (Schrader and Hill 1915). Sierrita Mountains, Pima supersystem, abundant in and near the mining districts (Ransome 1922). Silver Bell District (Kerr 1951).

PINAL COUNTY: Pioneer supersystem, Belmont District, Belmont Mine, as large crystals; Magma District—the Magma Mine is probably Arizona's premier pyrite locality, having yielded thousands of fine complex single crystals and groups, mostly from the A bed in the basal Martin Formation, as perfectly pyritohedral crystals up to 1 in. diameter, in soft clay gangue (Short et al. 1943; Mills and Eyrich 1966; Barnes and Hay 1983; Presmyk and Hay 2020); Resolution District, Resolution Mine, as an unusually strong pyrite halo that overprints the copper orebody (Hehnke et al. 2012). Mineral Creek supersystem, Ray District, as a widespread common gangue mineral associated with chalcopyrite (Ransome 1919); Galiuro Mountains, Bunker Hill (Copper Creek) District, Copper Creek, Childs-Aldwinkle Mine, as excellent crystals (Kuhn 1941). San Manuel District, San Manuel Mine, as the most abundant sulfide mineral (Schwartz 1947, 1949, 1958; Lovering 1948; Creasey 1959); Kalamazoo orebody (J. D. Lowell 1968).

SANTA CRUZ COUNTY: Patagonia Mountains, Palmetto District, Three-R Mine, as striated and twinned crystals up to 8 in. across and as large crystal aggregates; Querces District, Santo Niño Mine, as large crystal groups near the molybdenite bodies (Schrader and Hill 1915; Schrader 1917; J. W. Frondel and Wickman 1970). Washington Camp District, Duquesne, as a very large crystal (UA 5693) and crystal groups on prismatic quartz crystals. Four Metals District, Four Metals Mine, as striated crystals 2 cm on an edge (James Bleess, pers. comm., 1972). Trench District, Trench Mine (M. Hay, pers. comm., 2020). Flux District, Flux Mine, as striated cubes to 12 cm (Les Presmyk, pers. comm., 2020).

YAVAPAI COUNTY: Jerome supersystem, Verde District, Jerome, United Verde Mine, as one of the largest pyritic orebodies in the world (Fearing

1926; Lausen 1928; Schwartz 1938; C. A. Anderson and Creasey 1958; Moxham, Foote, and Bunker 1965). Other properties in the Jerome-Bradshaw Mountains area (Lindgren 1926).

#PYROCHLORE

The general formula is $A_{2-m}B_2X_{6-w}Y_{1-n}$, where A may be Na, Ca, Sr, Pb^{2+}, Sn^{2+}, Sb^{3+}, Y, U, Ag, Mn, Ba, Fe^{2+}, Bi^{3+}, Ce (and other rare-earth elements), Sc, Th, □, or H_2O; B may contain Ta, Nb, Ti, Sb^{5+}, W, V^{5+}, Sn^{4+}, Zr, Hf, Fe^{3+}, Mg, Al, or Si; X is typically O but can include subordinate OH and F; and Y is typically an anion, but can also be a vacancy, H_2O, or a very large (>> 1.0 Å) monovalent cation. The symbols m, w, and n represent parameters that indicate incomplete occupancy of the A, X, and Y sites. Pyrochlore is the name for a large group of minerals found in granite pegmatites; the exact identification of the species within the group is not known.

YAVAPAI COUNTY: White Picacho District, Outpost, Midnight Owl, and Picacho View Mines, where it is found in the intermediate quartz and perthite zones of peraluminous pegmatites (Jahns 1952).

PYROLUSITE

Manganese oxide: MnO_2. A member of the rutile group and a common and widespread manganese mineral formed under oxidizing conditions. Typically associated with other manganese minerals, such as manganite from which it forms by alteration, hausmannite, braunite, and psilomelane, as well as limonite, hematite, and goethite. Widely distributed in small amounts throughout Arizona and associated with epithermal veins in alkaline volcanic rocks, commonly associated with baryte and/or fluorite.

APACHE COUNTY: Sonsela Buttes District, Sonsela Buttes area, filling voids between pebbles or rock fragments in sandstone and conglomerate, with pyrolusite, and as nodules (Mayo 1955b).

COCHISE COUNTY: Tombstone District, in commercial quantities at the Oregon-Prompter, Lucky Cuss, Telephone, and Bunker Hill Mines (B. S. Butler, Wilson, and Rasor 1938; Rasor 1939; Romslo and Ravitz 1947; Havens et al. 1954; Hewett and Fleischer 1960). Warren District, Bisbee, Higgins Mine, with psilomelane and braunite (Palache and Shannon 1920). Texas Canyon area, as acicular crystals in vugs in the quartz-hübnerite veins that cut the Texas Canyon Quartz Monzonite and associated with the Adams Peak Leucogranite (J. R. Cooper and Silver 1964).

GILA COUNTY: Globe Hills District, with manganite and psilomelane, as the bulk of the gangue in the manganese-zinc-lead-silver deposits in the Globe Hills (N. P. Peterson 1962); as the variety polianite (UA 7756) at the Black Widow Mine. Banner supersystem, 79 District, 79 Mine, as numerous dendritic forms coating fractures in near-surface limestones (Stanley B. Keith 1972).

GRAHAM COUNTY: Fisher Hills District, northwest of Bowie, as microcrystals of the variety polianite (William Kurtz, UA x4139).

GREENLEE COUNTY: Copper Mountain supersystem, Morenci District, in black sooty masses with iron oxides in metamorphosed limestones (Lindgren 1905; Guild 1910).

LA PAZ COUNTY: Bouse Hills District, Dobbins claims, six miles east of Bouse. Southern Black Mountain District at a locality 2.5 miles west of Bouse (Bull. 181), with baryte and fluorite (Stanley B. Keith, pers. comm., 2019; Stanton B. Keith 1978). Plomosa Pass District, with baryte and fluorite, also in northeast-trending veins cutting Pliocene conglomerates (Stanton Keith 1978). Trigo Mountains, reported in the Silver District (E. D. Wilson 1933), as microcrystals (Robert Mudra, UA x185).

MARICOPA COUNTY: Big Horn Mountains, Aguila District, with manganite or wad (Bull. 181). White Picacho District, as crusts on lithium phosphate minerals in pegmatites (Jahns 1952). Black Vulture District, Black Vulture Mine, thirty-two miles south-southwest of Wickenburg, as replacements in limestone (Long, Batty, and Dean 1948).

MOHAVE COUNTY: Rawhide Mountains, Artillery District, in large deposits with wad (Head 1941; Lasky and Webber 1944, 1949); Stovall Mine, near Alamo Crossing, as good quality crystals (H 108492); Black Warrior Mine, six miles northeast of Alamo Crossing, with psilomelane (Long, Batty, and Dean 1948). Reported in veins four miles south of Hoover Dam (Bull. 181). Little Chemehuevi Valley, Pilot Rock District, Arizona manganese claims, in veins and shear zones with wad (Havens et al. 1947). Near the Colorado River, eighteen miles north of Parker Dam (Bull. 181). Artillery District, Black Jack, Price, and Priceless veins and Plancha bedded deposit, associated with a variety of oxidized manganese minerals (Mouat 1962).

PIMA COUNTY: Tucson Mountains, in quartz (UA 4486). Pima supersystem, Twin Buttes District, Twin Buttes Mine (William Hefferon, pers. comm., 3rd ed.). Santa Rosa District, Black Widow Mine, as lustrous striated crystals up to 8 mm (Les Presmyk, pers. comm., 2020).

PINAL COUNTY: East of Ripsey Wash near Kelvin, as fibrous material, pseudomorphous after manganite (Robert O'Haire, pers. comm., 1972). Pioneer supersystem, Magma District, in the outcrops of the Magma vein

(Bull. 181). Near Winkelman, Steamboat Mountain District, at several claims (Dean, Snedden, and Agey 1952). Black Hills, as microcrystals (William Kurtz, UA x4175).

YAVAPAI COUNTY: Box Canyon District, Mistake Mine, as radiating acicular pyrolusite crystals and replacing ramsdellite (Wilkinson et al. 1983).

PYROMORPHITE

Lead phosphate chloride: $Pb_5(PO_4)_3Cl$. A member of the apatite group and an uncommon mineral formed in the oxidized portion of lead deposits. Typically associated with cerussite, limonite, smithsonite, anglesite, malachite, wulfenite, vanadinite, and other oxidized minerals.

COCHISE COUNTY: Tombstone District, as small crystals associated with wulfenite (B. S. Butler, Wilson, and Rasor 1938). Warren District, Bisbee (Graeme 1993). Yellowstone District, about eight miles northeast of Benson, in quartz veins that cut gneiss, in minor amounts with chrysocolla, vauquelinite, and perite (Phelps Dodge Corp., pers. comm., 1972). Huachuca Mountains (UA 6254).

GILA COUNTY: Banner supersystem, 79 District, 79 Mine, sparingly as clear-yellow needle-like crystals on chrysocolla (Stanley B. Keith 1972).

GRAHAM COUNTY: Golondrina District, Golondrina property, in pyroclastic rocks, associated with chalcopyrite, chalcocite, and malachite; contains as much as 0.6 percent uranium (Granger and Raup 1962).

LA PAZ COUNTY: Bouse area, Iber-Plomosa Mine (3rd ed.).

MARICOPA COUNTY: White Picacho District, in pegmatites (Jahns 1952). Painted Rock District, Rowley Mine, as small green crystals up to 6 mm long (3rd ed.).

PIMA COUNTY: Cerro Colorado District, Cerro Colorado Mine (Bull. 181). South Comobabi Mountains, Cababi District, Mildren and Steppe claims (S. A. Williams 1963). Waterman District, Indiana-Arizona Mine, as small green to brown crystals (3rd ed.).

PINAL COUNTY: Mammoth District, Mammoth–St. Anthony Mine, as olive-green crystals on mottramite, with vanadinite (N. P. Peterson 1938a, 1938b). Slate District, Jackrabbit Mine, coating fault breccia and in vugs in silicified limestone (Hammer 1961).

SANTA CRUZ COUNTY: Patagonia Mountains, Trench District, Trench Mine, as encrustations (Bull. 181). Hardshell-Hermosa District, Hardshell Mine, as good yellow-green crystals (MM L589). Patagonia District,

Javalina prospect (Schrader and Hill 1915; Schrader 1917). Pajarito District, Sunset Mine Group, as microcrystals (Francis Sousa, UA x4611).

YAVAPAI COUNTY: Bradshaw Mountains, Black Canyon District, at a prospect on the Slate Creek property of Kalium Chemicals, as tiny transparent light-green barrel-shaped prismatic crystals, associated with plentiful mottramite and with botryoidal crusts of mimetite (William C. Berridge, pers. comm., 1973). Silver Mountain District, Fat Jack Mine, as green crusts (3rd ed.).

YUMA COUNTY: Castle Dome Mountains, Castle Dome District, in old mine workings, associated with wulfenite and grades into vanadinite (Blake 1881a; E. D. Wilson 1933).

PYROPE

Magnesium aluminum silicate: $Mg_3Al_2(SiO_4)_3$. A member of the garnet group that is typically formed as a metamorphic mineral in high-pressure basalts and gabbros; found in inclusions in serpentine ultramafic microbreccia diatremes and as pebbles in erosional remnants of the diatremes, especially in ant hills. It is also a rare component of calcsilicate skarns.

R050446   5 mm

FIGURE 3.205 Pyrope, Sunset Crater, Coconino County.

APACHE COUNTY: Navajo Reservation, Garnet Ridge SUM District, Garnet Ridge, as gem-quality pebbles (Helmstaedt and Doig 1972; Roden 1981). Buell Park District, Buell Park SUM diatreme, near Fort Defiance, in alluvium and agglomerate and porphyrocrysts in inclusions in igneous rock (Gregory 1917; Gavasci and Kerr 1968; Switzer 1977).

COCHISE COUNTY: Warren District, Bisbee, Gardner Mine, in skarn with magnetite and apatite (Graeme, Graeme, and Graeme 2015).

COCONINO COUNTY: Sunset Crater (R050446).

PYROPHYLLITE

Aluminum silicate hydroxide: $Al_2Si_4O_{10}(OH)_2$. An uncommon metamorphic mineral typically formed by hydrothermal alteration of feldspar in high-temperature, low-pH alteration assemblages. Commonly associated with quartz, rutile, kyanite, and andalusite; common as a contact-alteration mineral in metamorphic aureoles of peraluminous intrusions. Closely resembles talc.

COCHISE COUNTY: Warren District, at Warren, with baryte (UA 5978); common in the Sacramento Hill stock, Lavender pit (3rd ed.).

GILA COUNTY: South of Christopher Mountain, in Gordon Canyon, in a weathered zone in rhyolite beneath Mazatzal Quartzite (Donald L. Livingston, pers. comm., 1972).

LA PAZ COUNTY: Central Dome Rock Mountains, Sugarloaf Pyrophyllite District, near Quartzsite, Big Bertha Extension Mine, as well-crystallized pale-green tufts and sprays, on quartz and hematite (H 124764); Sugarloaf Peak area, where 200 to 250 ft. of pyrophyllite were encountered in a drill hole associated with natroalunite and pyrite (James D. Loghry, pers. comm., 1989; Goldsmith 2011). Granite Mountain area, K-D District, three miles southwest of Quartzsite, with dumortierite, andalusite, rutile, hematite, and kyanite (E. D. Wilson 1929).

MOHAVE COUNTY: Williams River, Cleopatra District, near Alamo, Cactus Queen Mine (Bull. 181). Reported in large quantities southeast of Yucca (Bull. 181).

PIMA COUNTY: At a locality fifty miles southwest of Ajo (UA 8865).

YUMA COUNTY: Alamo Springs District, near Alamo Springs, twenty-seven miles southeast of Quartzsite (Bull. 181).

pyroxene

A group name for minerals with the general formula $M2M1T_2O_6$, which specifies ranges of composition involving element substitutions. The T site may be filled using Si^{4+}, replaced by Al^{3+}, and then Fe^{3+} until the sum = 2.000. M1 may be filled using all Al^{3+} or Fe^{3+} in excess to fill T sites; if insufficient Al^{3+} or Fe^{3+} is available, then Ti^{4+}, Cr^{3+}, V^{3+}, Ti^{3+}, Zr^{4+}, Sc^{3+}, Mg^{2+}, Fe^{2+}, or Mn^{2+} may substitute until a sum of 1.000. M2 involves all Mg^{2+}, Fe^{2+}, or Mn^{2+} in excess of M1 sites, then Li^+, Ca^{2+}, or Na^+ may fill in until sum reaches 1.000 (see Morimoto 1988). In addition to variations in chemistry, the group includes species that differ to some extent in structural features. Individual species, where recognized, are listed individually. Common in mafic to intermediate igneous rocks, some of which consist almost entirely of pyroxene. Abundant in dark-colored volcanic rocks. Some species, especially diopside and hedenbergite, are typically formed in the contact-metamorphic environment.

PYRRHOTITE

Iron sulfide: Fe_7S_8. Formed as a high-temperature, early stage mineral in veins and as a primary mineral in some mafic igneous rocks, pegmatites, and contact-metamorphic deposits.

COCHISE COUNTY: Cochise supersystem, Johnson Camp District, Johnson Camp, sparse in a drill core taken near the Mammoth Mine (J. R. Cooper and Silver 1964).

GILA COUNTY: Southeastern Dripping Spring Mountains, Banner supersystem, Christmas District, Christmas Mine, common in the pyrrhotite-chalcopyrite zone of the lower Martin Formation (O'Carroll bed) orebody (Knoerr and Eigo 1963; Perry 1969; McCurry 1971). Sierra Ancha District, Workman Creek area, and Brush, Sorrel Horse, and Citation deposits in the Cherry Creek area, as a common disseminated constituent in hornfels and related metamorphic rocks (Granger and Raup 1969).

GREENLEE COUNTY: Copper Mountain supersystem, Morenci District, northern Morenci Mine, in an extensive contact-metamorphic assemblage (Moolick and Durek 1966).

MARICOPA COUNTY: White Picacho District, scattered in pegmatites, especially the coarse-grained interior portions (Jahns 1952).

MOHAVE COUNTY: Bunkerville District, near Littlefield, in mafic dikes with chalcopyrite and pentlandite (Bull. 181). Hualapai (Antler) District, Copper World Mine near Yucca, with sphalerite, chalcopyrite, and löllingite (Rasor 1946); Antler Mine, coated by covellite (Romslo 1948).

PIMA COUNTY: Santa Rita Mountains, Helvetia-Rosemont District, with pyrite (Schrader and Hill 1915); Busterville Mine, as blebs in sphalerite (Bull. 181). Sierrita Mountains, Pima supersystem, Sierrita District, in chalcopyrite ores (Bull. 181); Twin Buttes District, Twin Buttes Mine, as a primary mineral (Stanley B. Keith, pers. comm., 1973).

SANTA CRUZ COUNTY: Washington Camp District, in contact zones in Paleozoic limestones adjacent to the Washington Camp Quartz-Monzonite stock (Schrader 1917).

YAVAPAI COUNTY: Bradshaw Mountains, with gold ores (Bull. 181). Northeastern Bradshaw Mountains, Black Canyon District, Rainbow deposit near Turkey Creek station, in massive form (Lindgren 1926). White Picacho District, in pegmatites (Jahns 1952).

QUARTZ

Silicon oxide: SiO_2. Along with amorphous silica, quartz is by far the most abundant of the polymorphic crystalline forms of silica and, following the feldspar minerals, is the most widespread and abundant mineral in the Earth's sialic crust. Found in igneous, metamorphic, and sedimentary rocks; hydrothermal veins; and metasomatic and hot spring deposits. Only a few of the Arizona localities can be included here.

COCHISE COUNTY: Little Dragoon Mountains, Cochise supersystem, Johnson Camp District, Johnson Camp area, as well-formed crystals, some of which are large (Bull. 181); Russellville ghost town area, now the Johnson Mine, with inclusions of tiny yellow octahedral crystals of scheelite grown on phantoms within the quartz (S 105926). Huachuca Mountains, Hartford District, at Hamburg, as clear crystals twinned after the Japan law (C. Frondel 1962; H 83142; UA 9692) and as colorless, amethyst, and smoky crystals to 12 cm in Carr and Ash Canyons (Les Presmyk, pers. comm., 2020); Jack Wakefield Mine, as beautiful groups of clear crystals similar to those found at Hot Springs, Arkansas (S R15034). Courtland District (UA 8989) and Gleeson Ridge, Gleeson District, as prismatic crystals above the Defiance Mine and in the Tom Scott Mine (Stanley B. Keith, pers. comm., 2019).

GILA COUNTY: Diamond Point District, Diamond Point northeast of Payson, as clear gemmy doubly terminated crystals up to several inches that weather out of limestone over a wide area. They are commonly called *Arizona diamonds* (Peirce 1969a; Presmyk 1998). Globe Hills District, Old Dominion Mine, where crystals colored blue by associated chrysocolla line cavities in oxidized ores; some quartz is brilliant red from included finely divided hematite (Bull. 181). Green Valley District, as clear crystals up to 1 in. long at the Oxbow Mine (Bull. 181). Salt River asbestos and Seneca Districts, partly on San Carlos Apache Reservation, as tiger eye (silicified chrysotile) (Bull. 181).

GRAHAM COUNTY: Galiuro Mountains, Crystal Peak (UA 4424); Table Mountain, as twinned crystals up to 1 in. long (H 106631; UA 9521). Aravaipa supersystem, Stanley District, Stanley Butte, abundant as slender tapering clear, tan, or amethyst crystals up to 5 in. long, in places associated with and including andradite (S R12847); Grand Reef District, as good quality crystals in the veins of the Ten Strike Mine (Bull. 181). Found in the south end of the Santa Teresa Mountains as milky-white and gray crystals with fluorite and as dark smoky crystals with feldspar and epidote (Les Presmyk, pers. comm., 2020).

LA PAZ COUNTY: As large crystals in pegmatite in Precambrian granites over considerable areas (Bull. 181). Southern Plomosa Mountains, Kofa National Wildlife Reserve, Apache Chief District, Crystal Hill, southwest of Quartzsite, as large prismatic quartz crystals up to one foot long; also as a chatoyant variety from which gem-quality cats-eye cabochons have been produced (C. Frondel 1962; H 104852).

MARICOPA COUNTY: Northeast McDowell Mountains, Paraiso District, northeast of Scottsdale, rose quartz in a quarry and nearby Paraiso areas with muscovite, K-feldspar, and occasionally schorl in pegmatoid (Bull. 181); McDowell Sonoran Preserve (Gootee and Gruber 2015). Mazatzal Mountains, Four Peaks Amethyst District, Four Peaks, as gem-quality amethyst crystals of a rich red-violet color that rival the best Siberian material and as exceptional smoky quartz crystals (J. Lowell and Rybicki 1976; UA 5722). Near Lake Pleasant, as pseudomorphs of unknown mineral (UA 223). Along the Agua Fria River, seven miles west of New River, as tabular salmon-colored crystals to 4 cm as pseudomorphs after anhydrite (Les Presmyk, pers. comm., 2020).

MOHAVE COUNTY: Black Mountains, Oatman District, Moss Mine, as locally abundant rose quartz and as amethystine bands in colorless quartz of gold veins (Roedder, Ingram, and Hall 1963). Good quality specimens of rose quartz are reported from a locality forty miles northeast of Kingman (Bull. 181). Cerbat Mountains, McConnico District, northeast of Boulder Spring, as amethystine crystals in Precambrian granite; a crystal from this locality was reportedly sold to Tiffany & Co. (Guild 1910). Cerbat Mountains, Kingman Feldspar District, Kingman Feldspar Mine, in pegmatite, produced as a byproduct of feldspar mining (Heinrich 1960).

PIMA COUNTY: Tucson Mountains, near Sentinel Peak ("A" Mountain), as geodes in basalt flows (UA 6353); Contzen Pass area, with manganese oxide dendrites (UA 6324); unspecified mine dump, as crystals with copious chrysocolla inclusions (3rd ed.). Arivaca, as scepter crystals (UA 9361).

PINAL COUNTY: Galiuro Mountains, Bunker Hill (Copper Creek) District, Copper Creek Canyon, at the western end of the Aldwinkle and Longstreet claims, as Japan law twins and as individuals up to 2 in. long in a soft limonite matrix; smaller specimens are typically flattened on the prism, and the tips of crystals have smaller crystals perched on them. Some crystals show markedly flattened terminations that resemble the basal pinacoid. Red and green tourmaline is commonly enclosed in quartz crystals (William and Mildred Schupp, pers. comm., 1969; Gary M. Edson, pers. comm., 1972; UA 691, 3182). Picketpost District, in geodes in Middle Miocene, perlite-bearing rhyolite, famous for its Apache tears (UA 1813). Sacaton Mountains, Gila River Indian Reservation, as extensive outcrops of white vein quartz (E. D. Wilson 1969).

FIGURE 3.206 Quartz, Scarlet III claim, Santa Cruz County, 3.9 cm high.

Locality about three miles southwest of Pinal City ghost town, as hollow thin-walled crystals in sandstone (Kunz 1887).

SANTA CRUZ COUNTY: Patagonia Mountains, Duquesne, Washington Camp District, Holland Mine, as slender tapering crystals up to 12 in. long, some of which form Japan law twins; the individual crystals of one specimen are about 7 in. from tip to tip. These twins and slender crystals formed in a pocket with calcite, drusy siderite, and chlorite (UA 1096, 4459). This locality has probably produced the finest Japan law twins in the United States. Near Duquesne, as amethyst in pegmatite (Bull. 181); Belmont and Lead King properties, as a body 100 ft. wide containing crystals up to 2 ft. long (Schrader and Hill 1915). In geodes near Patagonia at Temporal Gulch (UA 9352), probably Ivanhoe District. Parker Canyon District, Sierra de Tordillo Mine (Bull. 181). Scarlet III claim near Ruby (Dick Morris, pers. comm., 2020).

YAVAPAI COUNTY: Bradshaw Mountains, Hassayampa District, Cash Mine, as clear crystals lining open veins, accompanied by crystals of adularia, calcite, and sulfide ore minerals (Bull. 181). Eureka supersystem, Bagdad District, Bagdad area, along the creek above the open-pit mine, as Japan law twins up to 2 in. long, commonly stained with iron oxides (Muntyan 2015a); Hillside District, near Hillside, as fern-like growths (H 106631). Date Creek, near Congress Junction, as amethyst scepters on milky quartz crystals up to 2 in. long, loose in the soil (ASDM, Hill collection). Mayer District, as Japan law twins (Robert Mudra, UA x646); about six miles east of Mayer, on a mine dump at the Yankee Boy Mine, as one group of multiply twinned crystals consisting of a central doubly terminated crystal about 5 mm long, to which are twinned (after the Japan law) two shorter individual crystals; other, smaller, simpler twins have also been collected (collection of Mrs. Donald C. Sonnenberg; Bideaux 1970). Silver Mountain District, Fat Jack Mine, near Crown King, as smoky crystals having many scepters, amethyst, phantoms, and crystals exhibiting various forms (Scovil and Wagner 1991). From Black Canyon City to the Verde Valley, as white and tan replacements after pseudohexagonal aragonite crystals to 5 cm (Les Presmyk, pers. comm., 2020).

AGATE

A term applied to chalcedony in which thin layers are typically accentuated by color differences. All gradations exist between chalcedony and agate, however, and many varieties of variegated and clouded agate have been given popular and commercial names. Moss agate is a variety in which manganese oxides form patterns. McMahan (2016) has many additional Arizona agate localities.

COCONINO AND MOHAVE COUNTIES: As nodules and geodes in the Kaibab Limestone (Guild 1910).

GRAHAM COUNTY: Deer Creek Fire Agate District, as locally gem-grade nodule fillings of fire agate in lithophysae in the vuggy zone of the Aravaipa tuff member (Krieger 1979).

GREENLEE COUNTY: Peloncillo Mountains, near Willow Springs Ranch (Dimick 1957). Round Mountain District, Bureau of Land Management, Round Mountain Rockhound Area (McMahan 2016).

MARICOPA COUNTY: Agua Fria River area, as plume agate and green and white fortification agate (3rd ed.). In an area of several miles from Gila Bend to the north (Richards 1956).

PIMA COUNTY: Tucson Mountains, near Sentinel Peak ("A" Mountain), as geodes of blue and white agate in basaltic rocks (Guild 1905). North of Pantano, as rare carnelian and blue-and-white-banded fortification agate (3rd ed.).

YAVAPAI COUNTY: Near Morgan City and Slow Springs Washes, as spherulitic nodules in lava (Bull. 181). On the Agua Fria River, west of New River station, as plume agate and mottled material (Richards 1956). A few miles north of Castle Hot Springs, as excellent gray, blue, pink, and violet material (Richards 1956).

CHALCEDONY

Cryptocrystalline quartz with fibrous microtexture formed as transparent to translucent crusts and coatings and as mammillary, botryoidal, nodular, and irregular masses; exhibits greasy or waxy luster. Commonly shows banding parallel to a free surface or to the surface on which it was deposited. Formed under low-temperature conditions by hydrothermal and weathering processes. Several variety names are applied to chalcedonic silica based mainly on color and textural characteristics.

APACHE AND NAVAJO COUNTIES: Several localities, as the principal constituent of petrified wood (see Petrified [Fossil] Wood below).

GILA COUNTY: Miami-Inspiration supersystem, Inspiration Mine, Live Oak Pit, colored blue or green from included chrysocolla or malachite and referred to as gem silica (Bull. 181).

GRAHAM COUNTY: Duncan area, Black Hills Fire Agate District, as an unusual and attractive type, locally termed fire agate, which exhibits a play of colors like that of fire opal (3rd ed.).

GREENLEE COUNTY: Copper Mountain supersystem, Shannon District, in limestone; also loose at Shannon Mountain (3rd ed.).

LA PAZ COUNTY: Chocolate Mountains, as excellent specimens (Bull. 181). Locality east of Parker, near milepost 87 of the Santa Fe Railway (Bull. 181).

MARICOPA COUNTY: South of Aguila (AM 31736). Saddle Mountain, as fire agate (3rd ed.).

PINAL COUNTY: Galiuro Mountains, in the pass between Little and Big Table Top Mountains, abundant as chalcedonic roses up to 12 in., and as other forms (3rd ed.). Ray District, Ray Mine, in the form of gem silica colored blue by chrysocolla (Les Presmyk, pers. comm., 2020).

SANTA CRUZ COUNTY: Southwestern Santa Rita Mountains, Grosvenor Hills, near the old village of Santa Cruz. Canelo Hills south of Sonoita (Bull. 181). Northwestern Santa Rita Mountains, Cottonwood Canyon, Tyndall District, Glove Mine, as pseudomorphs after calcite and as casts of hemimorphite crystals (H. J. Olson 1966; UA 700, 9367).

YAVAPAI COUNTY: Near Morgan City and Slow Spring Washes, as spherulitic nodules in lavas; much of the chalcedony is fluorescent (Bull. 181). Santa Maria Mountains, Saddle Mountain, about thirty miles northwest of Hassayampa, as abundant roses weathered from volcanic rocks (W. Rogers 1958).

YUMA COUNTY: Kofa Mountains, twenty miles southeast of Quartzsite, lining geodes in rhyolite (Walker 1957); west of the Kofa Mountains and north of the Castle Dome Mountains, about twenty-six miles south of Quartzsite, in geodes (Weight 1949).

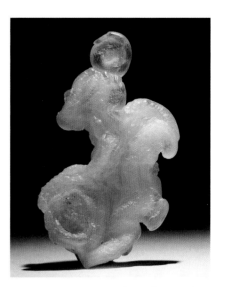

FIGURE 3.207 Quartz, variety chalcedony, Aquila, Maricopa County, 3.5 cm high.

CHERT, FLINT, AND JASPER

These are the massive fine-grained granular form of quartz. To most collectors, the difference between these materials is the color. Most cherts are fairly light in color and may be grayish-white, gray, yellowish, reddish, or brownish; flint is commonly darker, grays and black; and jasper is red, green, or yellow. They form in sedimentary rocks, especially limestones, in which they may be of primary or secondary origin. Very large beds of chert have formed in some limestones; the smaller bodies of chert assume a great variety of shapes. They are also found in metamorphic rocks such as jaspilite.

COCHISE AND PIMA COUNTIES: As abundant chert in some beds of Permian limestones, which are widely distributed in southeastern Arizona; also abundant in one bed of the Earp Formation, as small reddish jellybean-like bodies (Bull. 181; Gilluly 1956). Ajo District, New Cornelia Mine, as jasper associated with cuprite and copper and including shattuckite (UA 3431).

COCONINO AND MOHAVE COUNTIES: As abundant chert in the Permian Kaibab Limestone (Bull. 181).

GILA AND PINAL COUNTIES: As abundant black seams and lenses in the Mississippian Escabrosa Limestone (Stanley B. Keith, pers. comm., 2019).

MARICOPA COUNTY: Eighteen miles north of Cave Creek, as bright-red jasper (Sinkankas 1959).

CHRYSOPRASE

Chalcedony colored green by inclusions of nickel silicate (C. Frondel 1962). Other green or greenish-blue chalcedonies, such as those from the Live Oak and Keystone Mines in the Miami-Inspiration supersystem (Guild 1910), are not chrysoprase, because they are colored by chrysocolla. Most chrysoprase is a result of supergene weathering of nickel-rich serpentinite sources.

MOHAVE COUNTY: Reported on the western slopes of the northern Black Mountains, Pilgrim District (Bull. 181; Householder 1930).

LA PAZ COUNTY: Plomosa Mountains, north of Interstate 10—Raymond Perry had a chrysoprase mine in the 1960s. The chalcedony was colored green by the nickel mineral willemseite (Melchiorre 2017).

PETRIFIED (FOSSIL) WOOD

Fossil wood is petrified when replaced by cryptocrystalline quartz, largely chalcedony and jasper. As silica carried in underground waters replaced the original wood, even the most minute details of the original woody structures were commonly preserved. Petrified wood is the official Arizona state fossil.

APACHE AND NAVAJO COUNTIES: Petrified Forest National Monument is world-renowned for the abundance and quality of fossil wood in the Petrified Forest Member of the Chinle Formation, which extends over many square miles. Tree logs, trunks, limbs, and fragments are preserved, typically in beautiful colors. Near Nazlini Canyon, north of Ganado, and at numerous other localities (Gregory 1917).

COCONINO, MOHAVE, AND YUMA COUNTIES: Abundant along the banks of the Colorado River (Guild 1910; Wilson and Butler 1930).

PIMA COUNTY: Tucson Mountains area, where it is common in the Cretaceous Amole Formation (Donald L. Bryant, pers. comm., 1972). Empire Mountains, in middle Cretaceous sandstones of the upper Bisbee Group

(3rd ed.). Western Whetstone Mountains, Spring Water Canyon area, in sandy facies of the middle Cretaceous, Turney Ranch Formation of the upper Bisbee Group (Klute 1991; Lucas and Heckert 2005), where the petrified wood is locally associated with vanadinite and illite (collected from a single petrified log by Bob Downs).

SANTA CRUZ COUNTY: Santa Rita Mountains, Adobe Canyon, in Cretaceous clastic sedimentary rocks (3rd ed.).

QUEITITE

Zinc lead silicate sulfate: $Zn_2Pb_4(Si_2O_7)(SiO_4)(SO_4)$. A rare secondary mineral formed in the oxide zone of lead- and zinc-bearing deposits.

COCHISE COUNTY: Tombstone District, Lucky Cuss Mine, as a chalky-white calcian variety that replaces alamosite and forms masses of spherulitic nodules (3rd ed.).

PINAL COUNTY: Mammoth District, Mammoth—St. Anthony Mine (Robert Meyer, pers. comm., 2019).

QUETZALCOATLITE

Copper zinc tellurate hydroxide silver lead chloride: $Cu^{2+}_3Zn_6Te^{6+}_2O_{12}$ $(OH)_6 \cdot (Ag,Pb,\square)Cl$. A rare mineral formed by the oxidation of polymetallic mineral deposits rich in tellurium.

COCHISE COUNTY: Tombstone District, Old Guard Mine, in one tiny specimen, as small nodules of coarsely granular bright-blue crystals cemented with gold; the pockets may form a core to masses of dugganite rimmed with khinite (S. A. Williams 1978).

RAMEAUITE

Potassium calcium uranyl oxide hydroxide hydrate: $K_2Ca(UO_2)_6O_6$ $(OH)_4 \cdot 6H_2O$. A secondary uranium mineral associated with uraninite and other oxidized uranium minerals.

COCONINO COUNTY: South Rim of the Grand Canyon, Orphan District, in ores of the Orphan Mine (3rd ed.).

RAMMELSBERGITE

Nickel arsenide: $NiAs_2$. Formed at moderate temperatures in hydrothermal veins and found with other nickel and cobalt minerals.

COCONINO COUNTY: South Rim of the Grand Canyon, Orphan District, Orphan Mine, associated with uraninite and chalcocite (Gornitz 1986).

FIGURE 3.208 Ramsbeckite, 79 Mine, Gila County.

#RAMSBECKITE

Copper sulfate hydroxide hydrate: $Cu_{15}(SO_4)_4(OH)_{22} \cdot 6H_2O$. This secondary mineral forms from the oxidation of primary copper and zinc minerals.

GILA COUNTY: Banner supersystem, 79 District, 79 Mine, found as greenish-blue crusts and small tabular crystals on the sixth level and as coatings on the underside of sulfide-bearing rocks in the dump (R050295).

FIGURE 3.209 Ramsdellite, Mistake Mine, Yavapai County, 4 × 6 cm.

RAMSDELLITE

Manganese oxide: MnO_2. An orthorhombic polymorph of pyrolusite that is found in veins and bedded manganese deposits with various other oxidized manganese minerals.

MARICOPA COUNTY: Big Horn Mountains, Aguila District, Black Rock Mine, in hypogene veins in Precambrian crystalline gneisses (Hewett 1964).

MOHAVE COUNTY: Artillery Mountains, Artillery District, Black Jack, Price, and Priceless veins, in veinlets along fractures and joints, as small tabular to blocky crystals, with hollandite, psilomelane, cryptomelane, coronadite, pyrolusite, and lithiophorite; Plancha bedded manganese deposit (Mouat 1962; UA 9404).

PINAL COUNTY: Mammoth District, Mammoth–St. Anthony Mine, as pseudomorphs after groutite microcrystals (Bideaux 1980); Malpais Hill, as pseudomorphs (Thomssen 1983).

YAVAPAI COUNTY: Rich Hill District, locality east of Octave Mine and northeast of Wickenburg, as masses of equant to tabular crystals up to about 5 mm long (UA 10086). Box Canyon District, Mistake Mine, as crystals up to 1 cm long (Wilkinson et al. 1983).

RANCIÉITE

Calcium manganese oxide hydrate: $(Ca,Mn^{2+})_{0.2}(Mn^{4+},Mn^{3+})O_2 \cdot 0.6H_2O$. A rare mineral formed in the oxidized zone of mineral deposits, associated with limonite.

COCHISE COUNTY: Southeastern Dragoon Mountains, Courtland District, on the mine dumps at Courtland, in crystalline form on goethite (Hidemichi Hori, pers. comm., 1986).

PIMA COUNTY: Ajo District, Phelps Dodge Well No. 1, in vugs in volcanic rock with zeolites (William Thomas, pers. comm., 1988; William Hunt, UA x3229).

RANSOMITE

Copper iron sulfate hydrate: $CuFe^{3+}_2(SO_4)_4 \cdot 6H_2O$. A rare mineral originally found at Jerome, where it formed as a result of a mine fire. The United Verde Mine in Yavapai County is the type locality.

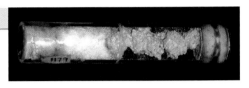

FIGURE 3.210 Ransomite, United Verde Mine, Yavapai County, 4.5 cm.

COCHISE COUNTY: Warren District, Bisbee, Cole and Campbell Mines, as a postmine mineral associated with römerite in warm (oxidizing) parts of pyritic ores (Graeme 1981).

YAVAPAI COUNTY: Black Hills, Jerome supersystem, Verde District, United Verde Mine, as crusts and small tufts of crystals formed by the burning of pyritic ore (Lausen 1928; Wood 1970).

#RASPITE

Lead tungstate: $Pb(WO_4)$. An uncommon secondary mineral found in the oxidized portions of tungsten-bearing base-metal deposits.

COCHISE COUNTY: Huachuca Mountains, Reef District, Reef Mine, as microtwinned and single monoclinic colorless to dark-brown crystals containing up to 25 percent tellurium in quartz openings associated with galena, gold, and stolzite (Walstrom 2013). Tombstone District, Grand Central Mine, as elongated prismatic to tabular pale-yellow to colorless crystals, associated with chlorargyrite, emmonsite, ottoite, quartz, and jarosite (R130514); a tellurium-rich raspite was also found at the Grand Central Mine (Andrade et al. 2014).

RAUVITE

Calcium uranyl vanadium oxide hydrate: $Ca(UO_2)_2V_{10}O_{28} \cdot 16H_2O$. A secondary mineral found in masses, as crusts and coatings, and as interstitial matter in sandstone, with other oxidized uranium-vanadium minerals, in Colorado Plateau–type sandstone-hosted uranium-copper deposits.

APACHE COUNTY: Cane Valley District, Monument No. 2 Mine (Weeks, Thompson, and Sherwood 1955; Finnell 1957; Witkind and Thaden 1963).

NAVAJO COUNTY: Monument Valley District, Monument No. 1 Mine and the adjoining Mitten No. 2 Mine (Witkind 1961; Witkind and Thaden 1963).

#RAYGRANTITE

Lead zinc sulfate silicate hydroxide: $Pb_{10}Zn(SO_4)_6(SiO_4)_2(OH)_2$. A very rare secondary mineral found in cavities in massive galena within the remnants of a galena-pyrite-chalcopyrite vein. It is a member of the iranite group. The Evening Star Mine is the type locality.

MARICOPA COUNTY: Big Horn Mountains, Big Horn District, Evening Star Mine, as very small clear bladed crystals always as fishtail twins in vugs in galena associated with anglesite, cerussite, lanarkite, leadhillite, mattheddleite, and other secondary lead minerals (Yang et al. 2016; R120151).

REALGAR

Arsenic sulfide: AsS. Found as a minor constituent in some gold, silver, and lead veins with orpiment, stibnite, and other arsenic minerals. It can also be formed by volcanic fumarolic action and some hot springs. Typically associated with orpiment, to which it readily alters.

PINAL COUNTY: In 1915, several pounds of realgar and orpiment were discovered at an unspecified locality near the junction of the Gila River and Hackberry Wash (Bull. 181),

YAVAPAI COUNTY: Bradshaw Mountains, Castle Creek District, near Castle Hot Springs (Bull. 181).

#RECTORITE

Sodium calcium aluminum silicate hydroxide hydrate: $(Na,Ca)Al_4(Si,Al)_8O_{20}(OH)_4·2H_2O$. A clay that forms as a low-temperature alteration mineral, 1:1 regular interstratification of a dioctahedral mica and a trioctahedral smectite.

COCHISE COUNTY: Northeastern Whetstone Mountains, Kartchner District, from Kartchner Caverns southwest of Benson (C. A. Hill 1999).

REEVESITE

Nickel iron carbonate hydroxide hydrate: $Ni_6Fe^{3+}_2(CO_3)(OH)_{16}\cdot4H_2O$. A secondary mineral found in altered meteorites and ultramafic rocks, a member of the hydrotalcite group.

COCHISE COUNTY: Southern Dragoon Mountains, Barrett Camp District, in the ghost town of Barrett Camp, in veins in crystalline dolomite, as greenish-yellow microcrystalline material that appears to be the product of leaching or dissolution of pecoraite; associated with earthy goethite (3rd ed.).

#REICHENBACHITE

Copper phosphate hydroxide: $Cu_5(PO_4)_2(OH)_4$. Found as a rare secondary copper mineral in oxide zones.

YAVAPAI COUNTY: Agua Fria District, from the Binghampton Mine with libethenite (R. Thomssen, pers. comm., 2000).

#RELIANCEITE-(K)

Potassium magnesium vanadate oxalate phosphate hydrate: $K_4Mg(V^{4+}O)_2(C_2O_4)(PO_3OH)_4(H_2O)_{10}$. A new organic mineral, the Rowley Mine is the type locality.

MARICOPA COUNTY: Painted Rock District, Rowley Mine, as very small pale blue-green elongated prisms associated with dendoraite-(NH_4), antipinite, fluorite, rowleyite, salammoniac, and others on a baryte-quartz matrix (A. R. Kampf, Cooper, Celestian, Ma, Marty 2021).

RHODOCHROSITE

Manganese carbonate: $Mn(CO_3)$. A member of the calcite group and a common gangue mineral in sulfide mineral deposits formed under a wide range of temperatures. It is also a secondary mineral in iron and manganese oxide deposits.

APACHE COUNTY: Springerville–St. Johns area, as nodules of iron oxide and rhodochrosite formed in small amounts in the Chinle Formation. In the northwestern part of Antelope Valley, these nodules weather out of the uppermost shale unit of the Chinle Formation and cover the shale slope (Sirrine 1958).

FIGURE 3.212 Rhodochrosite, with alabandite, Junction Mine, Bisbee, Cochise County, 4.3 cm.

COCHISE COUNTY: Tombstone District, Lucky Cuss Mine, as small grains in oxidized alabandite ore (Hewett and Rove 1930; B. S. Butler, Wilson, and Rasor 1938; Rasor 1939). Warren District, Bisbee, Higgins Mine, replacing dolomitic limestone, in drusy cavities in alabandite (Hewett and Rove 1930; Hewett and Fleischer 1960). Chiricahua Mountains, California supersystem, Leadville District, Humboldt Mine, with rhodonite, calcite, and quartz, associated with alabandite in lenses in a fissure vein that cuts limestone (Hewett and Rove 1930).

COCONINO COUNTY: In breccia pipes being mined for uranium (Wenrich and Sutphin 1988).

GILA COUNTY: Banner supersystem, Chilito District, London Range shaft. Globe Hills District, Ramboz (Silver Glance) deposit, with manganoan ankerite, the principal hypogene gangue mineral (Hewett and Fleischer 1960; N. P. Peterson 1962; Hewett, Fleischer, and Conklin 1963).

GREENLEE COUNTY: Ash Peak District, in a vein with quartz, calcite, and pyrite (Spencer and Welty 1989; Richter and Lawrence 1983).

PIMA COUNTY: Rincon Mountains, with pyrolusite (UA 7554). Pima supersystem, Twin Buttes District, Twin Buttes Mine (William Hefferon, pers. comm., 3rd ed.).

PINAL COUNTY: Pioneer supersystem, Magma District, Magma Mine, as massive material and as rare small crystals in the south vein (Barnes and Hay 1983).

SANTA CRUZ COUNTY: Patagonia Mountains, Trench District, Trench Mine, associated with alabandite, sphalerite, and galena (Schrader and Hill 1915; Schrader 1917; Hewett and Rove 1930). Northwestern Santa Rita Mountains, Cottonwood Canyon, Tyndall District, Glove Mine, with manganese oxide minerals (H. J. Olson 1966).

YAVAPAI COUNTY: White Picacho District, an alteration product of lithiophilite in pegmatites (London and Burt 1982a).

YUMA COUNTY: Kofa Mountains, Kofa District, North Star Mine, in veinlets with boulangerite, pyrite, chlorite, epidote, and fluorite (Hankins 1984). Sheep Tanks District, Sheep Tanks Mine, Booth Bonanza deposit, as brown veins in black-banded calcite veins (Cousins 1972).

RHODONITE

Calcium manganese silicate: $CaMn_3Mn(Si_5O_{15})$. A member of the pyroxenoid group that is commonly formed through hydrothermal processes, found in many manganese orebodies and pegmatites. It may also be formed in contact-metamorphic deposits through the alteration of rhodochrosite.

COCHISE COUNTY: Chiricahua Mountains, California supersystem, Leadville District, Humboldt Mine, with rhodochrosite and alabandite in lenses in a fissure vein in limestone (Hewett and Rove 1930).

PIMA COUNTY: Pima supersystem, Twin Buttes District, unspecified locality in the Twin Buttes area (UA 4496); Twin Buttes Mine (William Hefferon, pers. comm., 3rd ed.). Reported near Sasabe (Lee Hammonds, pers. comm., 1974). Baboquivari District, Baboquivari Mine, with manganese oxides (Stanton B. Keith 1974).

SANTA CRUZ COUNTY: South-central Patagonia Mountains, in quartz gangue in veins cutting Tertiary volcanic rocks (Schrader 1917). Salero District, Toluachi Group, in veins with rhodochrosite (Schrader and Hill 1915).

YAVAPAI COUNTY: Zannaropolis District, Zannaropolis Group, found with scheelite, epidote, pyrite, magnetite, garnet, calcite, and fluorite (Dale 1961).

YUMA COUNTY: Kofa Mountains, Kofa District, King of Arizona Mine, with galena, pyrite, and fluorite in veins in the Eichelberger exotic breccia (Hankins 1984).

RHODOSTANNITE

Copper iron tin sulfide: $Cu^{1+}(Fe^{2+}_{0.5}Sn^{4+}_{1.5})S_4$. Found as an alteration product of stannite.

COCHISE COUNTY: Warren District, Bisbee, Campbell Mine, as a component of the sulfide ores (Graeme 1993).

RHOMBOCLASE

Oxonium hydrate iron sulfate hydrate: $(H_5O_2)Fe^{3+}(SO_4)_2 \cdot 2H_2O$. A secondary mineral formed from the oxidation of pyrite as an encrustation on mine walls, associated with chalcanthite, römerite, and epsomite.

COCHISE COUNTY: Warren District, Bisbee, Copper Queen Mine, in porous crusts with römerite (Merwin and Posnjak 1937); Czar shaft, 400 level (Graeme 1981).

PINAL COUNTY: Pioneer supersystem, Magma District, Magma Mine, 1,000 ft. level, with voltaite and szomolnokite (3rd ed.).

RICHTERITE

Sodium calcium magnesium silicate hydroxide: $Na(NaCa)Mg_5Si_8O_{22}(OH)_2$. A member of the amphibole group, formed in igneous and contact-metamorphosed rocks.

COCONINO COUNTY: Canyon Diablo area, associated with krinovite, roed-derite, high albite, kosmochlor, and chromite in the Canyon Diablo octa-hedrite meteorite (Olsen and Fuchs 1968).

RICKARDITE

Copper telluride: $Cu_{3-x}Te_2$, where $x \approx 0.2$. A rare late-formed mineral found in veins with pyrite, tellurium, and other tellurides.

COCHISE COUNTY: Warren District, Bisbee, 1,400 ft. level of the Junction Mine, as small purple fragments in a sample of sulfide pulp; the identi-fication was based on the characteristic purple-red color and a positive qualitative test for tellurium (Crawford 1930); Campbell Mine (Graeme 1993). Tombstone District, Little Joe shaft, where it replaces empressite as patches of purple platelets in adularia gangue (3rd ed.).

RIEBECKITE

Sodium iron silicate hydroxide: $\square Na_2(Fe^{2+}_3Fe^{3+}_2)Si_8O_{22}(OH)_2$. Found in granitic igneous rocks and low-grade regionally metamorphosed schists.

COCONINO COUNTY: Southern San Francisco Mountains, at several places around Fremont Peak, in a blue to blue-gray fine-grained rhyolite with sanidine phenocrysts (Updyke 1977).

PIMA COUNTY: Northern Sierrita Mountains and southern Roskruge Mountains, as the asbestiform variety crocidolite (Robert O'Haire, pers. comm., 1972; UA 3065, 9212).

YUMA COUNTY: Castle Dome Mountains, as very fine-grained veins and impregnations in Mesozoic gneiss (Haxel et al. 2002).

ROALDITE

Iron nickel nitride: $(Fe,Ni)_4N$. Found as platelets in nickel-rich iron in iron meteorites.

COCONINO COUNTY: Meteor Crater near Winslow, in the Canyon Diablo iron meteorite (E. R. D. Scott, pers. comm., 1988).

ROBERTSITE

Calcium manganese oxide phosphate hydrate: $Ca_2Mn^{3+}_3O_2(PO_4)_3\cdot3H_2O$. A rare mineral previously reported only in certain pegmatites near Custer,

South Dakota, where it is associated with other dark-colored late-stage iron and manganese phosphate minerals.

YAVAPAI COUNTY: White Picacho District, altered lithiophilite in pegmatites (London and Burt 1982a).

RODALQUILARITE

FIGURE 3.213 Rodalquilarite, Grand Central Mine, Tombstone, Cochise County, FOV 2 mm.

Hydrogen iron tellurite chloride: $H_3Fe^{3+}_2(Te^{4+}O_3)_4Cl$. A very rare mineral found in hydrothermal deposits rich in tellurium.

COCHISE COUNTY: Tombstone District, where it was first found by R. W. Thomssen at the Joe Mine; common as highly perfect 1 mm crystals and as cleavage plates up to 2.5 cm across, with a distinctive oil-green to pistachio-green color inclining toward a bright cadmium yellow; formed on fractures in silicified and opalized pyritic shales, with emmonsite (to which it alters), sonoraite, and various other tellurides and tellurites.

The mineral that W. F. Hillebrand (1885) described as emmonsite from Tombstone (the type locality) was probably rodalquilarite. The physical description, optics, and chemical analysis are close to those of rodalquilarite. The emmonsite type locality should therefore be changed to Cripple Creek, Colorado.

ROEDDERITE

Potassium sodium magnesium silicate: $KNaMg_2(Mg_3Si_{12})O_{30}$. Found in iron meteorites.

COCONINO COUNTY: Canyon Diablo area, with krinovite in graphite nodules in the Canyon Diablo octahedrite meteorite; also associated with richterite, high albite, kosmochlor, and chromite (Olsen and Fuchs 1968).

ROMANÈCHITE

Barium hydrate manganese oxide: $(Ba,H_2O)_2(Mn^{4+},Mn^{3+})_5O_{10}$. A secondary oxide mineral that is found in many types of deposits and is probably more common than the literature suggests.

YAVAPAI COUNTY: Box Canyon District, Mistake Mine, with other manganese oxides (Wilkinson et al. 1983).

RÖMERITE

Iron sulfate hydrate: $Fe^{2+}Fe^{3+}_2(SO_4)_4 \cdot 14H_2O$. A secondary mineral commonly formed by the alteration of pyrite and typically associated with copiapite and other secondary sulfates. The louderbackite of Lausen (1928) was subsequently shown to be this species (Pearl 1950).

COCHISE COUNTY: Warren District, Bisbee, Copper Queen Mine, in porous crusts (Merwin and Posnjak 1937); Campbell shaft, as thick crusts on floors and walls; Junction shaft, as spongy crusts on the floor (Graeme 1981).

YAVAPAI COUNTY: Black Hills, Jerome supersystem, Verde District, United Verde Mine, as thin crusts on pyrite formed under fumarolic conditions resulting from burning pyritic ores (Lausen 1928; Wood 1970).

FIGURE 3.214 Rongibbsite, near Evening Star Mine, Maricopa County, FOV 1.3 mm.

#RONGIBBSITE

Lead aluminosilicate hydroxide: $Pb_2(Si_4Al)O_{11}(OH)$. A rare mineral found in the oxidized portions of polymetallic deposits. This is the first natural aluminosilicate with Pb as the only extra-framework cation. The Big Horn District locality is the type locality.

MARICOPA COUNTY: Big Horn Mountains, Big Horn District, unnamed prospect near the Evening Star Mine, as clear bladed crystals in divergent groups associated with wickenburgite, fornacite, creaseyite, and iranite in an oxidized quartz vein (Yang et al. 2012; R100031).

#ROQUESITE

Copper indium sulfide: $CuInS_2$. Found with other sulfides in high-temperature hydrothermal deposits.

COCHISE COUNTY: Warren District, Bisbee, Campbell Mine, 2000 ft. level in bornite sulfide ore (Schumer 2017).

ROSASITE

Copper zinc carbonate hydroxide: $CuZn(CO_3)(OH)_2$. A secondary mineral found with aurichalcite, brochantite, malachite, and other secondary minerals in the oxidized portions of copper-zinc-lead deposits.

COCHISE COUNTY: Tombstone District, Toughnut and Empire Mines, as bright-green mammillary spherules in siliceous linings of vugs and between hemimorphite crystals (B. S. Butler, Wilson, and Rasor 1938). Southeastern Dragoon Mountains, Gleeson District, Gleeson Ridge, with smithsonite and manganese oxides; Silver Bill and Mystery Tunnel Mines, as large specimens of botryoidal crusts, with smithsonite, aurichalcite, and hemimorphite on manganese oxides (Knudsen 1983). Warren District, Bisbee, common in gossans in underground workings, as dense spherules and stalactitic material of radiating blue-green fibers, with calcite, aurichalcite, and hemimorphite (Graeme 1981).

FIGURE 3.215 Rosasite, Silver Bill Mine, Cochise County, 8.2 cm.

GILA COUNTY: Southern Dripping Spring Mountains, Banner supersystem, 79 District, 79 Mine, in the oxidized portion as "deep blue-green velvety mats or warty crusts encrusting manganese oxides and smithsonite. Some of the mats are associated with aurichalcite, smithsonite, wulfenite, calcite, malachite, and mimetite"; formed contemporaneously with aurichalcite (Stanley B. Keith 1972).

MARICOPA COUNTY: Belmont Mountains, Osborne District, Tonopah-Belmont Mine, as balls of acicular crystals up to 4 mm in diameter (G. B. Allen and Hunt 1988).

PIMA COUNTY: Waterman Mountains, Waterman District, Silver Hill Mine (UA 6937, 5537). South Comobabi Mountains, Cababi District, Mildren and Steppe claims (S. A. Williams 1963). Pima supersystem, Mission-Pima District, San Xavier West Mine, intergrown with flat bladed calcite (Arnold 1964). Empire Mountains, Empire District, Chief Mine, as microcrystals with hemimorphite (Francis Sousa, UA x524).

SANTA CRUZ COUNTY: Santa Rita Mountains, Cottonwood Canyon, Tyndall District, Glove Mine (H. J. Olson 1966).

ROSCOELITE

Potassium vanadium aluminosilicate hydroxide: $KV^{3+}_2(Si_3Al)O_{10}(OH)_2$. A rare member of the mica group found in low-temperature ore deposits and in the oxidized portions of sedimentary uranium-vanadium ores.

APACHE COUNTY: Cane Valley District, Monument No. 2 Mine, with many other uranium-vanadium minerals (Gruner and Gardiner 1952).

COCHISE COUNTY: Warren District, Bisbee, Campbell orebody (Graeme 1993).

NAVAJO COUNTY: Monument Valley District, Mitten No. 2 Mine (Witkind and Thaden 1963); Monument No. 1 Mine (Holland et al. 1958).

ROSSITE

Calcium vanadium oxide hydrate: $Ca(VO_3)_2 \cdot 4H_2O$. A rare mineral associated with the oxidized uranium-vanadium deposits of the Colorado Plateau. Intimately associated with metarossite, to which it alters.

APACHE COUNTY: Lukachukai District, reported in the Mesa No. 1 Mine (Bull. 181); in mines in the district (Page, Stocking, and Smith 1956).

FIGURE 3.216 Rowleyite, Rowley Mine, Maricopa County, FOV 0.56 mm.

#ROWLEYITE

Ammonium-rich vanadate: $[Na(NH_4,K)_9Cl_4][V_2^{5+,4+}(P,As)O_8]_6 \cdot n[H_2O,Na,NH_4,K,Cl]$. A rare secondary mineral formed with a suite of low-temperature postmining minerals, some containing ammonia. The Rowley Mine is the type locality.

MARICOPA COUNTY: Painted Rock District, Rowley Mine, as very small equant dark-blue crystals with mottramite (Kampf et al. 2016).

ROZENITE

Iron sulfate hydrate: $Fe^{2+}(SO_4) \cdot 4H_2O$. A secondary sulfate commonly formed as a postmine efflorescence.

COCHISE COUNTY: Warren District, Bisbee, Campbell shaft, 2700 level as a powdery white efflorescence on pyrite with other minerals in postmine sulfate assemblages (Graeme 1981).

RUCKLIDGEITE

Lead bismuth telluride: $PbBi_2Te_4$. A mineral found in hydrothermal veins associated with gold, arsenopyrite, boulangerite, calaverite, and other hydrothermal minerals.

COCHISE COUNTY: Warren District, Bisbee, Campbell orebody, replacing granular pyrite in the massive sulfide ores (Graeme 1993).

RUIZITE

Calcium manganese silicate hydroxide hydrate: $Ca_2Mn^{3+}_2Si_4O_{11}(OH)_4 \cdot 2H_2O$. A rare mineral formed during retrograde metamorphism and oxidation of a contact-metamorphic calcsilicate assemblage developed in limestones. The Christmas Mine is the type locality.

GILA COUNTY: Southeastern Dripping Spring Mountains, Banner supersystem, Christmas District, Christmas Mine, as orange spherules of radial acicular crystals, in veinlets and on fracture surfaces in metamorphosed limestone that contain wollastonite, grossular, diopside, and vesuvianite; associated with junitoite, kinoite, apophyllite, smectite, xonotlite, and sepiolite (S. A. Williams and Duggan 1977).

FIGURE 3.217 Ruizite, Christmas Mine, Gila County, FOV 3 mm.

#RUTHERFORDINE

Uranium oxide carbonate: $(UO_2)CO_3$. A secondary uranium mineral that is formed by the alteration of uraninite.

LA PAZ COUNTY: Northern Plomosa Mountains, Rayvern District, Rayvern No. 2 Mine, with sharpite and carnotite (Melchiorre 2017).

RUTILE

Titanium oxide: TiO_2. A mineral that is widespread as an accessory mineral in granitic igneous rocks. It is also common in metamorphic rocks, as a product of the alteration of other titanium-bearing silicates, and as a detrital mineral in sands and sedimentary rocks. Cuproan rutile, or red rutile, is an important exploration guide to chalcopyrite-rich areas of porphyry copper deposits (S. A. Williams and Cesbron 1977). Rutile is a common accessory mineral in advanced argillic alteration assemblages and is variously associated with pyrophyllite, anatase, andalusite, and sillimanite.

FIGURE 3.218 Rutile, Santo Niño Mine, Santa Cruz County, 5.1 cm.

APACHE COUNTY: Monument Valley area, Garnet Ridge SUM District, as mineral fragments in a breccia dike that pierces the sedimentary rocks; common as needle-like crystals in garnet (Gavasci and Kerr 1968).

COCHISE COUNTY: Warren District, northwest of Bisbee, as an accessory mineral in granite; Sacramento Hill, in an argillically altered granite porphyry, as clusters of tiny crystals in pseudomorphs after biotite (Schwartz 1947).

GILA COUNTY: Miami-Inspiration supersystem, Miami District, Miami orebody, as a product of hydrothermal alteration of granite porphyry, and near the orebody, in schist (Schwartz 1947, 1958); Pinto Valley District, Castle Dome Mine, as a product of the clay mineral phase of alteration, recrystallized from titanite and ilmenite (N. P. Peterson, Gilbert, and Quick 1946; Schwartz 1947).

GILA AND PINAL COUNTIES: As an accessory mineral in the Madera diorite, Pinal Schist, and Solitude and Ruin Granites (N. P. Peterson 1962).

GREENLEE COUNTY: Copper Mountain supersystem, Morenci District, as clusters replacing ferromagnesian minerals in an intensely altered porphyry copper orebody, associated with a suite of hydrothermal alteration minerals (Reber 1916; Schwartz 1947, 1958).

LA PAZ COUNTY: Eastern Dome Rock Mountains, Middle Camp District, three miles southwest of Quartzsite, as small grains scattered throughout Precambrian schist and as well-formed crystals in fractures in quartz veins (E. D. Wilson 1929); also in kyanite-rich rocks (Marsh and Sheridan 1976).

MOHAVE COUNTY: Cerbat Mountains, Wallapai supersystem, Mineral Park District, in a hydrothermally altered, disseminated sulfide deposit in a porphyritic intrusive, as sagenite webs and in clusters of crystals; also replacing titanite (B. E. Thomas 1949).

PIMA COUNTY: Ajo District, as stumpy crystals that are an alteration product of biotite (Schwartz 1958; Hutton and Vlisidis 1960). Silver Bell District, as small acicular crystals associated with quartz and as small grains and prismatic crystals (Kerr 1951). Pima supersystem, Twin Buttes District, Twin Buttes Mine (William Hefferon, pers. comm., 3rd ed.); Esperanza District, in the New Year's Eve breccia pipe of the Esperanza pit as a tungsten-rich variety of rutile intergrown with sericite, orthoclase, and quartz (Kuck 1978).

PINAL COUNTY: Pioneer supersystem, Resolution District, as red rutile alteration product in the Precambrian diabase giant copper orebody (Stanley B. Keith, pers. comm., 2020). Mineral Creek supersystem, Ray District, Ray area, formed in hydrothermally altered quartz-monzonite porphyry, in pseudomorphs after "biotite," with chlorite, sericite, and hydromica (Schwartz 1952). San Manuel District, San Manuel Mine, in hydrothermally altered monzonite and quartz-monzonite porphyries, associated with chlorite pseudomorphs after biotite (Schwartz 1947, 1949, 1958); found in the propylitic, potassium silicate, and argillic phases of alteration (Creasey 1959). Sacaton Mountains, Sacaton (Casa Grande) District, S12, T5S R5E, as small irregular masses in felsite, with corundum and quartz (E. D. Wilson 1969). Bunker Hill (Copper Creek) District, Copper Creek, Childs-Aldwinkle Mine, as small single crystals, associated with molybdenite, quartz, and orthoclase crystals (3rd ed.). Rattler District, as red rutile in pan concentrates from copper-rich diabase (Tim Marsh, pers. comm., 2016).

SANTA CRUZ COUNTY: Patagonia Mountains, Querces District, Santo Niño Mine at Duquesne, as slender crystals and reticulated masses up to several inches in altering granite, associated with molybdenite (J. H. Courtright, pers. comm., 3rd ed.). Washington Camp District, as microcrystals (H. Peter Knudsen, UA x274). Santa Rita Mountains, east side of Madera Canyon in quartzite with andalusite, topaz, and lazulite (Marsh and Sheridan 1976).

YAVAPAI COUNTY: Black Hills, Jerome supersystem, Verde District, United Verde Mine, as well-developed crystals (Bull. 181). Bradshaw Mountains, Black Canyon District, Howard copper property, with tourmaline in the gangue. Eureka supersystem, Bagdad District, Bagdad, as stubby crystals or granular aggregates derived from titanite and biotite in an altered quartz-monzonite stock, associated with quartz and orthoclase and, in a few places, with sericite (C. A. Anderson 1950).

SABUGALITE

Hydrogen aluminum uranyl phosphate hydrate: $HAl(UO_2)_4(PO_4)_4 \cdot 16H_2O$. A rare secondary mineral formed in vein-type uranium deposits as an alteration product of uraninite and in Colorado Plateau–type sandstone-hosted uranium-copper-vanadium deposits.

APACHE COUNTY: Black Mesa Basin, Rough Rock District, Black Water Mine (Bull. 181).

COCONINO COUNTY: Cameron District, O'Jaco, Huskon No. 5, and Arrowhead Nos. 1, 3, and 7 claims (Holland et al. 1958).

#SALAMMONIAC

Ammonium chloride: $(NH_4)Cl$. Found as a sublimation product around volcanic fumaroles and in guano deposits.

MARICOPA COUNTY: Painted Rock District, Rowley Mine, occurs with rowleyite, antipinite, and struvite and is likely derived from bat guano (Kampf, Cooper, Nash, et al. 2017).

YUMA COUNTY: Reported from near Yuma (Bull. 181).

SALÉEITE

Magnesium uranyl phosphate hydrate: $Mg(UO_2)_2(PO_4)_2(H_2O)_{10}$. A secondary mineral of the autunite group associated with oxidized uranium deposits.

GILA COUNTY: Cherry Creek area, Sierra Ancha District, Sue Mine, where it is intimately associated with bassetite (Granger and Raup 1969); in weathered deposits in sandstone in the Dripping Spring Quartzite (Granger and Raup 1962).

#SAMARSKITE-(Y)

Yttrium iron niobium oxide: $YFe^{3+}Nb_2O_8$. A mineral found in granite pegmatites in small amounts with other rare-earth minerals such as monazite. The lighter rare-earth elements tend to enter the monazite structure, whereas the heavier elements enter the samarskite structure; thus, the two minerals are commonly found in the same pegmatite.

MOHAVE COUNTY: Aquarius Mountains, in granite pegmatites with allanite (Bull. 181). Reported northeast of Kingman (Bull. 181).

PIMA COUNTY: Sierrita Mountains, Sierrita-Esperanza supersystem, Esperanza District, New Year's Eve Mine, as shiny black crystals (Bull. 181).

YAVAPAI COUNTY: Black Hills, Jerome supersystem, Verde District, reported near Jerome (Bull. 181).

YUMA COUNTY: Fortuna District, in a small pegmatite at north-central S22 T10S R20W (Stanton B. Keith 1978).

SANIDINE

Potassium aluminum silicate: $K(AlSi_3O_8)$. A member of the alkali feldspar group that may be described as a structurally disordered orthoclase and is typically found in rocks formed at elevated temperatures. It is a constituent of alkali and felsic volcanic rocks such as rhyolites and trachytes, as transparent glassy crystals, occasionally chatoyant. The mineral is undoubtedly much more common in the state than suggested by the few localities noted below.

COCHISE COUNTY: Chiricahua National Monument, near Rustler's Park and on Flys Peak, in the groundmass and as porphyrocrysts to 1.5 cm, in welded rhyolite tuffs, associated with quartz and magnetite (Enlows 1955; Stanley B. Keith, pers. comm., 2019). Reported in the Warren District, as small phenocrysts in intrusive rhyolite (Nye 1968; 3rd ed.). Peloncillo Mountains, Cottonwood Canyon, common as phenocrysts, some showing a blue play of colors, in rhyolites (3rd ed.).

COCONINO COUNTY: Southern San Francisco Mountains, at several places around Fremont Peak, as phenocrysts in fine-grained rhyolite with riebeckite (Updyke 1977).

PIMA COUNTY: Northern Sierrita Mountains, Gunsight Mountain, as a barian variety, in a pulaskite dike (Bideaux, Williams, and Thomssen 1960). Batamote Mountains near Ajo, as clear round phenocrysts in rhyolite welded tuffs (3rd ed.).

SANTAFEITE

Calcium strontium sodium manganese iron vanadate hydroxide oxide hydrate: $(Ca,Sr,Na)_3(Mn^{2+},Fe^{3+})_2Mn^{4+}_2(VO_4)_4(OH,O)_5 \cdot 2H_2O$. A rare mineral found as a secondary coating on joint surfaces in limestone.

NAVAJO COUNTY: Monument Valley District, in the dumps of the Monument No. 1 Mine, as massive veinlets that cut sandstone in one large chunk of ore (3rd ed.).

SAPONITE

Calcium sodium magnesium iron aluminum silicate hydroxide hydrate: $(Ca,Na)_{0.3}(Mg,Fe)_3(Si,Al)_4O_{10}(OH)_2 \cdot 4H_2O$. A member of the smectite group, a hydrothermal mineral found in mineralized veins and in vesicular basalt.

MARICOPA COUNTY: Belmont Mountains, Osborne District, Tonopah-Belmont Mine, as smears along fractures in vein quartz (3rd ed.). Horseshoe Dam District, one mile south of Horseshoe Dam, in zeolites (D. M. Shannon 1983b).

PIMA COUNTY: Unspecified locality near Pantano (UA 6486).

SANTA CRUZ COUNTY: Northern Patagonia Mountains, Flux District, Flux Mine, in the open pit (David Shannon, pers. comm., 1988).

SAPPHIRINE

Magnesium aluminum silicate: $Mg_4(Mg_3Al_9)O_4[Si_3Al_9O_{36}]$. A relatively rare mineral formed in metamorphic rocks high in alumina and low in silica.

LA PAZ COUNTY: Granite Mountain area, K-D District, about three miles southwest of Quartzsite, as sparse irregular grains in schist, tentatively identified by optical means (E. D. Wilson 1929), associated with kyanite, pyrophyllite, rutile, magnetite, andalusite, hematite, sillimanite, and dumortierite (Stanton B. Keith 1978; Melchiorre 2013).

SAUCONITE

Sodium zinc aluminum silicate hydroxide hydrate: $Na_{0.3}Zn_3(Si,Al)_4O_{10}(OH)_2 \cdot 4H_2O$. A member of the smectite group found in oxidized zinc and copper deposits.

COCHISE COUNTY: Southeastern Dragoon Mountains, Gleeson District, Gleeson Ridge, Defiance and Silver Bill Mines, mixed with fraipontite (Foord, Taggert, and Conklin 1983). Tombstone Hills, Tombstone District (H 106306).

GILA COUNTY: Miami-Inspiration supersystem, Pinto Valley District, Castle Dome area, as purplish waxy lumps in manganiferous material (Bull. 181). Banner supersystem, 79 District, 79 Mine, associated with chrysocolla and hemimorphite in a stope system above the 470 ft. level (Stanley B. Keith 1972).

PINAL COUNTY: Pioneer supersystem, Magma District, Magma Mine, as soft waxy gouge-like material from near the lower limit of oxidation, associated with coronadite (Bull. 181).

scapolite

An aluminum silicate group with two end members, sodium-rich marialite ($Na_4Al_3Si_9O_{24}Cl$) and calcium-rich meionite [$Ca_4Al_6Si_6O_{24}(CO_3)$]. See specific species for occurrences.

SCAWTITE

Calcium silicate carbonate hydrate: $Ca_7(Si_3O_9)_2(CO_3) \cdot 2H_2O$. A rare late-stage mineral associated with calcium-rich tactites.

GILA COUNTY: Southeastern Dripping Spring Mountains, Banner supersystem, Christmas District, Christmas Mine, as tiny euhedral prisms projecting into vugs in grossular-diopside tactites (3rd ed.).

SCHEELITE

Calcium tungstate: $Ca(WO_4)$. A widespread mineral, it is the most common and important ore of tungsten. It is typically formed in high-temperature environments, including granite pegmatites, contact-metamorphic deposits formed in limestone, and quartz-rich high-temperature veins. In Arizona, the majority of scheelite occurrences are associated with quartz veins and aplopegmatites in or near peraluminous intrusions of various ages and commonly associated "wolframite."

COCHISE COUNTY: Cochise supersystem, Johnson Camp District, Johnson Camp area, near the Republic Mine, as crystals embedded in quartz crystals (Romslo 1949; J. R. Cooper 1957; Baker 1960; J. R. Cooper and Huff 1951; J. R. Cooper and Silver 1964). Bluebird District, about four miles north of Dragoon, in small quantities in quartz veins that cut granite (Guild 1910), associated with minor amounts of aquamarine and "wolframite." Teviston District, Cohen Tungsten Mine, ten miles east of Willcox, as orange crystals weighing up to 15 pounds, and as smaller crystals embedded in gray doubly termi-

FIGURE 3.219 Scheelite, Cohen Tungsten Mine, Cochise County, 2.6 × 3.5 cm.

nated quartz crystals (Bull. 181; Stanton B. Keith 1973; Dale, Stewart, and McKinney 1960), considered the best scheelite locality in Arizona. Northeastern Whetstone Mountains, Whetstone District, at the James and other mines, in quartz veins and as replacements in granite (Guild 1910; Dale, Stewart, and McKinney 1960). Huachuca Mountains, Reef District, Reef Mine, in a pegmatitic quartz vein associated with chalcopyrite, sphalerite, hübnerite, fluorite, galena, pyrite, and stolzite (Palache 1941a); Harper and other properties, relatively pure deposits occurring on or near the surface (E. D. Wilson 1941; Dale, Stewart, and McKinney 1960). Swisshelm District, in contact zones with siliceous, garnet-bearing rock (Dale, Stewart, and McKinney 1960). Chiricahua Mountains, California supersystem, Paradise area, in quartz veins in silicified limestones (Dale, Stewart, and McKinney 1960).

GILA COUNTY: Miami-Inspiration supersystem, Day Peaks District, on the lower east slope of Day Peaks, Lost Gulch area, as sparse isolated crystals with stolzite, in a mineralized zone in a fault that cuts diabase (Faick and Hildebrand 1958). Mazatzal Mountains, El Oso District, northeast of Four Peaks (Bull. 181).

GRAHAM COUNTY: Southwestern Pinaleño Mountains, south of Mt. Graham, Black Beauty District, SE quarter of S35 and SW quarter of S36, T8S R22E, with tourmaline from the Black Beauty Group (Dale 1959).

LA PAZ COUNTY: Trigo Mountains, Silver District, Gold Reef claims, in a sheared quartz vein (E. D. Wilson 1941). Little Harquahala District, Gold Dyke Group, originally found as sporadic nearly pure pockets of scheelite in quartz, but only sparse mineralization remains (Dale 1959). Livingston Hills, Livingston District, Colorado claims, as large crude crystals and masses over 5 cm, with "wolframite" (MM K128-135; Dale 1959; Melchiorre 2017); Jewel Anne Group, as masses up to 2.5 cm in quartz (MM L710-711); northwestern Granite Wash Mountains, Three Musketeers Mine, in S24, T6N R15W, as crystals up to 4 in. across, in white quartz; the crystals show brilliant blue-white fluorescence (David Shannon, pers. comm., 1971). Placer near Quartzsite (H 97745).

MARICOPA COUNTY: San Domingo District, near Morristown, on the upper San Domingo Wash, with powellite; northwest of Morristown, in several properties (Dale 1959). Cave Creek District, at an unspecified locality, with cuprotungstite and ferberite (Schaller 1932). Vulture Mountains, Flying Saucer District, Flying Saucer Group of claims, in S12, T6N R6W, with powellite as disseminations in granitic rocks (Dale 1959).

MOHAVE COUNTY: Hualapai Mountains, Boriana District, Boriana Mine, occurs with "wolframite" and chalcopyrite in narrow quartz veins (Hobbs 1944; Dale 1961) as crystals to 1.5 in. across; Maynard District, Telluride Chief Mine (E. D. Wilson 1941), and other properties, with "wolframite" in quartz veins (Bull. 181). Aquarius Mountains District, Boner Canyon, sparingly at the Williams Mine with "wolframite" (Hobbs 1944; Dale 1961). Greenwood District, in small amounts with "wolframite" (Dale 1961). Chemehuevi District, Mohave Mountains, Dutch Flat, in narrow quartz veins containing gold and "wolframite" (E. D. Wilson 1941).

PIMA COUNTY: Las Guijas Mountains, Las Guijas District, with "wolframite" in quartz veins (Dale, Stewart, and McKinney 1960). Mt. Benedict area, Nogales District, Reagan Mine, as minute crystals in the quartz near bands of "wolframite" (Schrader and Hill 1915). Sierrita Mountains, Pima supersystem, Twin Buttes District, Twin Buttes Mine, in contact-skarn zones (Stanley B. Keith, pers. comm., 1973). Gunsight District, as lenticular bodies in iron-stained milky quartz (E. D. Wilson 1941; Dale, Stewart, and McKinney 1960). Cababi District, eastern part of the Tohono O'odham (Papago) Reservation, about fifty miles west of Tucson, on several claims (Dale, Stewart, and McKinney 1960). Northern Santa Catalina Mountains, Marble Peak District, near Marble and Piety Peaks, found on tactite-marble contact or garnetized areas; some large crystals noted at Taylor X claims (Dale, Stewart, and McKinney 1960). Southernmost Pima County, east of Sasabe, San Luis Mountains, Easter District, southwest of Arivaca, in quartz veins in the Easter prospect (Dale, Stewart, and McKinney 1960).

PINAL COUNTY: Northern Santa Catalina Mountains, Oracle District, Campo Bonito area, Maudina and other properties (Guild 1910). Southern Tortilla Mountains, northwest of Mammoth, Gold Circle District,

Tarr, and Antelope Peak areas, with powellite or "wolframite" in quartz (E. D. Wilson 1941).

SANTA CRUZ COUNTY: Nogales District, south of Calabasas, with "wolframite" (Dale, Stewart, and McKinney 1960); at a small property with "wolframite" (Schrader 1917). Southern Patagonia Mountains, Querces District, four miles south of Duquesne, with molybdenite (Schrader and Hill 1915; E. D. Wilson 1941).

YAVAPAI COUNTY: Southern Bradshaw Mountains, Tip Top District, occurs sporadically with small quartz stringers or veins in schist with lead-zinc-silver-antimony mineralization (Dale 1961). Hassayampa District, Evelyn-Cordella, Lucky Bud, White Pearl, and other prospects, as sporadically disseminated masses up to 1 in. in iron-stained quartz (Dale 1961; 3rd ed.). White Picacho District, scattered sparsely through some quartz-rich zones (Jahns 1952); upper Santo Domingo and Little Santo Domingo Washes, disseminated in garnet-epidote schist (E. D. Wilson 1941). Silver Mountain District, reported to occur with massive "wolframite" in lenticular bodies of quartz in a fault zone (E. D. Wilson 1941).

YUMA COUNTY: Dome District, occurs sparsely in epidote-garnet zones and marble beds (Dale 1959). Kofa District, found as stringers, pods, and small grains with quartz, biotite schist, or hornblende schist (Dale 1959).

SCHIEFFELINITE

Lead tellurate hydroxide sulfate hydrate: $Pb_{10}Te^{6+}_{6}O_{20}(OH)_{14}(SO_4) \cdot 5H_2O$. A secondary mineral formed from other tellurium minerals under conditions of severe oxidation. Tombstone is the type locality.

COCHISE COUNTY: Tombstone District, on the dumps of the Joe shaft, as clusters of intergrown colorless or milk-white scales, with individuals up to 1 mm, in a large chunk of shattered quartz vein; tellurides have been converted to rodalquilarite, several unknown tellurites or tellurates, "girdite," and schieffelinite (S. A. Williams 1980).

SCHOEPITE

Uranyl oxide hydroxide hydrate: $(UO_2)_4O(OH)_6 \cdot 6H_2O$. A secondary mineral commonly associated with uraninite, from which it may form by alteration.

APACHE COUNTY: Cane Valley District, Monument No. 2 Mine, as a surficial alteration product of uraninite, with becquerelite and small amounts of fourmarierite (C. Frondel 1956).

COCONINO COUNTY: Cameron District, Black Point–Murphy Mine, sparsely distributed in Pleistocene gravels (Austin 1964).

YAVAPAI COUNTY: Black Rock District, Abe Lincoln Mine, northeast of Wickenburg, as the principal secondary uranium mineral, formed as a coating on pyrite grains (Raup 1954; Granger and Raup 1962).

#SCHOLZITE

Calcium zinc phosphate hydrate: $CaZn_2(PO_4)_2 \cdot 2H_2O$. A secondary mineral commonly found in zinc phosphate-bearing granite pegmatites and sediments.

MOHAVE COUNTY: Lead Pill District, Red Top Mine, as groups of colorless acicular crystals (T. Kampf, pers. comm., 2019; specimens collected by Jerry Baird in 2011).

FIGURE 3.220 Schorl, Castle Hot Springs, Maricopa County, 5.6 cm.

SCHORL

Sodium iron aluminum silicate borate hydroxide: $NaFe^{2+}_3Al_6(Si_6O_{18})(BO_3)_3(OH)_3(OH)$. Schorl is the most common member of the tourmaline group and is found in granite, granite pegmatites, high-temperature veins, some metamorphic rocks, and detrital in sediments. Many of these occurrences were found in the third edition under tourmaline. These occurrences may lack analysis of their specific species but may be assumed to be schorl until reported otherwise.

APACHE AND NAVAJO COUNTIES: Common as a heavy mineral in the basal member of the Monitor Butte Sandstone of the Chinle Formation of upper Triassic age, associated with apatite and zircon (R. G. Young 1964).

COCHISE COUNTY: Juniper Flat District, northwest of Bisbee, as nests of small prismatic crystals in muscovite (Graeme, Graeme, and Graeme 2015). Johnny Lyon Hills, Yellowstone District, as a common mineral at the contact zone of the Johnny Lyon Granodiorite (J. R. Cooper and Silver 1964).

COCONINO COUNTY: Grand Canyon Inner Gorge, Hermit Creek, as black crystals in pegmatites (Bull. 181).

GILA COUNTY: Miami-Inspiration supersystem, as a bluish-black variety widespread in the Pinal Schist (N. P. Peterson 1962).

LA PAZ COUNTY: Dome Rock Mountains, Cinnabar District, with magnetite and siderite in gangue in cinnabar veins (Bull. 181). Plomosa Mountains, Livingston District, Night Hawk and White Dike Mines, 5.7 miles south of Quartzsite, as black massive material associated with scheelite and quartz (Dale 1959).

MARICOPA COUNTY: Mazatzal Mountains District, in cinnabar veins (Lausen 1926). White Picacho District, in pegmatites (Jahns 1952). Phoenix Mountains District, Mummy Mountain, Phoenix (Bull. 181). Near Castle Hot Springs in schist (S. Celestian, pers. comm., 2020).

MOHAVE COUNTY: Wright Canyon area, Valentine District, near the Wright Creek Ranch, in several irregular pegmatite dikes in schist and granite, as large pod-like masses enclosing crystals of microcline, quartz, albite, fluorite, and titanite; also as abundant well-formed prismatic crystals in the cores of pegmatites (Schaller, Stevens, and Jahns 1962). Cerbat Mountains, Kingman Feldspar District, in small amounts with beryl in pegmatites (B. E. Thomas 1953).

PIMA COUNTY: Santa Rita Mountains, Santa Rita supersystem, Old Baldy District, Iron Mask Mine, with magnetite and siderite (Schrader 1917). Sierrita Mountains, Samaniego District, as vein-like masses with quartz in granite (Bull. 181). Little Ajo Mountains, widely distributed as brilliant black prisms coating joint planes in the Cardigan Gneiss and other rocks (Gilluly 1937). Pima supersystem, Twin Buttes District, Twin Buttes Mine (William Hefferon, pers. comm., 3rd ed.). Santa Catalina Mountains, as a common phase in the Lemmon Rock Leucogranite near Summerhaven (Force 1997).

PINAL COUNTY: Santa Catalina Mountains, Oracle District, in pegmatites in the Oracle Granite (Guild 1910; Ward 1931); in vein quartz in the Pinal Schist near post-Cambrian granite contacts (Ransome 1919). Galiuro Mountains, Bunker Hill (Copper Creek) District, as radiating groups of slender prismatic crystals, notably in the breccia pipe of the American Eagle Mine (Simons 1964); Childs-Aldwinkle Mine, as a gangue mineral in the breccia-pipe deposit (Kuhn 1941).

SANTA CRUZ COUNTY: Southern Patagonia Mountains, Washington Camp District, Washington Camp areas, abundant in contact-metamorphic deposits (Lehman 1978). Northwest Santa Rita Mountains, Tyndall District (Schrader 1917).

YAVAPAI COUNTY: Bradshaw Mountains, in pegmatites of the Bradshaw Granite, in lenses in schist, and scattered throughout the schist near granite contacts (Bull. 181). In veins of the Prescott District (Bull. 181).

Iron King District, Iron King Mine, as blue-gray prisms in quartz and dolomite (Lindgren 1926; Creasey 1952). In pegmatites of the Bagdad area (C. A. Anderson 1950).

SCHREIBERSITE

Iron nickel phosphide: $(Fe,Ni)_3P$. Found in many of the metallic meteorites found in Arizona.

APACHE COUNTY: Navajo meteorite, as veins of irregular width and varying extent in polished and etched samples (Roy and Wyant 1949).

COCONINO COUNTY: Canyon Diablo meteorite, found as cuneiform or lamellar skeleton crystals; continuous but irregular rims around troilite aggregates; or in metallic spheroids as white or pink inclusions intergrown with maghemite, nickel-rich iron, or troilite (Mead, Littler, and Chao 1965; Buchwald 1975).

PIMA COUNTY: Irwin-Ainsa (Tucson) meteorite, occurs with brezinaite, nickel-rich iron, and taenite (Bunch and Fuchs 1969).

SCHRÖCKINGERITE

Sodium calcium uranyl sulfate carbonate fluoride hydrate: $NaCa_3(UO_2)$ $(SO_4)(CO_3)_3F \cdot 10H_2O$. Formed as a late secondary mineral with other oxidized uranium minerals; also formed as a postmine mineral.

COCONINO COUNTY: Cameron District, Jack Daniels No. 1 Mine, coating fractures in sandstone, possibly of postmine origin (Austin 1964); Foley Bros. No. 5 Mine, filling fractures in logs undergoing oxidation (Austin 1964); Huskon No. 10 Mine, found with carnotite in buff-pinkish carbonaceous sandstone (Scarborough 1981).

NAVAJO COUNTY: Little John No. 1–3 Mines, occurs with uraninite, coffinite, zeunerite, and torbernite in gray medium to coarse-grained sandstone and in bentonitic mudstone (Scarborough 1981).

YAVAPAI COUNTY: Eureka supersystem, Hillside District, Hillside Mine, as 0.625 in. thick coatings on gypsum, with andersonite, bayleyite, and swartzite; johannite and uraninite are reported in the same mine (Axelrod et al. 1951).

SCHUBNELITE

Iron vanadate hydrate: $Fe^{3+}(V^{5+}O_4) \cdot H_2O$. A secondary mineral found with other vanadium and uranium minerals.

Cane Valley District, Monument No. 2 Mine, in Triassic fossilized wood; identified by Howard T. Evans Jr. (Haynes 1991).

#SCHULENBERGITE

Copper zinc sulfate hydroxide hydrate: $(Cu,Zn)_7(SO_4)_2(OH)_{10} \cdot 3H_2O$. A rare secondary mineral found in the oxidized portions of copper-zinc hydrothermal deposits, commonly postmining. Also found in oxidized copper-zinc slags.

YAVAPAI COUNTY: Old Dick District, Copper King Mine near Bagdad, with aurichalcite, smithsonite as very small blue plates and scales (R. Jenkins, pers. comm., 2016).

#SCHWERTMANNITE

Iron oxide hydroxide sulfate hydrate: $Fe^{3+}_{16}O_{16}(OH)_{9.6}(SO_4)_{3.2} \cdot 10H_2O$. A secondary mineral sometimes found in areas subject to acid mine drainage.

GREENLEE COUNTY: Copper Mountain supersystem, Gold Gulch District, Gold Gulch, found in manganese and iron oxide/hydroxide precipitates in conglomerate that forms from spring water (Walder et al. 2004).

SCOLECITE

Calcium aluminum silicate hydrate: $Ca(Si_3Al_2)O_{10} \cdot 3H_2O$. A member of the zeolite group found as seams and amygdule fillings in mafic volcanic rocks; less commonly, of hydrothermal origin in metamorphic rocks.

GRAHAM COUNTY: Black Point, about four miles below Geronimo on the Gila River, as highly fluorescent amygdule fillings up to 0.5 in. in diameter in basalt (Bull. 181).

PINAL COUNTY: Bunker Hill (Copper Creek) District, Copper Creek, as well-formed prisms up to 1 in. long, in brecciated zones in the Copper Creek Granodiorite, with calcite (3rd ed.).

SCORODITE

Iron arsenate hydrate: $Fe^{3+}(AsO_4) \cdot 2H_2O$. A relatively common mineral formed under oxidizing conditions in gossans associated with arsenic-bearing minerals. Scorodite is typically the first mineral to replace arsenopyrite. It may rarely be formed as a primary hydrothermal mineral.

COCONINO COUNTY: Grand Canyon National Park, Grandview District, Grandview (Last Chance) Mine, as light brown to gray mammillary masses with radiating internal structure, associated with metazeunerite and olivenite (Leicht 1971).

GILA COUNTY: Banner supersystem, 79 District, 79 Mine (Stanley B. Keith 1972).

MOHAVE COUNTY: Cerbat Mountains, Wallapai supersystem, as a common secondary mineral (Bull. 181); Mollie Gibson Mine (B. E. Thomas 1949), Golconda and Tennessee Districts, as a secondary oxidation product of arsenical ores (Stanley B. Keith, pers. comm., 2019).

YAVAPAI COUNTY: Mazatzal Mountains District, Stingy Lady Mine, as stains and coatings on arsenopyrite (Wrucke et al. 1983).

SCORZALITE

Iron aluminum phosphate hydroxide: $Fe^{2+}Al_2(PO_4)_2(OH)_2$. Formed in granite pegmatites associated with other phosphate minerals.

COCHISE COUNTY: Warren District, Bisbee, Lavender pit, Holbrook extension, as small amounts in variscite (Graeme, Graeme, and Graeme 2015).

MARICOPA AND YAVAPAI COUNTIES: White Picacho District, in pegmatites (London 1981).

#SCOTLANDITE

Lead sulfite: $Pb(S^{4+}O_3)$. A very rare secondary mineral found in the oxidized portions of hydrothermal lead deposits.

GRAHAM COUNTY: Aravaipa supersystem, Grand Reef District, Grand Reef Mine, as colorless elongated tabular crystals associated with aravaipaite and laurelite (R100043).

SELENIUM

Selenium: Se. A rare secondary mineral that can form as a sublimate from fumarolic gases, from the oxidation of selenium-bearing organic compounds in sedimentary uranium-vanadium deposits such as found on the Colorado Plateau, and from burning coal seams.

YAVAPAI COUNTY: Jerome supersystem, Verde District, United Verde Mine, as a coating of needle-like crystals on rock above the burning pyrite orebody; crystals are up to 2 cm long and are bounded by first- and second-order rhombohedrons (Palache 1934; H 92679, 92680). An analysis by F. A. Gonyer showed that the mineral contained no tellurium and only a trace of sulphur.

SENGIERITE

Copper uranyl vanadate hydroxide hydrate: $Cu_2(UO_2)_2(VO_4)_2(OH)_2 \cdot 6H_2O$. A rare secondary mineral formed from the alteration of uraninite.

COCHISE COUNTY: Warren District, Bisbee, Cole Mine, originally found in 1935 in one isolated pocket, associated with malachite, chalcocite, covellite, and chlorargyrite; later found to be more widely distributed in the mine; it "occurred in one of a series of five sulfide-oxide veins that extended from the 800 level to the 1,300 level of the Cole shaft" (Hutton 1957).

SEPIOLITE

Magnesium silicate hydroxide hydrate: $Mg_4Si_6O_{15}(OH)_2 \cdot 6H_2O$. Formed by the alteration of magnesian rocks and generally associated with serpentine or magnesite. It is a component of certain clays and found in some hydrothermal veins. A compact variety known as meerschaum was used to make smoking pipes.

COCHISE COUNTY: Warren District, Bisbee, Cole shaft and Southwest Mine (Graeme 1981).

GILA COUNTY: Southeastern Dripping Spring Mountains, Banner supersystem, Christmas District, Christmas Mine, occurs with ruizite (S. A. Williams and Duggan 1977).

GREENLEE COUNTY: Copper Mountain supersystem, Clifton area (UA 8847).

MARICOPA COUNTY: Locality forty-two miles north of Phoenix and two miles east of State Highway 69 (Bull. 181).

PIMA COUNTY: Pima supersystem, Mission-Pima District, San Xavier West Mine (UA 4164).

YAVAPAI COUNTY: In the basin of the Santa Maria River, west of the McCord Mountains, as crystalline fibrous material (Kauffman 1943).

serpentine

Serpentine is the name given to the magnesium silicates in the kaolinite-serpentine mineral group. Chrysotile, lizardite, and antigorite are polymorphs of magnesium silicate hydroxide, $Mg_3Si_2O_5(OH)_4$. Amesite is a magnesium silicate with aluminum. The fibrous habit of chrysotile is due to the rolled-up tube nature of its kaolinite-like sheet structure. Serpentine is of secondary origin, derived from the alteration of nonaluminous magnesian silicates, particularly olivine, amphibole, or pyroxene. In places, it is formed as large masses derived from peridotite or other mafic igneous rocks. Serpentine as antigorite is the principal matrix in the SUM diatremes. A common product of contact metamorphism of magnesian limestones, it is also found in hydrothermal veins. The most notable serpentine mineral in Arizona is chrysotile. It has been extensively mined as asbestos. Antigorite and lizardite are major constituents of the nonasbestiform serpentine rock associated with the asbestos ore. Extensive lists of asbestos localities are included in Funnell and Wolfe (1964) and in Bulletins 126 and 180 (E. D. Wilson 1928; Shride 1969). The exact species is not presently known for the following occurrences.

APACHE COUNTY: Garnet Ridge SUM District, Garnet Ridge, abundant in a boulder, pseudomorphous after an orthopyroxene; related to a breccia dike that pierces sedimentary rocks (Gavasci and Kerr 1968).

COCHISE COUNTY: Tombstone District, Lucky Cuss Mine, in altered limestone (B. S. Butler, Wilson, and Rasor 1938). Dos Cabezas Mountains, in metamorphosed limestones (Guild 1910). Little Dragoon Mountains, Johnson Camp area, Moore orebody, as a retrograde product of high-temperature metamorphism of forsterite, tremolite, and diopside (J. R. Cooper and Huff 1951; J. R. Cooper and Silver 1964).

COCONINO COUNTY: Grand Canyon region, Bass and Hance properties, where the Precambrian Bass Limestone has been altered adjacent to diabase sills (Selfridge 1936); cross-fiber asbestos is locally up to 4 in. long but is commonly shorter (Funnell and Wolfe 1964).

GILA COUNTY: Pinal and Mescal Mountains and Pinto Creek region (E. D. Wilson 1928). Sierra Ancha Mountains, at the head of Pocket Creek (Bateman 1923; Sampson 1924). Southeastern Dripping Spring Mountains, Banner supersystem, Christmas District, Christmas Mine, replacing dolomite and forsterite in the orebodies (Perry 1969). San Carlos Apache Reservation, Emsco and Bear Mines (Bromfield and Schride 1956). Globe Hills District, Old Dominion Mine, in Mescal Limestone near diabase sills (Lausen 1923).

GREENLEE COUNTY: Copper Mountain supersystem, Morenci District, northern part of the Morenci open-pit area, locally formed with other calcsilicate minerals in an extensive contact-metamorphic assemblage (Moolick and Durek 1966); on the ridge just west of Morenci, at the

Thompson Mine, as green-banded material associated with magnetite (Lindgren 1905; Reber 1916; Creasey 1959).

PIMA COUNTY: Pima supersystem, Twin Buttes District, Twin Buttes Mine, in metamorphic rocks (Stanley B. Keith, pers. comm., 1973).

YAVAPAI COUNTY: Copper Basin District, as an alteration product in ores in breccia pipes (Johnston and Lowell 1961).

SERPIERITE

Calcium copper zinc sulfate hydroxide hydrate: $Ca(Cu,Zn)_4(SO_4)_2(OH)_6 \cdot 3H_2O$. A rare secondary mineral found in the oxidized portions of copper-zinc orebodies.

COCHISE COUNTY: Middle Pass supersystem, Abril District, San Juan Mine, as microcrystals (Bruce Maier, UA x3086). Chiricahua Mountains, California supersystem, Leadville District, unnamed surface Cu-Zn-Pb-Ag prospect located in the center of the E half of S21, T17S R30E, as well-formed crystals with other copper and zinc minerals (Rolf Luetcke, pers. comm., 2015).

GILA COUNTY: Banner supersystem, 79 District, Brick or D. H. claims, northeast of Hayden, with ktenasite (William Hunt, UA x3715; R060475).

#SHANNONITE

Lead oxide carbonate: $Pb_2O(CO_3)$. A secondary mineral formed by the heating of cerussite. The Grand Reef Mine is the type locality.

GRAHAM COUNTY: Aravaipa supersystem, Grand Reef District, found at the Grand Reef Mine as white porcelaneous masses with plumbojarosite, hematite, quartz, litharge, massicot, hydrocerussite, and minium (Roberts et al. 1995).

MARICOPA COUNTY: Belmont Mountains, Osborne District, found in the main entrance to the Tonopah-Belmont Mine as the result of a mine fire (D. Shannon, pers. comm., 1992).

PINAL COUNTY: Pioneer supersystem, Black Prince (Olsen) Mine, from a fire (W. Hunt, pers. comm., 1990).

FIGURE 3.221 Shannonite, Tonopah-Belmont Mine, Maricopa County, FOV 2 mm.

#SHARPITE

Calcium uranyl carbonate hydrate: $Ca(UO_2)_3(CO_3)_4 \cdot 3H_2O$. A secondary uranium mineral formed from the alteration of uraninite.

LA PAZ COUNTY: Northern Plomosa Mountains, Rayvern District, Rayvern No. 2 Mine, with rutherfordine and carnotite (Melchiorre 2017).

SHATTUCKITE

Copper silicate hydroxide: $Cu_5(SiO_3)_4(OH)_2$. A secondary mineral formed as an alteration of other secondary copper minerals. The Shattuck Mine at Bisbee is the type locality.

COCHISE COUNTY: Warren District, Bisbee, Shattuck Mine, as pseudomorphs after malachite and as small blue spherules (Schaller 1915).

GREENLEE COUNTY: Morenci area, peripheral copper deposit in Gila Conglomerate (UA 1036).

LA PAZ COUNTY: New Water Mountains, Ramsey District, Eagle Eye Mine, as small spherical clusters of blue crystals (A. Roe 1980).

MARICOPA COUNTY: Belmont Mountains, south of Wickenburg, Osborne District, at several localities, including the Moon Anchor Mine and Potter-Cramer property, associated with several oxidized minerals (S. A. Williams 1968).

MOHAVE COUNTY: Rawhide Mountains, Rawhide District, Rawhide Mine, in oxidized ores with drusy quartz, alamosite, and luddenite (S. A. Williams 1982).

PIMA COUNTY: Ajo District, in veins at the New Cornelia Mine (Schaller and Vlisidis 1958; Newberg 1964; Mrose and Vlisidis 1966; Vlisidis and Schaller 1967; H. T. Evans and Mrose 1977; R050133). Santa Catalina Mountains, Molino Basin, with flecks of gold on chrysocolla-stained quartz (3rd ed.). Southern Roskruge Mountains, Roadside District, Roadside Mine, associated with fornacite and cerussite in mine dumps (3rd ed.).

PINAL COUNTY: Mammoth District, Mammoth–St. Anthony Mine (J. W. Anthony et al. 1995). San Manuel District, San Manuel Mine, as veinlets in chrysocolla (3rd ed.). Poston Butte (Florence) District, in a porphyry copper deposit (3rd ed.). Unspecified locality in the Tortolita Mountains (probably in the Suizo Mountains, Durham-Suizo District, Azurite Mine) (Bull. 181).

SHERWOODITE

Calcium aluminum vanadium oxide hydrate: $Ca_{4.5}AlV^{4+}_2V^{5+}_{12}O_{40}\cdot28H_2O$. A secondary oxidation product of lower-valence vanadium minerals commonly associated with selenium, metatyuyamunite, and melanovanadite.

APACHE COUNTY: Lukachukai District, Joleo Mine (AEC mineral collection).

SICKLERITE

Lithium manganese phosphate: $LiMn^{2+}(PO_4)$. A rare secondary mineral formed by the alteration of lithiophilite and triphylite in pegmatites.

YAVAPAI COUNTY: White Picacho District, as thin discontinuous rims on lithiophilite and triphylite (Jahns 1952); London and Burt (1982a) reported that most of this material is probably robertsite; White Ridge pegmatite (London and Burt 1982a).

SIDERITE

Iron carbonate: $Fe(CO_3)$. A member of the calcite group and typically found in bedded sedimentary rocks, where it may be an important ore mineral. It is also commonly formed as a hydrothermal vein mineral.

COCHISE COUNTY: Warren District, Bisbee, where boxwork siderites have proved to be a useful guide to ore deposits (Trischka, Rove, and Barringer 1929); Campbell Mine, as fine granular dark-brown stalactitic material, coated by small iridescent crystals, Gardner shaft, Hoatson shaft, and Junction shaft in small caves as stalactites and botryoidal masses "of incredible size and beauty" (Graeme 1981).

FIGURE 3.223 Siderite, Campbell shaft, Bisbee, Cochise County, 9.7 cm high.

GILA COUNTY: Dripping Spring Mountains, Banner supersystem, 79 District, McHur prospect, with wulfenite and vanadinite (Bull. 181).

GREENLEE COUNTY: Copper Mountain supersystem, Morenci District, as a weathering product found in massive form in highly mineralized limestone (Reber 1916).

LA PAZ COUNTY: Harquahala Mountains (UA 7586). Dome Rock Mountains, in cinnabar veins with tourmaline (H. Bancroft 1911); Harcuvar Mountains, Cunningham Pass area district, as nearly jet-black cleavable material with chalcopyrite (Bull. 181).

MOHAVE COUNTY: Cerbat Mountains, Wallapai supersystem, as a common gangue mineral (B. E. Thomas 1949). Gold Basin District (B. E. Thomas 1949), Hualapai (Antler) District, Antler Mine, as large botryoidal masses, some with an iridescent coating (3rd ed.).

PIMA COUNTY: Santa Rita Mountains, Old Baldy District, Iron Mask Mine, with magnetite and schorl (Bull. 181). Empire District, Hilton

Mine Group (Schrader and Hill 1915). South Comobabi Mountains, Cababi District, Mildren and Steppe claims (S. A. Williams 1963). Silver Bell District, as small crystals (C. A. Stewart 1912).

PINAL COUNTY: San Manuel District, San Manuel Mine (UA 8636).

SANTA CRUZ COUNTY: Santa Rita Mountains, Cottonwood Canyon, Tyndall District, Glove Mine, replacing limestone (H. J. Olson 1966). Patagonia Mountains, Mowry District, Mowry Mine (3rd ed.); Flux District, Flux Mine (Schrader 1917).

YAVAPAI COUNTY: Bradshaw Mountains, Lynx Creek, in veins with chlorite, pyrite, and tourmaline (Lindgren 1926); Turkey Creek District, Gold Note Group, as gangue in silver, gold, and lead ore (Lindgren 1926); Peck District, Peck and Swastika Mines, associated with silver and bromargyrite (Lindgren 1926). Weaver Mountains, near Yarnell, as large crystalline nodules (Bull. 181). Stonewall Mine near Prescott, as gangue associated with silver, chlorargyrite, and chalcanthite (Guild 1917). Black Hills District, at the Shea, Brindle Pup, and Mingus Mountain gold mines, in quartz veins with sulfides (Lindgren 1926).

SIDEROTIL

Iron sulfate hydrate: $Fe(SO_4) \cdot 5H_2O$. A secondary mineral formed as a product of the dehydration of melanterite.

COCHISE COUNTY: Warren District, Bisbee, Campbell shaft, as a cuprian variety replacing melanterite and Junction shaft as stalactites (Graeme 1981).

COCONINO COUNTY: In breccia pipes mined for uranium (Wenrich and Sutphin 1988).

SIEGENITE

Cobalt nickel sulfide: $CoNi_2S_4$. Found in hydrothermal veins, associated with other copper, nickel, and iron sulfides.

COCONINO COUNTY: Orphan District, Orphan Mine, on the South Rim of the Grand Canyon (3rd ed.)

MOHAVE COUNTY: Hack District, Hack Mine (3rd ed.).

SILLIMANITE

Aluminum silicate: Al_2SiO_5. Found in high-grade, thermally metamorphosed, argillaceous rocks and in the contact zones of peraluminous granites and granodiorites, where it is commonly associated with andalusite or cordierite.

COCHISE COUNTY: Bluebird District, about two miles southwest of Adams Peak, as acicular porphyroblasts up to 1 cm long in hornfels shale near the Texas Canyon stock (J. R. Cooper and Silver 1964).

COCONINO COUNTY: Grand Canyon National Park, abundant in the Inner Gorge 0.5 mile downstream from Monument Creek (Campbell 1936).

GILA AND PINAL COUNTIES: Pinal Ranch quadrangle, in Pinal Schist near post–Pinal Schist, muscovite-bearing peraluminous Solitude Granite contacts, where it is intimately intergrown with magnetite (N. P. Peterson 1962).

LA PAZ COUNTY: Near the eastern margins of the Dome Rock Mountains, three miles southwest of Quartzsite, K-D District, in Mesozoic schists, associated with dumortierite, kyanite, and andalusite (E. D. Wilson 1929; Melchiorre 2013). Granite Wash Mountains, Yuma King District, as a common metamorphic fiberform Late Cretaceous overprint on magnetite-chalcopyrite-garnet-actinolite skarns of Early Jurassic age (Stanley B. Keith, pers. comm., 2019).

MOHAVE COUNTY: Hualapai Mountains, Maynard District, in quartz veins that cut schist (Bull. 181). Cerbat Mountains, principally as the variety fibrolite in elongate masses of interlacing needles intergrown with "biotite," garnet, quartz, and feldspar in migmatite (B. E. Thomas 1953).

PIMA COUNTY: Santa Catalina Mountains, locally common in the metamorphic aureole of the peraluminous Wilderness Granite suite along its contact with Paleozoic rocks, Apache Group, and Pinal Schist (Force 1997; Bykerk-Kauffman 1990).

PINAL COUNTY: Gila River Indian Reservation, Sacaton Mountains, in metamorphosed sediments encased in granodiorite, associated with titanian andalusite, corundum, and cordierite (Bideaux, Williams, and Thomssen 1960).

YAVAPAI COUNTY: Santa Maria Mountains, Camp Wood District, near Camp Wood, as veins and nodules in schist near the peraluminous Lawler Peak Granite (Bull. 181); along Copper Creek near Bagdad, in the Hillside mica schist near the peraluminous Lawler Peak Granite (C. A. Anderson, Scholz, and Strobell 1955).

SILVER

Silver: Ag. Commonly found in the upper portions of silver-bearing deposits and in the zone of sulfide enrichment of copper deposits with chalcocite. It may also, but less commonly, be of primary origin, associ-

FIGURE 3.224 Silver, Stonewall Jackson Mine, Gila County, 8 cm.

ated with galena or tetrahedrite, and less commonly with nickel-cobalt-arsenide minerals.

COCHISE COUNTY: Tombstone District, as disseminated flakes at the Empire Mine, and as small masses of wire silver at the Flora Morrison Mine (Guild 1917; B. S. Butler, Wilson, and Rasor 1938). Warren District, Bisbee, notably in the Campbell and Cole Mines, in small amounts, commonly associated with, and in places coating, secondary chalcocite (H 94731, 94733; S 97831; UA 1249); Holbrook Mine, in a few places with halloysite-10Å (Schwartz and Park 1932); Copper Queen Mine and Lavender pit (Graeme 1981). Pearce Hill, Pearce District, Commonwealth Mine, with chlorargyrite, bromian chlorargyrite, jarosite, and acanthite (Endlich 1897).

GILA COUNTY: Globe Hills District, Continental Mine, as minute flakes in calcite associated with cuprite (N. P. Peterson 1962); Old Dominion Mine, as stout wires in the oxidized ores (Bull. 181); about four miles north of Globe, as fine specimens in placers (Ransome 1903b); Stonewall Jackson Mine, in siderite gangue with acanthite and chlorargyrite (Guild 1917; Presmyk 2015). Richmond Basin District, as one of the chief ore minerals, in fairly large masses—a recently found one weighed 407 pounds (Les Presmyk, pers. comm., 2020; N. P. Peterson 1962). Green Valley District, Silver Butte Mine at Payson, as wire silver in the oxidized ore (Lausen and Wilson 1925). Four Peaks District, as small dendritic pieces in calcite (MM K989).

GRAHAM COUNTY: Santa Teresa Mountains, Aravaipa supersystem, Grand Reef District, La Clede and Grand Reef Mines, with hematite (Frank Valenzuela, UA x572).

GREENLEE COUNTY: Copper Mountain supersystem, Morenci, Detroit, Arizona Copper, Shannon, Silver King, Gold Bar, Capote, Silver Bonanza, and other mines (3rd ed.).

MARICOPA COUNTY: White Picacho District, in pegmatites (Jahns 1952).

MOHAVE COUNTY: Wallapai supersystem, Tennessee District, Distaff Mine, as chunks weighing several pounds, with acanthite in the deeper workings (Bastin 1925); Tennessee District, Tennessee-Schuylkill Mine (Schrader 1909); Lucky Boy and Samoa Mines (B. E. Thomas 1949); Mineral Park District, Queen Bee Mine (Bastin 1925); Rural Mine (Bastin 1925); Stockton Hill District, Banner Group of claims, in solid chunks and as masses of wire silver (Schrader 1909); Buckeye Mine, noted for large masses of the solid mineral and for beautiful specimens of wire silver (Newhouse 1933; Schrader 1909).

PIMA COUNTY: Cerro Colorado District, Cerro Colorado Mine, partly as wire silver, with stromeyerite and tetrahedrite (Guild 1917; UA 2873). LeConte (1852) noted the presence of silver in a small mountain range about forty miles southeast of Tucson; it was associated with galena and "blende" (possibly the Empire Mountains, Empire District). Rincon Mountain foothills, Tanque Verde, as flakes "in considerable quantity" in copper ore (Engineering and Mining Journal 1892b).

PINAL COUNTY: Pioneer supersystem, Silver King District, as fine specimens in the Silver King Mine, where large masses of silver fill cracks in stromeyerite, bornite, and chalcopyrite; also in "beautiful filiform (specimens), the branches of which envelop individual chalcocite grains, some of the finer filaments even extending into fractures and cleavage cracks of the chalcocite" (Guild 1917; Presmyk 2015, S 65154, 65155; UA 9437, UA 12330; H 99620); Magma District, in the upper portions of the Magma Mine (UA 9524), as thin sheets in massive bornite (Les Presmyk, pers. comm., 2020). Galiuro Mountains, Saddle Mountain District, Little Treasure and Adjust Mines, as wire silver (C. P. Ross 1925b). Mineral Creek supersystem, Ray District, Ray Mine, on the west side of the Pearl Handle pit, with secondary copper minerals (Metz and Rose 1966). Mineral Hill District, Reymert Mine, as tiny blebs in chalcopyrite (K. S. Wilson 1984). Slate Mountains, Slate District, Jackrabbit Mine, as flakes and thin sheets in fractures (Hammer 1961). Tortolita Mountains, Owl Head District, Apache property, with chalcocite and chlorargyrite (Bull. 181).

SANTA CRUZ COUNTY: Patagonia Mountains, Palmetto District, Domino Mine, with crystallized cerussite and wulfenite (Schrader and Hill 1915); Flux District, World's Fair Mine, with tetrahedrite (UA 211); Hardshell District, Hardshell Mine, as a component of manganiferous silver ores (Koutz 1984). Southern slopes of the Santa Rita Mountains, as small crystals surrounded by magnetite in diorite (Schrader and Hill 1915). Santa Rita Mountains, Salero District, Eureka-Marble Mine (Blake 1904).

YAVAPAI COUNTY: Jerome supersystem, Verde District, United Verde Mine, as a thin layer of high-grade ore in gossan immediately above the sulfide orebody (C. A. Anderson and Creasey 1958). Northwestern Bradshaw Mountains, Groom Creek District, at several properties, for example, the Dos Oris Mine, with acanthite and chlorargyrite (Lindgren 1926); Monte Cristo District, Monte Cristo Mine, where it formed contemporaneously with nickelskutterudite, chalcopyrite, tennantite, enargite, and acanthite (Bastin 1922). Ticonderoga District, Arizona National Mine, as wire silver in cavities with acanthite (C. A. Anderson and Creasey 1958). Turkey Creek District, Goodwin properties, with chlorargyrite (Lindgren

1926). Black Canyon District, Thunderbolt Mine, with proustite (Lindgren 1926). Southern Bradshaw Mountains, Tip Top District, Tip Top Mine, with ruby silver and chlorargyrite (Lindgren 1926).

#SIMONKOLLEITE

Zinc hydroxide chloride hydrate: $Zn_5(OH)_8Cl_2 \cdot H_2O$. A rare secondary mineral found as a postmining weathering product of slag.

MARICOPA COUNTY: Painted Rock District, Rowley Mine, as colorless hexagonal tabular crystals found on a metallic mining artifact (Hazen et al. 2017; R130616).

SKUTTERUDITE

Cobalt arsenide: $CoAs_3$. This is the cobalt end member of a mineral series that forms a solid solution with nickelskutterudite, the nickel end member. Most minerals in the series are arsenic deficient; iron is a common substitutional element in the series. Formed in hydrothermal veins and associated with other cobalt and nickel minerals.

COCONINO COUNTY: Orphan District, Orphan Mine (Gornitz 2004).

GRAHAM COUNTY: Santa Teresa Mountains, Bluebird Cobalt District, Bluebird Mine, fifteen miles west of Fort Thomas, with cobaltite (3rd ed.)

YAVAPAI COUNTY: Northeastern Wickenburg Mountains Monte Cristo District, Monte Cristo Mine, with acanthite (Bastin 1922).

FIGURE 3.225 Smithsonite, 79 Mine, Gila County, 6 × 6 cm.

SMITHSONITE

Zinc carbonate: $Zn(CO_3)$. A secondary mineral commonly found in the oxidized zone of zinc deposits in limestone gangue. Formed from the alteration of sphalerite and commonly associated with hemimorphite, cerussite, malachite, azurite, anglesite, aurichalcite, rosasite, and other oxidized copper, lead, and zinc minerals.

COCHISE COUNTY: Tombstone District, Toughnut Mine, as tiny rhombohedral crystals (B. S. Butler, Wilson, and Rasor 1938). Gleeson District, Mystery and Silver Bill Mines, as blue to green encrustations and crystalline masses (Knudsen 1983; UA 6517). Middle Pass supersystem, Abril District, San Juan Mine (Trischka, Rove, and Barringer 1929). Huachuca Mountains, Hartford District (UA 1430).

Warren District, commonly associated with siderite in the typical box-work gossan areas (Trischka, Rove, and Barringer 1929).

COCONINO COUNTY: Grand Canyon National Park, Grandview District, Grandview (Last Chance) Mine, in small amounts, associated with malachite and azurite, lining vugs in clay in limestone (Leicht 1971).

GILA COUNTY: Banner supersystem, Chilito District, Curtin or Humphrey Mine, with cerussite and anglesite (Bull. 181); 79 District, 79 Mine, as a lustrous green cuprian variety, as crystals lining vugs, associated with wulfenite, hemimorphite, murdochite, and aurichalcite (Stanley B. Keith 1972).

GRAHAM COUNTY: Aravaipa supersystem, Iron Cap District, Head Center Mine (Simons 1964).

GREENLEE COUNTY: Copper Mountain supersystem, Shannon District, Shannon Mountain (Bull. 181).

LA PAZ COUNTY: Trigo Mountains, associated with cerussite and yellow lead oxide in cellular to crystalline masses (Bull. 181); Silver District, occurs as small crystals and porous encrustations in the oxidized zone of vein deposits in the southern portion of the district (E. D. Wilson 1933).

MARICOPA COUNTY: Belmont Mountains, Osborne District, Tonopah-Belmont Mine, found with paratacamite and boleite (Giester and Rieck 1996).

PIMA COUNTY: Empire District, Hilton Mines (Bull. 181). Santa Rita Mountains, Old Baldy District (Schrader 1917). Helvetia-Rosemont District, Omega Mine, as microcrystals (Brian Bond, UA x2675). Sierrita Mountains, Pima supersystem, Mission-Pima District, Daisy and Eagle-Picher Mines, as "dry bone ore" (Ransome 1922; Bull. 181; B. M. Williams 2018). Western Sierrita Mountains, Papago District, Yellow Bird Mine (UA 1363). Silver Bell District, as earthy mixtures of smithsonite and cerussite (Bull. 181). Waterman Mountains, Waterman District, Silver Hill Mine, as small rhombohedra, gray to pale-blue botryoidal masses (similar to material from the Kelly Mine in New Mexico), and "dry bone ore," associated with rosasite (UA 1186). Tucson Mountains, Amole District, Thunderbird Mine (UA 1710) and Gila Monster Mine (UA 2667). South Comobabi Mountains, Cababi District, Mildren and Steppe claims (S. A. Williams 1963).

PINAL COUNTY: Mammoth District, Mammoth–St. Anthony Mine, as blue rounded scalenohedra resembling rice grains, crusts, and porous masses (UA 6569); as crude crystals, in some places with wulfenite (Bull. 181; Bideaux 1980); and as green and white botryoidal hemispheres with hemimorphite (Les Presmyk, pers. comm., 2020).

SANTA CRUZ COUNTY: Southern Patagonia Mountains, Washington Camp District, Pride of the West Mine at Duquesne, associated with cerussite, anglesite, chrysocolla, and cuprite (Schrader and Hill 1915; Schrader 1917). Santa Rita Mountains, Tyndall District, at the Glove Mine in Cottonwood Canyon, abundant as "dry bone ore" encrustations in the intermediate zone of the replacement orebodies in limestone; also as rare gray balls (H. J. Olson 1966).

YAVAPAI COUNTY: Eureka supersystem, Hillside District, Hillside Mine, in the oxidation zone of sulfide-bearing veins in mica schist, associated with cerussite, anglesite, and hemimorphite (Axelrod et al. 1951).

YUMA COUNTY: Castle Dome Mountains, Castle Dome District, in channels and vugs, associated with hydrozincite, wulfenite, vanadinite, and mimetite (Batty et al. 1947).

#SODDYITE

Uranyl silicate hydrate: $(UO_2)_2SiO_4 \cdot 2H_2O$. A secondary mineral formed from the alteration of uraninite.

YAVAPAI COUNTY: Steel City Mine, with kasolite and uranium-rich thorite (M. E. Thompson, written comm., 1954, in C. Frondel 1958).

FIGURE 3.226 Sonoraite, Old Guard Mine, Tombstone, Cochise County, FOV 2.5 mm.

SONORAITE

Iron tellurite hydroxide hydrate: $Fe3^+(Te^{4+}O_3)(OH) \cdot H_2O$. A rare secondary mineral found in tellurium deposits.

COCHISE COUNTY: Tombstone District (R070745), Joe Mine, as distinctive yellow crystal fibers up to 2 mm long, in goethite gossan with emmonsite and rodalquilarite; also replaces emmonsite spicules and tubes (3rd ed.). Old Guard Mine (B. Murphy, pers. comm., 2020).

SPANGOLITE

Copper aluminum sulfate hydroxide chloride hydrate: $Cu_6Al(SO_4)(OH)_{12}Cl \cdot 3H_2O$. A rare secondary mineral associated with other oxidized copper minerals in the oxidized portions of copper deposits. It was first described in 1890 by S. L. Penfield on a specimen from an uncertain locality thought to be near Tombstone. Palache and Merwin (1909), however, subsequently argued that the type specimen was from Bisbee. A speci-

men at Yale University (No. 3482), which contains spangolite on cuprite with connellite and is identified as the
type specimen, is labeled from a locality near Tombstone and strongly resembles known Bisbee material.

COCHISE COUNTY: Warren District, Bisbee, Czar Mine,
in cuprite-azurite matrix with readily identifiable spots
of spangolite up to 0.625 in. diameter (C. Frondel 1949);

FIGURE 3.227 Spango-
lite, Southwest Mine,
Bisbee, Cochise County,
4.5 cm.

Copper Queen Mine (Ford 1914); Southwest Mine, with claringbullite,
atacamite, and paratacamite (Graeme 1993); some crystal cleavages are
up to 1 in. across (3rd ed.). Courtland District, Maid of Sunshine Mine, as
good crystals up to 1 mm long, with cyanotrichite, and as microcrystals
(Marvin Deshler, UA x1808).

GREENLEE COUNTY: Copper Mountain supersystem, Metcalf District, as
a scaly coating on chrysocolla from Metcalf Mine (Lindgren and Hillebrand 1904; Lindgren 1905).

SPERTINIITE

Copper hydroxide: $Cu(OH)_2$. A very rare secondary mineral formed by
the alteration of chalcocite.

COCHISE COUNTY: Warren District, Bisbee, as sparse pale blue-green encrustations with internal scaly structure, on a matrix of crystalline atacamite (Graeme 1993).

SPESSARTINE

Manganese aluminum silicate: $Mn^{2+}_3Al_2(SiO_4)_3$. A less
common member of the garnet group formed in skarns
and other manganese-rich associations of metamorphic
origin, also formed in peraluminous granite pegmatites,
where the garnet is commonly almandine-spessartine in
composition. The more orange garnets tend to be more
spessartine rich. The garnets found in the rhyolites at
several Arizona localities are almandine-spessartine in
composition and are listed as either one or the other.

COCHISE COUNTY: Warren District, Bisbee, Cole shaft, 700 level, 5 mm
crystals with azurite (Graeme 1993).

LA PAZ AND YUMA COUNTIES: Trigo Mountains, Castle Dome Mountains,
Neversweat Ridge, Cemetery Ridge, northern Plomosa Mountains,
common in ferromanganiferous metachert of the Orocopia Schist; with

FIGURE 3.228 Spessar-
tine, Aquarius Moun-
tains, Mohave County,
4.2 cm high.

magnetite and quartz (Haxel et al. 2002, 2021; Strickland, Singleton, and Haxel 2018).

MARICOPA COUNTY: Phoenix Mountains District, southwestern slope of Piestewa Peak (formerly Squaw Peak) in Phoenix, as microcrystals in schist, with manganese-rich andalusite (Thorpe 1980).

MOHAVE COUNTY: Near Wright Creek Ranch, fifteen miles south of Peach Springs, as a salmon-pink or wine-red accessory mineral in several pegmatite dikes in schist and granite (Schaller, Stevens, and Jahns 1962). Aquarius Mountains, Aquarius Mountains District, found with almandine (3rd ed.).

PIMA COUNTY: Santa Catalina Mountains, along the Mt. Lemmon Highway, as good quality crystals (Thomas Trebisky, UA x2279) and a major component of heavy mineral placers in Sabino Creek; as crystals up to 1 cm in aplo-granites and pegmatites in the Lemmon Rock Leucogranite complex on Mt. Lemmon. The garnet schlieren south of Summerhaven features line rock composed of orange garnets with manganese-rich whole rock chemistry suggesting end-member spessartine (Stanley B. Keith et al. 1980; Stanley B. Keith and Rasmussen 2019).

PINAL COUNTY: Northern Galiuro Mountains, southeast of Saddle Mountain, Lookout Mountain District, in rhyolite as highly perfect crystals, with pseudobrookite, topaz, and bixbyite (White 1992)

YAVAPAI COUNTY: White Picacho District, in nearly all the pegmatites, as grains up to 0.5 in. diameter (Jahns 1952).

SPHALERITE

Zinc sulfide: ZnS. This is the most abundant and important ore mineral of zinc. It is a widespread sulfide mineral common in the base-metal deposits in which it is formed under a wide range of conditions. Far more abundant in Arizona than is suggested by the listings below.

COCHISE COUNTY: Warren District, Bisbee, Campbell Mine, from which large quantities were mined, particularly during the years 1945 to 1949 (Trischka, Rove, and Barringer 1929; Hewett and Rove 1930; Schwartz and Park 1932; E. D. Wilson et al. 1950; Bain 1952); triboluminescent at the Junction Mine (Mitchell 1920; S 11794); and strongly fluorescent at the Copper Queen Mine (Les Presmyk, pers. comm., 2020). Tombstone District, Silver Thread and Sulfuret Mines, and less abundant elsewhere in the

FIGURE 3.229 Sphalerite, Iron Cap Mine, Graham County, 2.6 cm.

district (B. S. Butler, Wilson, and Rasor 1938; Rasor 1939). Little Dragoon Mountains, Cochise supersystem, Johnson Camp District, Johnson Camp area, with chalcopyrite in copper ores (Kellogg 1906; E. D. Wilson et al. 1950; J. R. Cooper 1957; J. R. Cooper and Silver 1964); Keystone and St. George deposits (Romslo 1949). Southeastern Dragoon Mountains, Gleeson District, as scattered bunches in pyritic ore, locally well crystallized (E. D. Wilson et al. 1951). Huachuca Mountains, Hartford District, State of Texas Mine (E. D. Wilson et al. 1951). Dragoon Mountains, Middle Pass supersystem, Abril District, Abril Mine, as the most abundant sulfide ore mineral (Perry 1964).

GILA COUNTY: Banner supersystem, 79 District, as an abundant ore mineral in the lower levels of the 79 Mine; as well-formed tetrahedral crystals (Kiersch 1949; E. D. Wilson et al. 1951; Stanley B. Keith 1972). Christmas District, Christmas Mine, as a moderately common sulfide in skarn (Knoerr and Eigo 1963; David Perry, pers. comm., 1967). Globe Hills District, where it is sparingly but widely distributed (N. P. Peterson 1962).

GRAHAM COUNTY: Aravaipa supersystem, Iron Cap District, as gem-quality green to yellow crystals at the Iron Cap Mine (UA 5829) and other properties (Denton 1947a; Simons and Munson 1963; W. E. Wilson 1988); Stanley District, Princess Pat Mine and Cold Spring prospect, with chalcopyrite, pyrite, or stibnite in calcite (C. P. Ross 1925a).

GREENLEE COUNTY: Copper Mountain supersystem, in large quantities in the deeper levels of many older mines (Lindgren 1905; Guild 1910; Reber 1916).

MOHAVE COUNTY: Wallapai supersystem, Tennessee and Golconda Districts, in most of the ores, especially at the Vanderbilt, Flores, Tennessee, Towne, and Keystone Mines (Haury 1947; B. E. Thomas 1949; E. D. Wilson et al. 1950); Ithaca Peak District, Ithaca Peak, in veins with chalcopyrite, pyrite, and galena in a quartz-monzonite stock (Eidel 1966). Union Pass District (Schrader 1909). McConnico District, near Yucca, as the most abundant sulfide in a lenticular orebody, with tremolite gangue and chalcopyrite, pyrrhotite, and minor amounts of löllingite (Rasor 1946).

PIMA COUNTY: Santa Rita Mountains, Santa Rita supersystem, Empire District, in many mines and prospects (Schrader and Hill 1915). Sierrita Mountains, Pima supersystem, as sulfide orebodies in limestone (Guild 1910); Paymaster Mine, Olive Camp (Ransome 1922; E. D. Wilson et al. 1951; Nye 1961); Mission-Pima District, San Xavier Mine (UA 7220);

Twin Buttes District, Twin Buttes Mine, in a drill hole in the arkose (Stanley B. Keith, pers. comm., 1973). Silver Bell District, occurs in small irregular amounts with pyrite and chalcopyrite in a gangue of garnet, quartz, calcite, and wollastonite (C. A. Stewart 1912).

PINAL COUNTY: Pioneer supersystem, Belmont District, Belmont Mine, and Magma District, Magma Mine, as fine dark-colored and green crystals up to 2.5 cm (Short et al. 1943; Barnes and Hay 1983); Silver King District, Silver King Mine, as the most abundant sulfide in the ore, and as cleavable masses described as being "held together by threads of native silver" (Guild 1917; Short et al. 1943; E. D. Wilson et al. 1950). Galiuro Mountains, Saddle Mountain District, Adjust, Saddle Mountain, and Little Treasure properties (Bull. 181). Mammoth District, Mammoth–St. Anthony Mine, on the lower levels but extensively altered to smithsonite and hemimorphite in the oxidized zone (N. P. Peterson 1938a, 1938b; Fahey, Daggett, and Gordon 1950; E. D. Wilson et al. 1950). Vekol District, Reward Mine, with pyrrhotite, malachite, chrysocolla, pyrite, and chalcopyrite (Denton and Haury 1946).

SANTA CRUZ COUNTY: Santa Rita and Patagonia Mountains, common in the copper and silver ores (Schrader and Hill 1915; Schrader 1917). Patagonia Mountains, Washington Camp District, as magnificent crystal groups at the Pride of the West Mine at Duquesne, where a single crystal measured nearly 1.5 in. diameter (Bull. 181); Callahan Lead-Zinc Mine (UA 1099) and Indiana Mine (Stanley B. Keith, pers. comm., 1973). Ruby District, Montana Mine, with galena (Guild 1910; E. D. Wilson et al. 1951); Idaho Mine (H. V. Warren and Loofbourrow 1932; R. Y. Anderson and Kurtz 1955). Santa Rita Mountains, Cottonwood Canyon, Tyndall District, Glove Mine (H. J. Olson 1966).

YAVAPAI COUNTY: Jerome supersystem, Verde District, United Verde Mine, in the pyritic ores (Fearing 1926; Lausen 1928; Moxham, Foote, and Bunker 1965); Copper Chief Mine, as streaks with pyrite and chalcopyrite (Lindgren 1926). Bradshaw Mountains, in most districts (Bull. 181). Tip Top District, Davis Mine, as an unusual golden-yellow variety (Bull. 181). Iron King District, Iron King Mine (Lindgren 1926; E. D. Wilson et al. 1950; Creasey 1952; C. A. Anderson and Creasey 1958).

SPHEROCOBALTITE

Cobalt carbonate: $Co(CO_3)$. This is a member of the calcite group found in cobalt-bearing veins and some uranium mineral deposits.

COCONINO COUNTY: Cameron District, associated with well-oxidized uranium occurrences (Austin 1964).

SPINEL

Magnesium aluminum oxide: $MgAl_2O_4$. A relatively common mineral formed under high-temperature conditions in igneous rocks, in regionally metamorphosed rocks and in contact-metamorphic rocks.

APACHE COUNTY: Unspecified locality near McNary (UA 3688). Garnet Ridge SUM District, Garnet Ridge, from an ultramafic diatreme on the Navajo Reservation (Wang et al. 2000)

COCHISE COUNTY: San Bernardino Valley, San Bernardino (Geronimo) volcanic field, in spinel lherzolite xenoliths in the basaltic and basanitic volcanic rocks of the Geronimo lavas (S. H. Evans and Nash 1979).

GILA COUNTY: At the south end of Peridot Mesa, Peridot District, as crystals in peridotitic inclusions in a nepheline basanite diatreme (Frey and Prinz 1978).

LA PAZ AND YUMA COUNTIES: Cemetery Ridge District, ferroan variety, common in several types of metasomatic schist and gneiss, with various combinations of hornblende, anorthite, epidote, diopside, clinochlore, tremolite, and olivine (Haxel et al. 2021).

MARICOPA AND YAVAPAI COUNTIES: White Picacho District, as greenish-black crystals of the ferroan variety in pegmatites (Jahns 1952).

MOHAVE COUNTY: About nine miles south of Hoover Dam, Hoover District, in road cuts along U.S. Highway 93, in lherzolite inclusions 1 to 2 cm in diameter, in basanite dikes (Garcia, Muenow, and Liu 1980).

SPIONKOPITE

Copper sulfide: $Cu_{39}S_{28}$. Found as a weathering product of primary copper sulfides such as djurleite in strata-bound red-bed copper deposits and in a serpentine-hosted magnetite-chromite orebody.

COCHISE COUNTY: Warren District, Bisbee, in the ores of the Campbell Mine (Graeme 1993).

SPIROFFITE

Manganese tellurite: $Mn^{2+}_2Te^{4+}_3O_8$. A very rare mineral; the Tombstone locality is one of only a few known in the world.

COCHISE COUNTY: Tombstone District, Joe Mine, very sparingly as large (0.5 in.) pale-pink deeply corroded crystals, commonly found in intensely silicified pyritic shales; the crystals are exposed by breaking the rock "across the grain" to reveal the isolated spiroffite nuggets within (3rd ed.).

SPODUMENE

Lithium aluminum silicate: $LiAlSi_2O_6$. An important ore of lithium formed in lithium-bearing granite pegmatites, where it is typically associated with quartz, albite, beryl, tourmaline, and lepidolite.

MARICOPA AND YAVAPAI COUNTIES: White Picacho District, in many of the pegmatites in the district (Jahns 1952, 1953; London and Burt 1978); Mitchell Wash, northeast of Morristown (UA 5625).

#SPURRITE

Calcium silicate carbonate: $Ca_5(SiO_4)_2(CO_3)$ Formed by high-temperature contact metamorphism of carbonate rock and mafic magma.

YAVAPAI COUNTY: Eureka supersystem, Bagdad (Bull. 181).

#SRILANKITE

Titanium zirconium oxide: Ti_2ZrO_6. Found as inclusions in other zirconium minerals and pyrope garnets.

APACHE COUNTY: Garnet Ridge SUM District, found in an ultramafic diatreme at Garnet Ridge serpentine ultramafic microbreccia (SUM) on the Navajo Reservation. It occurs as inclusions in pyrope crystals collected from surface concentrates. These pyropes also contain carmichaelite, rutile, spinel, minerals of the crichtonite group, and olivine (Wang et al. 2000).

STANNITE

Copper iron tin sulfide: Cu_2FeSnS_4. A primary mineral found in tin-bearing vein deposits of hydrothermal origin.

COCHISE COUNTY: Warren District, Bisbee, in the sulfide ores of the Campbell Mine (Graeme 1993).

MARICOPA COUNTY: Reported from the McDowell Mountains (3rd ed.).

STANNOIDITE

Copper iron zinc tin sulfide: $Cu_8(Fe,Zn)_3Sn_2S_{12}$. Formed in hydrothermal veins with chalcopyrite, quartz, and stannite.

COCHISE COUNTY: Warren District, Bisbee, Campbell Mine, as a coating on, or an infilling between, pyrite grains; associated with canfieldite (Graeme 1981); Junction Mine, as black grains in massive bornite and chalcopyrite (R160066).

STARKEYITE

Magnesium sulfate hydrate: $Mg(SO_4)\cdot4H_2O$. A water-soluble secondary mineral found as efflorescence in mine openings.

COCONINO COUNTY: In breccia pipes mined for uranium (Wenrich and Sutphin 1988).

PINAL COUNTY: San Manuel District, San Manuel Mine, one of several secondary sulfate minerals formed in drifts; on the 2,015 and 2,700 ft. levels with epsomite. The starkeyite may have formed when epsomite dehydrated after it was removed from the humid environment of the mine (material collected by Joseph Urban; J. W. Anthony and McLean 1976).

STAUROLITE

Iron aluminum silicate hydroxide: $Fe^{2+}_2Al_9Si_4O_{23}(OH)$. A relatively common mineral formed in schists during intermediate-grade metamorphism of argillaceous sediments. It is commonly associated with garnet, kyanite, andalusite, sillimanite, and tourmaline in mica schists, and in the aluminum-rich metasomatic zones in the contact aureoles of peraluminous intrusions.

COCONINO COUNTY: Grand Canyon National Park, Lone Tree Canyon, as brownish-red stout prismatic crystals, with garnet in metamorphic rocks (Campbell 1936).

MARICOPA COUNTY: Phoenix Mountains District, Piestewa Peak (formerly Squaw Peak) in Phoenix, as small 0.5 × 2.0 mm crystals in schist, with chloritoid, garnet, and biotite; contains 11 to 14 percent of the zincian end member (Thorpe and Burt 1980).

PIMA COUNTY: Santa Catalina Mountains, locally common in the metamorphic aureole of the peraluminous Wilderness Granite suite along its contact with Paleozoic rocks, Apache Group, and Pinal Schist, where it is associated with andalusite (Force 1997).

Bradshaw Mountains, in schist near contacts with the Crazy Basin peraluminous batholith (Bull. 181); as twinned crystals from Cleator (AM 35853).

STEIGERITE

Aluminum vanadate hydrate: $Al(VO_4)\cdot3H_2O$. A rare secondary mineral formed in sandstone with gypsum, corvusite, and secondary uranium and vanadium minerals.

APACHE COUNTY: Cane Valley District, Monument No. 2 Mine, with navajoite and tyuyamunite (Weeks, Thompson, and Sherwood 1955; Witkind and Thaden 1963; AM 19436).

STELLERITE

Calcium aluminum silicate hydrate: $Ca_4(Si_{28}Al_8)O_{72}\cdot28H_2O$. A member of the zeolite group that closely resembles stilbite and, like this mineral, is commonly found in cavities and vesicles in flow rocks, especially basalt.

COCHISE COUNTY: Unspecified locality near Douglas (UA 6337).

STEPHANITE

Silver antimony sulfide: Ag_5SbS_4. A late-forming primary mineral found in hydrothermal deposits, where it is associated with galena, tetrahedrite, and other silver minerals.

GILA COUNTY: Richmond Basin District, Mack Morris Mine, with stromeyerite (UA 1358).

SANTA CRUZ COUNTY: Patagonia Mountains, Flux District, Golden Rose Mine (Schrader 1917).

YAVAPAI COUNTY: Bradshaw Mountains, Tuscumbia District, Tuscumbia Mine (Bull. 181).

STERNBERGITE

Silver iron sulfide: $AgFe_2S_3$. A rare mineral found in silver ores, with the ruby silvers.

COCHISE COUNTY: Dos Cabezas Mountains, Mascot District, 150 ft. level of the Leroy Mine, as euhedral crystals up to 2 mm long enclosed in galena and sphalerite, in a silver-rich ore shoot (3rd ed.).

STETEFELDTITE

Silver antimony oxide hydroxide: $Ag_2Sb_2(O,OH)_7$. A rare secondary mineral regarded by some authorities as the silver analog of bindheimite; others believe it to be a mixture.

COCHISE COUNTY: East side of the Johnny Lyon Hills, Yellowstone District, in a prospect, as waxy white crusts on mercurian tetrahedrite (schwartzite) (3rd ed.).

LA PAZ COUNTY: Trigo Mountains, Silver District, Red Cloud Mine, as a pale-yellow powder or crust on wulfenite (Gary M. Edson, pers. comm., 1973; Edson 1980).

PIMA COUNTY: West side of Tucson Mountains, Saginaw Hill District, Snyder Hill Mine (3rd ed.).

STEVENSITE

Calcium sodium magnesium silicate hydroxide: $(Ca,Na)_xMg_{3-y}Si_4O_{10}$ $(OH)_2$. Formed by the hydrothermal alteration of igneous rocks and a member of the smectite group of minerals. The IMA lists status as questionable since further analysis indicates that the mineral may be a type of defect structure (see Faust, Hathaway, and Millot 1959).

COCHISE COUNTY: Warren District, Bisbee, Lavender pit, Holbrook extension, in the oxide ores, intimately intergrown with a chrysotile polytype (Graeme 1981).

STEWARTITE

Manganese iron phosphate hydroxide hydrate: $Mn^{2+}Fe^{3+}_2(PO_4)_2(OH)_2 \cdot 8H_2O$. A rare secondary mineral formed from the alteration of primary phosphate minerals in granite pegmatites.

MARICOPA AND YAVAPAI COUNTIES: White Picacho District, as numerous pale-yellow films, as finely crystallized aggregates cementing microbreccias of hureaulite and lithiophilite, and as thin subparallel fibers along fractures in other minerals, principally strengite and purpurite (Jahns 1952).

STIBICONITE

Antimony oxide hydroxide: $Sb^{3+}Sb^{5+}_2O_6(OH)$. A secondary mineral formed by the oxidation of stibnite and other antimony minerals. Its status as a distinct species is listed as questionable by the IMA due to a revised pyrochlore supergroup nomenclature (see Atencio et al. 2017).

COCHISE COUNTY: Warren District, Bisbee, 1,000 ft. level of the Cole Mine, with chalcocite (Graeme 1981; AM 28997). Johnny Lyon Hills, Yellowstone District, in a prospect on the bank of Tres Alamos Wash, associated with partzite, hydroxycalcioroméite, and bindheimite (3rd ed.).

LA PAZ COUNTY: Southern Dome Rock Mountains, Cinnabar District, on radiating blades of stibnite that are partly altered to cervantite and stibiconite, in veins (Bull. 181).

MARICOPA COUNTY: Belmont Mountains, Osborne District, Clifford claims, northeast of Tonopah, as an alteration product of massive stibnite (MM 1454).

PIMA COUNTY: South Comobabi Mountains, Cababi District, silver-lead claim, as a granular white powder in small pockets with lead oxides (S. A. Williams 1962).

PINAL COUNTY: Vekol Mountains, Vekol District, associated with galena and sphalerite (Bull. 181).

YAVAPAI COUNTY: Bradshaw Mountains near Black Canyon City, in a 2- to 3-ft vein (Bull. 181).

FIGURE 3.230 Stibnite, Dura Mine, Santa Cruz County, 4.5 × 3.5 cm.

STIBNITE

Antimony sulfide: Sb_2S_3. Typically formed in low-temperature hydrothermal veins with other antimony minerals. Stibnite is the most important ore mineral of antimony.

GILA COUNTY: Near Payson, in small amounts in some copper ores (Bull. 181); reported in the Mazatzal Mountains District, on Slate Creek (Bull. 181).

GRAHAM COUNTY: Stanley District, Cold Spring prospect, in contact-metamorphic ores (C. P. Ross 1925a).

LA PAZ COUNTY: Dome Rock Mountains, Cinnabar District, Yellow Devil Mine, eight miles southwest of Quartzsite, as radiating blades, with cervantite and stibiconite (Bull. 181; Melchiorre 2013).

MARICOPA COUNTY: Belmont Mountains, Osborne District, Clifford claims, northeast of Tonopah, as massive material partially altered to stibiconite (MM 1454).

MOHAVE COUNTY: Cerbat Mountains, Wallapai supersystem, Golconda District, Golden Gem and Vanderbilt Mines, with galena, sphalerite, and pyrite (Schrader 1909; B. E. Thomas 1949).

PIMA COUNTY: Pima supersystem, Mission-Pima District, Whitcomb property, near the Olivette Mine (UA 1718); Twin Buttes District, Twin Buttes Mine (UA 10764, x2653). Tucson Mountains, Amole District (UA 1296).

SANTA CRUZ COUNTY: Mt. Benedict area, Nogales District, Dura Mine (Schrader 1919).

YAVAPAI COUNTY: Bradshaw Mountains, Tip Top District, Seventy-Six and Ritha B Mines (C. P. Kortemeier 1984); Peck District, Swastika Mine (S 116870), as well as other properties (Bull. 181); Tuscumbia District, near Tuscumbia Mine, silver-bearing coarsely radial with quartz (Lindgren 1926); at Malley Hill Mine on Lynx Creek (Bull. 181); Walker District, Robinson property (Lindgren 1926); Turkey Creek District (Bull. 181).

STILBITE

Sodium calcium aluminum silicate hydrate: $(NaCa_4,Na_9)$ $(Si_{27}Al_9)O_{72}\cdot28H_2O$. A widespread member of the zeolite group and most commonly found in vesicles and cavities in basaltic lavas. Reclassified in 1997 by the IMA as a series with two species, stilbite-Ca and stilbite-Na, based on dominant cation. The specific species has not been identified for the following occurrences.

FIGURE 3.231 Stilbite, Sierrita Mine, Pima County, FOV 6 mm.

GILA COUNTY: Southeastern Dripping Spring Mountains, Banner supersystem, Christmas District, Christmas Mine, as the most common zeolite in hydrothermally altered granodiorite porphyry, in veinlets with associated sulfides; locally replaces granodiorite pervasively; also as radiating clusters of light pink to orange prismatic crystals that line vugs in diorite (David Perry, pers. comm., 1967, 1969). Miami-Inspiration supersystem, Pinto Valley District, Castle Dome Mine, sparse in altered quartz monzonite (N. P. Peterson, Gilbert, and Quick 1951; N. P. Peterson 1962).

GRAHAM COUNTY: Safford supersystem, Dos Pobres District, Dos Pobres Mine, as salmon-colored crystals to 1 cm in underground workings (Gibbs, pers. comm., 2014).

LA PAZ COUNTY: Southern Plomosa Mountains, Ramsey District, Moore claim, in S14, T3N R16W, with heulandite in basalt (William Hunt, pers. comm., 1985).

PIMA COUNTY: Sierrita Mountains, Esperanza District, Esperanza Mine, found in drill core in 1967 (3rd ed.); Sierrita District, Duval-Sierrita Mine, in gypsum (UA 9588). Ajo District, New Cornelia Mine, as yellow bladed crystals (W. J. Thomas and Gibbs 1983).

PINAL COUNTY: Suizo Mountains, Durham-Suizo District, Azurite Mine, with laumontite and copper silicates in quartz veins (Bideaux, Williams, and Thomssen 1960). Copper Creek area, in float boulders along road into Copper Creek (Ascher, pers. comm., 2015).

SANTA CRUZ COUNTY: Patagonia Mountains, Washington Camp District, on adularia (UA 8479).

YAVAPAI COUNTY: Burro Creek, as small clear crystals in tuff (William Hunt, pers. comm., 1985).

STILPNOMELANE

Potassium calcium sodium iron magnesium aluminum silicate hydroxide hydrate: $(K,Ca,Na)(Fe,Mg,Al)_8(Si,Al)_{12}(O,OH)_{36} \cdot nH_2O$. A metamorphic mineral widely distributed in iron ore deposits and certain schists.

YAVAPAI COUNTY: Agua Fria District, Binghampton Mine, as microcrystals (A. Roe 1980; William Hunt, UA x764).

STISHOVITE

Silicon oxide: SiO_2. Found as a product of the high-pressure shock waves produced by meteor impact. A high-density tetragonal polymorph of SiO_2, stishovite is notable for being the only polymorph in which silicon is in sixfold (octahedral) coordination with oxygen. Meteor Crater is the type locality.

COCONINO COUNTY: Meteor Crater, in shock impact–metamorphosed Coconino Sandstone, with coesite (Chao, Shoemaker, and Madsen 1962; Fahey 1964; Bohn and Stöber 1966; Gigl and Dachille 1968).

FIGURE 3.232 Stolzite, Reef Mine, Cochise County, FOV 3 mm.

STOLZITE

Lead tungstate: $Pb(WO_4)$. A member of the scheelite group, stolzite is a rare secondary mineral found in oxidized lead deposits with cerussite, wulfenite, vanadinite, mimetite, and limonite.

COCHISE COUNTY: Dragoon Mountains, Bluebird District, Boerich claims (3rd ed.); Primos Mine, as small highly complex pale-yellow crystals in cavities in a quartz vein containing scheelite, chalcopyrite, sphalerite, hübnerite, fluorite, and galena (Palache 1941a; H 101780, UA x1190). Huachuca Mountains, Reef District, Reef Mine, as pale-yellow crystals, 1 to 2 mm long, on the walls of cavities in a quartz vein, associated with scheelite, galena, chalcopyrite, pyrite, and limonite (Palache 1941a; H 94680; R130615). Northwest flank of the Dos Cabezas Mountains, Teviston District, as small white tablets up to 3 mm across, in quartz veins bearing scheelite, galena, and pyrite (3rd ed.). Warren District, Bisbee, Campbell shaft (Graeme 1981).

GILA COUNTY: Miami-Inspiration supersystem, Day Peaks District, Lost Gulch area, on the east flank of Day Peaks, as a molybdian variety (Dale 1961); Copper Cities District, locality about 0.75 mile southwest of the Copper Cities Mine, as imperfectly formed crystals in cavities in quartz and disseminated in limonite, associated with scheelite (Faick and Hildebrand 1958).

LA PAZ COUNTY: Livingston Hills, Livingston District, Livingston claims, southeast of Quartzsite (Bull. 181).

PIMA COUNTY: Helvetia-Rosemont District, Omega Mine, as small (less than 1 mm) crystals on rosasite (Vi Frazier, UA x2147).

YAVAPAI COUNTY: Bradshaw Mountains, Silver Mountain District, Fat Jack Mine, as rough cubic to platy crystals up to 5 cm across, with pyromorphite, mottramite, and quartz (Modreski and Scovil 1990; Scovil and Wagner 1991). Black Rock District, near the mouth of Amazon Wash at confluence with Hassayampa River, northeast of Wickenburg (Bull. 181).

#STRACZEKITE

Calcium potassium barium vanadium oxide hydrate: $(Ca,K,Ba)(V^{5+}, V^{4+})_8O_{20}\cdot 3H_2O$. A secondary mineral found in oxidized portions of vanadium or uranium-vanadium deposits.

APACHE COUNTY: Cane Valley District, Monument No. 2 Mine (H. T. Evans and Hughes 1990).

STRENGITE

Iron phosphate hydrate: $Fe^{3+}(PO_4)\cdot 2H_2O$. A member of the variscite group; forms a solid-solution series with the aluminum end member, variscite. Found under surface or near-surface conditions as a product of the alteration of iron-bearing phosphate minerals.

MARICOPA AND YAVAPAI COUNTIES: White Picacho District, as crusts and cavity fillings that are commonly small pinkish-lavender or deep-red felted aggregates, associated with purpurite (Jahns 1952).

YAVAPAI COUNTY: 7U7 Ranch near Bagdad, as small glassy pink crystals, with other phosphates (Peacor, Dunn, and Simmons 1984).

FIGURE 3.233 String-hamite, Christmas Mine, Gila County, FOV 2 mm.

STRINGHAMITE

Calcium copper silicate hydrate: $CaCu(SiO_4).H_2O$. An uncommon late-stage hydrothermal mineral known from only a few localities.

GILA COUNTY: Southeastern Dripping Spring Mountains, Banner supersystem, Christmas District, Christmas Mine, in tactite ores with kinoite and apophyllite; the crystals are a distinctive lavender color ("cornflower blue") but are minute and invariably form only on fractures in retrogressively altered tactites formerly rich in wollastonite (3rd ed.; R110018).

PIMA COUNTY: Pima supersystem, Twin Buttes District, Twin Buttes Mine, with kinoite, copper, and apophyllite, in wollastonite (William Hefferon, pers. comm., 3rd ed.).

STROMEYERITE

Copper silver sulfide: CuAgS. An uncommon secondary mineral found in zones of sulfide enrichment. Associated with argentian tetrahedrite and bornite, with which it is commonly intimately intergrown.

COCHISE COUNTY: Tombstone District (Romslo and Ravitz 1947), Empire and Toughnut Mines, where it was probably an important source of silver (Guild 1917; B. S. Butler, Wilson, and Rasor 1938). Warren District, Bisbee, Campbell Mine (Schwartz and Park 1932) and Cole Mine (Bideaux, Williams, and Thomssen 1960).

GILA COUNTY: Globe Hills District, in the ores of the Old Dominion Mine, occurs replacing bornite and covellite (Lausen 1923). Richmond Basin District, reported from Richmond Basin, Mack Morris Mine (Bull. 181).

PIMA COUNTY: Cerro Colorado District, Cerro Colorado (Heintzelman) Mine, associated with tetrahedrite and silver (Bull. 181). South Comobabi Mountains, Cababi District, Mildren and Steppe claims (S. A. Williams 1963).

PINAL COUNTY: Pioneer supersystem, Silver King District, Silver King Mine, as the most important silver mineral in the ores (Guild 1910, 1917); Magma District, Magma Mine, sparingly in the hypogene ores (Short et al. 1943; Hammer and Peterson 1968; H 108641). Galiuro Mountains, Bunker Hill (Copper Creek) District, Copper Creek, lower levels of the Bluebird Mine, associated with tennantite (Kuhn 1951).

SANTA CRUZ COUNTY: Santa Rita Mountains, Ivanhoe District, Ivanhoe Mine near Squaw Gulch, associated with proustite, polybasite, silver, and galena (Engineering and Mining Journal 1912).

STRONTIANITE

Strontium carbonate: $Sr(CO_3)$. A low-temperature hydrothermal mineral formed as a gangue mineral in veins with sulfides. It can also be found in veins in limestone with baryte, calcite, and celestine.

MARICOPA COUNTY: Sauceda Mountains, Sauceda District, Gila Bend celestite prospect, about fifteen miles south of Gila Bend, S26, T7S R5W, with celestine and gypsum (B. N. Moore 1935).

MOHAVE COUNTY: Locality ten miles south of Lake Mead City (Robert O'Haire, pers. comm., 1972).

#STRUVITE

Ammonium magnesium phosphate hydrate: $(NH_4)Mg(PO_4)\cdot6H_2O$. A mineral typically found in bird or bat guano deposits.

MARICOPA COUNTY: Painted Rock District, Rowley Mine, occurs with rowleyite, antipinite, and salammoniac and is likely derived from bat guano (Kampf, Cooper, Nash, et al. 2017).

STÜTZITE

Silver telluride: $Ag_{5-x}Te_3$, where $x = 0.24-0.36$. Found as replacement masses in hydrothermal deposits associated with other tellurides and sulfides.

COCHISE COUNTY: Warren District, Bisbee, in the sulfide ores of the Campbell Mine (Harris et al. 1984).

SULPHUR

Sulphur: S. Formed in various ways: as a result of volcanic activity; in the oxidized portions of sulfide deposits; as a product of mine-fire activity; as a result of reduction of gypsum and other sulfates; and by the partial oxidation of pyrite under special conditions, such as from pyritic waste in old mine dumps.

APACHE COUNTY: Cane Valley District, Monument No. 2 Mine, as rare crystals in silicified wood (Jensen 1958; Witkind and Thaden 1963).

COCHISE COUNTY: Tombstone District, fourth level of the Empire Mine, as small resinous amber-yellow crystals that resemble sphalerite and replace anglesite and galena in the Skipjack shaft fissure (B. S. Butler, Wilson, and Rasor 1938). Warren District, Bisbee, Campbell shaft, Junction shaft, and Lavender pit (Graeme 1981).

COCONINO COUNTY: San Francisco Mountains, in small amounts at Sunset Crater and other nearby craters (Guild 1910).

GILA COUNTY: Miami-Inspiration supersystem, Pinto Valley District, Castle Dome Mine, as small well-formed crystals in open spaces in veins, formed by the oxidation of galena and sphalerite (N. P. Peterson, Gilbert, and Quick 1951; N. P. Peterson 1962). Banner supersystem, 79 District, 79 Mine, as microcrystals in vugs in galena as an oxidation product of galena (Stanley B. Keith 1972).

MARICOPA COUNTY: Wickenburg Mountains, San Domingo District, Surprise claims, northeast of Morristown, as crystals in cavities in quartz, formed by the decomposition of pyrite (Bull. 181).

PIMA COUNTY: Sierrita Mountains, Sierrita District, Sierrita Mine (William Kurtz, UA x392). Ajo District, New Cornelia Mine, as bright-yellow inclusions in gypsum (Les Presmyk, pers. comm., 2020).

PINAL COUNTY: About 2.5 miles east of Winkelman, as tiny crystals lining small vugs in a quartz vein (Bull. 181). Mammoth District, Mammoth–St. Anthony Mine, in small amounts in the oxidized zone (Bull. 181).

SANTA CRUZ COUNTY: Patagonia Mountains, Washington Camp District, Duquesne area, as microcrystals on quartz (H. Peter Knudsen, UA x1040).

YAVAPAI COUNTY: Wickenburg Mountains, Black Rock District, Purple Passion Mine, found as microscopic crystals, usually associated with smoky quartz (B. Gardner and Davis 2000). Jerome supersystem, Verde District, United Verde Mine, deposited under solfataric conditions caused by the burning of the pyritic orebody (Bull. 181).

#SURITE

Lead calcium aluminum silicate carbonate hydroxide hydrate: $(Pb,Ca)_3Al_2(Si,Al)_4O_{10}(CO_3)_2(OH)_3 \cdot 0.3H_2O$. A very rare secondary mineral found in oxidized lead-zinc deposits.

FIGURE 3.234 Surite, Mammoth Mine, Pinal County.

PINAL COUNTY: Mammoth District, Mammoth–St. Anthony Mine, Tiger, as small spherical aggregates of white to colorless waxy platy crystals occasionally associated with mimetite (R110145 and R060794). Southeastern Dripping Spring Mountains, Banner supersystem, 79 District, Ben Hur prospect, as small spherical aggregates of platy crystals with willemite, mimetite, wulfenite, and other secondary lead minerals (Ruiz, pers. comm., 2014).

#SUSANNITE

Lead sulfate carbonate hydroxide: $Pb_4(SO_4)(CO_3)_2(OH)_2$. A rare secondary mineral formed in the oxidized zone of hydrothermal lead deposits. It is a trigonal dimorph of leadhillite.

PINAL COUNTY: Mammoth District, Mammoth–St. Anthony Mine, as colorless to snow-white rhombohedral crystals that have probably paramorphosed to leadhillite (Bideaux 1980; J. W. Anthony et al. 2003; R050642).

SVANBERGITE

Strontium aluminum sulfate phosphate hydroxide: $SrAl_3(SO_4)(PO_4)(OH)_6$. A member of the beudantite group, found in aluminous schists and may also be formed as a product of hydrothermal alteration of igneous rocks.

FIGURE 3.235 Svanbergite, Dome Rock Mountains, La Paz County, FOV 3 mm.

LA PAZ COUNTY: Northern Plomosa Mountains, Dutchman District, in an exotic block with pyrophyllite with hematite (UA 9725; collected by Richard L. Jones). Dome Rock Mountains (R. Gibbs, pers. comm., 2020).

PIMA COUNTY: South Comobabi Mountains, Cababi District, Mildren and Steppe claims (S. A. Williams 1963).

SWARTZITE

Calcium magnesium uranyl carbonate hydrate: $CaMg(UO_2)(CO_3)_3 \cdot 12H_2O$. A rare secondary mineral formed as efflorescences on the walls of mine workings. The Hillside Mine is the type locality.

YAVAPAI COUNTY: Eureka supersystem, Hillside District, Hillside Mine, with gypsum, schröckingerite, bayleyite, andersonite, johannite, and uraninite, as an efflorescence on the walls (Axelrod et al. 1951).

SWITZERITE

Manganese phosphate hydrate: $Mn^{2+}_3(PO_4)_2 \cdot 7H_2O$. Found in granite pegmatites associated with other phosphate minerals.

YAVAPAI COUNTY: Hillside area, 7U7 Ranch near Bagdad, associated with bermanite (Peacor, Dunn, and Simmons 1984).

SYLVANITE

Silver gold telluride: $AgAuTe_4$. Most commonly formed in low-temperature hydrothermal veins but also found in moderate- and high-temperature deposits.

COCHISE COUNTY: Warren District, Bisbee, among the sulfide minerals of the Campbell orebody (Criddle et al. 1983).

LA PAZ COUNTY: Central Granite Wash Mountains, Glory Hole District, True Blue Mine (Stanton B. Keith 1978).

SYLVITE

Potassium chloride: KCl. A member of the halite group commonly formed as a widespread precipitate from oceanic waters.

APACHE AND NAVAJO COUNTIES: Holbrook Basin Potash District, east-central Arizona, encountered in drill holes that delineate a northeast-trending potash zone beneath an area of about three hundred square miles, in Permian evaporites (Peirce 1969b). The log of a hole drilled in S24, T18N R25E, also listed carnallite, polyhalite, halite, anhydrite, and gypsum (H. Wesley Peirce, pers. comm., 1972).

#SYNCHYSITE

Calcium (cerium, neodymium, yttrium) fluorocarbonate: $Ca(Ce,Nd,Y)(CO_3)_2F$. A group of rare-earth fluorocarbonate minerals with three end members based on dominant cation: synchysite-(Ce), synchysite-(Nd), and synchysite-(Y). The specific species is not known for the following occurrence.

MOHAVE COUNTY: Cerbat Mountains, Kingman Feldspar District, Kingman feldspar quarry, found as an accessory mineral in a rare-earth element pegmatite with light-colored microcline (T. J. Brown 2010).

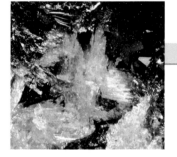

FIGURE 3.236 Szenicsite, Carlota Mine, Gila County, FOV 3 mm.

#SZENICSITE

Copper molybdate hydroxide: $Cu_3(MoO_4)(OH)_4$. A rare secondary mineral found in the oxidized portions of copper deposits. The Carlota Mine was the second occurrence of the mineral in the world.

GILA COUNTY: Carlota District, Carlota Mine, as very small green bladed crystals in an azurite-malachite vein with powellite and prosopite (W. E. Wilson and Jones 2015, R100179).

SZOMOLNOKITE

Iron sulfate hydrate: $Fe(SO_4) \cdot H_2O$. A water-soluble secondary mineral precipitated from highly acidic solutions derived from the breakdown of pyrite and associated with other sulfates.

COCHISE COUNTY: Warren District, Bisbee, Lavender pit, as brown warty crusts enclosing corroded pyrite grains, especially in intensely pyritized areas (Graeme 1981).

PINAL COUNTY: Pioneer supersystem, Magma District, Magma Mine, as clear colorless to tan crystalline material with rhomboclase and voltaite (3rd ed.).

TAENITE

Gamma nickel iron: γ-(Ni,Fe). Associated with nickel-rich iron, it is a major constituent of the iron meteorites of the state. Contains variable amounts (from about 27% to 65%) of nickel.

#TAKANELITE

Manganese calcium oxide hydrate: $(Mn^{2+},Ca)_{2x}(Mn^{4+})_{1-x}O_2 \cdot 0.7H_2O$. A secondary mineral that forms in various environments, including the oxide zone of a bedded manganese deposit in metamorphosed chert, the alteration of rhodochrosite in a manganese deposit, or a microbial catalysis of postmine groundwater.

GILA COUNTY: Globe Hills District, Pinal Creek, found as crusts containing a mixture of ranciéite and takanelite (Bilinski et al. 2002; Bargar et al. 2009).

TALC

Magnesium silicate hydroxide: $Mg_3Si_4O_{10}(OH)_2$. A layer-structured mica-like mineral formed by low-grade metamorphism of siliceous dolomites and by hydrothermal alteration of ultramafic rocks. It is commonly associated with serpentine.

APACHE COUNTY: Monument Valley, Garnet Ridge SUM District, in mineral fragments and in a serpentine ultramafic microbreccia (SUM) dike/diatreme that pierces Mesozoic sedimentary rocks (Gavasci and Kerr 1968).

COCHISE COUNTY: Gunnison Hills, as an alteration product of silty dolomite pebbles in Glance Conglomerate (J. R. Cooper and Silver 1964).

Cochise supersystem, Johnson Camp District, Little Dragoon Mountains, near Johnson Camp, in contact-metamorphosed dolomites around the Texas Canyon stock (J. R. Cooper 1957). Warren District, Bisbee, Boris shaft and Spray shaft (Graeme 1981). Dos Cabezas Mountains, Mascot District, at several mines, found with magnetite, pyrite, chalcopyrite, bornite, galena, epidote, chlorite, and garnet (Stanton B. Keith 1973).

COCONINO COUNTY: Grand Canyon National Park, in tremolite-talc dikes in the Vishnu Schist; the largest, at mile 83, is 80 m wide, containing tremolite nodules up to 15 cm across (M. D. Clark 1979).

GILA COUNTY: Dripping Spring Mountains, Christmas District, Christmas Mine, associated with tremolite and phlogopite in diopside hornfels in the footwall of the lower Martin orebody (Perry 1969).

GREENLEE COUNTY: Copper Mountain supersystem, Morenci District, northern part of the Morenci Mine, locally formed with other calcsilicate minerals in an extensive contact-metamorphic assemblage (Moolick and Durek 1966).

MARICOPA COUNTY: Vulture District, about ten miles south of Wickenburg, as massive green waxy material in pegmatite (Bull. 181).

PIMA COUNTY: Pima supersystem, Twin Buttes District, Twin Buttes Mine (William Hefferon, pers. comm., 3rd ed.).

PINAL COUNTY: Santa Catalina Mountains, Oracle District, Campo Bonito area (UA 8842).

YAVAPAI COUNTY: Eureka supersystem, reported as abundant, relatively pure material (Bull. 181).

YUMA COUNTY: Tank Mountains District, as light-green foliated masses (MM 6254–57).

TALMESSITE

Calcium magnesium arsenate hydrate: $Ca_2Mg(AsO_4)_2 \cdot 2H_2O$. A secondary mineral formed under oxidizing conditions.

MOHAVE COUNTY: Corkscrew District, Corkscrew Cave, on the Hualapai Indian Reservation, west side of Peach Springs Canyon, as a part of bright-white encrustations with green bands and spotty yellow patches, containing hörnesite, powellite, carnotite, conichalcite, calcite, and aragonite coating the walls of a cave (Wenrich and Sutphin 1989), found with tyuyamunite, claudetite, and others in breccia-pipe bodies (Onac, Hess, and White 2007).

#TANGEITE

Calcium copper vanadate hydroxide: $CaCu(VO_4)(OH)$. A rare secondary mineral found in the oxidized portion of copper deposits with vanadium. These occurrences include those for calciovolborthite from the third edition.

FIGURE 3.237 Tangeite, Monument No. 1 Mine, Navajo County, FOV 3 mm.

APACHE COUNTY: Monument Valley, Garnet Ridge U, in vein fillings and in cement of the Navajo Sandstone, along and near the contact with a breccia dike; associated with malachite, chrysocolla, tyuyamunite, limonite, pyrite, and chalcopyrite (Gavasci and Kerr 1968).

COCHISE COUNTY: Tombstone District, Gallagher vanadium property, as small (less than 0.3 mm) olive-green crystals on iron-stained quartz and anglesite in the Gallagher dump (Bowell, Luetcke, and Luetcke 2015). Warren District, as 0.5 mm yellow-green platy crystals associated with namibite and chrysocolla (J. McGlasson, pers. comm., 2021).

COCONINO COUNTY: Prospect Canyon (Ridenour) District, Ridenour Mine, with vésigniéite, naumannite, bromargyrite, tyuyamunite, and metatyuyamunite (Wenrich and Sutphin 1989).

LA PAZ COUNTY: New Water Mountains, Copper Queen Mine, as small aggregates of platy crystals (William Hunt, pers. comm., 1984).

MARICOPA COUNTY: Big Horn Mountains, Big Horn District, Evening Star Mine, as small yellow-green balls with conichalcite and plancheite in open cuts near Evening Star Mine (R. Jenkins, pers. comm., 2016).

NAVAJO COUNTY: Monument Valley District, Monument No. 1 Mine (Bull. 181).

YAVAPAI COUNTY: Big Bug Creek area, in NW quarter of S18, T8N R1W (Robert O'Haire, pers. comm., 1972).

#TANTALITE-(Fe)

Iron tantalate: $Fe^{2+}Ta_2O_6$. Often found in granite pegmatites, a member of the columbite-euxenite group.

PIMA COUNTY: Little Ajo Mountains, southwest of Ajo, San Antonio District, San Antonio Group, Valentine 1, 2, and 3 claims, as massive material in pegmatites associated with quartz, feldspar, and mica (MM 7865), which is identified as ferrotantalite.

#TAPIOLITE-(Fe)

Iron tantalate: $Fe^{2+}Ta_2O_6$. Often found in granite pegmatites, a member of the columbite-euxenite group.

YAVAPAI COUNTY: Southern Bradshaw Mountains, in stream gravels on Castle Creek (3rd ed.).

TEINEITE

Copper tellurite hydrate: $Cu^{2+}(Te^{4+}O_3)\cdot 2H_2O$. A very rare secondary mineral formed by oxidation of copper- and tellurium-bearing sulfides.

COCHISE COUNTY: Warren District, Bisbee, 1,200 ft. level of the Cole shaft, associated with cuprite, malachite, and graemite in a specimen collected from an ore car in 1959. One teineite crystal at least 2 cm long was totally replaced by graemite (S. A. Williams and Matter 1975; R100206); Shattuck shaft, as bright-blue oriented partial overgrowths on graemite pseudomorphs after teineite (Graeme 1993).

LA PAZ COUNTY: Dome Rock Mountains, in a small prospect, associated with goethite, gypsum, and graemite in small cavities in djurleite in quartz-tourmaline gangue; as at Bisbee, the graemite appears to replace teineite (S. A. Williams and Matter 1975).

TELLURITE

Tellurium oxide: TeO_2. Found in the oxidized zone of some hydrothermal deposits with tellurium minerals.

COCHISE COUNTY: Warren District, Bisbee, in the ores of the Campbell Mine (Graeme 1993).

TELLURIUM

Tellurium: Te. Elemental tellurium is rare and usually found in hydrothermal quartz veins with gold and telluride minerals.

COCHISE COUNTY: Tombstone District, as microscopic blebs in galena (B. S. Butler, Wilson, and Rasor 1938); with mackayite and emmonsite (Bideaux, Williams, and Thomssen 1960). Warren District, Bisbee, Campbell orebody (Graeme 1993).

LA PAZ COUNTY: Granite Wash Mountains, Glory Hole District, four miles north of Vicksburg (Bull. 181).

TELLUROBISMUTHITE

Bismuth telluride: Bi_2Te_3. Typically found in hydrothermal low-sulphur gold-quartz veins associated with gold, gold tellurides, bismuth, tetradymite, pyrite, and other minerals.

COCHISE COUNTY: Warren District, Bisbee, in the ores of the Campbell Mine (Graeme 1993).

LA PAZ COUNTY: La Cholla Gold Placers District, in the La Cholla placers as broken crystals in black sand concentrate (Melchiorre 2013).

TENNANTITE

Copper iron mercury nickel zinc arsenic sulfide: $Cu_6[Cu_4(Cu,Fe,Hg,Ni,Zn)_2]As_4S_{13}$. A widespread sulfosalt mineral formed in hydrothermal vein deposits with other sulfosalts and sulfides, especially in late-stage, arsenical, polymetallic mineralization related to porphyry copper deposits. Forms a complete solid-solution series with the antimony end member, tetrahedrite.

COCHISE COUNTY: Warren District, Bisbee, Campbell Mine, with tetrahedrite, chalcocite, bornite, enargite, and famatinite in Escabrosa Limestone, probably of hypogene origin (Schwartz and Park 1932).

GILA COUNTY: Northeast of Globe, Richmond Basin District, Helene vein, silver-bearing (Bull. 181). Mazatzal Mountains District, Ord Mine, as the mercurian variety (Faick 1958). Miami-Inspiration supersystem, Miami District, Miami Mine, with aikinite and enargite in veinlets that cut chalcopyrite (Legge 1939).

MOHAVE COUNTY: Cerbat Mountains, Wallapai supersystem, as a common constituent of the high-grade silver ores, typically associated with proustite in the Chloride and Stockton Hill Districts and in the Golconda and Tennessee Districts (Bastin 1925; B. E. Thomas 1949).

PIMA COUNTY: Ajo District, rare (Gilluly 1937). Pima supersystem, Mission-Pima District, Pima Mine, as a minor primary mineral (Himes 1972), in Mission and Eisenhower deposits (B. M. Williams 2018).

PINAL COUNTY: Pioneer supersystem, Magma District, Magma Mine (Barnes and Hay 1983), Tortilla Mountains, Ripsey District, Florence Lead-Silver Mine (S. A. Williams and Anthony 1970). Galiuro Mountains, Bunker Hill (Copper Creek) District, Bisbee, Childs-Aldwinkle Mine, with bornite and chalcopyrite (UA 543).

SANTA CRUZ COUNTY: Southern Patagonia Mountains, Washington Camp District, with digenite, johannsenite, palygorskite, and phlogopite (Lehman 1978).

YAVAPAI COUNTY: Wickenburg Mountains, Monte Cristo District, Monte Cristo Mine, argentiferous, with tetrahedrite, enargite, nickeline, and silver (Bastin 1922). Verde District, Jerome, United Verde Mine, as massive homogeneous material with high zinc content (C. A. Anderson and Creasey 1958). Iron King District, Iron King Mine, with sphalerite, galena, chalcopyrite, arsenopyrite, and pyrite (Creasey 1952).

TENORITE

Copper oxide: CuO. A relatively common mineral of secondary origin found in the oxidized zones of copper deposits, characteristically with chrysocolla, malachite, cuprite, limonite, and hematite. A study of many specimens labeled tenorite found them to be amorphous and mixtures of iron and manganese oxides. The only real tenorite found was closely associated with cuprite. Most localities below have not been confirmed.

COCHISE COUNTY: Warren District, Bisbee, as earthy material mixed with manganese oxides (Ransome 1904; Schwartz 1934; C. Frondel 1941). Dragoon Mountains, Courtland District, Maid of Sunshine and other mines (E. D. Wilson 1927b).

GILA COUNTY: Miami-Inspiration supersystem, sparingly in oxidized copper ores (Schwartz, 1921, 1934); Van Dyke Mine, with chrysocolla (N. P. Peterson 1962). Banner supersystem, 79 District, 79 Mine, associated with malachite and chrysocolla (Stanley B. Keith 1972).

GREENLEE COUNTY: Copper Mountain supersystem, Morenci District, may be related to "copper pitch ore" (Lindgren 1905; Bull. 181); Morenci Mine, Northwest extension, occurs with chrysocolla, brochantite, malachite, and neotocite (Enders 2000).

LA PAZ COUNTY: Yuma King District, Yuma King Mine, the main oxide mineral in the supergene zone, associated with azurite, chrysocolla, and malachite in oxidized Paleozoic skarns (Stanley B. Keith, pers. comm., 2019).

MOHAVE COUNTY: Emerald Isle District, Emerald Isle Mine, with chrysocolla (Bull. 181).

PIMA COUNTY: Pima supersystem, Twin Buttes District, Twin Buttes Mine, in the oxidized portions, especially in dissolution zones along the Twin Buttes fault, where it predates calcite deposition (Stanley B. Keith, pers. comm., 1973).

PINAL COUNTY: Mammoth District, Mammoth–St. Anthony Mine, as coal-black nodules surrounded by thin shells of chrysocolla (N. P.

Peterson 1938a, 1938b). San Manuel District, San Manuel orebody, in lesser quantities, associated with the more abundant chrysocolla, cuprite, malachite, and copper in the oxidized portions (L. A. Thomas 1966). Mineral Creek supersystem, Whitetail District, exotic copper deposit east of former Ray and in the east wall of the Pearl Handle Pit, with chrysocolla and malachite, cementing part of the White Tail Conglomerate (Phelps 1946; Clarke 1953); Sonora District, Black Copper Wash, in Holocene gravels with jarosite, goethite, malachite, and azurite (C. H. Phillips, Cornwall, and Rubin 1971).

YAVAPAI COUNTY: Weaver Mountains, Zonia District, Zonia Copper Mine, with malachite, cuprite, and chrysocolla (Kumke 1947).

TEPHROITE

Manganese silicate: $Mn^{2+}_2(SiO_4)$. A member of the olivine group, found in metamorphosed iron-manganese ore deposits and their associated skarns.

MOHAVE COUNTY: Rawhide Mountains, Alamo District, near Alamo Crossing (AM 34347).

TETRADYMITE

Bismuth tellurium sulfide: Bi_2Te_2S. An uncommon mineral found in gold-quartz veins and in contact-metamorphic deposits associated with tellurides and sulfides.

COCHISE COUNTY: Warren District, Bisbee, in the ores of the Campbell Mine (Graeme 1993). Little Dragoon Mountains, Tungsten King District, Tungsten King Mine, as exsolved phase in pockets of calcite with scheelite, pyrite, and galena (Stanton B. Keith 1973). Little Dragoon Mountains, Bluebird District, as an exsolution phenomenon in galena of the tungsten veins in the Texas Canyon stock (J. R. Cooper and Silver 1964). Little Dragoon Mountains, Precious Chemical Company property (UA 6323).

LA PAZ COUNTY: Reported near Vicksburg, but the exact locality is unknown (Bull. 181).

YAVAPAI COUNTY: Southern Bradshaw Mountains, Tip Top District, in small quantities at the Montgomery Mine (Guild 1910). Western Bradshaw Mountains, Tiger District, about two miles south of Bradshaw City, as bladed crystals in quartz, associated with pyrite (Genth 1890).

TETRAHEDRITE

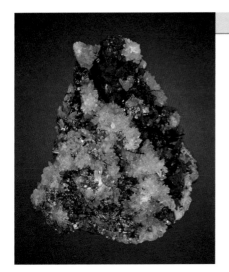

FIGURE 3.238 Tetrahedrite, Magma Mine, Pinal County, 7.9 cm high.

Copper iron mercury nickel zinc antimony sulfide: $Cu_6[Cu_4Fe,Hg,Ni,Zn_2]Sb_4S_{13}$. This is a common widespread sulfosalt mineral found in hydrothermal veins with other sulfosalts and sulfides. It forms a complete solid-solution series with the arsenic end member, tennantite. Crystals are typically complex and may be in zonal relationships with the isostructural minerals, freibergite and goldfieldite. More frequently found in antimony-rich portions of lead-zinc-silver districts, in contrast to tennantite, which forms in the arsenic-rich portions of the late stages of porphyry copper deposits.

COCHISE COUNTY: Tombstone District, notably at the Toughnut (UA 36), Lucky Cuss, and Ingersoll Mines, but present in most ores of the district, and containing silver (B. S. Butler, Wilson, and Rasor 1938; Rasor 1939; Romslo and Ravitz 1947). Warren District, widely distributed in primary copper sulfides (Graeme 1981). Pearce District, Commonwealth Mine, with proustite (L. A. Smith 1927). Cochise supersystem, Johnson Camp District, Johnson Camp area, in small quartz veins at the Republic Mine (J. R. Cooper and Silver 1964), and in fissure veins of the Moore orebody (J. R. Cooper and Huff 1951). Chiricahua Mountains, California supersystem, Ainsworth District, Humboldt Mine, argentiferous and associated with alabandite and other sulfides, rhodonite, and rhodochrosite (Hewett and Rove 1930).

GILA COUNTY: Globe Hills District, Old Dominion Mine, as crystals in cavities (Lausen 1923). Payson District, as the principal ore mineral at the Silver Butte Mine, where it reportedly carried considerable amounts of silver (Lausen and Wilson 1925). Banner supersystem, 79 District, on the dump of the 79 Mine (Stanley B. Keith 1972).

GRAHAM COUNTY: Aravaipa supersystem, Grand Reef District, Grand Reef Mine (Bull. 181).

MOHAVE COUNTY: Cerbat Mountains, Wallapai supersystem, common constituent of the high-grade silver ores (Bull. 181). Northern Hualapai Mountains, Hualapai (Antler) District, with galena (UA 148).

PIMA COUNTY: Santa Rita Mountains, Helvetia-Rosemont District, Silver Spur and Busterville Mines, with galena, pyrite, and chalcopyrite (Schrader and Hill 1915; C. A. Lee and Borland 1935); Greaterville District, Summit Mine (Schrader and Hill 1915). Pima supersystem,

Mission-Pima District, Paymaster Mine (Waller 1960; Nye 1961); Twin Buttes District, Twin Buttes Mine (William Hefferon, pers. comm., 3rd ed.). Cerro Colorado District, Cerro Colorado (Heintzelman) Mine, with stromeyerite and silver (Chaffee 1964; Bull. 181). Tucson Mountains, Amole District, very rare as spots on galena (Guild 1917).

PINAL COUNTY: Pioneer supersystem, Silver King District, Silver King Mine (Guild 1917); assays showed up to 3,000 ounces of silver per ton, but much of the value was thought to have been due to undetected stromeyerite; Magma District, abundant in the Magma Mine below the 900 ft. level (Short et al. 1943). Galiuro Mountains, Bunker Hill (Copper Creek) District, Bluebird Mine, where it is intimately associated with galena (Simons 1964); Childs-Aldwinkle Mine, associated with stromeyerite (Kuhn 1938; Simons 1964).

SANTA CRUZ COUNTY: Santa Rita Mountains, Salero District, Alto, Treasure Vault, and other mines, in copper-bearing ores with chalcopyrite (Schrader and Hill 1915); Wrightson District, as fine crystals at American Boy Mine (Schrader and Hill 1915); Mansfeld District, Armada Mine, as microcrystals with sphalerite (Peter Megaw, UA x2293). Patagonia Mountains, Flux District, World's Fair Mine, where it is argentiferous (UA 1520, 7307). Ruby District, Montana and Idaho Mines (H. V. Warren and Loofbourrow 1932); Oro Blanco District, Warsaw property (S 118120).

YAVAPAI COUNTY: Wickenburg Mountains, Monte Cristo District, Monte Cristo Mine, with enargite, argentiferous tennantite, nickeline, and silver (Bastin 1922). Verde District, Black Hills, United Verde Mine. Black Hills VMS District, Yeager and Shylock Mines (C. A. Anderson and Creasey 1958). Shea District, Shea Mine. Bradshaw Mountains, at several properties in the Walker, Hassayampa, Agua Fria, and other districts (Lindgren 1926). Turkey Creek District, Genth (1868) reported a "Fahlerz-like" mineral associated with quartz at the Goodwin Mine.

#TETRAHEDRITE-(Hg)

Copper mercury antimony sulfide: $Cu_6(Cu_4Hg_2)Sb_4S_{13}$. A member of the tetrahedrite subgroup of the tetrahedrite group, which contains significant amounts of mercury. It is the mercury analog of tetrahedrite-(Fe), tetrahedrite-(Ni), and tetrahedrite-(Zn).

MARICOPA COUNTY: Mazatzal Mountains District, Sunflower Group, as small lustrous black grains and masses associated with cinnabar, ankerite, and calcite (R. Gibbs, pers. comm., 2021).

TETRATAENITE

Iron nickel: FeNi. Found in several metallic meteorites, probably formed by the ordering of Fe and Ni atoms in taenite through very slow cooling.

COCONINO COUNTY: Meteor Crater near Winslow, sparsely in the Canyon Diablo meteorite (E. R. D. Scott, pers. comm., 1988).

#THALÉNITE-(Y)

Yttrium silicate fluoride: $Y_3Si_3O_{10}F$. A rare mineral formed in granite pegmatites. At the time of the third edition, thalénite was considered to be a hydroxide analog of fluorthalénite. Škoda et al. (2015), however, reexamined thalénite samples from various occurrences, including the Guy Hazen Group, and found that the mineral was identical to fluorthalénite, with fluorine strongly dominating the hydroxide. The IMA subsequently redefined the fluorine-dominant species as thalénite-(Y) and discredited fluorthalénite.

MOHAVE COUNTY: McConnico District, about five miles southwest of Kingman, in NE quarter of NW quarter of S15, T20N R17W, in a prospect in claims formerly known as the Guy Hazen Group (Pabst and Woodhouse 1964; Fitzpatrick and Pabst 1986); associated with microcline and "cyrtolite" (zircon, the crystal structure of which has been damaged by radioactivity) in a quartz outcrop, which is part of a pegmatite in coarse granitic rock; as rough grayish to beige-colored crystals up to several centimeters long that have been bleached white to a depth of up to 3 mm.

THAUMASITE

Calcium silicate hydroxide carbonate sulfate hydrate: $Ca_3Si(OH)_6(CO_3)$ $(SO_4) \cdot 12H_2O$. A very late-stage secondary mineral found in sulfide ore deposits, in contact-metamorphic zones, or where geothermal waters or seawaters react with basalts and tuffs (J. W. Anthony et al. 2003).

COCHISE COUNTY: Tombstone District, Lucky Cuss Mine, in small fissures and replacing limestone (B. S. Butler, Wilson, and Rasor 1938; Schaller 1939; UA 23, 615).

GRAHAM COUNTY: Aravaipa supersystem, Iron Cap District, Iron Cap Mine (W. E. Wilson 1988).

THÉNARDITE

Sodium sulfate: $Na_2(SO_4)$. Widespread in the arid Southwest, where it may be prevalent in dry lakes and playas; also formed as white encrustations on lavas. It is more abundant in Arizona than the few localities noted below would suggest.

COCHISE COUNTY: San Simon Basin, Bowie District, about seven miles northeast of Bowie, in tuff-bearing bedded lake deposits, associated with a variety of zeolite minerals (Regis and Sand 1967; Edson 1977).

COCONINO COUNTY: Near Sunset Crater, as tufts in ice caves and as coatings on basaltic lavas (Bull. 181; R040178).

PIMA COUNTY: Tucson Mountains, as ephemeral thin spotty crusts on andesitic volcanic rocks (3rd ed.).

PINAL COUNTY: Reported near Maricopa (Bull. 181).

YAVAPAI COUNTY: Camp Verde Salt District, Verde Salt Mine, in salt deposits of the Verde Valley lake beds, with halite, mirabilite, and glauberite (Silliman 1881; Blake 1890; Guild 1910; Peirce 1969b; J. R. Thompson 1983; R040183).

#THOMSONITE-Ca

Sodium calcium aluminum silicate hydrate: $NaCa_2$ $(Al_5Si_5)O_{20} \cdot 6H_2O$. A member of the zeolite group typically formed in cavities and vesicles in mafic igneous rocks, especially in amygdaloidal basalts.

COCHISE COUNTY: Warren District, Bisbee (Graeme 1981).

LA PAZ COUNTY: New Water Mountains, Ramsey District, with tangeite in a small prospect (Hunt 1983).

FIGURE 3.239 Thomsonite-Ca, Arnett Creek, Pinal County, FOV 3 mm.

MOHAVE COUNTY: Black Range, San Francisco District near Oatman, in amygdules in vesicular basalt (UA 8515, 8516).

PIMA COUNTY: Northern Santa Rita Mountains, Mt. Fagan, with clinochlore (variety penninite), pumpellyite, prehnite, epidote, and copper (Bideaux, Williams, and Thomssen 1960).

PINAL COUNTY: Northeast Suizo Mountains, just east of U.S. Highway 89 about midway between Tucson and Florence (3rd ed.), in a road cut in basalt. Arnett Creek, 5.5 miles south of Superior, as radiating fibrous material in amygdules in basalts (Wise and Tschernich 1975).

THORITE

Thorium silicate: $Th(SiO_4)$. Commonly contains substantial amounts of uranium, which substitutes for thorium. A primary mineral most commonly found in pegmatites.

LA PAZ COUNTY: Goodman District, Goodman Mine, near Quartzsite, in small amounts in a quartz vein that cuts a biotite schist (Adams and Staatz 1969).

MOHAVE COUNTY: Aquarius Mountains District, Uranium Basin claims, as the variety uranothorite, sparsely distributed in a vein at the contact between granite and pegmatite (Adams and Staatz 1969).

YAVAPAI COUNTY: Northeast Black Hills, Verde supersystem, Fairview and Becchetti claims, south side of the Verde River, in small amounts in veinlets that cut stockwork in green metavolcanic rocks, associated with quartz, limonite, hematite, siderite, dolomite, and magnetite (11 and 12 in J. C. Olson and Adams 1962; Adams and Staatz 1969). Steel City Mine, as uranium-rich variety with soddyite (M. E. Thompson, written comm., 1954, in C. Frondel 1958).

TILASITE

Calcium magnesium arsenate fluoride: $CaMg(AsO_4)F$. A rare secondary mineral formed in veins in limestone and dolomitic limestones, associated with braunite or hausmannite.

COCHISE COUNTY: Warren District, in an outcrop near the White Tail Deer Mine, as complex crystals up to 6 mm, with braunite, conichalcite, and calcite in veinlets in clean crystalline limestone (S. A. Williams 1970b; Bladh, Corbett, and McLean 1972).

#TIMROSEITE

Lead copper tellurate hydroxide: $Pb_2Cu_5(TeO_6)_2(OH)_2$. A rare secondary mineral found in the oxidized portion of lead-copper veins also containing tellurium.

COCHISE COUNTY: Huachuca Mountains, Reef District, Reef Mine, as a single specimen of very small green crystals intimately associated with mcalpineite and containing khinite-4O, gold, dugganite, and acanthite on quartz (Walstrom 2013).

TINZENITE

Calcium manganese aluminum borosilicate hydroxide: $Ca_2Mn^{2+}_4Al_4[B_2Si_8O_{30}](OH)_2$. A member of the axinite group that is commonly found in contact-metamorphic aureoles formed where intrusive rocks have invaded sediments, especially limestones. It is typically associated with other calcium silicate minerals.

COCHISE COUNTY: Huachuca Mountains, as yellow cleavage plates in massive quartz (H 95332).

TITANITE

Calcium titanium silicate: $CaTi(SiO_4)O$. This is a common accessory mineral in igneous rocks, pegmatites, and skarns. Titanite was called sphene in older geologic literature (generally pre-1980). An important early crystallizing accessory mineral, along with magnetite, in high-oxidation state magnetite or magnetite-sphene series granitic rocks. Magnetite-titanite assemblage is an important oxidation-state indicator in plutons associated with porphyry copper deposits.

FIGURE 3.240 Titanite, Big Lue Mountains, Greenlee County, FOV 3 mm.

COCHISE COUNTY: Tombstone District, as microscopic crystals in granodiorite and porphyritic rocks (B. S. Butler, Wilson, and Rasor 1938). Little Dragoon Mountains, Cochise supersystem, Johnson Camp District, near Johnson Camp, with wollastonite, grossular, epidote, and other contact-silicate minerals in the calcareous rocks, and as an accessory mineral with magnetite in the Texas Canyon Quartz Monzonite (J. R. Cooper 1957). Johnny Lyon Hills, as brilliant brown euhedral crystals, 1 to 2 mm in diameter (J. R. Cooper and Silver 1964). Mule Mountains, Juniper Flat District, as a common accessory mineral in the Juniper Flat Granite, northwest of Mule Pass Tunnel, in some places as well-formed crystals in vugs with chlorite and quartz (Graeme 1981).

GILA COUNTY: Miami-Inspiration supersystem, common accessory mineral in all the intrusions in the supersystem; Pinto Valley District, locally abundant in the Willow Spring Granodiorite (N. P. Peterson 1962); Castle Dome Mine, as a minor accessory mineral in Lost Gulch Quartz Monzonite and Gold Gulch Granodiorite (N. P. Peterson, Gilbert, and Quick 1946).

GRAHAM COUNTY: Mt. Turnbull, Aravaipa supersystem, Fisher District, in micropegmatite at the Fisher prospect (Bull. 181).

GREENLEE COUNTY: Copper Mountain supersystem, Morenci District, as an accessory mineral in Quartz Monzonite, associated with zircon (Schwartz 1947). Big Lue Mountains, Big Lue District, in road cuts along State Highway 78 (R. Gibbs pers. comm., 2018).

MARICOPA COUNTY: Belmont Mountains, Osborne District, Belmont pit, as small euhedral crystals in miarolitic cavities in the Belmont Granite, with epidote and fluorite (Gibbs and Turzi 2007).

MOHAVE COUNTY: Aquarius Mountains, Aquarius Mountains District, as small euhedral and long anhedral grains in pegmatite in a granite dike, associated with monazite, apatite, cronstedtite, and quartz; crosscuts and apparently replaces an earlier titanium-bearing mineral (Kauffman and Jaffe 1946). Cerbat Mountains, Wallapai supersystem, common as an alteration product in wall rocks associated with sulfide vein deposits, as long stringers and lenses replacing muscovite (B. E. Thomas 1949). Kingman, as black fragment (R050039). Near Wright Creek Ranch, Valentine District, Bluebird (Bountiful Beryl) prospect, in several irregular pegmatite dikes in schist and granite (Schaller, Stevens, and Jahns 1962).

PIMA COUNTY: East flank of the Quijotoa Mountains, Quijotoa District, Linda Lee claim, as tiny grains in spessartine-actinolite rock and in quartz monzonite (S. A. Williams 1960). Ajo District, as a primary constituent of all igneous rocks (Gilluly 1937). Silver Bell District, as an accessory mineral (Kerr 1951). Pima supersystem, as an accessory mineral in the Ruby Star granodioritic complex, along with magnetite and hornblende (E. Anthony 1986); Twin Buttes District, Twin Buttes Mine (William Hefferon, pers. comm., 3rd ed.). Northwestern Santa Catalina Mountains, as a common accessory mineral with biotite and magnetite in the Catalina Quartz-Monzonite pluton (Stanley B. Keith et al. 1980).

PINAL COUNTY: Mineral Creek supersystem, Ray District, a common accessory mineral, associated with magnetite, in all intrusions broadly related to the Ray porphyry copper system (Ransome 1919; Cornwall 1982). Central Dripping Spring Mountains, Troy supersystem, a common accessory mineral, along with magnetite, in all intrusive phases associated with porphyry copper mineralization (Ransome 1919; Cornwall 1982).

YAVAPAI COUNTY: Bradshaw Mountains, Big Bug District, in schist at the Butternut Mine (Lindgren 1926); Tiger District, Springfield Group of claims, as large crystals in Crown King Granodiorite (Lindgren 1926).

Eureka supersystem, Bagdad District, common accessory phase, along with magnetite, in all intrusions associated with the Bagdad porphyry copper system (C. A. Anderson, Scholz, and Strobell 1955).

TLAPALLITE

Calcium lead calcium copper oxide tellurite tellurate sulfate hydrate: $(Ca,Pb)_3CaCu_6O_2[Te^{4+}_3Te^{6+}O_{12}]_2(Te^{4+}O_3)_2(SO_4)_2 \cdot 3H_2O$. A rare mineral found with other tellurium minerals in the oxidized portion of two base-metal deposits.

COCHISE COUNTY: Tombstone District, a specimen labeled emmonsite, presumably from the Lucky Cuss Mine, was identified as tlapallite (S. A. Williams and Duggan 1978). This specimen, about 50 percent tlapallite, cemented a calcsilicate tactite gangue; Old Guard Mine, associated with khinite-4O, khinite-3T, dugganite, quetzalcoatlite, tenorite, chlorargyrite, chrysocolla, gold, and quartz (Cooper, Hawthorne, and Back 2008).

TOBERMORITE

Calcium silicate hydrate: $Ca_4Si_6O_{17}(H_2O)_2 \cdot (Ca \cdot 3H_2O)$. Found in retrograde contact-metamorphosed carbonate sediments. In 2015, the IMA redefined tobermorite as a supergroup of calcium silicate hydrates comprising two groups: tobermorite and a group of unclassified minerals. The tobermorite group consists of two species, tobermorite and kenotobermorite, which form a solid-solution series (Biagioni, Merlino, and Bonaccorsi 2015). The exact species is not known for the following occurrence.

GILA COUNTY: Southeastern Dripping Spring Mountains, Banner supersystem, Christmas District, Christmas Mine, as rare pearly-white plates plastered on fracture surfaces in metalimestone; in these rocks, the assemblage diopside grossular has reacted to form xonotlite, then tobermorite, zeolites, and apophyllite (Cesbron and Williams 1980).

TOCORNALITE

Silver mercury iodide: (Ag,Hg)I. A rare mineral found in hydrothermal silver deposits. The IMA lists status as questionable.

LA PAZ COUNTY: Trigo Mountains, Silver District, Padre Kino Mine, as massive green material, which darkens to black on exposure to light; associated with chlorargyrite (3rd ed.).

TODOROKITE

Sodium calcium potassium barium strontium manganese magnesium aluminum oxide hydrate: $(Na,Ca,K,Ba,Sr)_{1-x}(Mn,Mg,Al)_6O_{12}\cdot3\text{-}4H_2O$. A rare secondary mineral formed from the alteration of other manganese-bearing minerals.

COCHISE COUNTY: Southeasternmost Dragoon Mountains, Courtland District, in prospect pits north of Courtland, intimately mixed with birnessite; the mixtures form films on fractures in altered quartz monzonite (3rd ed.).

GREENLEE COUNTY: Copper Mountain supersystem, Morenci District, Morenci Mine, occurs in mixtures with cryptomelane, hollandite, and hausmannite (Enders 2000).

PINAL COUNTY: Mineral Hill District, Reymert Mine, among manganese oxide minerals, including psilomelane, hollandite, coronadite, and chalcophanite (K. S. Wilson 1984).

YUMA COUNTY: Kofa Mountains, Kofa District, North Star Mine, No. 2 vein, with psilomelane, bixbyite, groutite, and manganite (Hankins 1984); King of Arizona Mine, with groutite, chalcophanite, aurorite, and pyrolusite (Hankins 1984).

TOLBACHITE

Copper chloride: $CuCl_2$. As encrustations at fumaroles associated with basaltic lava flows.

COCHISE COUNTY: Warren District, Bisbee, Southwest Mine, as filmy brown crusts on masses of nantokite; alters to paratacamite under humid conditions (Graeme 1993).

TOPAZ

Aluminum silicate fluoride: $Al_2SiO_4F_2$. A widespread mineral typically associated with granites, pegmatites, and rhyolites. It is also formed in greisen and pneumatolytic bodies. It is a locally significant mineral in lithophysa in high-fluorine rhyolites, where it is commonly associated with spessartine garnet and as a rare accessory mineral in advanced argillic alteration zones associated with late-stage porphyry copper deposits and peraluminous granites.

COCHISE COUNTY: Dos Cabezas Mountains, DeBorde District, six miles east-southeast of Dos Cabezas and 0.75 mile southeast of the Cottonwood

Mine, on the William DeBorde property, as masses weighing up to several hundred pounds, with quartz and muscovite in coarse-grained granite (Bull. 181).

GILA COUNTY: Bread Pan District, Flying W Ranch, in a quartz-topaz-sodic-plagioclase rock (ongonite) and quartz-topaz (topazite) dikes in three areas northwest of the ranch: the Bread Pan claim, 0.5 km away; Dysart claim, 2.0 km away; and Spring Creek area, 1.3 km away; associated with muscovite, fluorite, tourmaline, and blue acicular beryl (W. T. Kortemeier 1986; W. T. Kortemeier and Burt 1988).

LA PAZ COUNTY: Northern Plomosa Mountains, Dutchman District, occurs as a major accessory mineral in highly argillized exotic blocks, hosted in Miocene trachyandesite agglomerate, associated with muscovite, hematite, svanbergite, and rutile (field stop 5 in Spencer and Pearthree 2015). Dome Rock Mountains, Stray Elephant Kyanite District, in rutile-bearing kyanite quartzite with lazulite and wagnerite in Middle Camp Quartz Monzonite (Marsh and Sheridan 1976), associated with andalusite and sillimanite (Melchiorre 2013).

MOHAVE COUNTY: Aquarius Mountains, Aquarius Mountains District, south of the Rare Metals Mine, in pegmatites, as euhedral crystals up to 2.5 in. long (Heinrich 1960). Kaiser Springs District, Negro Ed quadrangle, Negro Ed Mesa, as small crystals associated with spessartine garnets in cavities in a rhyolitic lava flow (Burt, Moyer, and Christiansen 1981); Lion Spring, with almandine in rhyolite (Stan Celestian, pers. comm., 2017).

PIMA COUNTY: Ajo Mountains, rare in the middle Tertiary age. Cardigan Peak pluton west of the Gibson Arroyo fault (Gilluly 1937). Pima supersystem, Twin Buttes District, Twin Buttes Mine (William Hefferon, pers. comm., 3rd ed.).

PINAL COUNTY: San Carlos Apache Reservation, Lookout Mountain District, southeast of Saddle Mountain, in the rhyolite-obsidian member of the Galiuro volcanics, with pseudobrookite, spessartine, and bixbyite (White 1992). Magma supersystem, Resolution District, Resolution copper deposit, with rutile, in late-stage highly argillized rocks (Hehnke et al. 2012).

TORBERNITE

Copper uranyl phosphate hydrate: $Cu(UO_2)_2(PO_4)_2 \cdot 12H_2O$. A member of the autunite group of layer-structured minerals whose water content varies between 8 and 12 percent; desiccation will produce the lower

FIGURE 3.241 Torbernite, King claims, Miami, Gila County, FOV 5 mm.

hydrated form, metatorbernite, and the inversion may or may not be reversible. A secondary mineral formed in the oxidized portions of uranium deposits. It is commonly associated with metatorbernite, metazeunerite, autunite, and other secondary uranium minerals. Also occurs with late-stage, oxide minerals in several porphyry copper deposits.

COCHISE COUNTY: Warren District, Bisbee, 770 level of the Junction Mine (Graeme, Graeme, and Graeme 2015)

COCONINO COUNTY: Cameron District, at several properties, including the Huskon claims (Scarborough 1981). Grand Canyon National Park, west of Maricopa Point, Orphan District, Orphan Mine, in a shear zone in Coconino Sandstone (Chenoweth 1986). Riverview District, Riverview Group of claims, in breccia-pipe deposit with uranophane and metatorbernite (Barrington and Kerr 1963; Wenrich and Sutphin 1989).

GILA COUNTY: Sierra Ancha District, Workman uranium deposit, and with bassetite and coffinite in the Dripping Spring Quartzite (Granger and Raup 1959). Lucky Boy District, Lucky Boy Mine, south of Globe, as small plates (MM 4903). Miami-Inspiration supersystem, Pinto Valley District, Castle Dome Mine (UA 6162); Inspiration District, Inspiration Mine, Live Oak pit (Stanley B. Keith, pers. comm., 2019). Globe Hills District, King claims south of Miami, as small green tablets (Gibbs, pers. comm., 2015).

GREENLEE COUNTY: Copper Mountain supersystem, Morenci District, Morenci Mine, as opaque crystals to 3 mm on native copper (Gibbs, pers. comm., 2015).

MOHAVE COUNTY: Hack District, Hack Canyon Mine, associated with metatorbernite, tyuyamunite, and coffinite (Wenrich and Sutphin 1989). Chapel District, Chapel prospect, with uranophane and autunite (Wenrich and Sutphin 1989).

NAVAJO COUNTY: Monument Valley District, Monument No. 1 Mine (Witkind and Thaden 1963) and Mitten No. 2 Mine (Witkind 1961).

PIMA COUNTY: Ajo Mountains, Ajo District, New Cornelia Mine, reportedly formed on copper (3rd ed.). Quijotoa Mountains, Quijotoa District, Linda Lee claims, with "gummite" and hematite in a vein in arkose (R. L. Robinson 1955). Silver Bell District, Silver Bell Mine, Oxide and El Tiro pits, as transparent emerald-green euhedral tabular crystals up to 1 cm; readily alters to apple-green opaque metatorbernite on exposure to air; associated with turquoise crystals and cacoxenite, both of which grow on torbernite (Kenneth W. Bladh, pers. comm., 1974).

PINAL COUNTY: Riverside District, one mile south of Kelvin, associated with turquoise (Charles L. Fair, pers. comm., 1975). Central Galiuro Mountains, Bunker Hill (Copper Creek) District, Old Reliable Mine, as transparent microcrystals with libethenite (Gibbs, pers. comm., 2015).

YAVAPAI COUNTY: Hillside area, 7U7 Ranch, in small amounts with bermanite and triplite in a pegmatite (Hurlbut 1936).

tourmaline

Minerals with a common structure and complex compositions, which may be represented by this general formula—sodium calcium magnesium iron aluminum lithium borate silicate hydroxide: $(Na,Ca)(Mg,Fe^{2+},Fe^{3+},Al,Li)_3Al_6(BO_3)Si_6O18(OH)_4$. Several species and varieties of tourmaline, which typically differ in color, have been identified. A widespread mineral group found in granite pegmatites, pneumatolytic veins, and granites. Also formed as products of metasomatism involving boron. Their resistance to mechanical and chemical weathering results in their accumulation in sands and in some sedimentary rocks. Tourmaline species described in this book include elbaite, dravite, magnesio-foitite, and schorl.

TREMOLITE

Calcium magnesium iron silicate hydroxide: $\square Ca_2(Mg_{5.0-4.5}Fe^{2+}_{0.0-0.5})Si_8O_{22}(OH)_2$. A member of the amphibole group, it is typically formed in contact-metamorphosed limestones and dolomites with other calcium silicate minerals. In Arizona it is a common gangue mineral associated with contact-metamorphic base-metal deposits.

COCHISE COUNTY: Warren District, as an abundant gangue mineral of the unoxidized pyritic ores (Graeme 1981). Tombstone District, Toughnut Mine, as long fibrous masses (B. S. Butler, Wilson, and Rasor 1938). Little Dragoon Mountains, in the metamorphosed dolomites (J. R. Cooper and Huff 1951; J. R. Cooper 1957; J. R. Cooper and Silver 1964).

COCONINO COUNTY: Grand Canyon National Park, in tremolite-talc ultramafic dikes in the Vishnu Schist; the largest, at mile 83, is 80 m wide, containing tremolite nodules up to 15 cm across (M. D. Clark 1979; shown on Billingsley 2000).

GILA COUNTY: Banner supersystem, Christmas District, Christmas Mine, associated with diopside, a prominent gangue mineral formed by contact metamorphism of dolomitic rocks (N. P. Peterson and Swanson 1956;

Perry 1969); Chilito and 79 Districts, 79 Mine, with andradite, diopside, and epidote in contact-metamorphosed limestones (Stanley B. Keith 1972).

GREENLEE COUNTY: Copper Mountain supersystem, Morenci District, in contact-metamorphosed limestones with other calcsilicate minerals (Lindgren 1905; Moolick and Durek 1966).

LA PAZ COUNTY: Northern Dome Rock Mountains, Dome Rock District, as an asbestiform variety in marbleized limestones (3rd ed.).

MARICOPA COUNTY: Northeastern Harquahala Mountains, Dushey Canyon, southwest of Aguila, replacing limestone (UA 5907). South-central Harquahala Mountains, eastern Harquahala District, in S14, T5N R10W, in a large outcrop of crystalline material containing crystals up to 4 cm long, in marble (David Shannon, pers. comm., 1988). Sierra Estrella Mountains, North End, in marble as an accessory mineral with diopside and zoisite (3rd ed.; Melchoirre 1992); Sunset Canyon, Yellow Flower Mine, as acicular rosettes in marble—the tremolite is fluorescent white under UV illumination (George V. Polman, pers. comm., 2020).

MOHAVE COUNTY: Hualapai (Antler) District, Copper World Mine near Yucca, as a gangue mineral in a lenticular metamorphosed volcanic massive sulfide, copper-zinc sulfide orebody (Rasor 1946) and at the nearby Antler Mine (Romslo 1948; More 1980).

PIMA COUNTY: Pima supersystem, Mission-Pima District, Pima Mine, abundant in limestone hornfels with diopside and grossular (Journeay 1959); Twin Buttes District, Twin Buttes Mine, in the tactites and skarns (Stanley B. Keith, pers. comm., 1973); Mission-Pima District, Mission Mine, in the hornfels host rock (Richard and Courtright 1959); Mineral Hill Mine and other properties, associated with mineral deposits of contact-metamorphic origin (Ransome 1922). Helvetia-Rosemont District, in contact-metamorphosed sedimentary rocks at several properties, including the Copper World, Leader, Narragansett, and Isle Royal Mines (Creasey and Quick 1955).

SANTA CRUZ COUNTY: Southern Patagonia Mountains, Washington Camp District, as gangue at the Pride of the West Mine (Schrader and Hill 1915); in contact-metamorphic deposits in limestone (Schrader 1917).

#TREVORITE

Nickel iron oxide: $NiFe^{3+}_2O_4$. Found in a nickeliferous serpentinite bodies, in ultramafic intrusions, and nickel sulfide deposits within gabbros that have intruded peridotite. It has also been found in meteorites.

COCONINO COUNTY: Meteor Crater area, Canyon Diablo meteorite, found in meteoritic iron oxide, which is composed of maghemite, goethite, and minor trevorite (Hofmann 1992).

TRIDYMITE

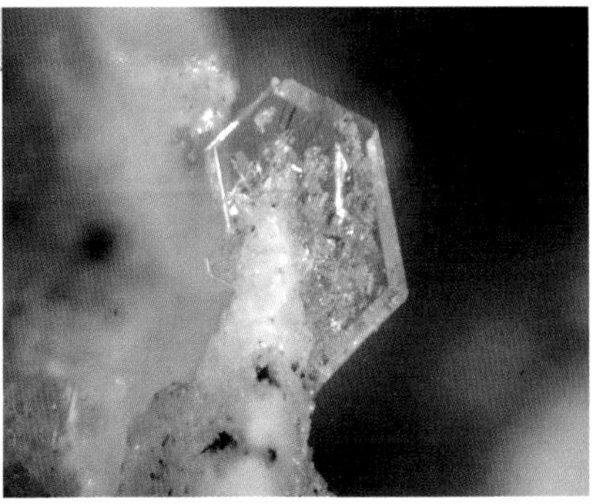

FIGURE 3.242 Tridymite, Big Lue Mountains, Greenlee County, FOV 2 mm.

Silicon oxide: SiO_2. A high-temperature polymorph of silica typically formed in felsic volcanic rocks. Present in many rhyolites in Arizona at a microscopic size.

GREENLEE COUNTY: Big Lue Mountains, Big Lue District, in road cuts along State Highway 78, about six miles west of the New Mexico border, as crystals up to 8 mm in light-colored volcanic rock, with pseudobrookite and titanite (William Hunt, pers. comm., 1989; R040143).

MOHAVE COUNTY: Cerbat Mountains, along the western edge of the Chloride District, with calcite in the matrix of pumice and andesite lithic fragments within a tuffaceous breccia found among stratiform volcanic rocks (B. E. Thomas 1953).

PIMA COUNTY: Roskruge Mountains, with cristobalite and clay in cavities in andesite (Bull. 181).

TRIPHYLITE

Lithium iron phosphate: $LiFe^{2+}(PO_4)$. A primary mineral formed in granite pegmatites. A complete solid-solution series that probably extends between the end members triphylite and lithiophilite, $Li(Mn^{2+},Fe^{2+})PO_4$.

YAVAPAI COUNTY: White Picacho District, with lithiophilite in the Midnight Owl and other pegmatites, as rough-faced equant to stubby prismatic crystals up to 6 in. long (Jahns 1952).

TRIPLITE

Manganese phosphate fluoride: $Mn^{2+}_2(PO_4)F$. A primary mineral formed in phosphate-rich granite pegmatites.

MARICOPA AND YAVAPAI COUNTIES: White Picacho District, in pegmatites, especially well-crystallized in the Midnight Owl pegmatite, as rough-faced tabular crystals up to 7 in. long (Jahns 1952).

YAVAPAI COUNTY: Hillside area, 7U7 Ranch, as a spherical aggregation about 2 ft. in diameter in a small pegmatite knot (Hurlbut 1936; Leavens 1967).

TRIPLOIDITE

FIGURE 3.243 Triploidite, 7U7 Ranch, Yavapai County, FOV 2 mm.

Manganese phosphate hydroxide: $Mn^{2+}_2(PO_4)(OH)$. Found in granite pegmatites as an uncommon alteration product of primary phosphate minerals.

MARICOPA AND YAVAPAI COUNTIES: White Picacho District, in pegmatites (London 1979; London and Burt 1982a). On a specimen from the Hillside District, with bermanite and phosphosiderite (Jim McGlasson, pers. comm., 2020).

TROILITE

Iron sulfide: FeS. Found in ultramafic rocks and in metallic meteorites.

COCONINO COUNTY: Meteor Crater, as a minor constituent of metallic spheroids, which were probably formed on impact of the Canyon Diablo meteorite; intergrown with schreibersite, formed interstitially along grain boundaries of the nickel-rich iron cores; also associated with maghemite and earthy iron oxide, both produced by weathering (Mead, Littler, and Chao 1965).

LA PAZ AND YUMA COUNTIES: Cemetery Ridge District, as small grains with other sulfides in magnetite in blocks of serpentinite and partially serpentinized harzburgite in the Orocopia Schist at Cemetery Ridge (Haxel et al. 2018).

NAVAJO COUNTY: Holbrook meteorite, as one of three iron phases (Gibson and Bogard 1978).

TSCHERMAKITE

Calcium magnesium aluminum silicate hydroxide: $\square Ca_2(Mg_3Al_2)(Si_6Al_2)$ $O_{22}(OH)_2$. Found in both ultramafic igneous rocks and metamorphic rocks, such as amphibolites.

PIMA COUNTY: Northern Santa Rita Mountains, Helvetia-Rosemont District, Blue Jay Mine, with graphite in metamorphosed limestone (3rd ed.).

TSUMEBITE

Lead copper phosphate sulfate hydroxide: $Pb_2Cu(PO_4)(SO_4)(OH)$. A very rare secondary mineral formed in the oxidized portion of base-metal deposits.

GILA COUNTY: Banner supersystem, 79 District, 79 Mine, as small euhedral crystals with pyromorphite on the lower fourth level (Callahan, pers. comm., 2014).

GREENLEE COUNTY: Copper Mountain supersystem, Morenci District, Morenci, with wulfenite in the oxidized portion of the Clay orebody (Bideaux, Williams, and Thomssen 1960; UA 2393, 2394, 5687).

PIMA COUNTY: Helvetia-Rosemont District, Central claim, as small blue crystals with azurite and wulfenite (Gibbs, pers. comm., 2014),

FIGURE 3.244 Tsumebite, 79 Mine, Gila County, FOV 1 mm.

PINAL COUNTY: Mammoth District, Mammoth–St. Anthony Mine, very rare as yellow-green spherules, with mimetite and wulfenite (Bideaux 1980); Collins vein, as sea-blue tabular crystals on quartz, associated with phosphohedyphane, goethite, hematite, cerussite, fluorite, and corkite (R070446); near the Pearl Mine as blue-green aggregates in a small prospect (Gibbs, pers. comm., 2016). Owl Head District, at a small prospect near the Big Mine as small blue crystals with mimetite and creaseyite (Gibbs, pers. comm., 2014).

#TUNGSTENITE

Tungsten sulfide: WS_2. Found in sulfide deposits replacing limestone with other tungsten minerals.

COCHISE COUNTY: Warren District, Bisbee, Campbell Mine, in bornite sulfide ore with kiddcreekite (Schumer 2017).

TUNGSTITE

Tungsten oxide hydrate: $WO_3 \cdot H_2O$. An uncommon mineral formed by the alteration of primary tungsten minerals such as "wolframite" and scheelite, with which it is commonly associated.

MARICOPA COUNTY: Cave Creek District, Gold Cliff Group, in S11, T6N R4E, associated with ferberite, cuprotungstite, fluorite, molybdenite, pyrite, and chalcopyrite (Dale 1959).

MOHAVE COUNTY: Hualapai Mountains, Boriana District, Boriana Mine (Chatman 1988; MM 1019).

TURQUOISE

Copper aluminum phosphate hydroxide hydrate: $CuAl_6(PO_4)_4(OH)_8 \cdot 4H_2O$. The aluminum-rich end member that forms a solid-solution series with the iron-rich end member chalcosiderite. Turquoise commonly contains some iron, which can help impart a greenish color. A widely distributed mineral commonly thought to be of secondary origin. Associated with limonite, clay minerals, and advanced argillic alteration in late stages of porphyry copper systems.

COCHISE COUNTY: Courtland District, mined at Turquoise Mountain, where it formed stringers up to a few inches wide and small nugget-like masses hosted in the Turquoise Granite and Bolsa Quartzite (Crawford and Johnson 1937; UA 9688). Warren District, Bisbee, 1,200 ft. level of the Cole shaft, as minute stringers in massive pyrite (UA 1243); Lavender pit, as large rich masses (Graeme 1981; UA 6669). Pearce District, reported as irregular patches and veinlets in kaolinized rhyolite (Guild 1910).

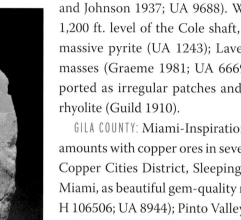

FIGURE 3.245 Turquoise, Ithaca Peak, Kingman, Mohave County, 7.5 cm high.

GILA COUNTY: Miami-Inspiration supersystem, in small amounts with copper ores in several deposits (Bull. 181); Copper Cities District, Sleeping Beauty Mine north of Miami, as beautiful gem-quality material (Jackson 1955; H 106506; UA 8944); Pinto Valley District, Castle Dome and Pinto Valley Mines, in substantial quantities (N. P. Peterson, Gilbert, and Quick 1946, 1951; N. P. Peterson 1962; W. W. Simmons and Fowells 1966; UA 1161, 1162). Salt River District, on Canyon Creek, one mile north of the Salt River, as small blebs and coatings on fine-grained beige-colored ortho-quartzite (R. T. Moore 1968).

GRAHAM COUNTY: Gila Mountains, Safford supersystem, Lone Star District, Lone Star porphyry copper deposit, intimately associated with jarosite and alunite in the oxidized zone (R. F. Robinson and Cook 1966).

GREENLEE COUNTY: Copper Mountain supersystem, Morenci District, as thin plates and nodules that are closely associated with a diabase dike system that crosses the Morenci orebody (Moolick and Durek 1966); Candelaria District, with alunite, wavellite, and conichalcite, in seams that cut the Candelaria breccia pipe (Bennett 1975).

MARICOPA COUNTY: Red Picacho District, east of Morristown (Bull. 181).

MOHAVE COUNTY: Cerbat Mountains, Wallapai supersystem, Ithaca Peak District, Ithaca Peak (Sterrett 1908; Crawford and Johnson 1937; B. E. Thomas 1949; Eidel, Frost, and Clippinger 1968), as gem-quality material in a porphyry body that cuts schist and gneiss (Schrader 1908; Guild 1910; NHM 86815). Turquoise Mountain, as gem-quality material in veins in gold-bearing quartz (Frenzel 1898).

PIMA COUNTY: Pima supersystem, Esperanza District, Esperanza Mine (Schmitt et al. 1959; Loghry 1972); Twin Buttes District, Twin Buttes Mine (William Hefferon, pers. comm., 3rd ed.). Silver Bell District, Silver Bell Mine, Oxide pit, as transparent tabular green crystals up to 1 mm, in subparallel aggregates; associated with torbernite and cacoxenite at the contact of an andesite dike (Kenneth W. Bladh, pers. comm., 1974; UA x215).

PINAL COUNTY: Mineral Creek supersystem, Riverside District, one mile south of Kelvin, associated with torbernite (Charles L. Fair, pers. comm., 1975).

YAVAPAI COUNTY: Reported in the county, but the specific locality was unknown (Bull. 181).

TYROLITE

Calcium copper arsenate carbonate hydroxide hydrate: $Ca_2Cu_9(AsO_4)_4$ $(CO_3)(OH)_8 \cdot 11H_2O$. A rare secondary mineral found in oxidized portions of copper deposits.

COCHISE COUNTY: Warren District, confirmed on a single specimen by Anthony J. Nikischer as sprays of 3 mm crystals on goethite with azurite and malachite. No information on the exact location at Bisbee was available (Graeme, Graeme, and Graeme 2015).

PINAL COUNTY: Superstition Mountains, Randolf District, as films on mercurian tetrahedrite known as the variety schwartzite (3rd ed.).

TYUYAMUNITE

Calcium uranyl vanadate hydrate: $Ca(UO_2)_2(VO_4)_2 \cdot$ 5-8H_2O. A widespread secondary mineral associated with carnotite, uranophane, and other uranium and vanadium minerals. Resembles carnotite but is less common.

APACHE COUNTY: Lukachukai District, Mesa No. 1 and No. 5 Mines, most common uranium ore mineral, coating sand grains or replacing calcite and carbon (Bull. 181; Chenoweth and Malan 1973). Carrizo Mountains, Black Rock Point and Cove Mesa Districts, disseminated in zones of dark-gray to black montroseite-bearing rock (Hershey 1958; Bull. 181). Chinle District, Black Mesa Basin, at numerous claims (Bull. 181). Cane Valley District, Monument No. 2 Mine, found in very rich cylindrical masses that have been called "rods" (Rosenzweig, Gruner, and Gardiner 1954; Mitcham and Evensen 1955; Weeks, Thompson, and Sherwood 1955; Finnell 1957;

FIGURE 3.246 Tyuyamunite, Ridenour Mine, Coconino County.

Evensen and Gray 1958; R. G. Young 1964; Witkind and Thaden 1963). Garnet Ridge U District near Dennehotso, sparsely disseminated in mineralized zones (Witkind and Thaden 1963; Gavasci and Kerr 1968).

COCHISE COUNTY: Warren District, Bisbee, Cole Mine, lining cracks and fissures in limestone country rock adjacent to massive chalcocite-covellite orebodies (Hutton 1957; UA 6158, 6239; S 96427); Campbell Mine (AM 21108).

COCONINO COUNTY: Cameron District, Black Point–Murphy Mine (Austin 1964). Prospect Canyon (Ridenour) District, Ridenour Mine, with vésigniéite, naumannite, bromargyrite, tangeite, and metatyuyamunite (Wenrich and Sutphin 1989).

MOHAVE COUNTY: Hack District, Hack Canyon Mine in breccia pipes (Bull. 181).

NAVAJO COUNTY: Monument Valley District, at several mines and claims, notably at the Monument No. 1 Mine, scattered irregularly with volborthite, azurite, and malachite, in brilliant blue-black corvusite-bearing ore (Evensen and Gray 1958; Holland et al. 1958; Witkind 1961; Witkind and Thaden 1963).

ULVÖSPINEL

Iron titanate: $Fe_2^{2+}TiO_4$. Found as a common constituent of titaniferous magnetite iron ores, in which it may form fine exsolution lamellae.

GILA COUNTY: San Carlos Apache Reservation, about three miles north-northwest of Coolidge Dam, Soda Spring vent, as skeletal-growth octahedral crystals up to 2 cm in a mugearite-type basalt with ferro-kaersutite, anorthoclase, and ferroan biotite (Caporuscio 1980).

UMOHOITE

Uranyl molybdate hydrate: $(UO_2)(MoO_4) \cdot 2H_2O$. A secondary mineral formed during the early stages of oxidation of primary uranium minerals.

COCONINO COUNTY: Cameron District, Alyce Tolino Mine, as a blue-black opaque isotropic mineral contained in sooty masses and carbonaceous trash replacements (Hamilton and Kerr 1959).

URANINITE

Uranium oxide: UO_2. Uraninite is commonly formed in hydrothermal veins. In Arizona, it is most notable in deposits in the sedimentary rocks of the Colorado Plateau, where it may be associated with coffinite and a wide variety of secondary uranium, vanadium, and copper minerals.

APACHE COUNTY: Lukachukai District, at several claims and mines, including Luki, Mesa 1¾, Mesa 2, Cove School, and Mesa 4½ (Joralemon 1952). Cane Valley District, Monument No. 2 Mine (Rosenzweig, Gruner, and Gardiner 1954; Mitcham and Evensen 1955; Jensen 1958; Witkind and Thaden 1963; R. G. Young 1964). Garnet Ridge U District near Dennehotso (Witkind and Thaden 1963).

COCHISE COUNTY: Warren District, reported in very small amounts in some of the underground workings (Graeme 1981).

COCONINO COUNTY: Cameron District (Austin 1964), Henry Sloan No. 1 Mine, with marcasite in calcite cement in sandstone bordering carbonaceous fossil wood (Austin 1964); Ramco No. 20-22 Mines (Peirce, Keith, and Wilt 1970); Huskon No. 5 Mine and related Huskon properties, with secondary uranium minerals, associated with petrified logs and halos around logs (Peirce, Keith, and Wilt 1970); Arrowhead Mine (Rosenzweig, Gruner, and Gardiner 1954; Holland et al. 1958); Boyd Teise No. 2 Mine, in flattened nodules with pyrite in sandstone (AEC mineral collection); Hosteen Nez claim, near Tuba City (Bull. 181; Peirce, Keith, and Wilt 1970). Vermilion Cliffs District, Sun Valley Mine, as replacements of rounded detrital grains with pyrite and sphalerite (Petersen, Hamilton, and Myers 1959; Petersen 1960). South Rim of the Grand Canyon, Orphan District, Orphan Mine, as the fairly pure mineral in disseminations and as veins and lenses up to several inches thick, in a nearly vertical circular or ellipsoidal breccia pipe developed primarily in the Permian Coconino Sandstone; associated with pyrite and other sulfides and sulfosalts of copper, lead, zinc, cobalt, nickel, and molybdenum; secondary minerals of uranium are common in the mine workings (Granger and Raup 1962; Gornitz and Kerr 1970).

GILA COUNTY: Sierra Ancha District in Dripping Spring Quartzite at the Black Bush, Hope, Little Joe, Lucky Stop, Red Bluff, Rock Canyon, Suckerite, Sue, Tomato Juice, and Workman deposits, as fissure and open-space fillings or as lenses and blebs in host rock (Granger and Raup 1969); Turquoise Mines, in T5N R15E, associated with pyrite, marcasite, and chalcopyrite (Granger and Raup 1959, 1962). Miami-Inspiration supersystem, Globe Hills District, Iron Cap Mine near Miami, with iron oxides (S 112722); Old Dominion Mine (3rd ed.).

MOHAVE COUNTY: Hack District, Hack Canyon Mine, mixed with chalcocite deposited in breccia zones and in some coarse-grained sandstones (Scarborough 1981; UA 6115, 6116, 6118).

NAVAJO COUNTY: Monument Valley District, Bootjack Mine (Chenoweth 1993); Monument No. 1 Mine (Evensen and Gray 1958; Witkind and Thaden 1963). Petrified Forest Wood District, near Holbrook, Ruth Group of claims, as replacement of carbonaceous matter (Gregg and Moore 1955). Cibecue District, Stinking Spring near Hunt (Rosenzweig, Gruner, and Gardiner 1954).

PIMA COUNTY: Eastern Rincon Mountains, Blue Rock District, Blue Rock claims, as sooty coatings in vugs, on fractures, and on pyrite, chalcopyrite, or fluorite grains (R. A. Miller 1958).

SANTA CRUZ COUNTY: Oro Blanco District, Annie Laurie prospect near Footes Spring (R. Y. Anderson and Kurtz 1955; Granger and Raup 1962). Pajarito District, Alamo Spring, with uranophane and autunite (Bull. 181). Santa Rita Mountains, Mansfeld District, Happy Jack Mine (G. M. Butler and Allen 1921).

YAVAPAI COUNTY: Eureka supersystem, Hillside District, Hillside Mine (Granger and Raup 1962).

URANOCIRCITE

Barium uranyl phosphate hydrate: $Ba(UO_2)_2(PO_4)_2 \cdot (10,12)H_2O$. A secondary mineral and uncommon member of the autunite group. Nuffield and Milne (1953) noted that the mineral's water content varies readily with humidity and temperature, while Walenta (1963) differentiated the species based on waters of hydration:

Uranocircite-I	$Ba(UO_2)_2(PO_4)_2 \cdot 12H_2O$
Uranocircite-II	$Ba(UO_2)_2(PO_4)_2 \cdot 10H_2O$
Metauranocircite-I	$Ba(UO_2)_2(PO_4)_2 \cdot 8H_2O$
Metauranocircite-II	$Ba(UO_2)_2(PO_4)_2 \cdot 6H_2O$

Because of limited characterization data and evidence of natural occurrences, uranocircite-I has not been approved by the IMA. Uranocircite-II, however, has been recognized as a valid species along with metauranocircite-II, which was renamed metauranocircite-I. The particular identity of the minerals from the following locations is not known.

COCONINO COUNTY: Cameron area, Cameron District, associated with metauranocircite-I in some of the uranium deposits of the area (Austin 1964).

GILA COUNTY: Sierra Ancha District, Sue Mine and Red Bluff Mine (Granger and Raup 1959).

URANOPHANE

Calcium uranyl silicate hydroxide hydrate: $Ca(UO_2)_2$ $(SiO_3OH)_2 \cdot 5H_2O$. This is one of the most common of the secondary uranium minerals. Uranophane has two polymorphs, which are designated by the suffixes -α and -β. The α dimorph forms over a wide range of neutral and alkaline solutions and is the more common occurrence (Barinova et al. 2003). The specific dimorph is not known for the following occurrences.

FIGURE 3.247 Uranophane, Muggins Mountains, Yuma County, FOV 3 mm.

APACHE COUNTY: Cane Valley District, Monument No. 2 and Cato Sells Mines (Witkind and Thaden 1963).

COCONINO COUNTY: Cameron District, in minor amounts in uranium ore in the Shinarump Conglomerate, in paleo-stream channels, and in the Chinle Formation in sandy portions of mounds, associated with uraninite, metatorbernite, and meta-autunite (Holland et al. 1958); Black Point–Murphy Mine, intimately mixed with uranophane-β in Pleistocene gravels (Austin 1964). Riverview District, Riverview Group of claims, in a breccia pipe with torbernite and metatorbernite (Wenrich and Sutphin 1989).

GILA COUNTY: Sierra Ancha District, in Dripping Spring Quartzite at the Fairview, Little Joe, and Red Bluff deposits (Granger and Raup 1969).

MOHAVE COUNTY: Chapel District, Chapel prospect, with autunite and torbernite (Wenrich and Sutphin 1989).

PIMA COUNTY: East flank of the Quijotoa Mountains, Quijotoa District, Linda Lee claims, as fracture coatings in a contact-metamorphic assemblage (S. A. Williams 1960). Rincon Mountains, Blue Rock District, Sure Fire No.1 claim, with autunite and fluorite (Granger and Raup 1962).

SANTA CRUZ COUNTY: Pajarito Mountains, Pajarito District, in a vein near Alamo Spring, associated with autunite and uraninite (Bull. 181); White Oak property, with kasolite and oxidized lead ore in shear zones in rhyolite (Granger and Raup 1962). Santa Rita Mountains, Duranium District, Duranium claims, in arkosic sandstone with kasolite and autunite (R. L. Robinson 1954). Nogales District.

YAVAPAI COUNTY: Weaver Mountains, Kirkland District, Peeples Valley Mine (Granger and Raup 1962).

YUMA COUNTY: Muggins Mountains, Muggins District (R. Jenkins, pers. comm., 2020).

URANOPHANE-β

Calcium uranyl silicate hydroxide hydrate: $Ca(UO_2)_2(SiO_3OH)_2 \cdot 5H_2O$. This is a secondary mineral, dimorphous with uranophane-α.

COCONINO COUNTY: Cameron District (Bollin and Kerr 1958).

GILA COUNTY: Sierra Ancha District, Red Bluff, Lucky Stop, and Hope deposits, in Dripping Spring Quartzite (Granger and Raup 1969).

#URANOPILITE

Uranyl sulfate oxide hydroxide hydrate: $(UO_2)_6(SO_4)O_2(OH)_6 \cdot 14H_2O$. A rare secondary uranium mineral.

COCHISE COUNTY: Warren District, Bisbee, Cole Mine, found as a post-mining efflorescence with gypsum and minor zippeite and johannite (Graeme, Graeme, and Graeme 2015).

NAVAJO COUNTY: Monument Valley District, Big Chief Mine (Shirley Wetmore, pers. comm., 1999).

URANOSPINITE

Calcium uranyl arsenate hydrate: $Ca(UO_2)_2(AsO_4)_2 \cdot 10H_2O$. A secondary member found in the oxidized zone of uranium-arsenic deposits and a member of the autunite group.

COCONINO COUNTY: Grand Canyon National Park, Orphan District, Orphan Mine (AM 33454).

#UREA

Carbonyl amide: $CO(NH_2)_2$. An organic mineral formed from bat guano and urine and stable only in arid conditions.

MARICOPA COUNTY: Painted Rock District, Rowley Mine, as small transparent elongated crystals with allantoin, aphthitalite, and natrosulfatourea in an area with abundant bat guano (Wilson 2020b).

#USHKOVITE

Hydrous magnesium iron phosphate hydroxide: $MgFe^{3+}_2(PO_4)_2(OH)_2 \cdot 8H_2O$. A secondary mineral found in granite pegmatites from the weathering of triplite.

COCHISE COUNTY: Warren District, Holbrook Mine, 300 level, as small pinkish crusts with cronstedtite and azurite (Jim McGlasson, pers. comm., 2020).

#UTAHITE

Magnesium copper zinc tellurate hydroxide hydrate: $MgCu^{2+}_4Zn_2Te^{6+}_3O_{14}(OH)_4 \cdot 6H_2O$. A very rare secondary mineral found in the oxidized zone of Cu-Zn-Te-bearing hydrothermal ore deposits.

COCHISE COUNTY: Tombstone District, Empire Mine, found on the 300 level as elongate brilliant-blue crystals to 3 mm as singles and divergent sprays associated with chlorargyrite, gold, dugganite, khinite-4O, khinite-3T, yafsoanite, quetzalcoatlite, and quartz (P. Megaw, pers. comm., 2017).

FIGURE 3.248 Utahite, Empire Mine, Tombstone, Cochise County, FOV 2 mm.

UVANITE

Uranyl vanadium oxide hydrate: $(UO_2)_2V^{5+}_6O_{17} \cdot 15H_2O(?)$. An uncommon mineral associated with the uranium-vanadium ores of the Colorado Plateau, where it probably forms by the alteration of uraninite and tyuyamunite. The IMA notes that the mineral is questionable but still approved.

APACHE COUNTY: Cane Valley District, Monument Valley, Monument No. 2 Mine (C. Frondel 1958).

UVAROVITE

Calcium chromium silicate: $Ca_3Cr_2(SiO_4)_3$. An uncommon member of the garnet group found with chromite in serpentines and granular limestones.

MOHAVE COUNTY: Central Cerbat Mountains, Wallapai supersystem, Tennessee District—an emerald-green garnet found in Precambrian migmatite just east of the staff house at the Tennessee Mine is believed to be this mineral (B. E. Thomas 1953).

UYTENBOGAARDTITE

Silver gold sulfide: Ag_3AuS_2. Formed in low-temperature silver-gold-quartz veins.

COCHISE COUNTY: Dos Cabezas Mountains, Teviston District, Comstock Mine, as small patches up to 0.5 mm in galena; the uytenbogaardtite is separated from the host galena by a phase whose composition, determined by microprobe analysis, is Ag_2PbS_2 (3rd ed.).

VAESITE

Nickel sulfide: NiS_2. A rare mineral, sometimes of primary origin, and also found as an alteration product of nickelskutterudite. It is a member of the pyrite group.

COCONINO COUNTY: Grand Canyon National Park, Orphan Mine, as brilliant tin-white crystals up to 2 mm in diameter perched on poorly formed yellow baryte crystals (3rd ed.).

VALLERIITE

Iron copper sulfide magnesium aluminum hydroxide: $2[(Fe,Cu)S] \cdot 1.53[(Mg,Al)(OH)_2]$. An uncommon mineral thought to be of high-temperature origin.

GILA COUNTY: Southeastern Dripping Spring Mountains, Banner supersystem, Christmas District, Christmas Mine, as rod-shaped grains in chalcopyrite in the pyrrhotite-chalcopyrite outer ore zone of orebodies, replacing dolomite (Perry 1969).

PIMA COUNTY: Pima supersystem, Mission-Pima District, Pima Mine, associated with magnetite (Pinch 1977); Twin Buttes District, Twin Buttes Mine (William Hefferon, pers. comm., 3rd ed.).

VANADINITE

Lead vanadate chloride: $Pb_5(VO_4)_3Cl$. A mineral commonly found in oxidized lead deposits and typically associated with wulfenite, cerussite, and descloizite. Vanadinite is widely distributed throughout Arizona with many outstanding localities having produced many beautiful specimens.

COCHISE COUNTY: Tombstone District, on calcite at the Tribute and Tombstone extension mines (B. S. Butler, Wilson, and Rasor 1938). Charleston District, Gallagher vanadium property near Charleston, as masses of brownish crystals with wulfenite, descloizite, and other arsenates (Bowell, Luetcke, and Luetcke 2015; S R 15016; UA 6903). Swisshelm Mountains, Swisshelm District, Scribner Mine near Elfrida (UA 5668, 9349). Whetstone Mountains, as brownish-red intergrown aggregates of barrel-shaped crystals coating cavities in silici-fied petrified wood from the Turney Ranch Member of the upper Bisbee Group (UA 15400).

COCONINO COUNTY: Bright Angel Creek, as hollow en-crustation pseudomorphs of brown vanadinite after cal-cite scalenohedra, in places only encrusting calcite (C. Frondel 1935; S 94280).

GILA COUNTY: Southern Dripping Spring Mountains, Banner supersystem, 79 District, 79 Mine (Stanley B. Keith 1972); near Hayden, with calcite and siderite (S 115636 00-05; UA 134); southwest and northeast work-ings of the Kullman-McCool (Finch, Barking Spider) Mine, as single crystals and as a replacement of wulfenite (K. D. Smith 1995); Christmas District, Santa Monica Camp, Premier Group, as coatings with wulfenite, lin-ing quartz vugs and cracks (C. P. Ross 1925b). Central Dripping Spring Mountains, Troy supersystem, C and B–Grey Horse District, C and B Vanadium Mine, found as bright red-orange, orange, and tan crystals or as flat hexagonal flower-like compos-ites (M. A. Allen and Butler 1921b; Crowley 1980). Globe Hills District, Lockwood claims; Clark and Stewart claims, near the Old Dominion Mine (Arizona Bureau of Mines 1916); Apache Mine, abundant as fine specimens of deep-red well-crystallized stocky prismatic crystals, char-acteristically having a background of dark mottramite on mottled quartz-ite of the Precambrian Pioneer Formation; crystals are commonly 1 to 2 mm long, although the best range up to 6 to 7 mm; crystals up to 2 cm have been reported (Guild 1910; W. E. Wilson 1971; UA 7092, 8988); Ruby Star Mine near the Apache Mine, as small red crystals up to 0.5 mm long (3rd ed.).

GREENLEE COUNTY: Copper Mountain supersystem, Granville District, north of Metcalf, as doubly terminated single crystals in a fracture in quartzite that may be the source of lead-vanadium-bearing boulders in the Haggin placers along the Coronado Trail (Bull. 181; Flagg 1942).

LA PAZ COUNTY: Trigo Mountains, Silver District, North Geronimo Mine (Pure Potential Mine), as bright-red crystals, some with wulfenite (W. E. Wilson and Godas 1996; Godas, Morris, and Hay 2015); Silver King Mine, as small red to orange crystals, many with fluorite (Parker 1966); Silver Clip and Princess Mines, as fine crystal aggregates (E. D. Wilson 1933; Stanton B. Keith 1978); as deep-red brilliant crystals at the Red Cloud Mine, where Silliman (1881) noted their great beauty and

FIGURE 3.249 Vanadi-nite, Apache Mine, Gila County, 4.8 cm.

extraordinary color (see also Hills 1890); these may have come from a nearby locality or been misidentified since very little if any vanadinite has been recently recovered from the Red Cloud Mine; Hamburg Mine (D. M. Shannon 1980; H 110870; R050171); Ronaldo Pacheco's Mine, as crystals of outstanding beauty (S 48793 00); also reported at Silver Glance and Mandeville Mines (Flagg 1942).

MARICOPA COUNTY: Vulture District, Collateral, Phoenix, Montezuma, and Frenchman Mines (Silliman 1881). Hieroglyphic Mountains, with wulfenite in veins (Silliman 1881). Painted Rock District, Rowley Mine near Theba, with wulfenite, mimetite, and on baryte and as pseudomorphs after wulfenite (W. E. Wilson and Miller 1974). White Picacho District, probably derived from galena (Jahns 1952). Baldy Mountain quadrangle, Pikes Peak District, Prince of Arizona Mine, with wulfenite (Flagg 1942).

MOHAVE COUNTY: Cerbat Mountains, Wallapai supersystem, Tennessee District, Aurora Mine; Golconda District, Climax Mine, at western end of Union Basin, as encrustations of crystals up to a quarter inch (Schrader 1909); Western (Western Union) Mine, as sheaf-like bundles and single doubly terminated crystals of the variety endlichite (Bull. 181; Schrader 1909; Hay 1997). Gold Basin District, Eldorado Mine, with cerussite and malachite (Schrader 1909).

PIMA COUNTY: Northern Tucson Mountains, Old Yuma District, Old Yuma Mine, as exceptionally beautiful specimens (Jenkins and Wilson 1920) and as pseudomorphs after wulfenite (C. Frondel 1935; AM 18716; UA 6294; NHM 1968, 1097, and others). Empire District, Total Wreck Mine (Schrader and Hill 1915; Schrader 1917; W. E. Wilson 2015). South Comobabi Mountains, Cababi District, Mildren and Steppe claims (S. A. Williams 1963). Pima supersystem, Twin Buttes District, Twin Buttes Mine (Stanley B. Keith, pers. comm., 1973). Helvetia-Rosemont District, as a secondary mineral in replacement deposits in metamorphosed limestones, associated with wulfenite, jarosite, and iron and manganese oxides (Schrader 1917).

PINAL COUNTY: Mammoth District, Mammoth–St. Anthony Mine, first noted by Genth (1887), as crystals up to one inch long, with descloizite and wulfenite; Ford Mine, once the largest producer of vanadium in Arizona, as orange-brown crystals with descloizite (W. E. Wilson 2020a). Oracle District, northeast side of the Santa Catalina Mountains, Bear Cat claims (Dale 1959); Royal Dane property, seven miles southeast of Oracle (Arizona Bureau of Mines 1916). Central Dripping Spring Mountains,

Troy supersystem, C and B–Grey Horse District, Grey Horse Mine (U.S. Vanadium mine), as skeletal crystals exceeding 2.5 cm in length, and in large groups embedded in calcite with descloizite (Newhouse 1934; A. Clark and Fleck 1980). Galiuro Mountains, Table Mountain District, Table Mountain Mine, as the arsenian variety endlichite, with wulfenite (Bull. 181; Simons 1964). Pioneer supersystem, Black Prince (Olsen) Mine, as doubly terminated crystals (Blake 1881b; Penfield 1886; AM 15129). Slate District, Jackrabbit Mine, as coarse crystalline masses with manganese oxides, in vugs in limestone (Hammer 1961).

SANTA CRUZ COUNTY: Southeastern Santa Rita Mountains, J. C. Holmes District, J. C. Holmes claims, near Temporal Gulch in S36, T21S R15E, as splendid orange crystals (Pellegrin 1911; Novak and Besse 1986). Northwestern Santa Rita Mountains, Cottonwood Canyon, Tyndall District, Glove Mine, with wulfenite and smithsonite (H. J. Olson 1966). Patagonia Mountains, Mowry District, Mowry Mine, with galena, cerussite, anglesite, bindheimite, and malachite (Schrader and Hill 1915). Flux District, Flux Mine (3rd ed.).

YAVAPAI COUNTY: Northern Bradshaw Mountains, Kirkland District, Kirkland gold mines, as fine specimens with descloizite (Bull. 181; Flagg 1942). Big Bug District, near the Silver Belt Mine (3rd ed.). Ticonderoga District, near Humboldt, as masses of 0.25 in. yellow-brown crystals (H 103031).

YUMA COUNTY: Castle Dome Mountains, Castle Dome District, in channels and vugs (E. D. Wilson 1933); Hull Mine, as small yellow to brown crystals (E. D. Wilson 1933); Puzzler Mine, as large green to brown crystals up to 3 cm long, some of which are barrel-shaped, associated with baryte, cerussite, anglesite, and wulfenite (E. D. Wilson 1933; Stanton B. Keith 1978; Domitrovic, Wilson, and Hay 1998; R050189). Reported in the Chocolate Mountains (Bull. 181). Muggins Mountains, Muggins District, Red Knob Mine, associated with wulfenite, cuprite, chalcedony, limonite, weeksite, opal, and carnotite (Honea 1959; Outerbridge et al. 1960).

VANDENDRIESSCHEITE

Lead uranyl oxide hydroxide hydrate: $Pb_{1.6}(UO_2)_{10}O_6(OH)_{11} \cdot 11H_2O$. A secondary mineral closely associated with uraninite, of which it is an oxidized alteration product.

APACHE COUNTY: Cane Valley District, Monument No. 2 Mine (3rd ed.).

VANMEERSSCHEITE

Uranium uranyl phosphate hydroxide hydrate: $U(UO_2)_3(PO_4)_2(OH)_6 \cdot 4H_2O$. A rare secondary uranium mineral found in altered granite pegmatites and in sedimentary uranium deposits.

COCONINO COUNTY: Cameron District, in some uranium prospects east of Cameron, as bright-yellow scaly crusts coating fractures that cut sandstone impregnated with carbon (3rd ed.).

VANOXITE

Vanadium oxide hydrate: $V_6O_{13} \cdot 8H_2O$. Found in the uranium-vanadium ores of the Colorado Plateau, where it formed massively, as cement in sandstone.

APACHE COUNTY: In the Lukachukai District, at King Tut Mesa, and on the Rattlesnake anticline, impregnating certain beds of the Salt Wash Member of the Morrison Formation, associated with carnotite (Masters 1955).

FIGURE 3.250 Variscite, Bisbee, Cochise County, 12.5 cm.

VARISCITE

Aluminum phosphate hydrate: $Al(PO_4) \cdot 2H_2O$. A mineral formed under near-surface conditions by the action of phosphatic waters on aluminous rocks.

APACHE COUNTY: Cane Valley District, Monument No. 2 Mine, as white to gray coatings and minute sprays of white crystals (Patrick Haynes, pers. comm., 1992).

COCHISE COUNTY: Warren District, Bisbee, Cole shaft, as unusual pale-green massive ferrian material (Sinkankas 1964; Graeme 1981).

VATERITE

Calcium carbonate: $Ca(CO_3)$. An uncommon mineral formed at low temperatures by the hydration of metamorphic calcsilicate rocks in the presence of carbon dioxide. It is a trimorph with calcite and aragonite.

MARICOPA COUNTY: In drill cores taken in basins of the Phoenix area (H. Wesley Peirce, pers. comm., 1989).

VAUQUELINITE

Copper lead chromate phosphate hydroxide: $CuPb_2(CrO_4)(PO_4)(OH)$. A rare mineral of secondary origin found in the oxidized portions of metalliferous deposits. It is typically associated with other chromates, including phoenicochroite and crocoite.

COCHISE COUNTY: Western Johnny Lyon Hills, Yellowstone District, about eight miles north of Benson, in quartz veins that cut gneiss, in minor amounts with pyromorphite, chrysocolla, and perite (Phelps Dodge Corp., pers. comm., 1972).

FIGURE 3.251 Vauquelinite, 79 Mine, Gila County, FOV 2 mm.

GILA COUNTY: Banner supersystem, 79 District, 79 Mine, as very small bright yellow-green crystals with tsumebite, wulfenite, and phosphohedyphane (R. Gibbs, pers. comm., 2020).

MARICOPA COUNTY: Wickenburg Mountains, east of Trilby Wash, at the Collateral, Chromate, Blue Jay, and Phoenix properties (Silliman 1881). South of Wickenburg in the Belmont Mountains in the Osborne District at several localities, including the Moon Anchor Mine, Rat Tail claim, and Potter-Cramer property, as a minor mineral in the oxidized zone of lead-zinc ore, associated with wickenburgite, willemite, mimetite, phoenicochroite, and hemihedrite (S. A. Williams 1968; S. A. Williams, McLean, and Anthony 1970; S. A. Williams and Anthony 1970); Tonopah-Belmont Mine, as greenish-brown to brown crystals up to 2 mm, with willemite and pyromorphite (G. B. Allen and Hunt 1988).

PINAL COUNTY: Tortilla Mountains, Ripsey District, Florence Lead-Silver Mine, in an oxide mineral assemblage that replaced galena, sphalerite, pyrite, and tennantite along a sheared and mineralized fault zone separating limestone and quartzite; associated with various other oxidized minerals, including wulfenite, hemihedrite, willemite, cerussite, minium, and mimetite (S. A. Williams and Anthony 1970).

SANTA CRUZ COUNTY: Patagonia Mountains, Hardshell-Hermosa District, Hardshell Mine, as apple-green tabular crystals (R060926).

VELIKITE

Copper mercury tin sulfide: Cu_2HgSnS_4. A rare sulfide mineral probably formed under hydrothermal conditions.

COCHISE COUNTY: Warren District, Bisbee, in the Campbell sulfide orebody (Graeme 1993).

VERMICULITE

Magnesium iron aluminum silicate hydroxide hydrate: $Mg_{0.7}(Mg,Fe,Al)_6$ $(Si,Al)_8O_{20}(OH)_4 \cdot 8H_2O$. A member of the smectite group of mica minerals, vermiculite has the property of expanding perpendicular to the micaceous layers when rapidly heated.

COCHISE COUNTY: Willcox Clay District, Willcox Playa, where it is, after illite and montmorillonite, the most abundant clay mineral in the younger playa sediments (Pipkin 1968).

LA PAZ COUNTY: Near Bouse (UA 8646).

MARICOPA COUNTY: Reported near Aguila and at a locality between Wickenburg and the Vulture Mountains (North and Jensen 1958; UA 8653). Bar FX Ranch, southwest of Wickenburg (E. D. Wilson and Roseveare 1949).

MOHAVE COUNTY: Hualapai Mountains, in a deposit fifteen miles southwest of Kingman (Engineering and Mining Journal 1940; E. D. Wilson 1944).

PIMA COUNTY: Tucson Mountains, possibly Amole District, in rocks metamorphosed by igneous intrusions (W. H. Brown 1939).

PINAL COUNTY: Near Oracle (R. T. Moore 1969b).

YUMA COUNTY: Formed by the alteration of biotite (UA 8651).

VÉSIGNIÉITE

Copper barium vanadate hydroxide: $Cu_3Ba(VO_4)_2(OH)_2$. A rare secondary mineral associated with oxidized uranium minerals.

APACHE COUNTY: Cane Valley District, Monument No. 2 Mine (Joseph Urban, pers. comm., 1967; UA 9501; NHM 1967, 148).

COCHISE COUNTY: Charleston and Bronco Hills, Charleston District, Gallagher vanadium property, identified in one sample with tangeite and vanadinite (Bowell, Luetcke, and Luetcke 2015).

COCONINO COUNTY: Prospect Canyon (Ridenour) District, Ridenour Mine, with naumannite, bromargyrite, tyuyamunite, tangeite, and metatyuyamunite (Wenrich and Sutphin 1989).

PINAL COUNTY: Central Galiuro Mountains, Bunker Hill (Copper Creek) District, Copper Creek, in a prospect pit 2 miles south of Mercer Ranch, as crude rounded pistachio-green crystals up to 1.5 mm long, in albitized quartz monzonite with cuprite and chrysocolla (3rd ed.).

VESUVIANITE

Calcium sodium aluminum magnesium iron silicate hydroxide fluoride oxide: $(Ca,Na)_{19}(Al,Mg,Fe)_{13}(SiO_4)_{10}(Si_2O_7)_4(OH,F,O)_{10}$. Typically formed in contact-metamorphosed limestones, where it is associated with other calcium silicate minerals.

COCHISE COUNTY: Warren District, around Sacramento Hill (Graeme 1981). Tombstone District, in the contact zone of Comstock Hill, occurs mainly as brownish to greenish crystalline masses and as excellent deep-green crystals with coarse crystalline calcite (B. S. Butler, Wilson, and Rasor 1938); Lucky Cuss Mine, with monticellite, hillebrandite, and thaumasite (B. S. Butler, Wilson, and Rasor 1938). Little Dragoon Mountains, Cochise supersystem, as pale-green square vertically striated prisms up to 2 mm long (Romslo 1949; J. R. Cooper and Huff 1951; J. R. Cooper 1957; J. R. Cooper and Silver 1964). Central Dragoon Mountains, Middle Pass supersystem, Stronghold District, northern end of Stronghold Canyon, as crystals up to 3 cm long in radiating prismatic clusters, with diopside, marialite, clinozoisite, and garnet (Rushing 1978).

GILA COUNTY: Southeastern Dripping Spring Mountains, Banner supersystem, Christmas District, Christmas Mine, in fibrous masses showing anomalous birefringence and color zoning; moderately common in skarn (C. P. Ross 1925b; David Perry, pers. comm., 1967); 79 District, 79 Mine, one of the minerals in the skarn deposits (Stanley B. Keith 1972).

GREENLEE COUNTY: Copper Mountain supersystem, Morenci District, northern part of Morenci, locally formed with other calcsilicate minerals, including garnet, tremolite, diopside, and epidote, in a large contact-metamorphosed zone (Moolick and Durek 1966).

PIMA COUNTY: Unspecified locality at the northern end of the Baboquivari Mountains (Bull. 181). In schist at the east end of the Coyote Mountains (UA 8095). Tucson Mountains, in contact-metamorphosed rocks (W. H. Brown 1939). Pima supersystem, Twin Buttes District, Twin Buttes Mine, in the skarns and tactites (Stanley B. Keith, pers. comm., 1973). Southern end of the Helvetia-Rosemont District (McNew 1981). Southern Baboquivari Mountains, Aguirre Peak District, Lesjimfre prospect, east of Aguirre Peak, near the head of Gallineta Wash, with scheelite and fluorite in a contact-metamorphic aureole adjacent to an east-west-striking pegmatite dike related to the Presumido Peak peraluminous granite (Stanton B. Keith 1974; Stanley B. Keith, pers. comm., 2016).

SANTA CRUZ COUNTY: Southern Patagonia Mountains, Washington Camp District, as lime-green crystal fragments in a wash, uphill behind the local store (Bideaux, Williams, and Thomssen 1960).

YUMA COUNTY: Locally abundant in metamorphosed limestones (Bull. 181). Gila Mountains, in a contact zone (E. D. Wilson 1933).

#VILLAMANÍNITE

Copper sulfide: CuS_2. An uncommon mineral found in hydrothermal ore deposits.

COCONINO COUNTY: South Rim of the Grand Canyon, Orphan District, Orphan Mine, as discrete zones in small euhedral zoned but complex Ni-Co-Fe-As-S minerals. Associated minerals include siegenite, violarite, polydymite, pyrite, and vaesite (Dr. Karen Wenrich, unpublished microprobe analysis, 1994).

#VIOLARITE

Iron nickel sulfide: $FeNi_2S_4$. An uncommon mineral found in hydrothermal ore deposits.

COCONINO COUNTY: South Rim of the Grand Canyon, Orphan District, Orphan Mine, as discrete zones in small euhedral zoned but complex Ni-Co-Fe-As-S minerals. Associated minerals include siegenite, polydymite, pyrite, and vaesite (Dr. Karen Wenrich, unpublished microprobe analysis, 1994).

MOHAVE COUNTY: Hack District, Hack No. 2 Mine, as discrete zones in small euhedral zoned but complex Ni-Co-Fe-As-S minerals (Dr. Karen Wenrich, unpublished microprobe analysis, 1994).

VIVIANITE

Iron phosphate hydrate: $Fe^{2+}_3(PO_4)_2 \cdot 8H_2O$. Found in the weathered portion of base-metal deposits and as an alteration product of primary iron-manganese phosphates in pegmatites.

MARICOPA AND YAVAPAI COUNTIES: White Picacho District, as finely crystalline bluish-gray films in triplite (Jahns 1952).

YAVAPAI COUNTY: Eureka supersystem, Hillside District, Hillside Mine, as thin blue coatings with montmorillonite and böhmite, on mica schist in old mine workings (3rd ed.).

VOGLITE

Calcium copper uranyl carbonate hydrate: $Ca_2Cu(UO_2)(CO_3)_4 \cdot 6H_2O$. A very rare secondary mineral formed as an alteration product of uraninite.

APACHE COUNTY: Garnet Ridge U District, Bluestone Mine, disseminated with tyuyamunite in the Navajo Sandstone adjacent to a serpentine dike

(Chenoweth and Malan 1973). Black Rock Point District, near the Red Mesa Trading Post (Bull. 181).

VOLBORTHITE

Copper pyrovanadate hydroxide hydrate: $Cu_3V_2O_7(OH)_2 \cdot 2H_2O$. A rare secondary mineral formed in the oxidized zones of base-metal deposits. It is also associated with certain uranium deposits on the Colorado Plateau.

APACHE COUNTY: Cane Valley District, Monument No. 2 Mine, as a barian variety (UA 1569).

COCONINO COUNTY: Prospect Canyon (Ridenour) District, Ridenour Mine, as splotches with malachite and tyuyamunite on light- to dark-gray dolomitic quartz sandstone and replacing a carbonate matrix composed originally of calcite and dolomite (Wenrich et al. 1990).

GILA COUNTY: Undisclosed locality north of Globe (Robert O'Haire, pers. comm., 1972).

NAVAJO COUNTY: (UA 9797); Monument Valley District, Mitten No. 2 Mine, as specks with tyuyamunite, azurite, and malachite scattered throughout the orebody (Witkind and Thaden 1963); Monument No. 1 Mine (Holland et al. 1958; Witkind 1961).

VOLKONSKOITE

Calcium chromium magnesium aluminum silicate hydroxide hydrate: $Ca_{0.3}(Cr,Mg)_2(Si,Al)_4O_{10}(OH)_2 \cdot 4H_2O$. This member of the smectite group is found in sedimentary rocks where it can fill voids left by the decomposition of organic matter. It is also found as a weathering product of serpentinite.

COCHISE COUNTY: Warren District, Bisbee, Campbell Mine, as waxy green masses with pyrite and quartz (Graeme 1993).

VOLTAITE

Potassium iron aluminum sulfate hydrate: $K_2Fe^{2+}_5Fe^{3+}_3Al(SO_4)_{12} \cdot 18H_2O$. Found in volcanic fumaroles, as secondary mineral in mineral deposits rich in pyrite, and as postmining efflorescence.

COCHISE COUNTY: Warren District, Bisbee, Copper Queen Mine, in irregular porous crusts several inches thick, associated with coquimbite, kornelite, copiapite, and rhomboclase (Merwin and Posnjak 1937).

FIGURE 3.252 Voltaite, Bisbee, Cochise County, FOV 3 mm.

COCONINO COUNTY: Sunset Crater National Monument, as small crystals and crusts with gypsum and jarosite in the fumarole deposits on Sunset Crater (S. L. Hanson, Falster, and Simmons 2008).

PINAL COUNTY: Pioneer supersystem, Magma District, Magma Mine, as crystals up to 0.25 in., associated with copiapite (3rd ed.).

YAVAPAI COUNTY: Black Hills, Jerome supersystem, Verde District, United Verde Mine, as black resinous cubo-octahedral crystals up to 5 mm, formed as a result of burning pyritic ore (C. A. Anderson 1927; Lausen 1928; Hutton 1959b; UA 64, 3074, and others; H 85680, 90536).

VOLYNSKITE

Silver bismuth telluride: $AgBiTe_2$. A rare hydrothermal mineral found with other tellurides in gold ores.

COCHISE COUNTY: Warren District, Bisbee, Campbell Mine, among the pyritic ores, with altaite, rhodostannite, and melonite (Graeme 1993).

#WAGNERITE

Magnesium phosphate fluoride: $Mg_2(PO_4)F$. Found as a rare accessory mineral in aluminosilicate metamorphic rocks.

LA PAZ COUNTY: Dome Rock Mountains, Stray Elephant Kyanite District, on the west side of the Dome Rock Mountains, in a kyanite quartzite with rutile, lazulite, andalusite, sillimanite, and topaz (Marsh and Sheridan 1976; Melchiorre 2013).

WAKEFIELDITE-(Y)

Yttrium vanadate: YVO_4. A rare mineral found in granite pegmatites.

MARICOPA COUNTY: White Tank Mountains, White Tank District, as good quality clear to milky-white plates that line voids in massive, vuggy, and corroded euxenite-(Y) (S. A. Williams, pers. comm., 1990).

WAVELLITE

Aluminum phosphate hydroxide hydrate: $Al_3(PO_4)_2(OH)_3 \cdot 5H_2O$. An uncommon but widespread mineral formed by low-grade metamorphism and as a product of hydrothermal alteration.

COCHISE COUNTY: Tombstone District, Gallagher and Manila Mines, as small white to clear spherical clusters up to 2 mm in diameter with jarosite (Rolf Luetcke, pers. comm., 2020).

GILA COUNTY: Miami-Inspiration supersystem, Pinto Valley District, Castle Dome Mine, where it is localized along fractures that cross the trend of ore veins (N. P. Peterson 1947; N. P. Peterson, Gilbert, and Quick 1951).

GREENLEE COUNTY: Copper Mountain supersystem, Candelaria District, north of Morenci Mine, Candelaria breccia pipe, with turquoise, alunite, and conichalcite (Bennett 1975).

MOHAVE COUNTY: Wallapai supersystem, Mineral Park District, Mineral Park Mine (Duval Pit), as microcrystals and spheroidal aggregates on quartz (Wilkinson, Roe, and Williams 1980; UA 10541).

FIGURE 3.253 Wavellite, Duval Pit, Kingman, Mohave County, 3.6 × 2.0 cm.

#WEDDELLITE

Calcium oxalate hydrate: $Ca(C_2O_4) \cdot 2H_2O$. Found in calcareous lake-bottom sediments and biological material, such as lichens and saguaros.

MARICOPA COUNTY: Found in decaying saguaro cacti in the Gila Bend Mountains (Garvie 2003).

PIMA COUNTY: Tucson area, Pusch Ridge in the Santa Catalina Mountains, as white to colorless spheroidal clusters of crystals found in the pith of living and dead saguaro cacti (R110123).

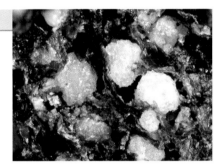

FIGURE 3.254 Weddellite, Pusch Ridge, Santa Catalina Mountains, Pima County, FOV 4 mm.

WEEKSITE

Potassium uranyl silicate hydrate: $(K)_2(UO_2)_2(Si_5O_{13}) \cdot 4H_2O$. A rare secondary mineral intimately associated with other oxidized uranium minerals.

YAVAPAI COUNTY: Near Congress Junction, as microcrystals (William Kurtz, UA x4477). Date Creek District, Anderson Mine, as bright-yellow coatings with chalcedony and carnotite in mid-Miocene tuffaceous sediments (Otton 1981); also as microcrystals with celestine and gypsum (William Hunt, pers. comm., 1985; R050330).

YUMA COUNTY: Muggins Mountains, Muggins District, Red Knob claims, as radial aggregates of fibrous to acicular crystals coating and intergrown

FIGURE 3.255 Weeksite, Anderson Mine, Yavapai County, FOV 2 mm.

with chalcedony, wulfenite, carnotite, vanadinite, cuprite, azurite, calcite, gypsum, and limonite (Honea 1959; Outerbridge et al. 1960).

WEISSITE

Copper telluride: $Cu_{2-x}Te$, where $x = 0$ to 0.33. Found in hydrothermal veins with pyrite, petzite, rickardite, and tellurium.

LA PAZ COUNTY: Dome Rock Mountains, in trace amounts with blebs of bornite in massive djurleite, in a quartz-tourmaline gangue (S. A. Williams and Matter 1975).

FIGURE 3.256 Wheatleyite, Rowley Mine, Maricopa County, FOV 2 mm.

#WHEATLEYITE

Sodium copper oxalate hydrate: $Na_2Cu(C_2O_4)_2 \cdot 2H_2O$. A rare mineral found in deposits of bat guano in contact with solutions containing copper.

MARICOPA COUNTY: Painted Rock District, Rowley Mine, as small pale blue-green bladed crystals with antipinite and rowleyite in bat guano (R160084, Kampf et al. 2016).

FIGURE 3.257 Whelanite, Christmas Mine, Gila County.

#WHELANITE

Copper calcium silicate carbonate hydroxide hydrate: $Cu_2Ca_6[Si_6O_{17}(OH)](CO_3)(OH)_3(H_2O)_2$. This mineral was described in the third edition as not being a valid mineral species; however, the mineral was approved by the IMA in 1977 (1977-006). A description was not published at that time due to the difficulty of solving the crystal structure. A full description of the mineral was published in 2012 based on material from the Bawana Mine in Utah (Kampf et al. 2012).

GILA COUNTY: Dripping Spring Mountains, Christmas District, Christmas Mine, moderately abundant as cleavable turquoise-blue prisms up to several centimeters long and as light-blue thin flat bladed fibers associated with hydroxyapophyllite, kinoite, and andradite (R050323).

WHERRYITE

Lead copper sulfate silicate hydroxide: $Pb_7Cu_2(SO_4)_4(SiO_4)_2(OH)_2$. A rare secondary mineral found in the oxidized portion of some lead deposits.

The Mammoth–St. Anthony Mine is the type locality. The species has been further substantiated by McLean and Anthony (1970).

MOHAVE COUNTY: Rawhide District, Rawhide Mine, as massive green material with chrysocolla (S. A. Williams, pers. comm., 1995; UA14970).

PINAL COUNTY: Mammoth District, Mammoth–St. Anthony Mine, in small vugs, with matlockite, leadhillite, hydrocerussite, paralaurionite, diaboleite, phosgenite, chrysocolla, anglesite, and cerussite (Fahey, Daggett, and Gordon 1950; R110089, R141161). Silver Reef Mountains, Silver Reef District, Silver Reef Mine, near Casa Grande (3rd ed.).

FIGURE 3.258 Wherryite, Tiger, Pinal County.

WHEWELLITE

Calcium oxalate hydrate: $Ca(C_2O_4)\cdot H_2O$. Found in sediments, plants, and urinary stones. Formation of whewellite sequesters high amounts of CO_2.

COCHISE COUNTY: Dos Cabezas Mountains, in a dead agave plant. It is found as very abundant minute needle-like colorless crystals in the cup-like depressions in the bases of dead agave in the southwestern deserts. In many places in Arizona, the material is so abundant that it sparkles in the brilliant desert sunlight (R070655).

WICKENBURGITE

Lead calcium aluminum silicate hydrate: $Pb_3CaAl_2Si_{10}O_{27}\cdot 4H_2O$. A rare secondary lead mineral formed from the oxidation of lead ores. Locally abundant in the Wickenburg area, Osborne District. The Potter-Cramer property is the type locality.

LA PAZ COUNTY: Silver District, Red Cloud and Padre Kino Mines (P. Bancroft and Bricker 1990).

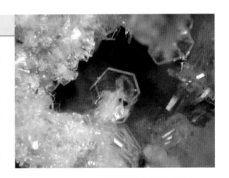

FIGURE 3.259 Wickenburgite, Evening Star Mine, Maricopa County, FOV 4 mm.

MARICOPA COUNTY: Belmont Mountains, Osborne District, Moon Anchor Mine and Potter-Cramer property, as transparent colorless crystals and rare salmon-pink crystals; also as dull-white granular to fine-grained masses; in the oxidized portion of galena-sphalerite veins, associated with many secondary minerals, such as phoenicochroite, crocoite, duftite, ajoite, vauquelinite, shattuckite, minium, and hemihedrite (S. A. Williams 1968; R060087, R060048). Big Horn Mountains, Big Horn District, Evening Star Mine (Gibbs 2011).

MOHAVE COUNTY: Rawhide District, Rawhide Mine, with luddenite, alamosite, and shattuckite, associated with drusy quartz (S. A. Williams 1982).

#WIDENMANNITE

Lead hydroxide uranyl carbonate: $Pb_2[(OH)_2(UO_2)(CO_3)_2]$. A rare secondary mineral found in oxidized lead deposits containing arsenic and uranium.

PIMA COUNTY: Redington District, Sure Fire No.1 claim, as very small yellow thin bladed crystals with malachite (R. Gibbs, pers. comm., 2021).

FIGURE 3.260 Willemite, Red Cloud Mine, La Paz County, FOV 6 mm.

WILLEMITE

Zinc silicate: Zn_2SiO_4. An uncommon mineral found in crystalline limestones, possibly formed by metamorphism of other zinc minerals, and more commonly as a secondary mineral in the oxidized portions of zinc deposits.

COCHISE COUNTY: Warren District, Bisbee, 1,500 ft. level of the Campbell Mine, as a fluorescent variety (Graeme 1981); Chiricahua Mountains, California supersystem, Hilltop District, Hilltop Mine, as small white- to rose-colored prisms in cavernous rock (Pough 1941; AM 21263; Stanton B. Keith 1973). Gunnison Hills, Cochise supersystem, Johnson Camp District, Texas-Arizona Mine, as minute glassy prisms that fluoresce straw-yellow, in metamorphosed limestones (J. R. Cooper 1957). Little Dragoon Mountains, Bluebird District, near the Little Fanny Mine, in a quartz vein (J. R. Cooper and Silver 1964). Southeasternmost Little Dragoon Mountains, Gleeson District, Defiance Mine, as microcrystals with wulfenite (William Kurtz, UA x4015).

GILA COUNTY: Globe Hills District, Apache Mine (Defiance Mine), abundant with vanadinite and descloizite (N. P. Peterson 1962); Irene vein (Liberty deposit), as granular masses with drusy crusts of hemimorphite (N. P. Peterson 1962).

GREENLEE COUNTY: Copper Mountain supersystem, Morenci District, Modoc Mountain, as small grayish crystals in garnet rock in the Modoc open cut (Lindgren 1905).

LA PAZ COUNTY: New Water Mountains, Ramsey District, Black Mesa Mine, in S16, T3N R17W, as brilliant nonfluorescing elongated hexagonal crystals, with aurichalcite and malachite on limonite (David Shannon, pers. comm., 1971). Trigo Mountains, Silver District, Red Cloud Mine, as minute prismatic crystals, commonly associated with wulfenite (Edson 1980); Padre Kino Mine, very abundant as both white acicular crystals and black botryoidal masses (3rd ed.).

MARICOPA COUNTY: South of Wickenburg, Belmont Mountains, Osborne District, at several localities, including the Moon Anchor Mine, Rat Tail claim, and Potter-Cramer property, as a secondary mineral in the oxidized zone of lead-zinc veins (S. A. Williams 1968); Tonopah-Belmont Mine (Robert O'Haire, pers. comm., 1970; G. B. Allen and Hunt 1988). Painted Rock District, Rowley Mine near Theba, as small white acicular crystals with wulfenite and as powdery masses mixed with fluorite (W. E. Wilson and Miller 1974).

PIMA COUNTY: Tucson Mountains, Old Yuma District, Gila Monster Mine, as glassy hexagonal crystals up to 3 mm in length that fluoresce bright yellowish-white under shortwave ultraviolet (Robbins 1994; UA 5691); Old Yuma Mine, as tiny opaque white to green hexagonal prisms (D. Jones 1983). Waterman District, Silver Hill Mine, near the head of the inclined surface tram, as small barrel-shaped crystals (Bull. 181; Stolburg 1988). South Comobabi Mountains, Cababi District, Mildren and Steppe claims (S. A. Williams 1963). Cimarron Mountains District, Paul Hinshaw property, in T11S R2E, as fluorescent material (E. D. Wilson and Roseveare 1949; Funnell and Wolfe 1964).

PINAL COUNTY: Mammoth District, Mammoth–St. Anthony Mine, as small colorless rhombs and bluish barrel-shaped crystals on wulfenite and vanadinite (Fahey 1955; Bideaux 1980; H 116242). Galiuro Mountains, Table Mountain District, Table Mountain Mine, in vugs with conichalcite, plancheite, and malachite (Bideaux, Williams, and Thomssen 1960; UA 6421; S 113802). Tortilla Mountains, Florence Lead District, Florence Lead-Silver Mine, with hemihedrite, wulfenite, cerussite, and other oxidized secondary minerals (S. A. Williams and Anthony 1970). Silver Reef Mountains, Silver Reef District, Silver Reef Mine, as dense tuffs of glassy acicular microcrystals, fluoresces bright yellowish-white under shortwave ultraviolet (Robbins 1994; UA 6855). Slate District, Jackrabbit Mine, as the most abundant zinc ore mineral (Hammer 1961). Mineral Mountain District, as microcrystals with hematite (Robert Mudra, UA x467).

#WILLEMSEITE

Nickel silicate hydroxide: $Ni_3Si_4O_{10}(OH)_2$. A secondary mineral found in nickel deposits.

LA PAZ COUNTY: Southern Plomosa Mountains, Black Beauty District, Perry Chrysoprase Mine, as inclusions in chalcedony, making the material chrysoprase-like, in thin veins cutting rhyolite (Melchiorre 2017).

WINSTANLEYITE

Titanium tellurite: $TiTe^{4+}_3O_8$. A rare mineral formed by the oxidation of other tellurium minerals. The Grand Central Mine dump is the type locality.

COCHISE COUNTY: Tombstone District, in the dumps of the Grand Central Mine, as simple cubes up to 0.5 mm on an edge, commonly yellow, but tan or cream in places; associated with pyrite, jarosite, chlorargyrite, and rodalquilarite in a quartz-adularia-opal rock formed by the alteration of a granodiorite (S. A. Williams 1979).

WITHERITE

Barium carbonate: $Ba(CO_3)$. An uncommon mineral formed in low-temperature hydrothermal veins with baryte and galena.

MOHAVE COUNTY: Hualapai Mountains, Hylstad deposit, in a vein along Moss Wash, SW quarter of S26, T19N R13W, with baryte (K. A. Phillips 1987).

PIMA COUNTY: Pima supersystem, Mission-Pima District, Mission Mine, as microcrystals (William Kurtz, UA x378). Quijotoa Mountains, as small tabular crystals on baryte (UA10655).

YUMA COUNTY: Castle Dome Mountains, Castle Dome District, De Luce Mine, as gangue in lead ores (Bull. 181).

WITTICHENITE

Copper bismuth sulfide: Cu_3BiS_3. An uncommon mineral of primary origin in vein deposits; less commonly formed during secondary enrichment of copper ores and typically associated with chalcocite.

COCHISE COUNTY: Warren District, as a primary mineral in some mines; as exsolution blebs in bornite from an unspecified mine (Julie De Azevedo Harlan, unpublished manuscript, 1966); Campbell orebody (Graeme 1993).

PINAL COUNTY: Pioneer supersystem, Magma District, Magma Mine, reported in the hypogene ores (Hammer and Peterson 1968).

WITTITE

Lead bismuth sulfide selenide: $Pb_8Bi_{10}(S,Se)_{23}$. Found in amphibolite at the type locality, Fahlan, Sweden, associated with nordströmite, friedrichite, magnetite, pyrite, and other minerals; also as a fumarolic sublimate.

COCHISE COUNTY: Dragoon Mountains, Middle Pass supersystem, Abril District, in an open-pit prospect northeast of Middlemarch Pass, as frayed prisms up to 10 mm long in grossularite-rich marble (3rd ed.).

WOLLASTONITE

Calcium silicate: $CaSiO_3$. A member of the pyroxenoid group and a common gangue mineral in mineral deposits of the high-temperature zones of magnesian skarns of contact-metamorphic origin.

COCHISE COUNTY: Tombstone District, Silver Thread and West Side Mines, as radiating fibrous masses (B. S. Butler, Wilson, and Rasor 1938). Little Dragoon Mountains, Cochise supersystem, Johnson Camp District, in garnetite skarn in contact with the Texas Canyon stock, commonly associated with vesuvianite (J. R. Cooper and Huff 1951; J. R. Cooper and Silver 1964); near Johnson Camp, with grossular, epidote, titanite, and other contact silicates in calcareous rocks (J. R. Cooper 1957). Dragoon Mountains, Middle Pass supersystem, Abril District, Abril Mine, abundant in contact-metamorphosed limestones (Perry 1964). Chiricahua Mountains, California supersystem, Leadville District, Paradise area, Gayleyville Group (Dale, Stewart, and McKinney 1960). Dos Cabezas Mountains, Apache Pass District, Siphon Canyon, south of Governors Peak, in high-temperature skarns of Escabrosa Limestone (UA 11800).

GILA COUNTY: Southeastern Dripping Spring Mountains, Banner supersystem, Christmas District, Christmas Mine, common and typically associated with chert nodules in marble of the Naco Formation, also locally present in skarn (C. P. Ross 1925b; Perry 1969).

LA PAZ COUNTY: Harcuvar District, Cobralla District, Cobralla Mine, where an entire bed of limestone is replaced adjacent to a copper-specularite skarn (H. Bancroft 1911).

MARICOPA COUNTY: Sierra Estrella Mountains, in NW quarter of S33, T2S R1E, as radiating fibrous masses associated with garnet in metamorphosed limestone (Robert O'Haire, pers. comm., 1971).

MOHAVE COUNTY: Hualapai (Antler) District, in gangue at the Antler Mine, a copper-zinc massive sulfide deposit (Romslo 1948).

PIMA COUNTY: Santa Rita Mountains, Helvetia-Rosemont District, where large masses of limestone are completely converted to wollastonite adjacent to the copper skarns (Creasey and Quick 1955). Sierrita Mountains, Pima supersystem, Mission-Pima District, Mineral Hill area, abundant throughout metamorphosed rocks (Mayuga 1942); Twin Buttes District, Twin Buttes Mine, in skarns and tactites (Stanley B. Keith, pers. comm., 1973). Silver Bell District, El Tiro Mine, locally abundant in contact-metamorphosed limestones (C. A. Stewart 1912; UA 4510, 11691; AM 31425); Atlas Mine area, in tactites associated with the Atlas copper-lead-zinc skarn deposit (Agenbroad 1962).

SANTA CRUZ COUNTY: Southern Patagonia Mountains, Washington Camp District, in contact-metamorphosed limestones near the Washington Camp pluton and associated copper-lead-zinc skarns (Schrader and Hill 1915).

#WOODHOUSEITE

Calcium aluminum sulfate phosphate hydroxide: $CaAl_3(SO_4)(PO_4)(OH)_6$. Created by the argillic alteration of hydrothermal vein and disseminated ore deposits.

PINAL COUNTY: Pioneer supersystem, Resolution District, as a component of advanced argillic alteration in the Resolution copper-molybdenum deposit (Hehnke et. al. 2012).

WOODWARDITE

Copper aluminum sulfate hydroxide hydrate: $(Cu_{1-x}Al_x)(SO_4)_{x/2}(OH)_2 \cdot nH_2O$, where ($x < 0.5$, $n < 3x/2$). A rare secondary mineral found with other uncommon secondary copper minerals such as cyanotrichite and langite.

COCHISE COUNTY: Southeasternmost Dragoon Mountains, Courtland District, Maid of Sunshine Mine, in fibrous to spherulitic form (identified by William Wise on material collected by David Shannon in 1991).

WULFENITE

Lead molybdate: $PbMoO_4$. A mineral generally considered to be of secondary origin. Wulfenite is widely distributed in Arizona in the oxidized portions of lead-bearing deposits, some of which have produced magnificently developed or colored crystals that are the equal of any in the world. Many collectors believe that the wulfenite from the Red Cloud Mine in La Paz County is unsurpassed in beauty because of its deep-red color, probably resulting from chromium substitution. In many districts, wulfenite color mimics that of its closely associated minerals, especially mimetite. In 2017, the Arizona state legislature passed a bill that was signed into law recognizing wulfenite as Arizona's state mineral. Only a few of the many Arizona localities are given here.

COCHISE COUNTY: Tombstone District, commonly present in large amounts with silver ores, as clusters and rosettes of crystals (B. S. Butler, Wilson, and Rasor 1938). Southeasternmost Dragoon Mountains, Gleeson District, Defiance Mine on Gleeson Ridge, as magnificent specimens in large amounts in a limestone cavern (Bideaux, Williams, and Thomssen 1960; J. R. Thompson 1980); Mystery Tunnel, Silver Bill and Tom Scott Mines (Knudsen 1983). Chiricahua Mountains, California supersystem, Hilltop District, Hilltop Mine, as groups of deep-yellow crystals on white calcite (D. K. Miller and Wilson 1983; H 97877; UA 2626). Pearce District, Commonwealth Mine, lining cavities with bromian chlorargyrite (Endlich 1897). Warren District, Bisbee, Campbell shaft between the 1,700 and 2,500 ft. levels, as small crystals associated with copper, malachite, cerussite, and mimetite; Holbrook shaft (Graeme 1981). Swisshelm Mountains, Swisshelm District, March Mine, as microcrystals (Carl Richardson, UA x1450).

FIGURE 3.261 Wulfenite, Hilltop Mine, Cochise County, 7.0 × 4.5 cm.

GILA COUNTY: Southern Dripping Spring Mountains, Banner supersystem, 79 District, 79 Mine, as transparent unflawed crystals up to 2 in. on an edge (Stanley B. Keith 1972); in the northeast workings of the Kullman-McCool (Finch, Barking Spider) Mine, as yellow to orange crystals, many coated with hemimorphite, vanadinite, and especially drusy quartz (K. D. Smith 1995); Christmas District, Santa Monica Camp, McHur and Premier properties; and Chilito District, London Range property, as coatings with vanadinite on spongy quartz or brecciated diabase (C. P. Ross 1925b; Bull. 181). Troy supersystem, C and B–Grey Horse District, C and B Vanadium Mine; and as brecciated plates in Martin Limestone at the Copper Chief property, three miles west of the C and B Vanadium Mine; and at the Grey Horse Mine, associated with mimetite, descloizite, and vanadinite. Less abundant in the

FIGURE 3.262 Wulfenite, 79 Mine, Gila County, 2.2 cm.

Pinto Valley District, Castle Dome Mine area, as small pointed prisms, a few of which are over 3 mm long (N. P. Peterson 1947; N. P. Peterson, Gilbert, and Quick 1951). Copper Cities District, Sleeping Beauty Mountain, two miles northwest of the Inspiration Mine, as a tungstenian variety (UA 8197).

GRAHAM COUNTY: Santa Teresa Mountains, Aravaipa supersystem, Grand Reef District, Silver Coin and Dogwater Mines (Simons 1964); Iron Cap District, Sinn Fein Mine, as yellow plates up to 5 mm on an edge (MM K090).

FIGURE 3.263 Wulfenite, Red Cloud Mine, La Paz County, 4.3 cm high.

LA PAZ COUNTY: Trigo Mountains, Silver District, most notably at the Red Cloud Mine, which many people regard as the world's premier wulfenite locality because of the remarkable crystals recovered from the mine in its 140-year history found in mineral collections throughout the world. Red Cloud Mine wulfenite crystals are brilliant orange-red and up to 2 in. on an edge (Blake 1881b; Silliman 1881; Foshag 1919; E. D. Wilson 1933; Fleischer 1959; Bideaux 1972; Edson 1980; W. E. Wilson 2008; UA 9949). In 1985, new finds at the Red Cloud Mine yielded superb modestly sized red-orange crystal groupings (W. E. Wilson 1985); Melissa Mine, as specimens with unusual forms (UA 4281); Padre Kino Mine, as bright-red transparent tabular crystals up to 1 cm on an edge, with unusual dipyramids with parallel stacking (Peter Megaw, UA x2976); Hamburg Mine and other properties (P. Bancroft and Bricker 1990); North Geronimo Mine (Pure Potential Mine) as crystals similar to Red Cloud specimens up to 2 cm (W. E. Wilson and Godas 1996).

FIGURE 3.264 Wulfenite, with mimetite, Rowley Mine, Maricopa County, 3.1 cm high.

MARICOPA COUNTY: Painted Rock District, Rowley Mine, as excellent bright orange to yellow crystals up to 2 cm on an edge, typically associated with mimetite and baryte (W. E. Wilson and Miller 1974; NHM 1970, 59); spectacular specimens were produced from this mine in the early 1980s; associated with microcrystals of boleite, diaboleite, and caledonite (W. E. Wilson 1986). White Picacho District, as a rare mineral in pegmatites (Jahns 1952).

MOHAVE COUNTY: Rawhide District, Rawhide Mine (Bladh 2019). Lost Basin District, Vanadinite Mine (Van-Wul), with vanadinite (Theodore, Blair, and Nash 1987, locality 310). Cerbat Mountains, Wallapai supersystem, Mineral Park District, Mineral Park Mine, as unusual modified crystals (Wilkinson, Roe, and Williams 1980).

PIMA COUNTY: Tucson Mountains, Old Yuma District, Old Yuma Mine, as deep-orange crystal groups associated with spectacular vanadinite, and with cerussite (Guild 1910, 1911; Newhouse 1934; D. Jones 1983; R050024). Empire Mountains, Empire District, Total Wreck Mine, with mottramite and Hilton Mine (Prince Mine) (W. E. Wilson 2015; UA 136, etc.; R. E. Birkholz, UA x1795). South Comobabi Mountains, Cababi District, Mildren and Steppe claims, with various secondary minerals formed from the oxidation of sulfide ores (S. A. Williams 1963). Silver Bell District, as brownish plates with fluorite (C. A. Stewart 1912), and as crystals showing obvious tetartohedrism (S. A. Williams 1966). Pima supersystem, Twin Buttes District, Twin Buttes Mine, associated with galena (Stanley B. Keith, pers. comm., 1973).

FIGURE 3.265 Wulfenite, Heavy Weight Mine, Pima County, FOV 2 mm.

PINAL COUNTY: Mammoth District, Mammoth–St. Anthony Mine, as light-yellow to bright-red crystals containing tungsten (Guild 1910; Newhouse 1934; N. P. Peterson 1938a, 1938b; Galbraith and Kuhn 1940; Palache 1941b; Fahey 1955; Fleischer 1959; Petersen, Hamilton, and Myers 1959; Bideaux 1980; H 101775; UA 10, 205; NHM 1961, 539); Ford Mine, as large orange crystals (W. E. Wilson 2020a). Pioneer supersystem, Black Prince (Oleson or Olson) Mine, with vanadinite (Bull. 181). Central Dripping Spring Mountains, C and B–Grey Horse District, Grey Horse Mine (U.S. Vanadium mine), four miles east of Kelvin (Newhouse 1934). Tortilla Mountains, Florence Lead District, Florence Lead-Silver Mine, with willemite, hemihedrite, vauquelinite, minium, mimetite, and other minerals, as oxidation products from the alteration of sphalerite and galena; adjacent country rocks contain chromium (S. A. Williams and Anthony 1970). Central Galiuro Mountains, Bunker Hill (Copper Creek) District, Copper Creek, Bluebird Mine, occurs with limonite, partly filling open spaces (Kuhn 1951). Table Mountain District, Table Mountain Mine, crusted by quartz and filling a breccia in silicified limestone (Stanley B. Keith, pers. comm., 2019). Bunker Hill (Copper Creek) District, Copper Creek, Childs-Aldwinkle Mine, as unusual white resinous stacked bipyramidal crystals (Ron Gibbs, pers. comm., 2019). Reymert supersystem, Mineral Mountain District, Gorilla claims, as dipyramidal crystals proximal to galena and as thin plates away from the galena (Stanley B. Keith, pers. comm., 2019).

FIGURE 3.266 Wulfenite, Glove Mine, Santa Cruz County, 7.1 cm.

SANTA CRUZ COUNTY: Santa Rita Mountains, Wrightson District, Gringo Mine, with gold (Schrader and Hill 1915); Tyndall District, Glove Mine, as remarkable crystal aggregates of various colors (typically light-yellow to pale-orange tones) and habits; some crystals measured up to 4 in. along the edge (H. J. Olson 1966; W. E. Wilson and Bideaux 1983; UA 2760, 15477; NHM 1957, 56); J. C. Holmes District, J. C. Holmes claims near Patagonia, with vanadinite, descloizite, and cerussite, on fracture planes in quartz vein filling (Pellegrin 1911). Patagonia Mountains, Palmetto District, as beautifully crystallized specimens associated with galena, cerussite, and silver (Schrader 1917).

YAVAPAI COUNTY: Black Rock District, Purple Passion Mine, as tabular to unusual fibrous crystals with fluorite (B. Gardner and Davis 2000). Eureka supersystem, Bagdad District, Mountain Spring deposit and vein west of Severing Gulch, as thin yellow plates (C. A. Anderson, Scholz, and Strobell 1955). Bradshaw Mountains, Silver Mountain District, Fat Jack Mine, as small crystals on quartz with stolzite (3rd ed.).

YUMA COUNTY: Muggins Mountains, Muggins District, Red Knob Mine, with weeksite, vanadinite, and cuprite (Honea 1959; Outerbridge et al. 1960). Castle Dome District, Hull and Puzzler Mines, as groups of tabular yellow or blocky orange crystals with mimetite, anglesite after galena pseudomorphs, or fluorite (Domitrovic, Wilson, and Hay 1998).

FIGURE 3.267 Wupatkiite, Grey Mountain, Coconino County, FOV 3 mm.

#WUPATKIITE

Cobalt aluminum sulfate hydrate: $CoAl_2(SO_4)_4 \cdot 22H_2O$. A rare secondary mineral found as a postmining encrustation. The Arizona occurrence is the type locality.

COCONINO COUNTY: Cameron District, in a shallow open cut eight miles east-southeast of Gray Mountain. It is found as a postmine encrustation which has formed a concrete-like crust on the walls of the open cut with pickeringite (S. A. Williams and Cesbron 1995).

WURTZITE

Zirconium sulfide: ZnS. A rare mineral found in hydrothermal veins with other sulfide minerals. It is a trimorph with matraite and sphalerite.

PINAL COUNTY: Mammoth District, Mammoth–St. Anthony Mine, where it is reported below the 900 ft. level (Bull. 181).

WÜSTITE

Iron oxide: FeO. A very rare mineral of the periclase group found as an alteration product of iron-bearing minerals, including some meteorites.

COCONINO COUNTY: In specimens of the Canyon Diablo meteorite as an alteration product of barringtonite and schreibersite (3rd ed.).

WYLLIEITE

Sodium manganese iron aluminum phosphate: $NaNaMn(Fe^{2+}Al)(PO_4)_3$. A primary mineral formed in granite pegmatites.

MARICOPA AND YAVAPAI COUNTIES: White Picacho District, in pegmatites (London 1981).

XENOTIME-(Y)

Yttrium phosphate: $Y(PO_4)$. Found as an accessory mineral in felsic igneous rocks, pegmatites, and gneiss.

GILA COUNTY: Two localities in the Diamond Butte quadrangle, in the hematite-rich portion of the feldspathic sandstone of the Dripping Spring Quartzite (Gastil 1954).

MOHAVE COUNTY: Aquarius Mountains District, Columbite prospect, as sparse crystals up to 1 inch across (Heinrich 1960); Wagon Bow No. 3 pegmatite, in small amounts with monazite-(Ce), uraninite, and niobian rutile (W. B. Simmons et al. 2012). Virgin Mountains, with monazite in Precambrian granite gneiss (E. J. Young and Sims 1961); near Mesquite, on the Nevada border (Clark County, Nevada), with monazite in granite augen gneiss (Overstreet 1967).

PIMA COUNTY: North side of the Rincon Mountains, with gadolinite in biotite gneiss (Bideaux, Williams, and Thomssen 1960).

XOCOMECATLITE

Copper tellurate hydroxide: $Cu_3(Te^{6+}O_4)(OH)_4$. A very rare secondary mineral formed from other tellurium minerals under conditions of severe oxidation.

COCHISE COUNTY: Tombstone District, Emerald Mine, where only one specimen was found on the mine dump; as pale-green spherules formed by oxidation on fracture surfaces; closely associated with khinite-4O and various other rare tellurates (S. A. Williams 1978).

XONOTLITE

Calcium silicate hydroxide: $Ca_6Si_6O_{17}(OH)_2$. Commonly found in contact-metamorphic rocks as a product of retrograde metamorphism.

GILA COUNTY: Southeastern Dripping Spring Mountains, Banner supersystem, Christmas District, Christmas Mine, associated with an assemblage of retrograde metamorphic minerals, including tobermorite, apachite, and gilalite (Cesbron and Williams 1980; UA10328).

YAFSOANITE

Calcium tellurium zinc oxide: $Ca_3Te^{6+}_2(ZnO_4)_3$. A rare secondary mineral formed by oxidation of primary tellurium-bearing ores.

COCHISE COUNTY: Tombstone District, Empire Mine, 400 ft. level, a specimen rich in various tellurites, including adamantine and white dodecahedra of yafsoanite, which thinly encrust some fracture surfaces (J. W. Anthony et al. 2003).

YARROWITE

Copper sulfide: Cu_9S_8. A rare secondary mineral found replacing other copper sulfides.

COCHISE COUNTY: Warren District, Bisbee, in the sulfide ores of the Campbell Mine (Graeme 1993).

YAVAPAIITE

Potassium iron sulfate: $KFe^{3+}(SO_4)_2$. A rare secondary mineral formed as a result of a mine fire at the United Verde Mine at Jerome. The United Verde Mine is the type locality.

YAVAPAI COUNTY: Jerome supersystem, Verde District, United Verde Mine, sparse as cement in rubble exposed in open-pit operations; also as rare short stumpy crystals; associated with voltaite, sulphur, and jarosite (Hutton 1959b; Graeber and Rosenzweig 1971; J. W. Anthony, McLean, and Laughon 1972).

YEDLINITE

Lead chromium chloride hydroxide: $Pb_6Cr(Cl,OH)_6$ $(OH,O)_8$. A very rare secondary mineral found in the oxidized portion of a polymetallic ore deposit. The Mammoth–St. Anthony Mine is the type locality.

PINAL COUNTY: Mammoth District, Mammoth–St. Anthony Mine, sparingly on a few specimens, associated with diaboleite, quartz, wulfenite, dioptase, phosgenite, and wherryite; as red-violet rhombohedral crystals up to 1 mm long (McLean et al. 1974; Wood, McLean, and Laughon 1974; UA11360; R050338).

R050338

FIGURE 3.268 Yedlinite, Tiger, Pinal County.

ZEUNERITE

Copper uranyl arsenate hydrate: $Cu(UO_2)_2(AsO_4)_2 \cdot 12H_2O$. A secondary mineral associated with other uranium minerals. Over time zeunerite may dehydrate to metazeunerite.

COCONINO COUNTY: Grand Canyon National Park, Grandview District, Grandview (Last Chance) Mine, as exceptional crystals, up to 0.25 in. on an edge, with brochantite, olivenite, scorodite, and chalcoalumite (Leicht 1971; UA2499). Orphan District, Orphan Mine (Gornitz 2004).

MOHAVE COUNTY: Hack District, Hack Canyon Mine (AM 26907).

ZINCITE

Zinc oxide: ZnO. Found in metamorphic zinc deposits and as a secondary mineral in zinc deposits.

COCHISE COUNTY: Tombstone District (P. Megaw, pers. comm., 2012).

MARICOPA COUNTY: Belmont Mountains, Osborne District, Tonopah-Belmont Mine, as dull red-brown earthy crusts, formed from sphalerite-bearing ore by a mine fire (William Hunt, pers. comm., 1990).

ZINCOBOTRYOGEN

Zinc iron sulfate hydroxide hydrate: $ZnFe^{3+}(SO_4)_2(OH) \cdot 7H_2O$. Formed in the oxidized zone of lead-zinc deposits.

COCHISE COUNTY: Warren District, Bisbee (Graeme 1993).

#ZINCOCOPIAPITE

Zinc iron sulfate hydroxide hydrate: $ZnFe^{3+}_4(SO_4)_6(OH)_2 \cdot 20H_2O$. A rare secondary mineral found in the oxidized portions of lead-zinc deposits in arid climates.

COCHISE COUNTY: Warren District, Bisbee, Higgins Mine, tunnel level. Found as a postmining mineral with copiapite, melanterite, and römerite on sphalerite and pyrite ore (Graeme, Graeme, and Graeme 2015).

ZINCZIPPEITE

Zinc uranyl sulfate oxide hydrate: $Zn(UO_2)_2(SO_4)O_2 \cdot 3.5H_2O$. A secondary uranium mineral, often postmining, formed as an efflorescence on mine walls and dumps. The Hillside Mine is the type locality.

YAVAPAI COUNTY: Eureka supersystem, Hillside District, Hillside Mine, as minute curved crystals and coatings; associated with natrozippeite and nickelzippeite, as well as schröckingerite, bayleyite, johannite, and gypsum (C. Frondel et al. 1976).

zinnwaldite

Potassium lithium iron aluminum silicate hydroxide fluoride: $K(Li,Fe,Al)_3(Si,Al)_4O_{10}(OH,F)_2$. An uncommon member of the mica group, typically associated with other lithium-bearing minerals, topaz, cleavelandite, beryl, tourmaline, fluorite, and monazite in tin veins and pegmatites. Zinnwaldite has been discredited as a species and is considered on the siderophyllite-polylithionite join, a dark mica containing lithium, in the mica group.

MARICOPA COUNTY: Rare in the pegmatites of the White Picacho District, as very dark cleavable crystals with golden-brown cleavage flakes; associated with spodumene and amblygonite (Jahns 1952).

SANTA CRUZ COUNTY: Patagonia Mountains, Querces District, Line Boy Mine (E. D. Wilson and Roseveare 1949).

ZIPPEITE

Potassium uranyl sulfate oxide hydroxide hydrate: $K_2[(UO_2)_4(SO_4)_2 O_2(OH)_2] \cdot 4H_2O$. A secondary mineral in oxidized uranium deposits and a postmine mineral formed on the walls of some mine workings and on dumps. Zippeite is a member of the zippeite group and is commonly associated with other secondary uranium minerals and gypsum. Several localities of zippeite group minerals are listed below because the exact species is not always known (C. Frondel et al. 1976).

COCHISE COUNTY: Warren District, Bisbee, Cole shaft, as postmining efflorescence with gypsum and uranopilite (Graeme, Graeme, and Graeme 2015).

COCONINO COUNTY: Cameron District, in minor amounts in uranium ore in the Shinarump Conglomerate and in the Chinle Formation, associated with uraninite, metatorbernite, and meta-autunite; Huskon No. 7 and No. 8 claims (Holland et al. 1958). Vermilion Cliffs District, Sun Valley Mine, associated with an unnamed uranyl phosphate as well as ilsemannite (Petersen, Hamilton, and Myers 1959). Additional uranium mines in the county have zippeite reported (Peirce, Keith, and Wilt 1970).

MOHAVE COUNTY: Hack District, Hack Canyon Mine, with other secondary uranium minerals (Scarborough 1981).

NAVAJO COUNTY: Holbrook District, Ruth claims (Bull. 181). Monument Valley District, Monument No. 1 and Mitten No. 2 Mines (Witkind 1961; Witkind and Thaden 1963).

YAVAPAI COUNTY: Eureka supersystem, Hillside District, Hillside Mine, with other zippeite group minerals (C. Frondel et al. 1976; UA 9286).

ZIRCON

Zirconium silicate: $Zr(SiO_4)$. Found as a common accessory mineral in igneous and metamorphic rocks. Because of its resistant nature, it is also present in sands and sedimentary rocks. Only some of the zircon occurrences as an accessory mineral are given here, as it is widespread in Arizona but generally very small and not of interest to collectors. Zircon is extremely important for age dating, however, as it is the most reliable mineral for uranium-lead radiometric age dating of igneous crystallization ages. Zircon age dates on the Elves Chasm Gneiss in the Upper Gorge of the Grand Canyon at about milepost 115 are 1.84 billion years old, which makes it the oldest rock in Arizona. Detrital zircons extracted from the Vishnu Schist are even older, at 3.8 billion years old. This makes zircon the oldest mineral in Arizona. Most of the geologic history of Arizona in the following chapter and on the county mineral district maps is based on radiometric dating of zircons.

FIGURE 3.269 Zircon, Belmont Pit, Maricopa County, FOV 2 mm.

APACHE COUNTY: Garnet Ridge SUM District, Garnet Ridge, as an accessory mineral in the Navajo Sandstone, associated with magnetite and tourmaline (Gavasci and Kerr 1968).

COCHISE COUNTY: Tombstone District, as microscopic grains in light-colored intrusive rocks (Silver and Deutsch 1963). Juniper Flat District, as small crystals in Pinal Schist with tourmaline, and in Juniper Flat per-aluminous granite northwest of Bisbee (Bull. 181).

GILA COUNTY: In the Pinal Schist, Madera diorite, and Ruin Granite (Ransome 1919), also a major accessory mineral in the Laramide plutons of the Globe-Miami region, where it was used for extensive uranium-lead dating (Seedorf et al. 2019).

GRAHAM COUNTY: Santa Teresa Mountains, with monazite in pegmatite (Robert O'Haire, pers. comm., 1972).

GREENLEE COUNTY: Copper Mountain supersystem, in granite (Reber 1916; N. P. Peterson, Gilbert, and Quick 1946).

LA PAZ, YUMA, AND MARICOPA COUNTIES: Northern Plomosa Mountains, Cemetery Ridge, southern Trigo Mountains, Castle Dome Mountains, and Neversweat Ridge, where numerous windows into the Orocopia Schist yielded abundant detrital zircons (Jacobson et al. 2017; Strickland, Singleton, and Haxel 2018; Seymour et al. 2018).

MARICOPA COUNTY: White Tank Mountains, Caterpillar tractor testing grounds, as crystals with euxenite-(Y) in a pegmatite (Phil Hooker, pers. comm., 1985). White Picacho District in pegmatites (London and Burt 1982a). Belmont Mountains, Belmont pit, as small well-formed crystals in miarolitic cavities of the Belmont Granite; thorium rich (Gibbs and Turzi 2007).

MOHAVE COUNTY: Cerbat Mountains, Kingman Feldspar District, Kingman feldspar quarry (Spar Consolidated mine), in granite pegmatite with allanite-(Ce), bastnäsite-(Ce), and "thorogummite" (S. L. Hanson et al. 2012). McConnico District, five miles southwest of Kingman at the Guy Hazen Group in a pegmatite with thalénite (Fitzpatrick and Pabst 1986). White Hills, Gold Basin District, in a gold placer claim as abundant lavender-colored sand (Theodore, Blair, and Nash 1987).

NAVAJO COUNTY: Monument Valley District, as one of the most common heavy minerals in uranium- and vanadium-bearing Shinarump Conglomerate; also common in the basal Monitor Butte Sandstone above the Shinarump Conglomerate, associated with apatite, baryte, and tourmaline (R. G. Young 1964).

PIMA COUNTY: Ajo District, as small, stout crystals in the New Cornelia Quartz Monzonite (W. J. Thomas and Gibbs 1983) and Cardigan Peak

Quartz Monzonite (Wilkinson et al. 2010). Santa Catalina Mountains, in mylonitic two-mica granite from Windy Point with monazite, where it was extensively dated by uranium-lead radiometric methods (Catanzaro and Kulp 1964; Fornash et al. 2013; Terrien 2012); northwest Santa Catalina Mountains, important accessory mineral in the Catalina Quartz-Monzonite pluton. Silver Bell District, as an accessory mineral (Kerr 1951).

PINAL COUNTY: Mineral Creek supersystem, Ray District, Ray Mine, as a common accessory mineral (R. W. Jones and Wilson 1983).

SANTA CRUZ COUNTY: Patagonia Mountains, in granite (Schrader 1913, 1917).

YAVAPAI COUNTY: Bradshaw Mountains, sparingly in the Bradshaw Granite (C. A. Anderson 1950). Kirkland–Copper Basin placers, reportedly in commercial quantities in the black sands (E. D. Wilson 1961). Jerome supersystem, Verde District, United Verde Mine, as very small crystals in black chloritic schist (C. A. Anderson and Creasey 1958).

ZOISITE

Calcium aluminum silicate hydroxide: $Ca_2Al_3[Si_2O_7][SiO_4]O(OH)$. Commonly found as a product of medium-grade thermal or contact metamorphism. The zoisite listed here may be clinozoisite as the two were not always differentiated in the older literature.

COCHISE COUNTY: Cochise supersystem, Johnson Camp District, Johnson Camp area, Republic Mine, sparse in contact-metamorphosed rocks containing other calcsilicate minerals; locally as the variety thulite, in vugs and as coatings on joint surfaces (J. R. Cooper 1957; J. R. Cooper and Silver 1964). Tombstone District, microscopically in igneous rocks (B. S. Butler, Wilson, and Rasor 1938). Warren District, Spray shaft, as fibrous gray groups with tremolite, talc, forsterite, and pyrite; Uncle Sam shaft, with epidote, quartz, pyrite, and chalcopyrite (Graeme 1981). Chiricahua Mountains, California supersystem, Hilltop District, in the area around Paradise and Hilltop, in altered limestone with scheelite, epidote, garnet, wollastonite, and calcite (Dale, Stewart, and McKinney 1960). In metamorphic rocks in central Cochise County at Middlemarch Canyon, central Dragoon Mountains, Middle Pass supersystem, Abril District, Four Canyon and Jordan Canyon (Gilluly 1956).

GILA COUNTY: Southeastern Dripping Spring Mountains, Banner supersystem, Christmas District, Christmas Mine, as the variety thulite, formed in metamorphosed diorite and in skarn (Perry 1969); 79 District, 79 Mine, as lovely pink crystals of thulite up to 1 in. long, and as coatings on fracture surfaces in metamorphosed limestones (3rd ed.).

GREENLEE COUNTY: Copper Mountain supersystem, Morenci District, Morenci Mine, as a product of contact metamorphism, found in highly altered granite (Reber 1916).

LA PAZ COUNTY: Dome Rock Mountains, Cinnabar District, in the wall rocks of cinnabar veins (Bull. 181).

MARICOPA COUNTY: North end of the Sierra Estrella Mountains with diopside, tremolite, and forsterite (Sommer 1982).

MOHAVE COUNTY: Cerbat Mountains, Wallapai supersystem, as an alteration product in the wall rocks of sulfide-bearing veins (B. E. Thomas 1949).

PIMA COUNTY: Baboquivari Mountains, Aguirre Peak District, Giant Mine, manganese-bearing zoisite with scheelite in porous quartzite (Dale, Stewart, and McKinney 1960); Quinlan District, Brown prospect, with epidote, garnet, and calcite in a quartzitic formation (Dale, Stewart, and McKinney 1960). Pima supersystem, Sierrita District, Sierrita Open-Pit Mine, as an alteration product of dioritic rocks (Roger Lainé, pers. comm., 1972).

PINAL COUNTY: San Manuel District, Kalamazoo and San Manuel orebodies, with epidote, chlorite, hydrobiotite, and secondary biotite (Schwartz 1947; Creasey 1959; J. D. Lowell 1968).

YAVAPAI COUNTY: Bradshaw Mountains, in scattered lenses in schist (Lindgren 1926); Mayer District, Blue Bell Mine, with "biotite," magnetite, chlorite, and quartz (Lindgren 1926). Eureka supersystem, Bagdad District at Bagdad, as an accessory mineral in bodies of titaniferous magnetite in the gangue of the copper deposits (Bull. 181).

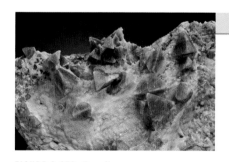

FIGURE 3.270 Zunyite, Big Bertha Extension Mine, Quartzsite, La Paz County, 7 cm.

ZUNYITE

Aluminum silicate hydroxide fluoride chloride: $Al_{13}Si_5$ $O_{20}(OH,F)_{18}Cl$. A rare mineral formed as a result of metamorphic or hydrothermal activity.

LA PAZ COUNTY: Dome Rock Mountains, Sugarloaf Pyrophyllite District, about five miles southwest of Quartzsite, south side of Sugarloaf Peak; Big Bertha Extension Mine, as transparent buff-colored crystals up to 2 cm on an edge, among the largest in the world (Bull. 181; Sprunger 1980; W. E. Wilson 1986; UA 5455; R050553; UAX050185).

PINAL COUNTY: San Manuel District, San Manuel Mine, in very small quantities in highly altered monzonite porphyry (Schwartz 1953). Pioneer supersystem, Resolution District, Resolution deposit, in drill core with topaz (Winant 2010).

4 Arizona Mineral Districts

During the early days of mining in the West, before state or county laws had been established, miners organized their mining areas into legal areas called mining districts. The phrase *mining district* means a geographically distinct area that is usually designated by a name related to a nearby natural feature or a recognizable mining area. Prior to 1872, these mining districts were worked under rules and regulations created by the miners. Geologic or production information about these areas was referenced to the named mining districts. These areas of mining activity were not precisely defined and had no surveyed boundaries. In the earlier editions of *Mineralogy of Arizona*, the mining district names were used with the locality information, and many of these old mining district names have been carried over and used for current mineral district names.

The mining districts were not geologically defined, and several different types of mineral deposits were frequently found within the same district. Therefore, to better provide geologic information for mineral exploration in Arizona, Stanley Keith and co-workers at the Arizona Bureau of Geology and Mineral Technology (now the Arizona Geological Survey) categorized the mineralized areas in Arizona according to geologic criteria of types of metals and ages of mineralization (Stanley B. Keith et al. 1983). These areas are called *mineral districts* to distinguish them from the legal mining districts of the early miners. The mineral districts are mineral systems related to the associated magma source, timing, and other geologic processes involved in the formation of the minerals, and the mineral district maps in appendix A have information for each district about the igneous rock association, the metals present, and the age of the deposit in million years ago (*Mega-annum*, or *Ma*). The mineral districts associated with mineral occurrences in chapter 3 are noted wherever possible.

For example, the Rowley Mine is a well-known wulfenite locality in the Painted Rock District. On the Maricopa County mineral district map, the Painted Rock District is given as MQA PbCu(AgMnAuMo) 18-15. This means that the Rowley Mine is associated with metaluminous quartz alkalic igneous rocks, has lead and copper as the major elements recovered with minor silver, manganese, gold, and molybdenum, and deposition of the original sulfides between 18 and 15 million years ago.

This chapter and the mineral district maps in appendix A give a mineral collector information about the geologic history of mineral formation in Arizona and a way to compare localities and find similar minerals in new localities. For example, after noting the information about the Rowley Mine, a search for similar districts in Maricopa County shows the Webb District is also MQA and formed 18 to 15 million years ago. Although the produced elements are different—CuAu(Ag)—the Webb area could potentially have oxide-zone copper minerals and other minerals that occur at the Rowley Mine.

Mineral Districts

A large amount of geologic information (such as age dates, geologic maps, and descriptive geologic information) has accumulated since the 1983 compilation, and the information continues to rapidly expand. The empirical relationship between magma-metal series and mineral deposits is better documented and understood (Stanley B. Keith et al. 1991; Wilt 1993, 1995; Stanley B. Keith and Swan 1996). Also, an ultradeep hydrothermal (UDH) process of mineral and petroleum formation has been identified that explains the deposition of some nonigneous mineral systems (Stanley B. Keith and Swan 2005; Rasmussen and Keith 2015; Keith, Spieth, and Rasmussen 2018). The criteria established by this research, plus the new geologic data, were used to produce the new and improved mineral district maps, which are presented on county maps in appendix A.

Despite the improvements in data and scientific understanding, there is still considerable variability in the precision of the various ages and classes for each mineral district. Some of the well-studied deposits, such as the porphyry copper deposits, are well dated and have abundant geochemical analyses with which to classify the district. Many mineral districts have not been studied in detail, however, and the metal associations are estimated based on mineral occurrences and production. The accuracy of geologic age information is also variable, and some ages of mineral districts are estimated based on geologic mapping and crosscutting relationships, without radiometric age date information on the deposit. A question mark is included with the map data to indicate this uncertainty. The outlines for each mineral district presented on the county maps, to the best of our knowledge at this time, represent the geographic area affected by deposition of a primary mineral suite at a particular time and related to the specific geochemistry of an igneous differentiation sequence. Within the area of the mineral system (mineral district), a similar suite of minerals is likely to occur.

Supersystems

Many geographic areas contain mineral districts that have metallogenic compositions and magmatism of the same age. These clusters of similar mineral districts occur in larger geographic areas that are designated as mineral supersystems. The presence of these supersystems implies that there could be a larger underlying batholithic-scale hydrothermal metal source in the middle crust underlying the individual plutons associated with the mineral system. The more obvious mineral supersystems are designated on the mineral district maps. Additional similar minerals could occur within these supersystems. A mineral collector might be able to find new localities for given mineral specimens within the supersystem. For example, the Red Cloud supersystem of La Paz and Yuma Counties contains many wulfenite occurrences. Additional supergene wulfenite occurrences could occur in the supergene zones of other districts in this supersystem.

Evolution of Arizona Mineral Systems

The succession of mountain-building episodes and intervening erosional periods resulted from plate tectonic movements of the North American continent and its interaction with adjoining oceanic and continental plates.

Orogenies

Most primary Arizona minerals were formed during five major orogenies, or mountain-building episodes, in the Proterozoic between 1825 and 541 Ma, and during five orogenies in the Mesozoic, 252–66 Ma, and Cenozoic, 66–10 Ma. A highly abbreviated summary of the progression of these orogenies and intervening rifting episodes is summarized in table 4.1.

A more detailed listing is found in table 4.2, at the end of this chapter. Three major orogenies in the Paleozoic, between 541 and 252 Ma, primarily affected the eastern United States but influenced the deposition of sedimentary rocks in Arizona. Other orogenies in the Paleozoic mainly affected adjoining states but also had an effect on Arizona sedimentation.

TABLE 4.1 Summary of mountain-building episodes, associated magmatism, age, resources, examples, and significant minerals

Orogeny	Phase	Age (Ma)	Type	Resources/ elements	Mineral district examples	Minerals
San Andreas	Late	2.6–0.012	MNA	olivine, gypsum	Peridot Mesa, Camp Verde	olivine; glauberite; gypsum
	Early	13–2.6	UDH	zeolites, salt	San Francisco volc., Luke salt	clinoptilolite, hectorite; halite
Galiuro	Latest	28–10	MNA	oil, He	Navajo volcanic field (Dineh-bi-Keyah)	phlogopite, oil
	Late	28–10	MQA	Mn, U, AuAg(CuPb)	Artillery Mtns., Oatman, Mammoth	Mn oxides, carnotite, gold, Tiger suite, specularite
	Middle	28–15	MAC	PbZnAg(AuCu) (Sn)	Ash Peak, Red Cloud, Aravaipa	silver, galena, sphalerite, cerussite, wulfenite (cassiterite)
	Early	30–21	MCA	Au(CuWAg)	Kofa, South Mountain, Gila Bend	gold, todorokite, chalcophanite, pyrolusite
		65–29	UDH	actinolite, garnet	Cemetery Ridge, Garnet Ridge	actinolite, serpentine group, pyrope garnet
	Late	70–35	PC	Au (PbMo)	Gold Basin, Vulture	gold, kyanite, wulfenite
		60–40	PCA	W	Bluebird	wolframite group, scheelite
Laramide	Middle	65–55	MCA	CuMoAg (W)	Ajo, Ray, San Manuel, Mineral Park, Pima, Bagdad, Silver Bell, Globe-Miami, Morenci	chalcopyrite, molybdenite, pyrite, bornite, epidote, garnet (powellite)
	Early	75–65	MAC	AgPbZn	Tombstone, Glove, Empire, Ruby, Salero	galena, sphalerite, alabandite, tetrahedrite, silver, enargite, cerussite, wulfenite, vanadinite
	Earliest	89–75	MQA	CuAuAg	Old Yuma, Mexican Hat, Golden Rule	gold, galena, cerussite, mottramite, wulfenite, vanadinite
	Late	93–89	MQA	U	Black Mountain (U)	carnotite, tyuyamunite, hewettite
Sevier	Middle	100–93	UDH	coal, fire clay	Dakota Sandstone, Deer Creek coal	coal, kaolinite
	Early	135–110	UDH	calcite flux	Paul Spur	calcite
	Earliest	155–135	MQA	UVCu (NiCo)	Orphan, Hermit, Arizona 1, EZ-2, Pigeon	pyrite, uraninite, bravoite, sphalerite, chalco-pyrite, galena
	Rifting	155–140	UDH	C, oil and gas?	Yellowbird, Crystal Cave Fm., Glance	graphite/graphene, black shales

		175–155	PC	Au	Nogales	gold
Nevadan	Late	175–155	PCA	W veins	Las Guijas, Juniper Flat	wolframite group, hübnerite, scheelite
		160–155	PAC	kyanite	Tung Hill	dumortierite, rutile, scheelite
	Middle, W AZ	173–155	MAC	PbZnAg	Comobabi Mtns, Cababi (Mildren-Steppe Mine)	galena, sphalerite, tetrahedrite (wulfenite)
		191–155	MQA	AuCu(AgWPb)	La Cholla, Sugarloaf (Big Bertha), Jaeger, Yuma King, Swansea, Planet	quartz, pyrite, gold, chalcopyrite, specular hematite, magnetite, tenorite
	Early, SE AZ	201–175	MAC	PbZnAg(CuAu)	Gleeson (South Turquoise), Courtland (North Turquoise), Hartford	galena, sphalerite, tetrahedrite, cerussite, wulfenite, vanadinite, azurite
		201–191	MQA	CuAu, PGE	Warren (Bisbee)	chalcopyrite, bornite, azurite, malachite, cuprite, copper, tenorite
	Early, N AZ	237–201	MQA	U-V-Cu (Ni-Co)	Orphan, Grandview, Monument Valley	uraninite, chalcopyrite, galena, torbernite; carnotite, montroseite
Ouachita	Pangaea assembly	318–271	UDH	salt, K; U; He	Holbrook salt, Holbrook potash	sylvite, carnallite, polyhalite, halite, gypsum, anhydrite
Antler (NV)	Island arcs	380–357	UDH	oolitic iron	Payson "diamond" quartz, Ranch Ck. Fe	Herkimer habit quartz, oolitic hematite
Post-Rodinia	Rifting	800–740	UDH	HC, black shale	Kwagunt Mbr. Chuar Grp.	kerogenous black shale
Grenville	Rodinia	1035–980	PC?	none	Peraluminous granites?	muscovite
Keweenawan	Rifting	1104–1035	MQA	asbestos; U (CuFe)	Sierra Ancha; Hope	chrysotile asbestos, lizardite; uraninite, pyrite, magnetite
Elzevirian	Island arcs	1360–1240	MQA	hematite	Apache / Chediski Fe, Troy Quartzite	hematite
Picuris	Late	1460–1370	PCA	W, Be, Li, Ce, Ta-Nb-Mn	White Picacho, Vulture, Wagon Bow, Tungstona	spodumene, lepidolite, FeMn phosphates, scheelite, beryl, wolframite group
	Late	1470–1450	PAC	amethyst	Four Peaks amethyst	amethyst, hematite, fluorapatite
	Early	1470–1420	MCA	Cu	Suspense	copper minerals

(continued)

TABLE 4.1 (*continued*)

Orogeny	Phase	Age (Ma)	Type	Resources/ elements	Mineral district examples	Minerals
Mazatzal	Late	1630–1610	PC	Au (Ag Pb Cu Bi)	Yellowstone	gold, galena, chalcopyrite, quartz, baryte, bismuth minerals, tourmaline, zircon, albite
	Late	1630–1590	PCA	W, Be, F, LREE	Black Beauty, Kingman Feldspar	scheelite, beryl, allanite-(Nd), bastnäsite, microcline
	Late	1680–1640	PAC	W	Garnet Mountain	scheelite, garnet, epidote
	Middle	1680–1630	MCA	Au	Roosevelt, Spring Creek, Prescott, Thumb Butte	gold
	Early	1702–1680	MAC	Be, F	Breadpan Fm.; Gordon Creek	acicular beryl, tourmaline, topaz; pyrophyllite
Yavapai	Late	1715–1700	PC	Au(AgCu)	Humbug, Cottonwood	gold, pyrite, galena, chalcopyrite
	Late	1715–1690	PCA	W(Be)	Boriana, Money Maker, Kingman Feldspar	scheelite, beryl, hübnerite, wolframite, microcline, quartz
	Early	1750–1720	MC	Zn-Cu-Ag VMS; Fe-Si BIF	VMS Jerome, Antler; Pikes Peak Iron	pyrite, chalcopyrite, sphalerite, galena, arsenopyrite, pyrrhotite; chert-hematite-magnetite
	Early	1770–1715	MCA	Hg; Au(Ag); MoCu; kyanite	Phoenix Mtns.; Groom Creek; Squaw Peak; Iron King VMS	cinnabar, kyanite, tourmaline; gold, quartz; molybdenite; galena, sphalerite, pyrite (chalcopyrite)
	Early	1750–1690	MAC	PbAg	Groom Creek, Ruth, Turkey Cr.?	galena, silver, gold
Mohave	Late	1820–1780	PC-PCA	none	none	muscovite, garnet, feldspar

Notes: Thick line = unconformity; MC = metaluminous calcic, MCA = metaluminous calc-alkalic, MAC = metaluminous alkali-calcic, MQA = metaluminous quartz alkalic, MNA = metaluminous nepheline alkalic, PC = peraluminous calcic, PCA = peraluminous calc-alkalic, PAC = peraluminous alkali-calcic, UDH = ultradeep hydrothermal, VMS = volcanogenic massive sulfide, BIF = banded iron formation, LREE = light rare-earth elements, Ma = mega-annum, or million years ago, Fm. = Formation, ? = uncertain minor commodities (enclosed in parentheses), semicolons separate different types of deposits

During mountain-building periods, Arizona was on the leading edge of a continental plate that was overriding a down-going subducting oceanic plate slab. The depth of the subducting plate typically becomes shallower through the evolution of each orogeny. As the subducting slab intersects different layers in the overlying mantle, the dehydrating melting zone intersects different metal-compositional layers in the mantle.

Melting occurred in mantle layers above the descending slab. The resulting magma rose into the crust like blobs in a lava lamp. When the stresses from plate tectonic motions opened cracks in the crust, hydrothermal fluids rose into open spaces, where the metals and ions in the fluids combined into minerals.

In a normally dipping subduction zone in Arizona, magmas are progressively more alkaline and potassium-rich with distance away from the trench (figure 4.1). Metal associations also change with distance from the trench. Near the trench, metaluminous magmas are more calcic (MC) and are associated with copper and/or zinc mineralization. Farther away from the trench, magmas are calc-alkalic (MCA) and are also associated with copper and/or gold mineralization, as well as including significant lead, barium, and silver minerals. Even farther away from the trench, magmas are more alkaline as alkalicalcic (MAC) and metallogeny become distinctly lead-zinc-silver biased with

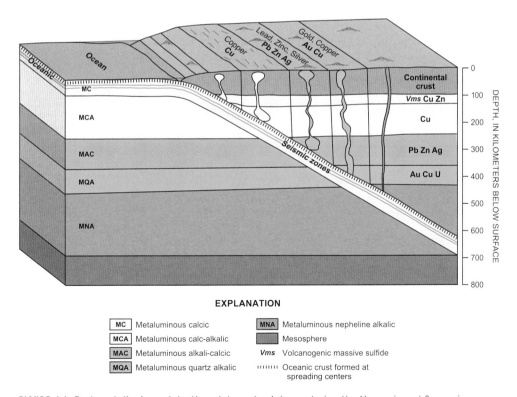

FIGURE 4.1 Eastward-dipping subducting plate under Arizona during the Mesozoic and Cenozoic.

continued amounts of barium and significant fluorine. At the farthest distance from the trench, magmatism is even more potassic and alkaline as quartz alkalic (MQA) and nepheline-alkalic (MNA) magmatism. The empirical metallogenic associations of these systems are copper, gold, silver, lead, and zinc rich, but the mineralization locally includes significant rare-earth elements, fluorine, manganese, vanadium, and uranium.

The metal and magmatic associations form distinctive belts parallel to the trench that are fixed in normal static-dip configurations. In a more dynamic, variable-dip situation, the magma-metallogenic belts migrate and overprint each other, as they did in Arizona during the Laramide orogeny. This is a major reason for the apparent complexity in Arizona's mineral district patterns.

In a plate tectonic setting of flattening subduction, the geographic pattern of metal deposits in a particular region starts with metaluminous quartz alkalic (MQA) magmatism, which is associated with copper-gold–rich deposits or uranium-rich deposits from the deepest mantle layers, and which is farthest from the trench associated with the subducting slab. These deposits are succeeded in time about 10 million years later with metaluminous alkali-calcic (MAC), lead-zinc-silver–rich deposits from the middle layers of the mantle. These deposits are associated with large volumes of volcanic rocks. Those deposits are succeeded about 10 million years later by metaluminous calc-alkalic (MCA) porphyry copper-silver–rich deposits from shallower layers of the mantle and are associated with no or very rare volcanic rocks.

As the down-going slab becomes nearly flat (figure 4.2), melting occurs in the aluminum-rich crust. This crustal melting produces tungsten and beryllium or gold-rich deposits associated with two-mica (muscovite and biotite) granites, pegmatites, and quartz veins. In cases of very flat subduction, large amounts of metagraywacke and serpentinite that were formerly deposited at the trench can be swept beneath the continental plate, becoming "attached" beneath the upper crustal plate. Because of their hydrous and low-density nature, serpentinites can be mobilized as diapiric pipes that rise into the upper crust as the serpentine ultramafic microbreccia (SUM) diatremes.

The orientation of the tectonic plates during the Proterozoic in Arizona was different from the orientation of the trench during the Mesozoic and Cenozoic. The Precambrian mountain building episodes added to the core of the North American continent on the southeastern edge (figure 4.3).

Erosional Periods

During flat subduction, continents are raised very high (similar to the Andes Mountains) and are then subjected to weathering and erosion. During ero-

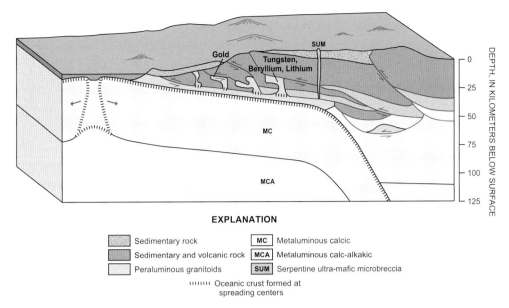

EXPLANATION

Sedimentary rock	**MC**	Metaluminous calcic
Sedimentary and volcanic rock	**MCA**	Metaluminous calc-alkakic
Peraluminous granitoids	**SUM**	Serpentine ultra-mafic microbreccia

ɪɪɪɪɪɪ Oceanic crust formed at
spreading centers

FIGURE 4.2 Flat-dipping subducting plate under Arizona crust and the metallic ore deposits that resulted from peraluminous magmas formed by crustal melting in different types of crust.

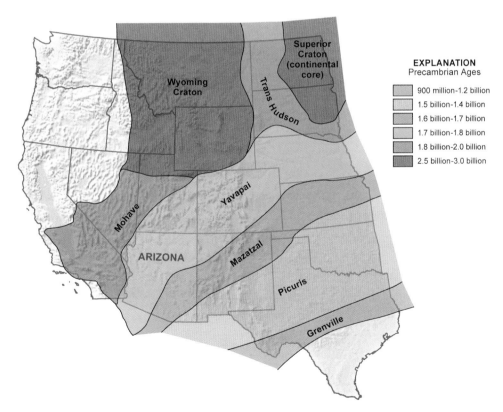

FIGURE 4.3 Stages in the growth of the North American continent during the Proterozoic (Late Precambrian) by addition of mountain-building belts along the southeastern margin of the Archean continental core, showing the southeastern limits of surface expression of rocks of that age.

sional periods between orogenic episodes, supergene enrichment produces beautiful oxide minerals, such as the malachite and azurite at Bisbee and the wulfenite at the mines in the Gleeson District of Cochise County.

Regional Variability of Mineral Systems

The complexity of Arizona ore deposits resulted from the juxtaposition of different magmatic sources through geologic time. Also, the different directions of plate tectonic movements opened different fracture orientations at different times, adding to this diversity. The level of emplacement of the mineral system also varied. For example, Laramide porphyry copper systems were emplaced at 3–5 km depths, but they are underlain at depth by two-mica granites that were emplaced at about 15–18 km depths at a later time. Later uplift can place these different levels adjacent to one another. Structural events have also raised or lowered deposits, influencing the level of erosion that exposes the mineral systems. In addition, low-angle faulting, especially regional thrust faults and later detachment faulting, juxtaposed different levels of crust.

Over geologic time, this complex geologic history has resulted in areas with different ages and types of deposits next to each other. Plutons with different metal characteristics overprint the earlier plutons and their ore deposits. This overprinting of ore deposits generally follows about 10 million years later and is usually offset from the previous ore deposit. The fractures and veins generally occurred in fracture systems of slightly different orientation, as the driving direction of plate tectonics changed through time.

Igneous Relationships of Mineral Systems

Most Arizona mineral occurrences have a direct or indirect relationship to igneous rocks. Examples of direct igneous relationships include minerals that are part of the igneous rock, such as magmatic hornblende crystals in an intrusive rock, or a late mineral precipitate in an igneous cavity, such as a zeolite-bearing vug in a volcanic rock or tourmaline in a pegmatite pocket. Examples of indirect igneous relationships mainly refer to hydrothermal emanations that have been expelled from crystallizing wet (hydrous) plutons, such as the well-known porphyry copper deposits. Some of Arizona's minerals are derived by later leaching and oxidation of an igneous rock source or ore deposit and redeposition as oxide minerals.

As magma rises, it cools and crystallizes some minerals into igneous rocks, starting with dark-colored iron- and magnesium-rich rocks, like gabbro, and continuing to differentiate into increasingly light-colored siliceous rocks, like granite. As lithostatic pressure is released by tectonic movement, each of these igneous rocks solidifies and releases a hot water-rich (hydrothermal) fluid that contains ore metals. This unmixing of the water from the metaluminous magma occurs in the upper crust at a depth of above 5 km. The unmixing of water from the more hydrous peraluminous magmas occurs deeper in the crust at a depth of about 5–15 km.

This step-wise release of hydrothermal fluids and accompanying pressure release is accompanied by fracturing of the surrounding host rocks, thus creating stockworks, disseminations, or fault and vein systems. If there is enough water in the hydrous magma, as shown by hornblende and biotite in the precursor igneous rock, then the final crystallization product is a metal deposit. As the metal-rich hydrothermal solutions become oversaturated, various minerals precipitate, and crystals are created in any open space. Expulsion of the hydrothermal metal system results in freezing of any remaining magma. Once released, a given hydrothermal plume will also undergo alteration and metal zoning.

Ultradeep Hydrothermal Episodes

When a subducting slab gets stuck and stops moving forward, there is a period of about 10 million years during which the tectonic plates reorganize. During plate reorganizations, there is no subduction-related igneous activity. As the large continental mass created by flat subduction at the end of the preceding orogeny bulged upward, it eventually cracked into rift zones. During this rifting, some hydrothermal fluids rose from sources at the crust-mantle boundary in a process similar to that making the present-day mud volcanoes, white smokers, and black smokers on the sea floor. These ultradeep hydrothermal (UDH) processes emit large volumes of brines and black muds. Some of the nonmetallic mineral districts in Arizona, such as the Holbrook salt and potash deposits, Luke salt, and Payson quartz "diamonds" (doubly terminated quartz crystals, as at Herkimer, New York), may have formed during these UDH episodes.

Magma-Metal Series Classification Used on Maps

The mineral districts on the county maps in appendix A are classified according to the magma-metal series classification (Stanley B. Keith et al. 1991). Categories in the system were defined by whole rock chemistry of associated igneous rocks. The lines between the categories were also determined based on the associated ore deposits and their production.

Aluminum Content

The magma-metal series classification used for the mineral district maps is divided into two primary megaseries: peraluminous (P) and metaluminous (M). The aluminum content is related to the source material for the magma, which depends on where melting occurs. If melting occurs in the mantle above a subduction zone and is induced by dehydration and metamorphism of the downgoing oceanic slab, then the result is metaluminous magmatism. If melting occurs in the crust as a result of continent-continent collision or very shallow subduction, then the result is peraluminous magmatism.

Peraluminous mineral districts are associated with igneous rocks that are silica- and aluminum-rich, with 65–78 percent silica. Peraluminous rocks originate from silica- and aluminum-rich crustal rocks. Minerals in peraluminous rocks commonly consist of muscovite, monazite, and/or cordierite and garnet. Less common are the aluminum-rich minerals andalusite, sillimanite, and corundum. Peraluminous magmatism crystallizes in the midcrust typically below 8 km (see figure 4.2) and lacks associated volcanics. Zircons in peraluminous rocks show xenocrystic cores and overgrowths on the cores that record various episodes of hydrous metamorphism and crystallization ages that can range over long periods.

As peraluminous magmatism originates as a melt of silica-rich crustal material, the initial melt composition is granodiorite, where silica contents begin at about 65 percent. Granodiorites differentiate into granite systems and related pegmatites. Outcrops of peraluminous magmatism constitute about 60 percent of the plutonic mass exposed in the mountain and desert province of Arizona.

Metaluminous mineral districts are associated with igneous rocks that are aluminum and silica poor with generally less than 65 percent silica. Metaluminous rocks generally contain hornblende or pyroxene as well as olivine in the more mafic phases. The presence of hornblende, pyroxene, or olivine in any amount conclusively indicates the metaluminous category. Accessory minerals in metaluminous rocks consist of sphene and relatively uncontaminated zircons. Metaluminous magmatism originates from various subaluminous layers in the mantle (see figure 4.1).

As metaluminous magmatism originates as a melt product of mafic mantle materials, the result is a full range of magmatic compositions with respect to silica content. Up to five differentiation stages may be present. In plutonic rocks, rock system stages sequentially consist of gabbro, diorite, granodiorite/quartz monzonite, granite, and locally quartz-feldspar porphyry. Outcrops of metaluminous magmatism constitute about 40 percent of plutonic exposure in the Basin and Range Province.

Alkalinity

Alkalinity is a result of the chemical composition of the Earth's layer where the magma originated. Metaluminous magmas originate from the mantle layer where they melted (see figure 4.1), and peraluminous magmas result from the composition of the crustal material where they melted. Metaluminous alkalinities depend on the depth and chemistry of the layer in the mantle that is melted.

Alkalinity categories of peraluminous series are identified by the position of the whole rock chemistry of the associated magmatism in a strontium versus calcium oxide (Sr vs. CaO) variation diagram (figure 4.4). Three peraluminous categories are present on the county mineral district maps: peraluminous calcic (PC—purple), peraluminous calc-alkalic (PCA—blue), and peraluminous alkali-calcic (PAC—turquoise).

Categories of metaluminous series are identified on the position of the whole rock chemistry of the associated magmatism on a potassium oxide versus silica oxide (K_2O vs. SiO_2) variation diagram (figure 4.5). The five metaluminous categories on the county mineral district maps are metaluminous calcic (MC—pale yellow), metaluminous calc-alkalic (MCA—bright yellow), alkali-calcic (MAC—orange), quartz alkalic (MQA—pink or red), and nepheline alkalic (MNA—maroon).

Alkalinity classes within the metaluminous category are shown on figure 4.6. The terminology is based on the Peacock (1931) alkali-lime index. Extensive compilation of whole rock geochemical data related to alkalinity, however, found the K_2O versus SiO_2 variation diagram to be more effective at separating the classes.

There is an empirical relationship between the mineralogical composition of the various mineral deposit types and their spatially and temporally associated igneous rocks. The Arizona data are a subset of a worldwide database that contains information for more than ten thousand mineral systems that have empirically matched combinations of petrochemistry, geochronological information, production data, and mineralogical information for the various mineral systems.

A magma-metal series distinction that is important for Arizona metallogeny is the boundary between alkali-calcic (orange on the maps) and calc-alkalic (yellow on the maps) series, as this is the boundary between copper-barren systems and economic porphyry copper systems. Production and geochemical data for alkali-calcic silver-dominated systems have silver to gold ratios from 40:1 to 50:1. This ratio is unique to alkali-calcic magma-metal series and is a major taxonomic criterion used to identify the MAC field on

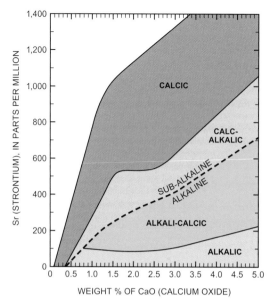

FIGURE 4.4 Variation diagram of weight percentage CaO versus Sr (in ppm) used to distinguish alkalinity of peraluminous igneous rocks.

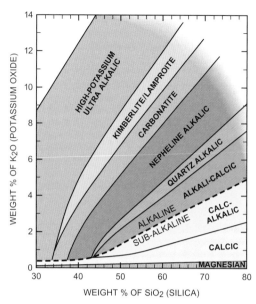

FIGURE 4.5 Variation diagram of weight percentage SiO$_2$ versus weight percentage K$_2$O of the whole rock geochemistry of unaltered igneous rocks, used to distinguish alkalinity of metaluminous rocks.

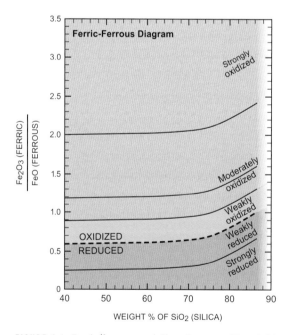

FIGURE 4.6 Ferric/ferrous variation diagram with weight percentage Fe$_2$O$_3$ divided by weight percentage FeO plotted against weight percentage SiO$_2$.

the alkalinity plots. These fields are color coordinated with the mineral system colors on the accompanying mineral district maps.

The MAC field was drawn using petrochemical data for plutons associated with lead-zinc-silver deposits such as the Tombstone District, and the MCA field was drawn using plutons associated with Arizona porphyry copper deposits such as the Ray District. The line between alkali-calcic and calc-alkalic fields also coincides with the line drawn between igneous rocks associated with epithermal silver deposits, such as in the Red Cloud Mine, and those associated with epithermal gold base-metal deposits, such as those in the Salt River Mountains.

Water Content

Water content is another important parameter in magma-metal series, as dry systems generally do not produce ore deposits. Most mineral districts on the maps are associated with hydrous magmatism. Hydration state is indicated by the mineralogy of the ferromagnesian suite. Dry or anhydrous magma series are dominated by anhydrous pyroxene and olivine ferromagnesian minerals. In contrast, wet (hydrous) magma series contain significant amounts of hydrous amphiboles, such as hornblende or biotite micas. Peraluminous magmas are invariably hydrous and can contain both muscovite and biotite.

Most of Arizona's interesting minerals are associated with oxide zones developed above various mineral systems associated with hydrous magma series. An exception is zeolites, found in vesicles of dry, mainly basaltic volcanism, and other vapor-phase minerals, such as topaz and cassiterite in rhyolitic cavities.

HYDROUS MINERAL SYSTEMS

Hornblende-bearing metaluminous magmatism is commonly associated with variously productive MC, MCA, MAC, and MQA districts with production coming from hydrothermal veins and skarns and disseminated porphyry mineralization. Hydrous magmatism is typically generated during periods of steep subduction (metaluminous magma series) or flat subduction (pegmatitic peraluminous magma series), and not during rifting.

ANHYDROUS MINERAL SYSTEMS

Vapor-phase zeolite occurrences in vesicular andesites and basalts are associated with dry pyroxene- or olivine-dominated magmatism. Minor amounts of felsic differentiates may also contain vapor-phase minerals, such as the topaz-bearing rhyolites near Kaiser Springs in Mohave County.

Anhydrous basaltic-rhyolitic magmatism is generated directly by depressurization of mantle materials during rifting, where water is absent or scarce. Rift-related magmatism generally occurred between the various orogenic episodes. Three major periods are recognized: the Precambrian diabase event, from 1100 to 1040 Ma; the late Jurassic to Early Cretaceous event, from 155 to 140 Ma; and the Basin and Range transtensional strain domains associated with the San Andreas orogeny, from 14 to 8 Ma.

Anhydrous magmatism that is rift-related exhibits less radiogenic non-crustal mantle isotopic signatures. In terms of surface geographic distribution, the transitional basalt-rhyolite volcanism is the most widespread volcanic event in Arizona. In terms of volume, however, the locally enormously thick metaluminous alkalic-calcic (MAC) ignimbrites of mid-Cenozoic age are probably the highest volume volcanics.

Oxidation State

A further subdivision of the magma-metal series classification is based on the oxidation state of the associated igneous rock. These subdivisions are based on the ilmenite-magnetite series distinction of Ishihara (1977). Chemically, oxidation state can be specified by ferric/ferrous ratios (weight percentage Fe_2O_3 divided by weight percentage FeO; figure 4.6).

The oxidation state controls the stability of early crystallizing accessory minerals, such as ilmenite, magnetite, and sphene. These minerals and the oxidation state determine which metals are incorporated into the structure of the early rock-forming minerals and which metals are expelled into the hydrothermal fluids to be deposited in the associated ore deposits.

For example, in reduced metaluminous calc-alkalic magmas where the ferric/ferrous ratios of the bulk igneous rock are less than 0.6, the crystallization of ilmenite will sequester copper in minerals such as ferro-amphibole and iron-rich biotite, thus releasing gold into the hydrothermal fluid. In contrast, in oxidized metaluminous calc-alkalic magmas where the ferric/ferrous ratio exceeds 0.9, crystallization of magnetite will sequester gold in magnetite, thus releasing copper into the hydrothermal fluid. A few reduced crustal areas in western, central, and eastern Arizona contain weakly reduced crust where the ferric/ferrous ratio is between 0.3 and 0.6, so the ore deposits in these areas are richer in gold but still contain a significant base-metal component (figure 4.7). On the maps, the oxidized crust deposits have an *o* after the magma-metal series class, and the weakly oxidized systems have a *wo* designation.

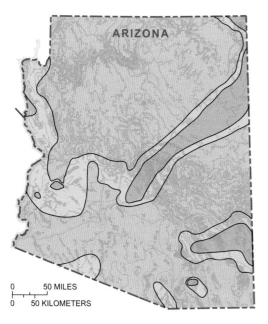

0 50 MILES
0 50 KILOMETERS

EXPLANATION
Ferric/Ferrous Ratio of Associated Plutons and Inferred Oxidation State of Mineral Systems

Oxidized–More than 0.9
Weakly oxidized–0.6-0.9
Weakly reduced–0.3-0.6

FIGURE 4.7 Oxidation state map of Arizona crust.

OXIDIZED SYSTEMS

Magmatic indicator minerals for oxidized magma series include magnetite, magnetite

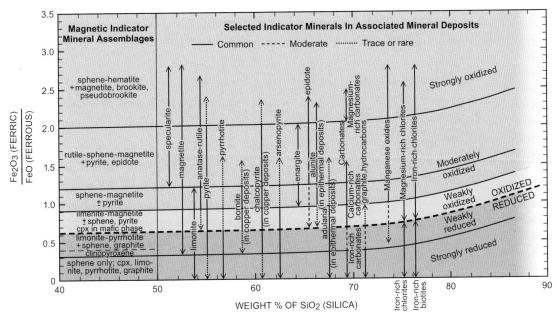

FIGURE 4.8 Empirical indicator mineral assemblages for oxidation states in plutons compared with empirical indicator minerals of oxidation states in associated mineral deposits on Fe_2O_3/FeO versus SiO_2 variation diagram.

plus sphene, brookite, rutile, specularite, and pyrite (left side of figure 4.8). The ferromagnesian suite contains magnesium-rich biotite and magnesium-rich amphibole.

Indicator minerals for hydrothermal systems derived from oxidized to strongly oxidized igneous rocks include the presence of enargite, hematite, brookite, anatase, helvine, bertrandite, and bastnäsite. Skarns associated with hydrothermalism sourced in oxidized plutons can include talc, tremolite/uralite, brucite, periclase, wollastonite, garnet, magnesite, forsterite, and uraninite (right side of figure 4.8). The hydrothermally altered areas associated with oxidized plutons contain alunite, pyrophyllite, magnesium-rich amphiboles, and magnesium-rich biotites. Late-stage diaspore, zunyite, alunite, and pyrophyllite can form extensive argillic and pyritic solfataras associated with volcanism and with plutonism related to oxidized rhyolitic or granitic differentiates.

WEAKLY OXIDIZED SYSTEMS

Magmatic indicator minerals for weakly oxidized igneous rocks are magnetite, ilmenite, sphene, iron-rich biotite, iron-rich amphibole, and clinopyroxene.

Mineral indicators for hydrothermal deposits are molybdenite, magnetite, scapolite, tetrahedrite-tennantite, phlogopite, epidote, rutile, andradite, iron-rich garnet, and magnesium-rich chlorite. Skarn indicator minerals for weakly oxidized igneous rocks can include magnesium-rich diopside, forsterite, hessite, altaite, pyrolusite, alabandite, manganite, orthoclase, ilvaite, wollastonite, anhydrite, johannsenite, and manganese-rich pyroxene. Chloride minerals that also indicate weakly oxidized systems include halite and sylvanite.

Wall-rock skarns are poorly to moderately developed in reduced to weakly oxidized classes. Where developed, skarns can display an iron-rich pyroxene, iron-rich biotite, chamosite, and manganese-rich pyroxenes, such as pyroxmangite and rhodonite. Andradite, a member of the garnet group, is absent or weakly developed. Stage 2 weakly oxidized skarns may develop abundant prograde wollastonite-calcite rocks, magnetite, and sulfides/sulfosalts.

WEAKLY REDUCED SYSTEMS

Magmatic mineral indicators for weakly reduced igneous rocks include ilmenite plus some magnetite, pyrrhotite, sphene, and graphite. Mineral indicators for weakly reduced mineral districts include ilmenite, graphite, iron-rich chlorite, ankerite, pyrrhotite, and wollastonite.

REDUCED SYSTEMS

Magmatic mineral indicators for reduced igneous rocks can include graphite, ilmenite, hercynite, titanomagnetite, pyrrhotite, rutile, olivine, clinopyroxene, and iron-rich hornblende and iron-rich biotite in gabbros and basalts. Mineral indicators for reduced mineral districts include ilmenite, graphite, hydrocarbon residues, petroleum, methane, cassiterite, annite, iron-rich chlorite, ankerite, pyrrhotite, stephanite, polybasite, pyrargyrite, miargyrite, adularia, rhodochrosite, hedenbergite, bournonite, jamesonite, boulangerite, arsenopyrite, illite, dickite, albite, scorodite, orpiment-realgar, mercury, stibnite, cinnabar, gold, rhodonite, pyroxmangite, blue calcite, and wollastonite.

Examples of dry reduced mineral systems are found in the Phoenix Mountains and the Mazatzal mercury districts. An example of a wet reduced district is the Green Valley gold district south of Payson, associated with the Gibson diorite complex.

Rock System Stages of Magma Series

A four- or five-step sequence of releases can occur during hydrothermal emissions from metaluminous differentiating magmas in the upper crust. This step-wise release of hydrothermal fluids and pressure reductions is accompanied by fracturing of the surrounding host rocks, thus creating stockworks, disseminations, or fault and vein systems. If there is enough water in the system to form a hydrous magma, then the final crystallization product can be a productive metal deposit. As the metal-rich hydrothermal solutions become oversaturated with metals, various ore minerals precipitate, and crystals are deposited in any open space. Expulsion of the hydrothermal metal system results in freezing of any remaining magma. Once released, a given hydrothermal plume will also undergo alteration and metal zoning.

Additional metal specializations are introduced based on the degree of differentiation of a given magma series at the rock system level. Rock systems are shown for metaluminous magma series on the K_2O versus SiO_2 diagram (figure 4.5).

Each hydrous igneous rock system releases its hot hydrothermal incompatible water component, which takes whatever metal is also acting incompatibly at that point in the differentiation sequence. In a hydrous metaluminous magma series, up to five rock systems and corresponding hydrothermal fluid releases may fractionate. An example is shown for the Arizona porphyry copper-related sequence in figure 4.9.

All rock system stages in both hydrous peraluminous and metaluminous systems can produce a hydrothermal fraction. For readability of the mineral district maps, however, only the MCAo systems that are related to the well-known porphyry copper-molybdenum deposits are shown with rock system stages. Mineral systems that are hydrothermal fractionates of that serial sequence are plotted on the map as MCAo2 for early stages, MCAo3 for the economic portion, and MCAo4 for later stages. The economically important porphyry copper-molybdenum systems are associated with the stage 3 biotite granodiorite rock system of hydrous oxidized calc-alkalic metaluminous magma-metal series (see figure 4.9). The much less-economic iron and minor copper-silver deposits are associated with earlier, Stage 2, hornblende-quartz diorite and biotite-hornblende granodiorite differentiates (MCAo2). Later, quartz-sericite-molybdenite-chalcopyrite veins and greisens, polymetallic veins and replacement deposits, and manganese-silver mineralization are associated with granite rock systems, quartz-feldspar porphyry intrusions of Stage 4 affinity (MCAo4). Occasionally, high-level epithermal quartz-alunite-pyrite veins can be associated with highly evolved quartz porphyry dikes of Stage 5 in a subvolcanic environment (MCAo5).

Exploration geologists can benefit from the mineral district maps by using combined geochemistry, mineralogy, production data, pluton geochemistry, and age dates to vector backward or forward toward the economically favorable Stage 3 part of the system.

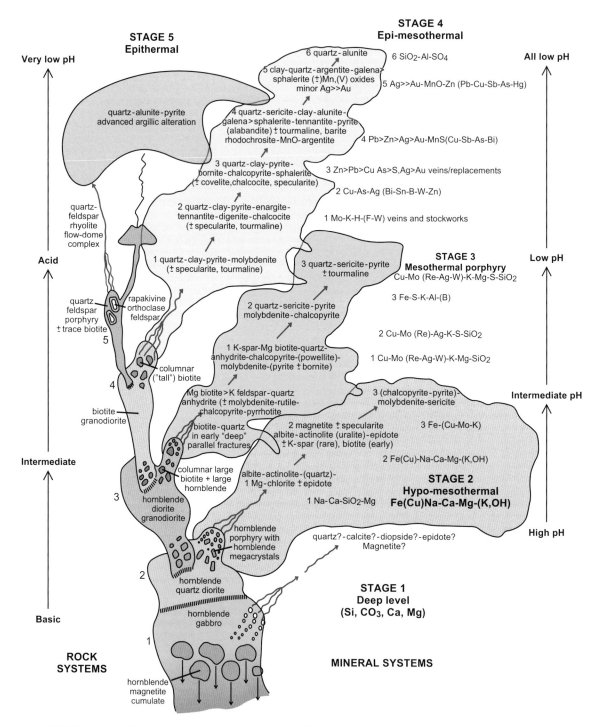

FIGURE 4.9 Fractionation chart for stages in the evolution of a porphyry copper system associated with a metaluminous calc-alkalic oxidized hydrous magma-metal series (MCAoh) supersystem.

Stages of Porphyry Copper Intrusions

Continued intrusions and movements on the main zone of weakness influenced fractionation of the mafic diorite to granodiorite into increasingly more felsic, quartz monzonite and granitic plutons. The stages in a typical MCA porphyry copper are shown on figure 4.9. The Stage 2 pluton is usually relatively mafic, with dark minerals of hornblende and biotite. The Stage 3 granodiorite is usually the more favorable source of the porphyry copper deposits. The later Stage 4 differentiates of quartz porphyry plugs, dikes, and breccias are generally associated with the most felsic differentiates and contain the molybdenite- and bornite-rich zones.

PORPHYRY COPPER DEPOSITS

The most productive ore deposits and source of many mineral species in Arizona are the porphyry copper deposits, such as those at Ajo, Bisbee, Globe-Miami, Mineral Park, Morenci, Ray, Mission-Pima, and Silver Bell. They are generally in the MCAo category and are associated with Stage 3 biotite granodiorite porphyries that contain large euhedral feldspar and biotite crystals set in a fine-grained groundmass of potassium feldspar, biotite, and quartz. The original sulfide copper minerals, including chalcopyrite, pyrite, and lesser amounts of bornite and molybdenite, commonly occur as magmatic or hydrothermally disseminated grains and fracture fillings.

Nonetheless, there are usually larger fractures or fault zones in these deposits that contain open spaces where larger mineral specimens have grown. Examples are breccias at Ajo and Sierrita-Esperanza in Pima County; breccia pipes at Copper Creek with molybdenite, adularia, and red rutile specimens; hydrothermal breccias at San Manuel in Pinal County; and breccia pipe at the Four Metals Mine with apatite and molybdenite specimens in Santa Cruz County. Locally, late-stage copper veins have yielded attractive specimens of the primary minerals. Most notable are the pyritohedral pyrite crystals and the brown and other barytes from the lower Martin Formation at the Magma Mine in Pinal County. Stage 2 of the sequence can also develop some specimen-grade material. Examples are the possible hypogene silver specimens associated with stromeyerite, bornite, chalcocite and chalcopyrite at the Silver King and Richmond Basin Mines that are associated with quartz-diorite intrusions.

Supergene Oxidation and Enrichment

The mineral district maps show the emplacement ages of the primary sulfide mineralization of a given mineral system. It is the later oxidation of these mineral systems, however, that produces the minerals that are popular with mineral collectors. Oxidation periods occur during the latest part of an orogeny, when flat subduction uplifts the deposits to expose them to weathering and erosion. The regional uplifts produced continental-scale unconformities on which typically mature continental sandstones accumulated. Examples are the Tapeats Sandstones and Bolsa Quartzites that accumulated on the Great Unconformity and the Mazatzal Quartzites and correlatives that were deposited on the unconformity between the Yavapai orogeny rocks and younger Mazatzal to Picuris orogeny rocks.

During the Eocene unconformity, Stage 3 portions (MCAo3) of the Laramide porphyry copper sequence underwent oxidation and supergene enrichment. Additional erosional episodes occurred between the Cenozoic orogenies (heavy dark lines on figure 4.10).

Oxide zones above the supergene enrichment zones appear on the surface as the famous iron hat or red thumbprint hematitic gossans (figure 4.11). The red gossan zones represent leach zones where copper is leached and reprecipitated at lower levels as secondary copper oxide minerals.

In the oxide zone of porphyry copper deposits, there is a crude zoning and/or paragenetic ordering in supergene copper minerals that grade downward in the following sequence:

1. Red barren hematitic gossan at the top
2. Blacker, so-called high-relief hematite that is cupriferous
3. Chrysocolla and possible tenorite
4. Copper carbonates, such as azurite and malachite
5. A potentially deeper, earlier zone of brochantite
6. Cuprite—native copper and possible tenorite zone

All these minerals commingle to various extents and may extend deeper along fractures into the other level assemblages. The various layers rest above and are younger than the next deeper layer.

All of the above oxide mineral assemblages rapidly grade down into the chalcocite supergene blanket, which occurs at the interface between the oxide zone and the underlying sulfide zone. Chalcocite enrichment results from acid dissolution of the underlying chalcopyrite-pyrite—dominated hypogene ores.

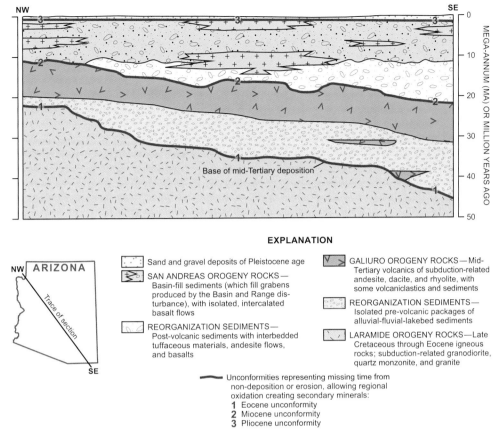

NW   SE

MEGA-ANNUM (MA) OR MILLION YEARS AGO

Base of mid-Tertiary deposition

EXPLANATION

NW ARIZONA

Trace of section

SE

Sand and gravel deposits of Pleistocene age

SAN ANDREAS OROGENY ROCKS—
Basin-fill sediments (which fill grabens
produced by the Basin and Range dis-
turbance), with isolated, intercalated
basalt flows

REORGANIZATION SEDIMENTS—
Post-volcanic sediments with interbedded
tuffaceous materials, andesite flows,
and basalts

GALIURO OROGENY ROCKS—Mid-
Tertiary volcanics of subduction-related
andesite, dacite, and rhyolite, with
some volcaniclastics and sediments

REORGANIZATION SEDIMENTS—
Isolated pre-volcanic packages of
alluvial-fluvial-lakebed sediments

LARAMIDE OROGENY ROCKS—Late
Cretaceous through Eocene igneous
rocks; subduction-related granodiorite,
quartz monzonite, and granite

Unconformities representing missing time from
non-deposition or erosion, allowing regional
oxidation creating secondary minerals:
1 Eocene unconformity
2 Miocene unconformity
3 Pliocene unconformity

FIGURE 4.10 Highly simplified NW-SE cross-section through Arizona's Basin and Range province showing the Eocene, Middle Miocene, and Pliocene unconformities with thick dark lines.

The rich chalcocite veins in the chalcocite enrichment zones were the main
targets of Arizona's early copper mining.

Pyrite content is critical as it supplies sulfur for sulfuric acid, which reacts
with chalcopyrite to precipitate as copper-rich chalcocite under more reducing
conditions at the oxide–primary sulfide interface. Acid leaching of the chalcopyrite component and downward percolation of the copper-bearing solutions
form the above reaction sequence. The paleo-water table provides a critical
reductive interface, which induces precipitation of the chalcocite component
to make the chalcocite blankets.

The upper oxide zones in the Ray, Twin Buttes, and Globe-Miami areas
produced gem-quality chrysocolla. Oxidized, phosphate-rich areas in the
Stage 4 sections have yielded lapidary and gem-quality turquoise, such as
at the former Sleeping Beauty Mine on the west side of Copper Cities Mine
in the Globe-Miami area. The lead-zinc-silver portions of Stage 4 porphyry
copper sequence have locally yielded high-quality specimens of aurichalcite,

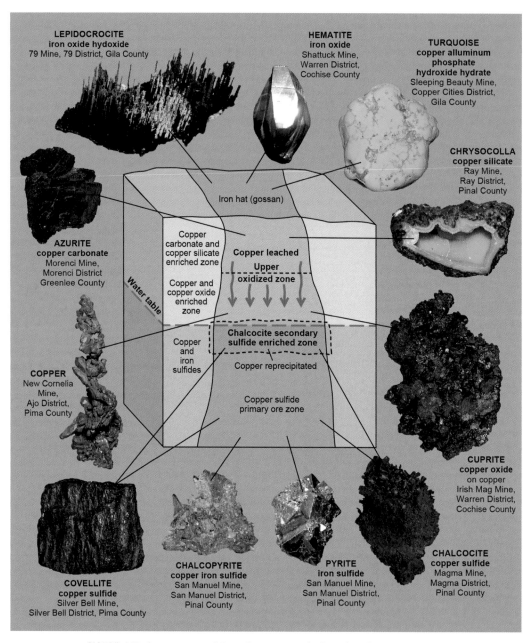

FIGURE 4.11 Supergene enrichment process and mineral zoning.

rosasite, smithsonite, hemimorphite, wulfenite, and mimetite. Of these, the
79 Mine is the most notable Stage 4 occurrence. It is the Stage 4 portions of
oxidized alkali-calcic systems, however, that have yielded the best museum-
quality wulfenite specimens. Red Cloud, Glove, and Defiance Mine wulfenites
are good examples. The same localities have yielded specimens of aurichalcite,
rosasite, cerussite, smithsonite, and hemimorphite.

Exotic Copper Deposits

The late Laramide porphyry copper deposits were uplifted and eroded, during which the chalcopyrite orebodies were weathered, and copper solutions were transported and redeposited as exotic copper deposits. It is likely this would have happened during or just before the early Galiuro orogeny.

Examples of exotic copper deposits near the porphyry copper deposits from which they were derived include the Table Mountain, Red Hills, and Copper Butte deposits from the Ray Mine in Pinal County; Emerald Isle from Mineral Park in Mohave County; and Ajo Cornelia from the New Cornelia Mine at Ajo in Pima County.

Minerals reported from these exotic copper deposits typically include azurite, malachite, chrysocolla, quartz, and gypsum. Less typical minerals reported from the Table Mountain District include quartz, austinite, conichalcite, wulfenite, vanadinite, willemite, dioptase, baryte, mimetite, mixite, plancheite, and duftite.

Summary

Since about 1,800 million years ago, mineralizing processes have been more or less continuous in Arizona. Primary minerals were generally emplaced during ten major orogenic periods and were oxidized during plate tectonic reorganization episodes between the orogenies. In addition, ultradeep hydrothermal processes were active between orogenic episodes and locally emplaced massive amounts of hydrothermal brines.

Arizona minerals result from many different metallogenic events and metal and magmatic sources. For mineral collectors and exploration geologists, the county mineral district maps found in appendix A can serve as a guide to mineral localities where similar mineral suites are likely to be found.

TABLE 4.2 Orogenies, ages, mineral districts, and mineralogy in Arizona through time (youngest at top, oldest at base)

Orogeny	Phase	Age	Age (Ma)	MMS class	Rocks, Strata, Magmatism	Resources	Mineral districts	Minerals
Modern	postmine	Holocene	0.0117–0	none	basaltic volcanism	tourism		gypsum, sulphur, alunite, alum grp., baryte, celestine, fluorite, hematite, opal, K alum, hydrokenoralstonite (ralstonite), voltaite, ferrohexahydrite, pickeringite
Modern	postmine	Holocene	0.0117–0	none	fumaroles	fumaroles	Sunset Crater	
Modern	postmine	Holocene	0.0117–0	none	hot springs / geothermal	geothermal energy	Indian Springs, Cactus Flat Hot Springs, Hooker's Hot Springs	gypsum, travertine, halite, silica, clays
Modern	postmine	Holocene	0.0117–0	none	postmine minerals	mineral specimens	(a) Jerome Mine fires; (b) Ray timbers (Cu); (c) sulfate efflorescences: 79 Mine, Flux; (d) carbonate efflorescences: Bisbee (SW Mine)	(a) Jerome mine-fire minerals ("jerometite," butlerite, lawsonite, guildite, ransomite, alunogen, copiapite, coquimbite, voltaite); (b) native copper; (c) copiapite, pickeringite, halotrichite, woodwardite, melanterite, gypsum; (d) aragonite, calcite, travertine/flowstone
Modern	postmine	Holocene	0.0117–0	none	rotting plants	mineral specimens	St. David, SE Gila Bend Mtns. rotting saguaros, San Francisco Mtns., north Flagstaff buried tree trunks	humboldtine (oxalate) mineral suite (whewellite, weddellite, monohydrocalcite, nitratine, glushinskite), flagstaffite
Modern	postmine	Holocene	0.0117–0	none	modern caves	tourism	Kartchner Caverns, Colossal Cave	calcite (cave fms.), nitrocalcite, hematite, brushite, hydroxylapatite, gypsum
Modern	postmine	Holocene	0.0117–0	none	lava tube cave	tourism	San Francisco Peaks ice cave	ice
Modern	postmine	Holocene	0.0117–0	none	fluvial sand and gravel	sand, gravel	Salt River, Orange Grove Road	sand, gravel
Modern	postmine	Holocene	0.0117–0	none	exotic copper	copper oxides	Kelvin, Tinhorn Wash, Mineral Creek	chrysocolla, malachite, tenorite
Modern	postmine	Holocene	0.0117–0	none	placer gold	gold	Greaterville, Gold Gulch, Mariquita, La Paz, Rich Hill	gold

Modern	postmine	Holocene	0.0117–0	none	placer heavy minerals	iron	Tortilla Mtns.	magnetite, ilmenite, sphene, apatite, zircon, diopside, quartz, biotite, feldspar
San Andreas (oblique breakup of SW America transform margin)	Basin and Range, transtensional domain	Quaternary	2.6–0.012	dry MNA	suballkaline basalt-rhyolite volcanism	cinders	San Francisco volcanic field: SP Crater and flow (70,000 ybp), Merriam Crater and flow (20,000 ybp), O'Leary rhyolite dome (Pleistocene)	pyroxene, plagioclase
San Andreas	transtensional domain	Quaternary	2.6–0.012	dry MNA	alkaline basalts and diatremes (San Carlos basanite, 0.6), Vulcan's throne (0.01), late San Bernardino tuff maars (1–0.2)	gem olivine, cinders	San Carlos maar, Hoover Dam, Vulcan's Throne, San Bernardino tuff maars	gem olivine, kaersutite, plagioclase, pyroxene, spinel
San Andreas	transtensional domain	Quaternary	2.6–0.012	NI UDH	saline lakes, Verde Fm.	salt	Luke salt; Camp Verde	halite, calcite pseudomorphs after glauberite
San Andreas	transtensional domain	Quaternary	2.6–0.012	NI UDH	Bowie chabazite	bedded zeolites	Bowie chabazite	chabazite, chabazite-Na (herschelite), erionite, clinoptilolite
San Andreas	transtensional domain	Quaternary	2.6–0.012	NI UDH	Silica hot springs (Quiburis Fm.)	diatomite	White Cliffs diatomite	diatomite
San Andreas	transtensional domain	Quaternary	2.6–0.012	NI UDH	shallow saline lakes and warm springs (St. David Fm.)	gypsum	St. David, Benson	gypsum
San Andreas	transtensional domain	Quaternary	2.6–0.012	NI GP	alluvial stream beds	placer gold	Rich Hill, Vulture, King Tut, La Paz–Quartzsite, Greaterville gold placers	placer gold in alluvium
San Andreas	transtensional domain	Quaternary	2.6–0.012	NI GP	alluvial stream beds	placer magnetite (Ti)	Tom Mix monument and Owl Head Buttes, Omega placer magnetite (ilmenite) sands	magnetite (ilmenite, titanite [sphene], apatite, zircon, diopside, quartz, biotite, and feldspar) sand
San Andreas	transtensional domain	Quaternary	2.6–0.012	NI	lake beds	clay	Willcox Playa	clay, illite?

(continued)

TABLE 4.2 (*continued*)

Orogeny	Phase	Age	Age (Ma)	MMS class	Rocks, Strata, Magmatism	Resources	Mineral districts	Minerals
San Andreas	transtensional domain	Quaternary	2.6–0.012	NI	Meteor Crater	tourism	Meteor Crater (Canyon Diablo meteorite), Tucson meteorite	native iron, diamond, stishovite, coesite, schreibersite, graphite, high albite, kosmochlor, rodderite, lonsdaleite, lawrencite, richterite, chromite
San Andreas	intra-San Andreas reorganization	Pliocene/Pleistocene unconformity	3–2	UDH	Hot springs, erosion, and oxidation	Silica Mn baryte fluorite, travertine, Li hot springs, veins	Metate, N. Plomosa MnBaF veins, Spectrum SiF veins	Mn oxides, baryte, fluorite, quartz/chalcedony
San Andreas	intra-San Andreas reorganization	Pliocene/Pleistocene unconformity	3–2	UDH	Hot springs, erosion, and oxidation	bedded zeolites	Big Sandy	clinoptilolite, mordenite, analcime, K-feldspar, quartz, opal, phillipsite, erionite
San Andreas	intra-San Andreas reorganization	Pliocene/Pleistocene unconformity	3–2	UDH	Hot springs, erosion, and oxidation	diatomite	White Cliffs	diatomite, silica
San Andreas	intra-San Andreas reorganization	Pliocene/Pleistocene unconformity	3–2	UDH	Hot springs, erosion, and oxidation	saponite bentonite	Burro Creek	saponite, bentonite, smectite, magnesite, opal
San Andreas	intra-San Andreas reorganization	Pliocene/Pleistocene unconformity	3–2	UDH	Hot springs, erosion, and oxidation	clay	Tolleson clay	illite?
San Andreas	intra-San Andreas reorganization	Pliocene/Pleistocene unconformity	3–2	UDH	Hot springs, erosion, and oxidation	exotic copper	Emerald Isle	chrysocolla, malachite, tenorite melaconite
San Andreas	Basin and Range, transtension domain	Middle and Late Neogene to Quaternary (Middle Miocene to Quaternary)	13–2.6	dry MNA, MQA, MAC, MCA	early subalkaline basalts and local andesites and rhyolites (Hickey basalts) (14–8), early San Bernardino basalt flows (3.2–1), San Francisco (5–0.01), Sentinel, and White Mtns. volcanic field	cinders	San Francisco volcanic field, San Bernardino (Geronimo) and Sentinel volc. fields	labradorite porphyrocrysts

San Andreas	Basin and Range, transtensional domain	Middle and Late Neogene to Quaternary (Middle Miocene to Quaternary)	13–2.6	dry MNA, MQA, MAC, MCA	Horseshoe Dam basalt (14.6)	amygdaloid zeolites	Horseshoe Dam zeolites	zeolites, gyrolite
San Andreas	Basin and Range, transtensional domain	Middle and Late Neogene to Quaternary (Middle Miocene to Quaternary)	13–2.6	dry MNA, MQA, MAC, MCA	Verde Fm. tuffs	bedded zeolites	Horseshoe clinoptilolite	clinoptilolite
San Andreas	Basin and Range, transtensional domain	Middle and Late Neogene to Quaternary (Middle Miocene to Quaternary)	13–2.6	dry MNA, MQA, MAC, MCA	hot springs MnBaF	hot springs, Mn-baryte-fluorite veins	Metate, Steamboat?, Plomosa Pass	Mn oxide, baryte, fluorite, silica
San Andreas	Basin and Range, transtensional domain	Middle and Late Neogene to Quaternary (Middle Miocene to Quaternary)	13–2.6	dry MNA, MQA, MAC, MCA	hot spring Li	hot spring Li	Lyles hectorite	hectorite
San Andreas	Basin and Range, transtensional domain	Middle and Late Neogene to Quaternary (Middle Miocene to Quaternary)	13–2.6	dry MNA, MQA, MAC, MCA	Hopi Buttes diatreme field	bentonite, U	Hopi Buttes, Cheto bentonite	bentonite (high Ca montmorillonite; schröckingerite)
San Andreas	Basin and Range, transtensional domain	Middle and Late Neogene to Quaternary (Middle Miocene to Quaternary)	13–2.6	dry MNA, MQA, MAC, MCA	Quiburis, Verde Fm.	gypsum	Winkelman, Camp Verde, and St. David gypsum	gypsum

(continued)

TABLE 4.2 (*continued*)

Orogeny	Phase	Age	Age (Ma)	MMS class	Rocks, Strata, Magmatism	Resources	Mineral districts	Minerals
San Andreas	Basin and Range, transtensional domain	Middle and Late Neogene to Quaternary (Middle Miocene to Quaternary)	13–2.6	dry MNA, MQA, MAC, MCA	saline brines enrichment	salt	Luke salt; Red Lake salt	halite, anhydrite
San Andreas	Basin and Range, transtensional domain	Middle and Late Neogene to Quaternary (Middle Miocene to Quaternary)	13–2.6	dry MNA, MQA, MAC, MCA	additional supergene enrichment	exotic Cu	Emerald Isle exotic Cu, Mineral Creek, Bisbee oxidation	chrysocolla, malachite, azurite, gypsum
Reorganization	reorganization	Middle Neogene (Middle Miocene to Late Miocene)	16–10	NI	Gila Congl, Big Dome Fm, Muddy Creek Fm, Laguna Mtns. congl.	gold placers	Laguna Mtns., King Tut, Vulture Au placers	gold nuggets
Reorganization	reorganization	Middle Miocene to Late Miocene	16–10	NI	Bear Canyon Fm., exotic landslide blocks in the above	gold-bearing exotic landslide blocks	Ocotillo (Calif.), Dutchman, Ariz.	gold, hematite, chrysocolla, chalcopyrite
Reorganization	reorganization	Middle Miocene to Late Miocene	16–10	NI	pre-Hickey (Verde) gravels	exotic Cu	Dundee-Arizona (E of Jerome), Cactus?, Carlota?, Emerald Isle?	chrysocolla, malachite, azurite, chalcocite, hematite
Reorganization	reorganization	Middle Miocene to Late Miocene	16–10	NI		Mn-baryte-Si veins	Steamboat Mtn.?	Mn oxides, baryte
Reorganization	reorganization	Middle Miocene to Late Miocene	16–10	NI	oxide zones of Galiuro orogeny PbAg and Au deposits	supergene Ag, mineral specimens	Commonwealth Mine, Tiger, Aravaipa, Red Cloud, Grand Reef, Castle Dome, Rowley	chlorargyrite, wulfenite, mimetite, descloizite, Tiger oxidation suite (caledonite)

Galiuro	Late	Late Paleogene to Middle Neogene (Late Oligocene to Late Miocene)	28–10	wet MQA	lacustrine Mn	Mn	Artillery Mtns.	Mn oxides, baryte, fluorite
Galiuro	Late	Late Oligocene to Late Miocene	28–10	wet MQA	rhyolitic tuffaceous sediments of Anderson Mine	U(V)	Anderson Mine	carnotite, weeksite, chalcedony
Galiuro	Late	Late Oligocene to Late Miocene	28–10	wet MQA	tuffaceous lacustrine sediments	Sr	Aguila	celestine
Galiuro	Late	Late Oligocene to Late Miocene	28–10	wet MQA	Batamote and Golden Deer volcanics (18–14)	zeolite amygdule linings	zeolites in amygdule vugs in volcanics	mordenite, ranciëite, heulandite, ferrierite, erionite, analcime, thomsonite, clinoptilolite
Galiuro	Late	Late Oligocene to Late Miocene	28–10	wet MQA	volcanics and local epizonal stocks (Oatman andesite, Moss pluton)	Au-Ag epithermal	Oatman, Allison Camp	fluorite, gold, baryte, wulfenite, Tiger suite, dioptase, adularia, quartz
Galiuro	Late	Late Oligocene to Late Miocene	28–10	wet MQA	Painted Rock volcanics, high-K rhyolite dikes (22–14)	epi-mesothermal AuAg (base metal, F)	Mammoth, Rowley, Rawhide	gold
Galiuro	Late	Late Oligocene to Late Miocene	28–10	wet MQA	alkaline mafic dikes; and sills (25–15?)	specularite CuAuAgF	Harquahala, Congress, Rich Hill?	specularite, Tiger anomalous suite (diaboleite etc.), linarite, plancheite, dioptase, gold
Galiuro	Late	Late Oligocene to Late Miocene	28–10	wet MQA	Mudersbach pluton (20)	CuAg	Mudersbach Mine	chalcopyrite, specularite
Galiuro	Late	Late Oligocene to Late Miocene	28–10	wet MQA	minette diatremes of Navajo volcanic field (~26–19), Sullivan Buttes latite (25–21); laccolith in Chuska Mtns.	oil, gas, helium	Navajo volcanic field minette diatremes, Sullivan Buttes latite, Dineh-bi-Keyah, helium, Four Corners	phlogopite

(continued)

TABLE 4.2 (continued)

Orogeny	Phase	Age	Age (Ma)	MMS class	Rocks, Strata, Magmatism	Resources	Mineral districts	Minerals
Galiuro	Middle	Late Paleogene and Early Neogene (Late Oligocene to Early Miocene)	28–15	wet MAC	ignimbritic volcanics and plutons	silica / fire agate, moss / plume agate	Deer Creek, Agua Fria River area, Willow Spring Ranch, Black Hills	silica/agate filling amygdules
Galiuro	Middle	Late Oligocene to Early Miocene	28–15	wet MAC	ignimbritic volcanics and plutons	zeolites filling amygdules	Malpais Hill, Horseshoe Dam, Blue Mtns., New Water Mtns.	zeolites filling amygdules, garronite, lévyne, erionite, analcime, chabazite, clinoptilolite, okenite
Galiuro	Middle	Late Oligocene to Early Miocene	28–15	wet MAC	NW Ash Peak rhyolite domes (20)	tin / topaz rhyolites	Ash Peak rhyolite domes, Upper Apsey Creek, Kaiser Springs at Burro Creek	cassiterite, topaz, brookite, pseudobrookite, hematite, garnet
Galiuro	Middle	Late Oligocene to Early Miocene	28–15	wet MAC	ignimbritic volcanics and plutons	epithermal Ag(AuFMn)	Silver (Red Cloud), Castle Dome, Ash Peak, Steeple Rock, Ramsey	silver, chlorargyrite, wulfenite, fluorite, proustite, argentite
Galiuro	Middle	Late Oligocene to Early Miocene	28–15	wet MAC	rhyolitic volcanics S of Superior	mesothermal PbZnAg (AuCu)	Stanley, Aravaipa	galena, sphalerite, silver, linarite, wulfenite, Aravaipa oxidative mineral suite (laurelite, grandreefite)
Galiuro	Early	Late Paleogene and Early Neogene (Oligocene to Early Miocene)	30–21	wet MCA	calc-alkalic volcanics and plutons	epithermal Au(CuWAg)	Kofa, South Mtn., Gila Bend Mtns., Mexican Hat	gold, todorokite, chalcophanite, aurorite, pyrolusite, adularia, quartz
Galiuro	Early	Oligocene to Early Miocene	30–21	wet MCA	calc-alkalic volcanics and plutons	epithermal hot spring HgSb(AuAg)	Cinnabar (Hg), Yellow Devil (Sb)	cinnabar, stibnite, tetrahedrite, quartz

Event	Process	Age	Range (Ma)		Description	Commodity	Localities	Minerals
Reorganization (change from flat subduction to steepening subduction)	erosion / supergene enrichment of Cu deposits	Middle to Late Paleogene (Oligocene)	35–28	none	fluvial fangls. (Whitetail Congl. and Helmet fangl., Locomotive fangl.)	exotic copper in fangl.	Ajo Cornelia (from Ajo), Copper Butte (from Ray)	chrysocolla, tenorite, malachite, gypsum
Reorganization (change from flat subduction to steepening subduction)	erosion / supergene enrichment of Cu deposits	Middle to Late Paleogene (Oligocene)	35–28	none	supergene copper oxide deposits above Stage 3 porphyry Cu	(a) barren leach zone; (b) supergene copper oxides; (c) supergene chalcocite enrichment blankets/zones; (d) oxidized skarns Cu	(a) Morenci, Ray, Globe-Miami, Silver Bell, Ajo; (b) Ray, Morenci, Miami-Inspiration, Copper Cities, Bagdad; (c) Ray, Silver Bell, Morenci, Castle Dome, Bagdad; (d) oxidized skarns: Christmas, Rosemont, Twin Buttes, Mission-Pima	(a) leach zone: hematite, goethite, tenorite; (b) Cu oxide zone: chrysocolla, azurite, malachite, conichalcite, native copper, cuprite; (c) supergene enriched zone: chalcocite, covellite, djurleite; (d) oxidized skarn: azurite, malachite, chrysocolla, diopside, celadonitic clay, kinoite suite
Reorganization	erosion / supergene enrichment	Oligocene	35–28	none	supergene Cu deposits above Stage 4 porphyry Cu	Stage 4 Cu oxides	Sunnyside, Four Metals, Red Mtn., 79 Mine, Stockton Hill (Wallapai), Saginaw Hill	Stage 4: turquoise, kaolinite, alunite, smithsonite, hemimorphite, anglesite, cerussite, luetheite, chenevixite, libethenite, lindgrenite, cornubite, pseudomalachite, tsumebite, cornwallite, cornetite
Reorganization	erosion / supergene enrichment	Oligocene	35–28	none	oxidation of Laramide PbZnAg deposits	Ag	Tombstone, Tyndall (Glove), Empire (Total Wreck), Richmond Basin, Cerro Colorado	chlorargyrite, native silver, cerussite, mimetite, pyromorphite, anglesite, wulfenite, aurichalcite, calcite, rosasite, vanadinite, mottramite
Reorganization	erosion / supergene enrichment	Oligocene	35–28	none	tuffaceous sediments in lake beds	U	Cardinal Ave., Mineta U shows	carnotite at Cochise
Reorganization	erosion / supergene enrichment	Oligocene	35–28	none	lake beds	clay	Pantano clay	high Al fire clay (kaolinite), copper, azurite, malachite, etc., and hematite chalcocite
Reorganization	erosion / supergene enrichment	Oligocene	35–28	none	reoxidation of Bisbee?	reoxidation of older Cu deposits	reoxidation Bisbee (Copper Queen and Cochise)	supergene mineral suite at Copper Queen (cuprite)

(continued)

TABLE 4.2 (*continued*)

Orogeny	Phase	Age	Age (Ma)	MMS class	Rocks, Strata, Magmatism	Resources	Mineral districts	Minerals
Reorganization	erosion / supergene enrichment	Oligocene	35–28	none	gold placer	paleoplacer gold	Vinegaroon?	gold
Reorganization	erosion / supergene enrichment	Oligocene	35–28	none	landslide block of porphyry copper mineralization	porphyry Cu mineralization in landslide block	Dynamite	mixed copper sulfides and oxides
Laramide (flattening subduction—to India collision with Asia)	Late	Paleogene (Paleocene to Eocene to Early Oligocene)	65–29	UDH	Orocopia Schist / hydrous metamorphism / dewatering	actinolite	tectonic windows: N. Plomosa Mtns., Cemetery Ridge, Neversweat Ridge, Yuma Wash	albite, graphite, actinolite, talc, tremolite, clinochlore, heazlewoodite, awaruite, serpentine grp.
Laramide	Late	Paleocene to Eocene to Early Oligocene	65–29	UDH	serpentine ultramafic diatremes (40–29)	pyrope garnet	serpentine ultramafic microbreccia diatremes: Garnet Ridge, Buell Park	2 serpentine grp., lawsonite, jadeite, pyrope garnet, Na garnet, rutile
Laramide	Late	Paleocene to Eocene to Early Oligocene	65–29	UDH	none	oil and gas?	Four Corners, Walker Creek, San Anido, Black Rock	oil, gas
Laramide	Late	Paleocene to Eocene to Early Oligocene	65–29	UDH	none	He		He
Laramide	Late	Late Cretaceous (Maastrichtian) to Late Paleogene (Eocene)	70–35	very wet PC	2-mica, garnet-muscovite hydrous granitic stocks, sills, dikes	Au disseminations and quartz veins	Gold Basin, Lost Basin, Fortuna, Vulture	gold, small wulfenite, kyanite, sillimanite, dumortierite
Laramide	Late	Early to Late Paleogene (Paleocene to Eocene)	60–40	very wet PCA	2-mica, garnet-muscovite hydrous granitic stocks, sills, dikes	W veins	Oracle (Wilderness Granite), Bluebird, Three Musketeers	wolframite grp., scheelite, beryl, kyanite, sillimanite, dumortierite

Laramide	Middle	Paleocene	65–55	wet MCA	granodiorite-quartz-monzonite porphyry stocks, NE to ENE-striking dike swarms	porphyry Cu-Mo-Ag (Au,Re)	porphyry Cu-Mo-Ag	chalcopyrite, molybdenite, pyrite, bornite
Laramide	Middle	Paleocene	65–55	wet MCA	granodiorite-quartz-monzonite porphyry stocks, NE to ENE-striking dike swarms	Fe(Cu)	Kelvin, Bagdad (Mammoth), Wooley breccia pipes	magnetite, Mg-Fe chlorite epidote, chlorapatite (minor chalcopyrite, molybdenite)
Laramide	Middle	Paleocene	65–55	wet MCA	granodiorite-quartz-monzonite porphyry stocks, NE to ENE-striking dike swarms	CuMo (AgAuRe)	Ajo, Ray, San Manuel, Mineral Park, Pima, Bagdad, Silver Bell, Globe-Miami, Morenci, Resolution, Magma vein; skarns: Christmas, Twin Buttes, Rosemont-Helvetia, Mission-Pima	chalcopyrite, molybdenite, pyrite, (phlogopite, clinochlore, K-feldspar, anhydrite), bornite, epidote, garnet; skarns: chalcopyrite, andradite-grossular garnet, diopside (magnetite bornite)
Laramide	Middle	Paleocene	65–55	wet MCA	granodiorite-quartz-monzonite porphyry stocks, NE to ENE-striking dike swarms	CuMoAg (Pb-ZnAuAsMnU) porphyry MoCu	79 Mine, Finch, Hardshell, Trench, Tennessee, Hillside, porphyry MoCu, Crown King, Pine Flat?	tennantite, chalcopyrite, bornite, enargite, molybdenite, rutile, adularia, pyrite, alunite, galena, sphalerite, alabandite, baryte, 79 Mine suite (wulfenite, vanadinite, auricalcite)
Laramide	Early	Late Cretaceous	75–65	wet MAC	quartz monzonite porphyry stocks; ash flows	Ag (AuPbZnSb)	epithermal pipes and veins: Silver King, Richmond Basin, Cerro Colorado	silver, tetrahedrite, proustite, galena
Laramide	Early	Late Cretaceous	75–65	wet MAC	quartz monzonite porphyry stocks; ash flows	Pb-Zn-Ag (CuAuTe)	mesothermal veins and replacements Tombstone, Tyndall (Glove), Washington Camp, Empire, Ruby, Salero	galena, sphalerite, alabandite, tetrahedrite, hessite, empressite, silver, enargite, altaite, Tombstone Te suite, wulfenite, vanadinite, mottramite, linarite
Laramide	Earliest	Late Cretaceous	89–75	wet MQA	volcanics	epithermal Cu-Au—Ag (Pb,Zn)	Old Yuma Mine, Golden Rule	gold, galena, cerussite, fornacite, mottramite, wulfenite, vanadinite

(continued)

TABLE 4.2 (continued)

Orogeny	Phase	Age	Age (Ma)	MMS class	Rocks, Strata, Magmatism	Resources	Mineral districts	Minerals
Laramide	Earliest	Late Cretaceous	89–75	wet MQA	small stocks (Mine Canyon)	mesothermal CuAg(Mo)	Mine Canyon	chalcopyrite, copper carbonates, oxides
Laramide	Earliest	Late Cretaceous (Coniacian to Santonian)	89–75	UDH	humic coal swamps (Wepo Fm.)	humic coal	Wepo Fm. coals in Black Mesa	coal
Laramide	Earliest	Late Cretaceous (Coniacian to Santonian)	89–75	UDH	Mulatto Tongue of Mancos = OAE 3 (87) Black Mesa	shale oil?	hydrocarbon-rich black shales? in Mulatto Tongue of Mancos Fm.	kerogenous shale?
Unconformity	Turonian unconformity	Late Cretaceous (Cenomanian to Turonian)	90–89.5	UDH or MQA?	erosion, no volcanism Gallup Sandstone			
Sevier (constant dip subduction)	Late	Late Cretaceous (mid to upper Turonian)	93–91	MQA, UDH	Toreva Fm.	U in sandstone V:U ratio = 1:1 to 5:1	Black Mtn. (U)	primary-secondary: carnotite; tyuyamunite, metatyuyamunite, hewettite, vanadium clay, metahewettite, melanovanadite, kerogenous material
Sevier	Middle	Late Cretaceous (Cenomanian to Turonian)	100–89	UDH	basal Mancos above Dakota Sandstone. In N AZ has OAE 2 (94.0–94.8)	coal	coal in Dakota Sandstone, Deer Creek coal, Pinetop coal	coal
Sevier	Middle	Cenomanian to Turonian	100–89	UDH		black shale hydrocarbons	basal Mancos Fm. at Black Mesa	kerogenous shale?
Sevier	Middle	Cenomanian to Turonian	100–89	UDH		fire clay	Pinetop kaolinite	kaolinite
Sevier	Early	Early Cretaceous (Aptian to Albian)	135–100	UDH?	Mural Limestone	calcite flux from Paul Spur	Paul Spur / Mural Limestone	calcite

Sevier	Early	135–100	Aptian to Albian	UDH?	Martyr Mbr. of Morita Fm. OAE 1b (~110) and 1a? (~125)	reoxidation of Bisbee	Bisbee-Cochise orebody	illite, alunite, hematite, chalcocite, covellite
Sevier	Early	135–100	Aptian to Albian	UDH?	reoxidation Bisbee (110–106)	(a) gossan; (b) CuAgAu oxide carbonate assemblage in limestones; (c) CuAg in red iron hat cap in Pinal Schist	(a) hematite gossan above Cochise orebody; (b) Cu oxide zones in Cochise, Copper Queen Mine (Bisbee), SW Mine; (c) Cochise chalcocite blanket orebody (Bisbee)	(a) limonite; (b) cuprite, tenorite, copper, malachite, azurite; (c) hematite, covellite, chalcocite
Reorganization	erosion surface	160–135	Late Jurassic to Early Cretaceous (Callovian to Barremian)	none	Glance Conglomerate and equivalents deposited on unconformity			secondary zones in Cochise orebody: malachite, chrysocolla, brochantite, copper oxides; secondary zones in limestone replacement deposits: azurite, malachite, cuprite, copper, chrysocolla, smithsonite, turquoise, red hematite, hemimorphite, calcite, aragonite, minor wulfenite
Reorganization	erosion surface	160–135	Callovian to Barremian	none	Glance Conglomerate	exotic Cu	exotic Cu deposits	
Reorganization	erosion surface	160–135	Callovian to Barremian	none	Glance Conglomerate	gold placers	gold paleoplacer? in Glance Congl. at Gold Gulch, Bisbee	gold flakes
Reorganization (from Nevadan subduction to Sevier subduction)	reorganization / failed rifting	154–132	Late Jurassic (Callovian to Oxfordian)	UDH (MQA)	U breccia pipes	U in breccia pipes: U-V-Cu (Ni-Co)	Orphan (146), Hermit (134), Arizona 1 (145), EZ-2 (147), Pigeon (154), Ridenour	pyrite, marcasite in caps; uraninite, bravoite, sphalerite, chalcopyrite, galena, siegenite, gypsum, calcite, baryte, kerogen; secondary: malachite, azurite, goethite, hematite, roscoelite, tyuyamunite, metatyuyamunite, volborthite, calciovolborthite, conichalcite, vésigniéite, naumannite, argentite

(continued)

TABLE 4.2 (continued)

Orogeny	Phase	Age	Age (Ma)	MMS class	Rocks, Strata, Magmatism	Resources	Mineral districts	Minerals
Reorganization / failed rifting	reorganization / failed rifting	Late Jurassic (Callovian to Oxfordian)	165–154	dry MAC; dry MQA; UDH	mafic alkaline volcanics (154) in basal McCoy Mtns. Fm. (Yellowbird Mbr.); Kerogenous shales; Onion Saddle Fm. (Callovian/Oxfordian); upper black shales of upper Crystal Cave Frm. (Callovian/Oxfordian); mafic volc.; (black shale N. Chiricahua Mtns.)	C, later (~80) turned to graphene by thrust burial of Maricopa thrust in Granite Wash Mtns.	Yellowbird graphene; Onion Saddle–Crystal Cave shale oil?	graphite, graphene, graphane, muscovite
Reorganization / failed rifting	reorganization / failed rifting	Late Jurassic (Callovian to Oxfordian)	165–154	dry MQA		sedimentary exhalative Mn	Cow Springs Mn	Mn oxides
Reorganization / failed rifting	reorganization / failed rifting	Late Jurassic (Callovian to Oxfordian)	165–154	dry MQA		CuAg	White Mesa CuAg	chrysocolla
Nevadan (arc assembly)	Late Nevadan	Late Jurassic	175-155	Very wet PC	Mt. Benedict aplites cut Mt. Benedict Quartz Monzonite	Au	Nogales	gold
Nevadan	Late Nevadan	Late Jurassic	175–155	very wet PCA	Las Guijas alaskite (174); Kitt Peak? (165–160); Juniper Flat gran. (175–171)	W veins	Las Guijas, Juniper Flat (175.2)	wolframite grp., hübnerite, scheelite, fluorite, muscovite, (minor galena, magnetite, chalcopyrite, gold)
Nevadan	Late Nevadan	Late Jurassic	160–155	very wet PAC?	Possibly leucogranites in western AZ	kyanite-W?	Tung Hill?, Dome Rock kyanite?	dumortierite, rutile, scheelite
Nevadan	Middle Nevadan W AZ	Middle and Late Jurassic	173–155	wet MAC	W AZ: volcanics, plutonics (Ko Vaya Super unit), Cargo Muchacho Super unit (173–161), Canelo Hills volcanics	PbZnAg	Comobabi Mtns., Cababi (Mildren-Steppe Mine)	galena, sphalerite, tetrahedrite (minor chalcopyrite, wulfenite, aurichalcite, rosasite, linarite), Mildren-Steppe mineral suite

Orogeny	Region	Age	Age range	Environment	Rock unit	Deposit type	Camp/Location	Minerals
Nevadan	Middle Nevadan W AZ	Late Jurassic	189–155	wet MQA	W AZ: Diablo alkali granite suite (163–158)	epithermal quartz veins: AuCu (AgWPb)	La Cholla, Granite Mtn. and Middle Camp, Copper Chief in northern Trigo	quartz, sericite, pyrite, gold, minor fluorite, tetrahedrite, chalcopyrite, cinnabar?
Nevadan	Middle Nevadan W AZ	Early and Middle Jurassic	189–163	wet MQA	Middle Camp Quartz-Monzonite suite (164–167.8)	disseminated pyritic gold: alunite, pyrophyllite, (PGE)	Sugarloaf (Big Bertha), Goodman	gold, specular hematite, quartz, alunite, pyrophyllite, kyanite, fluorite (chalcopyrite), sillimanite, zunyite (galena, tourmaline)
Nevadan	Middle Nevadan W AZ	Early and Middle Jurassic	189–170	wet MQA	Araz Wash syenodiorite suite (173.3–170.6), Dome Rock volcanic suite (189–169), Ko Vaya Super unit	quartz-calcite vein, Au (AgCu)	Jaeger	gold, pyrite, iron oxides (chalcopyrite), bornite, chalcocite
Nevadan	Early and Middle Nevadan SE AZ	Early Jurassic	191–175	wet MAC	SE AZ: Ox Frame Volcanics, Mt. Wrightson trachyandesite, Squaw Gulch monzonite, Cobre Ridge Tuff, Huachuca Quartz Monzonite (180–167), Gleeson Quartz Monzonite (191.3)	PbZnAg (CuAu)	Mesothermal: Gleeson (South Turquoise) (191.3), Hartford (180–167)	galena, sphalerite, (chalcopyrite), tetrahedrite, pyrite, cerussite, anglesite, smithsonite, chrysocolla, willemite, descloizite, turquoise, wulfenite, rosasite, aurichalcite
Nevadan	Early and Middle Nevadan SE AZ	Early Jurassic	201–191	wet MQA	monzonitic plutonic rocks, Sacramento stock complex at Bisbee (199–201); W AZ: Yuma King plutonic complex	porphyry CuAu, PGE	Warren (Bisbee), Yuma King	primary zone in replacement zones in Paleozoic carbonates: proximal chalcopyrite, bornite; more distal: altaite, galena, sphalerite, distal specularite

(continued)

TABLE 4.2 (continued)

Orogeny	Phase	Age	Age (Ma)	MMS class	Rocks, Strata, Magmatism	Resources	Mineral districts	Minerals
Nevadan	Early Nevadan N AZ	Late Triassic and Early Jurassic	237–182	wet MQA	tuffaceous continental clastics in Chinle Fm. (237–202), Petrified Forest Mbr. (216–211); monzonites to quartz-feldspar porphyries in adjacent Goodsprings, Nevada (217±4)	U-V-Cu (Ni-Co)	Orphan breccia pipe (182), Hack 3 pipes (190), Hack 2 pipes (186), Kanab (204), EZ1 pipe (208)	uraninite, chalcopyrite, galena, baryte, calcite, gypsum, marcasite, dolomite, siegenite, enargite, digenite, covellite, bornite, millerite, coffinite, gersdorffite, chalcocite, celestine; secondary: brochantite, malachite, torbernite, tyuyamunite, zeunerite, zippeite, bieberite
Nevadan	Early Nevadan N AZ	Late Triassic	237–201	wet MQA	monzonites to quartz-feldspar porphyries in nearby Goodsprings, Nevada (217±4)	2 U-Cu-V (C)	hosted in Chinle Fm. U deposits: Monument Valley, Petrified Forest, Cameron (~210); Promontory (225) hosted in lower Supai Fm. / Unit B1	in Shinarump Congl., (a) primary: uraninite, quartz, bornite, pyrite, sphalerite; (b) primary-secondary: carnotite, tyuyamunite, hewettite, navajoite, iron oxides, montroseite; (c) supergene: torbernite, uranophane, uranopilite, beta zippeite, johannite. (d) malachite, azurite, iron sulfates, gypsum
Nevadan	Early Nevadan N AZ	Late Triassic	237–201	wet MQA	Petrified Forest Mbr. of Chinle Fm.	petrified wood	Petrified Forest National Park and nearby ranches	petrified wood, groutite
Nevadan	Early Nevadan N AZ	Late Triassic	237–201	wet MQA		Fe, clay	Lyman Reservoir Fe in Chinle	specular hematite, peloidal hematite, kaolinite
Reorganization	erosion—start of Pangaea breakup	Latest Permian to Early Triassic	260–237	UDH MQA	breccia pipe emplacement	U-V-Cu (Ni-Co)	Pine Nut (255), Canyon breccia pipes (259)	uraninite, etc., in breccia pipes
Reorganization	erosion—start of Pangaea breakup	Latest Permian to Early Triassic	260–237	UDH MQA	Kaibab Limestone	Mn	Long Valley, Heber, Johnson & Hayden	Mn oxides / psilomelane, pyrolusite
Reorganization	erosion—start of Pangaea breakup	Latest Permian to Early Triassic	260–237	UDH (MQA)	Harrisburg Mbr., Kaibab Limestone	CuAg	Warm Springs	malachite, azurite, chrysocolla, chalcocite, chalcopyrite

Event	Tectonic process	Period	Age (Ma)	Designation	Stratigraphic notes	Products	Location	Minerals / products
Sonoma (S Great Basin) Permian Basin (P = 260–253)	final assembly of Pangaea	Permian	S = 260–250; P = 260–253	UDH (MQA)	potash emplacement; Toroweap Fm. UDH (285–270)	NaCl, potash, salt	Holbrook salt, potash	potash-bearing salts: sylvite, carnallite, polyhalite, salt (halite), gypsum, anhydrite
Alleghenian Ouachita = O	final assembly of Pangaea	Permian	O = 318–271	UDH		He, oil	Pinta Dome, Navajo Springs	He, carbonaceous dolomite
Reorganization	unconformity	Base of Permian (early Wolfcampian)	299–290	none/ UDH?	none: Red chert jelly bean congl. in Earp Fm., Wolfcamp black shale in Texas, Colina Limestone carbonaceous limestone	carbonaceous limestone and possible hydrothermal dolomite (Epitaph) replacing Colina Limestone	replacement of Upper Colina Limestone in Tombstone Hills by Epitaph Dolomite (Leonardian age)	tiny long c-axis doubly terminated quartz crystals
Ancestral Rockies	flat subduction along SW Laurentia toward NE	Pennsylvanian (Desmoinesian/ Missourian)	315–307	UDH	none: Horquilla Fm., Paradox grp. oil shows and production	oil, gas	S Paradox Basin; East Boundary Butte oil field	oil and gas
Reorganization	erosion	Early Pennsylvanian (Morrowan)	320–300	UDH	none: black shale basal Mbr. of Black Prince Limestone (323–315), Paradise Fm. (Chesteran)	shale hydrocarbons?	Winkelman oil show in Gila River Canyon	kerogenous black shale
Orogenic lull	shelf carbonate deposition	Mississippian (Viséan 331–327) (Kinderhook / Osage stage)	353–337	none	none: Redwall (Meramec or Viséan 345–341); Escabrosa (Osage or upper Tournaisian to upper Kinderhookian, 353–344)	High-calcium marble from Miss. limestone (hosts minor oil in NE AZ)	Twin Peaks cement, Clarkdale cement, Andrada marble, Superior marble	calcite, marble
Acadian (East Coast); Antler (Great Basin)	arc assemblies to W and E Laurentia	Late Devonian / earliest Mississippian	Antler = 380–357; Acadian = 375–325	UDH	Percha black shale and upper Martin Fm.	quartz	Payson "diamond quartz" at Diamond Point	Herkimer habit quartz
Acadian (East Coast)	arc assemblies to W and E Laurentia	Late Devonian / earliest Mississippian	Acadian = 375–325	UDH	McCracken, Elbert and Aneth Fm. oil shows and production	oil and gas	oil and gas in Walker Creek field	oil and gas, dolomite

(continued)

TABLE 4.2 (*continued*)

Orogeny	Phase	Age	Age (Ma)	MMS class	Rocks, Strata, Magmatism	Resources	Mineral districts	Minerals
Antler (Great Basin)	arc assemblies to W and E Laurentia	Late Devonian / earliest Mississippian	Antler = 380–357	UDH	3 oolitic iron	3 oolitic iron chamosite	3 Ranch Creek iron	3 oolitic hematite, chamosite
Erosion	erosion—unconformity	Late Ordovician (Silurian / Early Devonian)	480–385	none	none; erosion	none	none	none
Taconic (East Coast)	arc assembly East Coast; passive margin western US	Late Cambrian / Ordovician	485–430	none	none; erosion	none	none	none
Passive margin (west margin of Laurentia)	attempted breakup of Gondwana	Middle to Late Cambrian	515–485	UDH	alkaline magmatism in NM; hydrothermal activity? in Bright Angel Shale / Abrigo Fm.	Bolsa Quartzite, Troy Quartzite for silica flux for smelters	Chilito	quartz, glauconite shale
Pan-African / Avalonian	Gondwana assembly	Late Neo-Proterozoic to Early Cambrian	~530–506	UDH?	none; Sixtymile Fm., erosion	none	none	none
Rodinia breakup / rifting	Rodinia breakup / rifting	Early Neo-Proterozoic	950–510	UDH	Nankoweap Fm. (800–770 zircons), Chuar Grp. (800–742 detrital zircons), uppermost Walcutt Mbr. of upper Kwagunt Mbr. of Chuar Grp. (770–742±6)	Possible shale hydrocarbon?	carbon-rich shales in Grand Canyon	kerogenous black shale
Grenville	final assembly of Rodinia	Early Neo-Proterozoic	1035–980	none	none	none	none	none

Orogeny	Tectonic setting	Period	Dates	Deposit type	Geologic units	Commodity	Location	Minerals
Keweenawan failed rifting	failed rifting, Laurentia breakup—intra-orogenic	Late Mesoproterozoic	1104–1035	MQA	Diabase intrusions (1094–1080); sills in Sierra Ancha Mtns.; Cardenas lavas (1104)	serpentine asbestos	Sierra Ancha asbestos, Chrysotile (Salt River Canyon)	chrysotile asbestos, lizardite
Keweenawan failed rifting	failed rifting, Laurentia breakup—intra-orogenic	Late Mesoproterozoic	1104–1035	MQA	diabase granophyre (1038.5) and basaltic lavas	U (Cu) veins (1050)	uranium veins: Hope, Sierra Ancha supersys.	uraninite, pyrite and marcasite, pyrrhotite, molybdenite, galena, sphalerite, chalcopyrite, calcite, chlorite, nontronite
Keweenawan failed rifting	failed rifting, Laurentia breakup—intra-orogenic	Late Mesoproterozoic	1104–1035	MQA	diabase intrusions (1094–1080) sills in Sierra Ancha Mtns.	Fe contact skarns near diabase	Gentry Creek, Seneca, Fourth of July, Pine Ridge magnetite	specular hematite, magnetite, chrysotile, lizardite
Shawinigan	arc assembly	Late Mesoproterozoic	1240–1103	?	Red Rock Granite, NW Burro Mtns., NM (1220); possible megacrystic granites at Morenci	?	?	?
Elzevirian	arc	Late Mesoproterozoic	1360–1240	UDH Clinton ironstone type, syngenetic iron	tuff in Pioneer Shale (1330); Unkar Grp. (1256–<1167); tuff in Bass Limestone (1255); Dripping Spring Quartzite (1255); Troy Quartzite (1254); detrital zircon sources (1256±20); de- (1270–1250)	syn-sedimentary hematite in Mescal Limestone	Apache Iron, Chediski iron	hematite, apatite, muscovite, chert
Reorganization	erosion	Late Mesoproterozoic	1370–1360	none	none; unconformity under Apache Grp.	none	none	none
Picuris	Late (flat subduction)	Middle Mesoproterozoic	1460–1370	PCA	K-feldspar, rapaki-vine, megacrystic or porphyritic per-aluminous granites, pegmatites, greisens; White Picacho peg-matites (1376.5)	LCT: Be, Li, Ce, Ta-Nb, U; Li Be mica, Nb Ta, feldspar	White Picacho	White Picacho pegmatite suite (spodumene, lepidolite, scheelite, beryl, wolframite grp.)

(continued)

TABLE 4.2 (*continued*)

Orogeny	Phase	Age	Age (Ma)	MMS class	Rocks, Strata, Magmatism	Resources	Mineral districts	Minerals
Picuris	Late (flat subduction)	Middle Meso-proterozoic	1460–1370	PCA	Wagon Bow pegmatite (1460)	REE	REE Wagon Bow, Rare Metals?	REE Wagon Bow pegmatite suite (albite, biotite, cassiterite, epidote, goethite, hematite, microcline, monazite-Ce, muscovite, quartz, rutile (var. ilmenorutile), and xenotime-Y)
Picuris	Late (flat subduction)	Middle Meso-proterozoic	1460–1370	PCA	Oracle Granite (1434) and pegmatites, Ruin Granite (1436), Lawler Peak Granite (1411), Stockton Pass muscovite granite (1370)	W Be F	W Be F: Tungstona (1368), Tungsten King, Black Pearl, Oracle	scheelite, wolframite grp., fluorite, muscovite, microcline, gold
Picuris	Late (flat subduction)	Middle Meso-proterozoic	1460–1370	PCA	Whetstone alaskite complex	quartz/silica flux (W-U)	quartz/silica W U Whetstone	quartz/silica, scheelite
Picuris	Late	Early Meso-proterozoic	1470–1435	PAC	El Oso granites (1450), Pinaleño Granite	amethyst	Four Peaks amethyst in upper Mazatzal Quartzite	amethyst, hematite, fluorapatite
Picuris	Early (arc)	Early Meso-proterozoic	1470–1420	MCA?	Pinaleño Mtns. (1452)	Cu	Suspense	copper minerals
Reorganization	erosion	Early Meso-proterozoic	1550–1490	erosion	none; peneplanation/quartzites, Mazatzal Quartzite, Blackjack Quartzite (1494–1474)	none	none	none
Mazatzal	Late, flat subduction	Late Paleo-proterozoic	1630–1610	PC	aplopegmatites and garnet-muscovite soda episyenites that cut the Johnny Lyon Granodiorite	Au (Ag Pb Cu Bi)	Yellowstone	gold, quartz, baryte, bismuth minerals (nordströmite and friedrichite), stetefeldtite, schwartzite [mercurian tetrahedrite], tourmaline, zircon, fluorite
Mazatzal	Late, flat subduction	Late Paleo-proterozoic	1630–1590	PCA	peraluminous Pinaleño orthogneiss (1615), Cotton Center (1632)	W, Be, F, LREE	Black Beauty, Pinaleño Mtns., Beryl Hill?	scheelite, beryl, muscovite, allanite-(Nd), bastnäsite

Mazatzal	Late, flat subduction	1630–1590	PCA	Kingman pegmatite (1590)		Kingman Feldspar	microcline, bastnäsite-Ce, euxenite, allanite-(Nd), allanite-Ce
Mazatzal	Late flat subduction	1680–1690	PAC	Garnet Mtn.	W	Garnet Mtn.	scheelite, garnet, epidote
Mazatzal	Middle (arc)	1680–1630	MCA	Johnny Lyon Granodiorite, Madera Diorite, Beeline Granite, Maricopa batholith (1641), granite of Usery Mtns. (1644), Pinal Schist quartz porphyry (1634)	Au	Roosevelt, Spring Creek, Prescott, Thumb Butte	gold
Mazatzal	Early (arc)	1702–1680	MAC	basalt and rhyolite metavolcanics, schist, Buckhorn Granodiorite (1685), Soldier Camp Granites, Payson Granite (1703–1697), Green Valley Hills granophyre and Bear Flat alaskite (1697–1692), topazite ongonite	Be, F	Be veins in Breadpan District	acicular beryl, tourmaline, topaz
Mazatzal	Early (arc)	1702–1680	MAC		2 pyrophyllite	2 Gordon Creek pyrophyllite	2 pyrophyllite
Reorganization	erosion	1710–1680	none	none; peneplanation / erosional unconformity; Lower Mazatzal Quartzite, White Ledges Quartzite	none	none	none
Yavapai	Late, flat subduction	1715–1700	PC		Au(AgCu)	Humbug, Cottonwood	pyrite, galena, chalcopyrite, gold

(continued)

TABLE 4.2 (*continued*)

Orogeny	Phase	Age	Age (Ma)	MMS class	Rocks, Strata, Magmatism	Resources	Mineral districts	Minerals
Yavapai	Late, flat subduction	Late Paleoproterozoic	1715–1690	PCA	Antler peraluminous granites (1694); granites and pegmatites in upper Grand Canyon gorge (1710–1690)	W (Be) pegmatites and veins	Boriana W, Money Maker / North Star, H. S. Tungsten, Music Mtn. W	scheelite, beryl, hübnerite, wolframite grp, fluorite, microcline, quartz, muscovite, chalcopyrite, pyrite, apatite
Yavapai	Late, flat subduction	Late Paleoproterozoic	1715–1690	PCA	Crazy Basin peraluminous, 2-mica, quartz monzonite (~1700)	andalusite	andalusite-rich aluminosilicate alteration along east contacts of Crazy Basin pluton	andalusite-aluminosilicate alteration
Yavapai	Early (magmatic arc)	Late Paleoproterozoic	1750–1720	MC	mafic to intermediate volcanics, meta-andesite: Shea basalt, Brahma mafic schist (1750); felsic metavolcanics: Cleopatra-Deception Rhyolite (1738), Dick Rhyolite (1720), rhyolite metatuff at Antler Mine (1735), meta-amphibolite Antler Mine (1723), Rama felsic schist (1741)	Zn-Cu-Ag (Au,Pb) VMS	VMS Verde (Jerome), Old Dick (Bruce), Big Bug, Kay, Hualapai (Antler)	VMS mineral suite (pyrite, chalcopyrite, sphalerite, galena, minor cubanite, arsenopyrite, mackinawite, pyrrhotite; Mg chlorite in footwall)
Yavapai	Early	Late Paleoproterozoic	1750–1720	MC	cherty BIF (1728)	Fe-Si BIF	Pikes Peak Fe-Si belt in Yavapai Series (1728), Fe-Si deposits in Alder Group belt (Alder Creek, Pig Iron, Pittsburg-Tonto)	chert-hematite (magnetite)

Orogeny	Timing	Era	Age	Code	Rock unit	Deposit type	Deposit name	Minerals
Yavapai	Early	Late Paleoproterozoic	1750–1720	MC	Mn metapelites	Mn Si (Al,Zn)	Piestewa Mn	braunite I, kanonaite, piemontite, kyanite, andalusite, Zn-staurolite, Mn-Fe garnet
Yavapai	Early	Late Paleoproterozoic	1750–1720	MC	gabbro of Grapevine Gulch (1738–1740), Vishnu Schist (1740)	Co(As)	Grapevine cobalt	erythrite, cobaltian sulfides, arsenopyrite
Yavapai	Early	Late Paleoproterozoic	1750–1720	MC	tonalite of Cherry (1740)	Au (AgCu)	Cherry Creek	gold (chalcopyrite, bornite, sphalerite, galena)
Yavapai	Early (magmatic arc)	Late Paleoproterozoic	1770–1715	MCA	Phoenix Mtns. meta-rhyolites	meta-hot spring Hg	Phoenix-Mazatzal Mtns. Hg	cinnabar, kyanite, andalusite, pyrophyllite, tourmaline
Yavapai	Early	Late Paleoproterozoic	1755–1745	MCA	Spud Mountain Volcanics, Brady Butte Granodiorite	PbZnAg(AuCu), VMS	Iron King VMS	galena, sphalerite, chalcopyrite
Yavapai	Early	Late Paleoproterozoic	1770–1715	MCA	Alder Group metadacites (1730), Gibson Creek diorite complex (1730), Prescott Granodiorite	Au (Ag, base metal) mesothermal veins	Green Valley, Hassayampa, Groom Creek	gold, quartz, base-metal sulfides
Yavapai	Early	Late Paleoproterozoic	1770–1715	MCA	Squaw Peak intrusion	MoCu porphyry (1729–1738)	Squaw Peak MoCu	molybdenite, chalcopyrite
Yavapai	Early	Late Paleoproterozoic	1750–1690	MAC		PbAg	Groom Creek, Ruth, Turkey Creek	galena, sphalerite, silver, gold
Mohave / Penokean orogeny	Flat subduction and underplating of Mohave terrane?	Late Paleoproterozoic	1820?–1780?	P?; PCA?; PAC?	Undated pegmatites and other peraluminous phases in Elves Chasm complex; black quartzite (1770–2150); Turtle Mtns., CA?	none	none	muscovite, garnet, feldspar
Penokean orogeny	Early Mohave orogeny (magmatic volcanic arc)	Late Paleoproterozoic	1900–1820	MC	Elves Chasm hornblende tonalite orthogneiss (1840)	none	none	K-feldspar phenocrysts, hornblende, plagioclase

(continued)

TABLE 4.2 *(continued)*

Orogeny	Phase	Age	Age (Ma)	MMS class	Rocks, Strata, Magmatism	Resources	Mineral districts	Minerals
Mohave terrane		Paleoprotero-zoic	2300–1900	?	detrital zircons in Vishnu Schist and Del Rio Quartzite	none	none	detrital zircon
Algoman orog-eny in Wyoming craton	island arc accretion to S and E margins of Wyoming craton core	Late Archean	2720–2500	?	detrital zircons in Vishnu Schist and Del Rio Quartzite; plutonic belts mantle in SE parts of Wyo-ming core	none	none	detrital zircon
Core of Wyo-ming craton	formation of Wyoming craton core via amalga-mation of prim-itive island arcs (gray gneisses)	Early Archean	4000–3000	?	detrital zircons in Vishnu Schist and Del Rio Quartzite	none	none	detrital zircon

Notes: MMS = magma-metal series, MC = metaluminous calcic, MCA = metaluminous calc-alkalic, MAC = metaluminous alkali-calcic, MQA = metaluminous quartz alkalic, MNA = met-aluminous nepheline alkalic, PC = peraluminous calcic, PCA = peraluminous calc-alkalic, PAC = peraluminous alkali-calcic, UDH = ultradeep hydrothermal, VMS = volcanogenic massive sulfide, BIF = banded iron formation, LCT = lithium-cerium-tantalum pegmatites, NI = nonigneous, GP = gold placer, LREE = light rare earth elements, REE = rare earth elements, PGE = platinum group elements, OAE = oceanic anoxic event, Ma = mega-annum (or million years ago), ybp = years before present, Fm. = Formation, Congl.= Conglomerate, fangl. = fanglom-erate, Mtn. = Mountain, Mbr. = Member, Grp. = Group; numbers in parentheses are in Ma; parentheses enclosing elements or minerals indicate minor occurrences, ? indicates uncertain.

APPENDIX A

Maps of Arizona Mineral Districts

Appendix A consists of maps that display the mineral districts for each county, show the information discussed in detail in chapter 4, and help locate mineral occurrences referenced in chapter 3.

Mineral district maps delineated by Stanley B. Keith and Jan C. Rasmussen, with cartography by John Callahan.

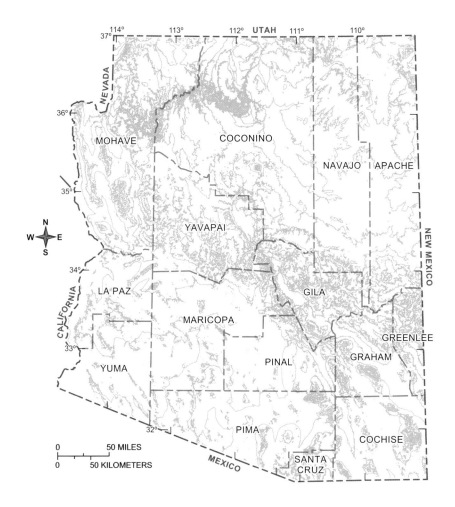

Apache County

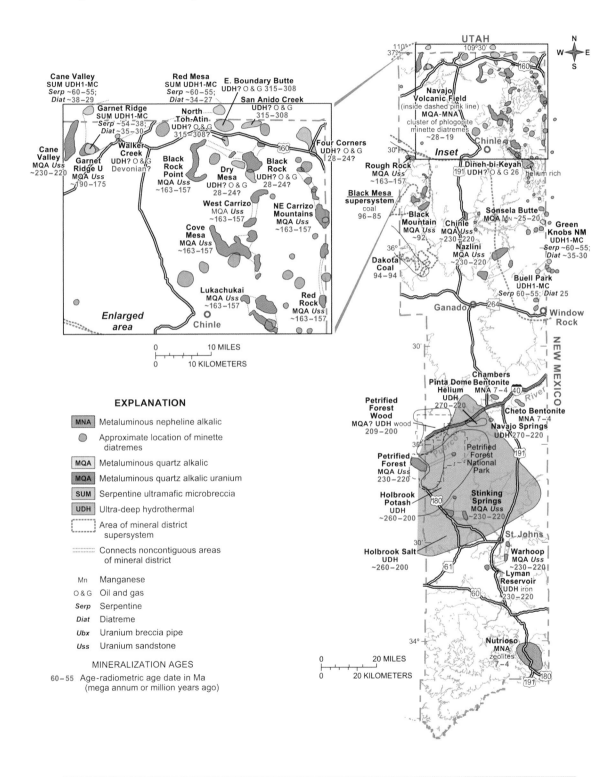

Cane Valley
SUM UDH1-MC
Serp ~60–55;
Diat ~38–29

Red Mesa
SUM UDH1-MC
Serp ~60–55;
Diat ~34–27

E. Boundary Butte
UDH? O & G 315–308

San Anido Creek
UDH? O & G
315–308

Garnet Ridge
SUM UDH1-MC
Serp ~54–38;
Diat ~35–30

North Toh-Atin
UDH? O & G
315–308?

Four Corners
UDH? O & G
28–24?

Cane Valley
MQA *Uss*
~230–220

Garnet Ridge U
MQA *Uss*
~190–175

Walker Creek
UDH? O & G
Devonian?

Black Rock Point
MQA *Uss*
~163–157

Dry Mesa
UDH? O & G
28–24?

Black Rock
UDH? O & G
28–24?

Rough Rock
MQA *Uss*
~163–157

Black Mesa supersystem
coal
96–85

Black Mountain
MQA *Uss*
~92

Chinle
MQA *Uss*
~230–220

Sonsela Butte
MQA *M*n ~25–20

Green Knobs NM
UDH1-MC
Serp ~60–55;
Diat ~35-30

West Carrizo
MQA *Uss*
~163–157

NE Carrizo Mountains
MQA *Uss*
~163–157

Nazlini
MQA *Uss*
~230–220

Cove Mesa
MQA *Uss*
~163–157

Buell Park
UDH1-MC
Serp 60–55; *Diat* 25

Dakota Coal
94–94

Lukachukai
MQA *Uss*
~163–157

Red Rock
MQA *Uss*
~163–157

Ganado

Window Rock

Enlarged area

Chinle

0   10 MILES
0   10 KILOMETERS

UTAH

Navajo Volcanic Field
(inside dashed pink line)
MQA-MNA
cluster of phlogopite
minette diatremes
~28–19

Inset

Chinle

Dineh-bi-Keyah
UDH? O & G 26 helium rich

NEW MEXICO

Chambers
Pinta Dome Bentonite Helium
UDH
270–220
MNA 7–4

Petrified Forest Wood
MQA? UDH wood
209–200

Cheto Bentonite
MNA 7–4

Navajo Springs
UDH 270–220

Petrified Forest
MQA *Uss*
230–220

Petrified Forest National Park

Holbrook Potash
UDH
~260–200

Stinking Springs
MQA *Uss*
~230–220

St. Johns

Holbrook Salt
UDH
~260–200

Warhoop
MQA *Uss*
~230–220

Lyman Reservoir
UDH iron
230–220

Nutrioso
MNA
zeolites
7–4

EXPLANATION

MNA	Metaluminous nepheline alkalic
●	Approximate location of minette diatremes
MQA	Metaluminous quartz alkalic
MQA	Metaluminous quartz alkalic uranium
SUM	Serpentine ultramafic microbreccia
UDH	Ultra-deep hydrothermal
⌐ ⌐	Area of mineral district supersystem
·········	Connects noncontiguous areas of mineral district
Mn	Manganese
O & G	Oil and gas
Serp	Serpentine
Diat	Diatreme
Ubx	Uranium breccia pipe
Uss	Uranium sandstone

MINERALIZATION AGES

60–55   Age-radiometric age date in Ma
(mega annum or million years ago)

0   20 MILES
0   20 KILOMETERS

Cochise County

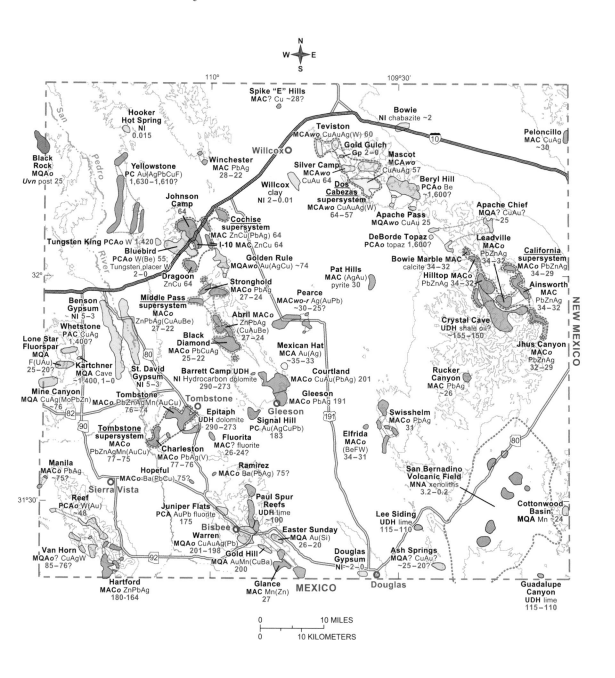

Spike "E" Hills
MAC? Cu ~28?

Bowie
NI chabazite ~2

Hooker
Hot Spring
NI
0.015

Teviston
MCAwo CuAuAg(W) 60

Peloncillo
MAC CuAg
~30

Willcox○

Gold Gulch
Gp 2-0

Mascot
MCAwo
CuAuAg 57

Black
Rock
MQAo
Uvn post 25

Yellowstone
PC Au(AgPbCuF)
1,630-1,610?

Winchester
MAC PbAg
28-22

Silver Camp
MCAwo
CuAu 64

Beryl Hill
PCAo Be
~1,600?

Johnson
Camp
64

Willcox
clay
NI 2-0.01

Dos
Cabezas
supersystem
MCAwo CuAuAg(W)
64-57

Apache Chief
MQA? CuAu?
~25

Cochise
supersystem
MAC ZnCu(PbAg) 64

Apache Pass
MQAwo CuAu 25

DeBorde Topaz
PCAo topaz 1,600?

Leadville
MACo
PbZnAg
34-32

Tungsten King PCAo W 1,420

I-10 MAC ZnCu 64

Bluebird
PCAo W(Be) 55;
Tungsten placer W
2-0

Golden Rule
MQAwo Au(AgCu) ~74

Pat Hills
MAC (AgAu)
pyrite 30

Bowie Marble MAC
calcite 34-32

California
supersystem
MACo PbZnAg
34-29

32°

Dragoon
ZnCu 64

Stronghold
MACo PbAg
27-24

Hilltop MACo
PbZnAg 34-32

Ainsworth
MAC
PbZnAg
34-32

Benson
Gypsum
NI 5-3

Middle Pass
supersystem
MACo
ZnPbAg(CuAuBe)
27-22

Pearce
MACwo-r Ag(AuPb)
~30-25?

Crystal Cave
UDH shale oil?
~155-150

Jhus Canyon
MACo
PbZnAg
32-29

Whetstone
PAC CuAg
1,400?

Abril MACo
ZnPbAg
(CuAuBe)
27-24

Lone Star
Fluorspar
MQA
F(UAu)
25-20?

Black
Diamond
MACo PbCuAg
25-22

Mexican Hat
MCA Au(Ag)
~35-33

Courtland
MACo CuAu(PbAg) 201

Rucker
Canyon
MAC PbAg
~26

Kartchner
MQA Cave
~1,400, 1-0

Barrett Camp UDH
NI Hydrocarbon dolomite
290-273

Mine Canyon
MQA CuAg(MoPbZn)
~76

St. David
Gypsum
NI 5-3

Gleeson
MACo PbAg 191

Tombstone
MACo PbZnAgMn(AuCu)
76-74

Tombstone

Gleeson
MACo PbAg 191

Swisshelm
MACo PbAg
31

Tombstone
supersystem
MACo
PbZnAgMn(AuCu)
77-75

Epitaph
UDH dolomite
290-273

Signal Hill
PC Au(AgCuPb)
183

Elfrida
MACo
(BeFW)
34-31

Charleston
MACo PbAg(V)
77-76

Fluorita
MAC? fluorite
26-24?

Manila
MACo PbAg
~75?

Hopeful
MACo-Ba(PbCu) 75?

Ramirez
MACo Ba(PbAg) 75?

San Bernadino
Volcanic Field
MNA xenoliths
3.2-0.2

Sierra Vista

Reef
PCAo W(Au)
~48

Paul Spur
Reefs
UDH lime
~100

31°30'

Juniper Flats
PCA AuPb fluorite
175

Lee Siding
UDH lime
115-110

Cottonwood
Basin
MQA Mn ~24

Bisbee

Easter Sunday
MQA Au(Si)
26-20

Warren
MQAo CuAuAg(Pb)
201-198

Douglas
Gypsum
NI ~2-0

Ash Springs
MQA? CuAu?
~25-20?

Van Horn
MQAo? CuAgW
85-76?

Gold Hill
MQA AuMn(CuBa)
200

Hartford
MACo ZnPbAg
180-164

Glance
MAC Mn(Zn)
27

MEXICO

Douglas

Guadalupe
Canyon
UDH lime
115-110

0   10 MILES

0   10 KILOMETERS

The earliest known image of the Czar Mine. The Copper Queen Mine can be seen to the right, with the smelter just below. Above and to the left is the small headframe of the Atlantic shaft, which first discovered the great Southwest and Atlantic orebodies, also to be mined from the Czar. The Copper Czar Mine, later called the "Czar," was the first vertical shaft at Bisbee, just 440 feet deep. The Czar operated from 1884 until 1944. Far and away, the Czar was the most prolific mineral-specimen producer at Bisbee. Photograph is from the Graeme/Larkin collection and text is courtesy of Richard Graeme.

Warren mineral district Copper Czar Mine enclosed headframe from February 1885.

EXPLANATION

MNA	Metaluminous nepheline alkalic
MQA	Metaluminous quartz alkalic
MQA	Metaluminous quartz alkalic uranium
MAC	Metaluminous alkali-calcic
MCA	Metaluminous calc-alkalic
PAC	Peraluminous alkali-calcic
PCA	Peraluminous calc-alkalic
PC	Peraluminous calcic

UDH	Ultra-deep hydrothermal
NI	Non-igneous
Gp	Gold placer
	Area of mineral district supersystem
	Connects noncontiguous areas of mineral district
o	Oxidized
wo	Weakly oxidized
r	Reduced

Ag	Silver	Pb	Lead	
Au	Gold	Si	Silica	
Ba	Barium	U	Uranium	
Be	Beryllium	V	Vanadium	
Cu	Copper	W	Tungsten	
F	Fluorine	Zn	Zinc	
Mn	Manganese	Au(Si)	Minor elements inside parentheses	
Mo	Molybdenum			
Ni	Nickel	*Uvn*	Uranium vein	

MINERALIZATION AGES

34–31 Age-radiometric age date in Ma
(mega annum or million years ago)

Coconino County

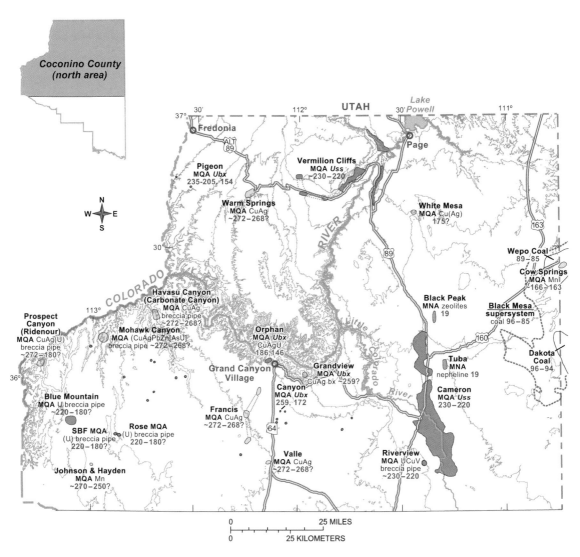

EXPLANATION

MNA	Metaluminous nepheline alkalic
MQA	Metaluminous quartz alkalic
MQA	Metaluminous quartz alkalic uranium
•	Uranium breccia pipe
[dashed box]	Area of mineral district supersystem
[dotted line]	Connects noncontiguous areas of mineral district

Ag	Silver	U	Uranium
As	Arsenic	V	Vanadium
Cu	Copper	Zinc	Zinc
Mn	Manganese	*Ubx*	Uranium breccia pipe
Pb	Lead	*Uss*	Uranium sandstone

MINERALIZATION AGES

230–220 Age-radiometric age date in Ma
(mega annum or million years ago)

The Orphan Mine was initially a copper mine established in a collapsed breccia pipe in 1906. Uranium was discovered in 1950, and the site was mined mainly for uranium from 1953 throught 1969. During the years of production, the mine produced over four million pounds of uranium oxide and significant quantities of copper and silver. The headframe in the 1959 photograph was constructed in 1959. The U.S. government assumed the ownership of the property in 1987, and the site is now reclaimed. Photograph is courtesy of Photo Archive, Arizona Geological Survey.

The Orphan Mine headframe on the south rim of the Grand Canyon less than two miles from Grand Canyon Village in the Orphan mineral district, 1959.

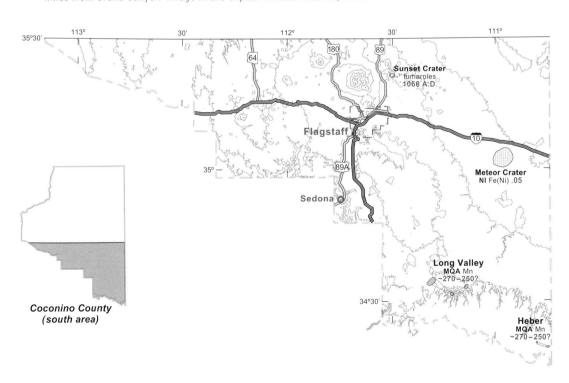

Coconino County
(south area)

Gila County

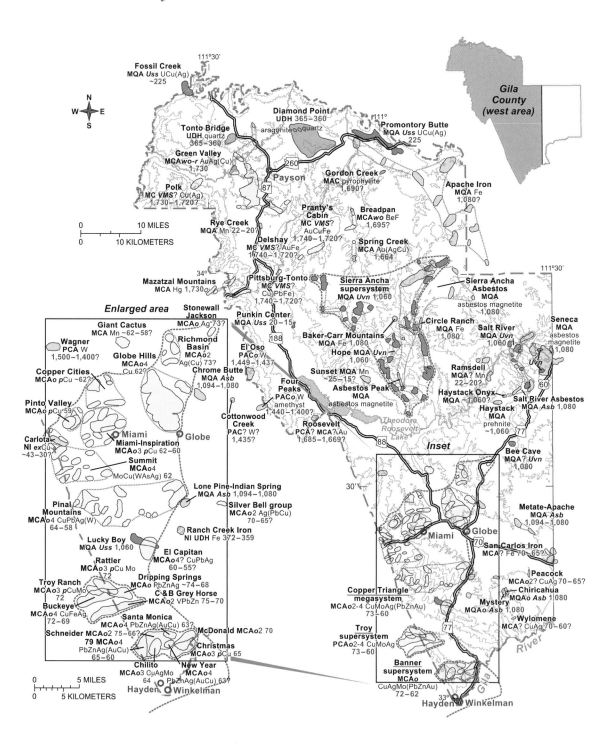

Fossil Creek
MQA *Uss* UCu(Ag)
~225

Diamond Point
UDH 365–360
aragonite quartz

Promontory Butte
MQA *Uss* UCu(Ag)
225

111°30'

111°

*Gila
County
(west area)*

Tonto Bridge
UDH quartz
365–360

Green Valley
MCA*wo-r* AuAg(Cu)
1,730

Payson

Gordon Creek
MAC pyrophyllite
1,690?

Apache Iron
MQA Fe
1,080?

Polk
MC *VMS*? Cu(Ag)
1,730–1,720?

Pranty's
Cabin
MC *VMS*?
AuCuFe

Breadpan
MCA*wo* BeF
1,695?

Rye Creek
MQA Mn 22–20?

Delshay
MC *VMS*? AuFe
1,740–1,720?

Spring Creek
MCA Au(AgCu)
1,664

Mazatzal Mountains
MCA Hg 1,730

Pittsburg-Tonto
MC *VMS*?
Cu(PbFe)
1,740–1,720?

Sierra Ancha
supersystem
MQA *Uvn* 1,060

Sierra Ancha
Asbestos
MQA
asbestos magnetite
1,080

111°30'

34°

Seneca
MQA
asbestos
magnetite
1,080

Enlarged area

Stonewall
Jackson
MCA*o* Ag 73?

Punkin Center
MQA *Uss* 20–15

Circle Ranch
MQA Fe
1,080

Salt River
MQA *Uvn*
1,060

Giant Cactus
MCA Mn ~62–58?

Richmond
Basin
Ag(Cu) 73?

El'Oso
PACo W
1,449–1,437

Baker-Carr Mountains
MQA Fe 1,080

Hope MQA *Uvn*
1,060

Wagner
PCA W
1,500–1,400?

Globe Hills
MCAo4
Cu.62?

Sunset MQA Mn
~25–15?

Ramsdell
MQA? Mn
22–20?

Uvn

Copper Cities
MCA*o* *p*Cu ~62?

Chrome Butte
MQA *Asb*
1,094–1,080

Four
Peaks
PACo W
amethyst
1,440–1,400?

Asbestos Peak
MQA
asbestos magnetite

Haystack Onyx
MQA ~1,060?

Salt River Asbestos
MQA *Asb* 1,080

Pinto Valley
MCA*o* *p*Cu 59?

Cottonwood
Creek
PAC? W?
1,435?

Roosevelt
PCA? MCA? Au
1,685–1,669?

Haystack
MQA
prehnite
~1,060

Carlota
NI *exCu*
~43–30?

Miami
Miami-Inspiration
MCA*o*3 *p*Cu 62–60

Globe

*Theodore
Roosevelt
Lake*

Inset

Bee Cave
MQA? *Uvn*
1,080

Summit
MCAo4
MoCu(WAsAg) 62

Pinal
Mountains
MCA*o*4 CuPbAg(W)
64–58

Lone Pine-Indian Spring
MQA *Asb* 1,094–1,080

Silver Bell group
MCA*o*2 Ag(PbCu)
70–65?

30'

Metate-Apache
MQA *Asb*
1,094–1,080

Lucky Boy
MQA *Uss* 1,060

Ranch Creek Iron
NI UDH Fe 372–359

Miami

Globe

San Carlos Iron
MCA? Fe 70–65?

El Capitan
MCA*o*4? CuPbAg
60–55?

Peacock
MCA*o*2? CuAg 70–65?

Rattler
MCA*o*3 *p*Cu Mo
72

Troy Ranch
MCA*o*3 *p*CuMo
72

Dripping Springs
MCA*o* PbZnAg ~74–68

C&B Grey Horse
MCA*o*2 VPbZn 75–70

Chiricahua
MQA*o* *Asb* 1080

Copper Triangle
megasystem
MCA*o*2-4 CuMoAg(PbZnAu)
73–60

Mystery
MQA*o* *Asb* 1,080

Buckeye
MCA*o*4 CuFeAg
72–69

Santa Monica
MCA*o*4 PbZnAg(AuCu) 63?

McDonald MCA*o*2 70

Troy
supersystem
PCA*o*2-4 CuMoAg
73–60

Wylomene
MCA? CuAg 70–60?

Schneider MCA*o*2 75–66?

79 MCA*o*4
PbZnAg(AuCu)
65–60

Christmas
MCA*o*3 *p*Cu 65

Chilito
MCA*o*3 CuAgMo
64

New Year
MCA*o*4
PbZnAg(AuCu) 63?

Banner
supersystem
MCA*o*
CuAgMo(PbZnAu)
72–62

33°

Hayden

Winkelman

Hayden Winkelman

River

Gila

N
W — E
S

0   10 MILES
0   10 KILOMETERS

0   5 MILES
0   5 KILOMETERS

260

87

188

88

70

77

60

77

The Old Dominion Mine operated from 1882 until 1931 with additional production in the district from small operations until 1957. This productive copper mine produced 800 million pounds of copper in the 75 years the Old Dominion Mine operated. Frederick Ransome was a U.S. Geological Survey geologist from 1897 to 1924. He took this 1901 photograph while writing his 1903 professional paper, *Geology of the Globe Copper District, Arizona*. Ransome headed the geology department at the University of Arizona from 1924 to 1927. A type mineral from the United Verde Mine in Jerome is named ransomite after Frederick Ransome. This photograph is from the U.S. Geological Survey collection.

Old Dominion Mine and smelter near Globe, Globe Hills mineral district, 1901.

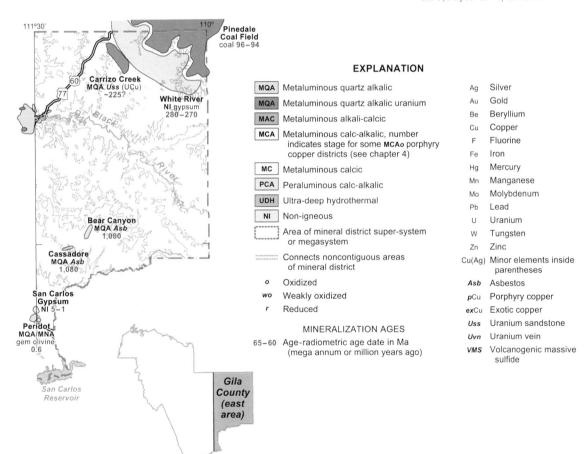

EXPLANATION

MQA	Metaluminous quartz alkalic
MQA	Metaluminous quartz alkalic uranium
MAC	Metaluminous alkali-calcic
MCA	Metaluminous calc-alkalic, number indicates stage for some **MCAo** porphyry copper districts (see chapter 4)
MC	Metaluminous calcic
PCA	Peraluminous calc-alkalic
UDH	Ultra-deep hydrothermal
NI	Non-igneous
[⸬]	Area of mineral district super-system or megasystem
⸱⸱⸱⸱⸱	Connects noncontiguous areas of mineral district
o	Oxidized
wo	Weakly oxidized
r	Reduced

MINERALIZATION AGES

65–60   Age-radiometric age date in Ma
(mega annum or million years ago)

Ag	Silver
Au	Gold
Be	Beryllium
Cu	Copper
F	Fluorine
Fe	Iron
Hg	Mercury
Mn	Manganese
Mo	Molybdenum
Pb	Lead
U	Uranium
W	Tungsten
Zn	Zinc
Cu(Ag)	Minor elements inside parentheses
Asb	Asbestos
pCu	Porphyry copper
exCu	Exotic copper
Uss	Uranium sandstone
Uvn	Uranium vein
VMS	Volcanogenic massive sulfide

Graham County

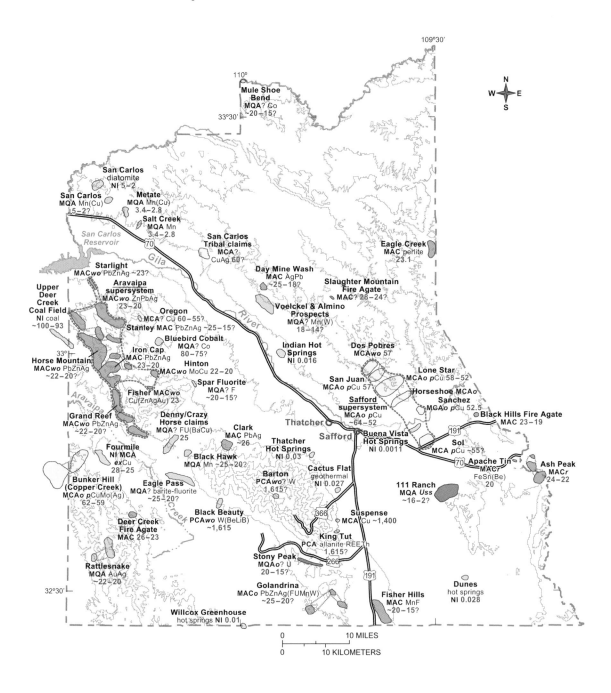

Mule Shoe Bend
MQA? Co
~20–15?

110°

33°30'

109°30'

San Carlos
diatomite
NI 5–2

San Carlos
MQA Mn(Cu)
5–2?

Metate
MQA Mn(Cu)
3.4–2.8

Salt Creek
MQA Mn
3.4–2.8

San Carlos
Reservoir

Gila River

70

San Carlos
Tribal claims
MCA?
CuAg 60?

Day Mine Wash
MAC AgPb
~25–18?

Eagle Creek
MAC perlite
23.1

Starlight
MACwo PbZnAg ~23?

Slaughter Mountain
Fire Agate
MAC? 28–24?

Aravaipa
supersystem
MACwo ZnPbAg
23–20

Voelckel & Almino
Prospects
MQA? Mn(W)
18–14?

Upper
Deer
Creek
Coal Field
NI coal
~100–93

Oregon
MCA? Cu 60–55?

Stanley MAC PbZnAg ~25–15?

Indian Hot
Springs
NI 0.016

Dos Pobres
MCAwo 57

Bluebird Cobalt
MQA? Co
80–75?

Iron Cap
MAC PbZnAg
23–20

Lone Star
MCAo pCu 58–52

33°

Horse Mountain
MACwo PbZnAg
~22–20?

Hinton
MACwo MoCu 22–20

San Juan
MCAo pCu 57

Horseshoe MCAo
Sanchez
MCAo pCu 52.5

Spar Fluorite
MQA? F
~20–15?

Safford
supersystem
MCAo pCu
~64–52

Black Hills Fire Agate
MAC 23–19

Fisher MACwo
Cu(ZnAgAu)? 23

Aravaipa

Grand Reef
MACwo PbZnAg
~22–20?

Denny/Crazy
Horse claims
MQA? FU(BaCu)
23

Clark
MAC PbAg
~26

Thatcher

Safford

Buena Vista
Hot Springs
NI 0.0011

191

Sol
MCA pCu ~55?

Fourmile
NI MCA
exCu
28–25

Black Hawk
MQA Mn ~25–20?

Thatcher
Hot Springs
NI 0.03

Apache Tin
MACr
FeSn(Be)
20

70

Ash Peak
MACr
24–22

Bunker Hill
(Copper Creek)
MCAo pCuMo(Ag)
62–59

Eagle Pass
MQA? barite-fluorite
~25–20?

Barton
PCAwo? W
1,615?

Cactus Flat
geothermal
NI 0.027

111 Ranch
MQA Uss
~16–2?

Suspense
MCA Cu ~1,400

366

Deer Creek
Fire Agate
MAC 26–23

Black Beauty
PCAwo W(BeLiB)
~1,615

King Tut
PCA allanite REETh
1,615?

Stony Peak
MQAo? U
20–15?

266

Rattlesnake
MQA AuAg
~22–20

Dunes
hot springs
NI 0.028

32°30'

Golandrina
MACo PbZnAg(FUMnW)
~25–20?

191

Fisher Hills
MAC MnF
~20–15?

Willcox Greenhouse
hot springs NI 0.01

0   10 MILES

0   10 KILOMETERS

N
W   E
S

The dark formation in the center background that projects up on both sides of Laurel Canyon gives the name Grand Reef to the mine. Tailings ponds show in the foreground, manager's house in center, mill and mine surface facilities in background next to the reef. The mine was discovered in 1890 and operated sporadically until all mining activity in the area ceased around 1958. The Grand Reef Mine produced lead, copper, silver, and zinc and is well known as a mineral-specimen location. Brilliant blue druses and crystals of the mineral linarite are well represented. Photograph is courtesy of Photo Archive, Arizona Geological Survey.

Grand Reef Mine in the Grand Reef mineral district, probably 1942.

EXPLANATION

MQA	Metaluminous quartz alkalic	Ag	Silver	Si	Silica
MQA	Metaluminous quartz alkalic uranium	Au	Gold	Sn	Tin
		B	Boron	Th	Thorium
MAC	Metaluminous alkali-calcic	Ba	Barium	U	Uranium
MCA	Metaluminous calc-alkalic	Be	Beryllium	W	Tungsten
PCA	Peraluminous calc-alkalic	Cu	Copper	Zn	Zinc
NI	Non-igneous	F	Fluorine	Mn(Cu)	Minor elements inside parenthesese
		Fe	Iron		
	Area of mineral district supersystem	Li	Lithium	REE	Rare earth elements
		Mn	Manganese	*p*Cu	Porphyry copper
	Connects noncontiguous areas of mineral district	Mo	Molybdenum	*ex*Cu	Exotic copper
		Pb	Lead	*Uss*	Uranium sandstone
o	Oxidized				
wo	Weakly oxidized		MINERALIZATION AGES		
r	Reduced	28–25	Age-radiometric age date in Ma (mega annum or million years ago)		

Greenlee County

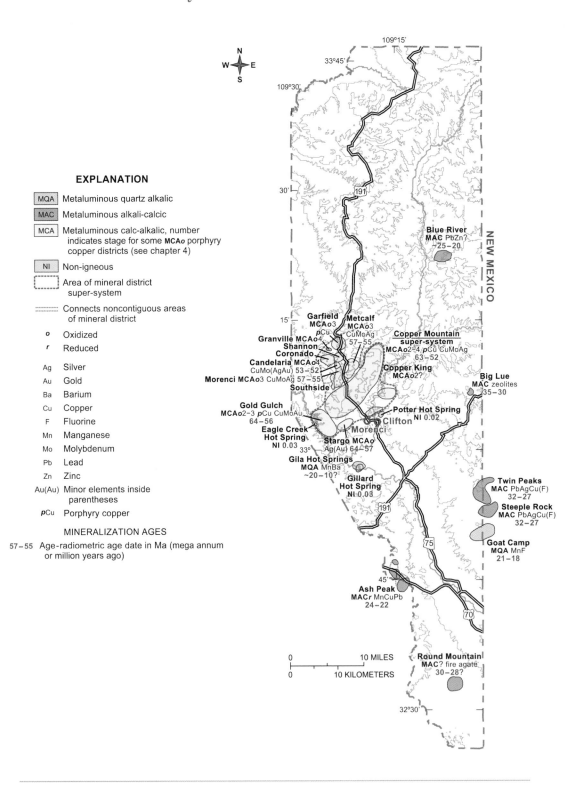

EXPLANATION

MQA	Metaluminous quartz alkalic
MAC	Metaluminous alkali-calcic
MCA	Metaluminous calc-alkalic, number indicates stage for some **MCA**o porphyry copper districts (see chapter 4)
NI	Non-igneous
⌐¬ (dashed box)	Area of mineral district super-system
········	Connects noncontiguous areas of mineral district

o	Oxidized
r	Reduced
Ag	Silver
Au	Gold
Ba	Barium
Cu	Copper
F	Fluorine
Mn	Manganese
Mo	Molybdenum
Pb	Lead
Zn	Zinc
Au(Au)	Minor elements inside parentheses
*p*Cu	Porphyry copper

MINERALIZATION AGES

57–55 Age-radiometric age date in Ma (mega annum or million years ago)

Map labels:
- Blue River MAC PbZn? ~25–20
- NEW MEXICO
- Garfield MCAo3 *p*Cu
- Metcalf MCAo3 CuMoAg
- Granville MCAo4
- Shannon
- Coronado
- Candelaria MCAo4 CuMo(AgAu) 53–52
- Morenci MCAo3 CuMoAg 57–55
- Southside
- 57–55
- Copper Mountain super-system MCAo2–4 *p*Cu CuMoAg 63–52
- Copper King MCAo2?
- Big Lue MAC zeolites ~35–30
- Gold Gulch MCAo2–3 *p*Cu CuMoAu 64–56
- Potter Hot Spring NI 0.02
- Clifton
- Morenci
- Eagle Creek Hot Spring NI 0.03
- Stargo MCAo Ag(Au) 64–57
- Gila Hot Springs MQA MnBa ~20–10?
- Gillard Hot Spring NI 0.03
- Twin Peaks MAC PbAgCu(F) 32–27
- Steeple Rock MAC PbAgCu(F) 32–27
- Goat Camp MQA MnF 21–18
- Ash Peak MAC*r* MnCuPb 24–22
- Round Mountain MAC? fire agate 30–28?

0   10 MILES
0   10 KILOMETERS

Mining in the Morenci mineral district in the 1880s utilized this narrow gauge railroad.

In 1882 the Coronado Railroad was extended from Clifton up to the base of the Coronado Incline. The incline brought ore down from the Coronado Mine, spanning a vertical distance of 1,200 feet via a steep inclined skipway. A short narrow gauge railway connected the upper end of the incline with the mine as shown in the photograph to the left. From 1872 to 1932 all mine operations in the Morenci area were underground. As a result of the low copper prices during the Great Depression, mining operation ceased for all Morenci-area mines in 1932. Mining started up again in 1937 utilizing open-pit technology. Today the Morenci Mine is two miles in length, almost as wide, and is the largest copper-producing mine in North America. Both of these photographs are courtesy Photo Archive, Arizona Geological Survey.

Panoramic view of the open pit Morenci Mine, 1957.

La Paz County

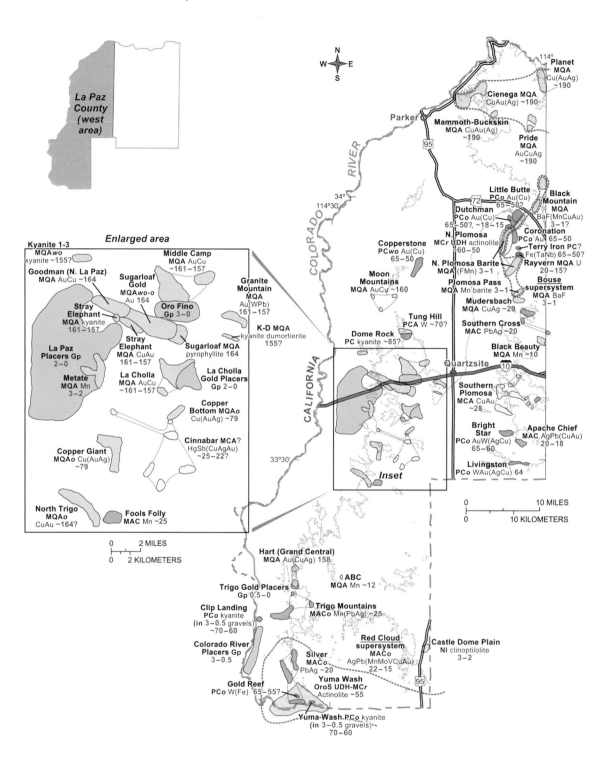

La Paz County (west area)

Enlarged area

Kyanite 1-3
MQA*wo*
kyanite ~155?

Goodman (N. La Paz)
MQA AuCu ~164

Stray Elephant
MQA kyanite 161–157

La Paz Placers Gp 2–0

Metate
MQA Mn 3–2

Sugarloaf Gold
MQA*wo-o*
Au 164

Stray Elephant
MQA CuAu 161–157

La Cholla
MQA AuCu ~161–157

Copper Giant
MQA*o* Cu(AuAg) ~79

North Trigo
MQA*o* CuAu ~164?

Middle Camp
MQA AuCu ~161–157

Oro Fino
Gp 3–0

Sugarloaf MQA pyrophyllite 164

La Cholla Gold Placers
Gp 2–0

Copper Bottom MQA*o*
Cu(AuAg) ~79

Cinnabar MCA?
HgSb(CuAgAu) ~25–22?

Fools Folly
MAC Mn ~25

Granite Mountain
MQA
Au(WPb)
161–157

K-D MQA
kyanite dumortierite 155?

0   2 MILES
0   2 KILOMETERS

114°
Planet MQA
Cu(AuAg) ~190

Cienega MQA
CuAu(Ag) ~190

Parker

Mammoth-Buckskin
MQA CuAu(Ag) ~190

Pride MQA
AuCuAg ~190

34°
114°30'

Little Butte
PCo Au(Cu) 65–50?

Dutchman
PCo Au(Cu) 65–50?, ~18–15

Black Mountain MQA
BaF(MnCuAu) 3–1?

N. Plomosa
MCr UDH actinolite ~60–50

Coronation
PCo Au 65–50

Terry Iron PC?
Fe(TaNb) 65–50?

Copperstone
PC*wo* Au(Cu) 65–50

Rayvern MQA U 20–15?

N. Plomosa Barite
MQA (FMn) 3–1

Bouse supersystem
MQA BaF 3–1

Moon Mountains
MQA AuCu ~160

Plomosa Pass
MQA Mn barite 3–1

Mudersbach
MQA CuAg ~20

Tung Hill
PCA W ~70?

Southern Cross
MAC PbAg ~20

Dome Rock
PC kyanite ~85?

Black Beauty
MQA Mn ~10

Quartzsite

Southern Plomosa
MCA CuAu ~28

Bright Star
PCo AuW(AgCu) 65–60

Apache Chief
MAC AgPb(CuAu) 20–18

Livingston
PCo WAu(AgCu) 64

CALIFORNIA

COLORADO RIVER

33°30'

Inset

0   10 MILES
0   10 KILOMETERS

Hart (Grand Central)
MQA Au(CuAg) 158

ABC
MQA Mn ~12

Trigo Gold Placers
Gp 0.5–0

Clip Landing
PCo kyanite (in 3–0.5 gravels) ~70–60

Colorado River Placers Gp 3–0.5

Gold Reef
PCo W(Fe) 65–55?

Trigo Mountains
MAC*o* Mn(PbAg) ~25

Red Cloud supersystem
MAC*o* AgPb(MnMoVCuAu) 22–15

Silver
MAC*o* PbAg ~20

Yuma Wash
OroS UDH-MC*r* Actinolite ~55

Castle Dome Plain
NI clinoptilolite 3–2

Yuma Wash PC*o* kyanite (in 3–0.5 gravels) 70–60

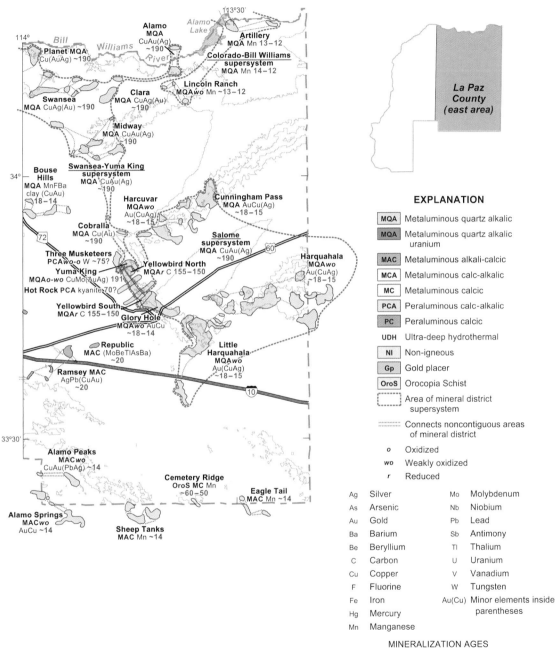

Alamo
MQA
CuAu(Ag)
~190

Alamo
Lake

113°30'

Artillery
MQA Mn 13–12

114°

Bill

Williams

Planet MQA
Cu(AuAg) ~190

River

Colorado-Bill Williams
supersystem
MQA Mn 14–12

Lincoln Ranch
MQA*wo* Mn ~13–12

Swansea
MQA CuAg(Au) ~190

Clara
MQA CuAg(Au)
~190

Midway
MQA CuAu(Ag)
~190

34°

Bouse
Hills
MQA MnFBa
clay (CuAu)
~18–14

Swansea-Yuma King
supersystem
MQA CuAu(Ag)
~190

Harcuvar
MQA*wo*
Au(CuAg)
~18–15

Cunningham Pass
MQA AuCu(Ag)
~18–15

Cobralla
MQA Cu(Au)
~190

72

Salome
supersystem
MQA CuAu(Ag)
~190

60

Three Musketeers
PCA*wo-o* W ~75?

Yuma King
MQA*o-wo* CuMo(AuAg) 191

Yellowbird North
MQA*r* C 155–150

Harquahala
MQA*wo*
Au(CuAg)
~18–15

Hot Rock PCA kyanite 70?

Yellowbird South
MQA*r* C 155–150

Glory Hole
MQA*wo* AuCu
~18–14

Republic
MAC (MoBeTlAsBa)
~20

Little
Harquahala
MQA*wo*
Au(CuAg)
~18–15

Ramsey MAC
AgPb(CuAu)
~20

10

33°30'

Alamo Peaks
MAC*wo*
CuAu(PbAg) ~14

Cemetery Ridge
OroS MC Mn
~60–50

Eagle Tail
MAC Mn ~14

Alamo Springs
MAC*wo*
AuCu ~14

Sheep Tanks
MAC Mn ~14

La Paz
County
(east area)

EXPLANATION

MQA	Metaluminous quartz alkalic
MQA	Metaluminous quartz alkalic uranium
MAC	Metaluminous alkali-calcic
MCA	Metaluminous calc-alkalic
MC	Metaluminous calcic
PCA	Peraluminous calc-alkalic
PC	Peraluminous calcic
UDH	Ultra-deep hydrothermal
NI	Non-igneous
Gp	Gold placer
OroS	Orocopia Schist

```
         Area of mineral district
         supersystem

.........  Connects noncontiguous areas
         of mineral district
```

o   Oxidized

wo Weakly oxidized

r   Reduced

Ag	Silver	Mo	Molybdenum
As	Arsenic	Nb	Niobium
Au	Gold	Pb	Lead
Ba	Barium	Sb	Antimony
Be	Beryllium	Tl	Thalium
C	Carbon	U	Uranium
Cu	Copper	V	Vanadium
F	Fluorine	W	Tungsten
Fe	Iron	Au(Cu)	Minor elements inside
Hg	Mercury		parentheses
Mn	Manganese		

MINERALIZATION AGES

18–14 Age-radiometric age date in Ma
       (mega annum or million years ago)

Maricopa County

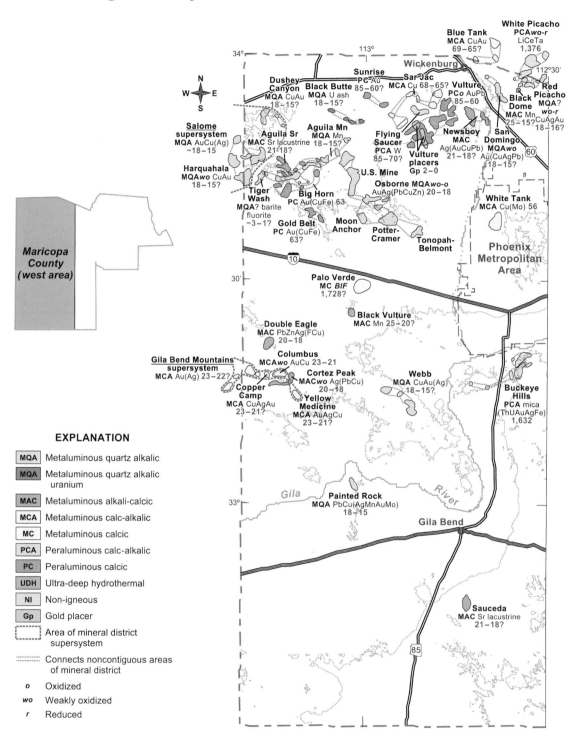

EXPLANATION

MQA	Metaluminous quartz alkalic
MQA	Metaluminous quartz alkalic uranium
MAC	Metaluminous alkali-calcic
MCA	Metaluminous calc-alkalic
MC	Metaluminous calcic
PCA	Peraluminous calc-alkalic
PC	Peraluminous calcic
UDH	Ultra-deep hydrothermal
NI	Non-igneous
Gp	Gold placer
⌐ ⌐	Area of mineral district supersystem
··········	Connects noncontiguous areas of mineral district
o	Oxidized
wo	Weakly oxidized
r	Reduced

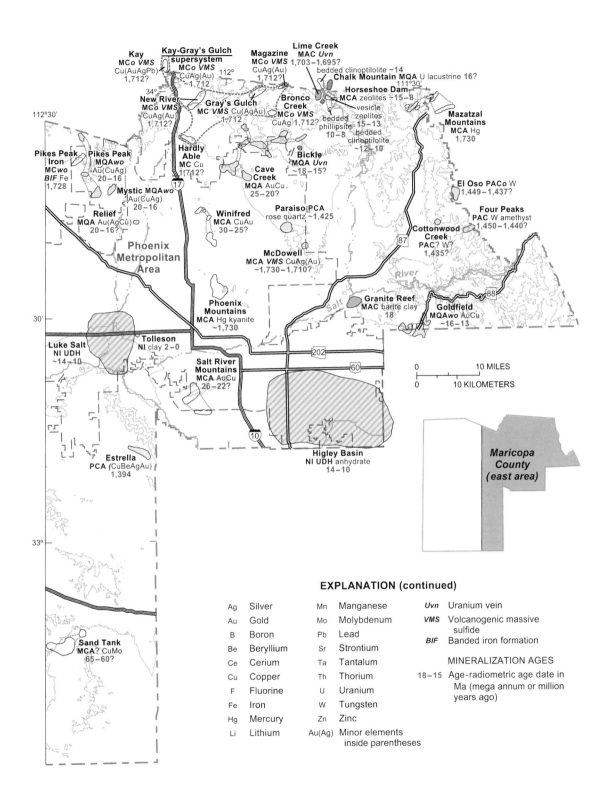

Kay
MCo *VMS*
Cu(AuAgPb)
1,712?

Kay-Gray's Gulch
supersystem
MCo *VMS*
CuAg(Au)
~1,712

Magazine
MCo *VMS*
CuAg(Au)
1,712?

Lime Creek
MAC *Uvn* 1,703–1,695?
bedded clinoptilolite ~14
Chalk Mountain MQA U lacustrine 16?

New River
MCo *VMS*
CuAg(Au)
1,712?

Gray's Gulch
MC *VMS* Cu(AgAu)
1,712

Bronco
Creek
MCo *VMS*
CuAg 1,712?

Horseshoe Dam
MCA zeolites ~15–8
vesicle
zeolites

bedded
phillipsite
15–13

Mazatzal
Mountains
MCA Hg
1,730

Pikes Peak
Iron
MC*wo*
BIF Fe
1,728

Pikes Peak
MQA*wo*
Au(CuAg)
20–16

Hardly
Able
MC Cu
1,712?

Bickle
MQA *Uvn*
~18–15?

bedded
clinoptilolite
~12–10

Mystic MQA*wo*
Au(CuAg)
20–16

Cave
Creek
MQA AuCu
25–20?

El Oso PAC*o* W
1,449–1,437?

Relief
MQA Au(AgCu)
20–16?

Winifred
MCA CuAu
30–25?

Paraiso PCA
rose quartz ~1,425

Four Peaks
PAC W amethyst
1,450–1,440?

Cottonwood
Creek
PAC? W?
1,435?

Phoenix
Metropolitan
Area

McDowell
MCA *VMS* CuAu(Au)
~1,730–1,710?

Granite Reef
MAC barite clay
18

Goldfield
MQA*wo* AuCu
~16–13

Phoenix
Mountains
MCA Hg kyanite
~1,730

Luke Salt
NI UDH
~14–10

Tolleson
NI clay 2–0

Salt River
Mountains
MCA AuCu
26–22?

Higley Basin
NI UDH anhydrate
14–10

Estrella
PCA (CuBeAgAu)
1,394

Sand Tank
MCA? CuMo
65–60?

*Maricopa
County
(east area)*

EXPLANATION (continued)

Ag	Silver	Mn	Manganese	*Uvn*	Uranium vein	
Au	Gold	Mo	Molybdenum	*VMS*	Volcanogenic massive sulfide	
B	Boron	Pb	Lead	*BIF*	Banded iron formation	
Be	Beryllium	Sr	Strontium			
Ce	Cerium	Ta	Tantalum		MINERALIZATION AGES	
Cu	Copper	Th	Thorium	18–15	Age-radiometric age date in	
F	Fluorine	U	Uranium		Ma (mega annum or million	
Fe	Iron	W	Tungsten		years ago)	
Hg	Mercury	Zn	Zinc			
Li	Lithium	Au(Ag)	Minor elements inside parentheses			

Mohave County

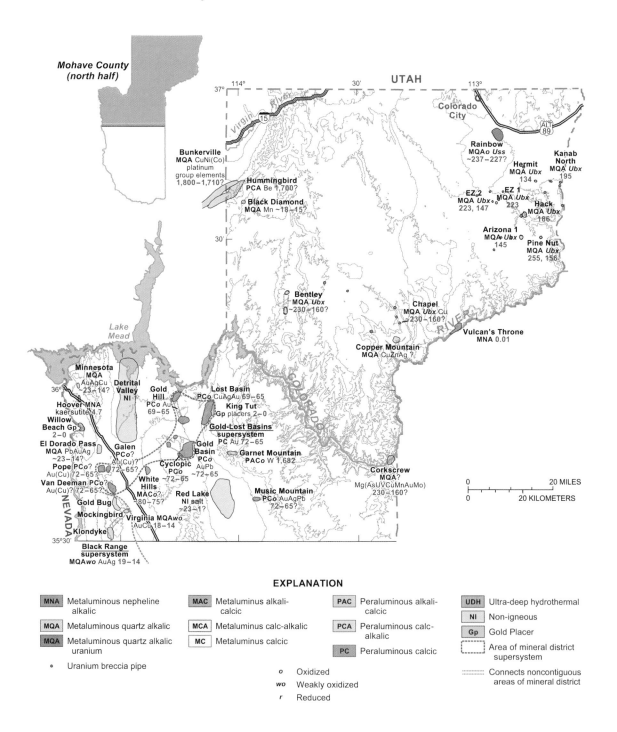

Mohave County (north half)

UTAH

37° 114° 30' 113°

Virgin River

Colorado City

15

ALT 89

Bunkerville
MQA CuNi(Co)
platinum
group elements
1,800–1,710?

Rainbow
MQAo *Uss*
~237–227?

Kanab North
MQA *Ubx*
195

Hermit
MQA *Ubx*
134

Hummingbird
PCA Be 1,700?

EZ 2
MQA *Ubx*
223, 147

EZ 1
MQA *Ubx*
223

Hack
MQA *Ubx*
186

Black Diamond
MQA Mn ~18–15?

30'

Arizona 1
MQA *Ubx*
145

Pine Nut
MQA *Ubx*
255, 156

Bentley
MQA *Ubx*
~230–160?

Chapel
MQA *Ubx* Cu
~230–160?

RIVER

Vulcan's Throne
MNA 0.01

Copper Mountain
MQA CuZnAg ?

Lake Mead

Minnesota
MQA
AuAgCu
~23–14?

36°

Detrital
Valley
NI

Gold
Hill
PCo Au
69–65

Lost Basin
PCo CuAgAu 69–65

King Tut
Gp placers 2–0

COLORADO

Hoover MNA
kaersutite 4.7

Willow
Beach Gp
2–0

El Dorado Pass
MQA PbAuAg
~23–14?

Galen
PCo?
Au(Cu)?
72–65?

Gold
Basin
PCo
AuPb
~72–65

Gold-Lost Basins
supersystem
PC Au 72–65

Garnet Mountain
PACo W 1,682

Corkscrew
MQA?
Mg(AsUVCuMnAuMo)
230–160?

Pope PCo?
Au(Cu) 72–65?

Cyclopic
PCo
~72–65

Van Deeman PCo?
Au(Cu)? 72–65?

White
Hills
MACo?
~80–75?

Red Lake
NI salt
~23–1?

Music Mountain
PCo AuAgPb
72–65?

Gold Bug

NEVADA

Mockingbird Virginia MQAwo
AuCu 18–14

Klondyke

35°30'

Black Range
supersystem
MQAwo AuAg 19–14

| | 0 | | 20 MILES |
| 0 | | 20 KILOMETERS | |

EXPLANATION

MNA	Metaluminous nepheline alkalic	**MAC**	Metaluminus alkali-calcic	**PAC**	Peraluminous alkali-calcic	**UDH**	Ultra-deep hydrothermal
MQA	Metaluminous quartz alkalic	**MCA**	Metaluminus calc-alkalic	**PCA**	Peraluminous calc-alkalic	**NI**	Non-igneous
MQA	Metaluminous quartz alkalic uranium	**MC**	Metaluminus calcic	**PC**	Peraluminous calcic	**Gp**	Gold Placer

• Uranium breccia pipe

o   Oxidized

wo   Weakly oxidized

r   Reduced

⬚ Area of mineral district supersystem

⋯⋯ Connects noncontiguous areas of mineral district

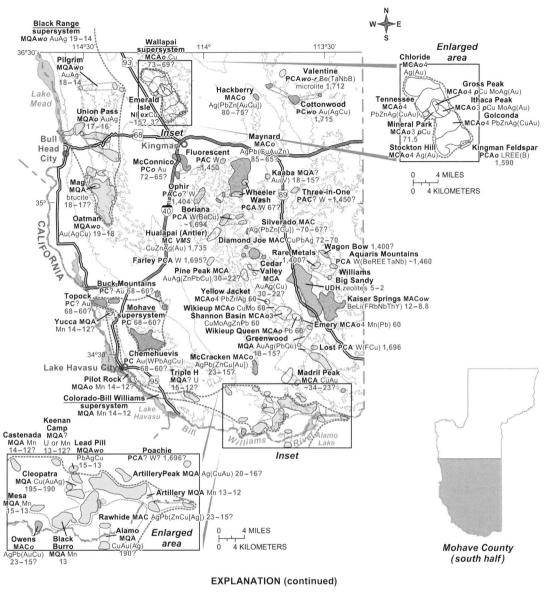

**Black Range
supersystem**
MQA*wo* AuAg 19–14
114°30'

Pilgrim
MQA*wo*
AuAg
18–14

Lake
Mead

Union Pass
MQAo AuAg
17–16?

Bull
Head
City

Mag
MQA
brucite
18–17?

Oatman
MQA*wo*
Au(AgCu) 19–18

Topock
PC? Au
68–60?

Yucca MQA
Mn 14–12?

Lake Havasu City

Pilot Rock
MQAo Mn 14–12?

**Colorado-Bill Williams
supersystem**
MQA Mn 14–12

**Keenan
Camp**
MQA?

Castenada
MQA Mn
14–12?

Lead Pill
MQA*wo*

Cleopatra
MQA Cu(AuAg)
195–190

Mesa
MQA Mn
15–13

Owens
MACo
AgPb(AuCu)
23–15?

**Black
Burro**
MQA Mn
13

**Wallapai
supersystem**
MCAo Cu
73–69?

**Emerald
Isle**
NI *exCu*
~15?, 3?

Kingman

McConnico
PCo Au
72–65?

Ophir
PACo? W
1,404

Boriana
PCA W(BeCu)
~1,694

Hualapai (Antler)
MC *VMS*
CuZnAg(Au) 1,735

Farley PCA W 1,695?

Buck Mountains
PC? Au 68–60?

**Mohave
supersystem**
PC 68–60?

Chemehuevis
PC Au(WPbAgCu)
68–60?

Lead Pill
MQA*wo*

Poachie
PCA? W? 1,696?

ArtilleryPeak MQA Ag(CuAu) 20–16?

Artillery MQA Mn 13–12

Rawhide MAC AgPb(ZnCu[Ag]) 23–15?

Alamo
MQA
CuAu(Ag)
190?

*Enlarged
area*

Hackberry
MACo
Ag(PbZn[AuCu])
80–75?

Valentine
PCA*wo-r* Be(TaNbB)
microlite 1,712

Cottonwood
PC*wo* Au(AgCu)
1,715

Maynard
MACo
AgPb(CuAuZn)
85–65?

Fluorescent
PAC W
~1,450

Kaaba MQA?
Au(V) 18–15?

**Wheeler
Wash**
PCA W 67?

Three-in-One
PAC? W ~1,450?

Silverado MAC
Ag(PbZn[Cu]) ~70–67?

Diamond Joe MAC CuPbAg 72–70

Rare Metals
~1,400?

**Cedar
Valley**
MCA
AuAg(Cu)
30–22?

Wagon Bow 1,400?

Aquarius Mountains
PCA W(BeREE TaNb) ~1,460

**Williams
Big Sandy**
UDH zeolites 5–2

Pine Peak MCA
AuAg(ZnPbCu) 30–22?

Yellow Jacket
MCAo4 PbZnAg 60

Wikieup MCAo CuMo 60

Shannon Basin MCAo3
CuMoAgZnPb 60

Wikieup Queen MCAo Pb 60

Greenwood
MQA AuAg(PbCu)
18–15?

McCracken MACo
AgPb(ZnCu[Au])

Kaiser Springs MACo*w*
BeLi(FRbNbThY) 12–8.8

Emery MCAo4 Mn(Pb) 60

Lost PCA W(FCu) 1,696

Triple H
MQA? U
15–12?

Madril Peak
MCA CuAu
~34–23?

Inset

*Enlarged
area*

Williams River

Alamo
Lake

Bill

Lake
Havasu

34°30'

*Enlarged
area*

Chloride
MCAo4
Ag(Au)

Tennessee
MCAo4
PbZnAg(CuAu)

Mineral Park
MCAo3 pCu
71.5

Stockton Hill
MCAo4 Ag(Au)

Gross Peak
MCAo4 pCu MoAg(Au)

Ithaca Peak
MCAo3 pCu MoAg(Au)

Golconda
MCAo4 PbZnAg(CuAu)

Kingman Feldspar
PCAo LREE(B)
1,590

0   4 MILES
0   4 KILOMETERS

0   4 MILES
0   4 KILOMETERS

Inset

*Mohave County
(south half)*

EXPLANATION (continued)

Ag	Silver	Li	Lithium	Th	Thorium	LREE Light rare earth elements
As	Arsenic	Mn	Manganese	U	Uranium	REE Rare earth elements
Au	Gold	Mo	Molybdenum	V	Vanadium	*p*Cu Porphyry copper
B	Boron	Nb	Niobium	W	Tungsten	*ex*Cu Exotic copper
Be	Beryllium	Pb	Lead	Zn	Zinc	*Ubx* Uranium breccia pipe
Co	Cobalt	Rb	Rubidium	W(FCu) Minor elements			*Uss* Uranium sandstone
Cu	Copper	Ta	Tantalum		inside parentheses		*VMS* Volcanogenic massive sulfide
F	Fluorine					

MINERALIZATION AGES

19–14   Age-radiometric age
date in Ma (mega
annum or million years
ago)

Navajo County

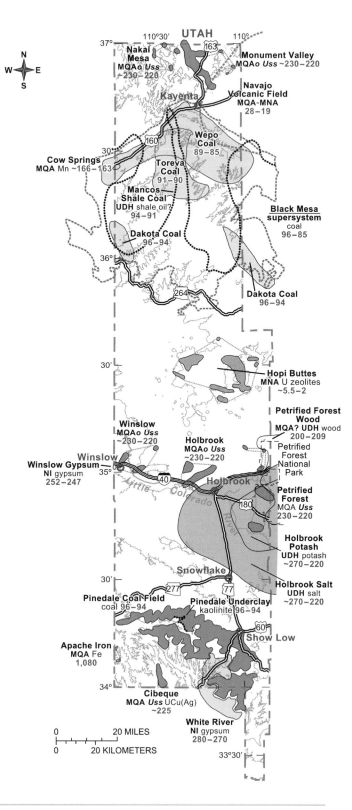

Map labels:

UTAH
110°30'   110°
37°
Nakai Mesa
MQAo *Uss* ~230–220
Monument Valley
MQAo *Uss* ~230–220
163
Navajo Volcanic Field
MQA–MNA 28–19
Kayenta
Wepo Coal 89–85
160
Cow Springs
MQA Mn ~166–163
30'
Toreva Coal 91–90
Mancos Shale Coal
UDH shale oil? 94–91
Black Mesa supersystem coal 96–85
Dakota Coal 96–94
36°
264
Dakota Coal 96–94
30'
Hopi Buttes
MNA U zeolites ~5.5–2
Petrified Forest Wood
MQA? UDH wood 200–209
Petrified Forest National Park
Winslow
MQAo *Uss* ~230–220
Holbrook
MQAo *Uss* ~230–220
Winslow Gypsum
NI gypsum 252–247
Winslow
35°
40
Little Colorado River
Holbrook
180
Petrified Forest
MQA *Uss* 230–220
Holbrook Potash
UDH potash ~270–220
Snowflake
30'
277   77
Holbrook Salt
UDH salt ~270–220
60
Show Low
Pinedale Coal Field coal 96–94
Pinedale Underclay kaolinite 96–94
Apache Iron
MQA Fe 1,080
34°
Cibeque
MQA *Uss* UCu(Ag) ~225
White River
NI gypsum 280–270
33°30'

0   20 MILES
0   20 KILOMETERS

The trees that silicified into petrified wood in Navajo and Apache Counties lived about 225 million years ago. The Colorado Plateau uplift of 60 million years ago and the subsequent erosion exposed the Chinle Formation, which contains the fossilized logs. In 1906 the Petrified Forest National Monument was established, and in 1964 the area was designated a national park. John Muir (1838–1914), an early advocate for wilderness conservation, is seen here visiting the park. This photograph is from the Holt-Atherton Special Collections.

Scottish American naturalist John Muir examining petrified wood in what is now the Petrified Forest National Park, 1906. Photograph taken by his daughter Helen Muir.

This quarry was mined from 1909 to 1914 for selenite, a variety of gypsum. This product was shipped to California for use in cement production. The original publication caption in a U.S. Geological Survey Bulletin *Gypsum Deposits of the United States* by Ralph Walter Stone stated that the ledge from which the gypsum was blasted is shown on the left. The ledge appears to be shown on the right of the photograph. Thanks to William Ascarza of *Mine Tales*. Photograph from the U.S. Geological Survey collection.

Gypsum quarry located about 3 miles west of Winslow, 1916.

Pima County

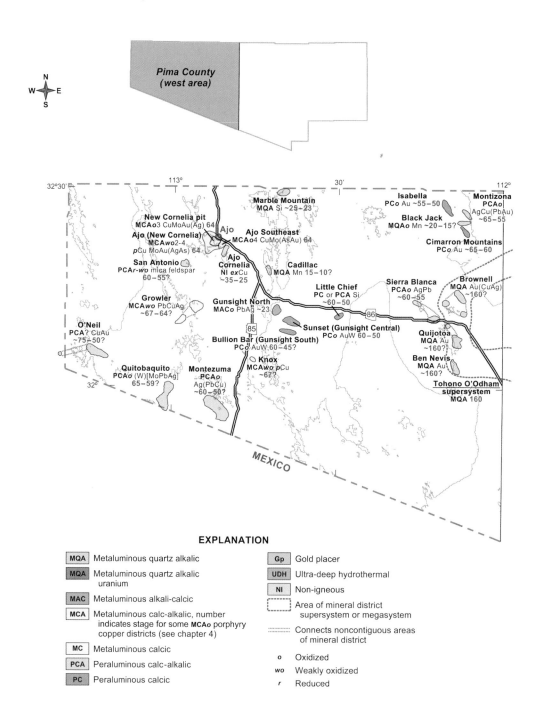

EXPLANATION

MQA	Metaluminous quartz alkalic
MQA	Metaluminous quartz alkalic uranium
MAC	Metaluminous alkali-calcic
MCA	Metaluminous calc-alkalic, number indicates stage for some **MCA**o porphyry copper districts (see chapter 4)
MC	Metaluminous calcic
PCA	Peraluminous calc-alkalic
PC	Peraluminous calcic

Gp	Gold placer
UDH	Ultra-deep hydrothermal
NI	Non-igneous
⌐ ¬	Area of mineral district supersystem or megasystem
··········	Connects noncontiguous areas of mineral district
o	Oxidized
wo	Weakly oxidized
r	Reduced

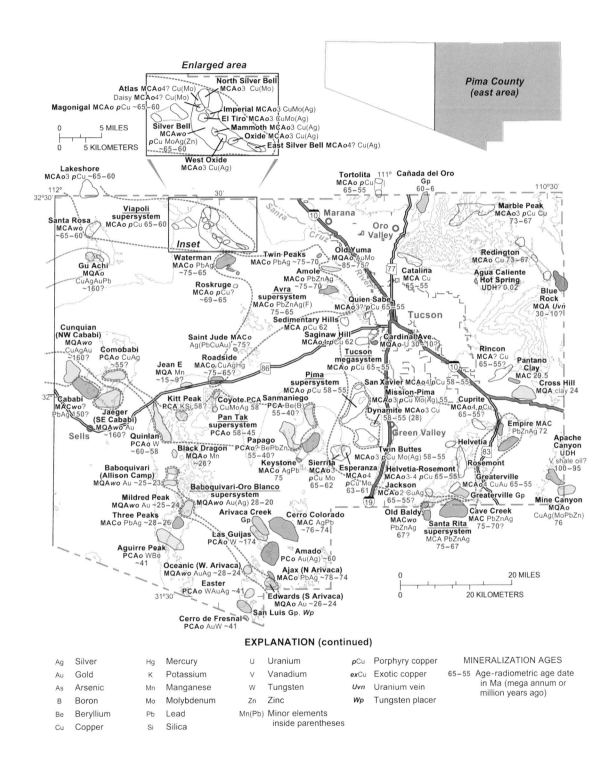

EXPLANATION (continued)

Ag	Silver	Hg	Mercury	U	Uranium	pCu	Porphyry copper		MINERALIZATION AGES
Au	Gold	K	Potassium	V	Vanadium	exCu	Exotic copper	65–55	Age-radiometric age date
As	Arsenic	Mn	Manganese	W	Tungsten	Uvn	Uranium vein		in Ma (mega annum or
B	Boron	Mo	Molybdenum	Zn	Zinc	Wp	Tungsten placer		million years ago)
Be	Beryllium	Pb	Lead	Mn(Pb)	Minor elements				
Cu	Copper	Si	Silica		inside parentheses				

Pinal County

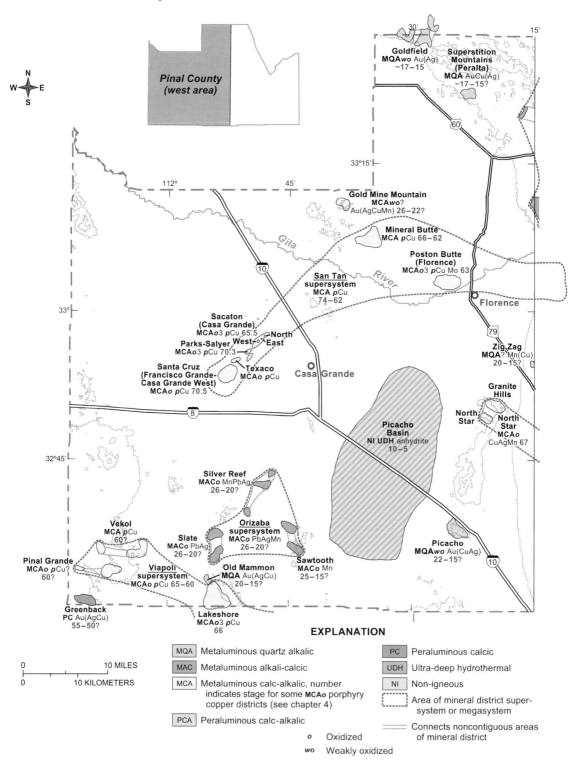

Pinal County (west area)

Goldfield
MQA*wo* Au(Ag)
~17–15

Superstition
Mountains
(Peralta)
MQA AuCu(Ag)
~17–15?

33°15'

Gold Mine Mountain
MCA*wo*?
Au(AgCuMn) 26–22?

Mineral Butte
MCA *p*Cu 66–62

Poston Butte
(Florence)
MCA*o*3 *p*Cu Mo 63

San Tan
supersystem
MCA *p*Cu
74–62

Florence

33°

Sacaton
(Casa Grande)
MCA*o*3 *p*Cu 65.5

North
West East
Parks-Salyer
MCA*o*3 *p*Cu 70.3

Zig Zag
MQA? Mn(Cu)
20–15?

Santa Cruz
(Francisco Grande-
Casa Grande West)
MCA*o* *p*Cu 70.5

Texaco
MCA*o* *p*Cu

Casa Grande

Granite
Hills

North Star

North
Star
MCA*o*
CuAgMn 67

Picacho
Basin
NI UDH anhydrite
10–5

32°45'

Silver Reef
MACo MnPbAg
26–20?

Vekol
MCA *p*Cu
60?

Slate
MACo PbAg
26–20?

Orizaba
supersystem
MACo PbAgMn
26–20?

Picacho
MQA*wo* Au(CuAg)
22–15?

Pinal Grande
MCA*o* *p*Cu?
60?

Viapoli
supersystem
MCA*o* *p*Cu 65–60

Old Mammon
MQA Au(AgCu)
20–15?

Sawtooth
MACo Mn
25–15?

Greenback
PC Au(AgCu)
55–50?

Lakeshore
MCA*o*3 *p*Cu
66

EXPLANATION

MQA	Metaluminous quartz alkalic
MAC	Metaluminous alkali-calcic
MCA	Metaluminous calc-alkalic, number indicates stage for some **MCAo** porphyry copper districts (see chapter 4)
PCA	Peraluminous calc-alkalic

PC	Peraluminous calcic
UDH	Ultra-deep hydrothermal
NI	Non-igneous

- - - - Area of mineral district super-system or megasystem

·········· Connects noncontiguous areas of mineral district

o Oxidized
wo Weakly oxidized

0   10 MILES
0   10 KILOMETERS

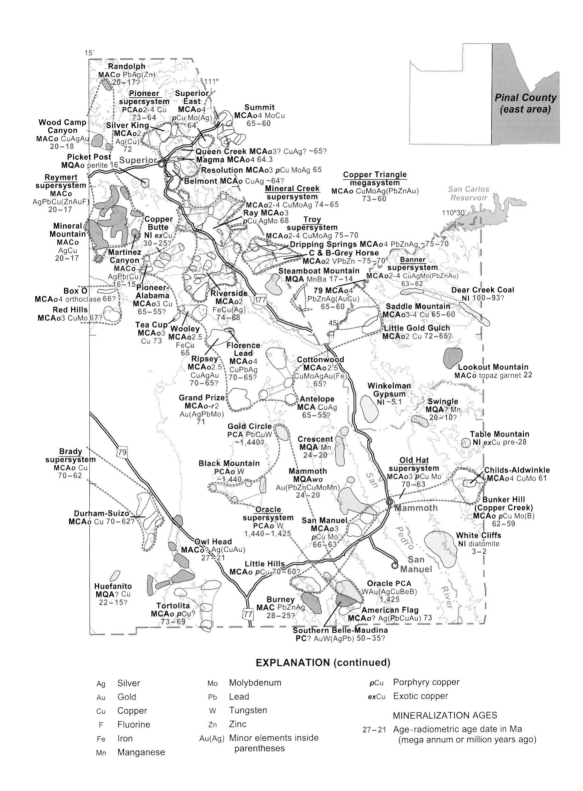

EXPLANATION (continued)

Ag	Silver	Mo	Molybdenum	*p*Cu	Porphyry copper	
Au	Gold	Pb	Lead	*ex*Cu	Exotic copper	
Cu	Copper	W	Tungsten			
F	Fluorine	Zn	Zinc	MINERALIZATION AGES		
Fe	Iron	Au(Ag)	Minor elements inside parentheses	27–21	Age-radiometric age date in Ma (mega annum or million years ago)	
Mn	Manganese					

Santa Cruz County

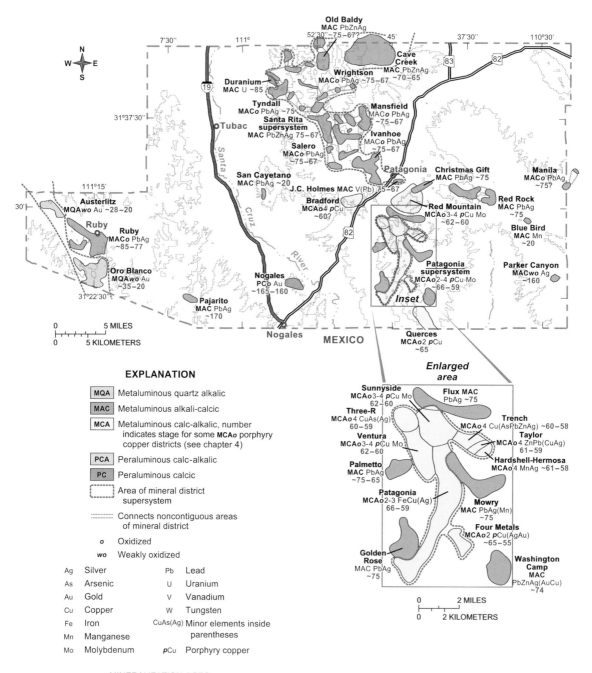

EXPLANATION

| MQA | Metaluminous quartz alkalic |

| MAC | Metaluminous alkali-calcic |

| MCA | Metaluminous calc-alkalic, number indicates stage for some **MCAo** porphyry copper districts (see chapter 4) |

| PCA | Peraluminous calc-alkalic |

| PC | Peraluminous calcic |

Area of mineral district supersystem

Connects noncontiguous areas of mineral district

o   Oxidized

wo   Weakly oxidized

Ag	Silver	Pb	Lead
As	Arsenic	U	Uranium
Au	Gold	V	Vanadium
Cu	Copper	W	Tungsten
Fe	Iron	CuAs(Ag)	Minor elements inside parentheses
Mn	Manganese		
Mo	Molybdenum	*p*Cu	Porphyry copper

MINERALIZATION AGES

75–65 Age-radiometric age date in Ma (mega annum or million years ago)

Ruby mineral district Montana Mine headframe as it appeared in the mid 1930s.

Spanish explorers discovered ore deposits in the mid-1700s near Montana Peak, where they did some placer mining for a couple of years. After the Gadsen Treaty of 1853 with Mexico, early mining entrepreneurs Charles Poston and Herman Ehrenberg rediscovered the Spanish workings in 1854. With the suppression of the Apache threat, mining activity in the Montana Peak area became active, and Montana Camp was established in 1877. Montana Camp was renamed Ruby in 1912 with the arrival of its post office, and the population of Ruby grew to 1,200 by the mid-1930s. The Montana Mine produced gold, silver, and zinc, and for several years in the 1920s was Arizona's greatest lead producer. Economic ore played out by 1939, and the Montana Mine closed in 1940. Photograph is courtesy of Photo Archive, Arizona Geological Survey.

Yavapai County

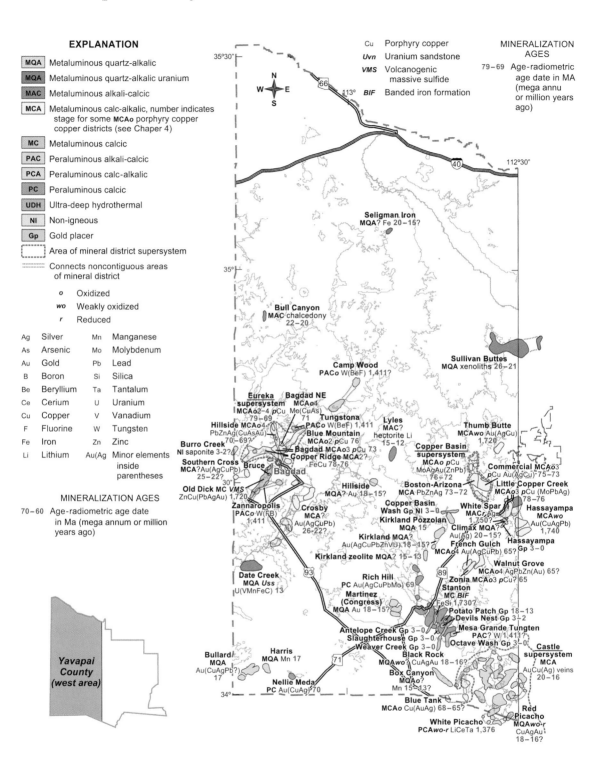

EXPLANATION

MQA Metaluminous quartz-alkalic

MQA Metaluminous quartz-alkalic uranium

MAC Metaluminous alkali-calcic

MCA Metaluminous calc-alkalic, number indicates stage for some MCAo porphyry copper copper districts (see Chaper 4)

MC Metaluminous calcic

PAC Peraluminous alkali-calcic

PCA Peraluminous calc-alkalic

PC Peraluminous calcic

UDH Ultra-deep hydrothermal

NI Non-igneous

Gp Gold placer

⌐ ¬ Area of mineral district supersystem
�
.... Connects noncontiguous areas of mineral district

o Oxidized

wo Weakly oxidized

r Reduced

Ag	Silver	Mn	Manganese
As	Arsenic	Mo	Molybdenum
Au	Gold	Pb	Lead
B	Boron	Si	Silica
Be	Beryllium	Ta	Tantalum
Ce	Cerium	U	Uranium
Cu	Copper	V	Vanadium
F	Fluorine	W	Tungsten
Fe	Iron	Zn	Zinc
Li	Lithium	Au(Ag	Minor elements inside parentheses

MINERALIZATION AGES

70–60 Age-radiometric age date in Ma (mega annum or million years ago)

Cu Porphyry copper

Uvn Uranium sandstone

VMS Volcanogenic massive sulfide

BIF Banded iron formation

MINERALIZATION AGES

79–69 Age-radiometric age date in MA (mega annu or million years ago)

Yavapai County (west area)

The Congress Mine in the Martinez minteral district with the town of Congress
in the foreground, 1898. Photograph taken by Phoenix photographer William Altenburgh.

The Congress Mine was established in the Date Creek Mountains of southwestern Yavapai County in 1884. This was the largest gold-producing mine in the Arizona Territory, with over 400,000 ounces of gold being produced in total and almost as many ounces of silver. Production declined after 1910, and the post office closed in 1938. Photograph courtesy of the Graeme/Larkin collection.

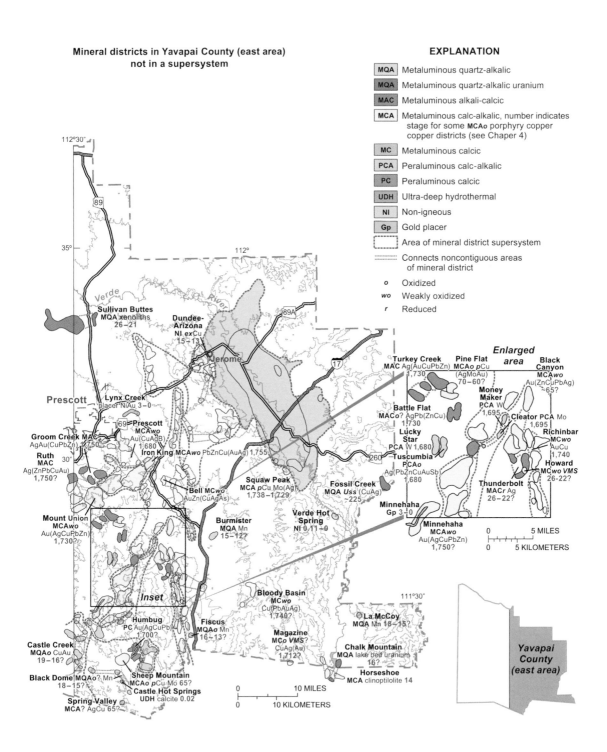

Mineral districts in Yavapai County (east area)
not in a supersystem

112°30"

89

35°

112°

Verde River

Sullivan Buttes
MQA xenoliths
26–21

Dundee-
Arizona
NI *exCu*
15–13

Jerome

89A

17

Turkey Creek
MAC Ag(AuCuPbZn)
1,730

Pine Flat
MCAo *pCu*
(AgMoAu)
70–60?

*Enlarged
area*

Black
Canyon
MCAwo
Au(ZnCuPbAg)
~65?

Money
Maker
PCA W
1,695

Cleator PCA Mo
1,695

Prescott

Lynx Creek
placer NiAu 3–0

69 Prescott
MCAwo
Au(CuAgB)
1,680

Battle Flat
MACo? AgPb(ZnCu)
1,730

Richinbar
MCwo
AuCu
1,740

Groom Creek MAC
AgAu(CuPbZn) 1,750

Ruth
MAC
Ag(ZnPbCuAu)
1,750?

30"

Iron King MCAwo PbZnCu(AuAg) 1,755

Lucky
Star
PCA W 1,680

Tuscumbia
PCAo
Ag(PbZnCuAuSb)
1,680

260

Howard
MCwo *VMS*
26–22?

Bell MCwo
AuZn(CuAgAs)

Squaw Peak
MCA *pCu* Mo(Ag)
1,738–1,729

Fossil Creek
MQA *Uss* (CuAg)
~225

Thunderbolt
MACr Ag
26–22?

Mount Union
MCAwo
Au(AgCuPbZn)
1,730?

Burmister
MQA Mn
15–12?

Verde Hot
Spring
NI 0.11–0

Minnehaha
Gp 3–0

Minnehaha
MCAwo
Au(AgCuPbZn)
1,750?

0   5 MILES

0   5 KILOMETERS

Inset

Bloody Basin
MCwo
Cu(PbAuAg)
1,740?

111°30"

La McCoy
MQA Mn 18–15?

Humbug
PC Au(AgCuPb)
1,700?

Fiscus
MQAo Mn
16–13?

Magazine
MCo *VMS*?
CuAg(Au)
1,712?

Chalk Mountain
MQA lake bed uranium
16?

Castle Creek
MQAo CuAu
19–16?

Black Dome MQAo? Mn
18–15?

Sheep Mountain
MCAo *pCu* Mo 65?

Castle Hot Springs
UDH calcite 0.02

Horseshoe
MCA clinoptilolite 14

Spring Valley
MCA? AgCu 65?

0   10 MILES

0   10 KILOMETERS

*Yavapai
County
(east area)*

Mineral districts in Yavapai County (east area) in supersystems

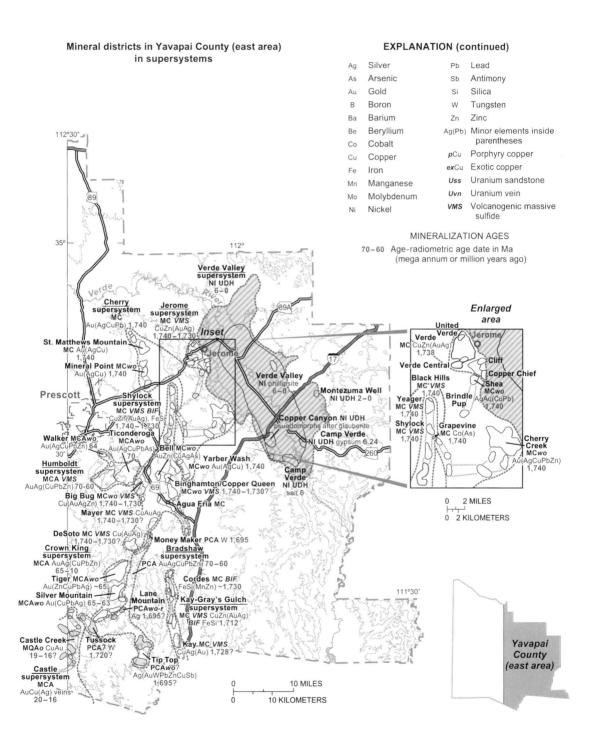

EXPLANATION (continued)

Ag	Silver	Pb	Lead
As	Arsenic	Sb	Antimony
Au	Gold	Si	Silica
B	Boron	W	Tungsten
Ba	Barium	Zn	Zinc
Be	Beryllium	Ag(Pb)	Minor elements inside parentheses
Co	Cobalt		
Cu	Copper	*p*Cu	Porphyry copper
Fe	Iron	*ex*Cu	Exotic copper
Mn	Manganese	*Uss*	Uranium sandstone
Mo	Molybdenum	*Uvn*	Uranium vein
Ni	Nickel	*VMS*	Volcanogenic massive sulfide

MINERALIZATION AGES

70–60 Age-radiometric age date in Ma (mega annum or million years ago)

Enlarged area

0   2 MILES
0   2 KILOMETERS

Yavapai County (east area)

111°30'

0   10 MILES
0   10 KILOMETERS

Yuma County

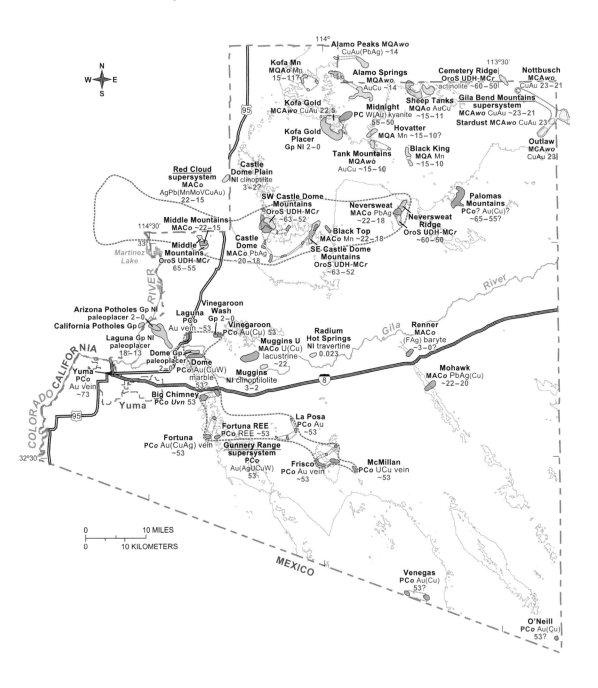

Alamo Peaks MQA*wo*
CuAu(PbAg) ~14

113°30'

Kofa Mn
MQA*o* Mn
15–11?

Alamo Springs
MQA*wo*
AuCu ~14

Cemetery Ridge|
OroS UDH-MC*r*
actinolite ~60–50|

Nottbusch
MCA*wo*
CuAu 23–21

Kofa Gold
MCA*wo* CuAu 22.5

Midnight
PC W(Au) kyanite
55–50

Sheep Tanks
MQA*o* AuCu

Gila Bend Mountains
supersystem
MCA*wo* CuAu ~23–21
Stardust MCA*wo* CuAu 23

Hovatter
MQA Mn ~15–10?

Kofa Gold
Placer
Gp NI 2–0

Tank Mountains
MQA*wo*
AuCu ~15–10

Black King
MQA Mn
~15–10

Outlaw
MCA*wo*
CuAu 23|

Red Cloud
supersystem
MACo
AgPb(MnMoVCuAu)
22–15

Castle
Dome Plain
NI clinoptilite
3–2?

SW Castle Dome
Mountains
OroS UDH-MC*r*
~63–52

Neversweat
MACo PbAg
~22–18

Neversweat
Ridge
OroS UDH-MC*r*
~60–50

Palomas
Mountains
PCo? Au(Cu)?
~65–55?

114°30'

Middle Mountains
MACo ~22–15

Middle
Mountains
OroS UDH-MC*r*
65–55

*Martinez
Lake*

Castle
Dome
MACo PbAg
~20–18

Black Top
MACo Mn ~22–18

SE Castle Dome
Mountains
OroS UDH-MC*r*
~63–52

RIVER

Vinegaroon
Wash
Gp 2–0

Arizona Potholes Gp NI
paleoplacer 2–0
California Potholes Gp

Laguna
PCo
Au vein ~53

Vinegaroon
PCo Au(Cu) 53

Radium
Hot Springs
NI travertine
0.023

Renner
MACo
(FAg) baryte
~3–0?

River

Gila

Laguna Gp NI
paleoplacer
18–13

Muggins U
MACo U(Cu)
lacustrine
~22

COLORADO CALIFORNIA

Dome Gp
paleoplacer
2–0

Dome
PCo Au(CuW)
marble
53?

Muggins
NI clinoptilolite
3–2

Mohawk
MACo PbAg(Cu)
~22–20

Yuma
PCo
Au vein
~73

Big Chimney
PCo *Uvn* 53

Yuma

La Posa
PCo Au
~53

Fortuna REE
PCo Au REE ~53

Fortuna
PCo Au(CuAg) vein
~53

Gunnery Range
supersystem
PCo
Au(AgUCuW)
53

Frisco
PCo Au vein
~53

McMillan
PCo UCu vein
~53

32°30'

MEXICO

Venegas
PCo Au(Cu)
53?

O'Neill
PCo Au(Cu)
53?

N
W E
S

| 0 | | 10 MILES |
| 0 | | 10 KILOMETERS |

114°

Southern Pacific crews are completing a rail bridge across the Colorado River to reach Yuma, Arizona. The Southern Pacific diamond stacker (coal burning) locomotive waits on the bridge and will be the first train to enter Arizona. Railroads significantly improved the economic viability of mining operations in Arizona. Photograph is from the Arizona Historical Society, Library & Archives.

The first train into Arizona, 1887.

EXPLANATION

MQA	Metaluminous quartz alkalic	Ag	Silver
MAC	Metaluminous alkali-calcic	Au	Gold
MCA	Metaluminous calc-alkalic	Ba	Barium
PAC	Peraluminous alkali-calcic	Be	Beryllium
PCA	Peraluminous calc-alkalic	Cu	Copper
PC	Peraluminous calcic	F	Fluorine
UDH	Ultra-deep hydrothermal	Mn	Manganese
NI	Non-igneous	Mo	Molybdenum
Gp	Gold placer		
OroS	Orocopia Schist		

Ag Silver
Au Gold
Ba Barium
Be Beryllium
Cu Copper
F Fluorine
Mn Manganese
Mo Molybdenum

Pb Lead
Si Silica
U Uranium
V Vanadium
W Tungsten
Au(Cu) Minor elements inside parentheses
REE Rare earth elements
Uvn Uranium vein

□ Area of mineral district supersystem

┈ Connects noncontiguous areas of mineral district

o Oxidized
wo Weakly oxidized
r Reduced

MINERALIZATION AGES
18–13 Age-radiometric age date in Ma (mega annum or million years ago)

APPENDIX B

Minerals Added since the 3rd Edition

#ALTERITE   indicates a type mineral for Arizona
Note: Many of the new Arizona species listed here are the result of changes in nomenclature. Those species do not occur with this name in the third edition.

#ABERNATHYITE
#ADANITE
#AERINITE
#AESCHYNITE-(Y)
#AGARDITE-(Nd)
#ALBITE
#ALLANITE-(Ce)
#ALLANITE-(Nd)
#ALLANTOIN
#ALTERITE
#ALUM-(K)
#ALUM-(Na)
#AMMINEITE
#ANILITE
#ANNABERGITE
#ANNITE
#ANORTHITE
#ANTIMONY
#ANTIPINITE
#APHTHITALITE
#ARCANITE

#ARSENDESCLOIZITE
#ARSENTSUMEBITE
#ARTHURITE
#ASHBURTONITE
#AUROSTIBITE
#AWARUITE
#AXINITE-(Mn)
#BABINGTONITE
#BACKITE
#BAIRDITE
#BANDYLITE
#BARLOWITE
#BARROISITE
#BASTNÄSITE-(Ce)
#BAYLISSITE
#BEAVERITE-(Cu)
#BECHERERITE
#BIPHOSPHAMMITE
#BOBMEYERITE
#BONATTITE
#BOOTHITE

#BREWSTERITE-Ba
#BRUSHITE
#BULTFONTEINITE
#BURCKHARDTITE
#BUSTAMITE
#CALCIOARAVAIPAITE
#CALDERÓNITE
#CARLOSBARBOSAITE
#CARMICHAELITE
#CARMINITE
#CERVELLEITE
#CHABAZITE-Ca
#CHABAZITE-Na
#CHALCONATRONITE
#CHALCOSTIBITE
#CHEVKINITE-(Ce)
#CLINOATACAMITE
#COBALTPENTLANDITE
#COLORADOITE
#COLUMBITE-(Fe)
#CORDEROITE
#CRICHTONITE
#CUMENGEITE
#CUPROAURIDE
#CUPROCOPIAPITE
#CUPRORIVAITE
#DAVIDBROWNITE-(NH$_4$)
#DAVIDITE-(Ce)
#DENDORAITE-(NH$_4$)
#DESTINEZITE
#DICKINSONITE-(KMnNa)
#EDDAVIDITE
#ERIOCHALCITE
#ERIONITE-K
#EULYTINE
#EUXENITE-(Y)
#FELSÖBÁNYAITE
#FERROHEXAHYDRITE
#FERRO-HORNBLENDE
#FERROSELITE

#FLAGGITE
#FLUORAPOPHYLLITE-(K)
#FUETTERERITE
#FURUTOBEITE
#GANOMALITE
#GEERITE
#GEORGEROBINSONITE
#GLAUCOCERINITE
#GLAUCODOT
#GLUSHINSKITE
#GORDAITE
#GRANDVIEWITE
#HALLOYSITE-10Å
#HEAZLEWOODITE
#HECHTSBERGITE
#HEDYPHANE
#HERBERTSMITHITE
#HOGANITE
#HOPEITE
#HOUSLEYITE
#HYALOTEKITE
#HYDROGLAUBERITE
#HYDROHONESSITE
#HYDROKENORALSTONITE
#HYDROXYAPOPHYLLITE-(K)
#HYDROXYCALCIOROMÉITE
#HYDROXYKENOELSMOREITE
#HYDROXYLPYROMORPHITE
#ICE
#IDAITE
#IIMORIITE-(Y)
#IKAITE
#IRON
#JACOBSITE
#JÔKOKUITE
#KALICINITE
#KAMITUGAITE
#KENHSUITE
#KIESERITE
#KINOSHITALITE

#KOLOVRATITE

#KRUPKAITE

#KUKSITE

#LANSFORDITE

#LAZARASKEITE

#LECHATELIERITE

#LIKASITE

#LIROCONITE

#LIUDONGSHENGITE

#LOVERINGITE

#MAGNESIOALTERITE

#MAGNESIO-FOITITE

#MARKASCHERITE

#MATTHEDDLEITE

#MAUCHERITE

#MCALPINEITE

#MEIONITE

#MERRILLITE

#METAURANOCIRCITE-I

#MILARITE

#MOGÁNITE

#MOLYBDITE

#MOLYBDOPHYLLITE

#MONAZITE-(Ce)

#MONOHYDROCALCITE

#MONTANITE

#MONTROYDITE

#MUNAKATAITE

#NAMIBITE

#NATROCHALCITE

#NATRODUFRÉNITE

#NATROSULFATOUREA

#NATROZIPPEITE

#NESQUEHONITE

#NICKELBOUSSINGAULTITE

#NICKELPHOSPHIDE

#NORDSTRANDITE

#OPAL

#ORCELITE

#ORTHOSERPIERITE

#OTAVITE

#OTTOITE

#PACEITE

#PARACOQUIMBITE

#PARKINSONITE

#PENTAHYDRITE

#PERROUDITE

#PETERSITE-(Ce)

#PHARMACOLITE

#PHILIPSBORNITE

#PHOSPHOHEDYPHANE

#PHOXITE

#PLANERITE

#PLUMBOTELLURITE

#POLYCRASE-(Y)

#POLYDYMITE

#POUGHITE

#PYROCHLORE

#RAMSBECKITE

#RASPITE

#RAYGRANTITE

#RECTORITE

#REICHENBACHITE

#RELIANCEITE-(K)

#RONGIBBSITE

#ROQUESITE

#ROWLEYITE

#RUTHERFORDINE

#SALAMMONIAC

#SAMARSKITE-(Y)

#SCHOLZITE

#SCHULENBERGITE

#SCHWERTMANNITE

#SCOTLANDITE

#SHANNONITE

#SHARPITE

#SIMONKOLLEITE

#SODDYITE

#SPURRITE

#SRILANKITE

#STRACZEKITE
#STRUVITE
#SURITE
#SUSANNITE
#SYNCHYSITE
#SZENICSITE
#TAKANELITE
#TANGEITE
#TANTALITE-(Fe)
#TAPIOLITE-(Fe)
#TETRAHEDRITE-(Hg)
#THALÉNITE-(Y)
#THOMSONITE-Ca
#TIMROSEITE
#TREVORITE
#TUNGSTENITE

#URANOPILITE
#UREA
#USHKOVITE
#UTAHITE
#VILLAMANÍNITE
#VIOLARITE
#WAGNERITE
#WEDDELLITE
#WHEATLEYITE
#WHELANITE
#WIDENMANNITE
#WILLEMSEITE
#WOODHOUSEITE
#WUPATKIITE
#ZINCOCOPIAPITE

APPENDIX C

Checklist of Arizona Minerals

# indicates minerals new to the fourth edition
BOLD type indicates an Arizona type mineral

- ❏ #ABERNATHYITE
- ❏ ACANTHITE
- ❏ ACTINOLITE
- ❏ ADAMITE
- ❏ #ADANITE
- ❏ AEGIRINE
- ❏ AENIGMATITE
- ❏ #AERINITE
- ❏ #AESCHYNITE-(Y)
- ❏ #AGARDITE-(Nd)
- ❏ AIKINITE
- ❏ AJOITE
- ❏ AKAGANEITE
- ❏ ALABANDITE
- ❏ ALAMOSITE
- ❏ #ALBITE
- ❏ #ALLANITE-(Ce)
- ❏ #ALLANITE-(Nd)
- ❏ #ALLANTOIN
- ❏ ALLOCLASITE
- ❏ ALLOPHANE
- ❏ ALMANDINE
- ❏ ALTAITE
- ❏ #ALTERITE
- ❏ #ALUM-(K)
- ❏ #ALUM-(Na)
- ❏ ALUNITE
- ❏ ALUNOGEN
- ❏ AMBLYGONITE
- ❏ AMESITE
- ❏ #AMMINEITE
- ❏ ANALCIME
- ❏ ANATASE
- ❏ ANDALUSITE
- ❏ ANDERSONITE
- ❏ ANDRADITE
- ❏ ANGLESITE
- ❏ ANHYDRITE
- ❏ #ANILITE
- ❏ ANKERITE
- ❏ #ANNABERGITE
- ❏ #ANNITE
- ❏ #ANORTHITE
- ❏ ANTHONYITE

- ❏ ANTHOPHYLLITE
- ❏ ANTIGORITE
- ❏ #ANTIMONY
- ❏ #ANTIPINITE
- ❏ ANTLERITE
- ❏ APACHITE
- ❏ #APHTHITALITE
- ❏ ARAGONITE
- ❏ ARAVAIPAITE
- ❏ #ARCANITE
- ❏ ARGENTOJAROSITE
- ❏ #ARSENDESCLOIZITE
- ❏ ARSENIC
- ❏ ARSENIOSIDERITE
- ❏ ARSENOLITE
- ❏ ARSENOPYRITE
- ❏ #ARSENTSUMEBITE
- ❏ #ARTHURITE
- ❏ ARTROEITE
- ❏ #ASHBURTONITE
- ❏ ATACAMITE
- ❏ AUGELITE
- ❏ AUGITE
- ❏ AURICHALCITE
- ❏ AURORITE
- ❏ #AUROSTIBITE
- ❏ AUSTINITE
- ❏ AUTUNITE
- ❏ #AWARUITE
- ❏ #AXINITE-(Mn)
- ❏ AZURITE
- ❏ #BABINGTONITE
- ❏ #BACKITE
- ❏ #BAIRDITE
- ❏ #BANDYLITE
- ❏ #BARLOWITE
- ❏ BARRINGERITE
- ❏ #BARROISITE
- ❏ BARYTE
- ❏ BASSANITE

- ❏ BASSETITE
- ❏ #BASTNÄSITE-(Ce)
- ❏ BAYLDONITE
- ❏ BAYLEYITE
- ❏ #BAYLISSITE
- ❏ #BEAVERITE-(Cu)
- ❏ #BECHERERITE
- ❏ BECQUERELITE
- ❏ BEIDELLITE
- ❏ BEMENTITE
- ❏ BERLINITE
- ❏ BERMANITE
- ❏ BERTRANDITE
- ❏ BERYL
- ❏ BEUDANTITE
- ❏ BEYERITE
- ❏ BIANCHITE
- ❏ BIDEAUXITE
- ❏ BIEBERITE
- ❏ BILINITE
- ❏ BINDHEIMITE
- ❏ #BIPHOSPHAMMITE
- ❏ BIRNESSITE
- ❏ BISMITE
- ❏ BISMUTH
- ❏ BISMUTHINITE
- ❏ BISMUTITE
- ❏ BIXBYITE
- ❏ BLÖDITE
- ❏ #BOBMEYERITE
- ❏ BOGDANOVITE
- ❏ BÖHMITE
- ❏ BOKITE
- ❏ BOLEITE
- ❏ BOLTWOODITE
- ❏ #BONATTITE
- ❏ #BOOTHITE
- ❏ BORNITE
- ❏ BOTALLACKITE
- ❏ BOTRYOGEN

- ❏ BOULANGERITE
- ❏ BOURNONITE
- ❏ BRACKEBUSCHITE
- ❏ BRANNERITE
- ❏ BRAUNITE
- ❏ BRAZILIANITE
- ❏ #BREWSTERITE-Ba
- ❏ BREZINAITE
- ❏ BROCHANTITE
- ❏ BROCKITE
- ❏ BROMARGYRITE
- ❏ BROOKITE
- ❏ BRUCITE
- ❏ #BRUSHITE
- ❏ #BULTFONTEINITE
- ❏ #BURCKHARDTITE
- ❏ #BUSTAMITE
- ❏ BUTLERITE
- ❏ BÜTSCHLIITE
- ❏ BUTTGENBACHITE
- ❏ CACOXENITE
- ❏ CALAVERITE
- ❏ #CALCIOARAVAIPAITE
- ❏ CALCITE
- ❏ #CALDERÓNITE
- ❏ CALEDONITE
- ❏ CALOMEL
- ❏ CANFIELDITE
- ❏ CANNIZZARITE
- ❏ CARBONATECYANOTRICHITE
- ❏ #CARLOSBARBOSAITE
- ❏ CARLSBERGITE
- ❏ #CARMICHAELITE
- ❏ #CARMINITE
- ❏ CARNALLITE
- ❏ CARNOTITE
- ❏ CASSITERITE
- ❏ CELADONITE
- ❏ CELESTINE
- ❏ CELSIAN

- ❏ CERUSSITE
- ❏ CERVANTITE
- ❏ #CERVELLEITE
- ❏ CESÀROLITE
- ❏ CESBRONITE
- ❏ #CHABAZITE-Ca
- ❏ #CHABAZITE-Na
- ❏ CHALCANTHITE
- ❏ CHALCOALUMITE
- ❏ CHALCOCITE
- ❏ CHALCOMENITE
- ❏ #CHALCONATRONITE
- ❏ CHALCOPHANITE
- ❏ CHALCOPHYLLITE
- ❏ CHALCOPYRITE
- ❏ CHALCOSIDERITE
- ❏ #CHALCOSTIBITE
- ❏ CHAMOSITE
- ❏ CHENEVIXITE
- ❏ #CHEVKINITE-(Ce)
- ❏ CHLORAPATITE
- ❏ CHLORARGYRITE
- ❏ CHLORITOID
- ❏ CHOLOALITE
- ❏ CHONDRODITE
- ❏ CHROMITE
- ❏ CHRYSOBERYL
- ❏ CHRYSOCOLLA
- ❏ CHRYSOTILE
- ❏ CINNABAR
- ❏ CLARINGBULLITE
- ❏ CLAUDETITE
- ❏ CLAUSTHALITE
- ❏ #CLINOATACAMITE
- ❏ CLINOBISVANITE
- ❏ CLINOCHLORE
- ❏ CLINOCLASE
- ❏ CLINOHEDRITE
- ❏ CLINOHUMITE
- ❏ CLINOPTILOLITE

- CLINOZOISITE
- CLINTONITE
- COBALTITE
- #COBALTPENTLANDITE
- COCONINOITE
- **COESITE**
- COFFINITE
- COHENITE
- COLEMANITE
- #COLORADOITE
- #COLUMBITE-(Fe)
- COLUSITE
- CONICHALCITE
- CONNELLITE
- COOKEITE
- COPIAPITE
- COPPER
- COQUIMBITE
- #CORDEROITE
- CORDIERITE
- CORKITE
- CORNETITE
- CORNUBITE
- CORNWALLITE
- **CORONADITE**
- CORRENSITE
- CORUNDUM
- CORVUSITE
- COSALITE
- COTUNNITE
- COVELLITE
- COWLESITE
- CRANDALLITE
- **CREASEYITE**
- CREDNERITE
- CREEDITE
- #CRICHTONITE
- CRISTOBALITE
- CROCOITE
- CRONSTEDTITE
- CRYPTOMELANE
- CUBANITE
- #CUMENGEITE
- CUMMINGTONITE
- CUPRITE
- #CUPROAURIDE
- #CUPROCOPIAPITE
- CUPROPAVONITE
- #CUPRORIVAITE
- CUPROTUNGSTITE
- CYANOTRICHITE
- DANALITE
- DANBURITE
- DARAPSKITE
- DATOLITE
- DAUBRÉELITE
- **#DAVIDBROWNITE-(NH$_4$)**
- #DAVIDITE-(Ce)
- DAVIDITE-(La)
- DELAFOSSITE
- **#DENDORAITE-(NH$_4$)**
- DESCLOIZITE
- #DESTINEZITE
- DEVILLINE
- DIABOLEITE
- DIADOCHITE
- DIAMOND
- DIASPORE
- #DICKINSONITE-(KMnNa)
- DICKITE
- DIGENITE
- DIOPSIDE
- DIOPTASE
- DJURLEITE
- DOLOMITE
- DOLORESITE
- DOMEYKITE
- DRAVITE
- DUFRÉNOYSITE
- DUFTITE

- ❏ DUGGANITE
- ❏ DUMONTITE
- ❏ DUMORTIERITE
- ❏ DYSCRASITE
- ❏ ECLARITE
- ❏ #EDDAVIDITE
- ❏ EDENITE
- ❏ EGLESTONITE
- ❏ ELBAITE
- ❏ EMMONSITE
- ❏ EMPLECTITE
- ❏ EMPRESSITE
- ❏ ENARGITE
- ❏ ENSTATITE
- ❏ EOSPHORITE
- ❏ EPIDOTE
- ❏ EPSOMITE
- ❏ #ERIOCHALCITE
- ❏ ERIONITE
- ❏ #ERIONITE-K
- ❏ ERYTHRITE
- ❏ ETTRINGITE
- ❏ EUCRYPTITE
- ❏ EUGENITE
- ❏ #EULYTINE
- ❏ #EUXENITE-(Y)
- ❏ FAIRBANKITE
- ❏ FAIRCHILDITE
- ❏ FAIRFIELDITE
- ❏ FAMATINITE
- ❏ FAYALITE
- ❏ #FELSÖBÁNYAITE
- ❏ FERBERITE
- ❏ FERGUSONITE
- ❏ FERNANDINITE
- ❏ FERRICOPIAPITE
- ❏ FERRIERITE
- ❏ FERRIMOLYBDITE
- ❏ FERRO-ACTINOLITE
- ❏ #FERROHEXAHYDRITE
- ❏ #FERRO-HORNBLENDE
- ❏ #FERROSELITE
- ❏ FERVANITE
- ❏ FIBROFERRITE
- ❏ FILLOWITE
- ❏ #FLAGGITE
- ❏ FLAGSTAFFITE
- ❏ FLUORAPATITE
- ❏ #FLUORAPOPHYLLITE-(K)
- ❏ FLUORITE
- ❏ FORNACITE
- ❏ FORSTERITE
- ❏ FOURMARIERITE
- ❏ FRAIPONTITE
- ❏ FREIBERGITE
- ❏ FREIESLEBENITE
- ❏ FRIEDRICHITE
- ❏ FROHBERGITE
- ❏ #FUETTERERITE
- ❏ #FURUTOBEITE
- ❏ GADOLINITE
- ❏ GAHNITE
- ❏ GALENA
- ❏ #GANOMALITE
- ❏ GARRONITE
- ❏ GEARKSUTITE
- ❏ GEDRITE
- ❏ #GEERITE
- ❏ GEIKIELITE
- ❏ #GEORGEROBINSONITE
- ❏ GERHARDTITE
- ❏ GERSDORFFITE
- ❏ GIBBSITE
- ❏ GILALITE
- ❏ GISMONDINE
- ❏ GLADITE
- ❏ GLAUBERITE
- ❏ #GLAUCOCERINITE
- ❏ #GLAUCODOT
- ❏ #GLUSHINSKITE

- GMELINITE
- GOETHITE
- GOLD
- GOLDFIELDITE
- GONNARDITE
- #GORDAITE
- GORMANITE
- GOSLARITE
- GRAEMITE
- GRANDREEFITE
- #GRANDVIEWITE
- GRAPHITE
- GREENOCKITE
- GROSSULAR
- GROUTITE
- GRUNERITE
- GUILDITE
- GYPSUM
- GYROLITE
- HALITE
- #HALLOYSITE-10Å
- HALOTRICHITE
- HARMOTOME
- HAUSMANNITE
- HAWLEYITE
- HAXONITE
- #HEAZLEWOODITE
- #HECHTSBERGITE
- HECTORITE
- HEDENBERGITE
- #HEDYPHANE
- HELVINE
- HEMATITE
- HEMIHEDRITE
- HEMIMORPHITE
- HENRYITE
- #HERBERTSMITHITE
- HERCYNITE
- HESSITE
- HETAEROLITE
- HEULANDITE
- HEWETTITE
- HEXAHYDRITE
- HIDALGOITE
- HILLEBRANDITE
- HINSDALITE
- HISINGERITE
- HOCARTITE
- HODRUŠITE
- #HOGANITE
- HOLLANDITE
- #HOPEITE
- HÖRNESITE
- #HOUSLEYITE
- HÜBNERITE
- HUMMERITE
- HUNTITE
- HUREAULITE
- #HYALOTEKITE
- HYDROBASALUMINITE
- HYDROBIOTITE
- HYDROCERUSSITE
- #HYDROGLAUBERITE
- #HYDROHONESSITE
- #HYDROKENORALSTONITE
- HYDROMAGNESITE
- HYDRONIUMJAROSITE
- #HYDROXYAPOPHYLLITE-(K)
- #HYDROXYCALCIOROMÉITE
- #HYDROXYKENOELSMOREITE
- HYDROXYLAPATITE
- #HYDROXYLPYROMORPHITE
- HYDROZINCITE
- #ICE
- #IDAITE
- #IIMORIITE-(Y)
- #IKAITE
- ILMENITE
- ILSEMANNITE
- ILVAITE

- ❏ INGODITE
- ❏ IODARGYRITE
- ❏ IRANITE
- ❏ IRIGINITE
- ❏ #IRON
- ❏ #JACOBSITE
- ❏ JADEITE
- ❏ JAHNSITE
- ❏ JALPAITE
- ❏ JAMESONITE
- ❏ JAROSITE
- ❏ JOHANNITE
- ❏ JOHANNSENITE
- ❏ #JÔKOKUITE
- ❏ JORDISITE
- ❏ JUNITOITE
- ❏ JURBANITE
- ❏ KAERSUTITE
- ❏ #KALICINITE
- ❏ #KAMITUGAITE
- ❏ KANONAITE
- ❏ KAOLINITE
- ❏ KASOLITE
- ❏ #KENHSUITE
- ❏ KËSTERITE
- ❏ KETTNERITE
- ❏ KHINITE
- ❏ KIDDCREEKITE
- ❏ #KIESERITE
- ❏ KINOITE
- ❏ #KINOSHITALITE
- ❏ KOECHLINITE
- ❏ #KOLOVRATITE
- ❏ KORNELITE
- ❏ KOSMOCHLOR
- ❏ KOSTOVITE
- ❏ KRENNERITE
- ❏ KRINOVITE
- ❏ #KRUPKAITE
- ❏ KTENASITE
- ❏ #KUKSITE
- ❏ KULANITE
- ❏ KURAMITE
- ❏ KUTNOHORITE
- ❏ KYANITE
- ❏ LANARKITE
- ❏ LANGITE
- ❏ #LANSFORDITE
- ❏ LAUMONTITE
- ❏ LAURELITE
- ❏ LAUSENITE
- ❏ LAUTITE
- ❏ LAWRENCITE
- ❏ LAWSONITE
- ❏ #LAZARASKEITE
- ❏ LAZULITE
- ❏ LEAD
- ❏ LEADHILLITE
- ❏ #LECHATELIERITE
- ❏ LEPIDOCROCITE
- ❏ LEUCITE
- ❏ LEUCOPHOSPHITE
- ❏ LEUCOSPHENITE
- ❏ LÉVYNE
- ❏ LIBETHENITE
- ❏ LIEBIGITE
- ❏ #LIKASITE
- ❏ LIME
- ❏ LINARITE
- ❏ LINDGRENITE
- ❏ LINNAEITE
- ❏ #LIROCONITE
- ❏ LITHARGE
- ❏ LITHIOPHILITE
- ❏ LITHIOPHORITE
- ❏ #LIUDONGSHENGITE
- ❏ LIZARDITE
- ❏ LÖLLINGITE
- ❏ LONSDALEITE
- ❏ #LOVERINGITE

- LUDDENITE
- LUDWIGITE
- LUETHEITE
- LUZONITE
- MACKAYITE
- MACKINAWITE
- MACPHERSONITE
- MACQUARTITE
- MAGHEMITE
- #MAGNESIOALTERITE
- MAGNESIOCHROMITE
- MAGNESIOCOPIAPITE
- #MAGNESIO-FOITITE
- MAGNESIO-HORNBLENDE
- MAGNESITE
- MAGNETITE
- MALACHITE
- MAMMOTHITE
- MANANDONITE
- MANGANBABINGTONITE
- MANGANITE
- MANJIROITE
- MARCASITE
- MARIALITE
- MARICOPAITE
- #MARKASCHERITE
- MASSICOT
- MATILDITE
- MATLOCKITE
- #MATTHEDDLEITE
- #MAUCHERITE
- MAWSONITE
- #MCALPINEITE
- MCKINSTRYITE
- #MEIONITE
- MELANOTEKITE
- MELANOVANADITE
- MELANTERITE
- MELILITE
- MELONITE
- MERCURY
- #MERRILLITE
- MESOLITE
- META-ALUNOGEN
- META-AUTUNITE
- METACINNABAR
- METAHEWETTITE
- METANOVÁČEKITE
- METAROSSITE
- METASIDERONATRITE
- METATORBERNITE
- METATYUYAMUNITE
- #METAURANOCIRCITE-I
- METAVOLTINE
- METAZEUNERITE
- MIARGYRITE
- MICROCLINE
- MICROLITE
- MIERSITE
- #MILARITE
- MILLERITE
- MIMETITE
- MINIUM
- MIRABILITE
- MIXITE
- #MOGÁNITE
- MOISSANITE
- MOLYBDENITE
- #MOLYBDITE
- MOLYBDOFORNACITE
- #MOLYBDOPHYLLITE
- #MONAZITE-(Ce)
- #MONOHYDROCALCITE
- #MONTANITE
- MONTEBRASITE
- MONTICELLITE
- MONTMORILLONITE
- MONTROSEITE
- #MONTROYDITE
- MOORHOUSEITE

- MORDENITE
- MORENOSITE
- MOTTRAMITE
- MROSEITE
- #MUNAKATAITE
- MURDOCHITE
- MUSCOVITE
- NACRITE
- #NAMIBITE
- NANTOKITE
- NATROALUNITE
- #NATROCHALCITE
- #NATRODUFRÉNITE
- NATROJAROSITE
- NATROLITE
- #NATROSULFATOUREA
- #NATROZIPPEITE
- NAUMANNITE
- NAVAJOITE
- NEKOITE
- NEKRASOVITE
- NELTNERITE
- NEOTOCITE
- NEPHELINE
- #NESQUEHONITE
- NEYITE
- #NICKELBOUSSINGAULTITE
- NICKELINE
- #NICKELPHOSPHIDE
- NICKELSKUTTERUDITE
- NICKELZIPPEITE
- NITER
- NITRATINE
- NITROCALCITE
- NOLANITE
- NONTRONITE
- #NORDSTRANDITE
- NORDSTRŐMITE
- NSUTITE
- NUKUNDAMITE

- OFFRETITE
- OKENITE
- OLIVENITE
- OMPHACITE
- #OPAL
- #ORCELITE
- ORPIMENT
- ORTHOCLASE
- #ORTHOSERPIERITE
- OSARIZAWAITE
- #OTAVITE
- #OTTOITE
- #PACEITE
- PALYGORSKITE
- PAPAGOITE
- #PARACOQUIMBITE
- PARAGONITE
- PARALAURIONITE
- PARAMELACONITE
- PARAMONTROSEITE
- PARARAMMELSBERGITE
- PARATACAMITE
- PARATELLURITE
- PARGASITE
- PARISITE-(Ce)
- #PARKINSONITE
- PARNAUITE
- PARSONSITE
- PASCOITE
- PAULKERRITE
- PEARCEITE
- PECORAITE
- PECTOLITE
- PEKOITE
- #PENTAHYDRITE
- PENTLANDITE
- PERICLASE
- PERITE
- PEROVSKITE
- #PERROUDITE

- #PETERSITE-(Ce)
- PETZITE
- #PHARMACOLITE
- PHARMACOSIDERITE
- #PHILIPSBORNITE
- PHILLIPSITE
- PHLOGOPITE
- PHOENICOCHROITE
- PHOSGENITE
- #PHOSPHOHEDYPHANE
- PHOSPHOSIDERITE
- PHOSPHURANYLITE
- #PHOXITE
- PICKERINGITE
- PIEMONTITE
- PIGEONITE
- PINALITE
- PINTADOITE
- PLANCHEITE
- #PLANERITE
- PLATINUM
- PLATTNERITE
- PLUMBOGUMMITE
- PLUMBOJAROSITE
- PLUMBONACRITE
- #PLUMBOTELLURITE
- PLUMBOTSUMITE
- POLLUCITE
- POLYBASITE
- #POLYCRASE-(Y)
- #POLYDYMITE
- POLYHALITE
- POLYLITHIONITE
- POSNJAKITE
- #POUGHITE
- POWELLITE
- PREHNITE
- PROSOPITE
- PROUSTITE
- PSEUDOBOLEITE
- PSEUDOBROOKITE
- PSEUDOGRANDREEFITE
- PSEUDOMALACHITE
- PUMPELLYITE
- PURPURITE
- PYRARGYRITE
- PYRITE
- #PYROCHLORE
- PYROLUSITE
- PYROMORPHITE
- PYROPE
- PYROPHYLLITE
- PYRRHOTITE
- QUARTZ
- QUEITITE
- QUETZALCOATLITE
- RAMEAUITE
- RAMMELSBERGITE
- #RAMSBECKITE
- RAMSDELLITE
- RANCIÉITE
- RANSOMITE
- #RASPITE
- RAUVITE
- #RAYGRANTITE
- REALGAR
- #RECTORITE
- REEVESITE
- #REICHENBACHITE
- #RELIANCEITE-(K)
- RHODOCHROSITE
- RHODONITE
- RHODOSTANNITE
- RHOMBOCLASE
- RICHTERITE
- RICKARDITE
- RIEBECKITE
- ROALDITE
- ROBERTSITE
- RODALQUILARITE

- ROEDDERITE
- ROMANÈCHITE
- RÖMERITE
- #RONGIBBSITE
- #ROQUESITE
- ROSASITE
- ROSCOELITE
- ROSSITE
- #ROWLEYITE
- ROZENITE
- RUCKLIDGEITE
- RUIZITE
- #RUTHERFORDINE
- RUTILE
- SABUGALITE
- #SALAMMONIAC
- SALÉEITE
- #SAMARSKITE-(Y)
- SANIDINE
- SANTAFEITE
- SAPONITE
- SAPPHIRINE
- SAUCONITE
- SCAWTITE
- SCHEELITE
- SCHIEFFELINITE
- SCHOEPITE
- #SCHOLZITE
- SCHORL
- SCHREIBERSITE
- SCHRÖCKINGERITE
- SCHUBNELITE
- #SCHULENBERGITE
- #SCHWERTMANNITE
- SCOLECITE
- SCORODITE
- SCORZALITE
- #SCOTLANDITE
- SELENIUM
- SENGIERITE

- SEPIOLITE
- SERPIERITE
- #SHANNONITE
- #SHARPITE
- SHATTUCKITE
- SHERWOODITE
- SICKLERITE
- SIDERITE
- SIDEROTIL
- SIEGENITE
- SILLIMANITE
- SILVER
- #SIMONKOLLEITE
- SKUTTERUDITE
- SMITHSONITE
- #SODDYITE
- SONORAITE
- SPANGOLITE
- SPERTINIITE
- SPESSARTINE
- SPHALERITE
- SPHEROCOBALTITE
- SPINEL
- SPIONKOPITE
- SPIROFFITE
- SPODUMENE
- #SPURRITE
- #SRILANKITE
- STANNITE
- STANNOIDITE
- STARKEYITE
- STAUROLITE
- STEIGERITE
- STELLERITE
- STEPHANITE
- STERNBERGITE
- STETEFELDTITE
- STEVENSITE
- STEWARTITE
- STIBICONITE

- STIBNITE
- STILBITE
- STILPNOMELANE
- STISHOVITE
- STOLZITE
- #STRACZEKITE
- STRENGITE
- STRINGHAMITE
- STROMEYERITE
- STRONTIANITE
- #STRUVITE
- STÜTZITE
- SULPHUR
- #SURITE
- #SUSANNITE
- SVANBERGITE
- SWARTZITE
- SWITZERITE
- SYLVANITE
- SYLVITE
- #SYNCHYSITE
- #SZENICSITE
- SZOMOLNOKITE
- TAENITE
- #TAKANELITE
- TALC
- TALMESSITE
- #TANGEITE
- #TANTALITE-(Fe)
- #TAPIOLITE-(Fe)
- TEINEITE
- TELLURITE
- TELLURIUM
- TELLUROBISMUTHITE
- TENNANTITE
- TENORITE
- TEPHROITE
- TETRADYMITE
- TETRAHEDRITE
- #TETRAHEDRITE-(Hg)
- TETRATAENITE
- #THALÉNITE-(Y)
- THAUMASITE
- THÉNARDITE
- #THOMSONITE-Ca
- THORITE
- TILASITE
- #TIMROSEITE
- TINZENITE
- TITANITE
- TLAPALLITE
- TOBERMORITE
- TOCORNALITE
- TODOROKITE
- TOLBACHITE
- TOPAZ
- TORBERNITE
- TREMOLITE
- #TREVORITE
- TRIDYMITE
- TRIPHYLITE
- TRIPLITE
- TRIPLOIDITE
- TROILITE
- TSCHERMAKITE
- TSUMEBITE
- #TUNGSTENITE
- TUNGSTITE
- TURQUOISE
- TYROLITE
- TYUYAMUNITE
- ULVÖSPINEL
- UMOHOITE
- URANINITE
- URANOCIRCITE
- URANOPHANE
- URANOPHANE-β
- #URANOPILITE
- URANOSPINITE
- #UREA

- #USHKOVITE
- #UTAHITE
- UVANITE
- UVAROVITE
- UYTENBOGAARDTITE
- VAESITE
- VALLERIITE
- VANADINITE
- VANDENDRIESSCHEITE
- VANMEERSSCHEITE
- VANOXITE
- VARISCITE
- VATERITE
- VAUQUELINITE
- VELIKITE
- VERMICULITE
- VÉSIGNIÉITE
- VESUVIANITE
- #VILLAMANÍNITE
- #VIOLARITE
- VIVIANITE
- VOGLITE
- VOLBORTHITE
- VOLKONSKOITE
- VOLTAITE
- VOLYNSKITE
- #WAGNERITE
- WAKEFIELDITE-(Y)
- WAVELLITE
- #WEDDELLITE
- WEEKSITE
- WEISSITE
- #WHEATLEYITE
- #WHELANITE
- WHERRYITE
- WHEWELLITE
- WICKENBURGITE
- #WIDENMANNITE
- WILLEMITE
- #WILLEMSEITE
- WINSTANLEYITE
- WITHERITE
- WITTICHENITE
- WITTITE
- WOLLASTONITE
- #WOODHOUSEITE
- WOODWARDITE
- WULFENITE
- #WUPATKIITE
- WURTZITE
- WÜSTITE
- WYLLIEITE
- XENOTIME-(Y)
- XOCOMECATLITE
- XONOTLITE
- YAFSOANITE
- YARROWITE
- YAVAPAIITE
- YEDLINITE
- ZEUNERITE
- ZINCITE
- ZINCOBOTRYOGEN
- #ZINCOCOPIAPITE
- ZINCZIPPEITE
- ZIPPEITE
- ZIRCON
- ZOISITE
- ZUNYITE

Arizona Type Minerals

Arizona has eighty-seven type mineral species (table D.1), that is, minerals that were first identified and described from an Arizona locality. Some are quite old, having been approved as early as 1885, before Arizona was granted statehood. Among Arizona's type minerals are some that are not minerals found naturally on earth. These include the suite of minerals formed during the underground fire in the United Verde Mine at Jerome and minerals found in meteorites discovered in Arizona.

Of the eighty-seven Arizona type minerals, twelve are either United Verde mine-fire minerals or found in meteorites. Of the remainder, twenty-six have been added since the third edition of *Mineralogy of Arizona* was published in 1995. These are preceded by # in the table below.

Five mineral districts have provided the bulk of Arizona's type minerals; in the Mammoth District, the Mammoth–St. Anthony Mine has ten type minerals; the mines of the Tombstone District have nine type minerals; the Warren District in the mines in and around Bisbee has eight type minerals; in the Aravaipa District, the Grand Reef Mine has seven; and last, in the Christmas District, the Christmas Mine has four type minerals.

Some minerals were described and characterized using material from more than one locality. There are seven minerals with an Arizona cotype locality.

United Verde Mine Fire

A mine fire that started in 1894 in the massive sulfide orebody in the underground workings of the United Verde Mine at Jerome burned for several decades, despite concerted efforts to extinguish it. The fire is thought to have been caused by spontaneous combustion of unstable sulfide minerals

TABLE D.1 Arizona Type Minerals

ajoite	1958	flagstaffite	1920	murdochite	1955
#allantoin	2020	#georgerobinsonite	2009	#natrosulfatourea	2020
#alterite	2018	gerhardtite	1885	navajoite	1954
andersonite	1951	gilalite	1980	papagoite	1960
antlerite	1889	graemite	1975	paramelaconite	1891
apachite	1980	grandreefite	1989	paulkerrite	1984
aravaipaite	1989	#grandviewite	2008	#petersite-(Ce)	2014
artroeite	1995	guildite	1928	#phoxite	2018
#backite	2013	haxonite	1971	pinalite	1989
bayleyite	1951	hemihedrite	1970	pseudograndreefite	1989
#bechererite	1996	henryite	1983	ransomite	1928
bermanite	1936	junitoite	1976	#raygrantite	2013
bideauxite	1970	jurbanite	1976	#relianceite-(K)	2021
#bobmeyerite	2012	khinite	1978	#rongibbsite	2010
brezinaite	1969	kinoite	1970	#rowleyite	2016
butlerite	1928	krinovite	1968	ruizite	1977
#calcioaravaipaite	1996	laurelite	1989	schieffelinite	1980
#carmichaelite	2000	lausenite	1928	#shannonite	1995
chalcoalumite	1925	#lazaraskeite	2019	shattuckite	1915
coesite	1960	#liudongshengite	2019	spangolite	1890
coronadite	1904	lonsdaleite	1967	stishovite	1962
creaseyite	1975	luddenite	1982	swartzite	1951
#davidbrownite-(NH$_4$)	2019	luetheite	1977	wherryite	1950
#dendoraite-(NH$_4$)	2021	macquartite	1980	wickenburgite	1968
dugganite	1978	#magnesioalterite	2020	winstanleyite	1979
#eddavidite	2018	mammothite	1985	#wupatkiite	1995
emmonsite	1885	maricopaite	1988	yavapaiite	1959
fairbankite	1979	#markascherite	2010	yedlinite	1974
#flaggite	2021	moissanite	1905	zinczippeite	1976

on exposure to air. Surface-stripping operations later exposed rocks above the fire area and revealed a suite of newly formed hydrated sulfate minerals. Of the eleven minerals formed in the mine fire, seven were previously unknown species. One of these, jeromite, was subsequently discredited, and another, selenium, was known earlier but not adequately described. Currently, the mine-fire-type minerals are butlerite, guildite, lausenite, ransomite, and yavapaiite.

The geology and mineralogy of the Jerome area have been described in detail by C. A. Anderson and Creasey (1958). Lausen (1928) reported on the fumarole-like formation of minerals deposited as a result of the fire. He

TABLE D.2 Arizona Cotype Minerals

bütschliite	1947
coconinoite	1966
cowlesite	1975
cryptomelane	1942
fairchildite	1947
kiddcreekite	1984
metatyuyamunite	1953

concluded that the addition of water and steam in attempts to extinguish the fires contributed to the formation of the new sulfates.

Arizona Meteorites

Meteorites have been found in almost every county in the state, and some are significant finds. Twenty-three minerals have been found in Arizona meteorites that occur nowhere else in Arizona. Meteorites recovered in Arizona have contributed five type mineral species: brezinaite, haxonite, krinovite, lonsdaleite, and moissanite. Two additional type minerals, coesite and stishovite, were created by the impact of the meteorite forming polymorphs of quartz from the high pressure and shock of impact. Four species were found in the analysis of the Canyon Diablo meteorite that created Meteor Crater in Coconino County, and one was found on analysis of the Tucson Ring meteorite found in Pima County.

The most famous meteorite to fall in Arizona must be the Canyon Diablo meteorite, which formed Meteor (Barringer) Crater. The meteorite was estimated to have been 80 feet in diameter and weighed more than 63,000 tons when it landed about fifty thousand years ago. It was shattered into innumerable pieces, and more than 30 tons of fragments have been collected from around the crater.

The Tucson Ring meteorite is the other well-known Arizona meteorite. It was found in the late 1700s south of Tucson and was used for many years as an anvil by a blacksmith. It weighs more than 1,500 pounds and is now in the Smithsonian Institution.

APPENDIX E

Doubtful and Discredited Arizona Minerals

Many mineral species that have been reported as occurring in Arizona are now considered doubtful. Some have poorly defined chemistry, and others are *field terms* used to describe a group of minerals when the specific species is unknown.

Some minerals were originally described as new species from Arizona only to be later found identical to existing species or to be mixtures of other species, have subsequently been discredited. Many of the discredited minerals that were included in the third edition have been removed and are discussed here.

Arsenosulvanite

Although it was described as a new species in 1941, it was shown in 1994 to be identical to colusite and was discredited in 2006. The third edition included one occurrence at Bisbee.

Arizonite

Palmer (1909) first described the mineral arizonite from a pegmatite in Mohave County associated with gadolinite. The material was given the formula $Fe_2Ti_3O_9$ and was thought to be an alteration product of ilmenite. Subsequent work by Overholt, Vaux, and Rodda (1950) concluded that it is a mixture of hematite, ilmenite, anatase, and rutile and that it is a weathering product of ilmenite. In 1994 the IMA discredited arizonite as a distinct species (Bideaux 1997). Arizonite was listed as questionable in the third edition.

Basaluminite

Recent work in 1997 showed that this mineral is identical to felsőbányaite, and it was subsequently discredited in 2006. There was one occurrence at Bisbee noted in the third edition.

Bisbeeite

Bisbeeite was originally proposed as a new species by Schaller in 1915 based on material from the Shattuck Mine in Bisbee. The mineral was described as a pseudomorph after shattuckite with the probable formula $CuSiO_3 \cdot H_2O$. Subsequent research suggested that bisbeeite might be a combination of plancheite and chrysocolla and advised that it be discredited as a distinct species. Bisbeeite is now considered equal to chrysocolla (Bideaux 1997). Bisbeeite was listed as a questionable species in the third edition, where several localities were listed.

Braunite II

Similar to braunite but with a doubled-cell parameter C, braunite II is not an approved IMA mineral name but is probably a valid species. It was originally found in manganese ores of the Kalahari manganese province, Republic of South Africa, which principally consists of braunite, bixbyite, cryptomelane, hematite, and pyrolusite (de Villiers and Herbstein 1967). It has been reported from the Warren District in Cochise County, and on the surface near the White Tail Deer Mine, where it is abundant as cores of braunite crystals up to 1 inch (Graeme 1993). A small amount of neltnerite is also present, suggesting that the three minerals are independent and stable phases.

Bursaite

This mineral had one occurrence listed in the third edition but has now been discredited. In 2008 it was noted to require a crystal structure determination.

Calciovolborthite

Known for many years, this mineral was found to be the same as tangeite and was discredited in 2006. There were several occurrences listed in the third edition, and they are now found in this edition under tangeite.

Cliftonite

A variety of graphite, with the formula of C, cliftonite occurs as octahedral grains in iron meteorites. Its connection to Arizona is unknown.

Cornuite

Considered a variety of chrysocolla, cornuite was described by A. F. Rogers (1917). It is known as the amorphous equivalent of chrysocolla and was given the formula $Cu_{2-x}Al_x(H_{2-x}Si_2O_5)(OH)_4 \cdot nH_2O$. It was listed in the third edition as a questionable species and has been reported from the Mildren and Steppe claims in Pima County.

Duhamelite

S. A. Williams (1981b) described duhamelite as a new mineral in material from a prospect at Lousy Gulch, in the Payson area, Gila County. The mineral was given the formula $Cu_4Pb_2Bi(VO_4)_4(OH)_3 \cdot 8H_2O$ and often occurs with a distinctive chartreuse color and an acicular habit. Subsequent analysis has determined that it is a bismuth-rich variety of mottramite, and it was discredited in 2003.

Dunhamite

This is a rare supergene lead tellurite that forms monoclinic gray crystals or granular masses. It is produced from altaite or other tellurides as a very early oxidation product.

COCHISE COUNTY: Tombstone District, Joe and Grand Central Mines, as a moderately abundant mineral that matches the original but scanty description, which was based on material that rimmed altaite in polished sections from the Hillside Mine, Organ Mountains, Doña Ana Mountains, New Mexico. The Tombstone mineral also rims altaite and is associated with rodalquilarite. No original material from the Doña Ana locality has been preserved, but old specimens from the Bob Root collection, obtained at the type locality, include a mineral that closely matches the Tombstone mineral.

Sid Williams discussed "dunhamite" in a paper describing four new minerals from Tombstone, Arizona (S. A. Williams 1979). He reiterated that a "tellurate de Plumb" was described with an incomplete analysis. In 1946 this species was named *dunhamite* without analyzing the material. Williams decried the species, noting that not even qualitative chemical work was done, and that the composition is "mere guesswork." The IMA discredited dunhamite in a mass discreditation of 130 species in 2006.

Eichbergite

Copper iron bismuth antimony sulfide: $(Cu,Fe)(Bi,Sb)_3S_5$. It was noted as a mineral of questionable validity in the third edition and reported in vein quartz from the Cortez prospect in Maricopa County with bismuthinite and neyite. In 2008 it was shown to be a mixture of jaskolskite and meneghinite and was rejected by the IMA.

Girdite

Girdite, $H_2Pb_3(Te^{4+}O_3)(Te^{6+}O_6)$, was originally approved as a new mineral in 1979 based on material from the Joe Mine at Tombstone in Cochise County. Later analysis of type material by Kampf, Mills, and Rumsey (2017) suggested that the description was based on the analysis of at least two different species, one of which may have been ottoite. Girdite was discredited in 2016.

Glauconite

Potassium sodium aluminum iron magnesium silicate hydroxide: $(K,Na)(Al,Fe^{3+},Mg)_2(Al,Si)_4O_{10}(OH)_2$. Glauconite is a mica mineral that needs more investigation and is not listed as an approved mineral by the IMA. Formed during marine diagenesis under restricted conditions in arenaceous sedimentary rocks, it is the typical coloring agent in *greensands*. It is believed to have formed by the alteration of other silicate minerals.

COCHISE COUNTY: North of the Swisshelm Mountains, in certain beds in the Cambrian Bolsa Quartzite (Bull. 181).

GREENLEE COUNTY: Copper Mountain (Morenci) supersystem, Morenci District, in shale of the Morenci Formation and in green shales above the Coronado Quartzite (Lindgren 1905).

MOHAVE COUNTY: Valley of the Big Sandy, east of Wikieup, as extensive sand beds of glauconite-coated analcime grains (E. D. Wilson 1944; Robert O'Haire, pers. comm., 1973).

PINAL COUNTY: Ash Creek area, east of Saddle Mountain, abundant in the shale facies of the Cambrian Abrigo Formation, where it locally replaces brachiopod fragments (Krieger 1968).

Gummite

A generic term for colorful hydrated uranium oxides whose true identity is unknown. Gummites commonly contain lead, thorium, and relatively large amounts of water. The term is similar in usage to *wad* and *limonite*. Gummites are typically formed by the alteration of uraninites. Gummite was not listed as a valid mineral in the third edition and has been reported in Arizona from the following localities.

APACHE COUNTY: Monument Valley, Monument No. 2 Mine, with other oxidized uranium and vanadium minerals (Mitcham and Evensen 1955).

PIMA COUNTY: Linda Lee claims, with torbernite and hematite in a vein cutting arkose (R. L. Robinson 1955).

Herschelite

Recognized as a species in the late 1800s, the validity of herschelite was questioned in 1962. In 1997, the Zeolite Subcommittee of the IMA renamed it chabazite-Na. Herschelite was listed as a valid species in the third edition and was reported from several localities. These are all listed under chabazite-Na in this edition.

Jeromite

The United Verde mine fire produced several unique minerals and compounds, which were studied by Lausen (1928). Lausen reported jeromite as "a coating on fragments of rock beneath iron hoods placed over vents from which sulfur dioxide gases are issuing" as a product of burning sulfide ores. It formed black globular opaque masses and has been described as an amorphous As-S-Se phase of variable composition. It was discredited by the IMA in 2007. It was listed in the third edition as a valid species, but only the one locality was listed.

Lewisite

This mineral name is now a synonym for hydroxycalcioroméite. It was listed in the third edition as a valid species from one locality in the Johnny Lion Hills of Cochise County. The nomenclature of the pyrochlore group was revised in 2010, and lewisite was reclassified as hydroxycalcioroméite, a member of the roméite group, in the pyrochlore supergroup.

Manandonite

This valid species was reported from a locality near Quartzsite in the third edition. Analysis by Dr. Robert Downs at the University of Arizona found it to be muscovite, so it is no longer listed as occurring in Arizona.

Oboyerite

The mineral was described and approved in 1979 from Tombstone in Cochise County. It was listed as a valid species from the one type locality in the third edition. Recent analysis revealed that the material is a combination of ottoite and plumbotellurite. Oboyerite was discredited by the IMA in 2019 (Miyawaki et al. 2019b).

Parakhinite

Described in 1978, parakhinite, $Pb^{2+}Cu^{2+}_3Te_6+O_6(OH)_2$, was found in the Emerald Mine at Tombstone, Cochise County. Subsequent research by Hawthorne et al. (2012) has determined that parakhinite is a polytype of khinite and was renamed khinite-3T. It was listed in the third edition from the type locality.

Partzite

Copper antimony oxide hydroxide: $Cu_2Sb_2(O,OH)_7$. A secondary mineral formed by the oxidation of antimony-bearing sulfides. Originally described in 1867 but discredited by IMA in 2016. Reinvestigation of the type material shows that it is constituted by a mixture of several phases, which include a member of the plumboroméite group and a chrysocolla-like amorphous phase. Partzite was reported as a valid mineral from the Johnny Lyon Hills in the third edition.

Thorogummite

This name has been used to describe heterogeneous mixtures of secondary, noncrystalline minerals, produced by the alteration, hydration, or metamictization of thorite. It was discredited in 2014. It was not listed in the third edition. An occurrence was reported in the Wallapai District at the Kingman Feldspar Mine as inclusions in allanite (T. J. Brown 2010).

APPENDIX F

Organic Minerals

The term *organic mineral* has many definitions because there is no agreement on the precise meaning of the term. Nonetheless, an organic mineral may commonly be defined as an organic compound formed by natural processes, with specific crystallographic and chemical properties, that has been recognized as a species by the International Mineralogical Association (IMA). Organic compounds are those that contain carbon, C. Chemists once thought that an organic compound could be created only through biological processes by living organisms. Then in the early 1800s, it was proved that organic compounds could be produced through inorganic processes. Some minerals containing carbon had already been classified as inorganic. These include carbides, carbonates, and pure carbon minerals such as graphite and diamond (Wenk and Bulakh 2016). Currently, there are three classes of organic minerals: hydrocarbons, salts of organic acids, and a group of miscellaneous minerals.

Hydrocarbon minerals (C-H system) are composed almost entirely of carbon and hydrogen. Examples include carpathite ($C_{24}H_{12}$) and idrialite ($C_{22}H_{14}$). These minerals may have formed from polycyclic aromatic hydrocarbon compounds that under burial eventually reached a temperature where pyrolysis occurred, followed by hydrothermal transport toward the surface where the minerals precipitated (M. Lee 1981).

Salts of organic acids (C-O-H system) are composed largely of carbon, oxygen, and hydrogen. Examples include the oxalate group of minerals containing $C_2O_4^{2-}$, formates containing $CHOO^-$, and acetates containing $C_2H_3O_2^-$. This is a large group of minerals, which includes the oxalates weddellite ($Ca(C_2O_4)\cdot2H_2O$), a mineral found in living saguaro cactus, and glushinskite ($Mg(C_2O_4)\cdot2H_2O$), a mineral formed where lichen grow on serpentinite.

TABLE F.1 Examples of organic minerals found in Arizona

Mineral	IMA approval	Type	Chemistry
#antipinite	2014	oxalate	$KNa_3Cu_2(C_2O_4)_4$
#davidbrownite-(NH_4)	2018	oxalate	$(NH_4)_5(V^{4+}O)_2(C_2O_4)[PO_{2.75}(OH)_{1.25}]_4 \cdot 3H_2O]$
#glushinskite	1980	oxalate	$Mg(C_2O_4) \cdot 2H_2O$
#hoganite	2001	acetate	$Cu(CH_3COO)_2 \cdot H_2O$
#lazaraskeite	2019	glycolate	$Cu(C_2H_3O_3)_2$
#paceite	2001	acetate	$CaCu(CH_3COO)_4 \cdot 6H_2O$
#phoxite	2018	oxalate	$(NH_4)2Mg_2(C_2O_4)(PO_3OH)_2(H_2O)_4$
#weddellite	1942	oxalate	$Ca(C_2O_4) \cdot 2H_2O$
#wheatleyite	1984	oxalate	$Na_2Cu(C_2O_4)_2 \cdot 2H_2O$
whewellite	1852	oxalate	$Ca(C_2O_4) \cdot H_2O$

TABLE F.2 Examples of inorganic minerals formed by biomineralization in Arizona

Mineral	IMA approval	Type	Chemistry
#aphthitalite	1813	sulfate	$K_3Na(SO_4)_2$
#brushite	1865	phosphate	$Ca(PO_3OH) \cdot 2H_2O$
#lansfordite	1888	carbonate	$Mg(CO_3) \cdot 5H_2O$
#monohydrocalcite	1964	carbonate	$Ca(CO_3) \cdot H_2O$
#nesquehonite	1890	carbonate	$Mg(CO_3) \cdot 3H_2O$
nitrocalcite	1835	nitrate	$Ca(NO_3)_2 \cdot 4H_2O$
#takanelite	1971	oxide	$(Mn^{2+},Ca)_{2x}(Mn^{4+})_{1-x}O_2 \cdot 0.7H_2O$

Miscellaneous organic minerals are those that do not fit the definitions above. These include H-C-N systems such as abelsonite and complex C-O-H systems such as flagstaffite. This is a catch-all category that includes anything different from the above two categories that is not a carbonate, carbide, or pure carbon.

Many new organic minerals have been approved recently by the IMA. These minerals are often oxalates as indicated in table F.1. In 2019, lazaraskeite, the first naturally occurring glycolate, was approved. The type locality is the Santa Catalina Mountains in Pima County. Recent new discoveries from the Rowley Mine in Maricopa County have included the oxalates davidbrownite-(NH_4) and phoxite.

Some carbonates and non-carbon-bearing minerals, such as sulfates and phosphates, have been formed through biological activity and are recognized as biominerals. Biomineralization can create organic minerals or inorganic minerals (Cuif, Dauphin, and Sorauf 2011). This is illustrated by comparing the calcite in a seashell created by a mollusk and calcite formed through hydrothermal processes. As another example, when bat guano or decaying vegetation interacts with descending surface waters, chemical compounds are put into solution that eventually may react with inorganic minerals, creating a *biologically influenced* mineral. Examples include the organic mineral antipinite, an oxalate, and the inorganic mineral brushite, a phosphate.

BIBLIOGRAPHY

Adams, J. W., and M. H. Staatz. 1969. "Rare Earths and Thorium." In *Mineral and Water Resources of Arizona*. Arizona Bureau of Mines Bulletin 180, 245–51. Tucson: University of Arizona.

Agenbroad, L. D. 1962. *The Geology of the Atlas Mine Area, Pima County, Arizona*. Master's thesis, University of Arizona.

Allen, G. B. 1985. *Economic Geology of the Big Horn Mountains of West-Central Arizona*. Arizona Bureau of Geology and Mineral Technology, Open-file Report 85-17. Tucson: Arizona Bureau of Geology and Mineral Technology.

Allen, G. B., and W. Hunt. 1988. "The Tonopah-Belmont Mine, Maricopa County, Arizona." *Mineralogical Record* 19:139-44.

Allen, J. E., and R. Balk. 1954. *Mineral Resources of Fort Defiance and Tohatchi Quadrangles, Arizona and New Mexico*. New Mexico Bureau of Mines and Mineral Resources Bulletin 36. Socorro: New Mexico Bureau of Mines and Mineral Resources.

Allen, M. A., and G. M. Butler. 1921a. *Fluorspar*. Arizona Bureau of Mines Bulletin 114. Tucson: University of Arizona.

Allen, M. A., and G. M. Butler. 1921b. *Vanadium*. Arizona Bureau of Mines Bulletin 115. Tucson: University of Arizona.

Anderson, C. A. 1927. "Voltaite from Jerome, Arizona." *American Mineralogist* 12:287–90.

Anderson, C. A. 1950. "Alteration and Metallization in the Bagdad Porphyry Copper Deposit, Arizona." *Economic Geology* 45:609–28.

Anderson, C. A. 1969. "Copper." in *Mineral and Water Resources of Arizona*. Arizona Bureau of Mines Bulletin 180, 117–56. Tucson: University of Arizona.

Anderson, C. A., and S. C. Creasey. 1958. *Geology and Ore Deposits of the Jerome Area, Yavapai County, Arizona*. U.S. Geological Survey Professional Paper 308. Washington, D.C.: Government Printing Office.

Anderson, C. A., E. A. Scholz, and J. D. Strobell Jr. 1955. *Geology and Ore Deposits of the Bagdad Area, Yavapai County, Arizona*. U.S. Geological Survey Professional Paper 278. Washington, D.C.: Government Printing Office.

Anderson, J. L., A. P. Barth, and E. D. Young. 1988. "Mid-crustal Cretaceous Roots of Cordilleran Metamorphic Core Complexes." *Geology* 16:366–69.

Anderson, R. Y., and E. B. Kurtz Jr. 1955. "Biochemical Reconnaissance of the Annie Laurie Uranium Prospect, Santa Cruz County, Arizona." *Economic Geology* 50:227–32.

Andrade, M. B., H. Yang, R. T. Downs, R. A. Jenkins, and I. Fay. 2014. "Te-rich Raspite, $Pb(W_{0.56}Te_{0.44})O_4$, from Tombstone, Arizona, U.S.A.: The First Natural Example of Te^{6+} Substitution for W^{6+}." *American Mineralogist* 99:1507–10.

Annis, David R., and Stanley B. Keith. 1986. "Petrochemical Variation in Post-Laramide Igneous Rocks in Arizona and Adjacent Regional Geotectonic and Metallogenic Implications." In *Frontiers in Geology and Ore Deposits of Arizona and the Southwest*, edited by Barbara Beatty and P. A. K. Wilkinson, 448–56. Arizona Geological Society Digest 16. Tucson: Arizona Geological Society.

Anthony, E. 1986. "Geochemical Evidence for Crustal Melting in the Origin of the Igneous Suite at the Sierrita Porphyry Copper Deposit, Southeastern." PhD diss., University of Arizona.

Anthony, J. W. 1951. "Geology of the Montosa-Cottonwood Canyons area, Santa Cruz County, Arizona." Master's thesis, University of Arizona.

Anthony, J. W., R. A. Bideaux, K. W. Bladh, and M. C. Nichols. 1990. *Handbook of Mineralogy*. Vol. 1, *Elements, Sulfides, Sulfosalts*. Tucson: Mineral Data.

Anthony, J. W., R. A. Bideaux, K. W. Bladh, and M. C. Nichols. 1995. *Handbook of Mineralogy*. Vol. 2, *Silica, Silicates (in Two Parts): Part 1*. Tucson: Mineral Data.

Anthony, J. W., R. A. Bideaux, K. W. Bladh, and M. C. Nichols. 1995. *Handbook of Mineralogy*. Vol. 2, *Silica, Silicates (in Two Parts): Part 2*. Tucson: Mineral Data.

Anthony, J. W., R. A. Bideaux, K. W. Bladh, and M. C. Nichols. 1997. *Handbook of Mineralogy*. Vol. 3. *Halides, Hydroxides, Oxides*. Tucson: Mineral Data.

Anthony, J. W, R. A. Bideaux, K. W. Bladh, and M. C. Nichols. 2000. *Handbook of Mineralogy*. Vol. 4, *Arsenates, Phosphates, Vanadates*. Tucson, Ariz.: Mineral Data.

Anthony, J. W., R. A. Bideaux, K. W. Bladh, and M. C. Nichols. 2003. *Handbook of Mineralogy*. Vol. 5, *Borates, Carbonates, Sulfates*. Tucson: Mineral Data.

Anthony, J. W., R. L. DuBois, and H. E. Krumlauf. 1955. "Gypsum." In *Mineral Resources of the Navajo-Hopi Indian Reservations, Arizona-Utah*, vol. 2, 78–93. Tucson: University of Arizona.

Anthony, J. W., and R. B. Laughon. 1970. "Kinoite, a New Hydrous Copper Calcium Silicate Mineral from Arizona." *American Mineralogist* 55:709–13.

Anthony, J. W., and W. J. McLean. 1976. "Jurbanite, a New Post-Mine Aluminum Sulfate Mineral from San Manuel, Arizona." *American Mineralogist* 61:1–4.

Anthony, J. W., W. J. McLean, and R. B. Laughon. 1972. "The Crystal Structure of Yavapaiite: A Discussion." *American Mineralogist* 57:1546.

Anthony, J. W., S. A. Williams, and R. A. Bideaux. 1977. *Mineralogy of Arizona*. Tucson: University of Arizona Press.

Anthony, J. W., S. A. Williams, and R. A. Bideaux. 1982. *Mineralogy of Arizona*. 2nd ed. Tucson: University of Arizona Press.

Anthony, J.W., S. A. Williams, R. A. Bideaux, and R. W. Grant. 1995. *Mineralogy of Arizona*. 3rd ed. Tucson: University of Arizona Press.

Aoki, K., K. Fujino, and M. Akaogi. 1976. "Titanochondrodite and Titanoclinohumite Derived from the Upper Mantle in the Buell Park Kimberlite, Arizona, USA." *Contributions to Mineralogy and Petrology* 56:243–53.

Arizona Bureau of Mines. 1916. *Mineralogy of Useful Minerals in Arizona, 1916–17.* Arizona Bureau of Mines Bulletin 41. Tucson: University of Arizona.

Arizona Geological Survey (AZGS). 2013a. Crystal Cavern files. ADMMR Mining Collection. Published January 10, 2013. http://docs.azgs.az.gov/OnlineAccessMine Files/C-F/CrystalcavernLapaz250.pdf.

Arizona Geological Survey (AZGS). 2013b. Sacaton Mine Files. ADMMR Mining Collection. Published July 24, 2013. http://docs.azgs.az.gov/OnlineAccessMineFiles/S -Z/SacatonminePinal441a.pdf.

Arizona Geological Survey (AZGS). 2013c. Tire Claims Files. ADMMR Mining Collection. Published July 24, 2013. http://docs.azgs.az.gov/OnlineAccessMineFiles/S -Z/TireYavapai1369-1.pdf.

Arizona Geological Survey (AZGS). 2015. Weatherby Beryl Mine Files. ADMMR Mining Collection. Published March 3, 2015. http://docs.azgs.az.gov/OnlineAccess MineFiles/S-Z/WeatherbyberylYavapai1287.pdf.

Arnold, L. C. 1964. "Supergene Mineralogy and Processes in the San Xavier Mine Area, Pima County, Arizona." Master's thesis, University of Arizona.

Atencio, D., M. B. Andrade, A. C. B. Neto, and V. P. Pereira. 2017. "Ralstonite Renamed Hydrokenoralstonite, Coulsellite Renamed Fluornatrocoulsellite, and Their Incorporation into the Pyrochlore Supergroup." *Canadian Mineralogist* 55:115–20.

Austin, S. R. 1957. "Recent Uranium Redistribution in the Cameron, Arizona Deposits." *Advances in Nuclear Engineering Proceedings of the Second Nuclear Engineering and Science Congress*, edited by J. R. Dunning and B. R. Prentice, 332–38. New York: Pergamon Press.

Austin, S. R. 1964. *Mineralogy of the Cameron Area, Coconino County, Arizona.* U.S. Atomic Energy Commission, Technical Information Service RME 99. Oak Ridge, Tenn.: Technical Information Service.

Axelrod, J. M., F. S. Grimaldi, C. C. Milton, and K. J. Murata. 1951. "The Uranium Minerals from the Hillside Mine, Yavapai County, Arizona." *American Mineralogist* 36:1–22.

Ayres, V. L. 1924. "Pyrite from Tucson, Arizona." *American Mineralogist* 9:91–92.

Back, M. E. 2018. *Fleischer's Glossary of Mineral Species.* Tucson: Mineralogical Record.

Bain, G. W. 1952. "The Age of the Lower Cretaceous from Bisbee, Arizona, Uraninite." *Economic Geology* 47:305–15.

Baker, A., III. 1960. "Chalcopyrite Blebs in Sphalerite at Johnson Camp, Arizona." *Economic Geology* 55:387–98.

Baldwin, E. J. 1971. "Environments of Deposition of the Moenkopi Formation in North-Central Arizona." PhD diss., University of Arizona.

Balk, R. 1954. "Petrology." In *Mineral Resources of Fort Defiance and Tohatchi Quadrangles, Arizona and New Mexico*, 192. New Mexico Bureau of Mines, Mineral Resources Bulletin 36. Socorro: New Mexico Bureau of Mines and Mineral Resources.

Ball, S. H., and T. M. Broderick. 1919. "Magmatic Iron Ore in Arizona." *Engineering and Mining Journal* 107:353–54.

Balla, J. C. 1962. "The Geology and Geochemistry of Beryllium in Southern Arizona." Master's thesis, University of Arizona.

Bancroft, H. 1910. "Rare Metals: Notes on the Occurrence of Cinnabar in Central Western Arizona." In *Contributions to Economic Geology, 1909: Part 1, Metals and Nonmetals Except Fuels*. U.S. Geological Survey Bulletin 430, 151–53. Washington, D.C.: U.S. Department of the Interior, Bureau of Mines.

Bancroft, H. 1911. "Reconnaissance of the Ore Deposits in Northern Yuma County, Arizona." U.S. Geological Survey Bulletin 451. Washington, D.C.: Government Printing Office.

Bancroft, P., and B. Bricker. 1990. "Arizona's Silver Mining District." *Mineralogical Record* 21:151–68.

Banerjee, A. K. 1957. "Structural and Petrological Study of the Oracle Granite." PhD diss., University of Arizona.

Bargar, J. R., C. C. Fuller, M. A. Marcus, A. J. Brearley, M. Perez De la Rosa, S. M. Webb, and W. A. Caldwell. 2009. "Structural Characterization of Terrestrial Microbial Mn Oxides from Pinal Creek, AZ." *Geochimica et Cosmochimica Acta* 73:889–910.

Barinova, A. V., R. K. Rastsvetaeva, G. A. Sidorenko, and I. A. Verin. 2003. "Crystal Structure of β-Uranophane from the Transbaikal Region and Its Relation to the Structure of the α Modification." *Crystallography Reports* 48 (1): 17–20.

Barnes, R., and M. Hay. 1983. "Famous Mineral Localities: The Magma Mine." *Mineralogical Record* 14:72–82.

Barra, F., J. Ruiz, R. Mathur, and S. Titley. 2003. "A Re-Os Study of Sulfide Minerals from the Bagdad Porphyry Cu-Mo Deposit, Northern Arizona, USA." *Mineralium Deposita* 38:585–96.

Barrington, J., and P. F. Kerr. 1961. "Breccia Pipe near Cameron, Arizona." *Geological Society of America Bulletin* 72:1661–74.

Barrington, J., and P. F. Kerr. 1962. "Alteration Effects at Tuba Dike, Cameron, Arizona." *Geological Society of America Bulletin* 73:101–12.

Barrington, J., and P. F. Kerr. 1963. "Collapse Features and Silica Plugs near Cameron, Arizona." *Geological Society of America Bulletin* 74:1237–58.

Barton, M., E. Seedorff, D. J. Maher, W. J. A. Stavast, R. J. Kamilli, T. S. Hayes, K. R. Long, and G. B. Haxel. 2007. *Laramide Porphyry Copper Systems and Superimposed Tertiary Extension: A Life Cycle Approach to the Globe-Superior-Ray Area, Arizona*. Arizona Geological Society, Ores and Orogenesis Symposium, Field Trip No. 4. Tucson: Arizona Geological Society.

Bastin, E. S. 1922. *Primary Native-Silver Ores near Wickenburg, Arizona, and Their Bearing on the Genesis of the Silver Ores of Cobalt, Ontario*. U.S. Geological Survey Bulletin 735. Washington, D.C.: Government Printing Office.

Bastin, E. S. 1925. *Origin of Certain Rich Silver Ores near Chloride and Kingman, Arizona*. U.S. Geological Survey Bulletin 750-B. Washington, D.C.: Government Printing Office.

Bateman, A. M. 1923. "An Arizona Asbestos Deposit." *Economic Geology* 18:663–83.

Bateman, A. M. 1929. "Some Covellite-Chalcocite Relationships." *Economic Geology* 24:424–39.

Batty, J. V., H. D. Snedden, G. M. Potter, and B. K. Shibler. 1947. *Concentration of Fluorite Ores from Arizona, California, Idaho, Montana, Nevada, and Wyoming*. U.S.

Bureau of Mines Report of Investigations RI-4133. Washington, D.C.: U.S. Department of the Interior, Bureau of Mines.

Bayliss, P. 1986. "Subdivision of the Pyrite Group, and a Chemical and X-ray Diffraction Investigation of Ullmannite." *Canadian Mineralogist* 24:27–33.

Beck, B. A. 2011. "The Forgotten Silver District Peck Mining District Yavapai County, Arizona." Abstract. *Minerals of Arizona Nineteenth Annual Symposium*, 1–4. Phoenix: Flagg Mineral Foundation.

Beckman, R. T., and W. H. Kerns. 1965. "Mercury in Arizona." In *Mercury Potential of the United States*. U.S. Bureau of Mines Information Circular 8252, 60–74. Washington, D.C.: U.S. Department of the Interior, Bureau of Mines.

Bennett, K. C. 1975. "Geology and Origin of the Breccias in the Morenci-Metcalf District, Greenlee County, Arizona." Master's thesis, University of Arizona.

Best, M. G. 1970. "Kaersutite-Peridotite Inclusions and Kindred Megacrysts in Basanitic Lavas, Grand Canyon, Arizona." *Contributions to Mineralogy and Petrology* 27:25–44.

Biagioni, C., S. Merlino, and E. Bonaccorsi. 2015. "The Tobermorite Supergroup: A New Nomenclature." *Mineralogical Magazine* 79:485–95.

Bideaux, R. A. 1970. "A Multiple Japan Law Quartz Twin." *Mineralogical Record* 1:33.

Bideaux, R. A. 1972. "The Collector: On Wulfenite." *Mineralogical Record* 3:148–50, 198–201.

Bideaux, R. A. 1973. "The Collector on Azurite." *Mineralogical Record* 4:34–35.

Bideaux, R. A. 1980. "Famous Mineral Localities: Tiger, Arizona." *Mineralogical Record* 11:155–81.

Bideaux, R. A. 1997. "Arizona's Lost Minerals—Arizonite and Bisbeeite." Abstract. *Fifth Annual Minerals of Arizona Symposium*, 2. Phoenix: Flagg Mineral Foundation.

Bideaux, R. A., S. A. Williams, and R. W. Thomssen. 1960. "Some New Occurrences of Minerals of Arizona." *Arizona Geological Society Digest* 3:53–56.

Bilbrey, J. H. 1962. *Cobalt, a Materials Survey*. U.S. Bureau of Mines Information Circular IC-8103. Washington, D.C.: U.S. Department of the Interior, Bureau of Mines.

Bilinski, H., R. Giovanoli, A. Usui, and D. Hanžel. 2002. "Characterization of Mn Oxides in Cemented Streambed Crusts from Pinal Creek, Arizona, USA, and in Hot-Spring Deposits from Yuno-Taki Falls, Hokkaido, Japan." *American Mineralogist* 87:580–91.

Billingsley, G. H. 2000. *Geologic Map of the Grand Canyon 30′ × 60′ Quadrangle, Coconino and Mohave Counties, Northwestern Arizona*. Reston, Va.: U.S. Geological Survey.

Blacet, P. M. 1969. "Gold Placer and Lode Deposits, Gold Basin—Lost Basin, Arizona." In *Geological Survey Research 1969*. U.S. Geological Survey Professional Paper 600-A, A1–A2. Washington, D.C.: Government Printing Office.

Bladh, K. W. 1973. "The Clay Mineralogy of Selected Fault Gouges." Master's thesis, University of Arizona.

Bladh, K. W. 2019. "Arizona Wulfenite." *Rocks and Minerals* 94:10–29.

Bladh, K. W., R. K. Corbett, and W. J. McLean. 1972. "The Crystal Structure of Tilasite." *American Mineralogist* 57:1880–84.

Blake, W. P. 1855. "Observations on the Extent of Gold Regions of California and Oregon." *American Journal of Science* 20:72–85.

Blake, W. P. 1865. "Iron Regions of Arizona." *American Journal of Science* 40:388.

Blake, W. P. 1866. *Annotated Catalogue of the Principal Mineral Species Hitherto Recognized in California and the Adjoining States and Territories, Being a Report to the California State Board of Agriculture.* Sacramento, Calif.: privately printed. Reprinted in J. R. Browne and J. W. Taylor 1867, 201–11.

Blake, W. P. 1881a. "On the Occurrence of Vanadates of Lead at the Castle Dome Mines in Arizona." *American Journal of Science* 22:410–11.

Blake, W. P. 1881b. "Vanadinite in Arizona." *American Journal of Science* 22:235.

Blake, W. P. 1890. "Mineralogical Notes." *American Journal of Science* 39:43–45.

Blake, W. P. 1896. "Gypsum in Arizona." *American Geologist* 18:394.

Blake, W. P. 1897. "The Fortuna Gold Mine." *Engineering and Mining Journal* 63:26.

Blake, W. P. 1898. "Wolframite in Arizona." *Engineering and Mining Journal* 65:21.

Blake, W. P. 1902. "The Geology of the Galiuro Mountains, Arizona, and of the Gold-Bearing Ledge Known as Gold Mountain." *Engineering and Mining Journal* 73:546.

Blake, W. P. 1903. "Tombstone and Its Mines." *American Institute of Mining, Metallurgical, and Petroleum Engineers (AIME) Transactions* 34:668–70.

Blake, W. P. 1904. "Mining in the Southwest." *Engineering and Mining Journal* 77:35.

Blake, W. P. 1905. "Iodobromite in Arizona." *American Journal of Science* 19:230.

Blake, W. P. 1909. *Minerals of Arizona: Their Occurrence and Association, with Notes on Their Composition, a Report to the Hon. J. H. Kibbey, Governor of Arizona.* Tucson: n.p.

Blake, W. P. 1910. "Manganese Ore in an Unusual Form." *American Institute of Mining, Metallurgical, and Petroleum Engineers (AIME) Transactions* 41:647–49.

Blanchard, R., and P. F. Boswell. 1930. "Limonite Types Derived from Bornite and Tetrahedrite." *Economic Geology* 25:557–80.

Boettcher, S. S., and S. Mosher. 1998. "Mid- to Late Cretaceous Ductile Deformation and Thermal Evolution of the Crust in the Northern Dome Rock Mountains, Arizona." *Journal of Structural Geology* 20 (6): 745–64.

Bohn, E., and W. Stöber. 1966. "Coesit und Stishovit als isolierte natürliche Mineralien." *Neues Jahrb. Min.*, 89–96.

Bollin, E. M., and P. F. Kerr. 1958. "Uranium Mineralization near Cameron, Arizona." In *Guidebook of the Black Mesa Basin, Northeastern Arizona: Ninth Field Conference*, edited by R. Y. Anderson and J. W. Harshbarger, 164–68. Socorro: New Mexico Geological Society.

Bowell, R., R. Luetcke, and M. R. Luetcke. 2015. "Mineralogy of the Gallagher Vanadium Property and Manila Mine, Cochise County, Arizona." *Rocks and Minerals* 90:338–51.

Bowen, N. L., and R. W. G. Wyckoff. 1926. "A Petrologic and X-ray Study of the Thermal Dissociation of Dumortierite from Clip, Arizona." *Journal of the Washington Academy of Science* 16:178–89.

Bowles, O. 1940. *Onyx Marble and Travertine.* U.S. Bureau of Mines Information Circular IC-6751R. Washington, D.C.: U.S. Department of the Interior, Bureau of Mines.

Bramlette, M. N., and E. Posnjak. 1933. "Zeolitic Alteration of Pyroclastics." *American Mineralogist* 18:167–71.

Brett, R., and R. A. Yund. 1964. "Sulfur-Rich Bornites." *American Mineralogist* 49:1084–98.

Briggs, D. F. 2018. *History of the Verde Mining District, Yavapai County, Arizona*. Arizona Geological Survey Contributed Report CR-18-D. Tucson: Arizona Geological Survey.

Britvin, S. N., V. D. Kolomensky, M. M. Boldyreva, A. N. Bogdanova, Y. L. Kretser, O. N. Boldyreva, and N. S. Rudashevsky. 1999. "Nickelphosphide (Ni,Fe)$_3$P—The Nickel Analog of Schreibersite." *Zapiski Vserossijskogo Mineralogicheskogo Obshchestva* 128 (3): 64–72.

Bromfield, C. S., and A. F. Schride. 1956. *Mineral Resources of the San Carlos Indian Reservation, Arizona*. U.S. Geological Survey Bulletin 1027-N. Washington, D.C.: Government Printing Office.

Brophy, G. P., and M. F. Sheridan. 1965. "Sulfate Studies IV: The Jarosite-Natrojarosite-Hydronium Jarosite Solid Solution Series." *American Mineralogist* 50:1595–1607.

Brown, T. J. 2010. "Geology and Geochemistry of the Kingman Feldspar, Rare Metals and Wagon Bow Pegmatites." Master's thesis, University of New Orleans.

Brown, W. H. 1939. "Tucson Mountains, an Arizona Basin and Range Type." *Geological Society of America Bulletin* 50:697–760.

Browne, J. F. 1958. "The Geology of the Cuprite Mine Area, Pima County." Tucson: Master's thesis, University of Arizona.

Browne, J. R., and J. W. Taylor. 1867. *Reports on the Mineral Resources of the United States*. Washington D.C.: Government Printing Office.

Brownell, G. M. 1959. "A Beryllium Detector for Field Exploration." *Economic Geology* 54:1103–14.

Brush, G. J. 1873. "On a Compact Anglesite from Arizona." *American Journal of Science* 5:421–22.

Bryant, D. G. 1968. "Intrusive Breccias Associated with Ore, Warren (Bisbee) Mining District, Arizona." *Economic Geology* 63:1–12.

Bryant, D. G., and H. E. Metz. 1966. "Geology and Ore Deposits of the Warren Mining District." In *Geology of the Porphyry Copper Deposits, Southwestern North America*, edited by S. R. Titley and C. L. Hicks, 189–203. Tucson: University of Arizona Press.

Buchwald, V. F. 1975. *Handbook of Iron Meteorites*. 3 vols. Berkeley: University of California Press.

Buerger, M. S. 1942. "The Unit Cell and Space Group of Claudetite, As$_2$O$_3$." Abstract. *American Mineralogist* 27:216.

Bunch, T. E., and L. H. Fuchs. 1969. "A New Mineral: Brezinaite, Cr$_3$S$_4$, and the Tucson Meteorite." *American Mineralogist* 54:1509–18.

Bunch, T. E., and K. Keil. 1969. "Mineral Compositions and Petrology of Silicate Inclusions in Iron Meteorites: Chemistry of Chromite in Non-chondritic Meteorites." *Meteoritics* 4:155–58.

Burchard, E. F. 1930. *Iron Ore on Canyon Creek, Fort Apache Indian Reservation*. U.S. Geological Survey Bulletin 821-C. Washington, D.C.: Government Printing Office.

Burchard, E. F. 1943. "Results of Exploration for Iron Ore in Far Western States." *Economic Geology* 38: 85-86.

Burke, E. A. J. 2006. "A Mass Discreditation of GQN Minerals." *Canadian Mineralogist* 44:1557–60.

Burt, D. M., T. C. Moyer, and E. H. Christiansen. 1981. "Garnet- and Topaz-Bearing Rhyolites from near Burro Creek, Mohave County, Western Arizona: Possible Exploration Significance." *Arizona Geological Society Digest* 13:1–4.

Butler, B. S. 1929. "Strontium Deposits near Aguila, Arizona." U.S. Department of the Interior Press Memo 31445. Washington, D.C.: U.S. Department of the Interior.

Butler, B. S., E. D. Wilson, and C. A. Rasor. 1938. *Geology and Ore Deposits of the Tombstone District, Arizona.* Arizona Bureau of Mines Bulletin 143. Tucson: University of Arizona.

Butler, G. M. 1928. "Corrections to Volume 13." *American Mineralogist* 13:594.

Butler, G. M., and M. A. Allen. 1921. *Uranium and Radium.* Arizona Bureau of Mines Bulletin 117. Tucson: University of Arizona.

Butler, J. R., and R. Hall. 1960. "Chemical Variations in Members of the Fergusonite-Formanite Series." *Mineralogical Magazine* 32:392–407.

Bykerk-Kauffman, A. 1990. "Structural Evolution of the Northeastern Santa Catalina Mountains, Arizona: A Glimpse of the Pre-Extension History of the Catalina Complex." PhD diss., University of Arizona.

Campbell, I. 1936. "On the Occurrence of Sillimanite and Staurolite in Grand Canyon." *Grand Canyon Natural History Association Bulletin* 5:17–22.

Campbell, I., and E. T. Schenk. 1950. "Camptonite Dikes near Boulder Dam, Arizona." *American Mineralogist* 35:671–92.

Cannaday, F. X. 1977. Memorandum from F. X. Cannaday to Kelsey L. Boltz, February 14, 1977. Arizona Geological Survey. http://docs.azgs.az.gov/SpecColl/2004-01/2004-01-0063.pdf.

Caporuscio, F. A. 1980. "Alkalic Lavas from the San Carlos, Arizona Province." Abstract. *Transactions, American Geophysical Union* 60 (46): 972.

Carter, N. L., and G. C. Kennedy. 1964. "Origin of Diamonds in the Canyon Diablo and Novo Urei Meteorites." *Journal of Geophysical Research* 69:2403–21.

Carter, T. L. 1911. "Gold Placers in Arizona." *Engineering and Mining Journal* 90:561.

Casebolt, L. L., J. M. Faurote, and D. M. Pillmore. 1986. "Underground Geologic Evaluation: Hack Canyon Breccia Pipe Uranium Ore Body, Mohave County, Arizona." Abstract. *Geological Society of America* 18:345.

Catanzaro, E. J., and J. L. Kulp. 1964. "Discordant Zircons from the Little Belt (Montana), Beartooth (Montana) and Santa Catalina (Arizona) Mountains." *Geochimica et Cosmochimica Acta* 28:87–124.

Cesbron, F. P. 1964. "Contribution à la minéralogie des sulfates de fer hydratés." *Bulletin de la Société Française de Minéralogie et de Cristallographie* 87:124–43.

Cesbron, F. P., and S. A. Williams. 1980. "Apachite and Gilalite, Two New Copper Silicates from Christmas, Arizona." *Mineralogical Magazine* 43:639–41.

Chaffee, M. A. 1964. "Dispersion Patterns as a Possible Guide to Ore Deposits in the Cerro Colorado District, Pima County, Arizona." Master's thesis, University of Arizona.

Chao, E. C. T., J. J. Fahey, J. Littler, and D. J. Milton. 1962. "Stishovite, SiO_2, a Very High Pressure New Mineral from Meteor Crater, Arizona." *Journal of Geophysical Research* 67:419–21.

Chao, E. C. T., E. M. Shoemaker, and B. M. Madsen. 1960. "First Natural Occurrence of Coesite." *Science* 132:220–22.

Chapman, T. L. 1947. *San Manuel Copper Deposit, Pinal County, Arizona*. U.S. Bureau of Mines Report of Investigations RI 4108. Washington, D.C.: U.S. Department of the Interior, Bureau of Mines.

Chatman, M. L. 1988. *Mineral Resources of the Wabayuma Peak Wilderness Study Area (AZ-020-037/043), Mohave County, Arizona*. Arizona Mineral Land Assessment. Open-File Report 5-88. Denver, Colo.: Intermountain Field Operations Center.

Chenoweth, W. L. 1967. "The Uranium Deposits of the Lukachukai Mountains." In *Guidebook of Defiance–Zuni–Mt. Taylor Region, Arizona and New Mexico: Eighteenth Field Conference*, edited by F. D. Trauger, 78–85. Socorro: New Mexico Geological Society.

Chenoweth, W. L. 1986. *The Orphan Lode Mine, Grand Canyon, Arizona: A Case History of a Mineralized Collapse-Breccia Pipe*. U.S. Geological Survey Open-File Report 86-510: 126. Denver, Colo.: U.S. Geological Survey.

Chenoweth, W. L. 1993. *Geology and Production History of the Bootjack Uranium Mine, Navajo County, Arizona*. Arizona Geological Survey Contributed Report 93-A. Tucson: Arizona Geological Survey.

Chenoweth, W. L., and R. C. Malan. 1973. "The Uranium Deposits of Northeastern Arizona." In *Guidebook of Monument Valley (Arizona, Utah, and New Mexico)*, edited by H. L. James, 139–49. New Mexico Geological Society Fall Field Conference Guidebook 24. Socorro: New Mexico Geological Society.

Clark, A., and G. Fleck. 1980. "The Grey Horse Mine, Pinal County, Arizona." *Mineralogical Record* 11:231–33.

Clark, M. D. 1979. "Geology of the Older Precambrian Rocks of the Grand Canyon, Part III. Petrology of Mafic Schists and Amphibolites." *Precambrian Research* 8:277–302.

Clarke, O. M., Jr. 1953. "Geochemical Prospecting for Copper at Ray, Arizona." *Economic Geology* 48:39–45.

Colchester, D. M., D. R. Klish, P. Leverett, and P. A. Williams. 2008. "Grandviewite, $Cu_3Al_9(SO_4)_2(OH)_{29}$, a New Mineral from the Grandview Mine, Arizona, USA." *Australian Journal of Mineralogy* 14:51–54.

Coleman, R. G., and M. Delevaux. 1957. "Occurrence of Selenium in Sulphides from Some Elementary Rocks of the Western United States." *Economic Geology* 5:499–527.

Conway, C., J. R. Hassemer, D. H. Jr. Knepper, J. A. Pitkin, RC. Jachens, and M. L. Chatman. 1990. *Mineral Resources of the Wabayuma Study Area, Mohave County, Arizona*. U.S. Geological Survey Bulletin 1737-E: E1–E52, map.

Cook, S. S. 1988a. "Petrologic Studies of the Supergene Copper Mineralization at the Santa Cruz Copper Deposit, Pinal County, Arizona." Abstract. *Geological Society of America Abstracts with Programs* 20:A336.

Cook, S. S. 1988b. "Supergene Copper Mineralization at the Lakeshore Mine, Pinal County, Arizona." *Economic Geology* 83:297–309.

Coombs, D. S., A. Alberti, T. Armbruster, G. Artioli, C. Colella, E. Galli, J. D. Grice, F. Liebau, J. A. Mandarino, H. Minato, E. H. Nickel, E. Passaglia, D. R. Peacor, S. Quartieri, R. Rinaldi, M. Ross, R. A. Sheppard, E. Tillmanns, and G. Vezzalini. 1997. "Recommended Nomenclature for Zeolite Minerals: Report of the Subcommittee on Zeolites of the International Mineralogical Association, Commission on New Minerals and Mineral Names." *Canadian Mineralogist* 35:1571–1606.

Cooper, J. R. 1957. "Metamorphism and Volume Losses in Carbonate Rocks near Johnson Camp, Cochise County, Arizona." *Geological Society of America Bulletin* 68:577–610.

Cooper, J. R. 1961. "Turkey-Track Porphyry—A Possible Guide for Correlation of Miocene Rocks in Southeastern Arizona." *Arizona Geological Society Digest* 4:17–35.

Cooper, J. R. 1971. *Mesozoic Stratigraphy of the Sierrita Mountains, Pima County, Arizona.* U.S. Geological Survey Professional Paper 658-D: D1–D42. Washington, D.C.: Government Printing Office.

Cooper, J. R. 1973. *Geologic Map of the Twin Buttes Quadrangle, Southwest of Tucson, Pima County, Arizona.* U.S. Geological Survey Miscellaneous Geologic Investigations Map I-745. Reston, Va.: U.S. Geological Survey.

Cooper, J. R., and L. C. Huff. 1951. "Geological Investigations and Geochemical Prospecting Experiment at Johnson, Arizona." *Economic Geology* 46 (7): 731–56.

Cooper, J. R., and L. T. Silver. 1964. *Geology and Ore Deposits of the Dragoon Quadrangle, Cochise County, Arizona.* U.S. Geological Survey Professional Paper 416. Washington, D.C.: U.S. Department of the Interior, Geological Survey.

Cooper, M. A., F. C. Hawthorne, and M. E. Back. 2008. "The Crystal Structure of Khinite and Polytypism in Khinite and Parakhinite." *Mineral Magazine* 72:763–70.

Cooper, M. A., N. A. Ball, F. C. Hawthorne, W. H. Paar, A. C. Roberts, and E. Moffatt. 2011. "Georgerobinsonite, $(Pb_4CrO_4)_2(OH)_2FCl$, A New Chromate Mineral from the Mammoth St. Anthony Mine, Tiger, Pinal County, Arizona: Description and Crystal Structure." *Canadian Mineralogist* 49: 865–76.

Cornwall, H. R. 1982. "Petrology and Chemistry of Igneous Rocks Ray Porphyry Copper District, Pinal County, Arizona." In *Advances in Geology of the Porphyry Copper Deposits, Southwestern North America,* edited by S. R. Titley, 259–73. Tucson: University of Arizona Press.

Cousins, N. B. 1972. "Relationship of Black Calcite to Gold and Silver Mineralization in the Sheep Tanks Mining District, Yuma County, Arizona." Master's thesis, Arizona State University.

Crawford, W. P. 1930. "Notes on Rickardite, a New Occurrence." *American Mineralogist* 15:272–73.

Crawford, W. P., and F. Johnson. 1937. "Turquoise Deposits of Courtland, Arizona." *Economic Geology* 32:511–23.

Creasey, S. C. 1952. "Geology of the Iron King Mine, Yavapai County, Arizona." *Economic Geology* 47:24–56.

Creasey, S. C. 1959. "Some Phase Relations in the Hydrothermally Altered Rocks of Porphyry Copper Deposits." *Economic Geology* 54:351–73.

Creasey, S. C., D. W. Peterson, and N. A. Gambell. 1983. *Geologic Map of the Teapot Mountain Quadrangle, Pinal County, Arizona.* U.S. Geological Survey Geologic quadrangle map GQ-1559. Washington, D.C.: U.S. Geological Survey.

Creasey, S. C., and G. L. Quick. 1955. *Copper Deposits of Part of Helvetia Mining District, Pima County, Arizona.* U.S. Geological Survey Bulletin 1027-F. Washington, D.C.: Government Printing Office.

Criddle, A. J., C. J. Stanley, J. E. Chisolm, and E. E. Fejer. 1983. "Henryite, A New Copper-Silver Telluride from Bisbee, Arizona." *Bulletin de Mineralogie* 106:511–17.

Crittenden, M. D., F. Cuttitta, H. J. Rose Jr., and M. Fleischer. 1962. "Studies of Manganese Oxide Minerals—VI. Thallium in Some Manganese Oxides." *American Mineralogist* 47: 1461–67.

Cross, C., and J. Holloway. 1974. "A Megacryst-Bearing Alkali Basalt, San Carlos, Arizona." Abstract. *Geological Society of America Abstracts with Programs* 6:437.

Crowley, J. A. 1980. "The C. and B. Mine, Gila County, Arizona." *Mineralogical Record* 11:213–18.

Cuif, Jean-Pierre, Yannicke Dauphin, and James E. Sorauf. 2011. *Biominerals and Fossils through Time*. Cambridge: Cambridge University Press.

Culin, F. L. 1917. *Gems and Precious Stones of Arizona*. Arizona Bureau of Mines Bulletin 48. Tucson: University of Arizona.

Cummings, J. B. 1946a. *Exploration of New Planet Iron Deposit, Yuma County, Arizona*. U.S. Bureau of Mines Rept. Inv. RI 3982. Washington, D.C.: U.S. Department of the Interior, Bureau of Mines.

Cummings, J. B. 1946b. *Exploration of the Packard Fluorspar Property, Gila County, Arizona*. U.S. Bureau of Mines Rept. Inv. RI 3880. Washington, D.C.: U.S. Department of the Interior, Bureau of Mines.

Dale, V. B. 1959. *Tungsten Deposits of Yuma, Maricopa, Pinal, and Graham Counties, Arizona*. U.S. Bureau of Mines Report of Investigations RI-5516. Washington, D.C.: U.S. Department of the Interior, Bureau of Mines.

Dale, V. B. 1961. *Tungsten Deposits of Gila, Yavapai, and Mohave Counties, Arizona*. U.S. Bureau of Mines Information Circular IC-8078. Washington, D.C.: U.S. Department of the Interior, Bureau of Mines.

Dale, V. B., L. A. Stewart, and W. A. McKinney. 1960. *Tungsten Deposits of Cochise, Pima and Santa Cruz Counties, Arizona*. U.S. Bureau of Mines Report of Investigations RI-5650. Washington, D.C.: U.S. Department of the Interior, Bureau of Mines.

Davis, D. G., and W. Breed. 1968. "Rock Fulgurites on the San Francisco Peaks, Arizona." *Plateau (Quarterly of the Museum of Northern Arizona)* 41:34.

Davis, R. E. 1955. "Geology of the Mary G Mine Area, Pima County, Arizona." Master's thesis, University of Arizona.

Davis, S. R. 1975. "The Hardshell Silver Deposit, Harshaw Mining District, Santa Cruz County, Arizona." Abstract. *Symposium of the New Mexico Geological Society and Arizona Geological Society: Abstracts with Programs*, 6. Silver City, N.Mex.: New Mexico Geological Society and Arizona Geological Society.

Day, D. T., and R. H. Richards. 1906. "Investigation of Black Sands from Placer Mines." In *Tin, Quicksilver, Platinum, Etc.: Tin in the Franklin Mountains, Tex.* U.S. Geological Survey Bulletin 285, 150–64. Washington, D.C.: Government Printing Office.

Dean, K. C., H. D. Snedden, and W. W. Agey. 1952. *Concentration of Oxide Manganese Ores from Vicinity of Winkelman, Pinal County, Arizona*. U.S. Bureau of Mines Report of Investigations RI-4848. Washington, D.C.: U.S. Department of the Interior, Bureau of Mines.

Deer, W. A., R. A. Howie, and J. Zussman. 1962. *Rock Forming Minerals*. Vol. 3, *Sheet Silicates*. London: Longman, Green.

De Kalb, C. 1895. "Onyx Marbles." *American Institute of Mining, Metallurgical, and Petroleum Engineers (AIME) Transactions* 25:562–63.

Denton, T. C. 1947a. *Aravaipa Lead-Zinc Deposits, Graham County, Arizona*. U.S. Bureau of Mines Report of Investigations RI 4007. Washington, D.C.: U.S. Department of the Interior, Bureau of Mines.

Denton, T. C. 1947b. *Old Reliable Copper Mine, Pinal County, Arizona*. U.S. Bureau of Mines Report of Investigations RI 4006. Washington, D.C.: U.S. Department of the Interior, Bureau of Mines.

Denton, T. C., and P. S. Haury. 1946. *Exploration of the Reward (Vekol) Zinc Deposit, Pinal County, Arizona*. U.S. Bureau of Mines Report of Investigations RI 3975. Washington, D.C.: Government Printing Office.

Denton, T. C., and C. A. Kumke. 1949. *Investigation of Snowball Fluorite Deposit, Maricopa County, Arizona*. U.S. Bureau of Mines Report of Investigations RI 4540. Washington, D.C.: U.S. Department of the Interior, Bureau of Mines.

DeRuyter, D. V. 1979. "Geology of the Granite Peak Stock Area, Whetstone Mountains, Cochise County, Arizona. Tucson." Master's thesis, University of Arizona.

de Villiers, P. R., and F. H. Herbstein. 1967. "Distinction between Two Members of the Braunite Group." *American Mineralogist* 52:20–30.

DeWitt, E., and J. Waegli. 1986. *Gold in the United Verde Massive Sulfide Deposit, Jerome, Arizona*. U.S. Geological Survey Open-File Report 85-0585; superseded by U.S. Geological Survey Bulletin 1857: D1–D26. Washington, D.C.: U.S. Department of the Interior, Geological Survey.

De Wolf, W. P. 1916. "Tungsten in Arizona." *Engineering and Mining Journal* 101:680.

Diery, H. D. 1964. "Petrography and Petrogenetic History of a Quartz Monzonite Intrusive, Swisshelm Mountains, Cochise County, Arizona." Master's thesis, University of Arizona.

Diller, J. S., and J. E. Whitfield. 1889. "Dumortierite from Harlem, N.Y., and Clip, Arizona." *American Journal of Science* 37:216–20.

Dimick, A. 1957. "Arizona's Peloncillo Agate." *Gems and Minerals* 242:24–27.

Dings, M. G. 1951. *The Wallapai Mining District, Cerbat Mountains, Mohave County, Arizona*. U.S. Geological Survey Bulletin 978-E. Washington, D.C.: Government Printing Office.

Dodd, P. H. 1955. "Examples of Uranium Deposits in the Upper Jurassic Morrison Formation of the Colorado Plateau." In *Contributions to the Geology of Uranium and Thorium by the United States Geological Survey and Atomic Energy Commission for the United Nations International Conference on Peaceful Uses of Atomic Energy, Geneva, Switzerland 1955*, edited by L. R. Page, H. E. Stocking, and H. B. Smith, 243–62. U.S. Geological Survey Professional Paper 300. Washington, D.C.: Government Printing Office.

Domitrovic, A. M., W. E. Wilson, and M. Hay. 1998. "Famous Mineral Localities: The Castle Dome District, Yuma County, Arizona." *Mineralogical Record* 29: 437–58.

Douglas, J. 1899. "The Copper Queen Mine, Arizona." *American Institute of Mining, Metallurgical, and Petroleum Engineers (AIME) Transactions* 29:511–46.

Dreyer, R. M. 1939. "Darkening of Cinnabar in Sunlight." *American Mineralogist* 24:457–60.

Duke, A. 1960. *Arizona Gem Fields*. 2nd ed. Yuma, Ariz.: self-published.

Dunn, P. J. 1978. "Cuprite Up Close." *Mineralogical Record* 9:259.

Dunn, P. J., J. D. Grice, and R. A. Bideaux. 1989. "Pinalite, a New Lead Tungsten Chloride Mineral from the Mammoth Mine, Pinal County, Arizona." *American Mineralogist* 74:934–35.

Eastlick, J. T. 1968. "Geology of the Christmas Mine and Vicinity, Banner Mining District, Arizona." In *Ore Deposits of the United States, 1933–1967*, edited by J. D. Ridge, 1191–1210. New York: American Institute of Mining, Metallurgical, and Petroleum Engineers.

Eckel, E. B. 1930. "Boxwork Siderite: An Analogous Occurrence of Silica and Siderite." *Economic Geology* 25:290–92.

Edson, G. M. 1977. "Some Bedded Zeolites, San Simon Basin, Southeastern Arizona." Master's thesis, University of Arizona.

Edson, G. M. 1980. "The Red Cloud Mine, Yuma, Arizona." *Mineralogical Record* 11:141–52.

Eidel, J. J. 1966. "The Crystallization and Mineralization of a Porphyry Copper Stock, Ithaca Peak, Mohave County, Arizona." *Economic Geology* 61:1305–6.

Eidel, J. J., J. E. Frost, and D. M. Clippinger. 1968. "Copper-Molybdenum Mineralization at Mineral Park, Mohave County, Arizona." In *Ore Deposits of the United States, 1933–1967*, edited by J. D. Ridge, 1258–81. New York: American Institute of Mining, Metallurgical, and Petroleum Engineers.

Elevatorski, E. A. 1978. *Arizona Industrial Minerals*. Arizona Department of Mineral Resources Mineral Report no. 2, MR-2. Phoenix: Arizona Department of Mineral Resources.

Elsing, M. J., and R. E. S. Heineman. 1936. *Arizona Metal Production*. Arizona Bureau of Mines Bulletin 140. Tucson: University of Arizona.

Enders, M. S. 2000. "The Evolution of Supergene Enrichment in the Morenci Porphyry Copper Deposit, Greenlee County, Arizona." PhD diss., University of Arizona.

Endlich, F. M. 1897. "The Pearce Mining District." *Engineering and Mining Journal* 63 (23): 571.

Engineering and Mining Journal. 1892a. "General Mining News." *Engineering and Mining Journal* 53 (16): 433.

Engineering and Mining Journal. 1892b. "General Mining News." *Engineering and Mining Journal* 53 (20): 527.

Engineering and Mining Journal. 1892c. "General Mining News." *Engineering and Mining Journal* 54 (14): 325.

Engineering and Mining Journal. 1892d. "General Mining News." *Engineering and Mining Journal* 54 (19): 445.

Engineering and Mining Journal. 1897. "General Mining News." *Engineering and Mining Journal* 64 (6): 163.

Engineering and Mining Journal. 1900. "White Gold." *Engineering and Mining Journal* 69:354.

Engineering and Mining Journal. 1904. "The Silverbell Camp, Arizona." *Engineering and Mining Journal* 77:639.

Engineering and Mining Journal. 1910. "A Tungsten Deposit in Western Arizona." *Engineering and Mining Journal* 90:1103.

Engineering and Mining Journal. 1912. "The Mining News." *Engineering and Mining Journal* 97:395.

Engineering and Mining Journal. 1915. "The Mining News." *Engineering and Mining Journal* 100:862.

Engineering and Mining Journal. 1918. "Editorial Correspondence." *Engineering and Mining Journal* 106:726.

Engineering and Mining Journal. 1922. "Onyx in Arizona." *Engineering and Mining Journal* 114:408.

Engineering and Mining Journal. 1926. "Large Marble Beds Reported in Arizona." *Engineering and Mining Journal* 121:416.

Engineering and Mining Journal. 1940. "News of the Industry." *Engineering and Mining Journal* 141:76.

Enlows, H. E. 1955. "Welded Tuffs of Chiricahua National Monument, Arizona." *Geological Society of America Bulletin* 66:1215–46.

Evans, H. T., and J. M. Hughes. 1990. "Crystal Chemistry of the Natural Vanadium Bronzes." *American Mineralogist* 75:508–21.

Evans, H. T., and M. E. Mrose. 1977. "The Crystal Chemistry of the Hydrous Copper Silicates, Shattuckite and Plancheite." *American Mineralogist* 62:491–502.

Evans, S. H., Jr., and W. P. Nash. 1979. "Petrogenesis of Xenolith-Bearing Basalts from Southern Arizona." *American Mineralogist* 64:249–67.

Evensen, C. G., and I. B. Gray. 1958. "Evaluation of Uranium Ore Guides, Monument Valley, Arizona and Utah." *Economic Geology* 53:639–62.

Eyde, T. H. 1978. *Arizona Zeolites*. Mineral Report 1. Phoenix: Arizona Department of Mineral Resources.

Eyde, T. H. 1986. "Field Trip to Selected Industrial Mineral Deposits of Arizona." *Arizona Geological Society Digest* 16:312–18.

Eyde, T. H., and J. C. Wilt. 1989. "Arizona Industrial Minerals: A Growing Industry in Transition." In *Geologic Evolution of Arizona*, edited by J. P. Jenney and S. J. Reynolds, 741–58. Arizona Geological Society Digest 17. Tucson: Arizona Geological Society.

Fahey, J. J. 1955. "Murdochite, a New Copper Lead Oxide." *American Mineralogist* 40:905–6.

Fahey, J. J. 1964. "Recovery of Coesite and Stishovite from Coconino Sandstone of Meteor Crater, Arizona." *American Mineralogist* 49:1643–47.

Fahey, J. J., E. B. Daggett, and S. G. Gordon. 1950. "Wherryite, a New Mineral from the Mammoth Mine, Arizona." *American Mineralogist* 35:93–98.

Faick, J. N. 1958. "Geology of the Ord Mine, Mazatzal Mountains Quicksilver District, Arizona." U.S. Geological Survey Bulletin 1042-R, 685–98. Washington, D.C.: Government Printing Office.

Faick, J. N., and F. A. Hildebrand. 1958. "An Occurrence of Molybdenian Stolzite in Arizona." *American Mineralogist* 43:156–59.

Fairbanks, E. E. 1923. "Mineragraphic Notes on Manganese Minerals." *American Mineralogist* 8:209.

Fanfani, L., A. Nunzi, and P. F. Zanazzi. 1971. "The Crystal Structure of Butlerite." *American Mineralogist* 56:751–57.

Farnham, L. L., and R. Havens. 1957. *Pikes Peak Iron Deposits, Maricopa County, Arizona*. U.S. Bureau of Mines Report of Investigations RI-5319. Washington, D.C.: U.S. Department of the Interior, Bureau of Mines.

Farnham, L. L., and L. A. Stewart. 1958. *Manganese Deposits of Western Arizona*. U.S. Bureau of Mines Information Circular IC 7843. Washington, D.C.: Government Printing Office.

Farrington, O. C. 1891. "On Crystallized Azurite from Arizona." *American Journal of Science* 41:300–307.

Faust, G. T., J. C. Hathaway, and G. Millot. 1959. "A Restudy of Stevensite and Allied Minerals." *American Mineralogist* 44:342–70.

Fearing, J. L., Jr. 1926. "Some Notes on the Geology of the Jerome District, Arizona." *Economic Geology* 21:757–73.

Fellows, M. L. 1976. "Composition of Epidote from Porphyry Copper Deposits." Master's thesis, University of Arizona.

Ferguson, C. A., and R. A. Trapp. 2001. *Stratigraphic Nomenclature of the Miocene Superstition Volcanic Field, Central Arizona*. Arizona Geological Survey Open-File Report 01-06. Tucson: Arizona Geological Survey.

Field, C. W. 1966. "Sulfur Isotopic Method for Discriminating between Sulfates of Hypogene and Supergene Origin." *Economic Geology* 61:1428–35.

Finch, W. I. 1967. *Geology of the Epigenetic Uranium Deposits in the United States*. U.S. Geological Survey Professional Paper 538. Washington, D.C.: Government Printing Office.

Finnell, T. L. 1957. "Structural Control of Uranium Ore at the Monument No. 2 Mine, Apache County, Arizona." *Economic Geology* 52:25–35.

Fischer, R. P. 1937. "Sedimentary Deposits of Copper, Vanadium-Uranium and Silver in Southwestern United States." *Economic Geology* 32:906–51.

Fitzpatrick, J., and A. Pabst. 1986. "Thalenite from Arizona." *American Mineralogist* 71:188–93.

Flagg, A. L. 1942. "Vanadium Report: Book I to Book VIII." Phoenix: Arizona Department of Mines and Mineral Resources, unpublished report.

Flagg, A. L. 1958. *Mineralogical Journeys in Arizona*. Scottsdale, Ariz.: Bitner.

Fleischer, M. 1959. "The Geochemistry of Rhenium, with Special Reference to Its Occurrence in Molybdenite." *Economic Geology* 54:1406–13.

Fleischer, M., and W. E. Richmond. 1943. "The Manganese Oxide Minerals: A Preliminary Report." *Economic Geology* 38:269–86.

Foord, E. E., J. E. Taggert, and N. M. Conklin. 1983. "Cuprian Fraipontite and Sauconite from the Defiance—Silver Bill Mines, Gleeson, Arizona." *Mineralogical Record* 14:131–32.

Foote, A. E. 1891. "A New Locality for Meteoric Iron with a Preliminary Notice of the Discovery of Diamonds in the Iron." *American Journal of Science* 42:413–41.

Force, E. R. 1997. *Geology and Mineral Resources of the Santa Catalina Mountains, Southeastern Arizona, a Cross-Sectional Approach*, with sections by D. M. Unruh and R. J. Kamilli. Monographs in Mineral Resource Science No. 1. Tucson: University of Arizona Center for Mineral Resources.

Ford, W. E. 1902. "On the Chemical Composition of Dumortierite from Clip, Arizona." *American Journal of Science* 14:426–30.

Ford, W. E. 1914. "New Occurrences of Spangolite." *American Journal of Science* 38:503–4.

Ford, W. E., and W. M. Beadley. 1915. "On the Identity of Footeite with Connellite Together with the Description of Two Occurrences of the Mineral." *American Journal of Science* 39:570–676.

Fornash, K. F., P. J. Patchett, G. E. Gehrels, and J. E. Spencer. 2013. "Evolution of Granitoids in the Catalina Metamorphic Core Complex, Southeastern Arizona: U–Pb, Nd and Hf Isotopic Constraints." *Contributions to Mineralogy and Petrology* 165:1295–1310.

Foshag, W. F. 1919. "Famous Mineral Localities: Yuma County, Arizona." *American Mineralogist* 4:149–50.

Frenzel, A. B. 1898. "A Turquoise Deposit in Mohave County, Arizona." *Engineering and Mining Journal* 66:697.

Frey, F. A., and M. Prinz. 1978. "Ultramafic Inclusions from San Carlos, Arizona: Petrologic and Geochemical Data Bearing on Their Petrogenesis." *Earth and Planetary Science Letters* 38:129–76.

Frondel, C. 1935. *Catalog of Mineral Pseudomorphs in the American Museum.* Bulletin of the American Museum of Natural History 67. New York: AMNH.

Frondel, C. 1941. "Paramelaconite: A Tetragonal Oxide of Copper." *American Mineralogist* 26:567–672.

Frondel, C. 1943. "Mineralogy of the Oxides and Carbonates of Bismuth." *American Mineralogist* 28:521–35.

Frondel, C. 1949. "Crystallography of Spangolite." *American Mineralogist* 34:181–87.

Frondel, C. 1956. "Mineral Composition of Gummite." *American Mineralogist* 41:539–68.

Frondel, C. 1958. *Systematic Mineralogy of Uranium and Thorium.* U.S. Geological Survey Bulletin 1064. Washington, D.C.: Government Printing Office.

Frondel, C. 1962. *The System of Mineralogy.* Vol. 3, *Silica Minerals.* New York: Wiley.

Frondel, C., and E. W. Heinrich. 1942. "New Data on Hetaerolite, Hydrohetaerolite, Coronadite and Hollandite." *American Mineralogist* 27:48–56.

Frondel, C., J. Ito, R. M. Honea, and A. M. Weeks. 1976. "Mineralogy of the Zippeite Group." *Canadian Mineralogist* 14:429–36.

Frondel, C., and U. B. Marvin. 1967a. "Lonsdaleite, a Hexagonal Polymorph of Diamond." *American Mineralogist* 52:1576.

Frondel, C., and U. B. Marvin. 1967b. "Lonsdaleite, a Hexagonal Polymorph of Diamond." *Nature* 214:587–89.

Frondel, C., and F. H. Pough. 1944. "Two New Tellurites of Iron: Mackayite and Blakeite, with New Data on Emmonsite and 'Durdenite.'" *American Mineralogist* 29:211–25.

Frondel, J. W. 1964. "Variation of Some Rare Earths in Allanite." *American Mineralogist* 49:1157–77.

Frondel, J. W., and F. E. Wickman. 1970. "Molybdenite Polytypes in Theory and Occurrence: II. Some Naturally-Occurring Polytypes of Molybdenite." *American Mineralogist* 55:1857–75.

Frost, R. L., P. A. Williams, W. Martens, P. Leverett, and J. T. Kloprogge. 2004. "Raman Spectroscopy of Basic Copper (II) and Some Complex Copper (II) Sulfate Minerals: Implications for Hydrogen Bonding." *American Mineralogist* 89:1130–37.

Fuchs, L. H. 1969. "The Phosphate Mineralogy of Meteorites." In *Meteorite Research,* edited by P. M. Millman, 683–95. Dordrecht: Reidel.

Funnell, J. E., and E. J. Wolfe. 1964. *Compendium on Nonmetallic Minerals of Arizona.* San Antonio, Tex.: Southwest Research Institute.

Gaft, M., R. Reisfeld, and G. Panczer. 2015. *Modern Luminescence Spectroscopy of Minerals and Materials.* 2nd ed. Heidelberg: Springer.

Galbraith, F. W. 1941. *Minerals of Arizona.* Arizona Bureau of Mines Bulletin 149. Tucson: University of Arizona.

Galbraith, F. W. 1947. *Minerals of Arizona.* 2nd ed. Arizona Bureau of Mines Bulletin 153. Tucson: University of Arizona.

Galbraith, F. W., and D. J. Brennan. 1959. *Minerals of Arizona.* 3rd ed. Physical Science Bulletin 4. Tucson: University of Arizona.

Galbraith, F. W., and D. J. Brennan. 1970. *Minerals of Arizona.* Arizona Bureau of Mines Bulletin 181. Tucson: University of Arizona.

Galbraith, F. W., and T. H. Kuhn. 1940. "A New Occurrence of Dioptase in Arizona." *American Mineralogist* 25:708–10.

Garcia, M. O., D. W. Muenow, and N. W. K. Liu. 1980. "Volatiles in Ti-rich Amphibole Megacrysts, Southwestern USA." *American Mineralogist* 65: 306–12.

Gardner, B., and E. Davis. 2000. "The Purple Passion Mine, Yavapai County, Arizona." *Mineralogical Record* 31:323–31.

Gardner, E. D. 1936. *Gold Mining and Milling in the Black Mountains, Western Mohave County, Arizona.* U.S. Bureau of Mines Information Circular IC 6901. Washington, D.C.: U.S. Department of the Interior, Bureau of Mines.

Garrels, R. M., and E. S. Larsen III. 1959. *Geochemistry and Mineralogy of the Colorado Plateau Uranium Ores.* U.S. Geological Survey Professional Paper 320. Washington, D.C.: Government Printing Office.

Garrison, F. L. 1907. "Notes on Minerals." *Proceedings of the Academy of Natural Science and Philosophy* 59:445.

Garvie, L. A. J. 2003. "Decay-Induced Biomineralization of the Saguaro Cactus (*Carnegiea gigantea*)." *American Mineralogist* 88:1879–88.

Garvie, L. A. J. 2016. "Mineralogy of Paloverde (*Parkinsonia microphylla*) Tree Ash from the Sonoran Desert: A Combined Field and Laboratory Study." *American Mineralogist* 101:1584–95.

Gastil, G. R. 1954. "An Occurrence of Authigenic Xenotime." *Journal of Sedimentary Petrology* 24:280–81.

Gastil, G. R. 1958. "Older Pre-Cambrian Rocks of the Diamond Butte Quadrangle, Gila County, Arizona." *Geological Society of America Bulletin* 69:1495–1514.

Gatehouse, B. M., I. E. Grey, and P. R. Kelly. 1979. "The Crystal Structure of Davidite." *American Mineralogist* 64:1010–17.

Gavasci, A. T., and P. F. Kerr. 1968. "Uranium Emplacement at Garnet Ridge." *Economic Geology* 63:859–76.

Genth, F. A. 1868. "Contributions to Mineralogy." *American Journal of Science* s2, 45 (135): 305–21.

Genth, F. A. 1887. "Contributions to Mineralogy." *Annals Phil. Soph. Soc.* 2:20.

Genth, F. A. 1890. "Contributions to Mineralogy." *American Journal of Science* s3, 40 (236): 114–20.

Getsinger, F. 1966. "Agate in Arizona." In *The Agates of North America*, edited by Hugh Leiper, 31. San Diego, Calif.: Lapidary Journal.

Gibbs, R. B. 1992. "Microminerals from the Big Lue Mountains, Greenlee County, Arizona and Grant County, New Mexico." Talk given at the Thirteenth Annual New Mexico Mineral Symposium, Socorro, New Mexico, November 14–15.

Gibbs, R. B. 2003. "An Occurrence of Creaseyite in the Tortolita Mountains, Pinal County, Arizona." *Minerals of Arizona, Eleventh Annual Symposium*. Phoenix: Flagg Mineral Foundation.

Gibbs, R. B. 2011. "Microminerals of the Evening Star Mine." Abstract. *Minerals of Arizona, Nineteenth Annual Symposium*, 20–21. Phoenix: Flagg Mineral Foundation.

Gibbs, R. B., and M. Ascher. 2012. "Through the 'Scope': Microminerals of the Tonopah-Belmont Mine, Big Horn Mountains, Maricopa County, Arizona." *Rocks and Minerals* 87:175–80.

Gibbs, R. B., and U. Turzi. 2007. "Minerals in the Miarolitic Cavities of the Belmont Granite, Belmont Mountains, Maricopa County, Arizona." *Minerals of Arizona, Fifteenth Annual Symposium*. Phoenix: Flagg Mineral Foundation.

Gibson, E. K., and D. D. Bogard. 1978. "Chemical Alterations of the Holbrook Chondrite Resulting from Terrestrial Weathering." *Meteorites* 13:277–89.

Giester, G., and B. Rieck. 1996. "Bechererite, $(Zn,Cu)_6Zn_2(OH)_{13}[(S,Si)(O,OH)_4]_2$, A Novel Mineral Species from the Tonopah-Belmont Mine, Arizona." *American Mineralogist* 81:244–48.

Gigl, P. D., and F. Dachille. 1968. "Effect of Pressure and Temperature on the Reversal Transitions of Stishovite." *Meteorites* 4:123–36.

Gilluly, J. 1937. *Geology and Ore Deposits of the Ajo Quadrangle, Arizona*. Arizona Bureau of Mines Bulletin 141. Tucson: University of Arizona.

Gilluly, J. 1942a. "Mineralization of the Ajo Copper District, Arizona." *American Mineralogist* 27:222–23.

Gilluly, J. 1942b. "The Mineralization of the Ajo Copper District, Arizona." *Economic Geology* 37:297–309.

Gilluly, J. 1946. *The Ajo Mining District, Arizona*. U.S. Geological Survey Professional Paper 209: 112. Washington, D.C.: Government Printing Office.

Gilluly, J. 1956. *General Geology of Central Cochise County, Arizona*. U.S. Geological Survey Professional Paper 281. Washington, D.C.: Government Printing Office.

Gleason, S. 1972. *Ultraviolet Guide to Minerals*. San Gabriel, Calif.: Ultra-Violet Products.

Godas, G., D. Morris, and M. Hay. 2015. "A Dream Pocket at the North Geronimo Mine, October 1996." *Mineralogical Record* 46:603–11.

Goldsmith, L. B. 2011. "Sugarloaf Peak Gold Project, T3N, R20W, Sec 3,4,5; T4N, R20W, Sec 29, 30, 31, 32,33,34, Quartzsite Area, La Paz County, Arizona." Unpublished report 43-101, prepared for Choice Gold, Vancouver, B.C.

Gootee, B. F., and D. G. Gruber. 2015. *Quartz Vein Investigation, McDowell Sonoran Preserve, Scottsdale, Maricopa County, Arizona*. Arizona Geological Survey, Open-File Report, OFR-15-03. Tucson: Arizona Geological Survey.

Gordon, S. G. 1923. "Recently Described Bisbeeite from the Grand Canyon Is Cyanotrichite." *American Mineralogist* 8:92–93.

Gornitz, V. 1986. "Uranium Mineralization at the Orphan Mine Breccia Pipe, Grand Canyon, Arizona." Abstract. *Geological Society of America Abstracts with Programs* 18:357.

Gornitz, V. 2004. "Grand Canyon Uranium: The Orphan Mine." *Mineral News* 20:627–41.

Gornitz, V., and P. F. Kerr. 1970. "Uranium Mineralization and Alteration, Orphan Mine, Grand Canyon, Arizona." *Economic Geology* 65:751–68.

Götze, J., M. Gaft, and R. Möckel. 2015. "Uranium and Uranyl Luminescence in Agate/ Chalcedony." *Mineralogical Magazine* 79 (4): 985–95.

Graeber, E. J., and A. Rosenzweig. 1971. "The Crystal Structures of Yavapaiite, $KFe(SO_4)_2$ and Goldichite, $KFe (SO_4)_2 \cdot 4 H_2O$." *American Mineralogist* 56:1917–33.

Graeme, R. W. 1981. "Famous Mineral Localities: Bisbee, Arizona." *Mineralogical Record* 12:258–319.

Graeme, R. W. 1993. "Bisbee Revisited." *Mineralogical Record* 24:421–36.

Graeme, R. W., III, R. W. Graeme IV, and D. L. Graeme. 2015. "An Update on the Minerals of Bisbee, Cochise County, Arizona." *Mineralogical Record* 46:627–41.

Granger, H. C. 1955. "Dripping Spring Quartzite, Arizona." In *Geologic Investigations of Radioactive Deposits—Semiannual Progress Report*, 187–90. U.S. Geological Survey TEI 540. Washington, D.C.: U.S. Geological Survey.

Granger, H. C., and R. B. Raup. 1959. *Uranium Deposits in the Dripping Spring Quartzite, Gila County, Arizona*. U.S. Geological Survey Bulletin 1049-P. Washington, D.C.: Government Printing Office.

Granger, H. C., and R. B. Raup. 1962. *A Reconnaissance Study of Uranium Deposits in Arizona*. U.S. Geological Survey Bulletin 1147-A. Washington, D.C.: Government Printing Office.

Granger, H. C., and R. B. Raup. 1969. *Geology of Uranium Deposits in the Dripping Spring Quartzite, Gila County, Arizona*. U.S. Geological Survey Professional Paper 595. Washington, D.C.: Government Printing Office.

Grant, R. W. 1982. *Checklist of Arizona Minerals*. Phoenix: Mineralogical Society of Arizona.

Grant, R. W. 1989. "The Blue Ball Mine, Gila County, Arizona." *Mineralogical Record* 20:447–50.

Grant, R. W. 2007. *Checklist of Arizona Minerals*. 2nd ed. Phoenix: Mineralogical Society of Arizona.

Grant, R. W., R. A. Bideaux, and S. A. Williams. 2006. *Minerals Added to the Arizona List 1995 to 2005*. Phoenix: Arizona Mineral and Mining Museum Foundation (now Flagg Mineral Foundation).

Gregg, C. C., and E. L. Moore. 1955. *Reconnaissance of the Chinle Formation in the Cameron-St. Johns Area Coconino, Navajo and Apache Counties, Arizona*. U.S. Atomic Energy Commission RME-51. Oak Ridge, Tenn.: Technical Information Service.

Gregory, H. E. 1916. "Garnet Deposits on the Navajo Reservation, Arizona and Utah." *Economic Geology* 11:223–30.

Gregory, H. E. 1917. *Geology of the Navajo Country—A Reconnaissance of Parts of Arizona, New Mexico, and Utah*. U.S. Geological Survey Professional Paper 93. Washington, D.C.: Government Printing Office.

Grout, F. F. 1946. "Microscopic Characters of Vein Carbonates." *Economic Geology* 41:475–502.

Gruner, J. W., and L. Gardiner. 1952. *Mineral Associations in the Uranium Deposits of the Colorado Plateau and Adjacent Regions with Special Emphasis on Those in the*

Shinarump Formation: Part 3, Annual Report July 1, 1951, to June 30, 1952. U.S. Atomic Energy Commission. Technical Information Service, RMO 566. Oak Ridge, Tenn.: U.S. Atomic Energy Commission, Technical Information Service.

Guild, F. N. 1905. "Petrography of the Tucson Mountains." *American Journal of Science* 20:313.

Guild, F. N. 1907. "The Composition of Molybdite from Arizona." *American Journal of Science* 23:455–56.

Guild, F. N. 1910. *The Mineralogy of Arizona.* Easton, Pa.: Chemical Publishing.

Guild, F. N. 1911. "Mineralogische Notizen." *Zeit. Krystal. und Mineralogical* 49:321–31.

Guild, F. N. 1917. "A Microscopic Study of the Silver Ores and Their Associated Minerals." *Economic Geology* 12:297–353.

Guild, F. N. 1920. "Flagstaffite, a New Mineral from Arizona." *American Mineralogist* 5:169–72.

Guild, F. N. 1921. "The Identity of Flagstaffite and Terpin Hydrate." *American Mineralogist* 6:133–35.

Guild, F. N. 1922. "Flagstaffite, a New Arizona Mineral, and Its Identity with Terpin Hydrate." Abstract. *Science* 55:543.

Guild, F. N. 1930. "The Relation of Pyrite to Wolframite." *American Mineralogist* 15:451–52.

Guild, F. N. 1934. "Microscopic Relations of Magnetite, Hematite, Pyrite and Chalcopyrite." *Economic Geology* 29:107–20.

Guild, F. N. 1935. "Piedmontite in Arizona." *American Mineralogist* 20:679–92.

Guillebert, C., and M. T. Le Bihan. 1965. "Contribution à l'étude structurale des silicates de cuivre: Structure atomique de la papagoite." *Bulletin de la Société Française de Minéralogie et de Cristallographie* 88:119–21.

Guiteras, J. R. 1936. *Gold Mining and Milling in the Black Canyon Area, Yavapai County, Arizona.* U.S. Bureau of Mines Information Circular IC-6905. Washington, D.C.: U.S. Department of the Interior, Bureau of Mines.

Hålenius, U., F. Hatert, M. Pasero, and S. J. Mills. 2016. "IMA Commission of New Minerals, Nomenclature and Classification (CNMNC) Newsletter 33: New Minerals and Nomenclature Modifications Approved in 2016." *Mineralogical Magazine* 80:1135–44.

Hålenius, U., F. Hatert, M. Pasero, and S. J. Mills. 2018a. "IMA Commission of New Minerals, Nomenclature and Classification (CNMNC) Newsletter 43: New Minerals and Nomenclature Modifications Approved in 2018." *Mineralogical Magazine* 82:779–85.

Hålenius, U., F. Hatert, M. Pasero, and S. J. Mills. 2018b. "IMA Commission of New Minerals, Nomenclature, and Classification (CNMNC) Newsletter 45: New Minerals and Nomenclature Modifications Approved in 2018." *Mineralogical Magazine* 82:1225–32.

Hall, R. B. 1978. *World Nonbauxite Aluminum Resources: Alunite.* U.S. Geological Survey Professional Paper 1076-A: A1–A35. Washington, D.C.: Government Printing Office.

Hamilton, P. 1884. *The Resources of Arizona.* 3rd ed. San Francisco: Bancroft.

Hamilton, P., and P. F. Kerr. 1959. "Umohoite from Cameron, Arizona." *American Mineralogist* 44:1248–60.

Hammer, D. F. 1961. "Geology and Ore Deposits of the Jackrabbit Area, Pinal County, Arizona." Master's thesis, University of Arizona.

Hammer, D. F., and D. W. Peterson. 1968. "Geology of the Magma Mine Area, Arizona." In *Ore Deposits of the United States, 1933–1967*, edited by J. D. Ridge, 1282–1310. New York: American Institute of Mining, Metallurgical, and Petroleum Engineers.

Handverger, P. A. 1963. "Geology of the Three R Mine, Palmetto Mining District, Santa Cruz County, Arizona." Master's thesis, University of Arizona.

Hankins, D. D. 1984. "Geologic Setting of Precious Metals Mineralization, King of Arizona District, Kofa Mountains, Yuma County, Arizona." Master's thesis, San Diego State University.

Hanson, H. S. 1966. "Petrography and Structure of the Leatherwood Quartz Diorite, Santa Catalina Mountains, Pima County, Arizona." PhD diss., University of Arizona.

Hanson, S. L., T. A. Brown, A. U. Falster, and W. B. Simmons. 2011. "'Allanite-(Nd)': A New Mineral from the Kingman Feldspar Mine in the Mojave Pegmatite District, Northwestern Arizona." Presented at the 38th Rochester Mineralogical Symposium, April 14–17, Rochester, N.Y., program notes, 20–22.

Hanson, S. L., A. U. Falster, and W. B. Simmons. 2001. "Mineralogy of Fumarole Deposits from Sunset Crater Volcano, Northern Arizona." In "Contributed Papers in Specimen Mineralogy: 27th Rochester Mineralogical Symposium," *Rocks and Minerals* 76:252–58.

Hanson, S. L., A. U. Falster, and W. B. Simmons. 2007. "Rare-Earth Mineralogy of the Mohave County Pegmatites, Northwestern Arizona." *Rocks and Minerals* 82:236–37.

Hanson, S. L., A. U. Falster, and W. B. Simmons. 2008. "Mineralogy of Fumarole Deposits: At Sunset Crater Volcano National Monument, Northern Arizona." *Rocks and Minerals* 83:534–44.

Hanson, S. L., A. U. Falster, W. B. Simmons, and T. A. Brown. 2012. "Allanite-(Nd) from the Kingman Feldspar Mine, Mojave Pegmatite District, Northwestern Arizona, USA." *Canadian Mineralogist* 50:815–24.

Harcourt, G. A. 1937. "The Distinction between Enargite and Famatinite (Luzonite)." *American Mineralogist* 22:517–25.

Harcourt, G. A. 1942. "Tables for the Identification of Ore Minerals by X-ray Powder Patterns." *American Mineralogist* 27:63–130.

Hardas, A. V. 1966. "Stratigraphy of Gypsum Deposits, South of Winkelman, Pinal County, Arizona." Master's thesis, University of Arizona.

Harness, C. L. 1942. *Strontium Minerals*. U.S. Bureau of Mines Information Circular IC 7200. Washington, D.C.: U.S. Department of the Interior, Bureau of Mines.

Harrer, C. M. 1964. *Reconnaissance of Iron Resources in Arizona*. U.S. Bureau of Mines Information Circular IC 8236. Washington, D.C.: U.S. Department of the Interior, Bureau of Mines.

Harris, D. C., A. C. Roberts., R. I. Thorpe, A. J. Criddle, and C. J. Stanley. 1984. "Kiddcreekite, a New Mineral Species from the Kidd Mine, Timmins, Ontario, and from the Campbell Orebody, Bisbee, Arizona." *Canadian Mineralogist* 22:227–32.

Hatert, F., S. J. Mills, M. Pasero, and P. A. Williams. 2013. "CNMNC Guidelines for the Use of Suffixes and Prefixes in Mineral Nomenclature, and for the Preservation of Historical Names." *European Journal of Mineralogy* 25:113–15.

Haury, P. S. 1947. *Examination of Lead-Zinc Mines in the Wallapai Mining District, Mohave County, Arizona.* U.S. Bureau of Mines Report of Investigations RI 4101. Washington, D.C.: U.S. Department of the Interior, Bureau of Mines.

Havens, R., S. J. Hussey, J. A. McAllister, and K. C. Dean. 1954. *Beneficiation of Oxide Manganese and Manganese-Silver Ores from Southern Arizona.* U.S. Bureau of Mines Report of Investigations RI 5024. Washington, D.C.: U.S. Department of the Interior, Bureau of Mines.

Havens, R., G. M. Potter, W. W. Agey, and R. R. Wells. 1947. *Concentration of Oxide Manganese Ores from the Lake Havasu District, California and Arizona.* U.S. Bureau of Mines Report of Investigations RI 4147. Washington, D.C.: U.S. Department of the Interior, Bureau of Mines.

Hawthorne, F. C., M. A. Copper, and M. E. Back. 2009. "Khinite-4*O* [=Khinite] and Khinite-3*T* [=Parakhinite]." *Canadian Mineralogist* 47:473–76.

Hawthorne, F. C., R. Oberti, G. E. Harlow, W. V. Maresch, R. F. Martin, J. C. Schumacher, and M. D. Welch. 2012. "Nomenclature of the Amphibole Supergroup." *American Mineralogist* 97:2031–48.

Haxel, G. B., G. S. Epstein, C. E. Jacobson, J. H. Wittke, K. G. Standlee, and S. R. Mulligan. 2021. "Geologic Map of Harzburgite and Associated Metasomatic Rocks in the Orocopia Schist Subduction Channel (Latest Cretaceous) at Cemetery Ridge, Southwest Arizona." *Arizona Geological Survey Contributed Report*, in press.

Haxel, G. B., S. M. Richard, R. M. Tosdal, and M. J. Grubensky. 2002. "The Orocopia Schist in Southwest Arizona: Early Tertiary Oceanic Rocks Trapped or Transported Far Inland." In *Contributions to Crustal Evolution of the Southwestern United States*, edited by Andrew Barth. Geological Society of America, Special Paper 365, 99–128. Boulder, Colo.: Geological Society of America.

Haxel, G. B., J. H. Wittke, G. S. Epstein, and C. E. Jacobson. 2018. "Serpentinization-Related Nickel, Iron, and Cobalt Sulfide, Arsenide, and Intermetallic Minerals in an Unusual Inland Tectonic Setting, Southern Arizona, USA." In *Tectonics, Sedimentary Basins, and Provenance: A Celebration of the Career of William R. Dickinson*, edited by R. V. Ingersoll, S. A. Graham, and T. F. Lawton, 65–86. Geological Society of America Special Paper 540. Boulder, Colo.: Geological Society of America.

Hay, M. 1997. "The Western Union Mine, Cerbat Mountains, Mohave County, Arizona." *Mineralogical Record* 28:167–74.

Hay, M., and G. Alexander. 2020. "The San Manual Mine, Pinal County, Arizona." *Mineralogical Record* 51:45–69.

Hay, M., and D. Morris. 2015. "Fluorite from the Oatman District, Mohave County, Arizona." *Mineralogical Record* 46:569–83.

Hayes, T. S. 2007. "Stop 4: Shallow Levels of Kelvin-Riverside Porphyry Copper System and Modern Supergene Enrichment and Exotic Copper System." In *Laramide Porphyry Copper Systems and Superimposed Tertiary Extension: A Life Cycle Approach to the Globe-Superior-Ray Area, Arizona*, 24–28. Arizona Geological Society Ores and Orogenesis Symposium Field Trip No. 4. Tucson: Arizona Geological Society.

Haynes, P. 1991. "Arizona News." *Mineral News*, January, 9.

Hazen, R. M., E. S. Grew, M. J. Origlieri, and R. T. Downs. 2017. "On the Mineralogy of the 'Anthropocene Epoch.'" *American Mineralogist* 102:595–611.

Head, R. E. 1941. *Physical Characteristics of Some Low-Grade Manganese Ores.* U.S. Bureau of Mines Report of Investigations RI 3560. Washington, D.C.: U.S. Department of the Interior, Bureau of Mines.

Headden, W. P. 1903. "Mineralogical Notes—Cuprodescloizite, Arizona." *Colorado Scientific Society Proceedings* 7:149–50.

Hehnke, C., G. Ballantyne, H. Martin, W. Hart, A. Schwarz, and H. Stein. 2012. "Geology and Exploration Progress at the Resolution Porphyry Cu-Mo Deposit, Arizona." In *Geology and Genesis of Major Copper Deposits and Districts of the World*, Special Publication 16, edited by J. W. Hedenquist, M. Harris, F. Camus, and R. H. Sillitoe, 147–66. Littleton, Colo.: Society of Economic Geologists.

Heidecker, E. J. 1978. "Precambrian Mineralization in Basalt Flows on Evaporites, Aztec Peak, Arizona." *Economic Geology* 73:106–8.

Heineman, R. E. S. 1930. "A Note on the Occurrence of Monazite in Western Arizona." *American Mineralogist* 15:536–37.

Heineman, R. E. S. 1931. "An Arizona Gold Nugget of Unusual Size." *American Mineralogist* 16:267–69.

Heineman, R. E. S. 1935. "Sugarloaf Butte Alunite." *Engineering and Mining Journal* 136:138–39.

Heinrich, E. W. 1960. *Some Rare-Earth Mineral Deposits in Mohave County, Arizona.* Arizona Bureau of Mines Bulletin 167. Tucson: University of Arizona.

Helmstaedt, H., and R. Doig. 1975. "Eclogite Nodules from Kimberlite Pipes of the Colorado Plateau—Samples of Subducted Franciscan-Type Oceanic Lithosphere." In *Physics and Chemistry of the Earth*, edited by L. H. Ahrens, A. R. Duncan, J. B. Dawson, and A. J. Erlank, vol. 9, 95–111. Oxford: Pergamon.

Henderson, R. 1941. "Sparkling Gems in the Aquarius Range." *Desert Magazine* 5:13–16.

Hershey, R. E. 1958. *Geology and Uranium Deposits of the Carrizo Mountains Area, Apache County, Arizona and San Juan County, New Mexico.* With a section on the Martin and Rattlesnake Incline Mines by V. A. Means and R. K. Labrecque. Atomic Energy Commission Report RME-117 (revised). Grand Junction, Colo.: U.S. Atomic Energy Commission.

Hess, F. L. 1909. "Note on a Wolframite Deposit in the Whetstone Mountains, Arizona." In *Contributions to Economic Geology, 1908: Part I, Metals and Nonmetals, Except Fuels.* U.S. Geological Survey Bulletin 380-D, 164–65. Washington, D.C.: Government Printing Office.

Hewett, D. F. 1925. "Carnotite Discovered near Aguila, Arizona." *Engineering and Mining Journal* 120:19.

Hewett, D. F. 1964. "Veins of Hypogene Manganese Oxide Minerals in the Southwestern United States." *Economic Geology* 59:1429–72.

Hewett, D. F. 1972. "Manganite, Hausmannite, Braunite: Features, Modes of Origin." *Economic Geology* 67:83–102.

Hewett, D. F., E. Callaghan, B. N. Moore, T. B. Nolan, W. W. Rubey, and W. T. Schaller. 1936. *Mineral Resources of the Region around Boulder Dam.* U.S. Geological Survey Bulletin 871. Washington, D.C.: Government Printing Office.

Hewett, D. F., and M. Fleischer. 1960. "Deposits of the Manganese Oxides." *Economic Geology* 55:55.

Hewett, D. F., M. Fleischer, and N. Conklin. 1963. "Deposits of the Manganese Oxides: Supplement." *Economic Geology* 58:1–51.

Hewett, D. F., and O. N. Rove. 1930. "Occurrence and Relations of Alabandite." *Economic Geology* 25:36–56.

Heyman, A. M. 1958. "Geology of the Peach-Elgin Copper Deposit, Helvetia District, Arizona." Master's thesis, University of Arizona.

Hibbs, D. E., P. Leverett, and P. A. Williams. 2002. "Buttgenbachite from Bisbee, Arizona, USA: A Single-Crystal X-ray Study." *Neues Jahrb. Min.—Monatshefte* 2002 (5): 225–40.

Hibbs, D. E., P. Leverett, and P. A. Williams. 2003. "A Single Crystal X-ray Study of a Sulphate-Bearing Buttgenbachite, $Cu_{36}Cl_{7.8}(NO_3)_{1.3}(SO_4)_{0.35}(OH)_{62.2}.5H_2O$, and a Re-examination of the Crystal Chemistry of the Buttgenbachite Connellite Series." *Mineralogical Magazine* 67:47–60.

Hill, C. A. 1999. "Mineralogy of Kartchner Caverns, Arizona." *Journal of Cave Karst Studies* 61:73–78.

Hill, J. M. 1914a. "Copper Deposits of the White Mesa District, Arizona." In *Copper Deposits near Superior Arizona*, edited by F. L. Ransome, 159–63. U.S. Geological Survey Bulletin 540-D. Washington, D.C.: Government Printing Office.

Hill, J. M. 1914b. "The Grand Gulch Mining Region, Mohave County, Arizona." *U.S. Geological Survey Bulletin* 580:39–58.

Hillebrand, W. F. 1885. "Emmonsite, a Ferric Tellurite." *Proceedings of the Colorado Scientific Society* 2:20–23.

Hillebrand, W. F. 1889a. "Analyses of Three Descloizites from New Localities." *American Journal of Science* 37:434–39.

Hillebrand, W. F. 1889b. "Antlerite, a Basic Cupric Sulfate." *U.S. Geologic Survey Bulletin* 55:54–55.

Hills, R. C. 1882. "Dioptase from Arizona." *American Journal of Science* 23:325.

Hills, R. C. 1890. "Informal Communication." *Proceedings of the Colorado Scientific Society* 3:257–58.

Himes, M. D. 1972. "Geology of the Pima Mine, Pima County, Arizona." Master's thesis, University of Arizona.

Hinkle, M. E., and G. S. Ryan. 1982. *Mineral Resource Potential of the Pusch Ridge Wilderness Area, Pima County, Arizona*. U.S. Geological Survey, Miscellaneous Field Studies Map 1356-B. Reston, Va.: U.S. Geological Survey.

Hobbs, S. W. 1944. "Tungsten Deposits in the Boriana District and the Aquarius Range, Mohave County, Arizona." In *Strategic Mineral Investigations, 1943, Tungsten, Boriana and Aquarius Range, Ariz.* U.S. Geological Survey Bulletin 940-I, 247–64. Washington, D.C.: Government Printing Office.

Hoelle, J. L. 1974. "Structural and Geochemical Analysis of the Catalina Granite, Santa Catalina Mountains, Arizona." Master's thesis, University of Arizona.

Hofmann, B. A. 1992. "Uranium Accumulation during Weathering of Canon Diablo Meteoritic Iron." *Meteoritics* 27:101–3.

Hoffman, V. J. 1963. "Heavy Mineral Distribution in Sands of the Tortolita Mountains Pediment, Southern Arizona." Master's thesis, University of Arizona.

Holden, E. F. 1922. "Ceruleofibrite, A New Mineral." *American Mineralogist* 7:80–83.

Holden, E. F. 1924. "Ceruleofibrite Is Connellite." *American Mineralogist* 9:55–56.

Holland, H. D., Jr., G. G. Witter, W. B. Head, and R. Petti. 1958. "The Use of Leachable Uranium in Geochemical Prospecting on the Colorado Plateau—2: The Distribution of Leachable Uranium in Surface Samples in the Vicinity of Ore Bodies." *Economic Geology* 53:190–209.

Honea, R. M. 1959. "New Data on Gastunite, an Alkali Uranyl Silicate." *American Mineralogist* 44:1047–56.

Honea, R. M. 1961. "New Data on Boltwoodite, an Alkali Uranyl Silicate." *American Mineralogist* 46:24.

Householder, E. R. 1930. "Geology of Mohave County, Arizona." Professional degree thesis, University of Missouri, School of Mines and Metallurgy.

Hovey, E. D. 1900. "Note on a Calcite Group from Bisbee, Arizona." *Bulletin of the American Museum of Natural History* 12:189–90.

Howell, K. K. 1977. "Geology and Alteration of the Commonwealth Mine, Cochise County, Arizona." Master's thesis, University of Arizona.

Hunt, W. 1983. "Calciovolborthite from the New Water Mountains, Yuma County, Arizona." *Mineralogical Record* 14:119.

Hurlbut, C. S., Jr. 1936. "A New Phosphate, Bermanite, Occurring with Triplite in Arizona." *American Mineralogist* 21:656–61.

Hurlbut, C. S., Jr., and L. F. Aristarain. 1968. "Bermanite, and Its Occurrence in Córdoba, Argentina." *American Mineralogist* 53:416–31.

Hutton, C. O. 1957. "Sengierite from Bisbee, Arizona." *American Mineralogist* 42:408–11.

Hutton, C. O. 1959a. "An Occurrence of Pseudomalachite at Safford, Arizona." *American Mineralogist* 44:1298–1301.

Hutton, C. O. 1959b. "Yavapaiite, an Anhydrous Potassium, Ferric Sulphate from Jerome, Arizona." *American Mineralogist* 44:1105–14.

Hutton, C. O., and A. C. Vlisidis. 1960. "Papagoite, a New Copper-Bearing Mineral from Ajo, Arizona." *American Mineralogist* 45:599–611.

Ipsen, A. O., and H. L. Gibbs. 1952. *Concentration of Oxide Manganese Ore from Doyle-Smith Claims, Northern Yuma County, Arizona.* U.S. Bureau of Mines Report of Investigations RI 4844. Washington, D.C.: U.S. Department of the Interior, Bureau of Mines.

Irvin, G. W. 1959. "Pyrometasomatic Deposits at San Xavier Mine." In *Southern Arizona Guidebook II, Combined with the 2nd Annual Arizona Geological Society Digest*, edited by L. A. Heindl, 195–97. Tucson: Arizona Geological Society.

Isachsen, Y. W., T. W. Mitcham, and H. B. Wood. 1955. "Age and Sedimentary Environments of Uranium Host Rocks, Colorado Plateau." *Economic Geology* 50:127–34.

Ishihara, S. 1977. "The Magnetite-Series and Ilmenite-Series Granitic Rocks." *Mining Geology* 27:293–305.

Jackson, D. 1955. "Turquoise in Eastern Arizona." *Gems and Minerals* 212:52–55.

Jacobson, C. E., J. K. Hourigan, G. B. Haxel, and M. Grove. 2017. "Extreme Latest Cretaceous-Paleogene Low-Angle Subduction: Zircon Ages from Orocopia Schist at Cemetery Ridge, Southwestern Arizona, USA." *Geology* 45:951–54.

Jahns, R. H. 1952. *Pegmatite Deposits of the White Picacho District, Maricopa and Yavapai Counties, Arizona.* Arizona Bureau of Mines Bulletin 162. Tucson: University of Arizona.

Jahns, R. H. 1953. "The Genesis of Pegmatites." *American Mineralogist* 38:563–98.

Jambor, J. L., V. A. Kovalenker, and A. C. Roberts. 2000. "New Mineral Names." *American Mineralogist* 85 (5–6): 873–77.

Jenkins, O. P., and E. D. Wilson. 1920. *A Geological Reconnaissance of the Tucson and Amole Mountains.* Arizona Bureau of Mines Bulletin 106. Tucson: Arizona Bureau of Mines.

Jensen, M. L. 1958. "Sulfur Isotopes and the Origin of Sandstone-Type Uranium Deposits." *Economic Geology* 53:598–616.

Johnson, D. H. 1963. "Mineralogy and Paragenesis of the Ore Deposit at the Monument No. 2 and Cato Sells Mines." In *Geology and Uranium-Vanadium Deposits of the Monument Valley Area, Apache and Navajo Counties, Arizona,* by I. J. Witkind and R. E. Thaden, 113–52. U.S. Geological Survey Bulletin 1103. Washington, D.C.: Government Printing Office.

Johnston, W. P., and J. D. Lowell. 1961. "Geology and Origin of Mineralized Breccia Pipes in Copper Basin, Arizona." *Economic Geology* 56:916–40.

Jones, D. 1983. "Famous Mineral Localities: The Old Yuma Mine." *Mineralogical Record* 14:95–107.

Jones, E. L., Jr. 1915. "Gold Deposits near Quartzsite, Arizona." In *Contributions to Economic Geology, 1915: Part I, Metals and Nonmetals Except Fuels.* U.S. Geological Survey Bulletin 620, 45–57. Washington, D.C.: Government Printing Office.

Jones, R. W. 1980. "The Grand Reef Mine, Graham County." *Mineralogical Record* 11:219–25.

Jones, R. W., and W. E. Wilson. 1983. "The Ray Mine." *Mineralogical Record* 14:311–22.

Joralemon, I. B. 1914. "The Ajo Copper District, Arizona." *Engineering and Mining Journal* 98:663.

Joralemon, I. B. 1952. "Age Cannot Wither Her Varieties of Geological Experience." *Economic Geology* 47:243–59.

Journeay, J. A. 1959. "Pyrometasomatic Deposits at Pima Mine." in *Southern Arizona Guidebook II, Combined with the 2nd Annual Arizona Geological Society Digest,* edited by L. A. Heindl, 198–99. Tucson: Arizona Geological Society.

Kampf, A. R., A. J. Celestian, B. P. Nash, and J. Marty. 2018. "Phoxite, IMA 2018-009." CNMNC Newsletter No. 43, June 2018. *Mineralogical Magazine* 82:779–85.

Kampf, A. R., A. J. Celestian, B. P. Nash, and J. Marty. 2020a. "Allantoin, IMA 2020-004a." CNMNC Newsletter 56, *European Journal of Mineralogy* 32 (4): 443–48.

Kampf, A.R., A. J. Celestian, B. P. Nash, and J. Marty 2020b. "Natrosulfatourea, IMA 2019-134." CNMNC Newsletter No. 55, *European Journal of Mineralogy* 32 (3): 367–71.

Kampf, A. R., M. A. Cooper, A. J. Celestian, C. Ma, J. Marty. 2021a. "Dendoraite-(NH4)." IMA 2020-103, CMNMC Newsletter 61, *Mineralogical Magazine* 85.

Kampf, A. R., M. A. Cooper, A. J. Celestian, C. Ma, J. Marty. 2021b. "Relianceite-(K)." IMA 2020-102, CMNMC Newsletter 61, *Mineralogical Magazine* 85.

Kampf, A. R., M. A. Cooper, B. P. Nash, T. E. Cerling, J. Marty, D. R. Hummer, A. J. Celestian, T. P. Rose, and T. J. Trebisky. 2016. "Rowleyite, IMA 2016-037." *Mineralogical Magazine* 50:1135–44.

Kampf, A. R., M. A. Cooper, B. P. Nash, T. E. Cerling, J. Marty, D. R. Hummer, A. J. Celestian, T. P. Rose, and T. J. Trebisky. 2017. "Rowleyite, [Na(NH$_4$,K)$_9$Cl$_4$][V$^{5+,4+}_2$(P,As)

$O_8]_6 \cdot n[H_2O,Na,NH_4,K,Cl]$, a New Mineral with a Microporous Framework Structure." *American Mineralogist* 102:1037–44.

Kampf, A. R., P. J. Dunn, and E. E. Foord. 1989. "Grandreefite, Pseudograndreefite, Laurelite, and Aravaipaite: Four New Minerals from the Grand Reef Mine, Graham County, Arizona." *American Mineralogist* 74:927–33.

Kampf, A. R., and E. E. Foord. 1995. "Artroeite, $PbAlF_3(OH)_2$, a New Mineral from the Grand Reef Mine, Graham County, Arizona: Description and Crystal Structure." *American Mineralogist* 80:179–83.

Kampf, A. R., and E. E. Foord. 1996. "Calcioaravaipaite, a New Mineral and Associated Lead Fluoride Minerals from the Grand Reef Mine, Graham County, Arizona." *Mineralogical Record* 27:293–300.

Kampf, A. R., R. M. Housley, G. R. Rossman, H. Yang, and R. T. Downs. 2020. "Adanite, IMA 2019-088." CNMNC Newsletter 53. *European Journal of Mineralogy* 32:209–13.

Kampf, A. R., S. J. Mills, A. J. Celestian, C. Ma, H. Yang, and B. Thorne. 2021. "Flaggite." IMA 2021-044, CNMNC Newsletter 63. *European Journal of Mineralogy* 33: in press.

Kampf A. R., S. J. Mills, S. Merlino, M. Pasero, A. M. McDonald, W. B. Wray, and J. R. Hindman. 2012. "Whelanite, $Cu_2Ca_6[Si_6O_{17}(OH)](CO_3)(OH)_3(H_2O)_2$, an (Old) New Mineral from the Bawana Mine, Milford, Utah." *American Mineralogist* 97:2007–15.

Kampf, A.R., S. J. Mills, and M. S. Rumsey. 2017. "The Discreditation of Girdite." *Mineralogical Magazine* 81:1125–28.

Kampf, A. R., J. J. Pluth, Y.-S. Chen, A. C. Roberts, and R. M. Housley. 2013. "Bobmeyerite, a New Mineral from Tiger, Arizona, USA, Structurally Related to Cerchiaraite and Ashburtonite." *Mineralogical Magazine* 77:81–91.

Kampf, A. R., I. M. Steele, and R. A. Jenkins. 2006. "Phosphohedyphane, $Ca_2Pb_3(PO_4)_3Cl$, the Phosphate Analog of Hedyphane: Description and Crystal Structure." *American Mineralogist* 91:1909–17.

Kampf, A. R., I. M. Steele, and T. A. Loomis. 2008. "Jahnsite-(NaFeMg), a New Mineral from the Tip Top Mine, Custer County, South Dakota: Description and Crystal Structure." *American Mineralogist* 93:940–45.

Karlstrom, K. E., and J. M. Timmons. 2012. "Many Unconformities Make One 'Great Unconformity.'" In *Grand Canyon Geology: Two Billion Years of Earth's History*, edited by J. M. Timmons and K. E. Karlstrom, 73–79. Geological Society of America Special Paper 489. Washington, D.C.: Geological Society of America.

Kartchner, W. E. 1944. "The Geology and Ore Deposits of the Harshaw District, Patagonia Mountains, Arizona." PhD diss., University of Arizona.

Kauffman, A. J., Jr. 1943. "Fibrous Sepiolite from Yavapai County, Arizona." *American Mineralogist* 28:512–20.

Kauffman, A. J., Jr., and H. W. Jaffe. 1946. "Chevkinite (Tscheffkinite) from Arizona." *American Mineralogist* 31:582–88.

Keith, Stanley B. 1969. "Limestone, Dolomite, and Marble." In *Mineral and Water Resources of Arizona*. Arizona Bureau of Mines Bulletin 180, 385–98. Tucson: University of Arizona.

Keith, Stanley B. 1972. "Mineralogy and Paragenesis of the 79 Mine Lead-Zinc-Copper Deposit." *Mineralogical Record* 3:247–64.

Keith, Stanley B. 1986. "Petrochemical Variations in Laramide Magmatism and Their Relationship to Laramide Tectonic and Metallogenic Evolution in Arizona and

Adjacent Regions." In *Frontiers in Geology and Ore Deposits of Arizona and the Southwest*, edited by B. Beatty, and P. A. K. Wilkinson, 89–101. Arizona Geological Society Digest 16. Tucson: Arizona Geological Society.

Keith, Stanley B., D. E. Gest, E. DeWitt, N. W. Toll, and B. A. Everson. 1983. *Metallic Mineral Districts and Production in Arizona*. Arizona Bureau of Geology and Mineral Technology, Geological Survey Branch, Bulletin 194. Tucson: University of Arizona.

Keith, Stanley B., D. P. Laux, J. Maughan, K. Schwab, S. Ruff, M. M. Abbott E. Swan, and S. Friberg. 1991. "Magma Series and Metallogeny: A Case Study from Nevada and Environs." In *Great Basin Symposium Field Trip 8: Magma Chemistry*, 404–93. Reno: Nevada Geological Society.

Keith, Stanley B., and J. C. Rasmussen. 2019. *Wilderness Granitic Suite along Catalina Highway, Petrologic and Tectonic Significance of the Wilderness Plutonic Suite: Products from a Massive Water-Flood Event of Hydrous Crustal Melting during High-Speed Flat Subduction in the Latest Laramide*. Arizona Geological Society Spring Field Trip 2019. Tucson: Arizona Geological Society.

Keith, Stanley B., S. J. Reynolds, P. E. Damon, M. Shafiqullah, D. E. Livingston, and P. D. Pushkar. 1980. "Evidence for Multiple Intrusion and Deformation within the Santa Catalina-Rincon-Tortolita Crystalline Complex, Southeastern Arizona." In *Cordilleran Metamorphic Core Complexes*, edited by M. D. Crittenden Jr., P. J. Coney, and G. H. Davis, 217–67. Geological Society of America Memoir 153. Boulder, Colo.: Geological Society of America.

Keith, Stanley B., V. Spieth, and J. C. Rasmussen. 2018. "Zechstein-Kupferschiefer Mineralization Reconsidered as a Product of Ultra-Deep Hydrothermal, Mud-Brine Volcanism." In *Contributions to Mineralization*, edited by Ali Al-Juboury, 23–66. http://www.DOI.org/10.5772/intechopen.72560.

Keith, Stanley B., and M. M. Swan. 1996. "The Great Laramide Porphyry Copper Cluster of Arizona, Sonora, and New Mexico: The Tectonic Setting, Petrology, and Genesis of a World Class Porphyry Metal Cluster." In *Geology and Ore Deposits of the American Cordillera*, edited by A. R. Coyner, and P. L. Fahey, 1667–1747. Reno: Geological Society of Nevada.

Keith, Stanley B., and M. M. Swan. 2005. "Hydrothermal Hydrocarbons." Goldschmidt Conference Abstracts, AAPG Datapages. http://www.searchanddiscovery.com/documents/abstracts/2005research_calgary/abstracts/extended/keith/keith.htm.

Keith, Stanley B., and J. C. Wilt. 1985. "Late Cretaceous and Cenozoic Orogenesis of Arizona and Adjacent Regions: A Strato-Tectonic Approach." In *Cenozoic Paleogeography of West-Central United States*, edited by R. M. Flores, and S. S. Kaplan, 403–37. Rocky Mountain Paleogeography Symposium 3. Denver, Colo.: Rocky Mountain Section, Society of Economic Paleontologists and Mineralogists.

Keith, Stanley B., and J. C. Wilt. 1986. "Laramide Orogeny in Arizona and Adjacent Regions: A Strato-Tectonic Synthesis." In *Frontiers in Geology and Ore Deposits of Arizona and the Southwest*, edited by B. Beatty, and P. A. K. Wilkinson, 502–54. Arizona Geological Society Digest 16. Tucson: Arizona Geological Society.

Keith, Stanton B. 1969. "Limestone, Dolomite, and Marble." In *Mineral and Water Resources of Arizona*. Arizona Bureau of Mines Bulletin 180, 385–98. Tucson: University of Arizona.

Keith, Stanton B. 1973. *Index of Mining Properties in Cochise County, Arizona.* Arizona Bureau of Mines Bulletin 187. Tucson: University of Arizona.

Keith, Stanton B. 1974. *Index of Mining Properties in Pima County, Arizona.* Arizona Bureau of Mines Bulletin 189. Tucson: University of Arizona.

Keith, Stanton B. 1975. *Index of Mining Properties in Santa Cruz County, Arizona.* Arizona Bureau of Mines Bulletin 191. Tucson: University of Arizona.

Keith, Stanton B. 1978. *Index of Mining Properties in Yuma County, Arizona.* Arizona Bureau of Mines Bulletin 192. Tucson: University of Arizona.

Keller, W. D. 1962. *Clay Minerals in the Morrison Formation of the Colorado Plateau.* U.S. Geological Survey Bulletin 1150. Washington, D.C.: Government Printing Office.

Kellogg, L. O. 1906. "Sketch of the Geology and Ore Deposits of the Cochise Mining District, Cochise County, Arizona." *Economic Geology* 1:651–59.

Kemp, J. F. 1905. "Secondary Enrichment in Ore Deposits of Copper." *Economic Geology* 1:11–25.

Kempton, P. D. 1983. "Petrography and Geochemistry of Amphibolite Peridotites from Geronimo Volcanic Field, SE Arizona." Abstract. *Geological Society of America Abstracts with Programs* 14:528.

Kempton, P. D., M. d. Blanchard, R. Dungan, R. Harmon, and J. Hoefs. 1984. "Petrogenesis of Alkalic Basalts from the Geronimo Volcanic Field: Geochemical Constraints on the Roles of Fractional Crystallization and Crustal Contamination." In *Proceedings of the ISeM Field Conference on Open Magmatic Systems*, 91–93. Dallas, Tex.: Institute for the Study of Earth and Man.

Kerr, P. F. 1951. "Alteration Features at Silverbell, Arizona." *Geological Society of America Bulletin* 62:451–80.

Khin, B. 1970. "Cornetite from Saginaw Hill, Arizona." *Mineralogical Record* 1:117–18.

Kiersch, G. A. 1947. "The Geology and Ore Deposits of the Seventy-Nine Mine Area, Gila County, Arizona." PhD diss., University of Arizona.

Kiersch, G. A. 1949. "Structural Control and Mineralization at the Seventy-Nine Mine, Gila County, Arizona." *Economic Geology* 44:24–39.

Kiersch, G. A. 1955. "Kaolin." In *Mineral Resources of the Navajo-Hopi Indian Reservations, Arizona-Utah.* Vol. 2, 72–76. Tucson: University of Arizona Press.

Kiersch, G. A., and W. D. Keller. 1955. "Bleaching Clay Deposits, Sanders–Defiance Plateau District, Navajo County, Arizona." *Economic Geology* 50:469–94.

Kinnison, J. E. 1966. "The Mission Copper Deposit, Arizona." In *Geology of the Porphyry Copper Deposits, Southwestern North America*, edited by S. R. Titley and C. L. Hicks, 281–87. Tucson: University of Arizona Press.

Kistner, D. J. 1984. "Fracture of a Volcanic Lithocap, Red Mountain Porphyry Copper Prospect, Santa Cruz County, Arizona." Master's thesis, University of Arizona.

Klute, M. A. 1991. "Sedimentology, Sandstone Petrofacies, and Tectonic Setting of the Late Mesozoic Bisbee Basin, Southeastern Arizona." PhD diss., University of Arizona.

Knoerr, A., and M. Eigo. 1963. "Arizona's Newest Copper Producer—The Christmas Mine." *Engineering and Mining Journal* 164:55–67.

Knudsen, H. P. 1983. "The Silver Bill Mine, Gleeson, Arizona." *Mineralogical Record* 14:127–31.

Koenig, B. A. 1978. "Oxidation-Leaching and Enrichment Zones of a Porphyry Copper Deposit—A Mineralogic and Quantitative Chemical Study." Master's thesis, University of Arizona.

Koenig, G. A. 1891. "On Paramelaconite and the Associated Minerals." *Proceedings of the Academy of Natural Sciences of Philadelphia* 43:284–91.

Kortemeier, C. P. 1984. "Geology of the Tip Top District, Yavapai County, Arizona." Master's thesis, Arizona State University.

Kortemeier, W. T. 1986. "Ongonite and Topazite in the Flying W Ranch Area, Tonto Basin, Arizona." Master's thesis, Arizona State University.

Kortemeier, W. T., and D. M. Burt. 1988. "Ongonite and Topazite Dikes in the Flying W Ranch Area, Tonto Basin, Arizona." *American Mineralogist* 73:507–23.

Korwin, A. 2013. "Escape at Four Peaks." *Lapidary Journal* 67 (7): 34–39.

Koutz, F. R. 1984. "The Hardshell Silver, Base-Metal, Manganese-Oxide Deposit." In *Gold and Silver Deposits of the Basin and Range Province, Western U.S.A.*, edited by Joe Wilkins Jr., 199–217. Arizona Geological Society Digest 15. Tucson: Arizona Geological Society.

Krieger, M. H. 1965. *Geology of the Prescott and Paulden Quadrangles, Arizona.* U.S. Geological Survey Professional Paper 467. Washington, D.C.: Government Printing Office.

Krieger, M. H. 1968. *Geologic Map of the Saddle Mountain Quadrangle, Pinal County, Arizona.* U.S. Geological Survey, Geologic Quadrangle Map GQ-671. Washington, D.C.: U.S. Geological Survey.

Krieger, M. H. 1979. *Ash-Flow Tuffs of the Galiuro Volcanics in the Northern Galiuro Mountains, Pinal County, Arizona.* U.S. Geological Survey Professional Paper 1104. Washington, D.C.: Government Printing Office.

Ksanda, C. J., and E. Henderson. 1939. "Identification of Diamond in the Canyon Diablo Iron." *American Mineralogist* 24:677–80.

Kuck, P. H. 1978. "The Behavior of Molybdenum, Tungsten, and Titanium in the Porphyry Copper Environment." PhD diss., University of Arizona.

Kuhn, T. H. 1938. "Childs-Aldwinkle Mine, Copper Creek, Arizona." In *Some Arizona Ore Deposits.* Arizona Bureau of Mines Bulletin 145, 127–30. Tucson: University of Arizona.

Kuhn, T. H. 1941. "Pipe Deposits of the Copper Creek Area, Arizona." *Economic Geology* 36:512–38.

Kuhn, T. H. 1951. "Bunker Hill District." In *Arizona Zinc and Lead Deposits.* Arizona Bureau of Mines Bulletin 158, 56–65. Tucson: University of Arizona.

Kumke, C. A. 1947. *Zonia Copper Mine, Yavapai County, Arizona.* U.S. Bureau of Mines Report of Investigations RI 4033. Washington, D.C.: U.S. Department of the Interior, Bureau of Mines.

Kunkely, H., and A. Vogler. 1995. "On the Origin of the Photoluminescence of Mercurous Chloride." *Chemistry and Physics Letters* 240:31–34.

Kunz, G. F. 1885. "On Remarkable Copper Minerals from Arizona." *Annals of the New York Academy of Sciences* 3:275–78.

Kunz, G. F. 1887. "Crystals of Hollow Quartz from Arizona." *American Journal of Science* 34:479.

Kunz, G. F. 1890. *Gems and Precious Stones of North America*. New York: Scientific Publishing.

Kunz, G. F. 1905. "Moissanite, a Natural Silicon Carbide." *American Journal of Science* 19:396.

Kunz, G. F., and O. W. Huntington. 1893. "On the Diamond in the Cañon Diablo Meteoric Iron and on the Hardness of Carborundum." *American Journal of Science* 46:470–73.

Ladoo, R. B. 1923. *Fluorspar Mining in the Western States*. U.S. Bureau of Mines Report of Investigations RI 2480. Washington, D.C.: U.S. Department of the Interior, Bureau of Mines.

Lang, J. 2001. "The Arizona Porphyry Province." In *Regional and System-Scale Controls on the Formation of Copper and/or Gold Magmatic-Hydrothermal Mineralization*. Special Publication 2, 53–76. Vancouver: Mineral Deposit Research Unit, University of British Columbia.

Langton, J. M., and S. A. Williams. 1982. "Structural, Petrological and Mineralogical Controls for the Dos Pobres Orebody: Lone Star Mining District, Graham County, Arizona." In *Advances in Geology of the Porphyry Copper Deposits, Southwestern North America*, edited by S. R. Titley, 335–52. Tucson: University of Arizona Press.

Larsen, E. S., and H. E. Vassar. 1925. "Chalcoalumite, a New Mineral from Bisbee, Arizona." *American Mineralogist* 10:79–83.

Larson, P. B. 1976. "The Metamorphosed Alteration Zone Associated with the Bruce Precambrian Volcanogenic Massive Sulfide Deposit, Yavapai County, Arizona." Master's thesis, University of Arizona.

Lasky, S. G., and B. N. Webber. 1944. *Manganese Deposits in the Artillery Mountains Region, Mohave County, Arizona*. U.S. Geological Survey Bulletin 936-R. Washington, D.C.: Government Printing Office.

Lasky, S. G., and B. N. Webber. 1949. *Manganese Resources of the Artillery Mountains Region, Mohave County, Arizona*. U.S. Geological Survey Bulletin 961. Washington, D.C.: Government Printing Office.

Laughlin, A. W., R. W. Charles, and M. J. Aldrich Jr. 1986. "Heteromorphism and Crystallization Paths of Katungites, Navajo Volcanic Field, Arizona, USA." In *Kimberlites and Related Rocks*, vol. 1, edited by J. Ross, A. L. Jaques, J. Ferguson, D. H. Green, S. Y. O'Reilly, R. V. Danchin, and A. J. A. Janse, 582–91. Geological Society Australia Special Publication 14. Carlton, Vic.: Blackwell.

Laughon, R. B. 1970. "New Data on Guildite." *American Mineralogist* 55:502–5.

Laughon, R. B. 1971. "The Crystal Structure of Kinoite." *American Mineralogist* 56:193–200.

Lausen, C. 1923. "Geology of the Old Dominion Mine, Globe, Arizona." Master's thesis, University of Arizona.

Lausen, C. 1926. "Tourmaline-Bearing Cinnabar Veins of the Mazatzal Mountains, Arizona." *Economic Geology* 21:782–91.

Lausen, C. 1927a. "The Occurrence of Olivine Bombs near Globe, Arizona." *American Journal of Science* 14:293–306.

Lausen, C. 1927b. "Piedmontite from the Sulphur Springs Valley, Arizona." *American Mineralogist* 12:283–87.

Lausen, C. 1928. "Hydrous Sulphates Formed under Fumerolic Conditions at the United Verde Mine." *American Mineralogist* 13:203–29.

Lausen, C. 1931a. *Geology and Ore Deposits of the Oatman and Katherine Districts, Arizona*. Arizona Bureau of Mines Bulletin 131. Tucson: University of Arizona.

Lausen, C. 1931b. "Gold Veins of the Oatman and Katherine Districts, Arizona." PhD diss., University of Arizona.

Lausen, C. 1936. "The Occurrence of Minute Quantities of Mercury in the Chinle Shales at Lee's Ferry, Arizona." *Economic Geology* 31:610–17.

Lausen, C., and E. D. Gardner. 1927. *Quicksilver (Mercury) Resources of Arizona*. Arizona Bureau of Mines Bulletin 122. Tucson: University of Arizona.

Lausen, C., and E. D. Wilson. 1925. *Gold and Copper Deposits near Payson, Arizona*. Arizona Bureau of Mines Bulletin 120. Tucson: University of Arizona.

Lawson, A. C. 1913. "The Gold in the Shinarump [Chinle] at Paria." *Economic Geology* 8:434–48.

Leavens, P. B. 1967. "Reexamination of Bermanite." *American Mineralogist* 52:1060–66.

LeConte, J. L. 1852. "Notice of Meteoric Iron in the Mexican Province of Sonora." *American Journal of Science* 13:289–90.

Lee, C. A., and G. C. Borland. 1935. "The Geology and Ore Deposits of the Cuprite Mining District." Master's thesis, University of Arizona.

Lee, M. 1981. *Analytical Chemistry of Polycyclic Aromatic Compounds*. Oxford: Elsevier Science.

Legge, J. A., Jr. 1939. "Paragenesis of the Ore Minerals of the Miami Mine, Arizona." Master's thesis, University of Arizona.

Lehman, N. E. 1978. "The Geology and Pyrometasomatic Ore Deposits of the Washington Camp-Duquesne District, Santa Cruz County, Arizona." PhD diss., University of Arizona.

Leicht, W. C. 1971. "Minerals of the Grandview Mine." *Mineralogical Record* 2:214–21.

Lewis, A. S. 1920. "Ore Deposits of Cave Creek District, Arizona." *Engineering and Mining Journal* 110:713.

Lewis, D. V. 1955. "Relationships of Ore Bodies to Dikes and Sills." *Economic Geology* 50:495–516.

Lindgren, W. 1903. "The Copper Deposits of Clifton, Arizona." *Engineering and Mining Journal* 75:705.

Lindgren, W. 1904. "The Genesis of Copper Deposits." *Engineering and Mining Journal* 78:987.

Lindgren, W. 1905. *The Copper Deposits of the Clifton-Morenci District, Arizona*. U.S. Geological Survey Professional Paper 43. Washington, D.C.: Government Printing Office.

Lindgren, W. 1926. *Ore Deposits of the Jerome and Bradshaw Mountains Quadrangle, Arizona*. U.S. Geological Survey Bulletin 782. Washington, D.C.: Government Printing Office.

Lindgren, W., and W. F. Hillebrand. 1904. "Minerals from the Clifton-Morenci District, Arizona." *American Journal of Science* 18:448–60.

Loghry, J. D. 1972. "Characteristics of Favorable Cappings from Several Southwestern Porphyry Copper Deposits." Master's thesis, University of Arizona.

London, D. 1979. "Occurrence and Alteration of Primary Lithium Minerals, White Picacho Pegmatites, Arizona." Master's thesis, Arizona State University.

London, D. 1981. "Lithium Mineral Stabilities in Pegmatites." PhD diss., Arizona State University.

London, D., and D. M. Burt. 1978. "Lithium Pegmatites of the White Picacho District, Maricopa and Yavapai Counties, Arizona." In *Guidebook to the Geology of Central Arizona*, edited by D. M. Burt and T. L. Péwé, 61–72. Arizona Bureau of Geology and Mineral Technology Special Paper 2. Tucson: Bureau of Geology and Mineral Technology.

London, D., and D. M. Burt. 1982a. "Alteration of Spodumene, Montebrasite, and Lithiophilite in Pegmatites of the White Picacho District, Arizona." *American Mineralogist* 67:97–113.

London, D., and D. M. Burt. 1982b. "Lithium Aluminosilicate Occurrences in Pegmatites and the Aluminosilicate Phase Diagram." *American Mineralogist* 67:483–509.

Long, W. J., J. V. Batty, and K. C. Dean. 1948. *Concentration of Miscellaneous Oxide Manganese Ores from Yavapai, Yuma, Maricopa, and Mohave Counties, Arizona*. U.S. Bureau of Mines Report of Investigations RI 4291. Washington, D.C.: U.S. Department of the Interior, Bureau of Mines.

Lovering, T. S. 1948. "Geothermal Gradients, Recent Climatic Changes, and Rate of Sulfide Oxidation in the San Manuel District, Arizona." *Economic Geology* 43:1–20.

Lovering, T. S., L. C. Huff, and H. Almond. 1950. "Dispersion of Copper from the San Manuel Copper Deposit, Pinal County, Arizona." *Economic Geology* 45:493–514.

Lowell, J., and T. Rybicki. 1976. "Mineralization of the Four Peaks Amethyst Deposit, Maricopa County, Arizona." *Mineralogical Record* 7:72–77.

Lowell, J. D. 1955. "Applications of Cross-Stratification Studies to Problems of Uranium Exploration, Chuska Mountains, Arizona." *Economic Geology* 50:177–85.

Lowell, J. D. 1956. "Occurrence of Uranium in Seth-La-Kai Diatreme, Hopi Buttes, Arizona." *American Journal of Science* 254:404–12.

Lowell, J. D. 1968. "Geology of the Kalamazoo Orebody, San Manuel District, Arizona." *Economic Geology* 63:645–54.

Lucas, S. G., and A. B. Heckert. 2005. "Distribution, Age and Correlation of Cretaceous Fossil Vertebrates from Arizona." In *Vertebrate Paleontology in Arizona*, edited by A. B. Heckert and S. G. Lucas, 104–9. New Mexico Museum of Natural History and Science Bulletin No. 29. Albuquerque: New Mexico Museum of Natural History and Science.

Ludwig, K. R. 1974. "Precambrian Geology of the Central Mazatzal Mountains, Arizona (Part I) and Lead Isotope Heterogeneity in Precambrian Igneous Feldspars (Part II)." PhD diss., California Institute of Technology.

Lurie, M. 2012. "Arizona Opal." *Lapidary Journal* 65 (8): 22–27.

Lynch, D. J. 1972. "Reconnaissance Geology of the Bernardino Volcanic Field, Cochise County, Arizona." Master's thesis, University of Arizona.

Lynch, D. J. 1976. "Basaltic Eruptions in the Bernardino Volcanic Field of Southeastern Arizona." *Geological Society of America Abstracts with Programs* 8:604–5.

Lynch, D. W. 1966. "The Economic Geology of the Esperanza Mine and Vicinity." In S. R. Titley and C. L. Hicks (eds.), *Geology of the Porphyry Copper Deposits, Southwestern North America*. Tucson: University of Arizona Press, 267–79.

Lynch, D. W. 1967. "The Geology of the Esperanza Mine and Vicinity, Pima County, Arizona." Master's thesis, University of Arizona.

Marsh, S. P., and D. M. Sheridan. 1976. *Rutile in Precambrian Sillimanite-Quartz Gneiss and Related Rocks, East-Central Front Range, Colorado*. U.S. Geological Society Professional Paper 959-G: G1–G17. Washington, D.C.: Government Printing Office.

Marshall, R. R., and O. Joensuu. 1961. "Crystal Habit and Trace Element Content of Some Galenas." *Economic Geology* 56:758–71.

Marvin, R. F., C. W. Naeser, and H. H. Mehnert. 1978. "Tabulation of Radiometric Ages—Including Unpublished K-Ar and Fission-Track Ages—For Rocks in Southeastern Arizona and Southwestern New Mexico." In *Land of Cochise—Southeastern Arizona*, edited by J. F. Callender, J. C. Wilt, R. E. Clemons, and H. L. James, 243–52. Albuquerque: New Mexico Geological Society.

Mason, B. 1968. "Kaersutite from San Carlos, Arizona, with Comments on the Paragenesis of This Mineral." *Mineralogical Magazine* 36:997–1002.

Masters, J. A. 1955. "Geology of the Uranium Deposits of the Lukachukai Mountains Area, Northeastern Arizona." *Economic Geology* 50:111–26.

Maucher, A., and G. Rehwald. 1961. *Bildkartei der Erzmikroskopie*. Frankfurt: Umschau Verlag.

Mayo, E. B. 1955a. "Copper." In *Mineral Resources of the Navajo-Hopi Indian Reservations, Arizona-Utah*. Vol. 1, *Metalliferous Minerals and Mineral Fuels*, 19–22. Tucson: University of Arizona Press.

Mayo, E. B. 1955b. "Manganese." In *Mineral Resources of the Navajo-Hopi Indian Reservations, Arizona-Utah*. Vol. 1, *Metalliferous Minerals and Mineral Fuels*, 39–47. Tucson: University of Arizona Press.

Mayuga, M. N. 1942. "The Geology and Ore Deposits of the Helmet Peak Area, Pima County, Arizona." PhD diss., University of Arizona.

Mazzi, F., and A. Pabst. 1962. "Reexamination of Cuprorivaite." *American Mineralogist* 47:409–11.

McCandless, T. E., and J. Ruiz. 1993. "Rhenium-Osmium Evidence for Regional Mineralization in Southwestern North America." *Science* 261 (3):1282–86.

McCarn, H. L. 1904. "The Planet Copper Mine." *Engineering and Mining Journal* 78:26.

McCauley, A., and D. C. Bradley. 2014. "The Global Age Distribution of Granitic Pegmatites." *Canadian Mineralogist* 52:183–90.

McCurry, W. G. 1971. "Mineralogy and Paragenesis of the Ores, Christmas Mine, Gila County, Arizona." Master's thesis, Arizona State University.

McGetchin, T. R., L. T. Silver, and A. A. Chodos. 1970. "Titanclinohumite: A Possible Mineralogical Site for Water in the Upper Mantle." *Journal of Geophysical Research* 75:255–59.

McKee, E. H., and C. A. Anderson. 1971. "Age and Chemistry of Tertiary Volcanic Rocks in Northcentral Arizona and Relation of the Rocks to the Colorado Plateaus." *Geological Society of America Bulletin* 82:2767–82.

McKee, E. D. 1982. *The Supai Group of the Grand Canyon*. U.S. Geological Survey Professional Paper 1173. Washington, D.C.: Government Printing Office.

McLean, W. J., and J. W. Anthony. 1970. "The Crystal Structure of Hemihedrite." *American Mineralogist* 55:1103–14.

McLean, W. J., and J. W. Anthony. 1972. "The Disordered Zeolite-like Structure of Connellite." *American Mineralogist* 57:426–38.

McLean, W. J. Bideaux R. A., and R. W. Thomssen. 1974. "Yedlinite, a New Mineral from the Mammoth Mine, Tiger, Arizona." *American Mineralogist* 59:1157–59.

McMahan, P. 2016. *Agates: The Pat McMahan Collection*. Cottonwood, Ariz.: McMahan Press.

McNew, G. E. 1981. "Tactite Alteration and Its Late-Stage Replacement in the Southern Half of the Rosemont Mining District." Master's thesis, University of Arizona.

Mead, C., W. J. Littler, and E. C. T. Chao. 1965. "Metallic Spheroids from Meteor Crater, Arizona." *American Mineralogist* 50:667–81.

Means, V. A., and R. K. Labrecque. 1958. "The Martin and Rattlesnake Incline Mines." In *Geology and Uranium Deposits of the Carrizo Mountains Area, Apache County, Arizona, and San Juan County, New Mexico*, edited by R. E. Hershey, 17–62. U.S. Atomic Energy Commission Report RME-117. Grand Junction, Colo.: U.S. Atomic Energy Commission, Grand Junction Operations Office.

Meeves, H. C. 1966. *Nonpegmatitic Beryllium Occurrences in Arizona, Colorado, New Mexico, Utah, and Four Adjacent States*. U.S. Bureau of Mines Report of Investigations RI 6828. Washington, D.C.: U.S. Department of the Interior, Bureau of Mines.

Meeves, H. C., C. M. Harrer, M. H. Salsbury, A. S. Konselman, and S. S. Shannon Jr. 1966. *Reconnaissance of Beryllium-Bearing Pegmatite Deposits in Six Western States*. U.S. Bureau of Mines Information Circular IC 8298. Washington, D.C.: U.S. Department of the Interior, Bureau of Mines.

Melchoirre, E. 1992. *Geology and Mineral Resources of the Sierra Estrella, Maricopa County, Arizona*. Arizona Geological Survey Open-File Report 92-15. Tucson: Arizona Geological Survey.

Melchiorre, E. 2008. *Octave Gold Mine: The Golden Queen of Rich Hill, Arizona*. San Bernardino, Calif.: Rock Doc.

Melchiorre, E. 2009. *Gold Atlas of Rich Hill, Arizona: A Comprehensive Guide to the Character and Distribution of Placer and Lode Gold from the Weaver II Mining District*. San Bernardino, Calif.: Rock Doc.

Melchiorre, E. 2010. "Geochemical Fingerprint of Placer Gold from the Middle Camp District, Quartzsite, AZ." In *2010 GSA Abstracts with Programs*, 105. Boulder, Colo.: Geological Society of America.

Melchiorre, E. 2011. *Gold Atlas of Quartzsite, Arizona*. Vol. 1, *Northern Dome Rock Mountains*. San Bernardino: Rock Doc Publications.284.

Melchiorre, E. 2013. *Gold Atlas of Quartzsite, Arizona*. Vol. 2, *Southern Dome Rock Mountains*. San Bernardino, Calif.: Rock Doc.

Melchiorre, E. 2017. *Gold Atlas of Quartzsite, Arizona*. Vol. 3, *Plomosa Mountains*. San Bernardino, Calif.: Rock Doc.

Melhase, J. 1925. "Asbestos Deposits of Arizona." *Engineering and Mining Journal* 120:805–10.

Merwin, H. E., and E. Posnjak. 1937. "Sulphate Encrustations in the Copper Queen Mine, Bisbee, Arizona." *American Mineralogist* 22:567–71.

Metz, R. A., and A. W. Rose. 1966. "Geology of the Ray Copper Deposit, Ray, Arizona." In *Geology of the Porphyry Copper Deposits, Southwestern North America*, edited by S. R. Titley and C. L. Hicks, 177–88. Tucson: University of Arizona Press.

Metzger, O. H. 1938. *Gold Mining and Milling in the Wickenburg Area, Maricopa and Yavapai Counties, Arizona.* U.S. Bureau of Mines Information Circular IC 6991. Washington, D.C.: U.S. Department of the Interior, Bureau of Mines.

Meyer, B. 2011. "The Micro Mineral Collector." *Pacific Northwest Friends of Mineralogy Newsletter* 41 (2): 15–26.

Michel, F. A., Jr. 1959. "Geology of the King Mine, Helvetia, Arizona." Master's thesis, University of Arizona.

Mielke, J. E. 1964. "Trace Elements Investigation of the Turkey Track Porphyry, Southeastern Arizona." Master's thesis, University of Arizona.

Miller, D. K., and W. E. Wilson. 1983. "Famous Mineral Localities: The Hilltop Mine." *Mineralogical Record* 14:121–26.

Miller, R. A. 1958. *Reconnaissance Geology of the Blue Rock Area, Pima-Cochise Counties, Arizona.* U.S. Atomic Energy Commission. RME-2077. Phoenix, Ariz.: U.S. Atomic Energy Commission, Division of Raw Materials.

Miller, R. J., F. Gray, J. R. Hassemer, W. F. Hanna, J. Brice III, and R. Schreiner. 1987. *Mineral Resources of the Lower Burro Creek Wilderness Study Area, Mohave and Yavapai Counties, Arizona.* U.S. Geological Survey Bulletin 1701-B: B1–B13.

Mills, J. W., and H. T. Eyrich. 1966. "The Rate of Unconformities in the Localization of Epigenetic Mineral Deposits in the United States and Canada." *Economic Geology* 61:1232–57.

Milton, C. 1944. "Stones from Trees." *Scientific Monthly* 59:421–23.

Milton, C., and J. Axelrod. 1947. "Fused Wood-Ash Stones: Fairchildite (n. sp.) $K_2CO_3 \cdot CaCO_3$, Buetschliite (n. sp.) $3K_2CO_3 \cdot 2CaCO_3 \cdot 6H_2O$, and Calcite, $CaCO_3$: Their Essential Components." *American Mineralogist* 32:607–24.

Missen, O., M. Rumsey, S. J. Mills, M. Weil, J. Najorka, J. Spratt, and U. Kolitsch. 2021. "Elucidating the Natural-Synthetic Mismatch of $Pb^{2+}Fe^{4+}O_3$: The Redefinition of Fairbankite $Pb^{2+}_{12}(Te_4O_3)_{11}(SO_4)$." *American Mineralogist* 106 (2): 309–16.

Mitcham, T. W., and C. G. Evensen. 1955. "Uranium Ore Guides, Monument Valley District, Arizona." *Economic Geology* 50:170–76.

Mitchell, G. J. 1920. "Vertical Extent of Copper Ore Minerals in the Junction Mine, Warren District, Arizona." *Engineering and Mining Journal* 109:1411–12.

Mitchell, G. J. 1921. "Rate of Formation of Copper Sulfate Stalactites." *Mining and Metallurgy* 170:30.

Mittlefehldt, D. W., T. J. McCoy, C. A. Goodrich, and A. Kracher. 1998. "Non-chondritic Meteorites from Asteroidal Bodies." In *Planetary Materials.* Vol. 36 of *Planetary Materials,* edited by J. J. Papike, 523–718. Washington, D.C.: Mineralogical Society of America.

Miyawaki, R., F. Hatert, M. Pasero, and S. J. Mills. 2019a. "IMA Commission on New Minerals, Nomenclature, and Classification (CNMNC) Newsletter 51." *Mineralogical Magazine* 83:757–61.

Miyawaki, R., F. Hatert, M. Pasero, and S. J. Mills. 2019b. "New Minerals and Nomenclature Modifications Approved in 2019." *Mineralogical Magazine* 83:615–20.

Mizer, J. D. 2018. "Early Laramide Magmatism in Southern Arizona: U-Pb Geochronology of Key Igneous Units and Implications for the Timing of Regional Porphyry Copper Mineralization." PhD diss., University of Arizona.

Modreski, P. J., and J. A. Scovil. 1990. "Stolzite, Wulfenite's Hidden Brother: Occurrence at the Fat Jack Gold Mine, Yavapai County, Arizona." Abstract. *Mineralogical Record* 21:99–100.

Mohon, J. P., Jr. 1975. "Comparative Geothermometry for the Monte Cristo Pegmatite, Yavapai County, Arizona." Master's thesis, University of Arizona.

Montoya, J. 1967. "A New Occurrence of Plancheite in Arizona." *Rocks and Minerals* 42:277.

Moolick, R. T., and J. J. Durek. 1966. "The Morenci District." In *Geology of the Porphyry Copper Deposits, Southwestern North America*, edited by S. R. Titley and C. L. Hicks, 221–31. Tucson: University of Arizona Press.

Moore, B. N. 1935. "Some Strontium Deposits of Southeastern California and Western Arizona." *American Institute of Mining, Metallurgical, and Petroleum Engineers (AIME) Transactions* 115:362–65.

Moore, B. N. 1936. "Celestite and Strontianite." In *Mineral Resources of the Region around Boulder Dam*, 151–54. U.S. Geological Survey Bulletin 871. Washington, D.C.: Government Printing Office.

Moore, P. B., and J. Ito. 1978. "Whiteite, a New Species and a Proposed Nomenclature for the Jahnsite-Whiteite Complex Series." *Mineralogical Magazine* 42:309–23.

Moore, R. T. 1968. *Mineral Deposits of the Fort Apache Indian Reservation, Arizona.* Arizona Bureau of Mines Bulletin 177. Tucson: University of Arizona.

Moore, R. T. 1969a. "Gold." In *Mineral and Water Resources of Arizona*. Arizona Bureau of Mines Bulletin 180, 156–67. Tucson: University of Arizona.

Moore, R. T. 1969b. "Vermiculite." In *Mineral and Water Resources of Arizona*. Arizona Bureau of Mines Bulletin 180, 462–64. Tucson: University of Arizona.

More, S. W. 1980. "The Geology and Mineralization of the Antler Mine and Vicinity, Mohave County, Arizona." Master's thesis, University of Arizona.

Morgan, W. C., and M. C. Tallmon. 1904. "A Peculiar Occurrence of Bitumen and Evidence as to Its Origin." *American Journal of Science* 18:363–77.

Morimoto, N., and A. Gyobu. 1971. "The Composition and Stability of Digenite." *American Mineralogist* 56:1889–1909.

Morimoto, N. 1988. "Nomenclature of Pyroxenes." *Mineralogy and Petrology* 39:55–76.

Morrison, S. M., K. J. Domanik, H. Yang, and R. T. Downs. 2016. "Petersite-(Ce), $Cu^{2+}_6Ce(PO_4)_3(OH)_6 \cdot 3H_2O$, A New Mixite Group Mineral from Yavapai County, Arizona, USA." *Canadian Mineralogist* 54:1505–11.

Moses, A. J., and L. M. Luquer. 1892. "Alabandite from Tombstone." *University of Columbia School Mines Quarterly* 13:236–39.

Moss, H. A. 1957. "The Nature of Carphosiderite and Allied Basic Sulphates of Iron." *Mineralogical Magazine* 31:407–12.

Mouat, M. M. 1962. "Manganese Oxides from the Artillery Mountains Area, Arizona." *American Mineralogist* 47:744–57.

Moxham, R. M., R. S. Foote, and C. M. Bunker. 1965. "Gamma-Ray Spectrometer Studies of Hydrothermally Altered Rocks." *Economic Geology* 60:653–71.

Moyer, T. C. 1982. "The Volcanic Geology of the Kaiser Spring Area, SE Mohave County, Arizona." Master's thesis, Arizona State University.

Mrose, M. E., H. J. Rose Jr., and J. W. Marinenko. 1966. "Synthesis and Properties of Fairchildite and Buetschliite: Their Relationship in Wood-Ash Stone Formation." Abstract. Geological Society of America Special Paper 101. Washington, D.C.: Geological Society of America.

Mrose, M. E., and A. C. Vlisidis. 1966. "Proof of the Formula of Shattuckite, Cu_5 $(SiO_3)_4(OH)_2$." Abstract. *American Mineralogist* 51:266–67.

Mueller, W. 2012. "Arizona Gemstones." *Rocks and Minerals* 87 (1): 64–69.

Mullens, R. L. 1967. "Stratigraphy and Environment of the Toroweap Formation (Permian) North of Ashfork, Arizona." Master's thesis, University of Arizona.

Muntyan, B. L. 2010. "The Hunt for Sierrita Mountain Aquamarine." In *Minerals of Arizona Eighteenth Annual Symposium*, 9–11. Phoenix: Flagg Mineral Foundation.

Muntyan, B. L. 2013. "The Rediscovery of Long-Lost Aquamarine from the Sierrita Mountains, Arizona." *Rocks and Minerals* 88:222–30.

Muntyan, B. L. 2015a. "The Bagdad Mine, Yavapai County, Arizona." *Mineralogical Record* 46:585–601.

Muntyan, B. L. 2015b. "Crystallized Gypsum Deposits of the San Pedro River Basin, Arizona." Talk given at Minerals of Arizona, Twenty-Third Annual Symposium, Phoenix, March 27–29.

Muntyan, B. L. 2017. "The Piedmont Mine: History, Minerals and Myths." *25th Annual Minerals of Arizona Symposium*, 7–12. Phoenix: Flagg Mineral Foundation.

Muntyan, B. L. 2019. "Arizona Pseudomorphs and Epimorphs." *Rocks and Minerals* 94:158–71.

Muntyan, B. L. 2020a, "Arizona Prehnite: A New Find from Gila County." *Rocks and Minerals* 95:432–39.

Muntyan, B. L. 2020b. "The Piedmont Mine Yavapai County, Arizona." *Mineralogical Record* 51:29–42.

Natkaniec-Nowak, L., M. Dumańska-Słowik, A. Gaweł, A. Łatkiewicz, J. Kowalczyk-Szpyt, A. Wolska, S. Milovská, J. Luptáková, and K. Ładoń. 2020. "Fire Agate from the Deer Creek Deposit (Arizona, USA)—New Insights into Structure and Mineralogy." *Mineralogical Magazine* 84:343–54.

Needham, A. B., and W. R. Storms. 1956. *Investigation of Tombstone District Manganese Deposits, Cochise County, Arizona*. U.S. Bureau of Mines Report of Investigations RI 5188. Washington, D.C.: U.S. Department of the Interior, Bureau of Mines.

Newberg, D. W. 1964. "X-ray Study of Shattuckite." *American Mineralogist* 49:1234–39.

Newberg, D. W. 1967. "Geochemical Implications of Chrysocolla-Bearing Alluvial Gravels." *Economic Geology* 62:932–56.

Newhouse, W. H. 1933. "Mercury in Native Silver." *American Mineralogist* 18:295–99.

Newhouse, W. H. 1934. "The Source of Vanadium, Molybdenum, Tungsten, and Chromium in Oxidized Lead Deposits." *American Mineralogist* 19 209–20.

North, O. S., and N. C. Jensen. 1958. "Vermiculite." In *Minerals Yearbook, 1954*, 1307–16. U.S. Bureau of Mines. Washington, D.C.: Government Printing Office.

Norton, J. J. 1965. "Lithium-Bearing Bentonite Deposit, Yavapai County, Arizona." U.S. Geological Survey Professional Paper 525-D: 163–66. Washington, D.C.: U.S. Geological Survey.

Novak, G., and W. W. Besse. 1986. "Vanadinite from the J. C. Holmes Claim, Santa Cruz County, Arizona." *Mineralogical Record* 17:111–15.

Nuffield, E. W., and L. H. Milne. 1953. "Studies of Radioactive Compounds: VI. Meta-uranocercite." *American Mineralogist* 38:476–88.

Nye, T. 1961. "Geology of the Paymaster Mine, Pima County, Arizona." *Arizona Geological Society Digest* 4:161–68.

Nye, T. 1968. "The Relationship of Structure and Alteration to Some Ore Bodies in the Bisbee (Warren) District, Cochise County, Arizona." PhD diss., University of Arizona.

O'Keefe, M., and J. O. Bovin. 1978. "The Crystal Structure of Paramelaconite, Cu_4O_a." *American Mineralogist* 63:180–85.

Olmstead, H. W., and D. W. Johnson. 1966. "Inspiration Geology." In *Geology of the Porphyry Copper Deposits, Southwestern North America*, edited by S. R. Titley and C. L. Hicks, 143–50. Tucson: University of Arizona Press.

Olsen, E., and L. Fuchs. 1968. "Krinovite, $NaMg_2CrSi_3O_{10}$, a New Meteorite Mineral." *Science* 161:786–87.

Olson, H. J. 1966. "Oxidation of a Sulfide Body, Glove Mine, Santa Cruz County, Arizona." *Economic Geology* 61:731–43.

Olson, J. C., and J. W. Adams. 1962. *Thorium and Rare Earths in the United States Exclusive of Alaska and Hawaii.* U.S. Geological Survey Mineral Investigations Resources Map MR-28. Washington, D.C.: U.S. Geological Survey.

Olson, J. C., and E. N. Hinrichs. 1960. Beryl-Bearing Pegmatites in the Ruby Mountains and Other Areas in Nevada and Northwestern Arizona. U.S. Geological Survey Bulletin 1082-D: 135–200. Washington, D.C.: Government Printing Office.

Omori, K., and P. F. Kerr. 1963. "Infrared Studies of Saline Sulfate Minerals." *Geological Society of America Bulletin* 74:709–34.

Onac, B. P., J. W. Hess, and W. B. White. 2007. "The Relationship between the Mineral Composition of Speleothems and Mineralization of Breccia Pipes: Evidence from Corkscrew Cave, Arizona, USA." *Canadian Mineralogist* 45:1177–88.

Otton, J. K. 1981. *Geology and Genesis of the Anderson Mine, a Carbonaceous Lacustrine Uranium Deposit, Western Arizona: A Summary Report.* U.S. Geological Survey, Open-File Report 81-780. Denver, Colo.: U.S. Geological Survey.

Outerbridge, W. F., M. H. Staatz, R. Meyrowitz, and A. M. Pommer. 1960. "Weeksite, a New Uranium Silicate from the Thomas Range, Juab County, Utah." *American Mineralogist* 45:39–52.

Overholt, J. L., G. Vaux, and J. L. Rodda. 1950. "The Nature of 'Arizonite.'" *American Mineralogist* 35:117–19.

Overstreet, W. C. 1967. *The Geologic Occurrence of Monazite.* U.S. Geological Survey Professional Paper 530. Washington, D.C.: Government Printing Office.

Pabst, A. 1938. "Crystal Structure and Density of Delafossite." Abstract. *American Mineralogist* 23:175–76.

Pabst, A. 1961. "X-ray Crystallography of Davidite." *American Mineralogist* 46:700–718.

Pabst, A., and R. W. Thomssen. 1959. "Davidite from the Quijotoa Mountains, Pima County, Arizona." Abstract. *Geological Society of America Bulletin* 70:1739.

Pabst, A., and C. D. Woodhouse. 1964. "Thalenite from Kingman, Arizona." Abstract. *Geological Society of America Special Paper* 82:269.

Page, L. R., H. E. Stocking, and H. B. Smith, comps. 1956. *Contributions to the Geology of Uranium and Thorium by the United States Geological Survey and Atomic Energy Commission for the United Nations International Conference on Peaceful Uses of Atomic Energy, Geneva, Switzerland, 1955.* U.S. Geological Survey Professional Paper 300. Washington, D.C.: Government Printing Office.

Palache, C. 1934. "Contributions to Crystallography: Claudetite, Minasragrite, Samsonite, Native Selenium, Iridium." *American Mineralogist* 19:194–205.

Palache, C. 1939a. "Antlerite." *American Mineralogist* 24:293–302.

Palache, C. 1939b. "Brochantite." *American Mineralogist* 24:463–81.

Palache, C. 1941a. "Crystallographic Notes: Cahnite, Stolzite, Zincite, Ultrabasite." *American Mineralogist* 26:429–36.

Palache, C. 1941b. "Diaboleite from Mammoth Mine, Tiger, Arizona." *American Mineralogist* 26:605–12.

Palache, C. 1950. "Paralaurionite." *Mineralogical Magazine* 29:341–45.

Palache, C., H. Berman, and C. Frondel. 1944. *The System of Mineralogy.* Vol. 1, 7th ed. New York: Wiley.

Palache, C., and L. W. Lewis. 1927. "Crystallography of Azurite from Tsumeb, Southwest Africa, and the Axial Ratio of Azurite." *American Mineralogist* 12:99–143.

Palache, C., and H. E. Merwin. 1909. "On Connellite and Chalcophyllite from Bisbee, Arizona." *American Journal of Science* 28:537–40.

Palache, C., and E. V. Shannon. 1920. "Higginsite, a New Mineral of the Olivenite Group." *American Mineralogist* 5:155–57.

Palmer, C. 1909. "Arizonite, Ferric Metatitanate." *American Journal of Science* 28:353–56.

Papish, J. 1928. "New Occurrences of Germanium." *Economic Geology* 23:660–70.

Parker, F. Z. 1966. "The Geology and Mineral Deposits of the Silver District, Trigo Mountains, Yuma County, Arizona." Master's thesis, San Diego State College.

Peacock, M. A. 1931. "Classification of Igneous Rock Series." *Journal of Geology* 39:54–67.

Peacor, D., P. J. Dunn., G. Schnorrer-Kohler, and R. A. Bideaux. 1985. "Mammothite, a New Mineral from Tiger, Arizona and Laurium, Greece." *Mineralogical Record* 15:117–20.

Peacor, D. R., P. J. Dunn, and W. B. Simmons. 1984. "Paulkerrite, a New Titanium Phosphate from Arizona." *Mineralogical Record* 15:303–7.

Peacor, D. R., P. J. Dunn, W. B. Simmons, F. J. Wicks, and M. Raudsepp. 1988. "Maricopaite, a New Hydrated Ca-Pb, Zeolite-like Silicate from Arizona." *Canadian Mineralogist* 26:309–13.

Peacor, D. R., R. C. Rouse, T. D. Coskren, and E. J. Essene. 1999. "Destinezite ("Diadochite"), $Fe_2(PO_4)(OH)–6H_2O$: Its Crystal Structure and Role as a Soil Mineral at Alum Cave Bluff, Tennessee." *Clays and Clay Minerals* 47:1–11.

Pearl, R. M. 1950. "New Data on Lossenite, Louderbackite, Zepharovichite, Peganite, and Sphaerite." *American Mineralogist* 35:1055–59.

Peirce, H. W. 1969a. "Gemstones, Quartz." In *Mineral and Water Resources of Arizona.* Arizona Bureau of Mines Bulletin 180, 362. Tucson: University of Arizona.

Peirce, H. W. 1969b. "Salines." In *Mineral and Water Resources of Arizona.* Arizona Bureau of Mines Bulletin 180, 417–24. Tucson: University of Arizona.

Peirce, H. W., and T. A. Gerrard. 1966. "Evaporite Deposits of the Permian Holbrook Basin, Arizona." Second Symposium on Salt. *Northern Ohio Geological Society* 1:1–10.

Peirce, H. W., Stanton B. Keith, and J. C. Wilt. 1970. *Coal, Oil, Natural Gas, Helium, and Uranium in Arizona*. Arizona Bureau of Mines Bulletin 182. Tucson: University of Arizona.

Pellegrin, A. L. 1911. "Rare Minerals in Southern Arizona." *Mining World* 34:450.

Penfield, S. L. 1886. "Crystallized Vanadinite from Arizona and New Mexico." *American Journal of Science* 32:441–43.

Penfield, S. L. 1890. "On Spangolite, a New Copper Mineral." *American Journal of Science* 39:370–78.

Percious, J. K. 1968. "Geochemical Investigation of the Del Bac Hills Volcanics, Pima County, Arizona." Master's thesis, University of Arizona.

Perry, D. V. 1964. "Genesis of the Contact Rocks at the Abril Mine, Cochise County, Arizona." Master's thesis, University of Arizona.

Perry, D. V. 1969. "Skarn Genesis at the Christmas Mine, Gila County, Arizona." *Economic Geology* 64:255–70.

Petereit, A. H. 1907. "Crystallized Native Copper from Bisbee, Arizona." *American Journal of Science* 23:232–33.

Petersen, R. G. 1960. "Detrital-Appearing Uraninite in the Shinarump Member of the Chinle Formation in Northern Arizona." *Economic Geology* 55:138–49.

Petersen, R. G., J. C. Hamilton, and A. T. Myers. 1959. "An Occurrence of Rhenium Associated with Uraninite in Coconino County, Arizona." *Economic Geology* 54:254–67.

Peterson, E. C. 1966. *Titanium Resources of the United States*. U.S. Bureau of Mines Information Circular IC 8290. Washington, D.C.: U.S. Department of the Interior, Bureau of Mines.

Peterson, N. P. 1938a. "Geology and Ore Deposits of the Mammoth Mining Camp, Pinal County, Arizona." PhD diss., University of Arizona.

Peterson, N. P. 1938b. *Geology and Ore Deposits of the Mammoth Mining Camp Area, Pinal County, Arizona*. Arizona Bureau of Mines Bulletin 144. Tucson: University of Arizona.

Peterson, N. P. 1947. "Phosphate Minerals in the Castle Dome Copper Deposit, Arizona." *American Mineralogist* 32:574–82.

Peterson, N. P. 1954. "Copper Cities Copper Deposit, Globe-Miami District, Arizona." *Economic Geology* 49:362–77.

Peterson, N. P. 1962. *Geology and Ore Deposits of the Globe-Miami District, Arizona*. U.S. Geological Survey Professional Paper 342. Washington, D.C.: Government Printing Office.

Peterson, N. P. 1963. *Geology of the Pinal Ranch Quadrangle, Arizona*. U.S. Geological Survey Bulletin. 1141-H: H1–H18. Washington, D.C.: Government Printing Office.

Peterson, N. P., C. M. Gilbert, and G. L. Quick. 1946. "Hydrothermal Alteration in the Castle Dome Copper Deposit, Arizona." *Economic Geology* 41:820–40.

Peterson, N. P., C. M. Gilbert, and G. L. Quick. 1951. *Geology and Ore Deposits of the Castle Dome Area, Gila County*. U.S. Geological Survey Bulletin 971. Washington, D.C.: Government Printing Office.

Peterson, N. P., and R. W. Swanson. 1956. *Geology of the Christmas Copper Mine, Gila County, Arizona*. U.S. Geological Survey Bulletin 1027-H. Washington, D.C.: Government Printing Office.

Phalen, W. C. 1914. *Celestite Deposits in California and Arizona.* U.S. Geological Survey Bulletin 540-T, 521–33. Washington, D.C.: Government Printing Office.

Phelps, H. D. 1946. *Exploration of the Copper Butte Mine, Mineral Creek Mining District, Pinal County, Arizona.* U.S. Bureau of Mines Report of Investigations RI 3914. Washington, D.C.: U.S. Department of the Interior, Bureau of Mines.

Phillips, C. H., H. R. Cornwall, and M. Rubin. 1971. "A Holocene Ore Body of Copper Oxides and Carbonates at Ray, Arizona." *Economic Geology* 66:495–98.

Phillips, K. A. 1987. *Arizona Industrial Minerals.* Mineral Report 4. 2nd ed. Phoenix: Arizona Department of Mines and Mineral Resources.

Phillips, K. A. 1988. *Gemstone Production in Arizona.* Mineral Report 5. Phoenix: Arizona Department of Mines and Mineral Resources.

Pierce, L., and P. R. Buseck. 1978. "Superstructuring in the Bornite-Digenite Series: A High-Resolution Electron Microscopy Study." *American Mineralogist* 63:1–15.

Pierce, W. G., and E. I. Rich. 1962. *Summary of Rock Salt Deposits in the United States as Possible Storage Sites for Radioactive Waste Materials.* U.S. Geological Survey Bulletin 1148. Washington, D.C.: Government Printing Office.

Pilkington, H. D. 1961. "A Mineralogic Investigation of Some Garnets from the Catalina Mountains, Pima County, Arizona." *Arizona Geological Society Digest* 4:117–22.

Pinch, W. W. 1977. "Rare Minerals Report." *Mineralogical Record* 8:51.

Pipkin, B. W. 1967. "Mineralogy of the 140-Foot Core from Willcox Playa, Cochise County, Arizona." Abstract. *American Association of Petroleum Geologists Bulletin* 51:478–79.

Pipkin, B. W. 1968. "Clay Mineralogy of the Willcox Playa and Its Drainage Basin, Cochise County, Arizona." PhD diss., University of Arizona.

Pirsson, L. V. 1891. "Mineralogical Notes." *American Journal of Science* 42:405–6.

Pogue, J. E. 1913. "On Cerussite Twins from the Mammoth Mine, Pinal County, Arizona." *American Journal of Science* 35:90–92.

Polman, G. V. 2021. "Potter-Cramer Mine, Maricopa County, Arizona." *Rocks and Minerals* 96:24–37.

Potter, G. M., and R. Havens. 1949. *Concentration of Oxide Manganese Ores from the Adams and Woody Properties, Coconino County, near Peach Springs, Arizona.* U.S. Bureau of Mines Report of Investigations RI 4439. Washington, D.C.: U.S. Department of the Interior, Bureau of Mines.

Potter, G. M., A. O. Ipsen, and R. R. Wells. 1946. *Concentration of Manganese Ores from Gila, Greenlee, and Graham Counties, Arizona.* U.S. Bureau of Mines Report of Investigations RI 3842. Washington, D.C.: U.S. Department of the Interior, Bureau of Mines.

Pough, F. N. 1941. "Occurrence of Willemite." *American Mineralogist* 26:92–102.

Prinz, M., D. G. Waggoner, and P. J. Hamilton. 1980. "Winonaites: A Primitive Achondritic Group Related to Silicate Inclusions in IAB Irons." Abstract. *Lunar and Planetary Science* 11:902–4.

Presmyk, L. 1998. "Diamond Point Quartz." Abstract. *Minerals of Arizona, Sixth Annual Symposium.* Phoenix: Flagg Mineral Foundation.

Presmyk, L. 2015. "The Arizona Silver Belt: Silver King to McMillenville." *Mineralogical Record* 46:507–27.

Presmyk, L., and M. Hay. 2020. "Treasures of the Magma Mine, Pinal County, Arizona." *Mineralogical Record* 51:135–43.

Quick, T. J., R. G. Corbett, and B. M. Manner. 1989. "Efflorescent Minerals Occurring in the Gorge of the Grand Canyon." *Geological Society of America Abstracts with Programs* 21 (4): 44–45.

Radcliffe, D., and W. B. Simmons Jr. 1971. "Austinite: Chemical and Physical Properties in Relation to Conichalcite." *American Mineralogist* 56:1359–65.

Ransome, F. L. 1903a. "The Copper Deposits of Bisbee, Arizona." *Engineering and Mining Journal* 75:444–45.

Ransome, F. L. 1903b. *Geology of the Globe Copper District, Arizona*. U.S. Geological Survey Professional Paper 12. Washington, D.C.: Government Printing Office.

Ransome, F. L. 1904. *The Geology and Ore Deposits of the Bisbee Quadrangle, Arizona*. U.S. Geological Survey Professional Paper 21. Washington, D.C.: Government Printing Office.

Ransome, F. L. 1914. *Copper Deposits near Superior, Arizona*. U.S. Geological Survey Bulletin 540-D. Washington, D.C.: Government Printing Office.

Ransome, F. L. 1916. "Quicksilver Deposits of the Mazatzal Range, Arizona." In *Contributions to Economic Geology*, U.S. Geological Survey Bulletin 620-F, 111–28. Washington, D.C.: Government Printing Office.

Ransome, F. L. 1919. *The Copper Deposits of Ray and Miami, Arizona*. U.S. Geological Survey Professional Paper 115. Washington, D.C.: Government Printing Office.

Ransome, F. L. 1922. *Ore Deposits of the Sierrita Mountains, Pima County, Arizona*. U.S. Geological Survey Bulletin 725-J. Washington, D.C.: Government Printing Office.

Ransome, F. L. 1923. *Geology of the Oatman Gold District, Arizona: A Preliminary Report*. U.S. Geological Survey Bulletin 743. Washington, D.C.: Government Printing Office.

Rasmussen, J. C., and S. B. Keith. 2015. "Magma-Metal Series Classification of Mineralization in the Vicinity of Yucca Mountain, Nevada." In *New Concepts and Discoveries*, vol. 2, edited by W. M. Pennell and L. J. Garside, 1131–52. Reno: Geological Society of Nevada.

Rasmussen, J. C., and C. Hoag. 2011. *Yuma King Project, Arizona*. NI 43-101 Technical Exploration Report, SRK consulting report prepared for Rare Green. Longmont, Colo.: Rare Green.

Rasor, C. A. 1938. "Bromyrite from Tombstone, Arizona." *American Mineralogist* 23:157–59.

Rasor, C. A. 1939. "Manganese Mineralization at Tombstone, Arizona." *Economic Geology* 34:790–803.

Rasor, C. A. 1946. "Loellingite from Arizona." *American Mineralogist* 31:406–8.

Raup, R. B. 1954. "Reconnaissance for Uranium in the United States, Southwest District." In *Geological Investigations of Radioactive Deposits*. U.S. Geological Survey TEI 440, 180–82. Oak Ridge, Tenn.: U.S. Atomic Energy Commission, Technical Information Service.

Reber, L. E., Jr. 1916. "The Mineralization at Clifton-Morenci." *Economic Geology* 11:528–73.

Reed, R. F., and W. W. Simmons. 1962. "Geological Notes on the Miami-Inspiration Mine." In *Guidebook of the Mogollon Rim Region (East-Central Arizona)*, edited by R. H. Weber and H. Wesley Peirce, 153–57. New Mexico Geological Society 13th Field Conference Guidebook. Albuquerque: New Mexico Geological Society.

Regis, A. J., and L. B. Sand. 1967. "Lateral Gradation of Chabazite to Herschelite in the San Simon Basin." Abstract. *Clays and Clay Minerals* 15 (1): 193.

Reiter, B. E. 1980. "Controls on Lead-Zinc Skarn Mineralization, Iron Cap Mine Area, Aravaipa District, Graham County, Arizona." Master's thesis, Arizona State University.

Reiter, B. E. 1981. "Controls on Lead-Zinc Skarn Mineralization, Iron Cap Mine Area, Aravaipa District, Graham County, Arizona." In *Arizona Geological Society Digest*, vol. 13, edited by C. Stone and J. P. Jenney, 117–25. Tucson: Arizona Geological Society.

Reynolds, S. J., S. M. Richard, G. B. Haxel, R. M. Tosdal, and S. E. Laubach. 1988. "Geologic Setting of Mesozoic and Cenozoic Metamorphism in Arizona." In *Metamorphism and Crustal Evolution of the Western United States: Rubey Volume 7*, edited by W. G. Ernst, 466–501. New York: Prentice Hall.

Reynolds, S. J., E. A. Scot, and R. T. O'Haire. 1985. "A Fluorite-Bearing Granite, Belmont Mountains, Central Arizona." *Arizona Bureau of Geology and Mineral Technology Fieldnotes* 15 (2): 3–5.

Reynolds, S. J., J. E. Spencer, S. E. Laubach, D. Cunningham, and S. M. Richard. 1989. *Geologic Map, Geologic Evolution, and Mineral Deposits of the Granite Wash Mountains, West-Central Arizona*. With addendum, "Geology of the Northwesternmost and Southwesternmost Granite Wash Mountains," by J. E. Spencer, S. M. Richard, and S. J. Reynolds. Arizona Geological Survey Open-File Report 89-04. Tucson: Arizona Geological Survey.

Rice, C. M., D. Atkin, J. F. W. Bowles, and A. J. Criddle. 1979. "Nukundamite, A New Mineral, and Idaite." *Mineralogical Magazine* 43:193–200.

Richard, K., and J. H. Courtright. 1959. "Some Geologic Features of the Mission Copper Deposit." In *Southern Arizona Guidebook II, Combined with the 2nd Annual Arizona Geological Society Digest*, edited by L. A. Heindl, 201–4. Tucson: Arizona Geological Society.

Richard, K., and J. H. Courtright. 1966. "Structure and Mineralization at Silver Bell, Arizona." In *Geology of the Porphyry Copper Deposits, Southwestern North America*, edited by S. R. Titley and C. L. Hicks, 157–63. Tucson: University of Arizona Press.

Richards, R. A. 1956. "Arizona Agate Hunting." *Rocks and Minerals* 31:279–80.

Richardson, C. A., D. A. Favorito, S. E. Runyon, E. Seedorff, D. J. Maher, M. D. Barton, and R. E. Greig. 2019. "Superimposed Laramide Contraction, Porphyry Copper Systems, and Cenozoic Extension, East-Central Arizona: A Road Log." In *Geologic Excursions in Southwestern North America*, edited by P. A. Pearthree, 337–67. Geological Society of America Field Guide 55. Boulder, Colo.: Geological Society of America.

Richmond, W. E. 1940. "Crystal Chemistry of the Phosphates, Arsenates and Vanadates of the type A2XO4(Z)." *American Mineralogist* 25:441–79.

Richmond, W. E., and M. Fleischer. 1942. "Cryptomelane, a New Name for the Commonest of the Psilomelane Minerals." *American Mineralogist* 27:607–10.

Richter, D. H., and V. A. Lawrence. 1983. *Mineral Deposit Map of the Silver City 1 × 2 Quadrangle, New Mexico and Arizona*. U.S. Geological Survey Map I-1310-B. Reston, Va.: U.S. Geological Survey.

Rieder, M., G. Cavazzini, Y. S. D'yakonov, V. A. Frank-Kamenetskij, G. Gottardi, S. Guggenheim, P. V. Koval, et al. 1998. "Nomenclature of the Micas." *Canadian Mineralogist* 36:905–12.

Robbins, M. 1983. *The Collector's Book of Fluorescent Minerals*. New York: Springer.

Robbins, M. 1994. *Fluorescence: Gems and Minerals under Ultraviolet Light*. Phoenix, Ariz.: Geoscience Press.

Roberts, A. C., J. A. R. Stirling, G. J. C. Carpenter, A. J. Criddle, G. C. Jones, T. C. Birkett, and W. D. Birch. 1995. "Shannonite, Pb_2OCO_3, a New Mineral from the Grand Reef Mine, Graham County, Arizona, USA." *Mineralogical Magazine* 59: 305–10.

Robinson, R. F., and A. Cook. 1966. "The Safford Copper Deposit, Lone Star Mining District, Graham County, Arizona." In *Geology of the Porphyry Copper Deposits, Southwestern North America*, edited by S. R. Titley and C. L. Hicks, 251–66. Tucson: University of Arizona Press.

Robinson, R. L. 1954. *Duranium Claims, Santa Cruz County, Arizona*. U.S. Atomic Energy Commission. Preliminary Reconnaissance Report PRR A-P-285. Washington, D.C.: U.S. Atomic Energy Commission.

Robinson, R. L. 1955. *Linda Lee Claims, Pima County, Arizona*. U.S. Atomic Energy Commission. Preliminary Reconnaissance Report PRR A-P-331. Washington, D.C.: U.S. Atomic Energy Commission.

Robinson, R. L. 1956. *Lucky Find Group, Maricopa County, Arizona*. U.S. Atomic Energy Commission. Preliminary Reconnaissance Report PRR A-96. Washington, D.C.: U.S. Atomic Energy Commission.

Roddy, M. S. 1986. "K-metasomatism and Detachment-Related Mineralization, Harcuvar Mountains, Arizona." Master's thesis, University of Arizona.

Roden, M. F. 1981. "Origin of Coexisting Minette and Ultramafic Breccia, Navajo Volcanic Field." *Contributions to Mineralogy and Petrology* 77:195–206.

Roe, A. 1980. "Micromounting in Arizona." *Mineralogical Record* 11:261–65.

Roe, R. R. 1976. "Geology of the Squaw Peak Porphyry Copper—Molybdenum Deposit, Yavapai County, Arizona." Master's thesis, University of Arizona.

Roedder, E., B. Ingram, and W. E. Hall. 1963. "Studies of the Fluid Inclusions III: Extractions and Quantitative Analysis of Inclusions in the Milligram Range." *Economic Geology* 58:353–74.

Rogers, A. F. 1913. "Delafossite, a Cuprous Metaferrite from Bisbee, Arizona." *American Journal of Science* 35:290–94.

Rogers, A. F. 1917. "A Review of the Amorphous Minerals." *Journal of Geology* 25: 515–41.

Rogers, A. F. 1922. "The Optical Properties and Morphology of Bisbeeite." *American Mineralogist* 7:153–54.

Rogers, A. F. 1930. "A Unique Occurrence of Lechatelierite or Silica Glass." *American Journal of Science* 19:195–202.

Rogers, W. 1958. "Saddle Mountain, Arizona." *Gems and Minerals* 254:26–28.

Romslo, T. M. 1948. *Antler Copper-Zinc Deposit, Mohave County, Arizona.* U.S. Bureau of Mines Report of Investigations RI 4214. Washington, D.C.: U.S. Department of the Interior, Bureau of Mines.

Romslo, T. M. 1949. *Investigation of Keystone and St. George Copper-Zinc Deposits, Cochise County, Arizona.* U.S. Bureau of Mines Report of Investigations RI 4504. Washington, D.C.: U.S. Department of the Interior, Bureau of Mines.

Romslo, T. M. 1950. *Investigation of the Lake Shore Copper Deposits, Pinal County, Arizona.* U.S. Bureau of Mines Report of Investigations RI 4706. Washington, D.C.: U.S. Department of the Interior, Bureau of Mines.

Romslo, T. M., and S. F. Ravitz. 1947. *Arizona Manganese-Silver Ores.* U.S. Bureau of Mines RI 4097. Washington, D.C.: U.S. Department of the Interior, Bureau of Mines.

Romslo, T. M., and C. S. Robinson. 1952. *Copper Giant Deposits, Pima County, Arizona.* U.S. Bureau of Mines Report of Investigations RI 4850. Washington, D.C.: U.S. Department of the Interior, Bureau of Mines.

Rose, A. W. 1970. "Zonal Relations of Wallrock Alteration and Sulfide Distribution at Porphyry Copper Deposits." *Economic Geology* 65:920–36.

Roseboom, E. H., Jr. 1966. "An Investigation of the System Cu-S and Some Natural Copper Sulfides between 25° and 700°C." *Economic Geology* 61:641–72.

Rosenzweig, A., J. W. Gruner, and L. Gardiner. 1954. "Widespread Occurrence and Character of Uraninite in the Triassic and Jurassic Sediments of the Colorado Plateau." *Economic Geology* 49:351–61.

Ross, C. P. 1925a. *Geology and Ore Deposits of the Aravaipa and Stanley Mining Districts, Graham County, Arizona.* U.S. Geological Survey Bulletin 763. Washington, D.C.: U.S. Department of the Interior, Bureau of Mines.

Ross, C. P. 1925b. *Geology of the Saddle Mountain and Banner Mining Districts.* U.S. Geological Survey Bulletin 771. Washington, D.C.: U.S. Department of the Interior, Bureau of Mines.

Ross, C. S. 1928. "Sedimentary Analcite." *American Mineralogist* 13:195–97.

Ross, C. S. 1941. "Sedimentary Analcite." *American Mineralogist* 26:627–29.

Ross, M. 1959. "Mineralogical Applications of Electron Diffraction, II: Studies of Some Vanadium Minerals of the Colorado Plateau." *American Mineralogist* 44:322–41.

Roy, S. K., and R. K. Wyant. 1949. *The Navajo Meteorite.* Geology Series of Field Museum of Natural History, vol. 7, no. 8. Chicago: Field Museum of Natural History.

Rubel, A. C. 1917–18. *Tungsten.* Arizona Bureau of Mines Bulletin 11. Tucson: University of Arizona.

Rushing, J. A. 1978. *Contact Metamorphism and Metasomatism of Paleozoic Rocks near Stronghold Canyon, Dragoon Mountains, Arizona.* Tucson: University of Arizona, M. S. Thesis: 96.

Sampson, E. 1924. "Discussion of an Arizona Asbestos Deposit." *Economic Geology* 19:386–88.

Sand, L. B., and A. J. Regis. 1966. "An Unusual Zeolite Assemblage, Bowie, Arizona." Abstract. *Geological Society of America Special Paper* 87:145–46.

Sandell, W. G., and D. T. Holmes. 1948. *Concentration of Oxide Manganese Ores from the Aguila District, Arizona.* U.S. Bureau of Mines Report of Investigations RI 4330. U.S. Department of the Interior, Bureau of Mines.

Sanders, J. 1911. "Hematite in Veins of Globe District." *Engineering and Mining Journal* 92:1191–92.

Santmyers, R. M. 1929. *Development of the Gypsum Industry by States.* U.S. Bureau of Mines Information Circular IC 6173. Washington, D.C.: U.S. Department of Commerce, Bureau of Mines.

Scarborough, R. B. 1981. *Radioactive Occurrences and Uranium Production in Arizona.* U.S. Department of Energy Report GJBX-143(81). Tucson: Arizona Bureau of Geology and Mineral Technology, Geological Survey Branch.

Schaller, W. T. 1905. "Dumortierite." In *Contributions to Mineralogy from the United States Geological Survey.* U.S. Geological Survey Bulletin 262, 91–120. Washington, D.C.: Government Printing Office.

Schaller, W. T. 1915. "Four New Minerals." *Journal of the Washington Academy of Science* 5:7.

Schaller, W. T. 1932. "Chemical Composition of Cuprotungstite." *American Mineralogist* 17:234–37.

Schaller, W. T. 1934. "Mottramite or Psittacinite—A Question of Nomenclature." *American Mineralogist* 16:180–81.

Schaller, W. T. 1939. "Corrections and Additions." *American Mineralogist* 24:346–47.

Schaller, W. T., R. E. Stevens, and R. H. Jahns. 1962. "An Unusual Beryl from Arizona." *American Mineralogist* 47:672–99.

Schaller, W. T., and A. C. Vlisidis. 1958. "Ajoite, a New Hydrous Aluminum Copper Silicate." *American Mineralogist* 43:1107–11.

Schlegel, D. M. 1957. "Gem Stones of the United States." In *Contributions to Economic Geology.* U.S. Geological Survey Bulletin 1042-G, 203–53. Washington, D.C.: Government Printing Office.

Schmidt, E. A. 1971. "A Structural Investigation of the Northern Tortilla Mountains, Pinal County, Arizona." PhD diss., University of Arizona.

Schmitt, H. A., D. M. Clippinger, W. J. Roper, and H. Toombs. 1959. "Disseminated Deposits at the Esperanza Copper Mine." In *Southern Arizona Guidebook II, Combined with the 2nd Annual Arizona Geological Society Digest,* edited by L. A. Heindl, 205. Tucson: Arizona Geological Society.

Schmitz, C. 1987. "Geology of the Black Pearl Mine Area, Yavapai County, Arizona." Master's thesis, Arizona State University.

Schrader, F. C. 1908. "Mineral Deposits of the Cerbat Range, Black Mountains, and Grand Wash Cliffs, Mohave County, Arizona." In *Contributions to Economic Geology 1907, Part I—Metals and Nonmetals, Except Fuels,* edited by C. W. Hayes and W. Lindgren. U.S. Geological Survey Bulletin 340, 53–83. Washington, D.C.: Government Printing Office.

Schrader, F. C. 1909. "Mineral Deposits of the Cerbat Range, Black Mountains, and Grand Wash Cliffs, Mohave County, Arizona." *U.S. Geological Survey Bulletin* 397:226.

Schrader, F. C. 1913. "Alunite in Patagonia, Arizona, and Bovard, Nevada." *Economic Geology* 8:752–67.

Schrader, F. C. 1914. "Alunite in Granite Porphyry near Patagonia, Arizona." U.S. *Geological Survey Bulletin* 540:347–50.

Schrader, F. C. 1917. "The Geologic Distribution and Genesis of the Metals in the Santa Rita-Patagonia Mountains, Arizona." *Economic Geology* 12:237–69.

Schrader, F. C. 1919. *Quicksilver Deposits of the Phoenix Mountains*. U.S. Geological Survey Bulletin 690-D. Washington, D.C.: Government Printing Office.

Schrader, F. C., and J. M. Hill. 1910. "Some Occurrences of Molybdenite in the Santa Rita and Patagonia Mountains, Arizona." In *Rare Metals: Notes on the Occurrence of Cinnabar in Central Western Arizona*, edited by F. C. Schrader and J. M. Hill, 154–63. U.S. Geological Survey Bulletin 430-D. Washington, D.C.: Government Printing Office.

Schrader, F. C., and J. M. Hill. 1915. *Mineral Deposits of the Santa Rita and Patagonia Mountains, Arizona*. U.S. Geological Survey Bulletin 582. Washington, D.C.: Government Printing Office.

Schultz, L. G. 1963. *Clay Minerals in Triassic Rocks of the Colorado Plateau*. U.S. Geological Survey Bulletin 1147-C. Washington, D.C.: Government Printing Office.

Schulze, D. J., and H. Helmstaedt. 1979. "Garnet Pyroxenite and Eclogite Xenoliths from the Sullivan Buttes Latite, Chino Valley, Arizona." In *The Mantle Sample: Inclusions in Kimberlites and Other Volcanics (Proceedings of the Second International Kimberlite Conference)*, vol. 2, edited by F. R. Boyd, and H. O. A. Meyer, 318–29. Washington, D.C.: American Geophysical Union.

Schulze, D. J., H. Helmstaedt, and R. M. Cassie. 1978. "Pyroxene-Ilmenite Intergrowths in Garnet Pyroxenite Xenoliths from a New York Kimberlite and Arizona Latite." *American Mineralogist* 63:258–65.

Schumer, B. N. 2017. "Mineralogy of Copper Sulfides in Porphyry Copper And Related Deposits." PhD diss., University of Arizona.

Schwartz, G. M. 1921. "Notes on Textures and Relationships in the Globe Copper Ores." *Economic Geology* 16:322–29.

Schwartz, G. M. 1928. "Experiments Bearing on Bornite-Chalcocite Intergrowths." *Economic Geology* 23:381–97.

Schwartz, G. M. 1934. "Paragenesis of the Oxidized Ores of Copper." *Economic Geology* 29:55–75.

Schwartz, G. M. 1938. "Oxidized Copper Ores of the United Verde Extension Mine." *Economic Geology* 33:21-33.

Schwartz, G. M. 1939. "Significance of Bornite-Chalcocite Microtextures." *Economic Geology* 34:399–418.

Schwartz, G. M. 1947. "Hydrothermal Alteration in the Porphyry Copper Deposits." *Economic Geology* 42:319–52.

Schwartz, G. M. 1949. "Oxidation and Enrichment in the San Manuel Copper Deposit, Arizona." *Economic Geology* 44:253–77.

Schwartz, G. M. 1952. "Chlorite-Calcite Pseudomorphs after Orthoclase Phenocrysts, Ray, Arizona." *Economic Geology* 47:665–72.

Schwartz, G. M. 1953. *Geology of the San Manuel Copper Deposit, Arizona*. U.S. Geological Survey Professional Paper 256.

Schwartz, G. M. 1956. "Argillic Alteration and Ore Deposits." *Economic Geology* 51:407–14.

Schwartz, G. M. 1958. "Alteration of Biotite under Mesothermal Conditions." *Economic Geology* 53:164–77.

Schwartz, G. M. 1959. "Hydrothermal Alteration." *Economic Geology* 54:161–83.

Schwartz, G. M. 1966. "The Nature of Primary and Secondary Mineralization in Porphyry Copper Deposits." In *Geology of the Porphyry Copper Deposits, Southwestern North America*, edited by S. R. Titley and C. L. Hicks, 41–50. Tucson: University of Arizona Press.

Schwartz, G. M., and C. F. Park Jr. 1932. "A Microscopic Study of Ores from the Campbell Mine, Bisbee, Arizona." *Economic Geology* 27:39–51.

Scott, E. R. D. 1971. "New Carbide (Fe,Ni)$_{23}$C$_6$, Found in Meteorites." *Nature Physical Science* 229:61–62.

Scovil, J., and L. Wagner Jr. 1991. "The Fat Jack Mine, Yavapai County, Arizona." *Mineralogical Record* 22:21–28.

Searls, F., Jr. 1950. "The Emerald Isle Copper Deposit." *Economic Geology* 45:175–76.

Seedorf, E., M. D. Barton, G. E. Gehrels, V. A. Valencia, D. A. Johnson, D. J. Maher, W. J. A. Stavast, and T. M. Marsh. 2019. "Temporal Evolution of the Laramide Arc: U-Pb Geochronology of Plutons Associated with Porphyry Copper Mineralization in East-Central Arizona." In *Geologic Excursions in Southwestern North America*, edited by P. A. Pearthree, 369–401. Geological Society of America Field Guide 55. Boulder, Colo.: Geological Society of America.

Selfridge, G. C., Jr. 1936. "An X-ray and Optical Investigation of the Serpentine Minerals." *American Mineralogist* 21:463–503.

Sell, J. D. 1961. "Bedding Replacement Deposit of the Magma Mine, Superior, Arizona." Master's thesis, University of Arizona.

Seymour, N. M., E. D. Strickland, J. S. Singleton, D. F. Stockli, and M. S. Wong. 2018. "Laramide Subduction and Metamorphism of the Orocopia Schist, Northern Plomosa Mountains, West-Central Arizona: Insights from Zircon U-Pb Geochronology." *Geology* 46:847–50.

Shafiqullah, M., P. E. Damon, D. J. Lynch, S. J. Reynolds, W. A. Rehrig, and R. H. Raymond. 1980. "K-Ar Geochronology and Geologic History of Southwestern Arizona and Adjacent Areas." In *Studies in Western Arizona*, edited by J. P. Jenney and C. Stone, 201–60. Arizona Geological Society Digest 12. Tucson: Arizona Geological Society.

Shakel, D. W. 1974. "The Geology of Layered Gneisses in Part of the Santa Catalina Forerange, Pima County, Arizona." Master's thesis, University of Arizona.

Shakel, D. W., L. T. Silver, and P. E. Damon. 1977. "Observations on the History of the Gneissic Core Complex, Santa Catalina Mountains, Southern Arizona." *Geological Society of America, Abstracts with Programs* 9:1169–70.

Shannon, D. M. 1980. "The Hamburg Mine and Vicinity, Yuma County, Arizona." *Mineralogical Record* 11:135–40.

Shannon, D. M. 1981. "What's New in Minerals? Flux Mine Cerussite." *Mineralogical Record* 12:117–18.

Shannon, D. M. 1983a. "Ferroan Dolomite from the Vekol Mine." *Mineralogical Record* 14:91.

Shannon, D. M. 1983b. "Zeolites and Associated Minerals from Horseshoe Dam, Arizona." *Mineralogical Record* 14:115–17.

Shannon, D. M. 1996. "Orthoserpierite from the Copper Creek District, Pinal County, Arizona." *Mineralogical Record* 27: 189–90.

Shannon, E. V., and F. A. Gonyer. 1927. "Natrojarosite from Kingman, Arizona." *Journal of the Washington Academy of Science* 17:536–37.

Shaw, V. E. 1959. *Extraction of Yttrium and Rare Earth Elements from Arizona Euxenite Concentrate*. U.S. Bureau of Mines Report of Investigations RI-5544. Washington, D.C.: U.S. Department of the Interior, Bureau of Mines.

Shawe, D. R. 1966. "Arizona–New Mexico and Nevada–Utah Beryllium Belts." In *Geological Survey Research 1966*, U.S. Geological Survey Professional Paper 550-C, 206–13. Washington, D.C.: Government Printing Office.

Sheppard, R. A. 1969. "Zeolites." In *Mineral and Water Resources of Arizona*. Arizona Bureau of Mines Bulletin 180, 464–67. Tucson: University of Arizona.

Sheppard, R. A. 1971. *Clinoptilolite, of Possible Economic Value in Sedimentary Deposits of the Conterminous United States*. U.S. Geological Survey Bulletin 1332-B: B1–B15. Washington, D.C.: U.S. Department of the Interior, Geological Survey.

Sheppard, R. A., and A. J. Gude III. 1973. *Zeolites and Associated Authigenic Silicate Minerals in Tuffaceous Rocks of the Big Sandy Formation, Mohave County, Arizona*. U.S. Geological Survey Professional Paper 830. Washington, D.C.: Government Printing Office.

Sherborne, J. E., Jr., W. A. Buckovic, D. B. Dewitt, T. S. Hellinger, and S. J. Pavlak. 1979. "Major Uranium Discovery in Volcaniclastic Sediments, Basin and Range Province, Yavapai County, Arizona." *American Association of Petroleum Geologists Bulletin* 63:621–46.

Sheridan, M. F., and C. F. Royse Jr. 1970. "Alunite: A New Occurrence near Wickenburg, Arizona." *American Mineralogist* 55:2016–22.

Short, M. N., F. W. Galbraith, E. N. Harshman, T. H. Kuhn, and E. D. Wilson. 1943. *Geology and Ore Deposits of the Superior Mining Area, Arizona*. Arizona Bureau of Mines Bulletin 151. Tucson: University of Arizona.

Shride, A. F. 1969. "Asbestos." In *Mineral and Water Resources of Arizona*. Arizona Bureau of Mines Bulletin 180, 303–11. Tucson: University of Arizona.

Sikka, D. B., W. Petruk, C. E. Nehru, and Z. Zhang. 1991. "Geochemistry of Secondary Copper Minerals from a Proterozoic Porphyry Copper Deposit, Malanjkhand, India." *Ore Geology Reviews* 6:257–90.

Silliman, B. 1866. "On Some Mining Districts of Arizona near Rio Colorado, with Remarks on the Climate, Etc." *American Journal of Science* 41:289–308.

Silliman, B. 1879. "Jarosite (with Gold)." *American Journal of Science* 17:73.

Silliman, B. 1881. "Mineralogical Notes." *American Journal of Science* 22:198–205.

Silver, L. T., and S. Deutsch. 1963. "Uranium-Lead Isotopic Variations in Zircons: A Case Study." *Journal of Geology* 71:721–58.

Simmons, W. B., S. L. Hanson, A. U. Falster, and K. L. Webber. 2012. "A Comparison of the Mineralogical and Geochemical Character and Geological Setting of Proterozoic REE-Rich Pegmatites of the North-Central and Southwestern US." *Canadian Mineralogist* 50:1695–1712.

Simmons, W. W., and J. E. Fowells. 1966. "Geology of the Copper Cities Mine." In *Geology of the Porphyry Copper Deposits, Southwestern North America*, edited by S. R. Titley and C. L. Hicks, 151–56. Tucson: University of Arizona Press.

Simons, F. S. 1964. *Geology of the Klondyke Quadrangle, Graham and Pinal Counties, Arizona*. U.S. Geological Survey Professional Paper 461. Washington, D.C.: Government Printing Office.

Simons, F. S., and E. Munson. 1963. "Johannsenite from the Aravaipa Mining District, Arizona." *American Mineralogist* 48:1154–58.

Singer, D. A., V. I. Berger, and B. C. Moring. 2008. *Porphyry Copper Deposits of the World: Database and Grade and Tonnage Models*. U.S. Geological Survey Open-File Report 2008-1155. Reston, Va.: U.S. Department of the Interior, Geological Survey.

Sinkankas, J. 1959. *Gemstones of North America*. Vol. 1. New York: Van Nostrand.

Sinkankas, J. 1964. *Mineralogy for Amateurs*. Princeton, N.J.: Van Nostrand.

Sinkankas, J. 1966. *Mineralogy: A First Course*. Princeton, N.J.: Van Nostrand.

Sinkankas, J. 1969. *Gemstones of North America*. Vol. 2. Princeton, N.J.: Van Nostrand.

Sinkankas, J. 1997. *Gemstones of North America*. Vol. 3. Tucson, Ariz.: Geoscience Press.

Sirrine, G. K. 1958. "Geology of the Springerville-St. Johns area, Apache County, Arizona." PhD diss., University of Texas, Austin.

Škoda, R., J. Plášil, R. Jonsson, R. Čopjaková, J. Langhof, and M. V. Galiová. 2015. "Redefinition of Thalenite-(Y) and Discreditation of Fluorthalenite-(Y): A Reinvestigation of Type Material from the Osterby Pegmatite, Dalarna, Sweden, and from Additional Localities." *Mineralogical Magazine* 79:965–83.

Smith, D. 1970. "Mineralogy and Petrology of the Diabasic Rocks in a Differentiated Olivine Diabase Sill Complex, Sierra Ancha, Arizona." *Contributions to Mineralogy and Petrology* 27:95–113.

Smith, D. S. 2013. *43-101 Technical Report on the Sugarloaf Peak Gold Project, La Paz County, Arizona*. Vancouver, B.C.: Riverside Resources.

Smith, K. D. 1995. "The Finch Mine and Vicinity, Gila County, Arizona." *Mineralogical Record* 26:439–48.

Smith, L. A. 1927. "The Geology of the Commonwealth Mine." Master's thesis, University of Arizona.

Smith, W. B. 1887. "Dioptase from Pinal County, Arizona." *Proceedings of the Colorado Scientific Society* 2:159–60.

Snyder, J. E. 1971. "Salt Mine Pseudomorphs." *Gems and Minerals*, June, 405.

Sommer, J. V. 1982. "Structural Geology and Metamorphic Petrology of the Northern Sierra Estrella, Maricopa County, Arizona." Master's thesis, Arizona State University.

Sousa, F. X. 1980. "Geology of the Middlemarch Mine and Vicinity, Central Dragoon Mountains, Cochise County, Arizona." Master's thesis, University of Arizona.

Spencer, J. E., and P. A. Pearthree. 2015. *Arizona Geological Society Field-Trip Guide: Northern Plomosa Mountains and Bouse Formation in Blythe Basin*. Arizona Geological Survey Open-File Report OFR-15-10. Tucson: Arizona Geological Survey.

Spencer, J. E., J. S. Singleton, E. Strickland, S. J. Reynolds, D. Love, D. A. Foster, and R. Johnson. 2018. "Geodynamics of Cenozoic Extension along a Transect across the Colorado River Extensional Corridor, Southwestern USA." *Lithosphere* 10:743–59.

Spencer, J. E., and J. W. Welty. 1989. "Mid-Tertiary Ore Deposits in Arizona." In *Geologic Evolution of Arizona*, edited by J. P. Jenney and S. J. Reynolds, 585–607. Arizona Geological Society Digest 17.

Sprunger, M. 1980. "Hematite from the Dome Rock Mountains near Quartzsite, Arizona." *Mineralogical Record* 11:227–29.

Staatz, M. H. 1985. *Geology and Description of the Thorium and Rare-Earth Veins in the Laughlin Peak Area, Colfax County, New Mexico.* U.S. Geological Survey Professional Paper 1049-E. Washington, D.C.: U.S. Department of the Interior, Geological Survey.

Staatz, M. H., W. R. Griffitts, and P. R. Barnett. 1965. "Differences in the Minor Element Composition of Beryl in Various Environments." *American Mineralogist* 50:1783.

Stephens, J. D., and R. A. Metz. 1967. "The Occurrence of Copper-Bearing Clay Minerals in Oxidized Portions of the Disseminated Copper Deposit at Ray, Arizona." Abstract. *Economic Geology* 62:876–77.

Stern, T. W., L. R. Stieff, H. J. Evans Jr., and A. H. Sherwood. 1957. "Doloresite, a New Vanadium Oxide Mineral from the Colorado Plateau." *American Mineralogist* 42:587–93.

Sterrett, D. B. 1908. "Turquoise-Mineral Park Area." *Mineral Resources of the United States: Part 2, Nonmetals*, 847–52. U.S. Geological Survey. Washington, D.C.: Government Printing Office.

Stewart, C. A. 1912. "The Geology and Ore Deposits of the Silverbell Mining District, Arizona." *American Institute of Mining, Metallurgical, and Petroleum Engineers (AIME) Transactions* 43:240–90.

Stewart, L. A. 1947. *Apache Iron Deposit, Navajo County, Arizona.* U.S. Bureau of Mines Report of Investigations RI 4093. Washington, D.C.: U.S. Bureau of Mines.

Stewart, L. A. 1956. *Chrysotile-Asbestos Deposits of Arizona.* U.S. Bureau of Mines Information Circular IC 7745, supp. to IC 7706. Washington, D.C.: U.S. Bureau of Mines.

Stewart, L. A., and P. S. Haury. 1947. *Arizona Asbestos Deposits, Gila County, Arizona.* U.S. Bureau of Mines Report of Investigations RI 4100. Washington, D.C.: U.S. Bureau of Mines.

Stewart, L. A., and A. J. Pfister. 1960. *Barite Deposits of Arizona.* U.S. Bureau of Mines Report of Investigations RI 5651. Washington, D.C.: U.S. Department of the Interior, Bureau of Mines.

Stimac, J. A., S. M. Richard, S. J. Reynolds, R. C. Capps, C. P. Kortemeier, M. J. Grubensky, G. B. Allen, F. Gray, and R. J. Miller. 1994. *Geologic Map and Cross Sections of the Big Horn and Belmont Mountains, West-Central Arizona.* Arizona Geological Survey Open-File Report 94-15. Tucson: Arizona Geological Survey.

Stolburg, C. S. 1988. "Osarizawaite from the Silver Hill Mine, Arizona." *Mineralogical Record* 19:311–13.

Strickland, E. D., J. S. Singleton, and G. B. Haxel. 2018. "Orocopia Schist in the Northern Plomosa Mountains, West-Central Arizona: A Laramide Subduction Complex Exhumed in a Miocene Metamorphic Core Complex." *Lithosphere* 10:723–42.

Strong, M. F. 1962. "Pink Marble in Arizona." *Gems and Minerals* 296:26-27.

Strunz, H., and B. Contag. 1965. "Evenkite, Idrialite, and Refikite." Abstract. *American Mineralogist* 50:2109–10.

Struthers, J. 1904. "The Production of the Minor Metals in 1903—Arsenic." *Engineering and Mining Journal* 77:74.

Sun, M.-S. 1954. "Titanclinohumite in Kimberlitic Tuff, Buell Park, Arizona." Abstract. *Geological Society of America Bulletin* 65:1311–12.

Sun, M.-S. 1961. "Differential Thermal Analysis of Shattuckite." *American Mineralogist* 46:67–77.

Sun, M.-S. 1963. "The Nature of Chrysocolla from Inspiration Mine, Arizona." *American Mineralogist* 48:649–58.

Switzer, G. S. 1977. "Composition of Garnet Xenocrysts from Three Kimberlite Pipes in Arizona and New Mexico." In *Mineral Sciences Investigations, 1974–1975*, edited by B. Mason, 1–21. Smithsonian Contributions to Earth Science 19. Washington, D.C.: Smithsonian Institution Press.

Taber, S., and W. T. Schaller. 1930. "Psittacinite from the Higgins Mine, Bisbee, Arizona." *American Mineralogist* 15:575–79.

Tainter, S. L. 1947a. *Amargosa (Esperanza) Molybdenum-Copper Property, Pima County, Arizona*. U.S. Bureau of Mines Report of Investigations RI 4016. Washington, D.C.: U.S. Department of the Interior, Bureau of Mines.

Tainter, S. L. 1947b. *Apex Copper Property, Coconino County, Arizona*. U.S. Bureau of Mines Report of Investigations RI 4013. Washington, D.C.: U.S. Department of the Interior, Bureau of Mines.

Tainter, S. L. 1947c. *Johnny Bull– Silver King Lead-Zinc Property, Cerbat Mountains, Mohave County, Arizona*. U.S. Bureau of Mines Report of Investigations RI 3998. Washington, D.C.: U.S. Department of the Interior, Bureau of Mines.

Tainter, S. L. 1948. *Christmas Copper Deposit, Gila County, Arizona*. U.S. Bureau of Mines Report of Investigations RI 4293. Washington, D.C.: U.S. Department of the Interior, Bureau of Mines.

Tait, K. T., V. Dicecco, N. A. Ball, F. C. Hawthorne, and A. R. Kampf. 2014. "Backite, $Pb_2Al(TeO_6)Cl$, a New Tellurate Mineral from the Grand Central Mine, Tombstone Hills, Cochise County, Arizona." *Canadian Mineralogist* 52:935–42.

Tarashchan, A. N., and G. Waychunas. 1995. "Interpretation of Luminescence Spectra in Terms of Band Theory and Crystal Field Theory." In *Advanced Mineralogy*. Vol. 2, *Methods and Instrumentations: Results and Recent Developments*, edited by A. S. Marfunin, 124–34. Berlin: Springer-Verlag.

Tarashchan, A. N., and G. Waychunas. 2015. "Interpretation of Luminescence Spectra in Terms of Band Theory and Crystal Field Theory." In *Late Jurassic Margin of Laurasia: A Record of Faulting Accommodating Plate Rotation*, edited by T. H. Anderson, A. N. Didenko, C. L. Johnson, A. I. Khanchuk, and J. H. MacDonald Jr., 189–221. Geological Society of America Special Paper 513.

Tenney, J. B. 1913. Unpublished report on 2,200 hand specimens and thin-section determinations. Phelps Dodge Corp. files. Northern Arizona University, Flagstaff.

Tenney, J. B. 1927. *History of Mining in Arizona*. Unpublished manuscript. Arizona Bureau of Mines, Tucson.

Tenney, J. B. 1936. *Mineral Survey of the State of Arizona*. Phoenix: Arizona State Planning Board.

Terrien, J. J. 2012. "The Role of Magmatism in the Catalina Metamorphic Core Complex, Arizona: Insights from Integrated Thermochronology, Gravity, and Aeromagnetic Data." PhD diss., Syracuse University.

Theodore, T. G., W. N. Blair, and J. T. Nash. 1987. *Geology and Gold Mineralization of the Gold Basin-Lost Basin Mining Districts, Mohave County, Arizona*. U.S. Geological Survey Professional Paper 1361. Washington, D.C.: Government Printing Office.

Thoenen, J. R. 1941. *Alunite Resources of the United States*. U.S. Bureau of Mines Report of Investigations RI 3561. Washington, D.C.: U.S. Department of the Interior, Bureau of Mines.

Thomas, B. E. 1949. "Ore Deposits of the Wallapai District, Arizona." *Economic Geology* 44 (8): 663–705.

Thomas, B. E. 1953. *Geology of the Chloride Quadrangle, Arizona*. Geological Society of America Bulletin 64. Boulder, Colo.: Geological Society of America.

Thomas, L. A. 1966. "Geology of the San Manuel Ore Body." In *Geology of the Porphyry Copper Deposits, Southwestern North America*, edited by S. R. Titley and C. L. Hicks, 133–42. Tucson: University of Arizona Press.

Thomas, W. J., and R. B. Gibbs. 1983. "Famous Mineral Localities: The New Cornelia Mine, Ajo, Arizona." *Mineralogical Record* 14:283–98.

Thompson, J. R. 1980. "The Defiance Mine and Vicinity, Cochise County." *Mineralogical Record* 11:203–9.

Thompson, J. R. 1983. "Camp Verde Evaporites." *Mineralogical Record* 14:85–90.

Thompson, W. A. 1980. "Chrysocolla Pseudomorphs from Ray, Arizona." *Mineralogical Record* 11:248–50.

Thomssen, R. W. 1983. "The Minerals of the Malpais Hills, Pinal County, Arizona." *Mineralogical Record* 14:109–13.

Thorpe, D. G. 1980. "Mineralogy and Petrology of Precambrian Metavolcanic Rocks, Squaw Peak, Phoenix, Arizona." Master's thesis, Arizona State University.

Thorpe, D. G., and D. M. Burt. 1980. "A Unique Chloritoid-Staurolite Schist from near Squaw Peak, Arizona." *Arizona Geological Society Digest* 12:193–200.

Throop, A. H. 1970. "The Nature and Origin of Black Chrysocolla at the Inspiration Mine, Arizona." Master's thesis, Arizona State University.

Throop, A. H., and P. R. Buseck. 1971. "Nature and Origin of Black Chrysocolla at the Inspiration Mine, Arizona." *Economic Geology* 66:1168–75.

Tosdal, R. M., and J. L. Wooden. 2015. "Construction of the Jurassic Magmatic Arc, Southeast California and Southwest Arizona." In *Late Jurassic Margin of Laurasia: A Record of Faulting Accommodating Plate Rotation*, edited by T. H. Anderson, A. N. Didenko, C. L. Johnson, A. I. Khanchuk, and J. H. MacDonald Jr., 189–221. Geological Society of America Special Paper 513. Boulder, Colo.: Geological Society of America.

Trebisky, T. J., and S. B. Keith. 1975. "Descloizite from the C and B Vanadium Mine." *Mineralogical Record* 6:109.

Trischka, C., O. N. Rove, and D. M. Barringer Jr. 1929. "Boxwork Siderite." *Economic Geology* 24:677–86.

Tschernich, R. W. 1992. *Zeolites of the World*. Phoenix, Ariz.: Geoscience Press.

Tsuji, K. S. 1984. "Silver Mineralization of the El Tigre Mine and Volcanic Resurgence in the Chiricahua Mountains, Cochise County, Arizona." Master's thesis, University of Arizona.

Uchida, H., B. Lavina, R. T. Downs, and J. Chesley. 2005. "Single-Crystal X-ray Diffraction of Spinels from the San Carlos Volcanic Field, Arizona: Spinel as a Geothermometer." *American Mineralogist* 90:1900–1908.

Updike, R. G. 1977. "The Geology of the San Francisco Peaks, Arizona." PhD diss., Arizona State University.

U.S. Geological Survey (USGS). 1981. "Deposit ID #10048258, Silver Flake Mine." Reported December 1981. MRDS database. https://mrdata.usgs.gov/mrds.

U.S. Geological Survey (USGS). 1990. "Deposit ID #10048120, Nelly James Mine." MRDS database. https://mrdata.usgs.gov/mrds.

U.S. Geological Survey (USGS). 1995. "Deposit ID #10062369, Paul Hinshaw Property." MRDS database. https://mrdata.usgs.gov/mrds.

U.S. Geological Survey (USGS). 2005a. "Deposit ID #10067567, Bluebird Mine." MRDS database. https://mrdata.usgs.gov/mrds.

U.S. Geological Survey (USGS). 2005b. "Deposit ID #10233632, Drury." MRDS database. https://mrdata.usgs.gov/mrds.

Vajdak, J. 1995. "New Mineral Finds in 1995." *Mineral News* 11–12:1.

Vajdak, J. 2002. "New Mineral Finds in the Second Half of 2001." *Mineral News* 18 (1): 5–7.

Van Tassel, R. 1958. "On Carphosiderite." *Mineralogical Magazine* 31:818–19.

Vega, L. A. 1984. "The Alteration and Mineralization of the Alum Wash Prospect, Mohave County, Arizona." Master's thesis, University of Arizona.

Verbeek, E. R. 1995. "Activators in Fluorescent Minerals." In *Ultraviolet Light and Fluorescent Minerals*, edited by T. S. Warren, S. Gleason, R. C. Bostwick, and E. R. Verbeek, 135–201. Rio, W.Va.: Gem Guides.

Vlisidis, A. C., and W. T. Schaller. 1967. "The Formula of Shattuckite." *American Mineralogist* 52:782–86.

Von Bernewitz, M. W. 1937. *Occurrence and Treatment of Mercury Ore at Small Mines*. Information Circular 6966. Washington, D.C.: U.S. Department of the Interior, Bureau of Mines.

Walder, I. F., D. Carr, J. S. Lee, and A. Williamson. 2004. "Similarities in Hydrogeochemical Processes between Ore Deposition Formation and Acid Mine Drainage: Example near a Porphyry Copper Mine." In *Proceedings of the International Mine Water Association Symposium 2004*, 133–40. Newcastle upon Tyne: University of Newcastle upon Tyne on behalf of the International Mine Water Association.

Walenta, K. 1963. "Über die Barium-Uranylphosphatmineralen Uranocircit I, Uranocircit II, Meta-Uranocircit I, Meta-Uranocircit II von Menzenschwand im südlichen Schwarzwald." *Jahresh. Geol. Landesamts Baden-Wurttemb* 6:113–35.

Walker, L. W. 1957. "The Geodes of Kofa." *Nature* 50:266–67.

Wallace, T., ed. 2012. *Collecting Arizona*. Denver, Colo.: Lithographie.

Waller, H. E., Jr. 1960. "The Geology of the Paymaster and Olivette Mining Areas, Pima County, Arizona." Master's thesis, University of Arizona.

Walstrom, R. E. 2012. "The Accidental Pocket." Talk given at the Northern California Mineralogical Association Meeting, El Dorado, Calif.

Walstrom, R. E. 2013. "New Tellurium Mineral Localities Huachuca Mountains, Hartford District, Cochise County, Arizona." Talk given at the Arthur Roe Memorial Micromount Symposium, Tucson Gem and Mineral Show, February 14–17.

Wang, L., E. J. Essene, and Y. Zhang. 1999. "Mineral Inclusions in Pyrope Crystals from Garnet Ridge, Arizona, USA: Implications for Processes in the Upper Mantle." *Contributions to Mineralogy and Petrology* 135:164–78.

Wang, L., R. C. Rouse, E. J. Essene, D. R. Peacor, and Y. Zhang. 2000. "Carmichaelite, a New Hydroxyl-Bearing Titanite from Garnet Ridge, Arizona." *American Mineralogist* 85:792–800.

Ward, G. W. 1931. "A Chemical and Optical Study of the Black Tourmalines." *American Mineralogist* 16:145–90.

Wardwell, H. R. 1941. "Geology of the Potts Canyon Mining Area near Superior, Arizona." Master's thesis, University of Arizona.

Warner, L. A., W. T. Holser, V. R. Wilmarth, and E. N. Cameron. 1959. *Occurrence of Nonpegmatite Beryllium in the United States*. U.S. Geological Survey Professional Paper 318. Washington, D.C.: Government Printing Office.

Warren, C. H. 1903. "Native Arsenic from Arizona." *American Journal of Science* 16:337–40.

Warren, H. V., and R. W. Loofbourrow. 1932. "The Occurrence and Distribution of the Precious Metals in the Montana and Idaho Mines, Ruby, Arizona." *Economic Geology* 27:578–85.

Watson, K. D., and D. M. Morton. 1969. "Eclogite Inclusions in Kimberlite Pipes at Garnet Ridge, Northeastern Arizona." *American Mineralogist* 54:267–85.

Webber, B. N. 1929. "Marcasite in the Contact Metamorphic Ore Deposits of the Twin Buttes District, Pima County, Arizona." *Economic Geology* 24:304–10.

Weeks, A. D., M. E. Thompson, and A. M. Sherwood. 1954. "Navajoite, a New Vanadium Oxide from Arizona." *Science* 119:326.

Weeks, A. D., M. E. Thompson, and A. M. Sherwood. 1955. "Navajoite, a New Vanadium Oxide from Arizona." *American Mineralogist* 40:207–12.

Weight, H. O. 1949. "Geodes and Palms in the Kofa Country." *Desert Magazine* 12:17–22.

Weitz, J. H. 1942. *High Grade Dolomite Deposits in the United States*. U.S. Bureau of Mines Information Circular IC 7226. Washington, D.C.: U.S. Department of the Interior, Bureau of Mines.

Wells, H. L., and S. L. Penfield. 1885. "Gerhardtite and Artificial Cupric Nitrates." *American Journal of Science* 30:50–57.

Wells, R. C. 1937. *Analysis of Rocks and Minerals from the Laboratory of the USGS, 1914–36*. U.S. Geological Survey Bulletin 878. Washington, D.C.: Government Printing Office.

Wells, R. L., and A. J. Rambosek. 1954. *Uranium Occurrences in Wilson Creek Area, Gila County, Arizona*. U.S. Atomic Energy Comm. RME 2005 (rev.). Washington, D.C.: U.S. Department of Commerce, Office of Technical Services.

Welty, J. W., S. J. Reynolds, S. B. Keith, D. E. Gest, R. A. Trapp, and E. DeWitt. 1985a. *Mine Index for Metallic Mineral Districts of Arizona*. Arizona Bureau of Geology and Mineral Technology Bulletin 196. Tucson: Arizona Bureau of Geology and Mineral Technology.

Welty, J. W., J. E. Spencer, G. B. Allen, S. J. Reynolds, and R. A. Trapp. 1985b. *Geology and Production of Middle Tertiary Mineral Districts in Arizona*. Arizona Bureau of Geology and Mineral Technology Open-File Report 85-1. Tucson: Arizona Bureau of Geology and Mineral Technology.

Wenk, Hans-Rudolf, and Andrey Bulakh. 2016. *Minerals: Their Constitution and Origin*. Cambridge: Cambridge University Press.

Wenrich, K. J. 2013. "TGMS Selenite Collecting Trip to St. David, Arizona, Saturday, April 13, 2013, Site Geological Discussion and Photos." *Tucson Gem and Mineral Society Newsletter* 7 (5): 7–8.

Wenrich, K. J., and H. B. Sutphin. 1988. "Recognition of Breccia Pipes in Northern Arizona." *Arizona Bureau of Geology and Mineral Technology Fieldnotes* 18:1–5, 11.

Wenrich, K. J., and H. B. Sutphin. 1989. *Lithotectonic Setting Necessary for Formation of a Uranium-Rich, Solution-Collapse Breccia-Pipe, Grand Canyon Region, Arizona.* U.S. Geological Survey Open-File Rept. 89-0173. Denver, Colo.: U.S. Geological Survey.

Wenrich, K. J., and H. B. Sutphin. 1994. "Grand Canyon Caves, Breccia Pipes and Mineral Deposits." *Geology Today* 10 (3): 97–104.

Wenrich, K. J., E. R. Verbeek, H. B. Sutphin, P. J. Modreski, B. S. Van Gosen, and D. E. Detra. 1990. *Geology, Geochemistry, and Mineralogy of the Ridenour Mine Breccia Pipe, Arizona.* U.S. Geological Survey. Open-File Report 90-0504. Denver, Colo.: U.S. Geological Survey.

Werner, Bob, and Tony Nikischer. 2016. "The Scheelite Fluorescence Analyzer Card." *Mineral News: The Mineral Collector's Newsletter* 32 (5): 1.

Wherry, E. T. 1915. "Notes on Wolframite, Beraunite, and Axinite." *U.S. National Museum Proceedings* 47:501–11.

White, J. S., Jr. 1974. "What's New in Minerals?" *Mineralogical Record* 5:233–36.

White, J. S., Jr. 1992. "An Occurrence of Bixbyite, Spessartine, Topaz and Pseudobrookite from Ash Creek near Hayden, Arizona." *Mineralogical Record* 23:487–92.

Wilkinson, W. H., Jr., R. W. Allgood, G. M. Allgood, and C. Williams. 1983. "Ramsdellite from the Mistake Mine." *Mineralogical Record* 14:333–35.

Wilkinson, W. H., Jr., A. Roe, and C. Williams. 1980. "Some Unusual Secondary Minerals from the Mineral Park Mine, Mohave County, Arizona." *Mineralogical Record* 112:243–45.

Wilkinson, W. H., Jr., R. J. Stegen, W. Brack, and D. P. Cox. 2010. *A Field Visit to the New Cornelia Open-Pit Copper Mine, Ajo, Arizona.* Arizona Geological Society, Fall 2010 Field Trip. Tucson: Arizona Geological Society.

Williams, B. M. 2018. *Playing in a Big Boy's Sandbox: ASARCO's Mission Mining Complex.* N.p.: self-published.

Williams, H. 1936. "Pliocene Volcanoes of the Navajo-Hopi Country." *Geological Society of America Bulletin* 47:111–72.

Williams, S. A. 1959. "Minerals in a Pulaskite-Melatutvetite Dike, Gunsight Mountain, Pima County, Arizona." *Mineral Explorer* 3.

Williams, S. A. 1960. "A New Occurrence of Allanite in the Quijotoa Mountains, Pima County, Arizona." *Arizona Geological Society Digest* 3:46–51.

Williams, S. A. 1961. "Gerhardtite from the Daisy Shaft, Mineral Hill Mine, Pima County, Arizona." *Arizona Geological Society Digest* 4:123.

Williams, S. A. 1962. "The Mineralogy of the Mildren and Steppe Mining Districts, Pima County, Arizona." PhD diss., University of Arizona.

Williams, S. A. 1963. "Oxidation of Sulfide Ores in the Mildren and Steppe Mining Districts, Pima County, Arizona." *Economic Geology* 58:1119–25.

Williams, S. A. 1966. "The Significance of Habit and Morphology of Wulfenite." *American Mineralogist* 51:1212–17.

Williams, S. A. 1968. "Wickenburgite, a New Mineral from Arizona." *American Mineralogist* 53:1433–38.

Williams, S. A. 1970a. "Bideauxite, a New Arizona Mineral." *Mineralogical Magazine* 37:637–40.

Williams, S. A. 1970b. "Tilasite from Bisbee, Arizona." *Mineralogical Record* 1:68-69.

Williams, S. A. 1974. "Cesbronite, a New Copper Tellurite from Moctezuma, Sonora." *Mineralogical Magazine* 39:744–46.

Williams, S. A. 1976. "Junitoite, a New Hydrated Calcium Zinc Silicate from Christmas, Arizona." *American Mineralogist* 61:1255–58.

Williams, S. A. 1977. "Luetheite, $Cu_2Al_2(AsO_4)_2(OH)_4.H_2O$, a New Mineral from Arizona Compared with Chenevixite." *Mineralogical Magazine* 41:27–42.

Williams, S. A. 1978. "Khinite, Parakhinite, and Dugganite, Three New Tellurates from Tombstone, Arizona." *American Mineralogist* 63:1016–19.

Williams, S. A. 1979. "Girdite, Oboyerite, Fairbankite, and Winstanleyite, Four New Tellurium Minerals from Tombstone, Arizona." *Mineralogical Magazine* 43:453–57.

Williams, S. A. 1980a. "Schieffelinite, a New Lead Tellurate-Sulfate from Tombstone, Arizona." *Mineralogical Magazine* 43:771–73.

Williams, S. A. 1980b. "The Tombstone District, Cochise County, Arizona." *Mineralogical Record* 11:251–56.

Williams, S. A. 1981a. "Choloalite, $CuPb(TeO_3)_2.H_2O$, a New Mineral." *Mineralogical Magazine* 44:55–57.

Williams, S. A. 1981b. "Duhamelite, $Cu_4Pb_2Bi(VO_4)_4(OH)_3.8H_2O$, a New Arizona Mineral." *Mineralogical Magazine* 44:151–52.

Williams, S. A. 1982. "Luddenite, a New Copper-Lead Silicate Mineral from Arizona." *Mineralogical Magazine* 46:363–64.

Williams, S. A., and J. W. Anthony. 1970. "Hemihedrite, a New Mineral from Arizona." *American Mineralogist* 55:1088–1102.

Williams, S. A., and R. A. Bideaux. 1975. "Creaseyite, $Cu_2Pb_2(Fe,Al)_2Si_5O_{17}.6H_2O$, a New Mineral from Arizona and Sonora." *Mineralogical Magazine* 40:227–31.

Williams, S. A., and F. P. Cesbron. 1977. "Rutile and Apatite: Useful Prospecting Guides for Porphyry Copper Deposits." *Mineralogical Magazine* 41:288–92.

Williams, S. A., and F. P. Cesbron. 1995. "Wupatkiite from the Cameron Uranium District, Arizona, a New Member of the Halotrichite Group." *Mineralogical Magazine* 59:553–56.

Williams, S. A., and M. Duggan. 1977. "Famous Mineral Localities: The Glove Mine." *Mineralogical Magazine* 41:429–32.

Williams, S. A., and M. Duggan. 1978. "Tlapallite, a New Mineral from Moctezuma, Sonora, Mexico." *Mineralogical Magazine* 42:183–86.

Williams, S. A., and M. Duggan. 1980. "La macquartite: Un nouveau silico-chromate de Tiger, Arizona." *Bulletin de Minéralogie* 103:530–32.

Williams, S. A., and B. Khin. 1971. "Chalcoalumite from Bisbee, Arizona." *Mineralogical Record* 2:126–27.

Williams, S. A., and P. Matter III. 1975. "Graemite, a New Bisbee Mineral." *Mineralogical Record* 6:32–34.

Williams, S. A., W. J. McLean, and J. W. Anthony. 1970. "A Study of Phoenicochroite—Its Structure and Properties." *American Mineralogist* 55:784–92.

Willis, C. W. 1915. *Directory of Arizona Minerals, 1915–16.* Arizona Bureau of Mines Bulletin 3. Tucson: University of Arizona.

Wilson, E. D. 1927a. *Arizona Gold Placers.* 2nd ed. Arizona Bureau of Mines Bulletin 124. Tucson: University of Arizona.

Wilson, E. D. 1927b. *The Geology and Ore Deposits of the Courtland-Gleeson Region, Arizona.* Arizona Bureau of Mines Bulletin 123. Tucson: University of Arizona.

Wilson, E. D. 1928. *Asbestos Deposits of Arizona.* Arizona Bureau of Mines Bulletin 126. Tucson: University of Arizona.

Wilson, E. D. 1929. "An Occurrence of Dumortierite near Quartzsite, Arizona." *American Mineralogist* 14:373–81.

Wilson, E. D. 1933. *Geology and Mineral Deposits of Southern Yuma County, Arizona.* Arizona Bureau of Mines Bulletin 134. Tucson: University of Arizona.

Wilson, E. D. 1941. *Tungsten Deposits of Arizona.* Arizona Bureau of Mines Bulletin 148. Tucson: University of Arizona.

Wilson, E. D. 1944. *Arizona Nonmetallics: A Summary of Past Production and Present Operations.* Arizona Bureau of Mines Bulletin 152. Tucson: University of Arizona.

Wilson, E. D. 1961. "Arizona Gold Placers." Part 1 of *Gold Placers and Placering in Arizona.* Arizona Bureau of Mines Bulletin 168. Tucson: University of Arizona.

Wilson, E. D. 1969. *Mineral Deposits of the Gila River Indian Reservation, Arizona.* Arizona Bureau of Mines Bulletin 179. Tucson: University of Arizona.

Wilson, E. D., and G. M. Butler. 1930. *Manganese Ore Deposits in Arizona.* Arizona Bureau of Mines Bulletin 127. Tucson: University of Arizona.

Wilson, E. D., J. B. Cunningham, and G. M. Butler. 1934. *Arizona Lode Gold Mines and Gold Mining.* Arizona Bureau of Mines Bulletin 137. Tucson: University of Arizona.

Wilson, E. D., G. R. Fansett, C. H. Johnson, and G. H. Roseveare. 1961. *Gold Placers and Placering in Arizona.* Arizona Bureau of Mines Bulletin 168. Tucson: University of Arizona.

Wilson, E. D., F. W. Galbraith, W. B. Loring, G. M. Fowler, T. H. Kuhn, and G. A. Kiersch. 1951. *Arizona Zinc and Lead Deposits, Part II.* Arizona Bureau of Mines Bulletin 158. Tucson: University of Arizona.

Wilson, E. D., W. G. Hogue, J. R. Cooper, S. C. Creasey, N. P. Peterson, C. A. Anderson, and M. G. Dings. 1950. *Arizona Zinc and Lead Deposits, Part I.* Arizona Bureau of Mines Bulletin 156. Tucson: University of Arizona.

Wilson, E. D., and G. H. Roseveare. 1949. *Arizona Nonmetallics: A Summary of Past Production and Present Operations.* 2nd ed. Arizona Bureau of Mines Bulletin 155. Tucson: University of Arizona.

Wilson, J. R. 1977. "Geology, Alteration, and Mineralization of the Korn Kob Mine Area, Pima County, Arizona." Master's thesis, University of Arizona.

Wilson, K. S. 1984. "The Geology and Epithermal Silver Mineralization of the Reymert Mine, Pinal County, Arizona." Master's thesis, University of Arizona.

Wilson, W. E. 1971. "Classic Locality: The Apache Mine." *Mineralogical Record* 2:252–58.

Wilson, W. E. 1977. "What's New in Minerals? Ray Chrysocolla." *Mineralogical Record* 8:58.

Wilson, W. E. 1985. "What's New in Minerals?" *Mineralogical Record* 16:497–500.

Wilson, W. E. 1986. "What's New in Minerals?" *Mineralogical Record* 17:210, 404–5.

Wilson, W. E. 1987a. "What's New in Gold?" *Mineralogical Record* 18:89.

Wilson, W. E. 1987b. "What's New in Minerals? The 79 Mine Aurichalcite Story." *Mineralogical Record* 18:299–301.

Wilson, W. E. 1988. "The Iron Cap Mine." *Mineralogical Record* 19:81–87.

Wilson, W. E. 2008. "The American Mineral Treasures Exhibition, Tucson Gem and Mineral Show 2008." *Mineralogical Record* 39:180–81.

Wilson, W. E. 2015. "The Total Wreck Mine, Pima County, Arizona." *Mineralogical Record* 46:485–503.

Wilson, W. E. 2020a. "The Ford Mine, Pinal County, Arizona." *Mineralogical Record* 51:169–80.

Wilson, W. E. 2020b. "The Rowley Mine, Painted Rock Mountains, Maricopa County, Arizona." *Mineralogical Record* 51:181–226.

Wilson, W. E., and R. A. Bideaux. 1983. "Famous Mineral Localities: The Glove Mine." *Mineralogical Record* 14:299–306.

Wilson, W. E., and G. Godas. 1996. "The North Geronimo Mine, La Paz County, Arizona." *Mineralogical Record* 27:363–72.

Wilson, W. E., and M. Hay. 2015. "The Flux Mine, Santa Cruz County, Arizona." *Mineralogical Record* 46:531–53.

Wilson, W. E., and E. Jones. 2015. "The Carlota Mine, Gila County, Arizona." *Mineralogical Record* 46:555–63.

Wilson, W. E., and D. K. Miller. 1974. "Minerals of the Rowley Mine." *Mineralogical Record* 5:10–30.

Wilt, J. C. 1993. "Geochemical Patterns of Hydrothermal Mineral Deposits Associated with Calc-Alkalic and Alkali-Calcic Igneous Rocks as Evaluated with Neural Networks." PhD diss., University of Arizona.

Wilt, J. C. 1995. "Correspondence of Alkalinity and Ferric/Ferrous Ratios of Igneous Rocks Associated with Various Types of Porphyry Copper Deposits." In *Porphyry Copper Deposits of the American Cordillera*, edited by F. W. Pierce, and J. G. Bolm, 180–200. Arizona Geological Society Digest 20. Tucson: Arizona Geological Society.

Wilt, J. C., and Stanley B. Keith. 1980. "Molybdenum in Arizona." *Fieldnotes from the State of Arizona, Bureau of Geology and Mineral Technology* 10 (3): 1–3, 7–9, 12.

Winant, A. R. 2010. "Sericitic and Advanced Argillic Mineral Assemblages and Their Relationship to Copper Mineralization, Resolution Porphyry Cu-(Mo) Deposit, Superior District, Pinal County, Arizona." Master's thesis, University of Arizona.

Winchell, R. E., and H. E. Wenden. 1968. "Synthesis and Study of Diaboleite." *Mineralogical Magazine* 36:933–39.

Wise, W. S., and R. W. Tschernich. 1975. "Cowlesite, a New Ca-zeolite." *American Mineralogist* 60:951–56.

Wise, W. S., and R. W. Tschernich. 1976. "The Chemical Compositions and Origin of the Zeolites Offretite, Erionite, and Levyne." *American Mineralogist* 61:853–63.

Witkind, I. J. 1961. *The Uranium-Vanadium Ore Deposit at the Monument No. 1–Mitten No. 2 Mine, Monument Valley, Navajo County, Arizona.* U.S. Geological Survey Bulletin 1107-C: 219–42. Washington, D.C.: Government Printing Office.

Witkind, I. J., and R. E. Thaden. 1963. *Geology and Uranium-Vanadium Deposits of the Monument Valley Area, Apache and Navajo Counties, Arizona.* U.S. Geological Survey Bulletin 1103. Washington, D.C.: Government Printing Office.

Wittke, J. H., G. B. Haxel, G. S. Epstein, and C. E. Jacobson. 2016. "Serpentinite-Hosted Nickel-, Iron-, and Cobalt-Sulfide, Arsenide, and Intermetallic Minerals in an Unusual Inland Tectonic Setting, Southwest Arizona." *Geological Society of America, Cordilleran Section, 112th Annual meeting, Abstracts with Programs* 48 (4): paper no. 25-2.

Wittke, J. H., and R. F. Holm. 1996. "The Association Basanitic Nephelinite: Feldspar Ijolite-Nepheline Monzosyenite at House Mountain Volcano, North-Central Arizona." *Canadian Mineralogist* 34:221–40.

Wood, M. M. 1963. "Metamorphic Effects of the Leatherwood Quartz Diorite, Santa Catalina Mountains, Pima County, Arizona." Master's thesis, University of Arizona.

Wood, M. M. 1970. "The Crystal Structure of Ransomite." *American Mineralogist* 55:729–34.

Wood, M. M., W. J. McLean, and R. B. Laughon. 1974. "The Crystal Structure and Composition of Yedlinite." *American Mineralogist* 59:1160–65.

Woodbridge, D. E. 1906. "Arizona and Sonora V: The Globe District." *Engineering and Mining Journal* 81:1229.

Wright, R. J. 1955. "Ore Controls in Sandstone Uranium Deposits of the Colorado Plateau." *Economic Geology* 50:135–55.

Wrucke, C. T. 1961. "Paleozoic and Cenozoic rocks in the Alpine-Nutrioso area, Apache County, Arizona." *U.S. Geological Survey Bulletin* 1121-H:H1–H26.

Wrucke, C. T., C. S. Bromfield, F. S. Simons, R. C. Greene, B. B. Houser, R. J. Miller, and F. Gray. 2004. *Geologic Map of the San Carlos Indian Reservation, Arizona*. U.S. Geological Survey, Geologic Investigations Series I-2780. Reston, Va.: U.S. Geological Survey.

Wrucke, C. T., S. P. Marsh, C. M. Conway, C. E. Ellis, D. M. Kulik, C. K. Moss, and G. L. Raines. 1983. *Mineral Resource Potential of the Mazatzal Wilderness and Contiguous Roadless Area, Gila, Maricopa, and Yavapai Counties, Arizona*. U.S. Geological Survey Misc. Field Studies MF-1573A. Reston, Va.: U.S. Geological Survey.

Yagoda, H. 1945. "The Localization of Copper and Silver Sulfide Minerals in Polished Sections by the Potassium Cyanide Etch Pattern." *American Mineralogist* 30: 51–64.

Yang, H., M. B. Andrade, R. T. Downs, R. G. Gibbs, and R. A. Jenkins. 2016. "Raygrantite, $Pb_{10}Zn(SO_4)_6(SiO_4)_2(OH)_2$, a New Mineral Isostructural with Iranite, from the Big Horn Mountains, Maricopa County, Arizona, USA." *Canadian Mineralogist* 54: 625–34.

Yang, H., and R. T. Downs. 2018. "Eddavidite." IMA 2018-010, CNMNC Newsletter No. 44. *Mineralogical Magazine* 82:1015–21.

Yang, H., R. T. Downs, S. H. Evans, R. A. Jenkins, and E. M. Bloch. 2010. "Rongibbsite, $Pb_2(Si_4Al)O_{11}(OH)$, a New Zeolitic Aluminosilicate Mineral with an Interrupted Framework from Maricopa County, Arizona, USA." *American Mineralogists* 98: 236–41.

Yang, H., R. B. Gibbs, S. H. Evans, R. T. Downs, and Z. Jibrin. 2018. "Alterite, IMA 2018-070." CNMNC Newsletter No. 45, October 2018, *Mineralogical Magazine* 82: 1232.

Yang, H., R. B. Gibbs, S. H. Evans, R. T. Downs, and Z. Jibrin. 2020. "Magnesioalterite, IMA 2020-050." CNMNC Newsletter No. 57, *Mineralogical Magazine* 84:791–94.

Yang, H., R. B. Gibbs, X. Gu, S. H. Evans, R. T. Downs, and Z. Jabrin. 2019. "Lazaraskeite." IMA 2018-137, CNMNC Newsletter No. 48. *European Journal of Mineralogy* 31:399–402.

Yang, H., R. A. Jenkins, R. M. Thompson, R. T. Downs, S. H. Evans, and E. M. Bloch. 2012. "Markascherite, $Cu_3(MoO_4)(OH)_4$, a New Mineral Species Polymorphic with Szenicsite, from Copper Creek, Pinal County, Arizona, U.S.A." *American Mineralogist* 97:197–202.

Young, E. J., and P. K. Sims. 1961. *Petrography and Origin of Xenotime and Monazite Concentrations, Central City District, Colorado.* U.S. Geological Survey Bulletin 1032-F: 273–97. Washington, D.C.: Government Printing Office.

Young, E. J., A. D. Weeks, and R. Meyrowitz. 1966. "Coconinoite, a New Uranium Mineral from Utah and Arizona." *American Mineralogist* 51:651–63.

Young, R. G., 1964. "Distribution of Uranium Deposits in the White Canyon— Monument Valley District, Utah-Arizona." *Economic Geology* 59:850–973.

PHOTOGRAPHY CREDITS

Chapter 1. Gemstones and Lapidary Materials of Arizona

Figure 1.1: Azurite-cementing breccia, San Xavier North Mine, Pima County, polished butterfly, 5.6 cm high, photo by Jeff Scovil, Bill Williams specimen.

Figure 1.2: Gem chrysocolla, Ray Mine, Pinal County, 14.3 ct faceted stone in ring with rough chrysocolla, photo by Sean Milliner, Somewhere Over the Rainbow specimens.

Figure 1.3: Pyrope, Navajo Nation, Apache County, 13.9 ct faceted stone with rough pyrope, photo by Sean Milliner, Somewhere Over the Rainbow specimens.

Figure 1.4: Olivine variety peridot, Peridot Mesa, San Carlos Apache Reservation, Gila County, 15 ct faceted stone, photo by Jeff Scovil, Paul Harter specimen.

Figure 1.5: Opal, Ruby, Santa Cruz County; largest cabochon is 27 × 33 mm, with slab of rough opal, photo by Jeff Scovil, Southern Skies Opal specimens.

Figure 1.6: Quartz variety amethyst, Four Peaks, Maricopa County, 80 ct faceted stone with rough amethyst, photo by Jeff Scovil, Somewhere Over the Rainbow specimens.

Figure 1.7: Quartz variety agate, Fourth of July Butte, Maricopa County, cut and polished nodule, 5.5 cm high, photo by Jeff Scovil, David Bunk specimen.

Figure 1.8: Fire agate, San Carlos Apache Reservation, Gila County, 17 ct polished stone, photo by Jeff Scovil, Paul Harter specimen.

Figure 1.9: Quartz variety petrified wood, Holbrook, Navajo County, polished slab 49 cm in diameter, photo by Jeff Scovil, Tellus Museum specimen.

Figure 1.10: Shattuckite, New Cornelia Mine, Pima County, 30 × 40 mm cabochon and rough shattuckite, photo by Ron Gibbs, Ron Gibbs specimens.

Figure 1.11: Turquoise cabochons: Sleeping Beauty Mine, Miami, Gila County, 31 mm high; Kingman, Mohave County, 45 mm high; and Bisbee, Cochise County, 39 mm high, photo by Wolfgang Mueller, Wolfgang Mueller specimens.

Figure 1.12: Campbellite cabochons, Bisbee, Cochise County, 25 × 35 mm and 30 × 40 mm, photo by Doug Graeme, Graeme Family Collection specimens.

Chapter 2. Fluorescent Minerals in Arizona

Figure 2.1a: Andersonite on gypsum, under shortwave UV, Hillside Mine, Yavapai County, 3.2 × 1.9 cm, photo by Harvey Jong, Harvey Jong specimen.

Figure 2.1b: Andersonite on gypsum, under white light, Hillside Mine, Yavapai County, 3.2 × 1.9 cm, photo by Harvey Jong, Harvey Jong specimen.

Figure 2.2a: Cerussite, under longwave UV, Grand Reef Mine, Graham County, 3 cm FOV, photo by Harvey Jong, Harvey Jong specimen.

Figure 2.2b: Cerussite, under white light, Grand Reef Mine, Graham County, 3 cm FOV, photo by Harvey Jong, Harvey Jong specimen.

Figure 2.3a: Powellite, under shortwave UV, Flying Saucer Group, Vulture Mountains, Maricopa County, 9.6 × 7.5 cm, photo by Harvey Jong, Harvey Jong specimen.

Figure 2.3b: Powellite, under white light, Flying Saucer Group, Vulture Mountains, Maricopa County, 9.6 × 7.5 cm, photo by Harvey Jong, Harvey Jong specimen.

Figure 2.4a: Scheelite, under shortwave UV, Johnson Camp, Cochise County, 2.5 cm, photo by Stan Celestian, Mardy Zimmermann specimen.

Figure 2.4b: Scheelite, under white light, Johnson Camp, Cochise County, 2.5 cm, photo by Stan Celestian, Mardy Zimmermann specimen.

Figure 2.5a: Calcite, Coals of Fire, under shortwave UV, Patagonia, Santa Cruz County, 15.0 × 8.5 cm, photo by Harvey Jong, Mardy Zimmermann specimen.

Figure 2.5b: Calcite, Coals of Fire, under white light, Patagonia, Santa Cruz County, 15.0 × 8.5 cm, photo by Harvey Jong, Mardy Zimmermann specimen.

Figure 2.6a: Calomel, under longwave UV, Saddle Mountain Mercury Mine, Maricopa County, 3.5 × 2.2 cm, photo by Harvey Jong, Harvey Jong specimen.

Figure 2.6b: Calomel, under shortwave UV, Saddle Mountain Mercury Mine, Maricopa County, 3.5 × 2.2 cm, photo by Harvey Jong, Harvey Jong specimen.

Figure 2.6c: Calomel, under white light, Saddle Mountain Mercury Mine, Maricopa County, 3.5 × 2.2 cm, photo by Harvey Jong, Harvey Jong specimen.

Figure 2.7a: Eucryptite, under shortwave UV, Midnight Owl Mine, Yavapai County, 8 × 5 cm, photo by George V. Polman, George V. Polman specimen.

Figure 2.7b: Eucryptite, under white light, Midnight Owl Mine, Yavapai County, 8 × 5 cm, photo by George V. Polman, George V. Polman specimen.

Figure 2.8a Fluorite, under longwave UV, Donna Anna workings, Dragoon Mountains, Cochise County, 6 cm FOV, photo by Harvey Jong, Harvey Jong specimen.

Figure 2.8b Fluorite, under shortwave UV, Donna Anna workings, Dragoon Mountains, Cochise County, 6 cm FOV, photo by Harvey Jong, Harvey Jong specimen.

Figure 2.8c Fluorite, under white light, Donna Anna workings, Dragoon Mountains, Cochise County, 6 cm FOV, photo by Harvey Jong, Harvey Jong specimen.

Figure 2.9a: Chalcedony epimorph of calcite in geode, under shortwave UV, Potts Canyon, Superior, Pinal County, 4.8 cm, photo by Stan Celestian, Stan Celestian specimen.

Figure 2.9b: Chalcedony epimorph of calcite in geode, under white light, Potts Canyon, Superior, Pinal County, 4.8 cm, photo by Stan Celestian, Stan Celestian specimen.

Figure 2.10a: Sphalerite, under longwave UV, Copper Queen Mine, Bisbee, Cochise County, 9 × 9 cm, photo by George V. Polman, George V. Polman specimen.

Figure 2.10b: Sphalerite, under white light, Copper Queen Mine, Bisbee, Cochise County, 9 × 9 cm, photo by George V. Polman, George V. Polman specimen.

Figure 2.10c: Sphalerite, showing triboluminescence, Campbell Shaft, Bisbee, Cochise County, 6.7 × 4.2 cm, photo by Harvey Jong, Raymond Grant specimen.

Figure 2.11a: Willemite, under shortwave UV, Nelly James Mine, Miller Canyon, Cochise County, 10 × 10 cm, photo by George V. Polman, George V. Polman specimen.

Figure 2.11b: Willemite, under white light, Nelly James Mine, Miller Canyon, Cochise County, 10 × 10 cm, photo by George V. Polman, George V. Polman specimen.

Figure 2.12a: Wickenburgite, under shortwave UV, Potter-Cramer Mine, Maricopa County, 11 × 8 cm, photo by George V. Polman, George V. Polman specimen.

Figure 2.12b: Wickenburgite, under white light, Potter-Cramer Mine, Maricopa County, 11 × 8 cm, photo by George V. Polman, George V. Polman specimen.

Figure 2.13a: Zunyite, under shortwave UV, Big Bertha Mine, Dome Rock Mountains, La Paz County, 4.4 × 4.7 cm, photo by Harvey Jong, Harvey Jong specimen.

Figure 2.13b: Zunyite, under white light, Big Bertha Mine, Dome Rock Mountains, La Paz County, 4.4 × 4.7 cm, photo by Harvey Jong, Harvey Jong specimen.

Figure 2.14a: Two-color fluorescence with intrinsic minerals, scheelite and powellite, under shortwave UV, Three Musketeers Mine, Granite Wash Mountains, La Paz County, 13.0 × 9.8 cm, photo by Harvey Jong, Mardy Zimmermann specimen.

Figure 2.14b: Two-color fluorescence with intrinsic minerals, scheelite and powellite, under white light, Three Musketeers Mine, Granite Wash Mountains, La Paz County, 13.0 × 9.8 cm, photo by Harvey Jong, Mardy Zimmermann specimen.

Figure 2.15a: Two-color fluorescence with impurity minerals, calcite and willemite, under shortwave UV, Case Grande, Pinal County, 7.6 × 7.6 cm, photo by George V. Polman, George V. Polman specimen.

Figure 2.15b: Two-color fluorescence with impurity minerals, calcite and willemite, under white light, Case Grande, Pinal County, 7.6 × 7.6 cm, photo by George V. Polman, George V. Polman specimen.

Figure 2.16a: Two-color fluorescence with intrinsic and impurity minerals, scheelite and calcite, under shortwave UV, Campo Bonito Mine, Santa Catalina Mountains, Pinal County, 15 × 14 cm, photo by George V. Polman, George V. Polman specimen.

Figure 2.16b: Two-color fluorescence with intrinsic and impurity minerals, scheelite and calcite, under white light, Campo Bonito Mine, Santa Catalina Mountains, Pinal County, 15 × 14 cm, photo by George V. Polman, George V. Polman specimen.

Figure 2.17a: Three-color fluorescence, calcite, fluorite, willemite, under shortwave UV, Black Rock Mine, La Paz County, 10 × 8 cm, Steve Hutchcraft photo—CC-BY-SA-3.0, via NaturesRainbows.com, Steve Hutchcraft specimen.

Figure 2.17b: Three-color fluorescence, calcite, fluorite, willemite, under white light, Black Rock Mine, La Paz County, 10 × 8 cm, Steve Hutchcraft photo—CC-BY-SA-3.0, via NaturesRainbows.com, Steve Hutchcraft specimen.

Figure 2.18a: Four-color fluorescence, fluorite, calcite, smithsonite, willemite, under midrange plus shortwave UV, Hogan Claim, Yavapai County, 17.8 cm wide, Eric Dawley photo—CC-BY-SA-3.0, via NaturesRainbows.com, Eric Dawley specimen.

Figure 2.18b: Four-color fluorescence, fluorite, calcite, smithsonite, willemite, under white light, Hogan Claim, Yavapai County, 17.8 cm wide, Eric Dawley photo—CC-BY-SA-3.0, via NaturesRainbows.com, Eric Dawley specimen.

Figure 2.19a: Five-color fluorescence, calcite, willemite, hydrozincite, sphalerite, smithsonite, under shortwave plus longwave UV, Miller Canyon, Cochise County, 10.8 × 8.9 cm, Eric Dawley photo—CC-BY-SA-3.0, via NaturesRainbows.com, Eric Dawley specimen.

Figure 2.19b: Five-color fluorescence, calcite, willemite, hydrozincite, sphalerite, smithsonite, under white light, Miller Canyon, Cochise County, 10.8 × 8.9 cm, Eric Dawley photo—CC-BY-SA-3.0, via NaturesRainbows.com, Eric Dawley specimen.

Chapter 3. Arizona Mineral Occurrences

Figure 3.1: Acanthite, Silver King Mine, Pinal County, 4.7 cm high, photo by Jeff Scovil, Les Presmyk specimen.

Figure 3.2: Adamite, Saginaw Hill, Pima County, FOV 2 mm, photo by Ron Gibbs, Bruce Murphy specimen.

Figure 3.3: Adanite, Tombstone, Cochise County, FOV 3.3 mm, R190033, RRUFF photo and specimen.

Figure 3.4: Agardite-(Nd), Narragansett Mine, Pima County, FOV 1.5 mm, photo by Ron Gibbs, Ron Gibbs specimen.

Figure 3.5: Ajoite, Magna Mine, Pinal County, FOV 4 mm, photo by Ron Gibbs, Ron Gibbs specimen.

Figure 3.6: Alamosite, Evening Star Mine, Maricopa County, FOV 3 mm, photo by Ron Gibbs, Joe Ruiz specimen.

Figure 3.7: Allantoin, Rowley Mine, Maricopa County, FOV 0.7 mm, photo by Tony Kampf, Natural History Museum of Los Angeles County specimen.

Figure 3.8: Alterite, near Cliff Dwellers Lodge, Coconino County, FOV 1.2 mm, R180005, RRUFF photo and specimen.

Figure 3.9: Alunogen, near Cliff Dwellers Lodge, Coconino County, FOV 3 mm, photo by Ron Gibbs, Ron Gibbs specimen.

Figure 3.10: Ammineite with mimetite, Rowley Mine, Maricopa County, FOV 3 mm, photo my Ron Gibbs, Ron Gibbs specimen.

Figure 3.11: Analcime, near Santa Maria River, Yavapai County, FOV 3 mm, photo by Ron Gibbs, UAX7129, University of Arizona Alfie Norville Gem and Mineral Museum specimen.

Figure 3.12: Anatase, Bagdad Mine, Yavapai County, FOV 2 mm, photo by Ron Gibbs, Ron Gibbs specimen.

Figure 3.13: Andradite, Grey Throne prospect, Graham County, 7.6 cm, photo by Jeff Scovil, Mark Hay specimen.

Figure 3.14: Anglesite, Flux Mine, Santa Cruz County, 5.2 cm, photo by Jeff Scovil, Mark Hay specimen.

Figure 3.15: Annite, near Mammoth, Pinal County, FOV 3 mm, photo by Ron Gibbs, Ron Gibbs specimen.

Figure 3.16: Antipinite, dark blue, Rowley Mine, Maricopa County, FOV 0.63 mm, photo by Tony Kampf, Natural History Museum of Los Angeles County specimen.

Figure 3.17: Antlerite, Antler Mine, Yavapai County, 2.5 cm high, photo by John Callahan, John Callahan specimen.

Figure 3.18: Apachite, Christmas Mine, Gila County, FOV 3 mm, photo by Ron Gibbs, Ron Gibbs specimen.

Figure 3.19: Apatite, Magna Mine, Copper Creek, Pinal County, FOV 2 mm, photo by Ron Gibbs, Ron Gibbs specimen.

Figure 3.20: Apophyllite, Christmas Mine, Gila County, FOV 2 mm, photo by Ron Gibbs, Ron Gibbs specimen.

Figure 3.21: Aragonite, Southwest Mine, Bisbee, Cochise County, 8 cm high, photo by Stephan Koch, Stephan Koch specimen.

Figure 3.22: Aravaipaite, Grand Reef Mine, Graham County, R100004, RRUFF photo and specimen.

Figure 3.23: Arsendescloizite, Tonopah-Belmont Mine, Maricopa County, FOV 0.4 mm, photo by Jerry Cone, Ron Gibbs specimen.

Figure 3.24: Arsenic, Double Standard Mine, Santa Cruz County, 7 cm, photo by Jeff Scovil, Mark Hay specimen.

Figure 3.25: Atacamite, Southwest Mine, Bisbee, Cochise County, 8.5 cm, photo by Jeff Scovil, Graeme Family Collection specimen.

Figure 3.26: Aurichalcite, 79 Mine, Gila County, 5.9 cm, photo by Jeff Scovil, Claude Yoder specimen.

Figure 3.27: Axinite-(Mn), Iron Cap Mine, Graham County, 4 × 4.5 cm, photo by Stephan Koch, Stephan Koch specimen.

Figure 3.28: Azurite, Bisbee, Cochise County, 8.4 cm, photo by Jeff Scovil, Paul Harter specimen.

Figure 3.29: Babingtonite, Iron Cap Mine, Graham County, R060093, RRUFF photo and specimen.

Figure 3.30: Bairdite, Empire Mine, Tombstone, Cochise County, FOV 3 mm, photo by Ron Gibbs, Michael Shannon specimen.

Figure 3.31: Baryte, San Manuel Mine, Pinal County, 5.4 cm, photo by Jeff Scovil, Mark Hay specimen.

Figure 3.32: Beaverite-(Cu), 79 Mine, Gila County, FOV 2 mm, photo by Ron Gibbs, Ron Gibbs specimen.

Figure 3.33: Bermanite, 7U7 Ranch, Yavapai County, FOV 3 mm, photo by Ron Gibbs, Jim McGlasson specimen.

Figure 3.34: Beryl, Palo Verde claim, Pima County, 6.1 cm high, photo by Jeff Scovil, Barbara Muntyan specimen.

Figure 3.35: Beyerite, Copperopolis, Yavapai County, FOV 3 mm, photo by Ron Gibbs, Ron Gibbs specimen.

Figure 3.36: Bobmeyerite, Mammoth–St. Anthony Mine, Pinal County, FOV 2 mm, photo by Ron Gibbs, Robert Jenkins specimen.

Figure 3.37: Boleite, Rowley Mine, Maricopa County, FOV 0.7 mm, photo by Jerry Cone, Ron Gibbs specimen.

Figure 3.38: Brewsterite-Ba, Trigo Mountains, La Paz County, FOV 7 mm, photo by Ron Gibbs, Ron Gibbs specimen.

Figure 3.39: Brochantite, Bisbee, Cochise County, 2 cm, photo by Jeff Scovil, Les Presmyk specimen.

Figure 3.40: Bromargyrite, Empire Mine, Tombstone, Cochise County, 6 × 3 cm, photo by Stephan Koch, Stephan Koch specimen.

Figure 3.41: Brookite, Bagdad Mine, Yavapai County, FOV 2 mm, photo by Ron Gibbs, Ron Gibbs specimen.

Figure 3.42: Butlerite, United Verde Mine, Yavapai County, R110160, RRUFF photo and specimen.

Figure 3.43: Calcite, Twin Buttes Mine, Pima County, 4.2 × 3.0 cm, photo by Kenny Don, John Callahan specimen.

Figure 3.44: Caledonite, Mammoth–St. Anthony Mine, Pinal County, 1 cm crystals, photo by Jeff Scovil, UA9675, University of Arizona Alfie Norville Gem and Mineral Museum specimen.

Figure 3.45: Calomel, Saddle Mountain Mine, Maricopa County, FOV 5 mm, photo by Ron Gibbs, Ron Gibbs specimen.

Figure 3.46: Carlosbarbosaite, Belmont Pit, Maricopa County, FOV 3 mm, photo by Ron Gibbs, Ron Gibbs specimen.

Figure 3.47: Carnotite, Anderson Mine, Yavapai County, FOV 3 mm, photo by Ron Gibbs, Ron Gibbs specimen.

Figure 3.48: Cassiterite, Wood Tin, Apache Tin claims, Graham County, 1.8 cm, photo by Ron Gibbs, UA12675, University of Arizona Alfie Norville Gem and Mineral Museum specimen.

Figure 3.49: Celadonite, Malpais Hill, Pinal County, FOV 6 mm, photo by Ron Gibbs, Ron Gibbs specimen.

Figure 3.50: Cerussite, Flux Mine, Santa Cruz County, 12.6 cm, photo by Jeff Scovil, Dick Morris specimen.

Figure 3.51: Chabazite-Ca, Well No. 1, Ajo, Pima County, FOV 3 mm, photo by Ron Gibbs, Ron Gibbs specimen.

Figure 3.52: Chalcanthite, Planet Mine, La Paz County, 10.2 cm, photo by Jeff Scovil, Conan Barker specimen.

Figure 3.53: Chalcoalumite, Bisbee, Cochise County, FOV 5 mm, photo by Ron Gibbs, Robert Jenkins specimen.

Figure 3.54: Chalcocite, Magma Mine, Pinal County, 2 mm crystals, photo by Ron Gibbs, Ron Gibbs specimen.

Figure 3.55: Chalcophanite, Cole shaft, Bisbee, Cochise County 1.3 cm, photo by Stan Celestian, Natural History Museum of Los Angeles County specimen.

Figure 3.56: Chalcopyrite, Holland Mine, Santa Cruz County, 8 cm, photo by Jeff Scovil, Mark Hay specimen.

Figure 3.57: Chalcosiderite, Silver Bell Mine, Pima County, FOV 7 mm, photo by Ron Gibbs, Ron Gibbs specimen.

Figure 3.58: Chenevixite, near Patagonia, Santa Cruz County, FOV 5 mm, photo by Ron Gibbs, Ron Gibbs specimen.

Figure 3.59: Chrysocolla, overgrown with quartz, Ray Mine, Pinal County, 3.2 × 2.5 cm, photo by Stephan Koch, Stephan Koch specimen.

Figure 3.60: Chrysotile, Sierra Anchas, Gila County, 12.8 cm high, photo by Jeff Scovil, Flagg Mineral Foundation specimen.

Figure 3.61: Cinnabar, San Juan Mine, Cochise County, 3.5 × 2.2 cm, photo by John Callahan, John Callahan specimen.

Figure 3.62: Clinoclase, Copper Queen Mine, Yavapai County, FOV 3 mm, photo by Ron Gibbs, Ron Gibbs specimen.

Figure 3.63: Clinohedrite, Christmas Mine, Gila County, FOV 3 mm, photo by Ron Gibbs, Robert Jenkins specimen.

Figure 3.64: Clinozoisite, Finch Mine, Gila County, FOV 3 mm, photo by Ron Gibbs, Ron Gibbs specimen.

Figure 3.65: Coconinoite, Sun Valley Mine, Coconino County, 2.8 × 1.0 cm, photo by John Callahan, John Callahan specimen.

Figure 3.66: Conichalcite, Bagdad Mine, Yavapai County, 11.1 cm, photo by Jeff Scovil, Evan Jones specimen.

Figure 3.67: Connellite, Bagdad Mine, Yavapai County, 3.5 × 5.0 cm, photo by Jeff Scovil, Robert Jenkins specimen.

Figure 3.68: Copper, New Cornelia Mine, Ajo, Pima County, 16 × 7 cm, photo by Stephan Koch, Stephan Koch specimen.

Figure 3.69: Corderoite, Saddle Mountain Mine, Maricopa County, FOV 3 mm, photo by Ron Gibbs, Ron Gibbs specimen.

Figure 3.70: Cornetite, Silver Bell Mine, Pima County, FOV 1.5 mm, photo by Ron Gibbs, Ron Gibbs specimen.

Figure 3.71: Cotunnite, Pusch Ridge, Santa Catalina Mountains, Pima County, FOV 1.5 mm, photo by Ron Gibbs, Warren Lazar specimen.

Figure 3.72: Covellite, Campbell Mine, Bisbee, Cochise County, 7 cm, photo by Ron Gibbs, UA12665, University of Arizona Alfie Norville Gem and Mineral Museum specimen.

Figure 3.73: Cowlesite, Arnett Creek, Pinal County, FOV 5 mm, photo by Ron Gibbs, Ron Gibbs specimen.

Figure 3.74: Creaseyite, Tonopah-Belmont Mine, Maricopa County, FOV 6 mm, photo by Ron Gibbs, Ron Gibbs specimen.

Figure 3.75: Cristobalite, Big Lue Mountains, Greenlee County, FOV 5 mm, photo by Ron Gibbs, Ron Gibbs specimen.

Figure 3.76: Cumengeite, Tonopah-Belmont Mine, Maricopa County, FOV 0.8 mm, photo by Jerry Cone, Ron Gibbs specimen.

Figure 3.77: Cuprite, Morenci Mine, Greenlee County, 1.2 cm high, photo by Jeff Scovil, Evan Jones specimen.

Figure 3.78: Cuprite, variety chalcotrichite, Bisbee, Cochise County, 8.2 cm, photo by Jeff Scovil, UA4008, University of Arizona Alfie Norville Gem and Mineral Museum specimen.

Figure 3.79: Cyanotrichite, Grandview (Last Chance) Mine, Coconino County, FOV 5 mm, photo by Thomas Finke, Stephan Koch specimen.

Figure 3.80: Davidbrownite-(NH4), Rowley Mine, Maricopa County, FOV 1.14 mm, photo by Tony Kampf, Natural History Museum of Los Angeles County specimen.

Figure 3.81: Delafossite, Bisbee, Cochise County, 2.3 cm, photo by Jeff Scovil, Marcus Origlieri specimen.

Figure 3.82: Descloizite, with vanadinite, Ben Hur Mine, Pinal County, 4 cm high, photo by Jeff Scovil, Mark Hay specimen.

Figure 3.83: Diaboleite, Mammoth–St. Anthony Mine, Pinal County, 2.6 cm high, photo by Jeff Scovil, Dave Bunk specimen.

Figure 3.84: Digenite, Campbell Mine, Bisbee, Cochise County, 2 cm, photo by Jeff Scovil, Evan Jones collection.

Figure 3.85: Dioptase, Mammoth–St. Anthony Mine, Pinal County, 2.1 cm high, photo by Jeff Scovil, Dave Bunk specimen.

Figure 3.86: Dolomite, Vekol Mine, Pinal County, 3.8 × 3.0 cm, photo by John Callahan, John Callahan specimen.

Figure 3.87: Dugganite, Empire Mine, Tombstone, Cochise County, R060498, RRUFF photo and specimen.

Figure 3.88: Dumortierite, Dome Rock Mountains, La Paz County, FOV 3 mm, photo by Ron Gibbs, Ron Gibbs specimen.

Figure 3.89: Eddavidite, Southwest Mine, Bisbee, Cochise County, R050381, RRUFF photo and specimen.

Figure 3.90: Eglestonite, Saddle Mountain Mine, Maricopa County, FOV 2 mm, photo by Ron Gibbs, Ron Gibbs specimen.

Figure 3.91: Epidote, near Mammoth, Pinal County, FOV 4 mm, photo by Ron Gibbs, Ron Gibbs specimen.

Figure 3.92: Epsomite, Southwest Mine, Bisbee, Cochise County, 18 cm, photo by Doug Graeme, Graeme Family Collection specimen.

Figure 3.93: Erionite, Well No. 1, Ajo, Pima County, FOV 1.8 mm, photo by Jerry Cone, Ron Gibbs specimen.

Figure 3.94: Eulytine, Flying Saucer Group, Vulture Mountains, Maricopa County, R130114, RRUFF photo and specimen.

Figure 3.95: Fairbankite, Tombstone, Cochise County, FOV 3 mm, R140375, RRUFF photo and specimen.

Figure 3.96: Ferrierite, New Cornelia Mine, Ajo, Pima County, FOV 3 mm, photo by Ron Gibbs, Ron Gibbs specimen.

Figure 3.97: Ferrimolybdite, Kingman, Mohave County, R050566, RRUFF photo and specimen.

Figure 3.98: Ferroselite, Well No. 1, Ajo, Pima County, FOV 2 mm, photo by Ron Gibbs, Ron Gibbs specimen.

Figure 3.99: Fluorite, Homestake Mine, Mohave County, 5.6 cm high, photo by Jeff Scovil, Dick Morris specimen.

Figure 3.100: Fornacite, Evening Star Mine, Maricopa County, FOV 5 mm, photo by Ron Gibbs, Ron Gibbs specimen.

Figure 3.101: Galena, Iron Cap Mine, Graham County, 5.4 cm high, photo by Jeff Scovil, Les Presmyk specimen.

Figure 3.102: Georgerobinsonite, Mammoth–St. Anthony Mine, Pinal County, R110148, RRUFF photo and specimen.

Figure 3.103: Gerhardtite, Tonopah-Belmont Mine, Maricopa County, R120082, RRUFF photo and specimen.

Figure 3.104: Gibbsite, Copper Queen Mine, Bisbee, Cochise County, 9.6 cm high, photo by Jeff Scovil, Brian Thurston specimen.

Figure 3.105: Gilalite, Christmas Mine, Gila County, FOV 7 mm, photo by Ron Gibbs, Ron Gibbs specimen.

Figure 3.106: Goethite, Bisbee, Cochise County, 11 cm, photo by Jeff Scovil, Les Presmyk specimen.

Figure 3.107: Gold, Quartzsite, La Paz County, 2.2 cm high, photo by Jeff Scovil, UA11400, University of Arizona Alfie Norville Gem and Mineral Museum specimen.

Figure 3.108: Graemite, Cole Mine, Bisbee, Cochise County, FOV 3 cm, photo by Richard Graeme IV, Graeme Family Collection specimen.

Figure 3.109: Grandviewite, Grandview (Last Chance) Mine, Coconino County, 1.5 cm, photo by Stephan Koch, Stephan Koch specimen.

Figure 3.110: Greenockite, San Juan Mine, Cochise County, 4.8 × 3.0 cm, photo by John Callahan, John Callahan specimen.

Figure 3.111: Gypsum, Duval Mine, Santa Cruz County, 4 × 6 cm, photo by Kenny Don, John Callahan specimen.

Figure 3.112: Halite, Verde Salt Mine, Yavapai County, 4.9 cm, photo by Stan Celestian, Natural History Museum of Los Angeles County Specimen.

Figure 3.113: Hedenbergite, Iron Cap Mine, Graham County, 5.2 × 4.1 cm, photo by Stephan Koch, Stephan Koch specimen.

Figure 3.114: Helvine, San Juan Mine, Cochise County, FOV 5 mm, photo by Ron Gibbs, Ron Gibbs specimen.

Figure 3.115: Hematite, Big Bertha Extension, Quartzsite, La Paz County, 4.4 cm, Jeff Scovil photo, Ed Huskinson specimen.

Figure 3.116: Hemihedrite, Rat Tail claim, Maricopa County, R141184, RRUFF photo and specimen.

Figure 3.117: Hemimorphite, 79 Mine, Gila County, 7 cm, photo by Jeff Scovil, Mark Hay specimen.

Figure 3.118: Herbertsmithite, Tonopah-Belmont Mine, Maricopa County, FOV 0.5 mm, photo by Jerry Cone, Ron Gibbs specimen.

Figure 3.119: Hetaerolite, Campbell shaft, Bisbee, Cochise County, 4 × 3 cm, photo by John Callahan, John Callahan specimen.

Figure 3.120: Heulandite, Malpais Hill, Pinal County, FOV 3 mm, photo by Ron Gibbs, Ron Gibbs specimen.

Figure 3.121: Hoganite, Holbrook Mine, Bisbee, Cochise County, 6 × 4 cm, photo by Stephan Koch, Stephan Koch specimen.

Figure 3.122: Hureaulite, 7U7 Ranch, Yavapai County, FOV 3 mm, photo by Ron Gibbs, Jim McGlasson specimen.

Figure 3.123: Hydroxylpyromorphite, Pusch Ridge, Santa Catalina Mountains, Pima County, FOV 2 mm, photo by Ron Gibbs, Warren Lazar specimen.

Figure 3.124: Iodargyrite, Rowley Mine, Maricopa County, FOV 4 mm, photo by Ron Gibbs, Ron Gibbs specimen.

Figure 3.125: Iranite, prospect near Evening Star Mine, Maricopa County, FOV 5 mm, photo by Ron Gibbs, Ron Gibbs specimen.

Figure 3.126: Jarosite, Morenci Mine, Greenlee County, FOV 3 mm, photo by Ron Gibbs, Ron Gibbs specimen.

Figure 3.127: Junitoite, Christmas Mine, Gila County, FOV 5 mm, photo by Ron Gibbs, Ron Gibbs specimen.

Figure 3.128: Jurbanite, San Manuel Mine, Pinal County, R130817, RRUFF photo and specimen.

Figure 3.129: Kenhsuite, Saddle Mountain Mine, Maricopa County, FOV 3 mm, photo by Ron Gibbs, Ron Gibbs specimen.

Figure 3.130: Kettnerite, Scorpion Hill, Vulture Mountains, Maricopa County, FOV 3 mm, photo by Ron Gibbs, Ron Gibbs specimen.

Figure 3.131: Khinite-3T, Empire Mine, Tombstone, Cochise County, R080124, RRUFF photo and specimen.

Figure 3.132: Kinoite, Christmas Mine, Gila County, 6 cm, photo by Jeff Scovil, Penny File specimen.

Figure 3.133: Ktenasite, DH Mine, Gila County, R090002, RRUFF photo and specimen.

Figure 3.134: Laumontite, Christmas Mine, Gila County, FOV 4 mm, photo by Ron Gibbs, Ron Gibbs specimen.

Figure 3.135: Laurelite, Grand Reef Mine, Graham County, 5 cm, photo by Jeff Scovil, Les Presmyk specimen.

Figure 3.136: Lausenite, United Verde Mine, Yavapai County, X070004, RRUFF photo and specimen.

Figure 3.137: Lazaraskeite, Pusch Ridge, Santa Catalina Mountains, Pima County, FOV 3 mm, photo by Ron Gibbs, Warren Lazar specimen.

Figure 3.138: Lead, Tubac, Santa Cruz County, 9.8 × 3.6 cm, photo by Jessie La Plante, John Callahan specimen.

Figure 3.139: Leadhillite, C and B Vanadium Mine, Gila County, 3 mm crystal, photo by Ron Gibbs, Ron Gibbs specimen.

Figure 3.140: Lévyne, Horseshoe Dam, Maricopa County, FOV 3 mm, photo by Ron Gibbs, Ron Gibbs specimen.

Figure 3.141: Libethenite, Old Reliable Mine, Pinal County, FOV 4 mm, photo by Thomas Finke, Stephan Koch specimen.

Figure 3.142: Linarite, Grand Reef Mine, Graham County, 3 cm, photo by Jeff Scovil, David Bunk specimen.

Figure 3.143: Lindgrenite, Childs-Aldwinkle Mine, Pinal County, FOV 3 mm, photo by Ron Gibbs, Mark Ascher specimen.

Figure 3.144: Liudongshengite, 79 Mine, Gila County, FOV 1.5 mm, R180016, RRUFF photo and specimen.

Figure 3.145: Luddenite, Evening Star Mine, Maricopa County, FOV 5 mm, photo by Ron Gibbs, Robert Jenkins specimen.

Figure 3.146: Luetheite, near Humboldt Mine, Santa Cruz County, FOV 3 mm, photo by Ron Gibbs, Ron Gibbs specimen.

Figure 3.147: Macquartite, Mammoth–St. Anthony Mine, Pinal County, FOV 3 mm, photo by Ron Gibbs, Robert Jenkins specimen.

Figure 3.148: Magnesioalterite, near Cliff Dwellers Lodge, Coconino County, R180015, RRUFF photo and specimen.

Figure 3.149: Malachite, Morenci Mine, Greenlee County, 8.1 cm, photo by Jeff Scovil, Les Presmyk specimen.

Figure 3.150: Mammothite, Rowley Mine, Maricopa County, FOV 0.6 mm, photo by Jerry Cone, Ron Gibbs specimen.

Figure 3.151: Manganite, Stovall Mine, Pinal County, 9 cm, photo by Jeff Scovil, Flagg Mineral Foundation specimen.

Figure 3.152: Marcasite, Antler Mine, Mohave County, 4.5 × 3.5 cm, photo by John Callahan, John Callahan specimen.

Figure 3.153: Maricopaite, Moon Anchor Mine, Maricopa County, R120147, RRUFF photo and specimen.

Figure 3.154: Markascherite, Childs-Aldwinkle Mine, Pinal County, FOV 3 mm, photo by Ron Gibbs, Robert Jenkins specimen.

Figure 3.155: Mercury, Saddle Mountain Mine, Maricopa County, FOV 3 mm, photo by Ron Gibbs, Ron Gibbs specimen.

Figure 3.156: Mesolite, near Clifton, Greenlee County, FOV 3 mm, photo by Ron Gibbs, Ron Gibbs specimen.

Figure 3.157: Metatorbernite, Silver Bell Mine, Pima County, 5.2 cm high, photo by Jeff Scovil, Les Presmyk specimen.

Figure 3.158: Microcline, Santa Nino Mine, Santa Cruz County, 6.1 cm, photo by Jeff Scovil, Barbara Muntyan specimen.

Figure 3.159: Miersite, on atacamite, Southwest Mine, Bisbee, Cochise County, FOV 1 cm, photo by Richard Graeme IV, Graeme Family Collection specimen.

Figure 3.160: Milarite, Belmont Pit, Maricopa County, FOV 3 mm, photo by Ron Gibbs, Ron Gibbs specimen.

Figure 3.161: Mimetite, Rowley Mine, Maricopa County, 2.2 cm, photo by Jeff Scovil, Paul Geffner specimen.

Figure 3.162: Minium, Rowley Mine, Maricopa County, 5.8 cm, photo by Jeff Scovil, Unique Minerals specimen.

Figure 3.163: Mixite, Table Mountain Mine, Pinal County, FOV 3 mm, photo by Ron Gibbs, UAX4591, University of Arizona Alfie Norville Gem and Mineral Museum specimen.

Figure 3.164: Molybdenite, Childs-Aldwinkle Mine, Pinal County, FOV 5 mm, photo by Ron Gibbs, Ron Gibbs specimen.

Figure 3.165: Mottramite, 79 Mine, Gila County, 6.1 cm, photo by Jeff Scovil, Claude Yoder specimen.

Figure 3.166: Munakataite, Tonopah-Belmont Mine, Maricopa County, R110005, RRUFF photo and specimen.

Figure 3.167: Murdochite, Mammoth Mine, Pinal County, R180001, RRUFF photo and specimen.

Figure 3.168: Muscovite, Santo Niño Mine, Santa Cruz County, 6.8 cm, photo by Jeff Scovil, Barbara Muntyan specimen.

Figure 3.169: Namibite, Copperopolis, Yavapai County, R100182, RRUFF photo and specimen.

Figure 3.170: Natrolite, Horseshoe Dam, Maricopa County, 7.2 × 5.3 cm, photo by Stephan Koch, Stephan Koch specimen.

Figure 3.171: Natrosulfatourea, Rowley Mine, Maricopa County, FOV 0.29 mm, photo by Tony Kampf, Natural History Museum of Los Angeles County specimen.

Figure 3.172: Navajoite, Monument No. 2 Mine, Apache County, 4.0 × 3.3 cm, photo by John Callahan, John Callahan specimen.

Figure 3.173: Nekoite, Iron Cap Mine, Graham County, FOV 3 mm, Ron Gibbs photo, Ron Gibbs specimen.

Figure 3.174: Nordstrandite, on azurite, Southwest Mine, Bisbee, Cochise County, 21.5 cm, photo by Richard Graeme IV, Graeme Family Collection specimen.

Figure 3.175: Olivenite, Copper Queen Mine, Yavapai County, FOV 3 mm, photo by Ron Gibbs, Ron Gibbs specimen.

Figure 3.176: Opal, Bagdad Mine, Yavapai County, 8.8 cm, photo by Jeff Scovil, Flagg Mineral Foundation specimen.

Figure 3.177: Orthoclase, Quail Wash, Maricopa County, 3 cm crystal, photo by Stan Celestian, Stan Celestian specimen.

Figure 3.178: Orthoserpierite, Childs-Aldwinkle Mine, Pinal County, FOV 5 mm, photo by Ron Gibbs, Ron Gibbs specimen.

Figure 3.179: Osarizawaite, Silver Hill Mine, Pima County, R060395, RRUFF photo and specimen.

Figure 3.180: Ottoite, Grand Central Mine, Tombstone, Cochise County, R130126, RRUFF photo and specimen.

Figure 3.181: Papagoite, New Cornelia Mine, Pima County, FOV 2 mm, photo by Ron Gibbs, Ron Gibbs specimen.

Figure 3.182: Paralaurionite, Mammoth–St. Anthony Mine, Pinal County, 4.1 cm, photo by Jeff Scovil, Mark Hay specimen.

Figure 3.183: Paramelaconite, Copper Queen Mine, Bisbee, Cochise County, 1.9 cm high, photo by Jeff Scovil, Evan Jones specimen.

Figure 3.184: Parnauite, Grandview (Last Chance) Mine, Coconino County, FOV 2 mm, photo by Ron Gibbs, Jim McGlasson specimen.

Figure 3.185: Parsonsite, near Fat Jack Mine, Yavapai County, FOV 5 mm, photo by Ron Gibbs, Ron Gibbs specimen.

Figure 3.186: Perroudite, Red Cloud Mine, La Paz County, FOV 4 mm, photo by Ron Gibbs, Ron Gibbs specimen.

Figure 3.187: Phillipsite, Malpais Hill, Pinal County, FOV 3 mm, photo by Ron Gibbs, Ron Gibbs specimen.

Figure 3.188: Phlogopite, Big Lue Mountains, Greenlee County, FOV 2 mm, photo by Ron Gibbs, Ron Gibbs specimen.

Figure 3.189: Phoenicochroite, Rat Tail claim, Maricopa County, FOV 6 mm, photo by Ron Gibbs, Ron Gibbs specimen.

Figure 3.190: Phosphohedyphane, Tonopah-Belmont Mine, Maricopa County, FOV 3 mm, photo by Ron Gibbs, Ron Gibbs specimen.

Figure 3.191: Phoxite, Rowley Mine, Maricopa County, FOV 0.68 mm, photo by Tony Kampf, Natural History Museum of Los Angeles County specimen.

Figure 3.192: Pickeringite, 79 Mine, Gila County, 4 cm, photo by Ron Gibbs, Ron Gibbs specimen.

Figure 3.193: Pinalite, Mammoth–St. Anthony Mine, Pinal County, FOV 1.5 mm, photo by Ron Gibbs, Robert Jenkins specimen.

Figure 3.194: Plancheite, Nugget Fraction Mine, Pinal County, 3 cm, photo by Stephan Koch, Stephan Koch specimen.

Figure 3.195: Plattnerite, 79 Mine, Gila County, FOV 3 mm, photo by Ron Gibbs, Ron Gibbs specimen.

Figure 3.196: Posnjakite, Childs-Aldwinkle Mine, Pinal County, FOV 2 mm, photo by Ron Gibbs, Ron Gibbs specimen.

Figure 3.197: Powellite, Carlota Mine, Gila County, FOV 5 mm, photo by Ron Gibbs, Ron Gibbs specimen.

Figure 3.198: Prehnite, Haystack Butte, Gila County, 13 × 10 cm, photo by Stephan Koch, Stephan Koch specimen.

Figure 3.199: Prosopite, Carlota Mine, Gila County, FOV 7 mm, photo by Ron Gibbs, Ron Gibbs specimen.

Figure 3.200: Pseudobrookite, Big Lue Mountains, Greenlee County, FOV 3 mm, photo by Ron Gibbs, Ron Gibbs specimen.

Figure 3.201: Pseudomalachite, Old Reliable Mine, Pinal County, FOV 5 mm, photo by Ron Gibbs, Ron Gibbs specimen.

Figure 3.202: Purpurite, Little Giant Mine, Yavapai County, FOV 3 mm, photo by Ron Gibbs, Ron Gibbs specimen.

Figure 3.203: Pyrite, Trench Mine, Santa Cruz County, 6.4 cm, photo by Jeff Scovil, Mark Hay specimen.

Figure 3.204: Pyromorphite, Hardshell Mine, Santa Cruz County, 1.5 cm, photo by Jeff Scovil, Frank Sousa specimen.

Figure 3.205: Pyrope, Sunset Crater, Coconino County, R050446, RRUFF photo and specimen.

Figure 3.206: Quartz, Scarlet III claim, Santa Cruz County, 3.9 cm high, photo by Jeff Scovil, Dick Morris specimen.

Figure 3.207: Quartz, variety chalcedony, Aquila, Maricopa County, 3.5 cm high, photo by Stan Celestian, Stan Celestian specimen.

Figure 3.208: Ramsbeckite, 79 Mine, Gila County, R050295, RRUFF photo and specimen.

Figure 3.209: Ramsdellite, Mistake Mine, Yavapai County, 4 × 6 cm, photo by Stephan Koch, Stephan Koch specimen.

Figure 3.210: Ransomite, United Verde Mine, Yavapai County, 4.5 cm, photo by John Callahan, John Callahan specimen.

Figure 3.211: Raygrantite, Evening Star Mine, Maricopa County, FOV 3 mm, photo by Ron Gibbs, Ray Grant specimen.

Figure 3.212: Rhodochrosite, with alabandite, Junction Mine, Bisbee, Cochise County, 4.3 cm, photo by Richard Graeme IV, Graeme Family Collection specimen.

Figure 3.213: Rodalquilarite, Grand Central Mine, Tombstone, Cochise County, FOV 2 mm, photo by Ron Gibbs, Jim McGlasson specimen.

Figure 3.214: Rongibbsite, near Evening Star Mine, Maricopa County, FOV 1.3 mm, photo by Jerry Cone, Jerry Cone specimen.

Figure 3.215: Rosasite, Silver Bill Mine, Cochise County, 8.2 cm, photo by Jeff Scovil, Evan Jones specimen.

Figure 3.216: Rowleyite, Rowley Mine, Maricopa County, FOV 0.56 mm, photo by Tony Kampf, Natural History Museum of Los Angeles County specimen.

Figure 3.217: Ruizite, Christmas Mine, Gila County, FOV 3 mm, photo by Ron Gibbs, Ron Gibbs specimen.

Figure 3.218: Rutile, Santo Niño Mine, Santa Cruz County, 5.1 cm, photo by Jeff Scovil, Barbara Muntyan specimen.

Figure 3.219: Scheelite, Cohen Tungsten Mine, Cochise County, 2.6 × 3.5 cm, photo by Kenny Don, John Callahan specimen.

Figure 3.220: Schorl, Castle Hot Springs, Maricopa County, 5.6 cm, photo by Stan Celestian, Stan Celestian specimen.

Figure 3.221: Shannonite, Tonopah-Belmont Mine, Maricopa County, FOV 2 mm, photo by Ron Gibbs, Ron Gibbs specimen.

Figure 3.222: Shattuckite, New Cornelia Mine, Pima County, 14.6 cm, photo by Jeff Scovil, Freeport McMoRan specimen.

Figure 3.223: Siderite, Campbell shaft, Bisbee, Cochise County, 9.7 cm high, photo by Jeff Scovil, Freeport McMoRan specimen.

Figure 3.224: Silver, Stonewall Jackson Mine, Gila County, 8 cm, photo by Jeff Scovil, Evan Jones specimen.

Figure 3.225: Smithsonite, 79 Mine, Gila County, 6 × 6 cm, photo by Stephan Koch, Stephan Koch specimen.

Figure 3.226: Sonoraite, Old Guard Mine, Tombstone, Cochise County, FOV 2.5 mm, photo by Bruce Murphy, Bruce Murphy specimen.

Figure 3.227: Spangolite, Southwest Mine, Bisbee, Cochise County, 4.5 cm, photo by Jeff Scovil, Graeme Family Collection specimen.

Figure 3.228: Spessartine, Aquarius Mountains, Mohave County, 4.2 cm high, photo by Jeff Scovil, Paul Harter specimen.

Figure 3.229: Sphalerite, Iron Cap Mine, Graham County, 2.6 cm, photo by Ray Grant, Ray Grant specimen.

Figure 3.230: Stibnite, Dura Mine, Santa Cruz County, 4.5 × 3.5 cm, photo by John Callahan, John Callahan specimen.

Figure 3.231: Stilbite, Sierrita Mine, Pima County, FOV 6 mm, photo by Ron Gibbs, Ron Gibbs specimen.

Figure 3.232: Stolzite, Reef Mine, Cochise County, FOV 3 mm, photo by Ron Gibbs, Ron Gibbs specimen.

Figure 3.233: Stringhamite, Christmas Mine, Gila County, FOV 2 mm, photo by Ron Gibbs, Ron Gibbs specimen.

Figure 3.234: Surite, Mammoth Mine, Pinal County, R110145, RRUFF photo and specimen.

Figure 3.235: Svanbergite, Dome Rock Mountains, La Paz County, FOV 3 mm, photo by Ron Gibbs, Ron Gibbs specimen.

Figure 3.236: Szenicsite, Carlota Mine, Gila County, FOV 3 mm, photo by Ron Gibbs, Ron Gibbs specimen.

Figure 3.237: Tangeite, Monument No. 1 Mine, Navajo County, FOV 3 mm, photo by Ron Gibbs, Jim McGlasson specimen.

Figure 3.238: Tetrahedrite, Magma Mine, Pinal County, 7.9 cm high, photo by Jeff Scovil, Les Presmyk specimen.

Figure 3.239: Thomsonite-Ca, Arnett Creek, Pinal County, FOV 3 mm, photo by Ron Gibbs, Ron Gibbs specimen.

Figure 3.240: Titanite, Big Lue Mountains, Greenlee County, FOV 3 mm, photo by Ron Gibbs, Ron Gibbs specimen.

Figure 3.241: Torbernite, King claims, Miami, Gila County, FOV 5 mm, photo by Ron Gibbs, Ron Gibbs specimen.

Figure 3.242: Tridymite, Big Lue Mountains, Greenlee County, FOV 2 mm, photo by Ron Gibbs, Ron Gibbs specimen.

Figure 3.243: Triploidite, 7U7 Ranch, Yavapai County, FOV 2 mm, photo by Ron Gibbs, Jim McGlasson specimen.

Figure 3.244: Tsumebite, 79 Mine, Gila County, FOV 1 mm, photo by Jerry Cone, Ron Gibbs specimen.

Figure 3.245: Turquoise, Ithaca Peak, Kingman, Mohave County, 7.5 cm high, photo by Jeff Scovil, SAS specimen.

Figure 3.246: Tyuyamunite, Ridenour Mine, Coconino County, R060983, RRUFF photo and specimen.

Figure 3.247: Uranophane, Muggins Mountains, Yuma County, FOV 3 mm, photo by Ron Gibbs, Robert Jenkins specimen.

Figure 3.248: Utahite, Empire Mine, Tombstone, Cochise County, FOV 2 mm, photo by Ron Gibbs, Michael Shannon specimen.

Figure 3.249: Vanadinite, Apache Mine, Gila County, 4.8 cm, photo by Jeff Scovil, Mark Hay specimen.

Figure 3.250: Variscite, Bisbee, Cochise County, 12.5 cm, photo by Ron Gibbs, UA7722, University of Arizona Alfie Norville Gem and Mineral Museum specimen.

Figure 3.251: Vauquelinite, 79 Mine, Gila County, FOV 2 mm, photo by Ron Gibbs, Ron Gibbs specimen.

Figure 3.252: Voltaite, Bisbee, Cochise County, FOV 3 mm, photo by Ron Gibbs, UAX6737, University of Arizona Alfie Norville Gem and Mineral Museum specimen.

Figure 3.253: Wavellite, Duval Pit, Kingman, Mohave County, 3.6 × 2.0 cm, photo by John Callahan, John Callahan specimen.

Figure 3.254: Weddellite, Pusch Ridge, Santa Catalina Mountains, Pima County, FOV 4 mm, photo by Ron Gibbs, Ron Gibbs specimen.

Figure 3.255: Weeksite, Anderson Mine, Yavapai County, FOV 2 mm, photo by Ron Gibbs, Ron Gibbs specimen.

Figure 3.256: Wheatleyite, Rowley Mine, Maricopa County, FOV 2 mm, photo by Ron Gibbs, Ron Gibbs specimen.

Figure 3.257: Whelanite, Christmas Mine, Gila County, R050323, RRUFF photo and specimen.

Figure 3.258: Wherryite, Tiger, Pinal County, R141161, RRUFF photo and specimen.

Figure 3.259: Wickenburgite, Evening Star Mine, Maricopa County, FOV 4 mm, photo by Ron Gibbs, Ron Gibbs specimen.

Figure 3.260: Willemite, Red Cloud Mine, La Paz County, FOV 6 mm, photo by Ron Gibbs, Ron Gibbs specimen.

Figure 3.261: Wulfenite, Hilltop Mine, Cochise County, 7.0 × 4.5 cm, photo by Stephan Koch, Stephan Koch specimen.

Figure 3.262: Wulfenite, 79 Mine, Gila County, 2.2 cm, photo by John Callahan, John Callahan specimen.

Figure 3.263: Wulfenite, Red Cloud Mine, La Paz County, 4.3 cm high, photo by Jeff Scovil, Les Presmyk specimen.

Figure 3.264: Wulfenite, with mimetite, Rowley Mine, Maricopa County, 3.1 cm high, photo by Jeff Scovil, Alex Schauss specimen.

Figure 3.265: Wulfenite, Heavy Weight Mine, Pima County, FOV 2 mm, photo by Ron Gibbs, Ron Gibbs specimen.

Figure 3.266: Wulfenite, Glove Mine, Santa Cruz County, 7.1 cm, photo by Jeff Scovil, UA12532, University of Arizona Alfie Norville Gem and Mineral Museum specimen.

Figure 3.267: Wupatkiite, Grey Mountain, Coconino County, FOV 3 mm, photo by Ron Gibbs, Ron Gibbs specimen.

Figure 3.268: Yedlinite: Tiger, Pinal County, R050338, RRUFF photo and specimen.

Figure 3.269: Zircon, Belmont Pit, Maricopa County, FOV 2 mm, photo by Ron Gibbs, Ron Gibbs specimen.

Figure 3.270: Zunyite, Big Bertha Extension Mine, Quartzsite, La Paz County, 7 cm, photo by Jeff Scovil, IKON Minerals specimen.

INDEX

Page references in *italic* type indicate topics discussed or noted within illustrations/graphics or captions. Page references that contain a "t" character indicate information found within a table.

Tennessee mine, 16
Teviston District, 34
Thomas Range, 23
Three Musketeers District, 44–45
Three Musketeers Mine, 44–45
Tiffany & Co., 19, 24
topaz, 23
Topaz Basin, 26
tourmaline, 23–24
travertine (calcite), 13
triboluminescence, 28, 37, 41
Trincheras Mountains (unidentifiable location), 15
Tucson, public mineral collections in, 8–9
tungsten ore, minerals associated with, 12, 32–33, 33, 34–35, 45, 538
turquoise, 7, 11, 24–25
Turquoise District, 24
Turquoise Mountain, 24

ultradeep hydrothermal (UDH) magmas, 532, 534–36t
ultraviolet (UV) light and fluorescence, 27–28, 30
United Verde Mine, 13–14
University of Arizona: Alfie Norville Gem and Mineral Museum, 8; Arizona Mining, Mineral and Natural Resources Education Museum, 9–10; mineralogists/scholars of note, 5, 6; RRUFF project, 4
uranyl ions in fluorescence, 30–31, 39–40
uvarovite garnet, 16

variscite, 25

Warren District, 40
water content and mineral systems: anhydrous, 545–46;

hydrous/hydrothermal systems, 541, 545, 546–47, 548–49, 550
watermelon tourmaline, 24
wavelength in UV fluorescence, 28
White Picacho District, 38
Wickenburg, Ariz., 33, 38, 43, 47
wickenburgite, 43
Wilcox, Ariz., 34
willemite, 41–42, 45, 46
Williams, Sidney, 6
Willis, Charles, 5–6
Winslow, Ariz., 22
wulfenite, 8, 532, 533

zebra agate, 21
zinc, 40, 41, 41–42, 47
zunyite, 43–44

ABOUT THE AUTHORS

Raymond W. Grant, co-author of the *Mineralogy of Arizona*, 3rd edition, has a PhD in geology and has retired after teaching geology at Mesa Community College in Mesa, Arizona, for thirty-one years. He is a past president of the Mineralogical Society of Arizona, past chair of the Flagg Mineral Foundation, and cochairman of the Minerals of Arizona Symposium for twenty-six years. He has recently started the Pinal Geology Museum in Coolidge, Arizona. His published works also include several articles on Arizona mineral localities and the *Checklist of Arizona Minerals.*

Ronald B. Gibbs has bachelor's degrees in geology and mining engineering and has retired after a career in the copper-mining industry. He has authored several articles on mineral localities, presented at several mineral symposia in Arizona and New Mexico, and co-authored papers describing new mineral species.

Harvey W. Jong received a bachelor's degree from MIT and a master's degree from the University of California, Santa Barbara; both degrees are in electrical engineering. After working in the microprocessor industry, he decided to pursue his passion for photography and started a digital media studio specializing in virtual reality. He is a founding member of the Earth Science Museum and currently serves as the president of this nonprofit organization, which is helping to reopen a mineral museum in Phoenix.

Jan Rasmussen received a PhD in economic geology from the University of Arizona. She was the associate curator of the University of Arizona Mineral Museum, later the curator of the Arizona Mining and Mineral Museum in

Phoenix, and has more than forty years' experience as an economic geologist working for the Arizona Geological Survey and several private companies. She has co-authored numerous books, open-file reports, and articles on Arizona geology.

Stanley Keith has a master's degree in geology from the University of Arizona and has worked for the Arizona Geological Survey before cofounding MagmaChem Exploration. He has worked on numerous exploration projects for mining companies and has advanced research in mineral deposits and their associated igneous rocks. Stan has authored many articles and technical papers and has presented at numerous symposia.

John Callahan earned a bachelor's degree in geology from Macalester College. He was employed for over thirty years as an illustrator for the U.S. Geological Survey, where he was able to observe the evolution and learn the craft of digital illustrating and publishing.